COLLECTED PAPERS OF
G. H. HARDY

LECTURE NOTES, c. 1937

The frontispiece is taken from a file of Hardy's Cambridge lectures on Ramanujan, intermediate between his Harvard lectures, 1937, 2, and his book *Ramanujan*, B9. See Comment on 1937, 2.

The remark that $2187 = 3^7$ is 'even rounder' than $1200 = 2^4 \cdot 3 \cdot 5^2$ is puzzling. It is reproduced in *Ramanujan*, Lecture III, but omitted from Hardy and Wright, *Theory of Numbers*, §22.14.

COLLECTED PAPERS
OF
G. H. HARDY

INCLUDING JOINT PAPERS WITH J. E. LITTLEWOOD
AND OTHERS

EDITED BY A COMMITTEE APPOINTED BY
THE LONDON MATHEMATICAL SOCIETY

VOLUME VII

CLARENDON PRESS · OXFORD
1979

Oxford University Press, Walton Street, Oxford OX2 6DP

OXFORD LONDON GLASGOW
NEW YORK TORONTO MELBOURNE WELLINGTON
IBADAN NAIROBI DAR ES SALAAM LUSAKA CAPE TOWN
KUALA LUMPUR SINGAPORE JAKARTA HONG KONG TOKYO
DELHI BOMBAY CALCUTTA MADRAS KARACHI

© *Oxford University Press* 1979

All rights reserved. No part of this publication may be reproduced, stored in a retrieval system, or transmitted, in any form or by any means, electronic, mechanical, photocopying, recording, or otherwise, without the prior permission of Oxford University Press

British Library Cataloguing in Publication Data
Hardy, Godfrey Harold
   Collected papers of G. H. Hardy
   Vol. 7
   1. Mathematics—Addresses, essays, lectures
   I. Littlewood, John Edensor   II. London
Mathematical Society
510'.8   QA7   77–30540
ISBN 0–19–853347–0

Printed and bound at William Clowes & Sons Limited, Beccles and London

EDITORIAL COMMITTEE

L. S. BOSANQUET
I. W. BUSBRIDGE
MARY L. CARTWRIGHT
*E. F. COLLINGWOOD
†H. DAVENPORT
‡T. M. FLETT
§H. HEILBRONN
∥A. E. INGHAM
R. RADO
R. A. RANKIN
**W. W. ROGOSINSKI
F. SMITHIES
†† E. C. TITCHMARSH
E. M. WRIGHT

died 25 October 1970
† died 9 June 1969
‡ died 13 February 1976
§ died 28 April 1975
∥ died 6 September 1967
*** died 23 July 1964*
†† died 18 January 1963

EDITORIAL NOTE

The completed work comprises seven volumes

THE contents of the separate volumes (except for part of Vol. VII) were planned by the Editorial Committee under the Chairmanship of H. Davenport. Davenport devoted much time and skill to the task of getting the project under way.

Vol. VII was left open-ended, and some further material has now been added. A list of Hardy's papers is printed at the end of each volume, based on that compiled by Titchmarsh (*Journal of the London Mathematical Society*, 25 (1950), 89–101). In the present volume the list has been expanded to include some additional Miscellaneous Papers (VII 2) and Questions from the *Educational Times* (VII 3), and also, at the end, a full list of Obituary Notices by Hardy (VII 4).

The section on Integral Equations and Integral Transforms has been divided into two sub-sections, headed VII (a) Fourier transforms, VII (b) Integral equations. Twenty-five book reviews by Hardy have been reproduced (VII 5); these are listed in the List of Contents for this volume, but are not in the list at the end. In VII 6 there is a list of Hardy's twelve books, and in VII 7 a list of writings about Hardy by other writers.

For convenience of reference, papers are numbered according to years, e.g. 1912, 4. When reference is made, in corrections and comments, to the pages of a paper, the numbers used are those of the original pagination and not the consecutive pages of this volume.

The allocation of work between the Editors, for the seven volumes, was as follows:

  Vol. I. Heilbronn and Davenport;
  Vol. II. Ingham and Rado;
  Vol. III. Flett (initially Rogosinski);
  Vol. IV. Wright and Cartwright;
  Vol. V. Smithies;
  Vol. VI. Bosanquet;
  Vol. VII. Busbridge and Rankin (initially Titchmarsh and Collingwood).
  Chairmen: Davenport (Vols. I–IV), Bosanquet (Vols. V–VII).

Finally, the Editors' grateful thanks are due to the publishing staff of the Oxford University Press, as well as the sub-editors, readers, photo-lithographers and compositors, for their unfailing cooperation in the production of this work.

ACKNOWLEDGEMENTS

The Archivist, The Bertrand Russell Archives, McMaster University, Ontario.

The Master and Fellows of Trinity College, Cambridge (for a letter from A. N. Whitehead).

The *Cambridge Review* Committee.

The Editors of *Eureka*.

The Editors of *Nature*.

Science (The American Association for the Advance of Science).

The Editor of *The Times Literary Supplement*.

CONTENTS OF VOLUME VII

1. INTEGRAL EQUATIONS AND INTEGRAL TRANSFORMS

(a) Fourier transforms

Introduction   3

1904, 6. Note on the function $\int_x^\infty e^{-\frac{1}{2}(x^2-t^2)}\,dt$.   10
Quarterly Journal of Mathematics, 35, 193–207.

1911, 9. Fourier's double integral and the theory of divergent integrals.   26
Transactions of the Cambridge Philosophical Society, 21, 427–51.

1912, 9. Notes on some points in the integral calculus XXXIII: Some cases of the inversion of the order of integration.   52
Messenger of Mathematics, 41, 102–9.

1913, 9. Notes on some points in the integral calculus XXXV: On an integral equation.   62
Messenger of Mathematics, 42, 89–93.

1917, 5. Notes on some points in the integral calculus XLII: On Weierstrass's singular integral, and on a theorem of Lerch.   68
Messenger of Mathematics, 46, 43–8.

1917, 9. Notes on some points in the integral calculus XLVI: On Stieltjes' 'problème des moments'.   75
Messenger of Mathematics, 46, 175–82.

1918, 6. Notes on some points in the integral calculus XLVII: On Stieltjes' 'problème des moments' (cont.).   84
Messenger of Mathematics, 47, 81–8.

1918, 8. Notes on some points in the integral calculus XLIX: On Mellin's inversion formula.   92
Messenger of Mathematics, 47, 178–84.

1920, 8. Notes on some points in the integral calculus LII: On some definite integrals considered by Mellin.   99
Messenger of Mathematics, 49, 85–91.

1921, 7. Notes on some points in the integral calculus LIV: Further notes on Mellin's inversion formulae.   106
Messenger of Mathematics, 50, 165–71.

1924, 8. Notes on some points in the integral calculus LVII: On Fourier transforms.   113
Messenger of Mathematics, 53, 135–42.

1925, 9. Notes on some points in the integral calculus LVIII: On Hilbert transforms.   122
Messenger of Mathematics, 54, 20–7.

1925, 10. Notes on some points in the integral calculus LIX: On Hilbert transforms (cont.).   131
Messenger of Mathematics, 54, 81–8.

1926, 11 (with S. Bochner). Notes on two theorems of Norbert Wiener.   139
Journal of the London Mathematical Society, 1, 240–4.

1927, 7. Notes on some points in the integral calculus LXII: A singular integral.   145
Messenger of Mathematics, 56, 10–16.

1927, 8. Notes on some points in the integral calculus LXIII: Some further applications of Mellin's inversion formula.   153
Messenger of Mathematics, 56, 186–92.

1929, 8. Notes on some points in the integral calculus LXVII: On the repeated integral which occurs in the theory of conjugate functions.   161
Messenger of Mathematics, 58, 53–8.

1930, 2 (with E. C. Titchmarsh). Self-reciprocal functions.   167
Quarterly Journal of Mathematics, 1, 196–231.

1931, 6 (with E. C. Titchmarsh). A note on Parseval's theorem for Fourier transforms.   203
Journal of the London Mathematical Society, 6, 44–8.

1932, 1 (with E. C. Titchmarsh). Formulae connecting different classes of self-reciprocal functions.   208
Proceedings of the London Mathematical Society (2), 33, 225–32.

1932, 2. On Hilbert transforms.   216
Quarterly Journal of Mathematics, 3, 102–12.

1933, 1 (with E. C. Titchmarsh). A class of Fourier kernels.   228
Proceedings of the London Mathematical Society (2), 35, 116–55.

1933, 4. A theorem concerning Fourier transforms.   268
Journal of the London Mathematical Society, 8, 227–31.

1935, 2. The resultant of two Fourier kernels.   274
Proceedings of the Cambridge Philosophical Society, 31, 1–6.

1937, 1. Ramanujan and the theory of transforms.   280
Quarterly Journal of Mathematics, 8, 245–54.

1937, 4. On a theorem of Paley and Wiener.   290
Proceedings of the Cambridge Philosophical Society, 33, 1–5.

1941, 4. A double integral.   295
Journal of the London Mathematical Society, 16, 89–94.

1947, 2. A double integral.   301
Journal of the London Mathematical Society, 22, 242–7.

(b) *Integral equations*

Introduction   309

1909, 3. On an integral equation.   313
Proceedings of the London Mathematical Society (2), 7, 445–72.

1925, 8 (with E. C. Titchmarsh). Solutions of some integral equations considered by Bateman, Kapteyn, Littlewood and Milne.   341
Proceedings of the London Mathematical Society (2), 23, 1–26, and Correction ibid., 24, xxxi–iii.

1929, 5 (with E. C. Titchmarsh). Solution of an integral equation.   372
Journal of the London Mathematical Society, 4, 300–4.

1930, 5 (with E. C. Titchmarsh). Additional note on certain integral equations.   378
Proceedings of the London Mathematical Society (2), 30, 95–106.

1932, 7 (with E. C. Titchmarsh). An integral equation.   390
Proceedings of the Cambridge Philosophical Society, 28, 165–73.

1936, 3 (with E. C. Titchmarsh). New solution of an integral equation.   399
Proceedings of the London Mathematical Society (2), 41, 1–15.

2. MISCELLANEOUS PAPERS

Introduction   417

(a) *Set theory*

1904, 2. A general theorem concerning absolutely convergent series.   421
Proceedings of the London Mathematical Society (2), 1, 285–90.

1904, 5. A theorem concerning the infinite cardinal numbers. 427
Quarterly Journal of Mathematics, 35, 87–94.

1904, 9. The cardinal number of a closed set of points. 435
Messenger of Mathematics, 33, 67–9.

1907, 1. The continuum and the second number class. 438
Proceedings of the London Mathematical Society (2), 4, 10–17.

1910, 2. The ordinal relations of the terms of a convergent sequence. 446
Proceedings of the London Mathematical Society (2), 8, 295–300.

(b) *Differential equations*

1912, 3. Some results concerning the behaviour at infinity of a real and continuous solution of an algebraic differential equation of the first order. 455
Proceedings of the London Mathematical Society (2), 10, 451–68.

1939, 1. A note on a differential equation. 473
Proceedings of the Cambridge Philosophical Society, 35, 652–3.

(c) *The Hardy–Weinberg Law*

1908, 10. Mendelian proportions in a mixed population. 477
Science (American Association for the Advancement of Science), new series 28, 49–50.

(d) *Elementary and expository notes*

1907, 8. Higher trigonometry. 481
Mathematical Gazette, 4, 13–14.

1907, 9. A curious imaginary curve. 483
Mathematical Gazette, 4, 14.

1907, 10. The line at infinity, etc. 484
Mathematical Gazette, 4, 14–15.

1910, 9. To find an approximation to the large positive root of the equation $e^{e^x} = 10^{10} x^{10} e^{10^{10} x^{10}}$. 485
Mathematical Gazette, 5, 333–4.

1915, 12. The definition of a complex number. 486
Mathematical Gazette, 8, 48–9.

1934, 6 (with J. E. Littlewood). A problem in elementary probability.   487
Mathematical Gazette, 18, 195.

1945, 4. A mathematical theorem about golf.   488
Mathematical Gazette, 29, 226–7.

(e) Addresses, invited lectures, etc.

1917, 11. Mr S. Ramanujan's mathematical work in England.   491
Journal of the Indian Mathematical Society, 9, 30–45. (Reprinted by Cambridge University Press.)

1918, 3. Sir George Stokes and the concept of uniform convergence.   505
Proceedings of the Cambridge Philosophical Society, 19, 148–56.

1922, 11. The theory of numbers.   514
British Association Report, 90, 16–24. (Reprinted in *Nature*, 110, 381–5.)

1925, 3. What is geometry?   519
Mathematical Gazette, 12, 309–16.

1926, 13. The case against the Mathematical Tripos.   527
Mathematical Gazette, 13, 61–71.

1929, 6. An introduction to the theory of numbers.   540
Bulletin of the American Mathematical Society, 35, 778–818.

1929, 12. Mathematical proof.   581
Mind, 38, 1–25.

1930, 7. Mathematics.   607
Oxford Magazine, 48, 819–21 (issue of 5 June).

1934, 7. The J-type and the S-type among mathematicians.   610
Nature, 134, 250.

1937, 2. The Indian mathematician Ramanujan.   612
American Mathematical Monthly, 44, 137–55.

1940, 2. Mathematics in war-time.   631
Eureka, Vol. I, No. 3, 5–8.

3. QUESTIONS FROM *THE EDUCATIONAL TIMES*

The volume and page numbers refer to *Mathematical Questions and Solutions from the 'Educational Times'* (Hodgson, London).

Introduction   637
1899, 1. Question 13848, 70, 43–4.   638
1899, 2. Question 13917, 70, 78–9.   640

1899, 2a. Question 13889, 71, 26.	641
1899, 2b. Question 13979, 71, 61–2.	642
1899, 2c. Question 13964, 71, 86–7.	643
1899, 3. Question 14124, 71, 100–1.	645
1899, 4. Question 14005, 71, 111–12.	646
1899, 5. Question 14155, 71, 117.	647
1900, 1a. Question 14089, 72, 49.	648
1900, 1b. Question 14090, 72, 57–9.	649
1900, 2. Question 14243, 72, 80–1.	650
1900, 2a. Question 14164, 72, 89–90.	652
1900, 2b. Question 14332, 72, 105–6.	653
1900, 3. Question 14271, 73, 36–7.	654
1900, 4. Question 14179, 73, 53–4.	655
1900, 5. Question 14317, 73, 61–3.	656
1900, 6. Question 14263, 73, 76.	658
1901, 5. Question 14496, 74, 37–8.	659
1901, 5a. Question 14511, 74, 42–3.	660
1901, 6. Question 14447, 74, 98–100.	662
1901, 7. Question 14467, 74, 111–12.	664
1901, 8. Question 14028, 74, 122–3.	665
1901, 9. Question 14369, 75, 135–6.	667
1902, 12. Questions 1423, 2316, 3941, 4794, (2), 1, 25.	669
1902, 13. Question 14851, (2), 1, 58–9.	670
1902, 14. Question 14055, (2), 2, 41–2.	672
1903, 10. Question 14988, (2), 3, 94–5.	674
1903, 10a. Question 9969, (2), 4, 44–5.	676
1903, 11. Question 14989, (2), 4, 69–70.	677
1903, 12. Question 15019, (2), 4, 75.	678
1903, 13. Question 15265, (2), 4, 109–10.	679
1904, 12a. Question 9515, (2), 5, 46.	681
1904, 13. Question 15300, (2), 5, 61.	682
1904, 13a. Question 15282, (2), 5, 89–92.	683
1904, 14. Question 15282 (Additional Note), (2), 5, 113–14.	685
1904, 15. Question 15361, (2), 5, 118.	687
1904, 16. Question 15125, (2), 6, 31.	688
1904, 17. Question 15357, (2), 6, 45.	689
1905, 15. Question 15686, (2), 8, 74.	690
1907, 11. Question 15498, (2), 12, 26–7.	691
1908, 8a. Question 16041, (2), 13, 42.	692
1908, 9. Question 16257, (2), 13, 79–80.	693

1909, 8. Question 16402, (2), 16, 66.   695
1917, 12. Question 14587, (3), 3, 85.   696

4. OBITUARY NOTICES BY G. H. HARDY
A full list of the notices is given at the end of the volume.

Introduction   701

O1.   S. Ramanujan.   702
Proceedings of the London Mathematical Society (2), 19 (1921), xl–lviii.

O4.   C. Jordan.   721
Proceedings of the Royal Society (A), 104 (1922), xxiii–vi.

O5.   G. Mittag-Leffler.   725
Journal of the London Mathematical Society, 3 (1928), 156–60.

O7.   J. W. L. Glaisher.   730
Messenger of Mathematics, 58 (1929), 159–60.

O8.   T. J. I'A Bromwich.   732
Journal of the London Mathematical Society, 5 (1930), 209–20.

O10.   R. E. A. C. Paley.   744
Journal of the London Mathematical Society, 9 (1934), 76–80.

O11.   E. W. Hobson.   749
Journal of the London Mathematical Society, 9 (1934), 225–37.

O14 (with H. Heilbronn).   E. Landau.   762
Journal of the London Mathematical Society, 13 (1938), 302–10.

O15.   W. H. Young.   771
Journal of the London Mathematical Society, 17 (1942), 218–37.

O18.   D. Hilbert.   791
Journal of the London Mathematical Society, 18 (1943), 191–2.

O19.   H. L. Lebesgue.   793
Nature, 152 (1943), 685.

O20 (with H. S. Carslaw).   J. R. Wilton.   795
Journal of the London Mathematical Society, 20 (1945), 58–64.

5. BOOK REVIEWS BY G. H. HARDY

Introduction   805

(a) Reviews from The Mathematical Gazette

R1.   O. Stolz & J. A. Gmeiner, *Theoretische Arithmetik (II)*.   808
Math. Gaz. 2 (1903), 312–13.

R2. O. Stolz & I. A. Gmeiner, *Einleitung in die Funktionentheorie (I)*. 810
    Math. Gaz. 3 (1905), 184–6.

R3. H. Liebmann, *N. J. Lobatschefskij's Imaginäre Geometrie und Anwendung der Imaginären Geometrie auf einige Integrale*. 812
    Math. Gaz. 3 (1905), 186–7.

R4. E. Borel, *Leçons sur les fonctions de variables réelles*. 814
René Baire, *Leçons sur les fonctions discontinues*.
E. Lindelöf, *Le calcul des residus et ses applications à la théorie des fonctions*.
    Math. Gaz. 3 (1905), 231–4.

R5. H. S. Carslaw, *Introduction to the infinitesimal calculus*. 817
Irving Fisher, *Kurze Einleitung in die Differential- und Integral Rechnung*.
    Math. Gaz. 3 (1905), 274–5.

R6. O. Stolz & J. A. Gmeiner, *Einleitung in die Funktionentheorie II*. 818
    Math. Gaz. 3 (1905), 304.

R7. W. F. Osgood, *A first course in the differential and integral calculus*. 819
W. Woolsey Johnson, *A treatise on the integral calculus, founded on the method of rates*.
    Math. Gaz. 4 (1907), 307–9.

R8. M. Bôcher, *An introduction to the study of integral equations*. 821
    Math. Gaz. 5 (1910), 208–9.

R9. E. Borel, *Leçons sur la théorie de la croissance*. 822
    Math. Gaz. 5 (1910), 210–11.

R10. G. Kowalewski, *Die complexen Veränd1ichen und ihre Funktionen*. 823
    Math. Gaz. 6 (1912), 345–6.

R11. Patricio Peñalver y Bachiller, *Etudio elemental de la prolongación analitica*. 824
    Math. Gaz. 6 (1912), 346.

R12. R. d'Adhemar, *Leçons sur les principes de l'analyse*. 825
    Math. Gaz. 6 (1912), 346–7.

R13. Charles Davison, *Higher algebra*. 826
S. Barnard & J. M. Child, *A new algebra*.
H. S. Carslaw, *An introduction to the infinitesimal calculus*.
W. M. Baker, *The calculus for beginners*.
W. A. Granville, *Elements of the differential and integral calculus*.
    Math. Gaz. 7 (1913), 21–4.

R14. Lorrain S. Hulburt, *Differential and integral calculus*. 829
    Math. Gaz. 7 (1914), 337.

R15. L. Zoretti, *Leçons de mathématiques générales.*   830
       Math. Gaz. 7 (1914), 338.

R16. M. Petrovitch, *Les spectres numériques.*   831
       Math. Gaz. 10 (1920), 77.

R17. J. L. S. Hatton, *The theory of the imaginary in geometry.*   832
       Math. Gaz. 10 (1920), 77–9.

R18. J. Hadamard, *The psychology of invention in the mathematical field.*   834
       Math. Gaz. 30 (1946), 111–15.

(b) *Reviews from* Nature

R19. E. W. Hobson, *The theory of functions of a real variable and the theory of Fourier's series.*   841
     H. S. Carslaw, *Introduction to the theory of Fourier's series and integrals and the mathematical theory of the conduction of heat.*
     J. Edwards, *A treatise on the integral calculus, with applications, examples and problems.*
       Nature, 109 (1922), 435–8.

R20. David Eugene Smith, *A source book in mathematics.*   844
       Nature, 126 (1930), 197–8.

R21. Richard Courant & Herbert Robbins, *What is mathematics?*   845
       Nature, 150 (1942), 673.

(c) *Reviews from* The Times Literary Supplement

R22. Bertrand Russell, *The principles of mathematics.*   851
       Times Lit. Supp. September 18, 1903, p. 263.

R23. T. L. Heath, *The thirteen books of Euclid's Elements.*   855
       Times Lit. Supp. November 11, 1909, pp. 420–1.

R24. A. N. Whitehead & B. Russell, *Principia mathematica.*   859
       Times Lit Supp. September 7, 1911, pp. 321–2.

(d) *Review from* The Cambridge Review

R25. Hastings Berkeley, *Mysticism in modern mathematics.*   864
       The Cambridge Review. May 26, 1910, Literary Supplement, pp. xiii–iv.

6. LIST OF OTHER WRITINGS

Introduction	869
List of books by G. H. Hardy	871

7. OBITUARY NOTICES OF G. H. HARDY AND OTHER WRITINGS CONCERNING HIS LIFE AND WORK

List of writings about Hardy	875
Arrangement of the Volumes	876
Complete list of Hardy's mathematical papers	878
Complete list of obituary notices by Hardy	897

1. INTEGRAL EQUATIONS AND INTEGRAL TRANSFORMS

(a) Fourier Transforms

INTRODUCTION TO THE PAPERS ON FOURIER TRANSFORMS

Hardy's interest in Fourier integrals goes back at least to his paper 1904, 6, which is concerned with a function self-reciprocal in the sine transform and the deduction of other self-reciprocal functions from it. His next paper on Fourier integrals (1911, 9) is more substantial; it is an investigation into sets of sufficient conditions for the truth of Fourier's double integral formula.

After 1911 there was a steady flow of papers on Fourier integrals and related topics. A major contribution to the theory is embodied in the papers 1917, 5; 1918, 8; 1920, 8; 1921, 7. These contain the theory of Mellin and Laplace transforms when one integral converges and the second is summable $(C, 1)$ or summable when interpreted in the sense

$$\lim_{\eta \to 0} \int_0^\infty e^{-\eta t} f(t)\, dt.$$

The last two papers of the group give various forms for Parseval's theorem for Mellin and Laplace transforms. Further applications of Parseval's theorem are given in 1927, 8.

Associated with this group of papers, because the same methods of proof are used, are the two on Stieltjes' moment problem: 1917, 9 and 1918, 6. These papers contribute little to the solution of the moment problem because Hardy dealt with it in the form: *Given an infinite sequence $\{c_n\}$ of real numbers, to find a function $f(x)$ such that*

$$\int_0^\infty x^n f(x)\, dx = c_n \quad (n = 0, 1, \ldots).$$

In the original problem as discussed by Stieltjes (**13**) and in most subsequent work, the problem has been to find a non-decreasing function $\Phi(x)$ such that

$$\int_0^\infty x^n\, d\Phi(x) = c_n \quad (n = 0, 1, 2, \ldots)$$

(or with the integrals over $(-\infty, \infty)$). In this form the problem is richer and much subsequent work has been done on it and on related problems. A full list of references is given in (**1**).

After 1921 the most important advance in the theory of Fourier transforms was due to Titchmarsh. In 1910, Plancheral (**9**) had discovered the complete reciprocity in the class L^2 as a corollary to a more general theory. In 1923 Titchmarsh (**15**) gave a direct proof of the L^2 theory, basing it on the theory of orthogonal functions, and he discovered the L^p ($1 < p \leq 2$) theory (see (**16**)). Hardy felt that a proof of the L^2 theory was required which was independent of orthogonal functions and he gave this in 1924, 8. Other contemporary proofs are given in Bochner (**2**) and F. Riesz (**11**).

When investigating general summability of integrals in 1911, 9, Hardy found it necessary to obtain, in sections I and II, sets of conditions sufficient to prove that

$$\int_0^\infty \phi(x)\,dx \int_0^\infty f(\lambda) \genfrac{}{}{0pt}{}{\cos}{\sin} \lambda x\,d\lambda = \int_0^\infty f(\lambda)\,d\lambda \int_0^\infty \phi(x) \genfrac{}{}{0pt}{}{\cos}{\sin} \lambda x\,dx. \tag{1}$$

If the Fourier (sine or cosine) transform of f is denoted by $T(f)$, (1) can be written

$$\int_0^\infty \phi \cdot T(f) = \int_0^\infty f \cdot T(\phi), \tag{2}$$

which is Parseval's formula. Without, I think, realising its theoretical importance, Hardy returned to (1) in 1912, 9, in which he gave further sets of conditions for its truth, his idea being to facilitate proofs involving the inversion of double integrals. It was not until the paper 1931, 6 that Hardy and Titchmarsh collected sets of conditions sufficient for the truth of (2) and for

$$\int_0^\infty fg = \int_0^\infty T(f) \cdot T(g),$$

but, oddly enough, this paper contains no reference to the earlier papers by Hardy. However, Titchmarsh included one theorem from 1911, 9 in his chapter on Parseval's formula in (**18**) (see Theorem 36).

In 1909, 2 (in Vol. V), Hardy gave the formulae relating Hilbert transforms (or conjugate functions) in the form

$$\pi g(y) = P \int_{-\infty}^\infty \frac{f(x)}{x-y}\,dx, \qquad -\pi f(y) = P \int_{-\infty}^\infty \frac{g(x)}{x-y}\,dx,$$

together with conditions for their validity. He then believed them to be new. Subsequently he discovered that the formulae had been known to Hilbert since at least 1904 and he therefore called them Hilbert transforms when he returned to the subject in 1925. Meanwhile he had used them in 1913, 9 in solving an integral equation and he had obtained some simple formulae resulting from pairs of Hilbert transforms.

In 1925, 9, 10, Hardy tried to develop the L^2 theory of Hilbert transforms *ab initio*, instead of deducing it from the corresponding theory for Fourier transforms. Unfortunately 1925, 9 contains a mistake which was pointed out to Hardy by G. N. Watson. It affects the analysis in both papers. In view of this, and because various other improvements could be made, Hardy rewrote the papers, the definitive version being 1932, 2. The papers 1927, 7 and 1929, 8, which deal with particular points arising from 1925, 9, 10, are unaffected by the mistake. The first of these deals with the singular integral encountered in the 1925, 9, 10 papers and it brings out the relationship between conjugate functions and Fourier series. If f is periodic and integrable and if the 'allied' series of the Fourier series of f is itself the Fourier series of a function g, then f and g are conjugate functions. In the second paper, 1929, 8, Hardy investigated conditions for the truth of the double integral formula

$$-\frac{1}{\pi^2} \int_{-\infty}^{\infty} \frac{dt}{t-x} \int_{-\infty}^{\infty} \frac{f(u)}{u-t} du = f(x)$$

for a particular x. Hardy's last paper on Hilbert transforms (1937, 4) gives the conditions to be satisfied by $f(x)$ if its conjugate $g(x)$ is to be an integrable function. A solution to this problem under very stringent conditions had already been found by Paley and Wiener. Hardy's solution is much more satisfactory.

In 1930, 2, Hardy's interest in self-reciprocal functions came to life again. This paper, written with Titchmarsh, investigates the general forms of functions which are self-reciprocal for the kernels $\cos xy$, $\sin xy$ and $(xy)^{\frac{1}{2}} J_\nu(x,y)$. The classes of such functions are denoted by R_c, R_s, R_ν respectively. In 1932, 1, Hardy and Titchmarsh went on to consider the kernels of linear transformations from R_μ to R_ν and, in particular, from R_c to R_s. These two papers (with corrections and improvements) form the main part of Titchmarsh (**18**), chap. IX. Self-reciprocal functions are, of course, eigenfunctions. If K is a 'Fourier kernel', so that there is a reciprocity

$$g(x) = \int_0^\infty f(y) K(xy)\, dy, \tag{3}$$

$$f(x) = \int_0^\infty g(y) K(xy)\, dy, \tag{4}$$

then the eigenfunction corresponding to an eigenvalue λ is a solution of

$$f(x) = \lambda \int_0^\infty f(y) K(xy)\, dy. \tag{5}$$

It follows from (3)–(5) that

$$f(x) = \lambda g(x), \qquad \lambda f(x) = g(x),$$

and hence that $\lambda^2 = 1$. Thus self- and skew-reciprocal functions are the only eigenfunctions of integral equations with Fourier kernels. This result is given by Hardy and Titchmarsh in 1933, 1.

Hardy returned at intervals to the problem of self-reciprocal functions. Thus in 1937, 1 the formula found for R_c functions in 1930, 2 is deduced from his investigation into some formulae of Ramanujan (see (10) and (11) below). In 1933, 4, when investigating a remark of Wiener that a pair of Fourier transforms, f and g, cannot both be small for large x, he used self-reciprocal functions for one proof of his theorems and a Phragmèn-Lindelöf argument for the other. A number of authors have dealt subsequently with self-reciprocal functions. See (8), where references to other papers will be found.

The general theory of Fourier kernels (see equations (3) and (4)) appeared first in 1933, 1. This very long paper was followed, in the same part of the *Proceedings of the London Mathematical Society*, by an equally long paper (19) by G. N. Watson on the same subject. Functions f and g related by equations (3) and (4) are now called Watson transforms. The characteristic property of a Fourier kernel is that, if

$$K_1(x) = \int_0^x K(t)\,dt, \tag{6}$$

then

$$\int_0^\infty K_1(ax)K_1(bx)\,\frac{dx}{x^2} = \min(a,b). \tag{7}$$

An alternative form for this condition is that, if

$$k(s) = \int_0^\infty K(x)x^{s-1}\,dx, \tag{8}$$

then
$$k(s)k(1-s) = 1. \tag{9}$$

Watson developed the theory of Fourier kernels in the space $L^2(0, \infty)$ and it is this theory which has proved fruitful. Although Watson's treatment was long and involved, short proofs were soon forthcoming (see (3), (5), (10), (17)) and subsequently in (6) the theory was extended to the space $L^p(0, \infty)$ $(1 < p \leq 2)$. What is most surprising is that neither of the papers 1933, 1, (19) contains a reference to an earlier paper by Temple, (14) II, published two years earlier (in the *Proceedings of the London Mathematical Society*) which includes part of Watson's theory as a special case. Watson refers in a footnote to (14) I. In the L^2 theory, condition (7), which implies $x^{-1}K_1(x) \in L^2(0, \infty)$, is the only condition on $K(x)$ which need be assumed.

Hardy and Titchmarsh (in 1933, 1) were concerned with the generalization of the classical theory of Fourier integrals and, in particular, with Fourier's integral formula. For this, much more stringent conditions have to be placed on $K(x)$ or on $k(s)$. Ten examples of Fourier kernels are listed in the paper and the theorems proved are sufficient to cover all of these. The first two theorems deal with summability and the rest with convergence; the last section is based on Watson's theory. The whole paper is a classical *tour de force* and also something of a museum piece. Titchmarsh, in (**18**), reproduced the last section and one other convergence theorem, devoting the rest of his chapter to Watson's theory. Hardy himself only made use of the last section in any subsequent paper and his next paper (1935, 2), which shows that the Watson transform of a Fourier kernel is itself a Fourier kernel, is entirely based on Watson's theory. In 1941, 4, again using Watson's theory, he showed that, if $K(x)$ is a Fourier kernel such that $K_1(x) = O(x^{\frac{1}{2}})$ for all $x > 0$, and if $H(x, y)$ is homogeneous of degree -1 and has derivatives H_x, H_y continuous except at $(0, 0)$, then

$$\int_0^\infty \int_0^\infty K(ax)K(by)H(x, y)\,dx\,dy = H(b, a),$$

the integral being a repeated Cauchy integral.

In 1937, 1, several of Hardy's interests were brought together. He was concerned to establish, under appropriate conditions, the following formulae of Ramanujan:

$$\int_0^\infty \left\{ \sum_0^\infty (-1)^n \frac{\phi(n)}{n!} t^n \right\} \cos xt\, dt = \sum_0^\infty (-1)^n \phi(-2n-1) x^{2n}, \tag{10}$$

$$\int_0^\infty \left\{ \sum_0^\infty (-1)^n \frac{\phi(n)}{n!} t^n \right\} \sin xt\, dt = \sum_1^\infty (-1)^{n-1} \phi(-2n) x^{2n-1}. \tag{11}$$

This is done in Theorem 1 by an argument of the classical type. Later in the paper, taking the case in which

$$\phi(2n+1) = 0, \quad n = 0, 1, 2, \ldots,$$

the formulae are used to show that the function

$$L(x) = \sum_0^\infty \frac{\chi(2n+1)}{n!} (-\tfrac{1}{2}x^2)^n,$$

where $\chi(u) = \chi(1-u)$, belongs to R_c.

Finally, analogues of Ramanujan's formulae are found for Watson transforms.

The following two papers do not fit into any systematic development of the subject.

1926, 11 (written with S. Bochner) deals with certain theorems in N. Wiener's memoir (**20**). Wiener had introduced the idea of a *nearly bounded function* (i.e. a function f such that, for some h, $\int_{\xi}^{\xi+h} f^2\,dx$ is bounded in ξ), and of *convergence almost in the mean* ($F(x, y)$ converges almost in the mean to $f(x)$ in $[a, b]$ as $y \to \infty$ if, for some $k > 0$,

$$\lim_{\eta \to \infty} \int_{\eta}^{\eta+k} dy \int_a^b \{F(x, y) - f(x)\}^2\,dx = 0),$$

and he had used them in the theory of almost periodic functions. By analysing Wiener's arguments, Hardy and Bochner isolated the kernel of his proofs and then gave a simpler proof under wider conditions.

1947, 2 was Hardy's last paper on Fourier transforms. He showed that, when f is $L^2(-\infty, \infty)$,

$$\iint e^{\alpha i x + \beta i y} \frac{f(x) - f(y)}{x - y}\,dx\,dy = -\pi i (\operatorname{sgn} \alpha + \operatorname{sgn} \beta) \int_{-\infty}^{\infty} e^{i(\alpha + \beta)x} f(x)\,dx$$

for all α, β, provided the integral on the left (taken over the whole (x, y) plane) is interpreted in some generalized sense.

Since Hardy's death (and, indeed, before it) the theory of Fourier transforms has developed in two main directions:

1. The development of the theory of functional analysis has provided a new setting for the theory of Fourier transforms. In particular Watson transforms of the class L^2 are now related to unitary transformations of the space $L^2(0, \infty)$. See references (3) and (4).

2. The theory of distributions (12) has introduced new classes of 'functions', and a new theory of Fourier transforms is being built up for these generalized functions which includes the theory of Fourier series as a special case. An elementary account is given in Lighthill (7).

REFERENCES

1. N. I. ARKHEIZER, *The Classical Moment Problems*, Oliver and Boyd, 1965.
2. S. BOCHNER, *Vorlesungen über Fouriersche Integrale*, Leipzig, 1932.
3. ——, Inversion formulae and unitary transformations, *Annals of Math.* 35, 111–15, 1934.
4. —— and K. CHANDRASEKHARAN, *Fourier Transforms*, Princeton, 1949.
5. I. W. BUSBRIDGE, On general transforms with kernels of the Fourier type, *Journal London Math. Soc.* 9, 179–87, 1934.
6. ——, A theory of general transforms for functions of the class $L^p(0, \infty)(1 < p \leq 2)$, *Quart. J. of Math.* (Oxford) 9, 148–60, 1938; 10, 11–12 and 13–27, 1939.
7. M. J. LIGHTHILL, *An Introduction to Fourier Analysis and Generalized Functions*, Cambridge, 1958.
8. B. M. MEHROTRA, A brief history of self-reciprocal functions, *Journal Indian Math. Soc.* (New Series) 1, 209–27, 1934.

9. M. Plancherel, Contribution à l'étude da la représentation d'une fonction arbitraire par des intégrales définies, *Rend. di Palermo* 30, 289–335, 1910.
10. ——, Sur les formules de réciprocité du type de Fourier, *Journal London Math. Soc.* 8, 220–6, 1933.
11. F. Riesz, Sur la formule d'inversion de Fourier, *Acta Szeged* 3, 235–41, 1927.
12. L. Schwartz, *Théorie des Distributions*, Hermann et Cie, 1950.
13. T. Stieltjes, Recherches sur les fractions continues, *Anns. Fac. Soc. Univ. Toulouse* 8, J1–J122; 9, A5–A47, 1894–5.
14. G. Temple, *Proc. London Math. Soc.* (2), 31, I 231–42, II 243–52, 1931.
15. E. C. Titchmarsh, Hankel Transforms, *Proc. Camb. Phil. Soc.* 21, 463–73, 1923.
16. ——, A contribution to the theory of Fourier transforms, *Proc. London Math. Soc.* (2), 23, 279–89, 1923.
17. ——, A proof of a theorem of Watson, *Journal London Math. Soc.* 8, 217–20, 1933.
18. ——, *Theory of Fourier Integrals*, Oxford, 1937.
19. G. N. Watson, General transforms, *Proc. London Math. Soc.* (2), 35, 156–99, 1933.
20. N. Wiener, On the representation of functions by trigonometrical integrals, *Math. Zeitschrift* 24, 575–616, 1945.

I.W.B.

NOTE ON THE FUNCTION $\int_x^\infty e^{\phi(x^2-t^2)}\,dt$.

By G. H. Hardy, Trinity College, Cambridge.

§ 1. IT is well-known* that

$$\int_0^\infty \frac{\sinh \mu x}{\sinh \tfrac{1}{2}x} \cos \alpha x\, dx = \frac{\pi \sin 2\mu\pi}{\cosh 2\alpha\pi + \cos 2\mu\pi},$$

$$\int_0^\infty \frac{\cosh \mu x}{\cosh \tfrac{1}{2}x} \cos \alpha x\, dx = \frac{2\pi \cosh \alpha\pi \cos \mu\pi}{\cosh 2\alpha\pi + \cos 2\mu\pi},$$

if α is real, and $0 \leq \mu < \tfrac{1}{2}$.

If we multiply both sides of each of these equations by $e^{-\beta\alpha^2}$ and integrate from $\alpha = 0$ to $\alpha = \infty$, with the help of the formula

$$\int_0^\infty e^{-\beta\alpha^2} \cos \alpha x\, d\alpha = \tfrac{1}{2}\sqrt{\left(\frac{\pi}{\beta}\right)} e^{-\frac{x^2}{4\beta}},$$

we obtain

$$(1) \quad \int_0^\infty \frac{\sinh \mu x}{\sinh \tfrac{1}{2}x} e^{-\frac{x^2}{4\beta}}\, dx$$

$$= 2\sqrt{(\beta\pi)} \sin 2\mu\pi \int_0^\infty \frac{e^{-\beta\alpha^2}\, d\alpha}{\cosh 2\alpha\pi + \cos 2\mu\pi},$$

and

$$(2) \quad \int_0^\infty \frac{\cosh \mu x}{\cosh \tfrac{1}{2}x} e^{-\frac{x^2}{4\beta}}\, dx$$

$$= 4\sqrt{(\beta\pi)} \cos \mu\pi \int_0^\infty \frac{e^{-\beta\alpha^2} \cosh \alpha\pi\, d\alpha}{\cosh 2\alpha\pi + \cos 2\mu\pi}.$$

In certain particular cases these formulæ take a symmetrical and very elegant form. If e.g. we make $\mu = \tfrac{1}{4}$ in (1), and put

$$x = 4\pi\xi, \quad \alpha = \tfrac{1}{2}\xi,$$

$$\frac{4\pi^2}{\beta} = a, \quad \tfrac{1}{4}\beta = b,$$

---

* See e.g. Harkness and Morley's *Introduction to the Theory of Analytic Functions*, pp. 226–229.

we obtain

$$(3) \quad a^{\frac{1}{4}} \int_0^\infty \frac{e^{-a\xi^2} d\xi}{\cosh \pi\xi} = b^{\frac{1}{4}} \int_0^\infty \frac{e^{-b\xi^2} d\xi}{\cosh \pi\xi},$$

if $ab = \pi^2$. Similarly, by making $\mu = \frac{1}{6}$, so that

$$\frac{\sinh \mu x}{\sinh \frac{1}{2} x} = \frac{\sinh \frac{1}{6} x}{\sinh \frac{1}{2} x} = \frac{1}{1 + 2\cosh \frac{1}{3} x},$$

and putting

$$x = 3\pi\xi, \quad \alpha = \tfrac{1}{2}\xi,$$
$$\frac{9\pi^2}{4\beta} = a, \quad \tfrac{1}{4}\beta = b,$$

we obtain

$$(4) \quad a^{\frac{1}{4}} \int_0^\infty \frac{e^{-a\xi^2} d\xi}{1 + 2\cosh \pi\xi} = b^{\frac{1}{4}} \int_0^\infty \frac{e^{-b\xi^2} d\xi}{1 + 2\cosh \pi\xi},$$

if $ab = (\tfrac{3}{4}\pi)^2$. If we make $\mu = \tfrac{1}{4}$ in (2), and put

$$x = 2\pi\xi, \quad \alpha = \tfrac{1}{2}\xi,$$
$$\frac{\pi^2}{\beta} = a, \quad \tfrac{1}{4}\beta = b,$$

we obtain

$$(5) \quad a^{\frac{1}{4}} \int_0^\infty \frac{\cosh \tfrac{1}{2}\pi\xi}{\cosh \pi\xi} e^{-a\xi^2} d\xi = b^{\frac{1}{4}} \int_0^\infty \frac{\cosh \tfrac{1}{2}\pi\xi}{\cosh \pi\xi} e^{-b\xi^2} d\xi,$$

if $ab = (\tfrac{1}{2}\pi)^2$.

§ 2. These formulæ at once suggest certain formulæ proved by Cauchy, of which

$$\sqrt{a}\{\tfrac{1}{2} + e^{-a^2} + e^{-4a^2} + e^{-9a^2} + \ldots\} = \sqrt{b}\{\tfrac{1}{2} + e^{-b^2} + e^{-4b^2} + e^{-9b^2} + \ldots\}$$

$(ab = \pi)$ is typical. These are, of course, special cases of the formulæ for the linear transformation of the Theta-functions.

We may transform (3), (4), (5) in such a way as to render their analogy with Cauchy's formulæ more apparent. For it is easy to prove that, if

$$I_n(a) = \int_0^\infty \frac{e^{-a\xi^2} \cos 2na\xi}{\cosh \pi\xi} d\xi,$$

$$I_0(a) - (-)^n e^{n^2 a} I_n(a) = \sum_{k=0}^{n-1} (-)^k e^{(k+\tfrac{1}{2})^2 a};$$

so that (3) takes the form

$$a^{\frac{1}{4}} \{e^{\frac{1}{2}a} - e^{\frac{3}{2}a} + e^{\frac{5}{2}a} - \ldots + (-)^{n-1} e^{(n-\frac{1}{2})^2 a} + (-)^n e^{n^2 a} I_n(a)\},$$
$$= b^{\frac{1}{4}} \{e^{\frac{1}{2}b} - e^{\frac{3}{2}b} + e^{\frac{5}{2}b} - \ldots + (-)^{n-1} e^{(n-\frac{1}{2})^2 b} + (-)^n e^{n^2 b} I_n(b)\}.$$

A similar transformation may be applied to (4) and (5). However, it seems that these formulæ cannot be themselves deduced from the theory of elliptic functions, but depend on the theory of the infinite products (such as the Double Gamma-function) out of which the elliptic functions may be built up. On the other hand I have been able to connect them with Cauchy's theory of *reciprocal functions* in a way which seems to me interesting.

§ 3. It follows from Fourier's double-integral theorem, that if

$$\psi(a) = \sqrt{\left(\frac{2}{\pi}\right)} \int_0^\infty \cos ax \, \phi(x) \, dx,$$

then

$$\phi(a) = \sqrt{\left(\frac{2}{\pi}\right)} \int_0^\infty \cos ax \, \psi(x) \, dx;$$

and if

$$\psi(a) = \sqrt{\left(\frac{2}{\pi}\right)} \int_0^\infty \sin ax \, \phi(x) \, dx,$$

then

$$\phi(a) = \sqrt{\left(\frac{2}{\pi}\right)} \int_0^\infty \sin ax \, \psi(x) \, dx,$$

provided $\phi(x)$ satisfies certain conditions. Cauchy calls $\phi(x)$ and $\psi(x)$ *reciprocal functions* of the first or of the second kind. Since

$$e^{-\frac{1}{2}a^2} = \sqrt{\left(\frac{2}{\pi}\right)} \int_0^\infty \cos ax \, e^{-\frac{1}{2}x^2} \, dx,$$

the function $e^{-\frac{1}{2}x^2}$ is its own reciprocal of the first kind. So far as I am aware, Cauchy gives no example of a function which is its own reciprocal of the *second* kind.

This property is, as I shall show, possessed by the function

$$\phi(x) = \int_x^\infty e^{\frac{1}{2}(x^2 - t^2)} \, dt.$$

§4. Before proving this it is essential to investigate the behaviour of $\phi(x)$ for very large values of x.

Now

$$\frac{1}{\sqrt{2}}\phi(x\sqrt{2}) = e^{x^2}\int_x^\infty e^{-t^2}dt = -\frac{1}{2x}\left\{xe^{x^2}\int_x^\infty \frac{1}{t}\frac{d}{dt}(e^{-t^2})dt\right\}$$

$$= \frac{1}{2x} - \tfrac{1}{2}e^{x^2}\int_x^\infty \frac{e^{-t^2}}{t^2}dt.$$

And as

$$e^{x^2}\int_x^\infty \frac{e^{-t^2}}{t^2}dt = -\tfrac{1}{2}e^{x^2}\int_x^\infty \frac{1}{t^3}\frac{d}{dt}(e^{-t^2})dt = \frac{1}{2x^3} - \tfrac{3}{2}e^{x^2}\int_x^\infty \frac{e^{-t^2}}{t^4}dt$$

and

$$e^{x^2}\int_0^\infty \frac{e^{-t^2}}{t^4}dt < \int_x^\infty \frac{dt}{t^4} = \frac{1}{3x^3}$$

it follows that

$$\frac{1}{\sqrt{2}}\phi(x\sqrt{2}) = \frac{1}{2x} + \frac{\rho}{x^3},$$

where ρ is a quantity whose modulus is less than $\tfrac{1}{2}$.

This process can evidently be repeated; in fact the asymtotic expansion of

$$e^{x^2}\int_x^\infty e^{-t^2}dt$$

is*

$$\frac{1}{2x} - \frac{1}{2^2 x^3} + \frac{1.3}{2^3 x^5} - \frac{1.3.5}{2^4 x^7} + \ldots.$$

It follows that

$$\int_0^\infty \sin\alpha x\, \phi(x)\, dx$$

is convergent. I shall now prove that the integral is equal to

$$\sqrt{\left(\frac{\pi}{2}\right)}\phi(\alpha).$$

---

* E. T. Whittaker, *Modern Analysis*, p. 166.

§ 5. If
$$\psi(\alpha) = \int_0^\infty \sin \alpha x \, \phi(x) \, dx,$$
$$e^{-\frac{1}{2}\alpha^2} \psi(\alpha) = \int_0^\infty e^{-\frac{1}{2}\alpha^2} \sin \alpha x \, \phi(x) \, dx.$$

We cannot differentiate this equation by the ordinary rule, since
$$\int_0^\infty \frac{\partial}{\partial \alpha} \left(e^{-\frac{1}{2}\alpha^2} \sin \alpha x \right) \phi(x) \, dx$$
is divergent. But we can obtain the value of the integral on the right by a procedure devised by M. de la Vallée-Poussin.*

If $\alpha > \beta > 0$,
$$e^{-\frac{1}{2}\alpha^2} \sin \alpha x - e^{-\frac{1}{2}\beta^2} \sin \beta x = \int_\beta^\alpha \frac{\partial}{\partial \alpha} \left(e^{-\frac{1}{2}\alpha^2} \sin \alpha x \right) d\alpha$$
$$= \int_\beta^\alpha e^{-\frac{1}{2}\alpha^2} (x \cos \alpha x - \alpha \sin \alpha x) \, d\alpha,$$

and so
$$e^{-\frac{1}{2}\alpha^2} \psi(\alpha) - e^{-\frac{1}{2}\beta^2} \psi(\beta)$$
$$= \lim_{X=\infty} \int_0^X \phi(x) \, dx \int_\beta^\alpha e^{-\frac{1}{2}\alpha^2} (x \cos \alpha x - \alpha \sin \alpha x) \, d\alpha$$
$$= \lim_{X=\infty} \int_\beta^\alpha e^{-\frac{1}{2}\alpha^2} d\alpha \int_0^X (x \cos \alpha x - \alpha \sin \alpha x) \, \phi(x) \, dx.$$

Now
$$\int_0^X (x \cos \alpha x - \alpha \sin \alpha x) \, e^{\frac{1}{2}x^2} dx \int_x^\infty e^{-\frac{1}{2}t^2} dt$$
$$= \int_0^X \cos \alpha x \, \frac{d}{dx} \left(e^{\frac{1}{2}x^2} \right) dx \int_x^\infty e^{-\frac{1}{2}t^2} dt$$
$$- \int_0^X \alpha \sin \alpha x \, e^{\frac{1}{2}x^2} dx \int_x^\infty e^{-\frac{1}{2}t^2} dt$$
$$= \left[\cos \alpha x \, e^{\frac{1}{2}x^2} \int_x^\infty e^{-\frac{1}{2}t^2} dt \right]_0^X + \int_0^X \cos \alpha x \, dx$$

---

* See Stolz, *Grundzüge*, III., p. 28,

(on integrating by parts)
$$= \frac{\sin \alpha X}{\alpha} + e^{\frac{1}{2}X^2} \cos \alpha X \int_X^\infty e^{-\frac{1}{2}t^2} dt - \sqrt{(\tfrac{1}{2}\pi)}.$$

Now
$$\left| \int_\beta^\alpha e^{-\frac{1}{2}\alpha^2} d\alpha \left[e^{\frac{1}{2}X^2} \cos \alpha X \int_X^\infty e^{-\frac{1}{2}t^2} dt \right] \right|$$
$$< (\alpha - \beta) e^{\frac{1}{2}X^2} \int_X^\infty e^{-\frac{1}{2}t^2} dt,$$

the limit of which for $X = \infty$ is 0; and[*]
$$\lim_{X=\infty} \int_\beta^\alpha \frac{\sin \alpha X}{\alpha} e^{-\frac{1}{2}\alpha^2} d\alpha = 0.$$

Hence
$$e^{-\frac{1}{2}\alpha^2} \psi(\alpha) - e^{-\frac{1}{2}\beta^2} \psi(\beta) = -\sqrt{(\tfrac{1}{2}\pi)} \int_\beta^\alpha e^{-\frac{1}{2}\alpha^2} d\alpha.$$

Now
$$e^{-\frac{1}{2}\beta^2} \psi(\beta) = \int_0^\infty e^{-\frac{1}{2}\beta^2} \sin \beta x \, \phi(x) \, dx,$$

and it follows from the fact that[†] for large values of x,
$$\phi(x) = \frac{1}{x} + \cdots,$$

that
$$\lim_{\beta=0} e^{-\frac{1}{2}\beta^2} \psi(\beta) = \frac{\pi}{2}.$$

Hence
$$e^{-\frac{1}{2}\alpha^2} \psi(\alpha) = \frac{\pi}{2} - \sqrt{(\tfrac{1}{2}\pi)} \int_0^\alpha e^{-\frac{1}{2}\alpha^2} d\alpha = \sqrt{(\tfrac{1}{2}\pi)} \int_\alpha^\infty e^{-\frac{1}{2}\alpha^2} d\alpha,$$

i.e.
$$\psi(\alpha) = \sqrt{\left(\frac{\pi}{2}\right)} e^{\frac{1}{2}\alpha^2} \int_\alpha^\infty e^{-\frac{1}{2}t^2} dt = \sqrt{\left(\frac{\pi}{2}\right)} \phi(\alpha).$$

Hence
$$\phi(x) = e^{\frac{1}{2}x^2} \int_x^\infty e^{-\frac{1}{2}t^2} dt$$

is its own reciprocal of the second kind.

---

[*] Kronecker, *Vorlesungen über Integrale*, p. 69.
[†] *Quarterly Journal*, Vol. XXXIV., p. 45.

§ 6. Cauchy proved that if $\phi(x)$, $\psi(x)$ were reciprocal functions of the first kind, and α, β were real quantities such that $\alpha\beta = \pi$, then

$$\sqrt{\alpha}\{\tfrac{1}{2}\phi(0) + \phi(\alpha) + \phi(2\alpha) + \ldots\}$$
$$= \sqrt{\beta}\{\tfrac{1}{2}\psi(0) + \psi(\beta) + \psi(2\beta) + \ldots\},$$

of course under certain conditions. The integral formulæ, which I gave in § 1, can be deduced from a similar relation which holds between two reciprocal functions of the second kind.

Suppose that $\phi(x)$ is a function of the complex variable $x = \xi + i\eta$, analytic within and on the boundary of the rectangle formed by the lines

$$\xi = 0, \; \xi = N\pi, \; \eta = 0, \; \eta = k > 0,$$

and that $\phi(\xi + i\eta)$ tends steadily to the limit 0 when ξ tends to ∞, for any value of η in $(0, k)$. And let us integrate

$$\int \frac{\phi(\lambda x)}{\cos \pi x} dx \quad (\lambda > 0)$$

round this rectangle, avoiding the poles by small semicircles in the usual way. In the limit, when the radii of the semicircles are indefinitely diminished, and N is indefinitely increased, we obtain the equation

$$(6) \quad P\int_0^\infty \frac{\phi(\lambda\xi)\, d\xi}{\cos \pi \xi} + i \sum_0^\infty (-)^n \phi\{(n+\tfrac{1}{2})\lambda\}$$
$$- \int_0^\infty \frac{\phi\{\lambda(\xi+ik)\}}{\cos \pi(\xi+ik)} d\xi - i\int_0^k \frac{\phi(\lambda i\eta)}{\cosh \pi\eta} d\eta = 0.$$

When k is indefinitely diminished the last integral tends to 0.

I shall suppose now that the first two derivates of $\phi(x)$ are continuous throughout the rectangle and vanish for $\xi = \infty$, and that the integrals

$$\int_0^\infty \phi\{\lambda(\xi+ik)\}\, e^{(2n+1)\pi\xi i}\, d\xi,$$

$$\int_0^\infty \phi'\{\lambda(\xi+ik)\}\, e^{(2n+1)\pi\xi i}\, d\xi,$$

$$\int_0^\infty \phi''\{\lambda(\xi+ik)\}\, e^{(2n+1)\pi\xi i}\, d\xi$$

are uniformly convergent for $n = 0, 1, 2, \ldots$, and all values of k in a small interval $(0, k')$. Then

$$\int_0^\infty \frac{\phi\{\lambda(\xi + ik)\}}{\cos \pi(\xi + ik)} d\xi$$

$$= 2 \sum_0^\infty (-)^n e^{-(2n+1)\pi k} \int_0^\infty \phi\{\lambda(\xi + ik)\} e^{(2n+1)\pi \xi i} d\xi.$$

Now

$$\int_0^\infty \phi\{\lambda(\xi + ik)\} e^{(2n+1)\pi \xi i} d\xi$$

$$= \frac{1}{(2n+1)\pi i} \left\{ -\phi(\lambda ik) - \lambda \int_0^\infty \phi'\{\lambda(\xi + ik)\} e^{(2n+1)\pi \xi i} d\xi \right\}$$

$$= -\frac{\phi(\lambda ik)}{(2n+1)\pi i}$$

$$+ \frac{\lambda}{(2n+1)^2 \pi^2} \left\{ -\phi'(\lambda ik) - \lambda \int_0^\infty \phi''\{\lambda(\xi + ik)\} e^{(2n+1)\pi \xi i} d\xi \right\}.$$

But

$$\sum \frac{(-)^n}{2n+1} e^{-(2n+1)\pi k}$$

and

$$\sum \frac{(-)^n}{(2n+1)^2} e^{-(2n+1)\pi k}$$

are continuous for $k = 0$.

And

$$\sum \frac{(-)^n}{(2n+1)^2} e^{-(2n+1)\pi k} \int_0^\infty \phi''\{\lambda(\xi + ik)\} e^{(2n+1)\pi \xi i} d\xi$$

is uniformly convergent in $(0, k')$, and therefore continuous for $k = 0$. Hence

$$\lim_{k=0} \int_0^\infty \frac{\phi\{\lambda(\xi + ik)\}}{\cos \pi(\xi + ik)} d\xi = 2 \sum_0^\infty (-)^n \int_0^\infty \phi(\lambda \xi) e^{(2n+1)\pi \xi i} d\xi.$$

§ 7. Equating the real and imaginary parts in (6), and making $k = 0$, we obtain

(7) $$P \int_0^\infty \frac{\phi(\lambda \xi) d\xi}{\cos \pi \xi} = 2 \sum_0^\infty (-)^n \int_0^\infty \phi(\lambda \xi) \cos(2n+1) \pi \xi d\xi,$$

and

$$(8) \quad \sum_0^\infty (-)^n \phi\{(n+\tfrac{1}{2})\lambda\} = 2\sum_0^\infty (-)^n \int_0^\infty \phi(\lambda\xi)\sin(2n+1)\pi\xi\, d\xi.$$

These formulæ may also be obtained by other* methods.

§ 8. If now $\psi(x)$ is the reciprocal of the second kind of $\phi(x)$,

$$\psi(a) = \sqrt{\frac{2}{\pi}} \int_0^\infty \sin ax\, \phi(x)\, dx,$$

and $\int_0^\infty \phi(\lambda\xi)\sin(2n+1)\pi\xi\, d\xi = \frac{1}{\lambda}\sqrt{\left(\frac{\pi}{2}\right)}\psi\left\{(2n+1)\frac{\pi}{\lambda}\right\}.$

And if $\lambda\mu = 2\pi$,

$$(9) \quad \sqrt{\lambda}\sum_0^\infty (-)^n \phi\{(n+\tfrac{1}{2})\lambda\} = \sqrt{\mu}\sum_0^\infty (-)^n \psi\{(n+\tfrac{1}{2})\mu\}.$$

If, for instance,

$$\phi(x) = \frac{1}{x^s} \quad (0 < s < 1),$$

$$\psi(a) = \sqrt{\frac{2}{\pi}} \int_0^\infty \frac{\sin ax}{x^s} dx$$

$$= \sqrt{\frac{2}{\pi}} \frac{\Gamma(1-s)\cos\tfrac{1}{2}s\pi}{a^{1-s}},$$

and the formula gives, after a little reduction,

$$(10) \quad \sum_0^\infty \frac{(-)^n}{(2n+1)^s} = \left(\frac{2}{\pi}\right)^{1-s} \cos\tfrac{1}{2}s\pi\, \Gamma(1-s) \sum_0^\infty \frac{(-)^n}{(2n+1)^{1-s}}.$$

This formula was originally found by Schlömilch† from the theory of Fourier series, and has since been generalised by various writers, notably Lipschitz.‡ Schlömilch also showed how to deduce the general formula (9).

It is to be observed that in the particular case in which $\phi(x) = x^{-s}$, the argument of § 6 requires to be modified slightly, owing to the fact that $\phi(x)$ becomes infinite at $x = 0$.

---

* See a paper 'On differentiation and integration of divergent series,' *Trans. Camb. Phil. Soc.*, XIX., p. 297.
† *Zeitschrift für Math. und Phys.*, iii., p. 130.
‡ *Crelle*, cv., p. 127.

§9. Now

$$(11) \quad \int_0^\infty \frac{e^{-k^2 x^2}}{1+x^2}\,dx = \sqrt{\pi} \cdot e^{k^2} \int_k^\infty e^{-t^2}\,dt.$$

For if we write r for k^2, and differentiate with respect to r, we find that the integral is a solution of

$$\frac{du}{dr} - u = -\tfrac{1}{2}\sqrt{\frac{\pi}{r}},$$

the solution of which is

$$u = Ce^r - \tfrac{1}{2}\sqrt{\pi}\,\frac{1}{D-1}\left(\frac{1}{\sqrt{r}}\right),$$

$\left(\text{where } D \equiv \dfrac{d}{dr}\right)$ or

$$= \tfrac{1}{2}\sqrt{\pi}\,e^r \int_r^K \frac{e^{-r}}{\sqrt{r}}\,dr,$$

where K is an arbitrary constant. To determine K we notice that for $r=0$, $u = \tfrac{1}{2}\pi$, so that $K = \infty$.*

But

$$\int_0^\infty \frac{e^{-a^2 x^2}}{\cosh \pi x}\,dx = \frac{4}{\pi}\int_0^\infty \sum_0^\infty \frac{(-)^n (2n+1)}{(2n+1)^2 + 4x^2}\,e^{-a^2 x^2}\,dx$$

$$= \frac{2}{\pi}\sum_0^\infty (-)^n \int_0^\infty \frac{e^{-(n+\frac{1}{2})^2 a^2 u^2}}{1+u^2}\,du$$

$$= \frac{2}{\pi}\sum_0^\infty (-)^n F(n+\tfrac{1}{2})\,a,$$

say. And

$$F(x) = \sqrt{\pi}\,e^{x^2}\int_x^\infty e^{-t^2}\,dt,$$

so that $F\left(\dfrac{x}{\sqrt{2}}\right)$ is its own reciprocal of the second kind. Hence

$$\sqrt{\alpha}\,\Sigma\,(-)^n F\!\left\{(n+\tfrac{1}{2})\frac{\alpha}{\sqrt{2}}\right\} = \sqrt{\beta}\,\Sigma\,(-)^n F\!\left\{(n+\tfrac{1}{2})\frac{\beta}{\sqrt{2}}\right\},$$

---

* See G. F. Meyer's edition of Dirichlet's lectures.

if $\alpha\beta = 2\pi$, or

$$\sqrt{\alpha} \, \Sigma \, (-)^n F\{(n+\tfrac{1}{2})\alpha\} = \sqrt{\beta} \, \Sigma \, (-)^n F\{(n+\tfrac{1}{2})\beta\},$$

if $\alpha\beta = \pi$. Thus, finally,

$$\sqrt{\alpha} \int_0^\infty \frac{e^{-\alpha^2 x^2}}{\cosh \pi x} dx = \sqrt{\beta} \int_0^\infty \frac{e^{-\beta^2 x^2}}{\cosh \pi x} dx,$$

if $\alpha\beta = \pi$.

§ 10. We can also prove this without the use of the equation (9). For

$$\int_0^\infty \frac{e^{-\alpha^2 \xi^2}}{\cosh \pi \xi} d\xi = 2 \sum_0^\infty (-)^n \int_0^\infty e^{-\alpha^2 \xi^2 - (2n+1)\pi \xi} d\xi,$$

and since

$$\int_0^\infty e^{-\alpha^2 \xi^2 - 2\lambda \xi} d\xi = e^{\left(\frac{\lambda}{\alpha}\right)^2} \int_{\frac{\lambda}{\alpha^2}}^\infty e^{-\alpha^2 t^2} dt$$

$$= \frac{1}{\alpha} e^{\left(\frac{\lambda}{\alpha}\right)^2} \int_{\frac{\lambda}{\alpha}}^\infty e^{-u^2} du = \frac{1}{\alpha \sqrt{\pi}} F\left(\frac{\lambda}{\alpha}\right),$$

$$\int_0^\infty \frac{e^{-\alpha^2 \xi^2}}{\cosh \pi \xi} d\xi = \frac{2}{\alpha \sqrt{\pi}} \sum_0^\infty (-)^n F\{(n+\tfrac{1}{2})\beta\},$$

where $\alpha\beta = \pi$; and therefore

$$\sqrt{\alpha} \int_0^\infty \frac{e^{-\alpha^2 \xi^2}}{\cosh \pi \xi} d\xi = \sqrt{\beta} \int_0^\infty \frac{e^{-\beta^2 \xi^2}}{\cosh \pi \xi} d\xi.$$

§ 11. In the same way it may be shown that (4) of § 1 is equivalent to

$$\sqrt{\alpha}\{F(\tfrac{2}{3}\alpha) - F(\tfrac{4}{3}\alpha) + F(\tfrac{8}{3}\alpha) - F(\tfrac{10}{3}\alpha) + \ldots\}$$
$$= \sqrt{\beta}\{F(\tfrac{2}{3}\beta) - F(\tfrac{4}{3}\beta) + F(\tfrac{8}{3}\beta) - F(\tfrac{10}{3}\beta) + \ldots\}$$

($\alpha\beta = \tfrac{3}{4}\pi$), a formula which may be found as in § 6 by means of the contour integral

$$\int \frac{\phi(\alpha x) \sin \pi x}{\sin 3\pi x} dx.$$

And (5) of § 1 is equivalent to

$$\sqrt{\alpha}\{F(\tfrac{1}{2}\alpha) + F(\tfrac{3}{2}\alpha) - F(\tfrac{5}{2}\alpha) - F(\tfrac{7}{2}\alpha) + \ldots\}$$
$$= \sqrt{\beta}\{F(\tfrac{1}{2}\beta) + F(\tfrac{3}{2}\beta) - F(\tfrac{5}{2}\beta) - F(\tfrac{7}{2}\beta) + \ldots\}$$

($\alpha\beta = \tfrac{1}{2}\pi$), which may be found by means of the integral

$$\int \frac{\phi(\alpha x)}{\cos \pi x} e^{-\tfrac{1}{2}\pi i x} dx.$$

§ 12. The formulæ (3), (4), and (5) of § 1, which were proved on the supposition that a was real, are easily extended to other values of a. For

$$a^{\tfrac{1}{2}} \int_0^\infty \frac{e^{-a\xi^2} d\xi}{\cosh \pi \xi} - b^{\tfrac{1}{2}} \int_0^\infty \frac{e^{-b\xi^2} d\xi}{\cosh \pi \xi}$$

is an analytic function of a throughout the region for which the real part of a is positive.

If we write $ae^{i\phi}$ ($a > 0$) for a, $be^{-i\phi}$ for b, where $ab = \pi^2$, and make ϕ increase from 0 to $\tfrac{1}{2}\pi$, we obtain ultimately

$$(12) \quad a^{\tfrac{1}{2}} e^{\tfrac{1}{8}\pi i} \int_0^\infty \frac{e^{-ai\xi^2} d\xi}{\cosh \pi \xi} - b^{\tfrac{1}{2}} e^{-\tfrac{1}{8}\pi i} \int_0^\infty \frac{e^{bi\xi^2}}{\cosh \pi \xi} d\xi = 0$$

($a, b > 0$, $ab = \pi^2$).

In the case in which a is the reciprocal of an odd integer n, the value of the first integral has been found by Kronecker.[*]

Integrate

$$\int \frac{e^{\tfrac{\pi i x^2}{n}} dx}{\cos \pi x}$$

round the contour bounded by

$$\xi = 0, \ \xi = \tfrac{1}{2}n, \ \eta = Y, \ \eta = -Y,$$

avoiding $x = \tfrac{1}{2}n$ by a small semicircle. The contribution of the sides $\eta = \pm Y$ vanishes when Y is made infinite. For consider

$$\int_0^{\tfrac{1}{2}n} \exp\left\{\frac{\pi i}{n}(\xi \pm iY)^2\right\} \sec \pi (\xi \pm iY) \, d\xi.$$

---

[*] *Crelle*, cv., p. 345.

The modulus of the subject of integration is

$$\frac{\sqrt{2}\exp\left(\mp\dfrac{2\pi\xi Y}{n}\right)}{\sqrt{(\cosh 2\pi Y + \cos 2\pi\xi)}},$$

which is less than a constant H for all values of ξ in $(0, \tfrac{1}{2}n)$ and all large values of Y. Hence we can choose ε so that

$$\left|\int_{\frac{1}{2}n-\varepsilon}^{\frac{1}{2}n}\right| < \sigma$$

for all large values of Y, however small be σ. And if $0 \leq \xi \leq \tfrac{1}{2}n - \varepsilon$ the modulus of the subject of integration is less than

$$K\exp\left(-\dfrac{2\varepsilon\pi Y}{n}\right),$$

where K is a constant; and this tends uniformly to 0 for $Y = \infty$.

The contribution of $\xi = 0$ is

$$-\int_{-\infty}^{\infty}\frac{e^{-\frac{\pi i\eta^2}{n}}}{\cosh \pi\eta}\,id\eta = -2i\int_{0}^{\infty}\frac{e^{-\frac{\pi i\eta^2}{n}}}{\cosh \pi\eta}\,d\eta,$$

when Y is made infinite. That of $\xi = \tfrac{1}{2}n$ is

$$P\int_{-\infty}^{\infty}\frac{e^{\frac{\pi i}{n}\left(\frac{n}{2}+i\eta\right)^2}}{\cos \pi\left(\frac{n}{2}+i\eta\right)}\,id\eta$$

$$= (-)^{\frac{n+1}{2}}e^{\frac{n\pi i}{4}}P\int_{-\infty}^{\infty}e^{-\frac{\pi i\eta^2}{n}}\frac{e^{-\pi\eta}}{\sinh \pi\eta}\,d\eta$$

$$= (-)^{\frac{n-1}{2}}2e^{\frac{n\pi i}{4}}\int_{0}^{\infty}e^{-\frac{\pi i\eta^2}{n}}\,d\eta$$

$$= (-)^{\frac{n-1}{2}}e^{\frac{(n-1)\pi i}{4}}\sqrt{n}.$$

That of the small semicircle round $x = \tfrac{1}{2}n$ is

$$-(-)^{\frac{n+1}{2}}ie^{\frac{n\pi i}{4}}.$$

Hence

$$(13) \quad \int_0^\infty \frac{e^{-\frac{\pi i \eta^2}{n}}}{\cosh \pi \eta} d\eta = \frac{1}{2i}\left\{(-)^{\frac{n-1}{2}} e^{\frac{(n-1)\pi i}{4}} \sqrt{n} - 2i \Sigma (-)^{k+1} \varepsilon_k e^{(k+\frac{1}{2})^2 \frac{\pi i}{n}}\right\},$$

where the summation extends from $k = 0$ to $k = \frac{1}{2}(n-1)$, and $\varepsilon_k = \frac{1}{2}$ if $k = \frac{1}{2}(n-1)$, $= 1$ otherwise.

Thus

$$(14) \quad \int_0^\infty \frac{\cos \frac{\pi \eta^2}{n}}{\cosh \pi \eta} d\eta = \frac{1}{2}(-)^{\frac{n-1}{2}} \sin \frac{(n-1)\pi}{4} \sqrt{n} - \Sigma (-)^{k+1} \varepsilon_k \cos\left\{(k+\frac{1}{2})^2 \frac{\pi}{n}\right\},$$

$$(15) \quad \int_0^\infty \frac{\sin \frac{\pi \eta^2}{n}}{\cosh \pi \eta} d\eta = \frac{1}{2}(-)^{\frac{n-1}{2}} \cos \frac{(n-1)\pi}{4} \sqrt{n} + \Sigma (-)^{k+1} \varepsilon_k \sin\left\{(k+\frac{1}{2})^2 \frac{\pi}{n}\right\}.$$

If $n = 1$ we find

$$(16) \quad \int_0^\infty \frac{\cos \pi \eta^2}{\cosh \pi \eta} d\eta = \frac{1}{2\sqrt{2}}, \quad \int_0^\infty \frac{\sin \pi \eta^2}{\cosh \pi \eta} d\eta = \frac{1}{2}\left(1 - \frac{1}{\sqrt{2}}\right),$$

$$(17) \quad \int_0^\infty \frac{e^{\pm \pi i \eta^2}}{\cosh \pi \eta} d\eta = \frac{1 \pm i(\sqrt{2}-1)}{2\sqrt{2}};$$

and as

$$\frac{1+i(\sqrt{2}-1)}{1-i(\sqrt{2}-1)} = \frac{-2+2\sqrt{2}+2i(\sqrt{2}-1)}{4-2\sqrt{2}} = \frac{1+i}{\sqrt{2}} = e^{\frac{1}{4}\pi i},$$

this affords a verification of equation (12).

It is evident that from (12) and (13) we can deduce the values of

$$\int_0^\infty \frac{\cos n\pi \eta^2}{\cosh \pi \eta} d\eta, \quad \int_0^\infty \frac{\sin n\pi \eta^2}{\cosh \pi \eta} d\eta,$$

for all odd integral values of n.

§ 13. If n is even and $= 2m$, the argument is the same except that $x = \dfrac{n}{2}$ is not a pole, and the contribution of $\xi = \dfrac{n}{2}$ is

$$\int_{-\infty}^{\infty} \frac{e^{\frac{\pi i}{n}\left(\frac{n}{2}+i\eta\right)^2}}{\cos\pi\left(\frac{n}{2}+i\eta\right)} i\, d\eta$$

$$= (-)^m \, ie^{\frac{n\pi i}{4}} \int_{-\infty}^{\infty} e^{-\frac{\pi i\eta^2}{n}} \frac{e^{-\pi\eta}}{\cosh\pi\eta} d\eta$$

$$= (-)^m \, 2ie^{\frac{n\pi i}{4}} \int_{0}^{\infty} e^{-\frac{\pi i\eta^2}{n}} d\eta$$

$$= (-)^m \, ie^{\frac{(n-1)\pi i}{4}} \sqrt{n}.$$

Therefore

(18) $\displaystyle\int_0^\infty \frac{e^{-\frac{\pi i\eta^2}{n}}}{\cosh\pi\eta} d\eta = \tfrac{1}{2}(-)^{\frac{1}{2}n} e^{\frac{(n-1)\pi i}{4}} \sqrt{n} - \sum_{k=0}^{\frac{1}{2}n-1} (-)^{k+1} e^{(k+\frac{1}{2})^2\frac{\pi i}{n}},$

(19) $\displaystyle\int_0^\infty \frac{\cos\frac{\pi\eta^2}{n}}{\cosh\pi\eta} d\eta = \tfrac{1}{2}(-)^{\frac{1}{2}n} \cos\frac{(n-1)\pi}{4} \sqrt{n}$
$\qquad\qquad - \displaystyle\sum_{k=0}^{\frac{1}{2}n-1} (-)^{k+1} \cos\left\{(k+\tfrac{1}{2})^2 \frac{\pi}{n}\right\},$

(20) $\displaystyle\int_0^\infty \frac{\sin\frac{\pi\eta^2}{n}}{\cosh\pi\eta} d\eta = -\tfrac{1}{2}(-)^{\frac{1}{2}n} \sin\frac{(n-1)\pi}{4} \sqrt{n}$
$\qquad\qquad + \displaystyle\sum_{k=0}^{\frac{1}{2}n-1} (-)^{k+1} \sin\left\{(k+\tfrac{1}{2})^2 \frac{\pi}{n}\right\}.$

Thus we can find the values of

$$\int_0^\infty \frac{\cos\frac{\pi\eta^2}{n}}{\cosh\pi\eta} d\eta, \quad \int_0^\infty \frac{\sin\frac{\pi\eta^2}{n}}{\cosh\pi\eta} d\eta,$$

$$\int_0^\infty \frac{\cos n\pi\eta^2}{\cosh\pi\eta} d\eta, \quad \int_0^\infty \frac{\sin n\pi\eta^2}{\cosh\pi\eta} d\eta,$$

for all integral values of n.

Mr. Hardy, Note on an integral function.

CORRECTIONS

p. 193, §1, (2), *p.* 195, §3, *line* 7, *p.* 199, *line* 3 *up*, and *p.* 202, *line* 2. The upper limit ∞ of an integral is missing.

p. 202, *line* 5 *up*. $F(n+\tfrac{1}{2})\alpha$ should be $F\{(n+\tfrac{1}{2})\alpha\}$.

XVI. *Fourier's Double Integral and the Theory of Divergent Integrals.*

By G. H. Hardy, M.A., F.R.S.

[*Received* Sept. 1, 1910. *Read* Oct. 31, 1910.]

I. *Introduction.*

1. Fourier's "double-integral theorem" is expressed by the equation

$$\int_0^\infty dx \int_{-\infty}^\infty f(\lambda) \cos x (\lambda - \xi) \, d\lambda = \tfrac{1}{2}\pi \{ f(\xi + 0) + f(\xi - 0) \} \quad \ldots\ldots\ldots\ldots(1).$$

Here ξ is a constant which may, without loss of generality, be taken to be zero, as appears at once if we make the substitution $\lambda - \xi = \mu$, and then write λ for μ and $f(\lambda)$ for $f(\mu + \xi)$.

It has long been known that the following conditions (which I shall call the "classical" conditions) are sufficient for the validity of the equation (1):

(i) $f(\lambda)$ *is integrable and absolutely integrable*[*] *in any finite interval;*

(ii) $f(\xi + 0), f(\xi - 0)$ *exist;*

(iii) $f(\lambda)$ *is monotonic on each side of* $\lambda = \xi$ (*more generally, of limited total fluctuation in some interval including* $\lambda = \xi$);

(iv) *the integral* $\displaystyle\int_{-\infty}^\infty |f(\lambda)| \, d\lambda$

is convergent.

For condition (iii) we may substitute

(iii a) *the integrals*

$$\int_\xi^{\xi+\delta} \left| \frac{f(\lambda) - f(\xi + 0)}{\lambda - \xi} \right| d\lambda, \quad \int_{\xi-\delta}^\xi \left| \frac{f(\xi - 0) - f(\lambda)}{\xi - \lambda} \right| d\lambda$$

are convergent[†].

2. These results are deduced from a consideration of Dirichlet's integral

$$\int_{-\infty}^\infty f(\lambda) \frac{\sin w\lambda}{\lambda} d\lambda \quad \ldots\ldots\ldots\ldots\ldots\ldots\ldots\ldots(2),$$

---

[*] If a function is *limited* in an interval, its integrability throughout that interval involves its absolute integrability, whereas the converse is not true. Thus the function $f(\lambda)$ which is equal to 1 or -1, according as λ is rational or irrational, is absolutely integrable but not integrable in any finite interval. On the other hand many functions are integrable, in intervals in which they are *unlimited*, without being absolutely integrable.

[†] See Hobson, *Theory of functions of a real variable*, pp. 758 *et seq.* Prof. Hobson states all these results for the more general case in which the integrals concerned exist only as Lebesgue integrals. In this paper I do not consider this generalisation, which has no particular relevance to the theorems that I have in view.

which is known to have the limit
$$\tfrac{1}{2}\pi \{f(+0) + f(-0)\},$$
as $w \to \infty$, if $f(\lambda)$ is subject to conditions similar to those laid down in § 1. In a paper published in 1908* Prof. Hobson has shown that this last result is still correct if the condition (iv) is replaced by the less stringent condition that
$$\int_0^\infty \left|\frac{f(\lambda)}{\lambda}\right| d\lambda, \quad \int_{-\infty}^0 \left|\frac{f(\lambda)}{\lambda}\right| d\lambda$$
are convergent.

It is, however, not permissible to replace condition (iv) by this condition, in enunciating conditions sufficient for the validity of Fourier's result. Fourier's theorem is in fact deduced from Dirichlet's by the transformation
$$\int_0^\infty dx \int_{-\infty}^\infty f(\lambda) \cos \lambda x \, d\lambda = \lim_{w \to \infty} \int_0^w dx \int_{-\infty}^\infty f(\lambda) \cos \lambda x \, d\lambda$$
$$= \lim_{w \to \infty} \int_{-\infty}^\infty f(\lambda) \, d\lambda \int_0^w \cos \lambda x \, dx$$
$$= \lim_{w \to \infty} \int_{-\infty}^\infty f(\lambda) \frac{\sin w\lambda}{\lambda} d\lambda,$$

and Prof. Hobson's condition is not sufficient to justify the inversion of the order of integration, or even to ensure the convergence of
$$\int_{-\infty}^\infty f(\lambda) \cos \lambda x \, d\lambda.$$
This is easily shown by an example.

Let
$$a_1 < b_1 < a_2 < b_2 < a_3 < \ldots\ldots$$
be a sequence of numbers increasing above all limit, and let
$$f(\lambda) = 1 \; (a_n < \lambda < b_n), \quad f(\lambda) = 0 \; (b_n < \lambda < a_{n+1}).$$
Then
$$\int^\infty \frac{f(\lambda)}{\lambda} d\lambda$$
is convergent if
$$\sum^\infty \log\left(\frac{b_n}{a_n}\right)$$
is convergent; and
$$\int^\infty f(\lambda) \cos x\lambda \, d\lambda$$
is certainly not convergent unless
$$\sum^\infty \{\sin(b_n x) - \sin(a_n x)\}$$
is so; and it is easy to choose a_n and b_n so that the first of these series is convergent but not the second. Thus if
$$a_n = n^2, \quad b_n = n^2 + 1$$
the series are
$$\sum^\infty \log\left(1 + \frac{1}{n^2}\right), \quad 2 \sin \tfrac{1}{2}x \sum^\infty \cos(n^2 + \tfrac{1}{2})x,$$
and the latter series is in general oscillatory.

* *Proc. Lond. Math. Soc.* vol. VI. p. 372.

3. Quite recently there has appeared a paper by Pringsheim* in which he generalises the conditions to be imposed upon $f(\lambda)$, in so far as they relate to the behaviour of $f(\lambda)$ for large values of λ, considerably further than had been effected by any previous writer. In the first place he shows that condition (iv) may be replaced by

(iv a) $f(\lambda)$ *is monotonic for* $\lambda > \Lambda$ *or* $\lambda < -\Lambda$, *and tends to zero as* $\lambda \to \infty$ *or* $\lambda \to -\infty$; *or, more generally,* $f(\lambda)$ *is of limited total fluctuation in the infinite intervals* $(\Lambda, \infty), (-\infty, -\Lambda)$.

In the second place he introduces a function $F(\lambda)$ defined by

$$F(\lambda) = \sum_0^\infty c_\nu \cos(q_\nu \lambda + r_\nu),$$

where *either* $c_\nu = 0$ for $\nu > n$ (so that the series is finite), *or* $\Sigma |c_\nu|$ is convergent, $q_\nu \to \infty$, and $F(\lambda)$ possesses a derivative integrable and absolutely integrable in any finite interval. Writing $f(\lambda) F(\lambda)$ in the place of $f(\lambda)$ he shows that

$$\int_0^\infty dx \int_{-\infty}^\infty f(\lambda) F(\lambda) \cos(\lambda - \xi) x\, d\lambda = \tfrac{1}{2}\pi \{f(\xi+0) + f(\xi-0)\} F(\xi) \quad\ldots\ldots\ldots(3)$$

if $f(\lambda)$ is still subject to the conditions (i), (ii), (iii) or (iii a), and (iv) or (iv a), provided that when we chose the condition (iv a), and $F(\lambda)$ is not a mere constant, we impose the additional condition

(v)   *the integrals* $\int^\infty \left|\dfrac{f(\lambda)}{\lambda}\right| d\lambda, \quad \int_{-\infty} \left|\dfrac{f(\lambda)}{\lambda}\right| d\lambda$

are convergent†.

Further, he shows that the formula (3) remains valid if we replace its right-hand side by

$$\tfrac{1}{2}\pi \lim_{\epsilon \to 0} \{f(\xi+\epsilon) + f(\xi-\epsilon)\} F(\xi+\epsilon)$$

and suppose only that this limit exists and that $f(\lambda) F(\lambda)$, and not necessarily $f(\lambda)$ itself, is subject to the conditions (i) and (iii).

Finally he shows that, when $F(\lambda)$ reduces to a finite sum, we can dispense with condition (v), provided we regard Fourier's integral as a *principal value* (in respect to the integration with regard to x).

It should be observed that these conditions of Pringsheim's are not valuable merely on account of their greater theoretical generality. There are perfectly simple and obvious forms of $f(\lambda)$, such as

$$f(\lambda) = \frac{\sin \lambda}{\lambda},$$

with which no theorems previously established were sufficient to deal.

4. Generalisations of the ordinary form of Fourier's integral theorem, of a different type, have been considered by Sommerfeld‡ and by the present writer.

The essence of these generalisations is that they show that, if we agree to regard Fourier's integral as *summable* only (by one method or another) and not necessarily convergent, we can

---

* Math. Annalen, Bd. 68, S. 367.

† If $f(\lambda)$ is ultimately monotonic, the signs of the absolute value may be omitted; but they must be retained if all that we know is that $f(\lambda)$ is of limited total fluctuation.

‡ Sommerfeld, "Die willkürlichen Funktionen in der Math. Physik," *Inaugural-Dissertation*, Königsberg, 1901; Hardy, "Further researches in the theory of divergent series and integrals," *Camb. Phil. Trans.* vol. XXI. p. 39.

dispense entirely with all conditions which have regard to the behaviour of $f(\lambda)$ near $\lambda = \xi$, except the conditions (i) and (ii). That is to say we can omit the conditions (iii) or (iii a), or the conditions relating to the function $f(\lambda) F(\lambda)$ substituted for them by Pringsheim. Similarly, we shall find, we can omit the condition, used by Pringsheim, that $F(\lambda)$ possesses an integrable and absolutely integrable derivative.

In the paper in the *Cambridge Philosophical Transactions* referred to above I considered three definitions of the generalised or summable integral, viz. those expressed by the equations

$$G \int_0^\infty f(x)\,dx = \lim_{\delta \to 0} \int_0^\infty e^{-\delta x} f(x)\,dx \quad \ldots\ldots\ldots\ldots(4),$$

$$G \int_0^\infty f(x)\,dx = \lim_{\delta \to 0} \int_0^\infty e^{-(\delta x)^2} f(x)\,dx \quad \ldots\ldots\ldots\ldots(5),$$

$$G \int_0^\infty f(x)\,dx = \lim_{x \to \infty} \frac{1}{x} \int_0^x dt \int_0^t f(u)\,du \quad \ldots\ldots\ldots\ldots(6).$$

But I considered (as Sommerfeld had done) only the case in which the interval of integration with respect to λ is finite. The result was to establish the formula

$$G \int_0^\infty dx \int_\beta^\gamma f(\lambda) \cos(\lambda - \xi) x\, d\lambda = \tfrac{1}{2}\pi \{f(\xi + 0) + f(\xi - 0)\}$$

(where $\beta < \xi < \gamma$) under the conditions that

(i) $f(\lambda)$ *is integrable and absolutely integrable in* (β, γ),

(ii) $f(\xi + 0)$ *and* $f(\xi - 0)$ *exist*.

In the present paper I propose to complete and generalise these results by

(*a*) adopting a more general form of the definition of the summable integral,

(*b*) supposing the interval of integration with respect to λ infinite,

(*c*) supposing $f(\lambda)$ to have one of the more general forms considered by Pringsheim.

II. *Some Properties of the Summable Integral.*

5. The integral
$$\int_a^\infty f(x)\,dx \quad \ldots\ldots\ldots\ldots(1)$$

will be said to be *summable* $(C\,1)$ to sum s if

$$\frac{1}{x} \int_a^x dt \int_a^t f(u)\,du \to s \quad \ldots\ldots\ldots\ldots(2)$$

as $x \to \infty$. We may suppose $a > 0$. An alternative and equivalent definition is given by the formula

$$\int_a^x \left(1 - \frac{t}{x}\right) f(t)\,dt \to s \quad \ldots\ldots\ldots\ldots(3).$$

Similarly the integral is said to be *summable* $(C\,r)$ if

$$\frac{r!}{x^r} \left(\int_a^x dt\right)^{r+1} f(t)\,dt \to s$$

or

$$\int_a^x \left(1 - \frac{t}{x}\right)^r f(t)\,dt \to s.$$

More general types of definition, modelled on the above, and analogous to those given by Riesz* for summable series, are easily framed; and r need not necessarily be integral. But in what follows we shall only have occasion to use the simplest form—viz. that expressed by (2) or (3).

We shall in the main be concerned with a different type of definition, a generalisation of that expressed by (4) and (5) of § 4.

Let $\phi(x)$ be a function of x subject to the following conditions:

(i) $\phi(x)$ has at most a finite number of maxima and minima†;

(ii) $\phi''(x)$ is continuous and ultimately positive;

(iii) the integral $\int^\infty \phi(x)\,dx$ is convergent;

(iv) $\phi(0) \neq 0$.

We note in passing that these properties imply a number of others. Thus $\phi'(x)$ is ultimately negative and increases steadily to zero as $x \to \infty$, and $\phi(x)$ is ultimately positive and decreases steadily to zero. Also
$$\int^\infty \phi'(x)\,dx$$
is convergent, and so, as $\phi'(x)$ is ultimately monotonic, we must have
$$x\phi'(x) \to 0.$$
Finally $\int_a^x t\phi''(t)\,dt = x\phi'(x) - a\phi'(a) - \phi(x) + \phi(a) \to \phi(a) - a\phi'(a),$
so that
$$\int^\infty x\phi''(x)\,dx$$
is convergent.

Then we shall say that (1) is *summable* (ϕ), to sum s, if
$$\frac{1}{\phi(0)} \int_a^\infty \phi(\delta x) f(x)\,dx \to s$$
as $\delta \to 0$ by positive values: this condition implies the convergence of the integral on the left-hand side for all positive values of δ.

6. The following properties of summable integrals will be used in the sequel.

(i) *If the integral* (1) *is convergent and equal to s, it is summable* $(C1)$ *to sum* $s_+^\ddagger$.

This result is very easily extended to the more general definitions: but it is not necessary for our present purpose to enter into this.

(ii) *If the integral* (1) *is convergent, and has the value s, it is summable* (ϕ), *and has the sum s.*

For the integral $\int_a^\infty \phi(\delta x) f(x)\,dx$

is uniformly convergent for $0 \leq \delta \leq \delta_0$.§

---

* *Comptes Rendus*, 5 July, 1909.

† This is not a consequence of condition (ii). Consider, for example, the function $\phi(x) = x^5 \sin(1/x)$.

‡ Hardy, *Quarterly Journal*, vol. xxxv. p. 54; C. N. Moore, *Trans. Amer. Math. Soc.* vol. viii. p. 312.

§ Bromwich, *Infinite Series*, pp. 434–5.

(iii) *If (a) the integral* (1) *is summable* ($C1$),

(b) $\phi(\delta x) \int_a^x f(t)\, dt \to 0$ *for all positive values of* δ,

then the integral (1) *is summable* (ϕ) *to sum* s*.

Let
$$f_1(x) = \int_a^x f(t)\, dt, \quad f_2(x) = \int_a^x f_1(t)\, dt,$$

so that
$$\frac{1}{x} f_2(x) \to s.$$

Then
$$\int_a^X \phi(\delta x) f(x)\, dx = \phi(\delta X) f_1(X) - \delta \phi'(\delta X) f_2(X) + \delta^2 \int_a^X \phi''(\delta x) f_2(x)\, dx.$$

Now $\int_a^\infty \phi''(\delta x) f_2(x)\, dx$ is convergent (absolutely), by comparison with $\int_a^\infty x \phi''(\delta x)\, dx$; and

$$\delta \phi'(\delta X) f_2(X) = \left\{ \frac{1}{X} f_2(X) \right\} \{\delta X \phi'(\delta X)\} \to 0.$$

Using condition (b) we see that
$$\int_a^\infty \phi(\delta x) f(x)\, dx = \delta^2 \int_a^\infty \phi''(\delta x) f_2(x)\, dx;$$

the convergence of each integral having been established incidentally. The integral on the right-hand side is equal to

$$\delta^2 \int_a^\infty (s + \epsilon_x) x \phi''(\delta x)\, dx = s \{\phi(\delta a) - \delta a \phi'(\delta a)\} + \delta^2 \int_a^\infty x \phi''(\delta x) \epsilon_x\, dx,$$

where $\epsilon_x \to 0$ as $x \to \infty$: and it is easily shown (by a type of proof so familiar that it is hardly worth repeating) that the limit of the right-hand side, when $\delta \to 0$, is $s \phi(0)$.

(iv) *If (a) the integral* (1) *is summable* (ϕ) *to sum* s,

(b) $xf(x) \to 0$ *as* $x \to \infty$,

then (1) *is convergent, and has the value* s†.

In the first place
$$\phi(0) - \phi(\delta x) = -\delta x \phi'(\xi),$$
where $0 < \xi < \delta x$; and so
$$|\phi(0) - \phi(\delta x)| < K \delta x.$$

Hence
$$\left| \int_a^X \{\phi(0) - \phi(\delta x)\} f(x)\, dx \right| < K\delta \int_a^X x |f(x)|\, dx.$$

Again, if $F(X)$ denotes the upper limit of $x|f(x)|$ for $x \geq X$, we have

$$\left| \int_X^\infty \phi(\delta x) f(x)\, dx \right| \leq \frac{F(X)}{X} \int_X^\infty \phi(\delta x)\, dx = F(X) \Phi(\delta X) / \delta X,$$

where
$$\Phi(x) = \int_x^\infty \phi(t)\, dt.$$

---

* Cf. Moore, *loc. cit.*; Bromwich, *Math. Annalen*, Bd. 65, S. 367.

† Cf. Tauber, *Monatshefte für Math.* Bd. 8, S. 273; Landau, *ibid.* Bd. 18, S. 8; Bromwich, *Infinite Series*, p. 251. It has been shown by Mr Littlewood, with the aid of more elaborate analysis, that the condition (b) may be replaced by the less stringent condition that $|xf(x)| < K$, as in theorem (vi) below. See *Proc. Lond. Math. Soc.* vol. IX. p. 434: the theorem there proved is the analogous theorem for series, but the same method is applicable to integrals.

Suppose that $X = 1/\delta$. Then

$$\left| \phi(0) \int_a^X f(x)\,dx - \int_a^\infty \phi(\delta x) f(x)\,dx \right|$$

$$< \frac{K}{X} \int_a^X x |f(x)|\,dx + F(X)\Phi(1) \to 0 ;$$

and so
$$\int_a^X f(x)\,dx \to s$$

as $X \to \infty$; which proves the theorem.

(v) *The necessary and sufficient condition that the integral* (1), *when summable* ($C\,1$), *should also be convergent, is that*

$$\frac{1}{x} \int_a^x t f(t)\,dt \to 0$$

as $x \to \infty$.

This follows at once from the equations

$$\frac{f_2(x)}{x} = \frac{1}{x} \int_a^x f_1(t)\,dt = f_1(x) - \frac{1}{x} \int_a^x t f(t)\,dt.$$

(vi) *If* (i) *the integral* (1) *is summable* ($C\,1$) *to sum* s,

    (ii) $|xf(x)| < K$,

then the integral (1) *is convergent, and has the value* s.

Let
$$g(x) = xf(x), \quad G(x) = \int_a^x g(t)\,dt.$$

Then
$$f_1(x) = \int_a^x f(t)\,dt = \int_a^x \frac{1}{t} G'(t)\,dt$$

$$= \frac{G(x)}{x} - \frac{G(a)}{a} + \int_a^x G(t)\frac{dt}{t^2},$$

and so
$$\frac{1}{x}\int_a^x f_1(t)\,dt = f_1(x) - \frac{G(x)}{x}$$

$$= -\frac{G(a)}{a} + \int_a^x G(t)\frac{dt}{t^2}.$$

Hence, as the left-hand side tends to a limit as $x \to \infty$, the integral

$$\int_a^\infty G(t)\frac{dt}{t^2}$$

is convergent. I shall now prove that this cannot be the case unless

$$\frac{1}{x} G(x) \to 0 :$$

from which it will follow, by (v), that the integral (1) is convergent.

If this last relation does not hold, it must be possible to find a positive number K such that
$$G(x) > K_1 x,$$

or $G(x) < -K_1 x$, for values of x surpassing all limit. Let us adopt the first hypothesis: we may clearly suppose $K_1 < K$. And let X be a value of x for which the inequality above written is satisfied. Let

$$X_1 = \left(1 - \frac{K_1}{2K}\right) X.$$

Then, for $X_1 \leq x \leq X$, we have

$$|G(x) - G(X)| = \left|\int_x^X t f(t) dt\right|$$
$$< K(X - x)$$
$$\leq K(X - X_1);$$

and so
$$G(x) \geq G(X) - |G(x) - G(X)|$$
$$> \tfrac{1}{2} K_1 X.$$

Thus
$$\int_{X_1}^X G(t) \frac{dt}{t^2} > \tfrac{1}{2} K_1 X \int_{X_1}^X \frac{dt}{t^2}$$
$$= \tfrac{1}{2} K_1 \left(\frac{X}{X_1} - 1\right)$$
$$= \tfrac{1}{2} K_1^2 / (2K - K_1)$$
$$= K_2,$$

say. And this inequality is plainly inconsistent with the convergence of the integral $\int^\infty G(t) \frac{dt}{t^2}$. Thus the theorem follows*.

III. *Theorems relating to the inversion of the order of integration in a repeated infinite integral.*

7. This section will be devoted to the formulation of a variety of sets of sufficient conditions for the truth of the equations

$$\int_0^\infty \phi(x) dx \int_0^\infty f(\lambda) \frac{\cos}{\sin} \lambda x \, d\lambda = \int_0^\infty f(\lambda) d\lambda \int_0^\infty \phi(x) \frac{\cos}{\sin} \lambda x \, dx \quad \ldots\ldots\ldots\ldots(1).$$

We shall suppose throughout that $\phi(x)$ and $f(\lambda)$ are subject to the following condition:

Condition of integrability. *Each of the functions ϕ, f is integrable and absolutely integrable throughout any finite integral*—or, as we shall say, **regularly integrable** throughout any finite interval.

8. It will be convenient to begin by stating certain lemmas. These lemmas are proved by Pringsheim†, but, as the proofs are very short, it seems worth while to repeat them here.

Lemma A. *The integrals*

$$\int_\Lambda^\infty f(\lambda) \frac{\sin \lambda \xi}{\lambda} d\lambda, \quad \int_\Lambda^\infty f(\lambda) \frac{\cos \lambda \xi}{\lambda} d\lambda \quad (\Lambda > 0)$$

---

* For similar theorems relating to *series*, see *Proc. L. M. S.* vol. VIII. pp. 301 et seq.
† *Math. Annalen*, loc. cit.

will tend to zero, as $\xi \to \infty$, if $\psi(\lambda) = f(\lambda)/\lambda$ satisfies any one of the following conditions—

(a) $\int_\Lambda^\infty |\psi(\lambda)| d\lambda$ *is convergent;*

(b) $\psi(\lambda)$ *tends steadily to zero as $\lambda \to \infty$;*

(c) $\psi(\lambda)$ *is of limited total fluctuation in the interval (Λ, ∞);*

—*and, in the case of conditions (b) and (c), it is sufficient to suppose them satisfied for values of λ greater than some definite value Λ'.*

In the first place, it is well known* that
$$\int_\Lambda^{\Lambda'} \psi(\lambda) \genfrac{}{}{0pt}{}{\sin}{\cos} \lambda\xi \, d\lambda \to 0$$
as $\xi \to \infty$, for any finite value of Λ'; and all that we have to do is to consider whether we may replace Λ' by ∞.

Now
$$\int_\Lambda^\infty \psi(\lambda) \genfrac{}{}{0pt}{}{\sin}{\cos} \lambda\xi \, d\lambda = \int_\Lambda^{\Lambda'} + \int_{\Lambda'}^\infty,$$
if only the last integral is convergent. And it is sufficient to show that this is so and that we can make
$$\left|\int_{\Lambda'}^\infty\right| < \epsilon$$
by choice of Λ', independently of ξ. For then, Λ' being fixed, we can choose ξ_0 so that
$$\left|\int_\Lambda^{\Lambda'}\right| < \epsilon$$
for $\xi \geqq \xi_0$; and the truth of the lemma will follow.

(a) If this condition is satisfied,
$$\left|\int_{\Lambda'}^\infty \psi(\lambda) \genfrac{}{}{0pt}{}{\sin}{\cos} \lambda\xi \, d\lambda \right| \leqq \int_{\Lambda'}^\infty |\psi(\lambda)| \, d\lambda,$$
and the result follows at once.

(b) We have (Λ' being large enough)
$$\int_{\Lambda'}^\infty = \lim_{\Lambda'' \to \infty} \int_{\Lambda'}^{\Lambda''} = \lim \psi(\Lambda') \int_{\Lambda'}^{\Lambda'''} \genfrac{}{}{0pt}{}{\sin}{\cos} \lambda\xi \, d\lambda,$$
where $\Lambda' < \Lambda''' < \Lambda''$: that the limit exists and the integral is convergent follows at once from Dirichlet's test†.

Also
$$\left|\int_{\Lambda'}^\infty\right| \leqq \frac{2}{\xi} |\psi(\Lambda')|,$$
and the result of the lemma follows immediately.

(c) In this case we may write
$$\psi(\lambda) = \psi_1(\lambda) - \psi_2(\lambda),$$
where ψ_1 and ψ_2 are subject to (b).

---

* See, e.g., Hobson, *Theory of functions of a real variable*, pp. 672 et seq.
† Bromwich, *Infinite Series*, p. 430.

Lemma B. *If $f(\lambda)$ satisfies any one of the conditions imposed upon $f(\lambda)/\lambda$ in Lemma A, then*
$$\int_\Lambda^\infty f(\lambda) \frac{\sin \lambda \xi}{\lambda} d\lambda \to 0$$
as $\xi \to 0$.

This will certainly be the case if the integral is uniformly convergent in an interval including $\xi = 0$. If condition (a) is satisfied this is obviously the case. If condition (b) is satisfied we have
$$\int_{\Lambda'}^\infty f(\lambda) \frac{\sin \lambda \xi}{\lambda} d\lambda = f(\Lambda') \int_{\Lambda'}^{\Lambda''} \frac{\sin \lambda \xi}{\lambda} d\lambda,$$
which is numerically less than
$$\pi \, |f(\Lambda')|$$
for all values of ξ; and the result follows.

Lemma C. *If*
$$\int^\infty \frac{|f(\lambda)|}{\lambda} d\lambda$$
is convergent, then
$$\int_\Lambda^\infty f(\lambda) \frac{\sin \lambda \xi}{\lambda} d\lambda \to 0, \quad \int_\Lambda^\infty f(\lambda) \frac{\cos \lambda \xi}{\lambda} d\lambda \to \int_\Lambda^\infty \frac{f(\lambda)}{\lambda} d\lambda,$$
as $\xi \to 0$.

For each integral is plainly uniformly convergent in an interval including $\xi = 0$. In so far as the sine-integral is concerned, this lemma includes case (a) of Lemma B.

9. **Theorem I.** *If the integrals*
$$\int^\infty |\phi(x)| \, dx, \quad \int^\infty |f(\lambda)| \, d\lambda$$
are convergent, then
$$\int_0^\infty \phi(x) \, dx \int_0^\infty f(\lambda) \genfrac{}{}{0pt}{}{\cos}{\sin} \lambda x \, d\lambda = \int_0^\infty f(\lambda) \, d\lambda \int_0^\infty \phi(x) \genfrac{}{}{0pt}{}{\cos}{\sin} \lambda x \, dx.$$

This may be deduced at once from the existence of the double integral
$$\int_0^\infty \int_0^\infty |\phi(x)| \, |f(\lambda)| \left| \genfrac{}{}{0pt}{}{\cos}{\sin} \lambda x \right| dx \, d\lambda.$$
In view of our subsequent results, however, it is more convenient to proceed as follows.

(a) In virtue of the "condition of integrability," we have
$$\int_{x_0}^X \int_{\lambda_0}^\Lambda = \int_{\lambda_0}^\Lambda \int_{x_0}^X$$
for any finite values of the limits, zero included.

(b) The integral
$$\int_0^\infty \phi(x) \genfrac{}{}{0pt}{}{\cos}{\sin} \lambda x \, dx$$
is uniformly convergent for all values of λ. Hence
$$\int_0^\infty \int_0^\Lambda = \int_0^\Lambda \int_0^\infty,$$

however great be Λ; and it only remains to show that

$$\int_0^\infty \int_\Lambda^\infty$$

is convergent, and tends to zero as $\Lambda \to \infty$: for then

$$\int_0^\infty \phi(x)\,dx \int_0^\infty f(\lambda) \genfrac{}{}{0pt}{}{\cos}{\sin} \lambda x\,d\lambda = \lim_{\Lambda \to \infty} \int_0^\infty \int_0^\Lambda = \lim_{\Lambda \to \infty} \int_0^\Lambda \int_0^\infty$$

$$= \int_0^\infty f(\lambda)\,d\lambda \int_0^\infty \phi(x) \genfrac{}{}{0pt}{}{\cos}{\sin} \lambda x\,dx,$$

by the definition of the latter repeated integral.

Now
$$\int_\Lambda^\infty f(\lambda) \genfrac{}{}{0pt}{}{\cos}{\sin} \lambda x\,d\lambda$$

is uniformly convergent, and so continuous, throughout any finite interval of values of x; and so

$$\phi(x) \int_\Lambda^\infty f(\lambda) \genfrac{}{}{0pt}{}{\cos}{\sin} \lambda x\,d\lambda$$

is regularly integrable throughout any such interval. Also

$$\left| \phi(x) \int_\Lambda^\infty \right| < |\phi(x)| \int_0^\infty |f(\lambda)|\,d\lambda;$$

and so
$$\int_0^\infty \phi(x)\,dx \int_\Lambda^\infty f(\lambda) \genfrac{}{}{0pt}{}{\cos}{\sin} \lambda x\,d\lambda$$

is convergent. It is moreover plainly less in absolute value than

$$\int_0^\infty |\phi(x)|\,dx \int_\Lambda^\infty |f(\lambda)|\,d\lambda$$

and so tends to zero as $\Lambda \to \infty$. Thus the theorem is established.

10. THEOREM II. *The inversion of the order of integration, in the case of the* **cosine** *integral, is also legitimate under the following conditions:*

(i) $$\int_0^\infty |\phi(x)|\,dx$$

is convergent;

(ii) $\phi(x)$ *tends steadily to a limit* $\phi(+0)$ *as* $x \to 0$;

(iii) $f(\lambda)$ *tends steadily to zero as* $\lambda \to \infty$, *or, more generally, is of limited total fluctuation in an interval* (Λ, ∞).

(a) Precisely as in the proof of Theorem I, we show that

$$\int_0^\Lambda \int_0^\infty = \int_0^\infty \int_0^\Lambda.$$

All that remains, therefore, is to shew that

$$\int_0^\infty \int_\Lambda^\infty$$

is convergent, and tends to zero as $\Lambda \to \infty$.

(b) The problem of establishing the truth of the last statement may be divided into three, viz. that of establishing the same property in the cases of the integrals

$$(1) \qquad \int_0^{x_0} \int_\Lambda^\infty,$$

$$(2) \qquad \int_{x_0}^{X} \int_\Lambda^\infty,$$

$$(3) \qquad \int_X^\infty \int_\Lambda^\infty.$$

And, in the first place, it is obviously possessed by the integral (2), inasmuch as

$$\int_\Lambda^\infty f(\lambda) \cos \lambda x \, d\lambda$$

is uniformly convergent throughout any interval of the type $0 < x_0 \leq x \leq X$.

Again, it follows at once from the second theorem of the mean that

$$\left| \int_\Lambda^\infty f(\lambda) \cos \lambda x \, d\lambda \right| < 2 \, |f(\Lambda)| / X \qquad (x \geq X);$$

and from this, and condition (i), it follows that the property in question is possessed by the integral (3). There remains (1), the treatment of which is slightly more difficult.

We observe first that if $\bar{x} > 0$ then

$$\int_{\bar{x}}^{x_0} dx \int_\Lambda^\infty f(\lambda) \cos \lambda x \, d\lambda = \int_\Lambda^\infty f(\lambda) \frac{\sin \lambda x_0 - \sin \lambda \bar{x}}{\lambda} d\lambda.$$

Making $\bar{x} \to 0$, and using Lemma B, we see that

$$\int_0^{x_0} dx \int_\Lambda^\infty f(\lambda) \cos \lambda x \, d\lambda$$

is convergent and may be calculated by inverting the order of integration.

Now $\phi(x)$ is monotonic near $x = 0$, and we may suppose x_0 so small that it is monotonic throughout $(0, x_0)$. It then follows from the second theorem of the mean that

$$\int_0^{x_0} \phi(x) \, dx \int_\Lambda^\infty f(\lambda) \cos \lambda x \, d\lambda$$

is convergent and equal to

$$\phi(+0) \int_0^{x_1} dx \int_\Lambda^\infty f(\lambda) \cos \lambda x \, d\lambda + \phi(x_0) \int_{x_1}^{x_0} dx \int_\Lambda^\infty f(\lambda) \cos \lambda x \, d\lambda$$

$$= \phi(+0) \int_\Lambda^\infty f(\lambda) \frac{\sin \lambda x_1}{\lambda} d\lambda + \phi(x_0) \int_\Lambda^\infty f(\lambda) \frac{\sin \lambda x_0 - \sin \lambda x_1}{\lambda} d\lambda,$$

where $0 < x_1 < x_0$. And as (Lemma B) the integral

$$\int_\Lambda^\infty f(\lambda) \frac{\sin \lambda \xi}{\lambda} d\lambda$$

is uniformly convergent in an interval of values of ξ including $\xi = 0$, it follows that

$$\int_0^{x_0} \phi(x) \, dx \int_\Lambda^\infty f(\lambda) \cos \lambda x \, d\lambda \to 0$$

as $\Lambda \to \infty$. Thus the proof of Theorem II is completed*.

11. THEOREM III. *The same result holds of the* **sine** *integral if the additional condition is satisfied that*

$$\int^\infty \left| \frac{f(\lambda)}{\lambda} \right| d\lambda$$

is convergent.

The proof is the same as that of Theorem II except that at the end we are left with an expression of the type

$$\phi(+0) \int_\Lambda^\infty f(\lambda) \frac{1 - \cos \lambda x_1}{\lambda} d\lambda + \phi(x_0) \int_\Lambda^\infty f(\lambda) \frac{\cos \lambda x_1 - \cos \lambda x_0}{\lambda} d\lambda,$$

and that we use Lemma C in the place of Lemma B.

12. We have now to consider the question of the inversion of the order of integration in the more general integral

$$\int_0^\infty \phi(x) \, dx \int_{-\infty}^\infty f(\lambda) \frac{\cos}{\sin} a\lambda \cos(\lambda - \xi) x \, d\lambda \quad (a \geq 0).$$

Putting $\lambda - \xi = \mu$, and writing $F(\mu)$ for $f(\mu + \xi)$, we obtain

$$\cos a\xi \int_0^\infty \phi(x) \, dx \int_{-\infty}^\infty F(\mu) \frac{\cos}{\sin} a\mu \cos \mu x \, d\mu \mp \sin a\xi \int_0^\infty \phi(x) \, dx \int_{-\infty}^\infty F(\mu) \frac{\sin}{\cos} a\mu \cos \mu x \, d\mu,$$

so that the question is reduced to the same question for the original integral, with $\xi = 0$. It may then be expressed in the form

$$\tfrac{1}{2} \int_0^\infty \phi(x) \, dx \int_{-\infty}^\infty f(\lambda) \frac{\cos}{\sin} (x + a) \lambda \, d\lambda \pm \tfrac{1}{2} \int_0^\infty \phi(x) \, dx \int_{-\infty}^\infty f(\lambda) \frac{\cos}{\sin} (x - a) \lambda \, d\lambda.$$

We now put $x + a = y$ or $x - a = y$, and, using our previous results, we obtain the theorem which follows.

THEOREM IV. *The equation*

$$\int_0^\infty \phi(x) \, dx \int_{-\infty}^\infty f(\lambda) \frac{\cos}{\sin} a\lambda \cos(\lambda - \xi) x \, d\lambda = \int_{-\infty}^\infty f(\lambda) \frac{\cos}{\sin} a\lambda \, d\lambda \int_0^\infty \phi(x) \cos(\lambda - \xi) x \, dx$$

is true if

(i) f and ϕ *satisfy the condition of integrability,*

(ii) $$\int_0^\infty |\phi(x)| \, dx$$

is convergent, and (iii) **either**

---

* I have established a more general theorem dealing with the case in which neither of the integrals

$$\int^\infty |\phi(x)| \, dx, \quad \int^\infty |f(\lambda)| \, d\lambda$$

is convergent. The theorem here proved is sufficiently general for the purposes of this paper.

(iii a) $$\int_{-\infty}^{\infty} |f(\lambda)|\, d\lambda$$

is convergent, **or**

(iii b) *the following three conditions are satisfied—*

    (iii bα) $\phi(x)$ *tends steadily to a limit* $\phi(a)$ *as* $x \to a$,

    (iii bβ) $f(\lambda)$ *tends steadily to zero as* $\lambda \to \infty$ *or* $-\infty$,

    (iii bγ) *the integrals* $\int^{\infty}\left|\dfrac{f(\lambda)}{\lambda}\right|d\lambda,\ \int_{-\infty}\left|\dfrac{f(\lambda)}{\lambda}\right|d\lambda$

are convergent—where however it is to be observed that (iii bβ) *may be replaced by the more general condition that* $f(\lambda)$ *is of limited total fluctuation in the intervals* $(-\infty, -\Lambda), (\Lambda, \infty)$, *and that* (iii bγ) *may be dispensed with when* $a = 0$.

IV. *The limit of the integral* $\displaystyle\int_0^{\infty} \phi(\delta x)\, dx \int_{-\infty}^{\infty} f(\lambda) \genfrac{}{}{0pt}{}{\cos}{\sin} a\lambda \cos(\lambda - \xi)x\, d\lambda$.

13. If $\phi(x)$ satisfies the first three conditions of § 5, viz. that

    (i) $\phi(x)$ has at most a finite number of maxima and minima,

    (ii) $\phi''(x)$ exists, and is ultimately positive,

    (iii) the integral $\displaystyle\int_0^{\infty} \phi(x)\, dx$

is convergent, it is clear that $\phi(\delta x)$, where δ is any positive number, satisfies the conditions of the preceding theorems.

Hence, if $f(\lambda)$ satisfies one or other of the sets of conditions stated in Theorem IV, we have the right to invert the order of integration in the integral written at the head of this paragraph.

Put $\quad\lambda - \xi = \mu,\quad f(\mu + \xi) \genfrac{}{}{0pt}{}{\cos}{\sin} a(\mu + \xi) = \psi(\mu).$

Then the integral reduces to

$$\int_0^{\infty} \phi(\delta x)\, dx \int_{-\infty}^{\infty} \psi(\mu) \cos \mu x\, d\mu = \int_{-\infty}^{\infty} \psi(\mu)\, d\mu \int_0^{\infty} \phi(\delta x) \cos \mu x\, dx.$$

We may write this in the form

$$\frac{1}{\delta} \int_{-\infty}^{\infty} \psi(\mu) \Delta\left(\frac{\mu}{\delta}\right) d\mu,$$

where $\qquad \Delta(\mu) = \displaystyle\int_0^{\infty} \phi(x) \cos \mu x\, dx.$

We shall have occasion to use the following lemma.

Lemma. *The integral* $\displaystyle\int_{-\infty}^{\infty} |\Delta(\mu)|\, d\mu$

is convergent.

In fact, integrating twice by parts, we obtain

$$\Delta(\mu) = -\frac{1}{\mu} \int_0^{\infty} \phi'(x) \sin \mu x\, dx = \frac{1}{\mu^2} \int_0^{\infty} \phi''(x)(1 - \cos \mu x)\, dx.$$

From this equation, and the fact that $\Delta(\mu)$ is plainly a continuous function of μ, the truth of the lemma follows at once.

It should be observed that the convergence of the integral
$$\int_{-\infty}^{\infty} \Delta(\mu) d\mu = \int_{-\infty}^{\infty} d\mu \int_0^{\infty} \phi(x) \cos \mu x \, dx$$
is an immediate consequence of the ordinary forms of Fourier's integral-theorem: the value of the integral is in fact $\pi \phi(0)$. In the particular case (which in point of fact includes all the most interesting cases) in which $\phi'(x)$ is monotonic not merely ultimately but for all positive values of x, it is clear that
$$\Delta(\mu) = -\frac{1}{\mu} \int_0^{\infty} \phi'(x) \sin \mu x \, dx > 0;$$
and then the convergence of $\int_{-\infty}^{\infty} \Delta(\mu) d\mu$ of course implies its absolute convergence.

We observe also that
$$|\Delta(\mu)| < K/\mu^2.$$

14. Theorem V. *If*

(i) $\phi(x)$ *is subject to the first three conditions of § 5,*

(ii) $f(\lambda)$ *satisfies the condition of integrability,*

(iii) $f(\xi+0)$ *and* $f(\xi-0)$ *exist,*

then will
$$\int_0^{\infty} \phi(\delta x) dx \int_{\Lambda_1}^{\Lambda_2} f(\lambda) \genfrac{}{}{0pt}{}{\cos}{\sin} a\lambda \cos \lambda(x-\xi) d\lambda,$$
where $a \geqq 0$, $\Lambda_1 < \xi < \Lambda_2$, *tend, as* $\delta \to 0$, *to the limit*
$$\tfrac{1}{2} \pi \phi(0) \genfrac{}{}{0pt}{}{\cos}{\sin} a\xi \{f(\xi+0) + f(\xi-0)\}.$$

Further, if condition (iii) *is replaced by the condition, sometimes more general, that* $\psi(\xi+0)$ *and* $\psi(\xi-0)$ *exist, where*
$$\psi(\lambda) = f(\lambda) \genfrac{}{}{0pt}{}{\cos}{\sin} a\lambda,$$
the limit will still have the value
$$\tfrac{1}{2} \pi \phi(0) \{\psi(\xi+0) + \psi(\xi-0)\}.$$

Finally, if the inequalities $\Lambda_1 < \xi < \Lambda_2$ *are not satisfied, the result remains true if the final formula is modified in the manner usual in the theory of Fourier's series**.

Putting
$$\lambda - \xi = \mu, \quad f(\mu+\xi) \genfrac{}{}{0pt}{}{\cos}{\sin} a(\mu+\xi) = \psi(\mu),$$
$$\Lambda_1 - \xi = M_1, \quad \Lambda_2 - \xi = M_2,$$
we reduce the integral to the form
$$\int_0^{\infty} \phi(\delta x) dx \int_{M_1}^{M_2} \psi(\mu) \cos \mu x \, d\mu,$$
which may be transformed into
$$\frac{1}{\delta} \int_{M_1}^{M_2} \psi(\mu) \Delta\left(\frac{\mu}{\delta}\right) d\mu.$$

---
* The result is zero if ξ falls outside (Λ_1, Λ_2), etc.

The problem therefore reduces itself to proving that

$$\frac{1}{\delta}\int_0^\Lambda \psi(\lambda)\,\Delta\left(\frac{\lambda}{\delta}\right)d\lambda \to \tfrac{1}{2}\pi\phi(0)\,\psi(+0)$$

as $\delta \to 0$. Since

$$\frac{1}{\delta}\int_0^\Lambda \Delta\left(\frac{\lambda}{\delta}\right)d\lambda = \int_0^{\Lambda/\delta}\Delta(\mu)\,d\mu \to \tfrac{1}{2}\pi\phi(0),$$

it is sufficient to prove that

$$\frac{1}{\delta}\int_0^\Lambda \chi(\lambda)\,\Delta\left(\frac{\lambda}{\delta}\right)d\lambda \to 0,$$

where $\chi(\lambda) = \psi(\lambda) - \psi(+0)$. I shall prove first that the integral tends to zero when the lower limit 0 is replaced by any positive number λ_0. That this is so follows in fact at once from the inequality

$$\frac{1}{\delta}\left|\int_{\lambda_0}^\Lambda \chi(\lambda)\,\Delta\left(\frac{\lambda}{\delta}\right)d\lambda\right| < K\delta\int_{\lambda_0}^\Lambda \frac{|\chi(\lambda)|}{\lambda^2}d\lambda,$$

which is itself an immediate consequence of the remark at the end of the last paragraph.

We can suppose λ_0 so chosen that $|\chi(\lambda)| < K$ for $0 < \lambda \leq \lambda_0$. Then

$$\frac{1}{\delta}\int_0^{\lambda_0}\chi(\lambda)\,\Delta\left(\frac{\lambda}{\delta}\right)d\lambda = \int_0^{\lambda_0/\delta}\chi(\delta\mu)\,\Delta(\mu)\,d\mu = \left(\int_0^M + \int_M^{\lambda_0/\delta}\right)\chi(\delta\mu)\,\Delta(\mu)\,d\mu.$$

The second integral is less than

$$K\int_M^{\lambda_0/\delta}|\Delta(\mu)|\,d\mu,$$

which may be made less than ϵ by choice of M, independently of δ. And when M is fixed we can choose δ_0 so that $|\chi(\delta\mu)| < \epsilon$ for $0 < \delta \leq \delta_0$, $0 < \mu \leq M$. Then

$$\left|\int_0^M \chi(\delta\mu)\,\Delta(\mu)\,d\mu\right| < \epsilon\int_0^M |\Delta(\mu)|\,d\mu,$$

and the proof of the theorem is completed.

It is interesting to consider an example in which the result is not true. Let

$$\phi(x) = 1 \ (0 \leq x \leq X), \ = 0 \ (x > X).$$

Then

$$\int_0^\infty \phi(\delta x)\,dx\int_0^\Lambda f(\lambda)\cos\lambda x\,d\lambda = \int_0^{X/\delta}dx\int_0^\Lambda f(\lambda)\cos\lambda x\,d\lambda$$

$$= \int_0^\Lambda f(\lambda)\frac{\sin(\lambda X/\delta)}{\lambda}d\lambda,$$

and it is known that the continuity of $f(\lambda)$ is not a sufficient condition to ensure that this integral shall tend to a limit as $\delta \to 0$.

15. THEOREM VI. *The result of Theorem V remains true for* $\Lambda_1 = -\infty$, $\Lambda_2 = \infty$ *provided* **either**

(α) $\int_{-\infty}^\infty |f(\lambda)|\,d\lambda$ *is convergent,*

or (β) *the following conditions are satisfied,*

(β 1) $f(\lambda)$ *tends steadily to zero as* $\lambda \to -\infty$ *or* ∞ (*or, more generally, is of limited total fluctuation in the intervals* $(-\infty, \Lambda_1)$, $(\Lambda_2, -\infty)$),

(β 2) *the integrals* $\quad \int_{-\infty} \left|\frac{f(\lambda)}{\lambda}\right| d\lambda, \quad \int^{\infty} \left|\frac{f(\lambda)}{\lambda}\right| d\lambda$

are convergent.

If $a = 0$ *the condition* (β 2) *may be dispensed with.*

We have only to show that Λ_1 and Λ_2 may be chosen so that

$$\int_0^\infty \phi(\delta x)\,dx \int_{-\infty}^{\Lambda_1} f(\lambda) \genfrac{}{}{0pt}{}{\cos}{\sin} a\lambda \cos(\lambda - \xi)x\,d\lambda \to 0,$$

$$\int_0^\infty \phi(\delta x)\,dx \int_{\Lambda_2}^{\infty} f(\lambda) \genfrac{}{}{0pt}{}{\cos}{\sin} a\lambda \cos(\lambda - \xi)x\,d\lambda \to 0,$$

as $\delta \to 0$. Take the second relation, for example. Putting $\lambda - \xi = \mu$ we reduce it to

$$\int_0^\infty \phi(\delta x)\,dx \int_{M_2}^\infty \psi(\mu) \cos \mu x\,d\mu \to 0,$$

or, as the conditions of Theorem IV are satisfied, to

$$\frac{1}{\delta} \int_{M_2}^\infty \psi(\mu) \Delta\left(\frac{\mu}{\delta}\right) d\mu \to 0.$$

We can suppose M_2 chosen so that

$$|\psi(\mu)| < K \qquad (\mu \geqq M_2).$$

Then $\quad \dfrac{1}{\delta} \displaystyle\int_{M_2}^\infty \psi(\mu) \Delta\left(\dfrac{\mu}{\delta}\right) d\mu = \int_{M_2/\delta}^\infty \psi(\delta\lambda) \Delta(\lambda)\,d\lambda,$

which is in absolute value less than

$$K \int_{M_2/\delta}^\infty |\Delta(\lambda)|\,d\lambda$$

and so tends to zero with δ.

16. We can now state

THEOREM VII. *If*

(i) $\phi(x)$ *is subject to the conditions of* § 5,

(ii) $f(\lambda)$ *is subject to the conditions* (α) *or* (β),

(iii) $f(\xi + 0)$, $f(\xi - 0)$ *exist, or more generally,* $\psi(\xi + 0)$, $\psi(\xi - 0)$ *exist, where*

$$\psi(\lambda) = f(\lambda) \genfrac{}{}{0pt}{}{\cos}{\sin} a\lambda;$$

then will the integral $\quad \displaystyle\int_0^\infty dx \int_{-\infty}^\infty f(\lambda) \genfrac{}{}{0pt}{}{\cos}{\sin} a\lambda \cos(\lambda - \xi)x\,d\lambda$

be summable (ϕ), *and its sum will be*

$$\tfrac{1}{2}\pi \genfrac{}{}{0pt}{}{\cos}{\sin} a\xi \{f(\xi + 0) + f(\xi - 0)\},$$

or

$$\tfrac{1}{2}\pi \{\psi(\xi + 0) + \psi(\xi - 0)\}.$$

V. Summability by Cesàro's method.

17. It will be convenient at this stage to prove

Theorem VIII. *If the conditions of Theorem VII are satisfied (except (i)) then the same result holds in respect of summability (C 1).*

We have to prove that

$$\frac{1}{X}\int_0^X dx \int_0^x dt \int_{-\infty}^{\infty} f(\lambda) \genfrac{}{}{0pt}{}{\cos}{\sin} a\lambda \cos(\lambda - \xi) t\, d\lambda \to \tfrac{1}{2}\pi \genfrac{}{}{0pt}{}{\cos}{\sin}(a\xi)\{f(\xi+0)+f(\xi-0)\}$$

as $X \to \infty$. It is clear that the generality of the investigation is in no way affected by supposing $\xi = 0$.

The triple repeated integral is then the same as

$$\frac{1}{X}\int_0^X (X-x)\, dx \int_{-\infty}^{\infty} f(\lambda) \genfrac{}{}{0pt}{}{\cos}{\sin} a\lambda \cos \lambda x\, d\lambda = \int_{-\infty}^{\infty} f(\lambda) \genfrac{}{}{0pt}{}{\cos}{\sin} a\lambda\, d\lambda \frac{1}{X}\int_0^X (X-x)\cos \lambda x\, dx$$

$$= \int_{-\infty}^{\infty} \psi(\lambda)\left(\frac{\sin w\lambda}{w\lambda}\right)^2 w\, d\lambda,$$

where $\psi(\lambda) = f(\lambda) \genfrac{}{}{0pt}{}{\cos}{\sin} a\lambda$ and $w = \tfrac{1}{2}X$. We have therefore to determine the limit of this integral as $w \to \infty$. I shall prove first that, as $w \to \infty$,

$$\int_0^{\Lambda} \psi(\lambda)\left(\frac{\sin w\lambda}{w\lambda}\right)^2 w\, d\lambda \to \tfrac{1}{2}\pi \psi(+0).$$

Let $\psi(\lambda) = \psi(+0) + \chi(\lambda);$

then $\chi(\lambda) \to 0$ as $\lambda \to 0$. And it is clear that what we have to prove is that

$$\int_0^{\Lambda} \chi(\lambda)\left(\frac{\sin w\lambda}{w\lambda}\right)^2 w\, d\lambda \to 0.$$

Choose λ_0 so that $|\chi(\lambda)| < \epsilon \quad (0 < \lambda \leq \lambda_0);$

then

$$\left|\int_0^{\lambda_0} \chi(\lambda)\left(\frac{\sin w\lambda}{w\lambda}\right)^2 w\, d\lambda\right| < \epsilon \int_0^{\infty}\left(\frac{\sin w\lambda}{w\lambda}\right)^2 w\, d\lambda = \tfrac{1}{2}\pi\epsilon.$$

Also

$$\left|\int_{\lambda_0}^{\Lambda} \chi(\lambda)\left(\frac{\sin w\lambda}{w\lambda}\right)^2 w\, d\lambda\right| < \frac{1}{w\lambda_0^2}\int_0^{\Lambda} |\chi(\lambda)|\, d\lambda;$$

and from these two inequalities our conclusion follows at once.

In order to complete the proof of the theorem we have only to prove that

$$\int_{\Lambda}^{\infty} \psi(\lambda)\left(\frac{\sin w\lambda}{w\lambda}\right)^2 w\, d\lambda \to 0.$$

We can suppose Λ so chosen that

$$|\psi(\lambda)| < K \quad (\lambda \geq \Lambda),$$

and then
$$\left|\int_\Lambda^\infty \psi(\lambda)\left(\frac{\sin w\lambda}{w\lambda}\right)^2 w\,d\lambda\right| < K\int_\Lambda^\infty \left(\frac{\sin w\lambda}{w\lambda}\right)^2 w\,d\lambda = K\int_{w\Lambda}^\infty \left(\frac{\sin u}{u}\right)^2 du,$$
which plainly tends to zero as $w \to \infty$. Thus
$$\int_0^\infty \psi(\lambda)\left(\frac{\sin w\lambda}{w\lambda}\right)^2 w\,d\lambda \to \tfrac{1}{2}\pi\psi(+0);$$
and the truth of the theorem follows immediately*.

VI. *The introduction of Cauchy's Principal Value.*

18. The condition relating to the convergence of the integrals
$$\int_{-\infty}^{\ } \left|\frac{f(\lambda)}{\lambda}\right| d\lambda, \quad \int^\infty \left|\frac{f(\lambda)}{\lambda}\right| d\lambda$$
was introduced in Theorem III. In fact, if the condition is *not* satisfied, it is not generally true that
$$\int_0^\ \phi(x)\,dx \int_\Lambda^\infty f(\lambda) \sin \lambda x\, d\lambda$$
is convergent: although this is true of the corresponding integral involving $\cos \lambda x$.

Suppose, for example, that
$$f(\lambda) = \frac{1}{\log \lambda}$$
so that the condition is not satisfied. Then it is not hard to show that, as $x \to 0$,
$$\int_\Lambda^\infty \frac{\cos \lambda x}{\log \lambda} d\lambda \sim \tfrac{1}{2}\pi \frac{1}{x\{\log(1/x)\}^2},$$
$$\int_\Lambda^\infty \frac{\sin \lambda x}{\log \lambda} d\lambda \sim \frac{1}{x\{\log(1/x)\}},$$
the first, but not the second, function, being integrable down to $x = 0$.

If the condition is not satisfied, however, we can impose an additional condition on $\phi(x)$, in the neighbourhood of $x = 0$, so that†
$$P \int_{-x_0}^\infty \phi(x)\,dx \int_\Lambda^\infty f(\lambda) \sin \lambda x\, d\lambda$$
$(x_0 > 0)$ shall still be convergent and equal to
$$\int_\Lambda^\infty f(\lambda)\,d\lambda \int_{-x_0}^\infty \phi(x) \sin \lambda x\, dx,$$
as I shall now proceed to prove.

---

* It is not difficult to prove that, if $f(\lambda)$ is of limited total fluctuation in an interval including $\lambda = \xi$, then
$$\left|\int_{-\infty}^\infty \psi(\lambda) \cos(\lambda - \xi)x\, d\lambda\right| < K/x$$
for large values of x. It then follows from Theorem (v) of § 6 that Fourier's double-integral is convergent in the ordinary sense, so that his integral-theorem becomes a corollary of Theorem VIII. Similarly we can use Mr Littlewood's extension of Theorem (iv) of § 6 to exhibit Fourier's integral-theorem as a corollary of Theorem VII. Analogous remarks apply to Dirichlet's form of Fourier's series theorem: cf. Littlewood, *loc. cit.*, and Hardy, *Proc. Lond. Math. Soc.* vol. VIII. p. 308.

† For a detailed exposition of the theory of Cauchy's "Principal Values" I may refer to four papers in the *Proc. Lond. Math. Soc.* (*Old Series*, vol. XXXIV. p. 16 and p. 55 and vol. XXXV. p. 81, and *New Series*, vol. VII. p. 181).

19. Since
$$\int_{x_0}^{\infty}\int_{\Lambda}^{\infty} = \int_{\Lambda}^{\infty}\int_{x_0}^{\infty},$$

it is clear that what we have to prove is that

$$P\int_{-x_0}^{x_0}\int_{\Lambda}^{\infty} = \int_{\Lambda}^{\infty}\int_{-x_0}^{x_0}.$$

The principal value will be convergent, and it will be legitimate to calculate its value by a change in the order of integration, if and only if the same is true of the ordinary integral

$$\int_0^{x_0} x\chi(x)\,dx \int_{\Lambda}^{\infty} f(\lambda)\sin\lambda x\,d\lambda$$

where
$$x\chi(x) = \phi(x) - \phi(-x).$$

We suppose $\chi(x)$ continuous and monotonic near $x = 0$.

Now
$$\int_0^{x_0}\int_{\Lambda}^{\Lambda'}$$

is convergent, and may be calculated by inversion, for any finite value of Λ'. Hence what we have to prove is that

$$\int_0^{x_0}\int_{\Lambda'}^{\infty}$$

is convergent and tends to zero as $\Lambda' \to \infty$.

Now, if $f(\lambda)$ is monotonic, $\left|\int_{\Lambda'}^{\infty} f(\lambda)\sin\lambda x\,d\lambda\right| < \dfrac{2}{x}|f(\Lambda')|$

and so
$$\int_0^{x_0} x\,dx \int_{\Lambda'}^{\infty} f(\lambda)\sin\lambda x\,d\lambda$$

is convergent. Hence, by the second mean-value theorem,

$$\int_0^{x_0} x\chi(x)\,dx \int_{\Lambda'}^{\infty} f(\lambda)\sin\lambda x\,d\lambda$$

is convergent and equal to

$$\chi(0)\int_0^{x_1} x\,dx \int_{\Lambda'}^{\infty} f(\lambda)\sin\lambda x\,d\lambda + \chi(x_0)\int_{x_1}^{x_0} x\,dx \int_{\Lambda'}^{\infty} f(\lambda)\sin\lambda x\,d\lambda,$$

where $0 < x_1 < x_0$. These repeated integrals are convergent, and may be calculated by inversion. The only point in this statement that requires proof is that the first of them may be calculated by inversion. We have, however small $\bar{x}$,

$$\int_{\bar{x}}^{x_1} x\,dx \int_{\Lambda'}^{\infty} f(\lambda)\sin\lambda x\,d\lambda = \int_{\Lambda'}^{\infty}\int_{\bar{x}}^{x_1} = \int_{\Lambda'}^{\infty} \frac{f(\lambda)}{\lambda^2}\{\theta(x_1) - \theta(\bar{x})\}\,d\lambda,$$

where
$$\theta(x) = \sin\lambda x - \lambda x\cos\lambda x.$$

We have therefore to show that

$$\bar{x}\int_{\Lambda'}^{\infty} \frac{f(\lambda)}{\lambda}\cos\lambda\bar{x}\,d\lambda - \int_{\Lambda'}^{\infty} \frac{f(\lambda)}{\lambda^2}\sin\lambda\bar{x}\,d\lambda$$

tends to zero as $\bar{x} \to 0$. This is obvious in the case of the second integral. In the case of the first we have

$$\bar{x}\int_{\Lambda'}^{\infty} = \bar{x}\int_{\Lambda'}^{\Lambda''} + \bar{x}\int_{\Lambda''}^{\infty}.$$

The second integral is numerically less than

$$\bar{x} \cdot \frac{|f(\Lambda'')|}{\Lambda''} \cdot \frac{2}{\bar{x}},$$

which may be made less than ϵ by choice of Λ'', independently of $\bar{x}$; and the first term may then be made less than ϵ by choice of $\bar{x}$.

Hence
$$\int_0^{x_0} x\,\chi(x)\,dx \int_{\Lambda'}^{\infty} f(\lambda) \sin \lambda x\,d\lambda = \chi(0)\int_{\Lambda'}^{\infty} \frac{f(\lambda)}{\lambda^2}\,\theta(x_1)\,d\lambda$$
$$+ \chi(x_0)\int_{\Lambda'}^{\infty} \frac{f(\lambda)}{\lambda^2}\{\theta(x_0) - \theta(x_1)\}\,d\lambda,$$

and, by what is practically a repetition of the argument which immediately precedes, we can prove that this tends to zero as $\Lambda' \to \infty$.

20. THEOREM IX. *If the condition that the integrals*

$$\int_{-\infty}^{\infty} \left|\frac{f(\lambda)}{\lambda}\right| d\lambda, \quad \int^{\infty} \left|\frac{f(\lambda)}{\lambda}\right| d\lambda$$

be dropped, in Theorem III, the result of that theorem will in general not be correct; but if we impose upon $\phi(x)$ the additional condition that

$$\chi(x) = \frac{1}{x}\{\phi(x) - \phi(-x)\}$$

is continuous and monotonic near $x = 0$, it will still be true that

$$P\int_{-x_0}^{\infty} \phi(x)\,dx \int_{-\infty}^{\infty} f(\lambda) \sin \lambda x\,d\lambda = \int_{-\infty}^{\infty} f(\lambda)\,d\lambda \int_{-x_0}^{\infty} \phi(x) \sin \lambda x\,dx,$$

where $x_0 > 0$, and P is the sign of Cauchy's Principal Value.

THEOREM X. *In Theorem IV we may drop condition* (iii bγ), *if we introduce the additional condition that*

$$\chi(x) = \frac{\phi(a+x) - \phi(a-x)}{x}$$

is continuous and monotonic near $x = 0$, and insert the sign of the principal value before the outer integration with regard to x.

21. We have now to consider how far these modifications affect the work of § IV. In the first place, the effect on the proof of Theorem V is obviously *nil*. Secondly, the missing condition affects the proof of Theorem VI only in so far as it is required to justify a certain inversion of integrations—its effect on which we have already discussed. Hence

THEOREM XI. *Theorem VI remains valid if modified as Theorem IV was modified above.*

VII. *The introduction of a function defined by an infinite series.*

22. Let
$$F(\lambda) = \sum_{0}^{\infty} c_\nu \cos(q_\nu \lambda + r_\nu),$$
where (q_ν) is an ascending sequence of positive numbers, tending to ∞ with ν, and
$$\sum |c_\nu|$$
is convergent. I shall now prove that if we assume that $\phi(x)$ and $f(\lambda)$ satisfy the conditions of Theorem IV, and in addition that $\phi(x)$ is monotonic, or has at most a finite number of maxima and minima, then we may invert the order of integration in the integral
$$\int_0^\infty \phi(x)\,dx \int_{-\infty}^\infty f(\lambda)\,F(\lambda+y)\cos\lambda x\,d\lambda$$
(for any value of y).

This will be proved if we can justify the following series of inversions:

$$\int_0^\infty \phi(x)\,dx \int_{-\infty}^\infty f(\lambda) \cos\lambda x\,d\lambda \sum_0^\infty c_\nu \cos\{q_\nu(\lambda+y)+r_\nu\}$$

$$= \int_0^\infty \sum_0^\infty \int_{-\infty}^\infty \quad \ldots\ldots\ldots\ldots\ldots\ldots\ldots\ldots\ldots(A)$$

$$= \sum_0^\infty \int_0^\infty \int_{-\infty}^\infty \quad \ldots\ldots\ldots\ldots\ldots\ldots\ldots\ldots\ldots(B)$$

$$= \sum_0^\infty \int_{-\infty}^\infty \int_0^\infty \quad \ldots\ldots\ldots\ldots\ldots\ldots\ldots\ldots\ldots(C)$$

$$= \int_{-\infty}^\infty \sum_0^\infty \int_0^\infty \quad \ldots\ldots\ldots\ldots\ldots\ldots\ldots\ldots\ldots(D)$$

$$= \int_{-\infty}^\infty \int_0^\infty \sum_0^\infty \quad \ldots\ldots\ldots\ldots\ldots\ldots\ldots\ldots\ldots(E)$$

(A) The equation
$$\int_{-\infty}^\infty f(\lambda)\cos\lambda x\,d\lambda \sum_0^\infty c_\nu \cos\{q_\nu(\lambda+y)+r_\nu\} = \sum_0^\infty \int_{-\infty}^\infty$$
will be true if

(i) the series is uniformly convergent throughout any finite interval of values of λ (as is certainly the case),

(ii)
$$\sum_0^\infty \int_\Lambda^\infty$$
is convergent and tends to zero as $\Lambda \to \infty$,

(iii)
$$\sum_0^\infty \int_{-\infty}^{-\Lambda}$$
is convergent and tends to zero as $\Lambda \to \infty$. It is clearly enough to justify the assertion (ii). Its truth is practically obvious if $f(\lambda)$ satisfies condition (iii a) of Theorem IV: we shall therefore suppose that $f(\lambda)$ is subject to the set of conditions (iii b).

We can choose ν_0 so that $q_{\nu_0} > x$, and it is enough to consider the series

$$\sum_{\nu_0}^{\infty} \int_{\Lambda}^{\infty}.$$

But $\int_{\Lambda}^{\infty}$ is numerically less than the product of $|c_\nu|$ and

$$\frac{1}{2}\left|\int_{\Lambda}^{\infty} f(\lambda) \cos\{(q_\nu + x)\lambda + q_\nu y + r_\nu\}\, d\lambda\right| + \frac{1}{2}\left|\int_{\Lambda}^{\infty} f(\lambda) \cos\{(q_\nu - x)\lambda + q_\nu y + r_\nu\}\, d\lambda\right|$$

and so less than

$$|f(\Lambda)|\left(\frac{1}{q_\nu + x} + \frac{1}{q_\nu - x}\right).$$

It follows that our series is convergent and numerically less than

$$|f(\Lambda)| \sum_{\nu_0}^{\infty} |c_\nu| \left(\frac{1}{q_\nu + x} + \frac{1}{q_\nu - x}\right);$$

and so tends to zero as $\Lambda \to \infty$.

(B) The equation

$$\int_0^{\infty} \sum_0^{\infty} \int_{-\infty}^{\infty} = \sum_0^{\infty} \int_0^{\infty} \int_{-\infty}^{\infty}$$

will certainly be true if

(i)
$$\int_0^X \sum_0^{\infty} \int_{-\infty}^{\infty} = \sum_0^{\infty} \int_0^X \int_{-\infty}^{\infty}$$

for any finite X, and

(ii)
$$\sum_0^{\infty} \int_X^{\infty} \int_{-\infty}^{\infty}$$

is convergent and tends to zero as $X \to \infty$. As above I confine myself to the case in which $f(\lambda)$ satisfies the conditions (iii b), leaving the easier case in which it satisfies (iii a) to the reader.

It is easily proved, by a slight modification of the argument used under (A) above, that

$$\sum_0^{\infty} \int_{-\infty}^{\infty}$$

is uniformly convergent for $0 \leq x \leq X$. In the first place, this is obviously true of

$$\sum_0^{\infty} \int_{\Lambda_1}^{\Lambda_2}$$

for any finite values of Λ_1 and Λ_2; for the series may be compared with

$$\sum_0^{\infty} |c_\nu| \int_{\Lambda_1}^{\Lambda_2} |f(\lambda)|\, d\lambda.$$

We have therefore only to justify the assertion for

$$\sum_0^{\infty} \int_{\Lambda}^{\infty}.$$

Choose ν_0 so that $q_{\nu_0} > X$. Then we need only consider

$$\sum_{\nu_0}^{\infty} \int_{\Lambda}^{\infty}$$

and the result follows at once from the analysis given under (A). It should be observed that each term of the series may become infinite for one special value of x, viz. $x = q_\nu$. But as this only happens, in the interval $(0, X)$, to a finite number of terms, no difficulty is caused thereby.

Next, as regards (ii), we observe first that
$$\sum_0^\infty \int_X^\infty \int_{\Lambda_1}^{\Lambda_2}$$
is convergent and tends to zero as $X \to \infty$, for any finite values of Λ_1 and Λ_2. For the series may be compared with
$$\sum_0^\infty |c_\nu| \int_X^\infty |\phi(x)| dx \int_{\Lambda_1}^{\Lambda_2} |f(\lambda)| d\lambda,$$
which has certainly this property. We need therefore only establish the property for a series of the type
$$\sum_0^\infty \int_X^\infty \int_\Lambda^\infty .$$

Now
$$\int_X^\infty \int_\Lambda^\infty = \int_\Lambda^\infty \int_X^\infty,$$
and
$$\left| \int_X^\infty \phi(x) \cos \lambda x \, dx \right| < \frac{2|\phi(X)|}{\lambda}*,$$
and so
$$\left| \int_\Lambda^\infty \int_X^\infty \right| < 2|\phi(X)| \int_\Lambda^\infty \frac{|f(\lambda)|}{\lambda} d\lambda.$$
Hence
$$\left| \sum_0^\infty \int_X^\infty \int_\Lambda^\infty \right| < 2|\phi(X)| \sum_0^\infty |c_\nu| \int_\Lambda^\infty \frac{|f(\lambda)|}{\lambda} d\lambda$$
and the truth of (ii) is established.

(C) The legitimacy of this inversion of the order of integration has already been established.

(D) In order to prove that
$$\sum_0^\infty \int_{-\infty}^\infty \int_0^\infty = \int_{-\infty}^\infty \sum_0^\infty \int_0^\infty$$
we must prove that (i)
$$\sum_0^\infty \int_{\Lambda_1}^{\Lambda_2} \int_0^\infty = \int_{\Lambda_1}^{\Lambda_2} \sum_0^\infty \int_0^\infty$$
for any finite values of Λ_1 and Λ_2, and

(ii)
$$\sum_0^\infty \int_\Lambda^\infty \int_0^\infty$$
is convergent and tends to zero as $\Lambda \to \infty$.

The truth of (i) follows at once from the fact that the general term of the series
$$\sum_0^\infty \int_0^\infty$$
is numerically less than
$$|c_\nu| \int_0^\infty |\phi(x)| dx.$$

* Provided $\phi(x)$ is monotonic for $x \geqq X$, a condition certainly satisfied if X is large enough.

The truth of (ii) follows at once from the inequalities

$$\left| \int_0^\infty \phi(x) \cos \lambda x \, dx \right| < \frac{K}{\lambda},$$

$$\left| \int_\Lambda^\infty f(\lambda) \cos \{q_\nu (\lambda + y) + r_\nu\} \, d\lambda \int_0^\infty \phi(x) \cos \lambda x \, dx \right| < K \int_\Lambda^\infty \frac{|f(\lambda)|}{\lambda} \, d\lambda.$$

(E) Finally, in this inversion there is nothing to justify, the two limit operations applying to entirely distinct factors of the subject of integration and summation.

Hence we deduce

THEOREM XII. *If*
$$F(\lambda) = \sum_0^\infty c_\nu \cos (q_\nu \lambda + r_\nu),$$

where (q_ν) is an ascending sequence of positive numbers whose limit is infinity, and Σc_ν is an absolutely convergent series: if further $\phi(x)$, besides satisfying the conditions of Theorem IV, has at most a finite number of maxima and minima; and $f(\lambda)$ is subject to the conditions of Theorem IV (including (iii bγ)); then will

$$\int_0^\infty \phi(x) \, dx \int_{-\infty}^\infty f(\lambda) F(\lambda + y) \cos \lambda x \, d\lambda = \int_{-\infty}^\infty f(\lambda) F(\lambda + y) \, dy \int_0^\infty \phi(x) \cos \lambda x \, dx.$$

THEOREM XIII. *We have, under similar conditions,*

$$\int_0^\infty \phi(x) \, dx \int_{-\infty}^\infty f(\lambda) F(\lambda) \cos (\lambda - \xi) x \, d\lambda = \int_{-\infty}^\infty f(\lambda) F(\lambda) \, d\lambda \int_0^\infty \phi(x) \cos (\lambda - \xi) x \, dx.$$

23. THEOREM XIV. *Let $F(\lambda)$ be defined as in Theorem XII. Let $\phi(x)$ and $f(\lambda)$ be subject to the conditions of Theorem VII. Then will the integral*

$$\int_0^\infty dx \int_{-\infty}^\infty f(\lambda) F(\lambda) \cos (\lambda - \xi) x \, d\lambda$$

be summable (ϕ), and its sum will be

$$\tfrac{1}{2} \pi F(\xi) \{f(\xi + 0) + f(\xi - 0)\},$$

or, more generally,
$$\tfrac{1}{2} \pi \{\psi(\xi + 0) + \psi(\xi - 0)\},$$

where
$$\psi(\lambda) = f(\lambda) F(\lambda).$$

In fact no substantial modification is required in the argument of Section IV in consequence of this more general hypothesis as to the structure of the function $\psi(\lambda)$.

CORRECTIONS

p. 430, line 13. The inner integral should be $\int_\beta^\gamma$.

p. 433, line 3 up. K should be K_1.

p. 434, line 6 up. 'finite integral' should be 'finite interval'.

NOTES ON SOME POINTS IN THE INTEGRAL CALCULUS.

By *G. H. Hardy.*

XXXIII.

Some cases of the inversion of the order of integration.

1. I HAVE discussed elsewhere,* for a special purpose, the question of the inversion of the order of integration expressed by the formula

$$(1) \quad \int_{-\infty}^{\infty} \phi(x)\, dx \int_{-\infty}^{\infty} f(\lambda) \genfrac{}{}{0pt}{}{\cos}{\sin} \lambda x\, d\lambda$$
$$= \int_{-\infty}^{\infty} f(\lambda)\, d\lambda \int_{-\infty}^{\infty} \phi(x) \genfrac{}{}{0pt}{}{\cos}{\sin} \lambda x\, dx.$$

I proved there that the formula holds under a variety of conditions, which I will re-state.

THEOREM 1. *The inversion is legitimate if* (i) ϕ *and* f *are regularly integrable†in any finite intervals,* (ii) *the integrals*

$$\int_{-\infty}^{\infty} |\phi|\, dx, \quad \int_{-\infty}^{\infty} |f|\, d\lambda$$

are convergent.‡

This theorem is of course only a very special case of a general theorem due to de la Vallée-Poussin.§

THEOREM 2. *The inversion is legitimate if* (i) ϕ *and* f *are regularly integrable in any finite intervals,* (ii) *the integral*

$$\int_{-\infty}^{\infty} |\phi(x)|\, dx$$

is convergent, (iii) $\phi(x)$ *tends steadily to limits* $\phi(+0), \phi(-0)$

---

* "Fourier's double integral and the theory of divergent integrals," *Camb. Phil. Trans.*, vol. xxi., p. 427. I refer to this paper as "F. I."
† *I.e.*, integrable and absolutely integrable; see "F. I.," p. 427 (footnote) and p. 434.
‡ "F. I.," p. 436.
§ Bromwich, *Infinite Series*, p. 457.

as $x \to 0$ *by positive or negative values*, (iv) $f(\lambda)$ *tends steadily to zero as* $\lambda \to \infty$ *or* $\lambda \to -\infty$, (v) *the integrals*

$$\int^{\infty} \frac{f(\lambda)}{\lambda} d\lambda, \quad \int_{-\infty} \frac{f(\lambda)}{\lambda} d\lambda$$

are convergent.*

With regard to the conditions of this theorem it was to be remarked (*a*) that condition (v) is unnecessary in the case of the cosine integral, and (*b*) that condition (iv) may be replaced by the more general condition that *$f(\lambda)$ is of limited total fluctuation in intervals* $(-\infty, -l)$, (l, ∞)—a condition which will certainly be satisfied if $f'(\lambda)$ *exists and is absolutely integrable up to* ∞ *and down to* $-\infty$.

2. These sets of conditions were general enough for the end that I had in view. But the question is one which arises frequently in analysis, and I find that there are simple and interesting applications which require more general conditions. I propose, therefore, in this and a subsequent note, to state and illustrate some additional theorems. The conditions of these theorems are framed with an eye to applications, and make no pretence to a *maximum* of generality.

3. In the theorems which follow I suppose, for the sake of simplicity of statement, that the range of integration with respect to each variable is $(0, \infty)$. There is, of course, no difficulty in modifying the enunciations so as to apply to a range infinite both ways. The first two theorems apply only to the integral which contains $\sin \lambda x$.

THEOREM 3. *The inversion is legitimate, in the case of the sine-integral, if* (i) $\phi(x)$ *is regularly integrable throughout any finite interval*, (ii) $f(\lambda)$ *is regularly integrable throughout any finite interval which does not include* $\lambda = 0$, (iii) $f(\lambda)$ *may be expressed, near* $\lambda = 0$, *in the form* $\lambda^{-1-s} F(\lambda)$, *where* $0 \leq s < 1$ *and* $F(\lambda)$ *tends steadily to a limit* $F(+0)$ *as* $\lambda \to 0$, (iv) *the integrals*

$$\int_0^\infty x^s |\phi| \, dx, \quad \int_{\lambda_0}^\infty |f| \, d\lambda \quad (\lambda_0 > 0)$$

are convergent.

---

* "F. I.," p. 437: the theorem there is stated for intervals of integration $(0, \infty)$ instead of $(-\infty, \infty)$. It is, of course, to be understood that $\phi(x)$ and $f(\lambda)$ need only be monotonic for sufficiently small values of x and sufficiently large values of λ.

THEOREM 4. *The inversion is legitimate if* (i) $f(\lambda)$ *satisfies conditions* (ii) *and* (iii) *of Theorem* 3, (ii) $\phi(x)$ *satisfies similar conditions* (*with r instead of s*), (iii) *the integrals*

$$\int_{x_0}^{\infty} x^s |\phi| \, dx \quad (x_0 > 0), \qquad \int_{\lambda_0}^{\infty} \lambda^r |f| \, d\lambda \quad (\lambda_0 > 0)$$

are convergent.

It will be noted that if (*e.g.*) $f(\lambda)$ is of the form $F(\lambda)/\lambda$ in the neighbourhood of the origin (so that $s=0$), the condition imposed on $\phi(x)$ in these theorems, as regards its behaviour at infinity, is identical with that of Theorem 1.

These theorems are extensions of Theorem 1. As an extension of Theorem 2, we have the following theorem (applying to both the sine and cosine integrals).

THEOREM 5. *The inversion is legitimate if* (i) ϕ *satisfies conditions* (i) *and* (ii) *of Theorem* 2, (ii) f *satisfies conditions* (i) *and* (iv) *of Theorem* 2,* (iii) ϕ *can be expressed, near* $x=0$, *in the form* $x^{-s} \psi(x)$, *where* $0 \leq s < 1$, *and* $\psi(x)$ *tends steadily to a limit* $\psi(+0)$ *as* $x \to 0$, (iv) *the integral*

$$\int^{\infty} \frac{f(\lambda)}{\lambda^{1-s}} \, d\lambda$$

s convergent.

These three theorems can all be proved by a modification of the arguments used in my former paper. I shall content myself with showing this in the case of Theorem 5.

4. We proceed exactly as in the proof of Theorem 2,† until we come to the last stage, where we have to prove that

$$(2) \qquad \int_0^{x_0} \frac{\psi(x)}{x^s} dx \int_l^{\infty} f(\lambda) \cos \lambda x \, d\lambda$$

is convergent and tends to zero as $l \to \infty$. I prove this first in the special case in which $\psi(x)$ is replaced by unity. We have

$$(3) \qquad \int_{\xi}^{x_0} \frac{dx}{x^s} \int_l^{\infty} f(\lambda) \cos \lambda x \, d\lambda = \int_l^{\infty} f(\lambda) \, d\lambda \int_{\xi}^{x_0} \frac{\cos \lambda x}{\lambda^s} dx,$$

---

* Naturally only in so far as *positive* values of x and λ are concerned.
† "F. I.," p. 438. The same argument applies to the sine-integral.

for $0 < \xi < x_0$. This equation will still hold for $\xi = 0$ if
$$\int_l^\infty f(\lambda)\, d\lambda \int_0^\xi \frac{\cos \lambda x}{x^s}\, dx$$
is convergent and tends to zero as $\xi \to 0$, or if this is true of

(4) $$\int_l^\infty \frac{f(\lambda)}{\lambda^{1-s}}\, d\lambda \int_0^{\lambda\xi} \frac{\cos u}{u^s}\, du.$$

Now $\int_0^{\lambda\xi} \frac{\cos u}{u^s}\, du$ is a continuous function of λ and ξ, and ess in absolute value than an absolute constant K. It follows hat (4) is convergent and uniformly convergent throughout in interval of values of ξ including the value 0. It therefore tends to zero as $\xi \to 0$. Hence

(5) $$\int_0^{x_0} \frac{dx}{x^s} \int_l^\infty f(\lambda) \cos \lambda x\, d\lambda = \int_l^\infty f(\lambda)\, d\lambda \int_0^{x_0} \frac{\cos \lambda x}{x^s}\, dx.$$

It follows, by the second mean value theorem, that the integral (2) is convergent and equal to

$$\psi(+0) \int_0^{x_1} \frac{dx}{x^s} \int_l^\infty f(\lambda) \cos \lambda x\, d\lambda + \psi(x_0) \int_{x_1}^{x_0} \frac{dx}{x^s} \int_l^\infty f(\lambda) \cos \lambda x\, d\lambda*$$
$$= \psi(+0) \int_l^\infty \frac{f(\lambda)}{\lambda^{1-s}}\, d\lambda \int_0^{\lambda x_1} \frac{\cos u}{u^s}\, du + \psi(x_0) \int_l^\infty \frac{f(\lambda)}{\lambda^{1-s}}\, d\lambda \int_{\lambda x_1}^{\lambda x_0} \frac{\cos u}{u^s}\, du,$$

where $0 < x_1 < x_0$. But this is plainly less in absolute value than a constant multiple of $\int_l^\infty \frac{f(\lambda)}{\lambda^{1-s}}\, d\lambda$,† and so tends to zero as $l \to \infty$. This completes the proof of Theorem 5.

It might be thought that there was room, in the case of the *sine* integral, for a further generalisation of Theorem 2, in which $\phi(x)$ or $f(\lambda)$ should behave, near $x = 0$ or $\lambda = 0$, like x^{-1-r} or λ^{-1-s}. A little consideration shows that, for practical purposes, there is only one such case of importance. If $\phi(x)$ had the form suggested, we should have to impose on $f(\lambda)$ a condition, viz., the convergence of $\int^\infty \lambda^r f(\lambda)\, d\lambda$, which cannot possibly be satisfied unless $\int^\infty f(\lambda)\, d\lambda$ is convergent. And when $f(\lambda)$ has the form suggested it usually happens, in cases of interest, that $\int^\infty f(\lambda)\, d\lambda$ is convergent,

---

* If ψ has an ordinary discontinuity for $x = x_0$ (as is consistent with the conditions), we must replace $\psi(x_0)$ by $\psi(x_0 - 0)$.

† $f(\lambda)$, being ultimately monotonic, is of course ultimately of constant sign we may suppose this sign positive.

so that it is hardly necessary to frame a general theorem to meet this case. The exception to these remarks arises when $s = 0$, so that $f(\lambda)$, near the origin, is of the form $F(\lambda)/\lambda$, where $F(\lambda)$ is monotonic. This case is of some importance. Suppose, for example, that the subject of integration is

$$e^{-x}\frac{\sin \lambda x}{\lambda}.$$

Integration, first with respect to x, gives

$$\int_0^\infty \frac{d\lambda}{1+\lambda^2} = \tfrac{1}{2}\pi.$$

Integration, first with respect to λ, gives

$$A\int_0^\infty e^{-x}\,dx = A,$$

where
$$A = \int_0^\infty \frac{\sin u}{u}\,du.$$

Thus, if the inversion of the order of integration can be justified, we see that $A = \tfrac{1}{2}\pi$. But, as neither of the integrals

$$\int_0 \frac{d\lambda}{\lambda}, \quad \int^\infty \frac{d\lambda}{\lambda}$$

is convergent, we cannot justify the inversion either by Theorem 2 or by Theorem 3. This case is met by

THEOREM 6. *The inversion is legitimate, in the case of the sine integral, if the conditions of Theorem 2 or of Theorem 5 are satisfied, except that, near $\lambda = 0$,*

$$f(\lambda) = F(\lambda)/\lambda,$$

where $F(\lambda)$ tends steadily to a limit as $\lambda \to 0$.*

The reasoning by which this result is established is of precisely the same character as that already used, and I need not write out a proof.

5. The theorems which precede may all be generalised by supposing the integrals (1) to contain, instead of $\cos \lambda x$ or $\sin \lambda x$, a general function $\theta(\lambda x)$ subject to appropriate restrictions. Thus Theorem 1 holds if $\theta(u)$ is any con-

tinuous function whose modulus has a finite upper limit; and Theorem 5 holds if we suppose in addition that

$$\int^u \theta(v)\,dv$$

oscillates at most finitely. This includes (for $s=0$) the corresponding generalisation of Theorem 2. Finally, Theorems 3, 4, and 6 hold if $\theta(u)$ vanishes to the first order for $u=0$ (*i.e.*, if $\theta(u) = u\Theta(u)$, where Θ is continuous for $u=0$). If, more generally, we suppose $\theta(u) = u^t\Theta(u)$, we must suppose r and s less than t.† All these conclusions follow without any serious change in the arguments we have used.

6. I proceed now to give some illustrations of the use of Theorems 1–5. The only difficulty is to make a selection from the large number that suggest themselves.

(α) Let

$$I(s) = \int_0^\infty x^{s-1}\cos x\,dx, \quad J(s) = \int_0^\infty x^{s-1}\sin x\,dx,$$

$$K(s) = \int_0^\infty \frac{x^{s-1}}{1+x}\,dx.$$

Then

$$\Gamma(1-s)I(s) = \int_0^\infty e^{-x}dx \int_0^\infty \lambda^{s-1}\cos\lambda x\,d\lambda$$

$$= \int_0^\infty \frac{\lambda^{s-1}d\lambda}{1+\lambda^2} = \tfrac{1}{2}\int_0^\infty \frac{\mu^{\frac{1}{2}s-1}d\mu}{1+\mu} = \tfrac{1}{2}K(\tfrac{1}{2}s)$$

if $0 < s < 1$. The inversion here is justified by Theorem 2. Again

$$\Gamma(1-s)J(s) = \int_0^\infty e^{-x}dx \int_0^\infty \lambda^{s-1}\sin\lambda x\,d\lambda$$

$$= \int_0^\infty \frac{\lambda^s d\lambda}{1+\lambda^2} = \tfrac{1}{2}\int_0^\infty \frac{\mu^{\frac{1}{2}(s-1)}}{1+\mu}d\mu = \tfrac{1}{2}K\{\tfrac{1}{2}(1+s)\}$$

if $-1 < s < 1$. The inversion here is justified by Theorem 2 if $0 < s < 1$, by Theorem 3 if $-1 < s < 0$, and by Theorem 6 if $s = 0$. Finally

$$\Gamma(s)\Gamma(1-s) = \int_0^\infty e^{-x}dx \int_0^\infty e^{-\lambda x}\lambda^{s-1}d\lambda = \int_0^\infty \frac{\lambda^{s-1}}{1+\lambda}d\lambda = K(s),$$

if $0 < s < 1$. Here we may appeal to de la Vallée-Poussin's

---

* Or $\lambda \to -0$ in the case of Theorem 2.
† These statements seem sufficiently general for ordinary purposes; there would of course be no difficulty in extending them further.

standard theorem. We thus obtain the values of $I(s)$, $J(s)$, and $K(s)$, in all cases in which they are convergent, in terms of gamma-functions: if we use the formula

$$\Gamma(s)\Gamma(1-s) = \pi\operatorname{cosec} s\pi,$$

or evaluate $K(s)$ independently, we obtain the ordinary forms of the values of the integrals.

(β) In this example I shall assume that we know the values of the integrals

$$\int_0^\infty e^{-x} x^{r-1} \frac{\cos}{\sin} \lambda x\, dx = \frac{\Gamma(r)}{(1+\lambda^2)^{\frac{1}{2}r}} \frac{\cos}{\sin}(r \arctan \lambda),$$

where $r > 0$ for the cosine integral, and $r > -1$ for the sine integral.

Now let us take

$$e^{-x} x^{r-1} \lambda^{s-1} \cos \lambda x,$$

where $r > 0$, $0 < s < 1$, and integrate from 0 to ∞ with respect to each variable. Integrating first with respect to λ, we obtain

$$\Gamma(s)\cos\tfrac{1}{2}s\pi \int_0^\infty e^{-x} x^{r-s-1} dx = \Gamma(s)\Gamma(r-s)\cos\tfrac{1}{2}s\pi,$$

provided $r > s$. Integrating first with respect to x, we obtain

$$\Gamma(r)\int_0^\infty \frac{\lambda^{s-1}}{(1+\lambda^2)^{\frac{1}{2}r}} \cos(r \arctan \lambda)\, d\lambda$$

$$= \Gamma(r)\int_0^{\frac{1}{2}\pi} (\cos\phi)^{r-s-1}(\sin\phi)^{s-1} \cos r\phi\, d\phi,$$

again provided $r > s$. Hence we are led to the formula

$$\int_0^{\frac{1}{2}\pi} (\cos\phi)^{r-s-1}(\sin\phi)^{s-1} \cos r\phi\, d\phi = \frac{\Gamma(s)\Gamma(r-s)}{\Gamma(r)} \cos\tfrac{1}{2}s\pi,$$

holding for $r > s$, $0 < s < 1$.* The inversion of the order of integration is justified by Theorem 5, since $\int^\infty \lambda^{s-r-1} d\lambda$ is convergent if $r > s$.

If we take

$$e^{-x} x^{r-1} \lambda^{s-1} \sin \lambda x$$

as the subject of integration, where $r > s$, $-1 < s < 1$, we obtain, in the same way, the formula

$$\int_0^{\frac{1}{2}\pi} (\cos\phi)^{r-s-1}(\sin\phi)^{s-1} \sin r\phi\, d\phi = \frac{\Gamma(s)\Gamma(r-s)}{\Gamma(r)} \sin\tfrac{1}{2}s\pi.$$

If $0 < s < 1$ (in which case also $r > 0$), the inversion is justified by Theorem 5. If $-1 < s < 0$ and $r > 0$, it is justified by Theorem 3; and if $s = 0$, $r > 0$ by Theorem 6. Finally, if $-1 < s < r \leq 0$, it is justified by Theorem 4, since

$$\int_0^\infty e^{-x} x^{r-s-1} dx, \quad \int_0^\infty \lambda^{s-r-1} d\lambda$$

are convergent.†

(γ) Let us take as our subject of integration

$$\lambda^{s-1} \operatorname{sech} \pi x \cos \lambda x \quad (0 < s < 1).$$

Using Theorem 2, we obtain

$$\tfrac{1}{2} \int_0^\infty \frac{\lambda^{s-1}}{\cosh \tfrac{1}{2}\lambda} d\lambda = \Gamma(s) \cos \tfrac{1}{2} s\pi \int_0^\infty \frac{x^{-s}}{\cosh \pi x} dx.$$

If we put $\lambda = 2\xi$ on the left-hand side and $\pi x = \xi$ on the right-hand side, and observe that

$$\tfrac{1}{2} \int_0^\infty \frac{\xi^{s-1}}{\cosh \xi} d\xi = \Gamma(s) \left(\frac{1}{1^s} - \frac{1}{3^s} + \frac{1}{5^s} - \ldots \right) = \Gamma(s) \eta(s),$$

say, we obtain the formula

$$\eta(1-s) = \Gamma(s) (\tfrac{1}{2}\pi)^{-s} \sin \tfrac{1}{2} s\pi \, \eta(s).\ddagger$$

7. There are still a variety of interesting questions to consider. We have supposed so far either (i) that $f(\lambda)$ is absolutely integrable up to ∞ or (ii) that $f(\lambda)$ is ultimately monotonic. We must consider next the case in which $f(\lambda)$ is the product of $F(\lambda)$ by an oscillating factor such as $\cos a\lambda$ or $\sin a\lambda$, $F(\lambda)$ being subject to (ii), but not to (i). There are also interesting cases in which neither $f(\lambda)$ nor $\phi(x)$ is absolutely integrable up to ∞. I shall return to these questions in another note.

---

* This formula was first given by Kummer; see Dirichlet-Meyer, *Bestimmte Integrale*, p. 224.

† The theorems give between them the *exact* ranges of r and s, for which the transformations are valid. The ultimate formulæ hold for wider ranges; in fact, the inequality $s < 1$ is superfluous. But this shows, not that the theorems do not give complete information about the inversion of integrations, but that the ultimate formulæ hold in cases in which the inversion is *not* legitimate. There is no difficulty in extending the formulæ to their full range by means of elementary reduction formulæ.

‡ See Bromwich, *Infinite Series*, p. 494. The formula is originally due to Schlömilch: see his *Compendium der höheren Analysis*, vol. ii., p. 286. It is of course analogous to the functional equation satisfied by $\zeta(s)$.

CORRECTIONS

p. 104, *last line of Theorem* 5. The first word is 'is'.
——, *formula* (3). λ^s should be x^s.
p. 105, *lines* 6, 7, 8. The first words are 'less', 'that', 'an'.

NOTES ON SOME POINTS IN THE INTEGRAL CALCULUS.

By *G. H. Hardy*.

XXXV.

On an integral equation.

§1. In a recent note* Mr. Bateman gives a formula for the solution of the equation

$$(1) \quad f(x) = \frac{1}{\pi} \int_0^\infty \tfrac{1}{2} \log \left(\frac{t+x}{t-x}\right)^2 \phi(t)\, dt \quad (x > 0),$$

viz.,
$$\phi(x) = \int_0^\infty \sin xt\, \chi(t)\, dt,$$

where $\chi(t)$ is determined from the equation

$$f(x) = \int_0^\infty \frac{\sin xt}{t} \chi(t)\, dt.$$

This solution has the obvious disadvantage that it demands the previous solution of another integral equation. A more effective solution can be deduced from results that I have given elsewhere.† I proved that if $f(x)$ is a function whose first two derivatives are continuous, and the integral

$$\int^\infty \frac{f(x) \log x}{x}\, dx$$

is convergent, then the equation

$$(2) \quad \tfrac{1}{2}\pi f(x) = P \int_0^\infty \frac{x}{t^2 - x^2} g(t)\, dt$$

is satisfied by

$$(3) \quad \tfrac{1}{2}\pi g(x) = - P \int_0^\infty \frac{t}{t^2 - x^2} f(t)\, dt.‡$$

It is important in applications to extend the result to certain simple cases in which $f(x)$ is discontinuous. It is easy to see that it still holds if we can divide the interval $(0, \infty)$ into a finite number of parts $(0, a)$, (a, b), ..., (k, ∞)

---

* *Messenger*, vol. xli., p. 180.
† *Proc. Lond. Math. Soc.*, vol. vii., pp. 181 *et seq.*
‡ *l.c.*, p. 204. The actual inversion formula is (b) of §25. It may be deduced from Theorem C (pp. 203-4) by supposing ϕ an odd function and making certain elementary transformations.

inside each of which f and its first two derivatives are continuous, while at the points of division they have at most ordinary discontinuities.

Now let us suppose $\phi(t)$ continuous, and write
$$\int_0^x \phi(t)\,dt = g(x).$$
Then
$$\int_0^{x-\epsilon} \tfrac{1}{2}\log\left(\frac{t+x}{t-x}\right)^2 \phi(t)\,dt = \log\left(\frac{2x-\epsilon}{\epsilon}\right) g(x-\epsilon)$$
$$+ 2\int_0^{x-\epsilon} \frac{x}{t^2-x^2} g(t)\,dt,$$
and, provided $g(t)/t \to 0$ as $t \to \infty$,
$$\int_{x+\epsilon}^\infty \tfrac{1}{2}\log\left(\frac{t+x}{t-x}\right)^2 \phi(t)\,dt = -\log\left(\frac{2x+\epsilon}{\epsilon}\right) g(x+\epsilon)$$
$$+ 2\int_{x+\epsilon}^\infty \frac{x}{t^2-x^2} g(t)\,dt.$$
But, as $g'(x)$ is continuous,
$$\lim_{\epsilon \to 0}\left\{\log\left(\frac{2x-\epsilon}{\epsilon}\right) g(x-\epsilon) - \log\left(\frac{2x+\epsilon}{\epsilon}\right) g(x+\epsilon)\right\} = 0.$$

Hence (1) may be transformed into (2), and, if f satisfies the conditions stated, a solution is given by (3). It is then not difficult to verify that g has a continuous derivative, and $g(t)/t \to 0$ as $t \to \infty$.* Thus if $f(x)$ is any function subject to the conditions of my former paper, a solution of (i) is given by
$$\phi(x) = -\frac{2}{\pi}\frac{d}{dx} P\int_0^\infty \frac{t}{t^2-x^2} f(t)\,dt.$$

The solution (3) is obviously not the only solution of (2), for we can add any constant to $g(t)$ without affecting the value of
$$P\int_0^\infty \frac{x}{t^2-x^2} g(t)\,dt.$$
It is not difficult to impose restrictions on g which are sufficient to ensure the uniqueness of the solution; but I do not propose to enter into this at present.

---

* The necessary apparatus will be found in my papers on Cauchy's Principal Values, of which the paper already quoted is the last.

§ 2. If g is to be continuous without exception, f must vanish for $x = 0$, as when
$$f(x) = \sin ax, \qquad g(x) = -\cos ax.$$
In this case, we have
$$P\int_0^\infty \frac{x \cos at}{t^2 - x^2} dt = -\tfrac{1}{2}\pi \sin ax, \qquad P\int_0^\infty \frac{t \sin at}{t^2 - x^2} dt = \tfrac{1}{2}\pi \cos ax.$$
If f does not vanish for $x = 0$, g will have a logarithmic infinity for $x = 0$. Suppose, for example, that
$$f = 1 \quad (0 \leq t \leq 1), \qquad f = 0 \quad (t > 1).$$
An elementary calculation gives
$$g(x) = -\frac{1}{2\pi} \log\left(1 - \frac{1}{x^2}\right)^2;$$
so that
$$P\int_0^\infty \frac{x}{t^2 - x^2} \log\left(1 - \frac{1}{t^2}\right)^2 dt = \begin{matrix} -\pi^2 & (0 < x < 1) \\ 0 & x > 1 \end{matrix}.$$
For $x = 1$, the value of the integral is $-\tfrac{1}{2}\pi^2$.

§ 3. Next let us suppose
$$f(x) = x \quad (0 < x \leq a), \qquad f(x) = a \quad (x > a).$$
In this case the conditions for the truth of the equations (2) and (3) are not satisfied, and the integral on the right-hand side of (3) is, in fact, divergent. We can, however, obtain a solution by first differentiating (1) with respect to x. This (assuming the legitimacy of the differentiation, which is easily verified afterwards) gives
$$f'(x) = \frac{2}{\pi} P\int_0^\infty \frac{t}{t^2 - x^2} \phi(t)\, dt.$$
As $f' = 1$ $(0 < x < a)$ and $f' = 0$ $(x > a)$, we obtain
$$\phi(x) = -\frac{2}{\pi} P\int_0^a \frac{x\, dt}{t^2 - x^2} = \frac{1}{2\pi} \log\left(\frac{x+a}{x-a}\right)^2.$$
We thus obtain Mr. Bateman's elegant formula
$$\int_0^\infty \log\left(\frac{t+x}{t-x}\right)^2 \log\left(\frac{t+a}{t-a}\right)^2 dt = \begin{matrix} 4\pi^2 x & (0 \leq x \leq a) \\ 4\pi^2 a & (x > a) \end{matrix}^*.$$

---

* In the *Proc. Lond. Math. Soc.* (old series), vol. xxxv, p. 86, I gave the formulæ
$$\int_0^\infty \log\left(1 - \frac{a^2}{x^2}\right)^2 \log\left(1 - \frac{\beta^2}{x^2}\right)^2 dx = \begin{matrix} 2\pi^2 a & (0 \leq a \leq \beta) \\ 2\pi^2 \beta & (0 \leq \beta \leq a) \end{matrix}.$$
These formulæ are inaccurate; the right-hand side should be multiplied by 2.

§4. On p. 180 of the same note Mr. Bateman has occasion to discuss the question of the inversion of the order of integration in the integral

$$(4) \qquad \int_0^\infty \frac{1 - J_0(x)}{x} dx \int_z^\infty \frac{\sin \lambda x}{\sqrt{(\lambda^2 - z^2)}} d\lambda.$$

He states correctly that the theorems which I gave in Note XXXIII. do not cover this case, as the function $\{1 - J_0(x)\}/x$ is not integrable in the infinite interval.

In my note I supposed throughout that the integral $\int |\phi| dx$ was convergent over the whole interval $(0, \infty)$; but I stated at the end that I had proved theorems applicable when this is not the case. It seems worth while that I should now state the simplest of these theorems and show how it, in conjunction with Theorem 2 of my previous note, enables us to deal with the integral (4). The theorem is as follows:

The inversion is legitimate if (i) ϕ *and* f *are regularly integrable in any finite intervals,* (ii) $\phi(x)$ *tends steadily*[*] *to a limit* $\phi(+0)$ *as* $x \to 0$, (iii) $f(\lambda)$ *tends steadily to a limit* $f(+0)$ *as* $\lambda \to 0$, (iv) $\phi(x)$ *tends steadily to zero as* $x \to \infty$, (v) $f(\lambda)$ *tends steadily to zero as* $\lambda \to \infty$, *and* (vi) *the integrals*

$$\int^\infty \frac{\phi(x)}{x} dx, \qquad \int^\infty \frac{f(\lambda)}{\lambda} d\lambda$$

are convergent.

Condition (vi) *may be omitted in the case of the cosine integral.*

The proof of this theorem follows very much the same lines as those of my earlier theorems.

In order to apply the theorems to the integral (4), we divide it into the two parts

$$\int_0^\xi \int_z^\infty + \int_\xi^\infty \int_z^\infty = I_1 + I_2.$$

In I_1 we write

$$\phi(x) = \{1 - J_0(x)\}/x \quad (0 < x \leq \xi), \qquad \phi(x) = 0 \quad (x > \xi),$$

and $f(\lambda) = 1/\sqrt{(\lambda^2 - z^2)}$; and the legitimacy of the inversion follows at once from Theorem 2. It should be observed that it does not follow immediately from classical theorems, since the integral with respect to λ is not uniformly convergent in the neighbourhood of $x = 0$.

---

[*] Of course this condition applies only to an arbitrary small interval to the right of $x = 0$. Similar remarks apply to the next three conditions.

We now divide I_2 into the two parts

$$\int_\xi^\infty \frac{dx}{x} \int_z^\infty \frac{\sin \lambda x}{\sqrt{(\lambda^2-z^2)}} d\lambda - \int_\xi^\infty \frac{J_0(x)}{x} dx \int_z^\infty \frac{\sin \lambda x}{\sqrt{(\lambda^2-z^2)}} d\lambda.$$

The theorem of this section applies to the first part; and, since the integral

$$\int^\infty \frac{|J_0(x)|}{x} dx$$

is convergent, Theorem 2 applies to the second.* Thus the inversion is completely justified.

§5. I conclude this note by indicating a very simple and elegant proof of the formulæ

$$\int_z^1 \frac{\arccos x}{\sqrt{(x^2-z^2)}} dx = \int_1^{1/z} \frac{\operatorname{arg\,cosh} x}{\sqrt{\{(1/z)^2-x^2\}}} dx = -\tfrac{1}{2}\pi \log z$$

where $0 < z < 1$, given by Mr. Bateman. Let us take the first integral, for example. If we denote its value by $\phi(z)$, we have, on writing $x = y/a$, and then x again for y,

$$\phi(z/a) = \int_z^a \frac{\arccos(x/a)}{\sqrt{(x^2-z^2)}} dx,$$

and so, differentiating with respect to a,

$$-(z/a^2) \phi'(z/a) = \frac{1}{a} \int_z^a \frac{x\,dx}{\sqrt{\{(x^2-z^2)(a^2-x^2)\}}} = \frac{\pi}{2a},$$

or
$$\phi'(z/a) = -\tfrac{1}{2}\pi a/z.$$

The result now follows at once from the fact that

$$|\phi(z)| < K \int_z^1 \frac{dx}{\sqrt{(x^2-z^2)}} \to 0$$

as $z \to 1$.

The result is equivalent to the formula

$$\int_a^{\frac{1}{2}\pi} \frac{\theta \cos\theta\,d\theta}{\sqrt{(\cos^2\alpha - \cos^2\theta)}} = \tfrac{1}{2}\pi \log(1+\cos\alpha)$$

given by Legendre.†

---

* Naturally we take $\phi(x)=0$ if $x<\xi$, and $f(\lambda)=0$ if $\lambda<z$.
† *Exercises de Calcul Intégral* (vol. i., Supplément, p. 19).

NOTES ON SOME POINTS IN THE INTEGRAL CALCULUS.

By *G. H. Hardy.*

XLII.

On Weierstrass's singular integral, and on a theorem of Lerch.

1. THEOREM. *If*

(i) x, α, and $f(x)$ are real;
(ii) $f(x)$ is summable in any finite interval of values of x;
(iii) the integral
$$\int_{-\infty}^{\infty} e^{-Ax^2} f(x)\, dx$$

is convergent for some value of A (and so for all greater values):

then

$$(1.1) \quad \lim_{\alpha \to 0} \frac{1}{\sqrt{(\pi\alpha)}} \int_{-\infty}^{\infty} e^{-(x-\xi)^2/\alpha} f(x)\, dx = f(\xi)$$

for almost all values of ξ, and in particular wherever $f(\xi)$ is continuous; and the limit is equal to

$$(1.2) \quad \tfrac{1}{2}\{f(\xi+0) + f(\xi-0)\}$$

whenever this expression has a meaning.

The first accurate discussion of the formula (1.1) is that of Weierstrass*. Weierstrass supposes $f(x)$ uniformly continuous and bounded. The formula has been discussed since by Hobson† and Lebesgue‡. Hobson supposes $f(x)$ to be summable in any finite interval, and

$$(1.3) \qquad \int_{-\infty}^{\infty} |f(x)|\, dx$$

to be convergent; but confines himself to values of ξ for which the expression (1.2) has a meaning. In a note added to his paper, and attributed to the suggestion of Bromwich, he replaces the condition that (1.3) should be convergent by the much less stringent condition that

$$(1.4) \qquad |f(x)| < e^{A|x|} §$$

for some value of A and all sufficiently large values of $|x|$.

The question whether (1.1) holds for almost all values of ξ was first discussed by Lebesgue. Lebesgue proved that this is so if $f(x)$ is summable in any finite interval, and (1.3) is convergent, or if $f(x)$ is summable in some finite interval including ξ, and bounded outside it. The last condition may naturally be replaced by the more general condition (1.4).

The work of Hobson and Lebesgue thus includes everything stated in my theorem except that I employ a very much more general condition as to the behaviour of $f(x)$ at infinity. The generalisation which I make is however a very material one. For it is plain that, unless my condition (iii) is satisfied, there can be no question of the existence of the limit in question. My conditions are therefore *necessary and sufficient*.

2. Hobson and Lebesgue both discuss Weierstrass's integral as a special case of a 'singular integral' of a very general

---

* K. Weierstrass, 'Ueber die analytische Darstellbarkeit sogenannter willkurlichen Functionen reeller Argumente', *Berliner Sitzungsberichte* (1885), pp. 633–639, 789–805 (*Werke*, vol. iii., pp. 1–37). See also E. Borel, *Leçons sur les fonctions de variables réelles*, pp. 51 et seq.

† E. W. Hobson, 'On a general convergence theorem, and the theory of the representation of a function by series of normal functions', *Proc. London Math. Soc.*, ser. 2, vol. vi. (1908), pp. 349–395 (especially pp. 367–370).

‡ H. Lebesgue, 'Sur les intégrales singulières', *Annales de la Faculté des Sciences de Toulouse*, ser. 3, vol. i. (1909), pp. 25–118 (especially pp. 90–92).

§ The condition is stated in the form

$$|f(x)| < |x|^p e^{A|x|},$$

but the simpler form adopted here asserts neither more nor less.

With this condition, Hobson's result includes everything proved by Weierstrass. With the more stringent condition it did not, since Weierstrass's conditions do not imply the convergence of (1.3).

type. I shall therefore prove the theorem *ab initio* and without reference to their work. I begin by recalling a fundamental theorem of Lebesgue, that if $f(x)$ is summable then

(2.1) $\quad \phi(x) = \int_0^x |f(\xi+t) + f(\xi-t) - 2f(\xi)| \, dt = o(x),$

when $x \to 0$, for almost all values of ξ; and I shall prove that (1.1) is true for all such values of ξ.

It is obvious that, in proving this, we may suppose without loss of generality that $\xi = 0$ and $f(\xi) = 0$. Further we may write

$$f_1 = f \ (x>0), \quad f_1 = 0 \ (x<0),$$
$$f_2 = 0 \ (x>0), \quad f_2 = f \ (x<0),$$

so that

$$f = f_1 + f_2,$$

and discuss the contributions of f_1 and f_2 to the integral separately. And when we have made these formal simplifications we have to prove that *if*

(2.2) $\quad \phi(x) = \int_0^x |f(t)| \, dt = o(x)$

when $x \to 0$, *then*

(2.3) $\quad J(\alpha) = \int_0^\infty e^{-x^2/\alpha} f(x) \, dx = o(\sqrt{\alpha})$

when $\alpha \to 0$.

3. I observe first that *if*

$$\int_0^\infty e^{-Ax^2} f(x) \, dx$$

is convergent, and $x_0 \geq 0$, *then*

(3.1) $\quad \left| \int_{x_0}^\infty e^{-x^2/\alpha} f(x) \, dx \right| < K e^{-x_0^2/2\alpha},$

where K is independent of x_0 and α, for all sufficiently small values of α. To prove this, choose α_0 so that $2A\alpha_0 < 1$. Then

$$\int_0^\infty e^{-x^2/2\alpha} f(x) \, dx$$

is uniformly convergent for $0 < \alpha \leq \alpha_0$. There is therefore a constant K such that

$$\left| \int_{x_0}^{x_1} e^{-x^2/2\alpha} f(x) \, dx \right| < K,$$

for $0 \leq x_0 \leq x_1$ and $0 < \alpha \leq \alpha_0$. But

$$\int_{x_0}^{\infty} e^{-x^2/\alpha} f(x)\, dx = \lim_{X \to \infty} \int_{x_0}^{X} e^{-x^2/\alpha} f(x)\, dx$$

$$= \lim_{X \to \infty} e^{-x_0^2/2\alpha} \int_{x_0}^{x_1} e^{-x^2/2\alpha} f(x)\, dx,$$

where $x_0 \leq x_1$. From this the truth of (3.1) follows immediately.

Now choose δ so that

(3.2) $\qquad \phi(x) < \eta x = \dfrac{\epsilon x}{\sqrt{\pi}}$

for $0 < x \leq \delta$; and write

$$F(x) = \int_0^x f(t)\, dt,$$

so that

(3.3) $\qquad |F(x)| \leq \phi(x).$

Dividing the range of integration in $J(\alpha)$ into the two parts $(0, \delta)$ and (δ, ∞), and integrating the first part of the integral by parts, we obtain

$$J(\alpha) = e^{-\delta^2/\alpha} F(\delta) + \frac{2}{\alpha} \int_0^\delta x e^{-x^2/\alpha} F(x)\, dx + \int_\delta^\infty e^{-x^2/\alpha} f(x)\, dx.$$

It follows from (3.1), (3.2), and (3.3) that

$$|J(\alpha)| < e^{-\delta^2/\alpha} \phi(\delta) + K e^{-\delta^2/2\alpha} + \frac{2\eta}{\alpha} \int_0^\delta x^2 e^{-x^2/\alpha}\, dx$$

$$< e^{-\delta^2/\alpha} \phi(\delta) + K e^{-\delta^2/2\alpha} + \tfrac{1}{2}\epsilon \sqrt{\alpha};$$

since

$$\frac{2\eta}{\alpha} \int_0^\infty x^2 e^{-x^2/\alpha}\, dx = 2\eta \sqrt{\alpha} \int_0^\infty u^2 e^{-u^2}\, du = \tfrac{1}{2}\epsilon \sqrt{\alpha}.$$

But, when δ has been fixed, we have obviously

$$e^{-\delta^2/\alpha} \phi(\delta) + K e^{-\delta^2/2\alpha} < \tfrac{1}{2}\epsilon \sqrt{\alpha}$$

for all sufficiently small values of α; and so

$$|J(\alpha)| < \epsilon \sqrt{\alpha}.$$

This proves the theorem, in so far as it asserts that (1) holds almost everywhere and in particular at all points of continuity.

In order to prove the last clause of the theorem we have only to replace (2.1) by

(3.4) $\quad \phi(x) = \displaystyle\int_0^x |f(\xi + t) + f(\xi - t) - f(\xi + 0) - f(\xi - 0)|\, dt,$

to suppose that $\phi(x) = o(x)$, and to make a few trivial changes in the preceding argument.

4. It was proved by Lerch* that *if $f(x)$ is continuous and*

(4.1)
$$\int_0^1 x^n f(x)\, dx = 0$$

for all positive integral values of n, then $f(x)$ is identically zero. Lerch bases his proof on Weierstrass's theorem concerning the representation of a continuous function of a series of polynomials. Another proof of a more direct character had been indicated earlier by Stieltjes† and was made explicit and rigorous by Landau‡. A third proof was given independently by Phragmén.§

It has been proved, more generally, by Hobson,|| that if $f(x)$ is any summable function, and (10) holds for all positive integral values of n, then $f(x) = 0$ except at a set of measure zero. Hobson, like Lerch, uses Weierstrass's theorem. The result may, as has been remarked by Lebesgue,¶ be deduced at once from the theorem of § 1; for, since the expansion

$$e^{-(x-\xi)^2/\alpha} = 1 - \frac{(x-\xi)^2}{\alpha} + \frac{(x-\xi)^4}{2!\, \alpha^2} - \cdots$$

is uniformly convergent, we have

$$\int_0^1 e^{-(x-\xi)^2/\alpha} f(x)\, dx = 0$$

for all positive values of α. An alternative method would be to observe that, since the power series for $\cos 2n\pi x$ and $\sin 2n\pi x$ are uniformly convergent, all the Fourier constants of $f(x)$ must be zero, and then to use Lebesgue's extension of Fejér's theorem on Fourier series.

5. It should be observed that Lerch states his theorem in the form: *if*

$$J(a) = \int_0^\infty e^{-ax} f(x)\, dx$$

and

$$J(a) = 0$$

---

* M. Lerch, 'Sur un point de la théorie des fonctions génératrices d'Abel', *Acta Mathematica*, vol. xxvii. (1903), pp. 339–351.
† *Correspondance d' Hermite et de Stieltjes*, vol. ii., pp. 337–339.
‡ E. Landau, 'Ueber die Approximation einer stetigen Funktion durch eine ganze rationale Funktion', *Rendiconti di Palermo*, vol. xxv. (1908), pp. 337–345. I am indebted to this paper for my references to Stieltjes and Phragmén.
§ E. Phragmén, 'Sur une extension d'un théorème classique de la théorie des fonctions', *Acta Mathematica*, vol xxviii. (1904), pp. 351–368 (pp. 361–363).
|| E. W. Hobson, 'The fundamental lemma of the calculus of variations', *Proc. London Math. Soc.*, ser. 2, vol. xi. (1911), pp. 17–28.
¶ *l.c.*, p. 104.

for
$$a = \alpha + n\beta \quad (\beta > 0; \; n = 0, 1, 2, \ldots),$$
then $J(a)$ is identically zero and $f(x)$ is 'in general' zero.

Lerch naturally supposes $f(x)$ integrable in Riemann's sense. If we suppose merely that $f(x)$ is summable and the integral convergent, then it is easy to show, by Lerch's argument, that $f(x) = 0$ except at a set of measure zero, and in particular at all points of continuity.

We may suppose without loss of generality that $\alpha > \beta$. Then $a > 0$, and, if we write
$$F(x) = \int_0^x f(t)\,dt$$
it is easy to show that
$$F(x) = o(e^{ax})$$
and
$$J(a) = a\int_0^\infty e^{-ax} F(x)\,dx.^*$$

If now we put $e^{-\beta x} = y$ and
$$\psi(y) = y^{(\alpha-\beta)/\beta} F\left(\frac{1}{\beta} \log \frac{1}{y}\right),$$
then $\psi(y)$ is continuous for $0 \leq y \leq 1$ and
$$\int_0^1 y^n \psi(y)\,dy = 0$$
for all positive integral values of n. It follows that F is identically zero, and so that f is zero except at a set of measure zero.

Some very beautiful applications of Lerch's theorem have been made recently by Ramanujan.†

I intend in some succeeding notes to give some applications of the theorems discussed in this note to various interesting problems concerning definite integrals, in particular to the 'problème des moments' of Stieltjes and certain 'inversion formulæ' of Mellin and other writers. I have therefore thought it best to discuss the theorems in some detail, even where I have but little to add to them.

---

* The proofs depend merely on partial integration, and I leave them to the reader.
† S. Ramanujan, 'Some definite integrals connected with Gauss's sums', *Messenger*, vol. xliv. (1915), pp. 75–81; 'On certain infinite series', *ibid.*, vol. xlv. (1916), pp. 11–15.

CORRECTIONS

p. 46, *line* 7 *up*. (1) should be (1.1).
p. 47, *line* 11. (10) should be (4.1).

NOTES ON SOME POINTS IN THE INTEGRAL CALCULUS.

By *G. H. Hardy*.

XLVI.

On Stieltjes' 'problème des moments'.

1. The *problème des moments* is that of finding a function $f(x)$ such that

(1.1) $$\int_0^\infty x^n f(x)\, dx = c_n$$

for $n = 0, 1, 2, \ldots$, the c's being given constants. The problem was first attacked by Stieltjes[*], who supposed $f(x)$ *positive*, and obtained the complete solution[†] in this case by means of some singularly beautiful analysis based on the theory of continued fractions.

Stieltjes proved in particular that there cannot be two different continuous functions $f_1(x)$ and $f_2(x)$, each of which satisfies the equations (1.1) and the inequalities

(1.2) $$0 \leq f(x) < e^{-k\sqrt{x}},$$

where k is a positive constant; and Borel showed that (1.2) may be replaced by

(1.3) $$|f(x)| < e^{-k\sqrt{x}}.$$

---

[*] T. J. Stieltjes, 'Recherches sur les fractions continues', *Annales de la Faculté des Sciences de Toulouse*, ser. 1, vol. viii. (1894), J., pp. 1–122, and vol. ix. (1895), A., pp. 1–47. See also E. Le Roy, 'Sur les séries divergentes et les fonctions définies par un développement de Taylor', *ibid.*, ser. 2, vol. ii. (1900), pp. 317–431: H. Lebesgue, 'Sur les intégrales singulières', *ibid.*, ser. 3, vol. i. (1909), pp. 25–111 (pp. 101 *et seq.*): E. Borel, *Leçons sur les séries divergentes* (1901), ch. 3: O. Perron, *Die Lehre von den Kettenbrüchen* (1913), ch. 9

[†] It is necessary, in order to obtain such a solution, to develop a theory of 'Stieltjes integrals' of the form $\int x^n d\psi(x)$, where $\psi(x)$ is a monotonic but not necessarily continuous function. I shall not be concerned in this note with such generalisations of the notion of an integral.

Thus no continuous function which satisfies (1.3) can also satisfy

$$(1.4) \qquad \int_0^\infty x^n f(x)\, dx = 0$$

for $n = 0, 1, 2, \ldots$.

This last theorem is certainly untrue if enunciated without some limitation on the order of $f(x)$. Thus Stieltjes gave the examples

$$f(x) = e^{-x^{\frac{1}{4}}} \sin x^{\frac{1}{4}}, \quad f(x) = e^{-(\log x)^2} \sin(2\pi \log x),$$

both of which satisfy (1.4) for all positive integral values of n. I may observe in passing that, since

$$\int_0^\infty e^{-x^\mu \cos \alpha} \sin(x^\mu \sin \alpha)\, dx = \frac{1}{\mu} \Gamma\left(\frac{n+1}{\mu}\right) \sin \frac{(n+1)\alpha}{\mu}$$

if $\mu > 0$ and $0 < \alpha < \frac{1}{2}\pi$, the function

$$f(x) = e^{-x^\mu \cos \mu \pi} \sin(x^\mu \sin \mu \pi)$$

satisfies the equations (1.4) for every value of μ less than $\frac{1}{2}$. It is therefore impossible that (1.3) should be replaced by any condition

$$|f(x)| < e^{-kx^\mu},$$

where $\mu < \frac{1}{2}$*.

It has been suggested by Lebesgue† that it should be possible to prove some at any rate of Stieltjes' results, and in particular this theorem concerning the uniqueness of the solution, by methods independent of the theory of continued fractions. But the argument indicated by Lebesgue applies only to functions such that the integral

$$\int_0^\infty |f(x)|\, e^{Ax^2}\, dx$$

is convergent for every value of A.

In this note I suppose that the integral

$$(1.5) \qquad \int_0^\infty |f(x)|\, e^{k\sqrt{x}}\, dx$$

is convergent for some positive k, and I give two different proofs of the uniqueness of the solution. Both are indepen-

---

* It is clear that Stieltjes (though he does not state the result explicitly) was aware that (1.2) could not be replaced by any inequalities of the form

$$0 \leq f(x) < e^{-kx^\mu} \quad (\mu < \tfrac{1}{2}).$$

It is probable that he knew the example that I have just given.
† *Loc. cit*, p. 104.

dent of the theory of continued fractions. In each of the proofs I obtain incidentally a formula for the solution which is valid whenever a solution satisfying the condition exists. One of these formulæ is that given by Le Roy*, who established its validity in certain special cases. I also consider some other forms of the solution valid under narrower restrictions.

I impose no condition on $f(x)$, except at infinity, save that of being summable. The value of $f(x)$ can of course be altered arbitrarily at the points of any set of measure zero, without affecting the values of the integrals (1.1). Two functions which differ only at the points of such a set will be regarded as identical; and when I say that 'the solution is unique' I must be understood to mean that any two solutions are the same for almost all values of x, that is to say except at the points of a set of measure zero.

2. Suppose that $f(x)$ is summable, that the equations (1.1) are satisfied, and that the integral (1.5) is convergent. Then the integral

$$\int_0^\infty |f(x)| \Sigma \frac{|sx|^n}{2n!} dx = \int_0^\infty |f(x)| \cosh \sqrt{|sx|}\, dx,$$

is convergent for $|s| < k$, so that

$$F(s) = \sum_0^\infty \frac{c_n (-s)^n}{2n!} = \int_0^\infty f(x) \sum_0^\infty \frac{(-sx)^n}{2n!} dx$$

$$= \int_0^\infty f(x) \cos \sqrt{(sx)}\, dx$$

for $|s| < k$. The integral just written is plainly an analytic function of s, regular for all positive values of s†.

We have therefore

$$F\left(\frac{t^2}{u}\right) = \int_0^\infty f(x) \cos\left\{t \sqrt{\left(\frac{x}{u}\right)}\right\} dx,$$

for all positive values of t and u. Multiply by e^{-t}, and integrate from $t = 0$ to $t = \infty$. Since

---

* *Loc. cit.*, p. 387.

† The integral is uniformly convergent throughout any finite domain which lies inside the parabola whose polar and Cartesian equations are

$$r = k^2 \operatorname{cosec}^2 \tfrac{1}{2}\theta, \quad y^2 = 4k^2(x + k^2).$$

On this parabola the imaginary part of $\sqrt{s}$ is equal to k.

$$\int_0^\infty e^{-t} dt \int_0^\infty |f(x)| dx$$

is convergent, change of the order of integration is legitimate. We have therefore

$$\int_0^\infty e^{-t} F\left(\frac{t^2}{u}\right) dt = \int_0^\infty f(x) dx \int_0^\infty e^{-t} \cos\left\{t \sqrt{\left(\frac{x}{u}\right)}\right\} dt$$
$$= \int_0^\infty \frac{u f(x)}{x+u} dx.$$

Thus
$$\chi(u) = \frac{1}{u} \int_0^\infty e^{-t} F\left(\frac{t^2}{u}\right) dt = \int_0^\infty \frac{f(x)}{x+u} dx$$

for all positive values of u.

3. The integral last written is plainly an analytic function of u, regular except at points of the negative real axis, which is generally a *coupure*, and the equation

(3.1) $$\chi(u) = \int_0^\infty \frac{f(x)}{x+u} dx$$

holds for all such values of u. It was shown by Stieltjes* that, if $f(x)$ is continuous,

(3.2) $$\frac{1}{2\pi i} \lim_{\varepsilon \to 0} \{\chi(-\xi - i\epsilon) - \chi(-\xi + i\epsilon)\} = f(\xi),$$

when ϵ tends to zero through positive values.

I shall now prove that, *if $f(x)$ is any summable function,* (3.2) *holds for almost all values of* ξ, and in particular for all values of ξ for which

(3.3) $$X(t) = \int_0^t |f(\xi + \tau) + f(\xi - \tau) - 2f(\xi)| d\tau = o(t),$$

as in the case for almost all values of ξ.

We have to prove that

(3.4) $$\int_0^\infty \frac{\epsilon f(x) dx}{(x-\xi)^2 + \epsilon^2} \to \pi f(\xi) ;$$

and it is obvious that we may replace the limits of integration by $\xi - \delta$ and $\xi + \delta$, where δ is any positive number, and that we may suppose, without real loss of generality, that $f(\xi) = 0$.

---

* The idea is due to Hermite. See his memoir 'Sur quelques points de la théorie des fonctions', *Journal für Math.*, vol xci. (1881), pp. 54–78; and his *Cours professé à la Faculté des Sciences*, ed. 4 (1888), pp. 76 *et seq.*, 154 *et seq.*

We have then
$$\int_{\xi-\delta}^{\xi+\delta} \frac{\epsilon f(x)\,dx}{(x-\xi)^2+\epsilon^2} = \int_0^\delta \frac{\epsilon \psi(t)}{t^2+\epsilon^2}\,dt,$$
where
$$\psi(t) = f(\xi+t) + f(\xi-t).$$
Let
$$\Psi(t) = \int_0^t \psi(\tau)\,d\tau,$$
so that
$$|\Psi(t)| \leq X(t) = o(t).$$

Given any positive η, we can choose δ so that
$$X(t) < \eta t$$
for $0 < t \leq \delta$. Then
$$\int_0^\delta \frac{\epsilon \psi(t)}{t^2+\epsilon^2}\,dt = \frac{\epsilon}{\delta^2+\epsilon^2}\Psi(\delta) + \int_0^\delta \frac{2\epsilon t}{(t^2+\epsilon^2)^2}\Psi(t)\,dt.$$

The first integral tends to zero with ϵ, and the second is less than
$$2\eta \int_0^\infty \frac{\epsilon t^2\,dt}{(t^2+\epsilon^2)^2} = 2\eta \int_0^\infty \frac{w^2\,dw}{(w^2+1)^2}.$$

It follows that
$$\int_0^\delta \frac{\epsilon \psi(t)}{t^2+\epsilon^2}\,dt \to 0;$$
and therefore that (3.4) is true.

4. Returning to the analysis of § 2, we see that

(4.1) $\quad f(x) = \dfrac{1}{2\pi i}\lim\{\chi(-x-i\epsilon) - \chi(-x+i\epsilon)\}$

for almost all values of x, χ being the analytic function defined by the formula

(4.2) $\quad \chi(u) = \dfrac{1}{u}\int_0^\infty e^{-t} F\!\left(\dfrac{t^2}{u}\right) dt,$

and $F(s)$ the function defined by

(4.3) $\quad F(s) = \sum_0^\infty \dfrac{c_n(-s)^n}{2n!}.$

It follows that $f(x)$ is determined, except at a set of measure zero, by the equations (1.1) and by the condition that the integral (1.5) is convergent. The solution of the problem

---

* This is the condition from which Lebesgue has drawn so many important consequences.

is therefore, under this condition, and *a fortiori* under the condition (1.3), unique. We have thus proved:

THEOREM A. *There can be at most one solution $f(x)$ of the problème des moments, such that the integral*

$$\int_0^\infty |f(x)| e^{k\sqrt{x}} dx$$

is convergent for a positive value of k. This solution of the problem is, if it exists, given by the formulæ (4.1), (4.2), *and* (4.3).

5. Suppose, for example, that c_n is of the form

$$c_n = 2n! \int_0^1 \lambda^n \theta(\lambda) d\lambda.$$

Then
$$F(s) = \int_0^1 \frac{\theta(\lambda)}{1+s\lambda} d\lambda,$$

and
$$\chi(u) = \int_0^\infty e^{-t} dt \int_0^1 \frac{\theta(\lambda) d\lambda}{t^2 \lambda + u} = \int_0^1 \theta(\lambda) d\lambda \int_0^\infty \frac{e^{-t} dt}{t^2 \lambda + u}$$

$$= \tfrac{1}{2} \int_0^1 \theta(\lambda) \frac{d\lambda}{\sqrt{\lambda}} \int_0^\infty \frac{e^{-\sqrt{(\omega/\lambda)}}}{\sqrt{\omega}} \frac{d\omega}{\omega+u}.$$

Now
$$\frac{1}{2\pi i} \lim_{\varepsilon \to 0} \int_0^\infty \frac{e^{-\sqrt{(\omega/\lambda)}}}{\sqrt{\omega}} \frac{2i\varepsilon}{\varepsilon^2 + (\omega-x)^2} d\omega = \frac{e^{-\sqrt{(x/\lambda)}}}{\sqrt{x}};$$

and it is easily deduced that

$$\frac{1}{2\pi i} \lim_{\varepsilon \to 0} \{\chi(-x-i\varepsilon) - \chi(-x+i\varepsilon)\} = \frac{1}{2\sqrt{x}} \int_0^1 e^{-\sqrt{(x/\lambda)}} \theta(\lambda) \frac{d\lambda}{\sqrt{\lambda}}.$$

Thus the solution, if it exists, is

$$f(x) = \frac{1}{2\sqrt{x}} \int_0^1 e^{-\sqrt{(x/\lambda)}} \theta(\lambda) \frac{d\lambda}{\sqrt{\lambda}};$$

and it is easily verified that this function is in fact a solution. For

$$\int_0^\infty x^n f(x) dx = \tfrac{1}{2} \int_0^1 \theta(\lambda) \frac{d\lambda}{\sqrt{\lambda}} \int_0^\infty x^{n-\frac{1}{2}} e^{-\sqrt{(x/\lambda)}} dx$$

$$= 2n! \int_0^1 \lambda^n \theta(\lambda) d\lambda = c_n.$$

A similar method may be applied in more general cases, for example when

$$c_n = 2n! \int_0^{a_1} \lambda_1^n \theta_1(\lambda_1) d\lambda_1 \int_0^{a_2} \lambda_2^n \theta_2(\lambda_2) d\lambda_2 \ldots.$$

6. It is sometimes more convenient to use a different form of the solution, valid only under narrower conditions. Let us suppose that there are numbers k and α, where
$$k > 0, \quad \alpha > \tfrac{1}{2},$$
such that the integral
$$\int_0^\infty |f(x)| e^{kx^\alpha} dx$$
is convergent. Let
$$E_\alpha(y) = \sum_0^\infty \frac{(-y)^n}{\Gamma(\beta n + 1)},$$
where $\beta = 1/\alpha$. Then it is known* that
$$|E_\alpha(y)| < K e^{|y|^\alpha},$$
where K is a constant, for all values of y. Hence
$$\int_0^\infty |f(x)| \sum_0^\infty \frac{|sx|^n}{\Gamma(\beta n + 1)} dx < K \int_0^\infty |f(x)| e^{(|s| x)^\alpha} dx,$$
so that the integral on the left-hand side is certainly convergent for sufficiently small values of $|s|$, in fact for $|s|^\alpha < k$. It follows that
$$F(s) = \sum_0^\infty \frac{c_n(-s)^n}{\Gamma(\beta n + 1)} = \int_0^\infty f(x) E_\alpha(sx) dx$$
for sufficiently small values of $|s|$. But†

(6.1) $$|E_\alpha(y)| < \frac{K}{|y|}$$

at all points of a certain region of the type
$$|y| \geq |y_0|, \quad -\delta \leq am . y \leq \delta,$$
which includes the part of the real axis which extends to positive infinity. It follows that the formula
$$F(s) = \int_0^\infty f(x) E_\alpha(sx) dx$$
is valid for all positive values of s. Further, we have
$$\chi(u) = \frac{1}{u} \int_0^\infty e^{-t} F\left(\frac{t^\beta}{u}\right) dt = \frac{1}{u} \int_0^\infty f(x) dx \int_0^\infty e^{-t} E_\alpha\left(\frac{x t^\beta}{u}\right) dt$$
$$= \int_0^\infty \frac{f(x)}{x + u} dx;$$

---

* G. Mittag-Leffler, 'Sur la représentation analytique d'une branche uniforme d'une fonction monogène (Cinquième note)', *Acta Mathematica*, vol. xxix. (1905), pp. 101–181. See also A. Wiman, 'Ueber den Fundamentalsatz in der Theorie der Funktionen $E_\alpha(x)$', *ibid.*, pp. 191–201; and 'Ueber die Nullstellen der Funktionen $E_\alpha(x)$', *ibid.*, pp. 217–234.
In my notation $E_\alpha(y) = E_{1/\alpha}(-y)$.
† Mittag-Leffler and Wiman, *l.c. supra*.

and $\chi(u)$ is regular for all values of u other than negative values. Repeating the arguments of §§ 3—4, we obtain

THEOREM B. *If there is a solution of the problème des moments such that*

$$\int_0^\infty |f(x)| e^{kx^\alpha} dx \quad (k > 0,\ \alpha > \tfrac{1}{2})$$

is convergent, it is given, for almost all values of x, by the formulæ

$$f(x) = \frac{1}{2\pi i} \lim_{\varepsilon \to 0} \{\chi(-x - i\varepsilon) - \chi(-x + i\varepsilon)\},$$

$$\chi(u) = \frac{1}{u} \int_0^\infty e^{-t} F\left(\frac{t^\beta}{u}\right) dt,$$

$$F(s) = \sum_0^\infty \frac{c_n(-s)^n}{\Gamma(\beta n + 1)}, \quad \beta = \frac{1}{\alpha}.$$

It is well to point out explicitly how this method fails when $\alpha < \tfrac{1}{2}$. In this case the function $E_\alpha(y)$ is an integral function of real order less than $\tfrac{1}{2}$, and (to put the matter roughly) becomes exponentially infinite (for appropriate values of $|y|$) when $y \to \infty$ in *any* direction*; so that there is no such relation as (6.1).

7. It is easy to deduce, from Theorem B, that, if

$$c_n = \Gamma(\beta n + 1) \int_0^1 \lambda^n \theta(\lambda) d\lambda,$$

where $0 < \beta < 2$, the solution of the *problème des moments* is given by

$$f(x) = \alpha x^{\alpha - 1} \int_0^1 e^{-(x/\lambda)^\alpha} \lambda^{-\alpha} \theta(\lambda) d\lambda,$$

where $\alpha = 1/\beta$. More generally, we can, by methods similar to those of this section, obtain the solution for

$$c_n = \Gamma(\beta_1 n + \gamma_1) \Gamma(\beta_2 n + \gamma_2) \ldots \int_0^{a_1} \lambda_1^n \theta_1(\lambda_1) d\lambda_1 \int_0^{a_2} \lambda_2^n \theta_2(\lambda_2) d\lambda_2 \ldots,$$

where $\beta_1, \beta_2, \ldots$ are positive numbers whose sum is less than 2.

A different form of the solution has been given by Le Roy. This I shall discuss in a second note on the subject.

---

* See A. Wiman, 'Sur une extension d'un théorème de M. Hadamard', *Arkiv för Matematik*, vol. ii. (1905), No. 14, pp. 1–5.

CORRECTIONS

p. 176, *line* 1. Insert 'non-null' before 'continuous'.
——, *line* 10. Insert factor x^n in integrand.
p. 177, §2, *lines* 5 *and* 8. $|s|<k$ should be $|s|<k^2$.

NOTES ON SOME POINTS IN THE INTEGRAL CALCULUS.

By *G. H. Hardy.*

XLVII.

On Stieltjes' 'problème des moments' (continued).

1. LE ROY, in his memoir quoted in Note XLVI,* gave, as a formal solution of the *problème des moments*, the formula

$$(1.1) \quad f(x) = \int_0^\infty J_0\{2\sqrt{(xy)}\} \sum_0^\infty \frac{(-1)^n c_n}{(n!)^2} y^n \, dy;$$

and showed that this formula really gives a solution for certain special forms of c_n.

I shall now prove

THEOREM C. *If there is a solution $f(x)$ of the problem, such that*

$$(1.2) \quad \int_0^\infty |f(x)| \, e^{k\sqrt{x}} \, dx$$

is convergent for some positive value of k, then

$$(1.3) \quad f(x) = G \int_0^\infty J_0\{2\sqrt{(xy)}\} \phi(y) \, dy$$

for almost all values of x; $\phi(y)$ being the analytic function defined, for sufficiently small values of y, by the equation

$$(1.4) \quad \phi(y) = \sum \frac{(-1)^n c_n}{(n!)^2} y^n,$$

and the 'generalised integral'

$$G \int_0^\infty F(y) \, dy$$

being defined as meaning

$$\lim_{\delta \to 0} \int_0^\infty e^{-\delta y} F(y) \, dy.$$

We have

$$\phi(y) = \sum_0^\infty \frac{(-1)^n c_n}{(n!)^2} y^n = \int_0^\infty f(x) \sum_0^\infty \frac{(-1)^n (xy)^n}{(n!)^2} \, dx$$

$$= \int_0^\infty f(x) J_0\{2\sqrt{(xy)}\} \, dx,$$

if term by term integration be legitimate. Now

---
* Vol. xlvi., p. 175.

$$\sum_0^\infty \frac{|s|^n}{(n!)^2} < K e^{\sqrt{|s|}},$$

for all values of $|s|$, so that

$$\int_0^\infty |f(x)| \sum_0^\infty \frac{|xy|^n}{(n!)^2} dx$$

is convergent if $|y| < k$. Thus term by term integration is legitimate for sufficiently small values of $|y|$. It follows, as in §2 of Note XLVI, that

$$\phi(y) = \int_0^\infty f(x) J_0\{2\sqrt{(xy)}\} dx$$

for all positive values of y, $\phi(y)$ being the analytic function defined for sufficiently small values of y by the series (1.4).

We have therefore

$$\int_0^\infty e^{-\delta y} J_0\{2\sqrt{(wy)}\} \phi(y) dy$$
$$= \int_0^\infty e^{-\delta y} J_0\{2\sqrt{(wy)}\} dy \int_0^\infty f(x) J_0\{2\sqrt{(xy)}\} dx$$
$$= \int_0^\infty f(x) dx \int_0^\infty e^{-\delta y} J_0\{2\sqrt{(wy)}\} J_0\{2\sqrt{(xy)}\} dy,$$

if this integral is convergent and the change of the order of integration legitimate: and this is certainly so, since the Bessel's functions are bounded and the double integral

$$\int_0^\infty |f(x)| dx \int_0^\infty e^{-\delta y} dy$$

is convergent. Now

$$\int_0^\infty e^{-\delta y} J_0\{2\sqrt{(wy)}\} J_0\{2\sqrt{(xy)}\} dy$$
$$= 2\int_0^\infty e^{-\delta \eta^2} J_0(2\eta \sqrt{w}) J_0(2\eta \sqrt{x}) \eta\, d\eta = \frac{1}{\delta} e^{-(x+w)/\delta} I_0\left\{\frac{2\sqrt{(xw)}}{\delta}\right\},*$$

where $$I_0(t) = J_0(it),$$

so that

(1.51) $$\qquad 0 < I_0(t) < K e^t$$

for all positive values of t, and

(1.52) $$\qquad I_0(t) = \frac{e^t}{\sqrt{(2\pi t)}} \left\{1 + O\left(\frac{1}{t}\right)\right\}$$

when t is large.

---

* This formula is due to H. Weber, 'Über einige bestimmte Integrale', *Journal für Mathematik*, vol. lxix., 1869, pp. 1–36.

2. We have thus to consider the limit, when $\delta \to 0$, of the integral
$$\mathbf{I}(\delta) = \frac{1}{\delta} \int_0^\infty f(x) \, e^{-(x+w)/\delta} \, I_0 \left\{ \frac{2\sqrt{(xw)}}{\delta} \right\} dx.$$

We write this integral in the form
$$\mathbf{I}(\delta) = \frac{1}{\delta} \int_0^{x_0} + \frac{1}{\delta} \int_{x_0}^{x_1} + \frac{1}{\delta} \int_{x_1}^\infty = \mathbf{I}_1(\delta) + \mathbf{I}_2(\delta) + \mathbf{I}_3(\delta),$$
where $0 < x_0 < w < x_1$.

In $\mathbf{I}_1(\delta)$ we use the crude inequality (1.51), which gives
$$e^{-(x+w)/\delta} I_0 \left\{ \frac{2\sqrt{(xw)}}{\delta} \right\} < K e^{-(\sqrt{w}-\sqrt{x})^2/\delta} < K e^{-(\sqrt{w}-\sqrt{x_0})^2/\delta},$$
so that
$$|\mathbf{I}_1(\delta)| < \frac{K}{\delta} e^{-(\sqrt{w}-\sqrt{x_0})^2/\delta} \int_0^{x_0} |f(x)| \, dx \to 0.$$
Similarly
$$|\mathbf{I}_3(\delta)| < \frac{K}{\delta} e^{-(\sqrt{x_1}-\sqrt{w})^2/\delta} \int_{x_1}^\infty |f(x)| \, dx \to 0.$$

We have therefore only to obtain the limit of $\mathbf{I}_2(\delta)$.

§3. Writing $x = \xi^2$, $w = \omega^2$, we have
$$\mathbf{I}_2(\delta) = \frac{2}{\delta} \int_{\sqrt{x_0}}^{\sqrt{x_1}} \xi f(\xi^2) \, e^{-(\xi^2+\omega^2)/\delta} \, I_0 \left(\frac{2\xi\omega}{\delta} \right) d\xi.$$
Also, by (1.52),
$$I_0 \left(\frac{2\xi\omega}{\delta} \right) = \frac{e^{2\xi\omega/\delta}}{2\sqrt{(\pi\xi\omega/\delta)}} \{1 + O(\delta)\}.$$

The term $O(\delta)$ may be neglected. For it contributes only
$$O \left\{ \sqrt{\delta} \int_{\sqrt{x_0}}^{\sqrt{x_1}} \xi |f(\xi^2)| \, d\xi \right\} = o(1).$$
Thus we need consider only
$$I(\delta) = \frac{1}{\sqrt{(\delta\pi\omega)}} \int_{\sqrt{x_0}}^{\sqrt{x_1}} \sqrt{\xi} f(\xi^2) \, e^{-(\xi-\omega)^2/\delta} \, d\xi,$$
and the limit of $I(\delta)$ is $f(\omega^2) = f(w)$ for almost all values of w.*

We have therefore
$$\lim_{\delta \to 0} \int_0^\infty e^{-\delta y} J_0 \{2\sqrt{(wy)}\} \phi(y) \, dy = f(w)$$

---

* H. Lebesgue, *loc. cit.* (Note XLVI), p. 91.

for almost all values of w; and the proof of Theorem C is completed.*

4. Theorem C asserts that the integral

$$(4.1) \qquad \int_0^\infty J_0\{2\sqrt{(xy)}\}\,\phi(y)\,dy$$

is summable in a certain manner, to sum $f(x)$, for almost all values of x. Another theorem of a similar character is

THEOREM D. *Under the same conditions as those of Theorem C, the integral* (4.1) *is summable* $(C, 1)$, *to sum* $f(x)$, *for almost all values of x; that is to say*

$$\lim_{Y \to \infty} \frac{1}{Y} \int_0^Y (Y-y) J_0\{2\sqrt{(xy)}\}\,\phi(y)\,dy = f(x),$$

for almost all values of x.

The proof of this theorem is much like several of the other proofs of this note and Note XLVI; and I therefore omit it.

5. If, as in Theorem B, we suppose $f(x)$ to be subject to some more stringent condition, we can obtain alternative forms of the solution.

Suppose in particular that

$$\int_0^\infty |f(x)|\,e^{kx}\,dx \qquad (k>0)$$

is convergent. Then

$$\phi(s) = \Sigma \frac{(-1)^n c_n s^n}{n!} = \int_0^\infty f(x) \Sigma \frac{(-1)^n (sx)^n}{n!}\,dx = \int_0^\infty f(x) e^{-sx}\,dx,$$

if term by term integration is legitimate, which is certainly the case if $|s| < k$; and $\phi(s)$ is plainly an analytic function of s, regular for all values of s whose real part is greater than $-k$.

---

* It was proved by Sommerfeld ('Die willkürlichen Functionen in der mathematischen Physik', *Dissertation*, Königsberg, 1891) that

$$\lim_{\delta \to 0} \int_0^\infty e^{-(\delta y)^2} J_n(xy)\,y\,dy \int_\beta^\gamma J_n(ya)\,af(a)\,da = f(x)$$

at any point of continuity of $f(x)$ which falls inside (β, γ). The result proved here is more special in that $n=0$, but more general in other respects.

Further results concerning this integral, and the corresponding integral with $e^{-\delta y}$, will be found in Sommerfeld's dissertation and in my paper 'Further researches in the theory of divergent series and integrals' (*Trans. Camb. Phil. Soc.*, vol. xxi., 1908, pp. 1–48). I take this opportunity of pointing out that a number of the formulæ of the latter paper (§§ 31–34) are incorrect owing to the omission of the factor a before $f(a)$: see (e.g.) equation (8), p. 43. The error does not affect the validity of the arguments used, but the results are stated inaccurately, and a number of small corrections are required.

Let us write, as usual, $s = \sigma + it$, and suppose that $\kappa > -k$ and $c > 0$. Then

$$\int_{\kappa-i\infty}^{\kappa+i\infty} e^{\delta s^2 + cs} \phi(s)\, ds = \int_{\kappa-i\infty}^{\kappa+i\infty} e^{\delta s^2 + cs}\, ds \int_0^\infty f(x) e^{-sx}\, dx$$

$$= \int_0^\infty f(x)\, dx \int_{\kappa-i\infty}^{\kappa+i\infty} e^{\delta s^2 + (c-x)s}\, ds,$$

the convergence and equality of all the integrals which occur being secured by the convergence of the integral

$$\int_0^\infty |f(x)|\, e^{-\kappa x}\, dx \int_{-\infty}^\infty e^{-\delta t^2}\, dt.$$

But

$$\int_{\kappa-i\infty}^{\kappa+i\infty} e^{\delta s^2 + (c-x)s}\, ds = 2i \int_0^\infty e^{-\delta t^2} \cos(x-c)t\, dt = i\sqrt{\left(\frac{\pi}{\delta}\right)} e^{-(x-c)^2/4\delta}.$$

Hence

$$\int_{\kappa-i\infty}^{\kappa+i\infty} e^{\delta s^2 + cs} \phi(s)\, ds = i\sqrt{\left(\frac{\pi}{\delta}\right)} \int_0^\infty f(x)\, e^{-(x-c)^2/4\delta}\, dx;$$

and so[*]

$$\lim_{\delta \to 0} \frac{1}{2\pi i} \int_{\kappa-i\infty}^{\kappa+i\infty} e^{\delta s^2 + cs} \phi(s)\, ds = f(c)$$

for almost all values of c.

We have thus proved

THEOREM E. *If there is a solution of the problem for which*

$$\int_0^\infty |f(x)|\, e^{kx}\, dx \quad (k > 0)$$

is convergent, it is given, for almost all values of x, by the formula

$$f(x) = \frac{1}{2\pi i}\, G \int_{\kappa-i\infty}^{\kappa+i\infty} \phi(s)\, e^{xs}\, ds,$$

where $\kappa > -k$, $\phi(s)$ is the analytic function defined, for sufficiently small values of s, by the series

$$\Sigma \frac{(-1)^n c_n}{n!} s^n,$$

---

[*] G. H. Hardy, 'Notes on some points in the Integral Calculus', No. XLII (*On Weierstrass's singular integral*), *Messenger of Mathematics*, vol. xlvi., pp. 43–48. See also Lebesgue, *loc. cit.*

and the generalised integral
$$G \int_{-\infty}^{\infty} \psi(t)\, dt$$
is defined as meaning
$$\lim_{\delta \to 0} \int_{-\infty}^{\infty} e^{-\delta t^2} \psi(t)\, dt.$$

We have also

THEOREM F. *The integral*
$$\frac{1}{2\pi i} \int_{\kappa-i\infty}^{\kappa+i\infty} \phi(s)\, e^{xs}\, ds,$$
is summable $(C, 1)$, *with sum* $f(x)$, *for almost all values of* x; *that is to say*
$$\frac{1}{4\pi i T} \int_{\kappa-iT}^{\kappa+iT} (T-|t|)\, \phi(s)\, e^{xs}\, ds \to f(x),$$
as $T \to \infty$, *for almost all values of* x.

6. In Theorems C—F we have proved that, in certain circumstances, an integral is summable to sum $f(x)$. In order to prove the *convergence* of the integrals in question we must naturally assume more, as in the theory of the convergence of Fourier's series. It is not difficult to see that any of the standard hypotheses as to the behaviour of $f(x)$, in the immediate neighbourhood of the value of x considered, suffice to secure the convergence of the integral in the ordinary sense.

It is instructive to consider the application of Theorems A to F in two special cases.

(a) Suppose $\quad c_n = 1/(n+1)$.

If we follow the solution of Theorem A, we find
$$F(s) = \sum_{0}^{\infty} \frac{(-s)^n}{(n+1)\, 2n!} - \frac{2}{s}(\sqrt{s}\sin\sqrt{s} + \cos\sqrt{s} - 1),$$
$$\chi(u) = \frac{2}{\sqrt{u}} \int_0^{\infty} e^{-t} \sin\frac{t}{\sqrt{u}} \frac{dt}{t} - 2\int_0^{\infty} e^{-t}\left(1 - \cos\frac{t}{\sqrt{u}}\right) \frac{dt}{t^2}$$
$$= \log\left(1 + \frac{1}{u}\right),$$
$$f(x) = 1 \quad (0 < x < 1), \quad f(x) = 0 \quad (x > 1).$$

In order to apply the method of Theorem B we take $\alpha = 1$. We find
$$F(s) = \sum_{0}^{\infty} \frac{(-s)^n}{(n+1)!} = \frac{1-e^{-s}}{s},$$

$$\chi(u) = \int_0^\infty e^{-t}(1-e^{-t/u})\frac{dt}{t} = \log\left(1+\frac{1}{u}\right).$$

In the solution of Theorems C and D we have

$$\phi(y) = \sum_0^\infty \frac{(-y)^n}{n!(n+1)!} = \frac{J_1(2\sqrt{y})}{\sqrt{y}},$$

$$f(x) = \int_0^\infty J_0\{2\sqrt{(xy)}\}\frac{J_1(2\sqrt{y})}{\sqrt{y}}\,dy$$

$$= 2\int_0^\infty J_0(2u\sqrt{x})J_1(2u)\,du,$$

$$f(x) = 1 \quad (0 < x < 1), \quad f(x) = 0 \quad (x > 1).$$

Finally, in the solution of Theorems E and F, we have

$$\phi(s) = \frac{1-e^{-s}}{s},$$

$$f(x) = \frac{1}{2\pi i}\int_{\kappa-i\infty}^{\kappa+i\infty}\frac{1-e^{-s}}{s}e^{xs}\,ds$$

$$= \frac{1}{\pi}\int_0^\infty \frac{\sin xt}{t}\,dt - \frac{1}{\pi}\int_0^\infty \frac{\sin(x-1)t}{t}\,dt,$$

which gives the same result.

(b) Suppose $c_n = 1$ for all values of n. Then it will be found that all our solutions fail. For example, in the last solution, we obtain

$$\phi(s) = e^{-s},$$

$$\frac{1}{2\pi i}\int_{\kappa-i\infty}^{\kappa+i\infty} e^{\delta s^2 + xs}\phi(s)\,ds = \frac{1}{2\sqrt{(\pi\delta)}}e^{-(x-1)^2/4\delta},$$

which tends to infinity if $x = 1$ and to zero otherwise. Thus there is no solution satisfying our conditions, as is easily proved independently.

It should be observed that, so soon as we introduce Stieltjes' generalisation of the notion of an integral, this example admits an obvious solution. For if $\psi(x) = 0$ for $0 < x < 1$, and $\psi(x) = 1$ for $x \geq 1$, we have

$$\int_0^1 x^n\,d\psi(x) = \int_0^\infty x^n\,d\psi(x) = 1$$

for all values of n.

The methods used in this note are no doubt capable of adaptation to prove the uniqueness, under appropriate conditions, of any solution by means of a Stieltjes' integral. They would however require as a preliminary the proof of certain general theorems concerning passage to a limit under the sign of Stieltjes' integration.

I observe in conclusion that, if $f(x)$ is *positive*, the condition that
$$\int_0^\infty f(x)\, e^{k\sqrt{x}}\, dx$$
is convergent for some positive k is easily shown to be equivalent to the condition that
$$c_n < 2n!\, K^n$$
for some value of K.

CORRECTIONS

p. 82, *line* 1. $e^{\sqrt{|s|}}$ should be $e^{2\sqrt{|s|}}$.
———, *line* 4. $|y|<k$ should be $|y|<\tfrac{1}{4}k^2$.

p. 86, *line* 10. $\dfrac{1}{4\pi i T}$ should be $\dfrac{1}{2\pi i T}$.

NOTES ON SOME POINTS IN THE INTEGRAL CALCULUS.

By *G. H. Hardy*.

XLIX.

On Mellin's inversion formula.

1. THE relationship between the two functions f and ϕ, expressed by the reciprocal equations

$$f(s) = \int_0^\infty \phi(x) x^{s-1} dx \quad\quad\quad (1)$$

and

$$\phi(x) = \frac{1}{2\pi i} \int_{a-i\infty}^{a+i\infty} f(s) x^{-s} ds \quad\quad\quad (2),$$

was (so far as I am aware) first observed by Riemann in his famous memoir on the distribution of primes.* Riemann deduces (2) from (1) by a purely formal process based on Fourier's integral theorem; and the first accurate discussion of the question was that of Mellin,† with whose name the formulæ are generally associated.

Mellin bases his discussion on a somewhat restrictive hypothesis as to the nature of $\phi(x)$, viz. that it is an analytic function regular (except at the origin and infinity) throughout an angular region including the positive real axis; and the formulæ are so interesting that a more general discussion seems called for. Such a discussion will also form a natural sequel to some recent notes of this series, which deal with problems of a very similar character.‡

---

* B. Riemann, 'Ueber die Anzahl der Primzahlen unter einer gegebener Grösse', *Werke*, (ed. 1876), pp. 136–144 (p. 140). Riemann's formulæ are not quite the same as (1) and (2), but embody the same idea.

† Hj. Mellin, 'Ueber die fundamentale Wichtigkeit des Satzes von Cauchy für die Theorien der Gamma- und der hypergeometrischen Funktionen', *Acta Societatis Fennicae*, vol. xxi., 1896, no. 1, pp. 1–115, §§ 14–29: 'Abriss einer einheitlichen Theorie der Gamma- und der hypergeometrischen Funktionen', *Math. Annalen*, vol. lxviii., 1910, pp. 305–337, § 8.

‡ Note XLII, 'On Weierstrass's singular integral'; Notes XLVI and XLVII, 'On Stieltjes' *problème des moments*': *Messenger of Mathematics*, vol. xlvi., 1917, pp. 43–48, 175–182; vol. xlvii., 1918, pp. 81–88.

2. In discussing the integral (2) I shall take as fundamental not the ordinary idea of convergence of an infinite integral but that of *summability*. I shall say that the integral

$$\frac{1}{2\pi i}\int_{a-i\infty}^{a+i\infty} f(s)\,x^{-s}\,ds = \frac{1}{2\pi}\int_{-\infty}^{\infty} f(a+it)\,x^{-a-it}\,dt$$

is *summable* $(C, 1)$ if

$$\frac{1}{2\pi T}\int_{-T}^{T}(T-|t|)f(a+it)\,x^{-a-it}\,dt$$

tends to a limit when $T \to \infty$ *. It is easily verified that this definition satisfies the 'condition of consistency', *i.e.* that a convergent integral is summable to a sum equal to its value.

THEOREM A. *Suppose that* $\phi(x)$ *is integrable (in the sense of Lebesgue* †) *in every interval* $0 < \delta \leq x \leq \Delta$; *and that*

$$\int_0 x^{\alpha-1}\phi(x)\,dx, \qquad \int x^{\beta-1}\phi(x)\,dx,$$

where $\alpha < \beta$, *are convergent* ‡, *so that*

$$\int_0^\infty x^{s-1}\phi(x)\,dx,$$

where $s = \sigma + it$, *is uniformly convergent throughout any strip* $\alpha < \alpha' \leq \sigma \leq \beta' < \beta$, *and represents an analytic function* $f(s)$ *regular in the strip. Then the integral*

$$\frac{1}{2\pi i}\int_{a-i\infty}^{a+i\infty} f(s)\,x^{-s}\,ds \quad (\alpha < a < \beta)$$

is summable $(C, 1)$, *to sum* $\phi(x)$, *for almost all positive values of* x. *In particular this is true at all points of continuity of* $\phi(x)$: *more generally, the integrable is summable to sum*

$$\tfrac{1}{2}\{\phi(x-0)+\phi(x+0)\}$$

whenever this expression has a meaning.

---

* On p. 86 of Note XLVII, in Theorem F, $4\pi i T$ should be $2\pi i T$.

† I use 'integrable' rather than 'summable' in order to avoid using one word in two quite different senses As Prof. Young has justly remarked, Lebesgue's idea has superseded the idea of 'Riemann integrability' so completely that it is quite unnecessary to mark the distinction any longer by the use of different words.

‡ Not necessarily absolutely.

We have

$$\frac{1}{2\pi i T}\int_{a-iT}^{a+iT}(T-|t|)x^{-s}f(s)\,ds$$

$$=\frac{1}{2\pi T}\int_{-T}^{T}(T-|t|)\,dt\int_{c}^{\infty}y^{a+it-1}\phi(xy)\,dy$$

$$=\frac{1}{2\pi T}\int_{0}^{\infty}y^{a-1}\phi(xy)\,dy\int_{-T}^{T}(T-|t|)y^{it}\,dt,$$

the inversion of the order of integration being permissible in virtue of the uniform convergence of the integral with respect to y. But

$$\int_{-T}^{T}(T-|t|)y^{it}\,dt = 2\int_{0}^{T}(T-t)\cos(t\log y)\,dt$$

$$= 2\,\frac{1-\cos(T\log y)}{(\log y)^2};$$

and so

$$\frac{1}{2\pi i T}\int_{a-iT}^{a+iT}(T-|t|)x^{-s}f(s)\,ds$$

$$=\frac{1}{2\pi T}\int_{0}^{\infty}y^{a-1}\phi(xy)\left\{\frac{\sin(\tfrac{1}{2}T\log y)}{\tfrac{1}{2}\log y}\right\}^2 dy.$$

We write the integral on the right-hand side in the form

$$\frac{1}{2\pi T}\left(\int_{0}^{\delta}+\int_{\delta}^{\Delta}+\int_{\Delta}^{\infty}\right) = J_1 + J_2 + J_3,$$

say; and we begin by considering J_3.

Let

$$\int_{y}^{\infty} w^{a-1}\phi(xw)\,dw = \Phi(y),$$

so that $\Phi(y)\to 0$ when $y\to\infty$. Then

$$\pi J_3 = -\frac{1}{T}\int_{\Delta}^{\infty}\Phi'(y)\,\frac{1-\cos(T\log y)}{(\log y)^2}\,dy$$

$$=\frac{\Phi(\Delta)}{T(\log\Delta)^2}\{1-\cos(T\log\Delta)\} + \int_{\Delta}^{\infty}\Phi(y)\,\frac{\sin(T\log y)}{y(\log y)^2}\,dy$$

$$-\frac{2}{T}\int_{\Delta}^{\infty}\Phi(y)\,\frac{1-\cos(T\log y)}{y(\log y)^3}\,dy.$$

Of these three terms, the second (which is obviously the most important) is in absolute value less than

$$\mu \int_\Delta^\infty \frac{dy}{y(\log y)^2} = \frac{\mu}{\log \Delta},$$

where μ is the upper bound of $|\Phi|$ in (Δ, ∞); and this tends to zero when $\Delta \to \infty$, uniformly in T. It is plain that the same is true of the other two terms; and we can therefore choose Δ so that

$$|J_3| < \epsilon$$

for all values of T greater than unity. A similar argument shews that δ may be chosen so that

$$|J_1| < \epsilon$$

for all values of T greater than unity.

On the other hand, if we write $y = e^u$, $\delta = e^\omega$, $\Delta = e^\Omega$, we obtain

$$J_2 = \frac{1}{2\pi T} \int_\omega^\Omega e^{au} \phi(xe^u) \left(\frac{\sin \frac{1}{2} Tu}{\frac{1}{2}u}\right)^2 du.$$

But this is an integral of Fejér's type. It tends to $\phi(x)$ almost always, and in particular whenever x is a point of continuity; and to $\frac{1}{2}\{\phi(x-0) + \phi(x+0)\}$ whenever this expression has a meaning. From this, and our previous results about J_1 and J_3, our theorem follows.

3. We can obtain practically the same result by using other definitions of summability, and in particular that employed for the solution of Stieltjes' problem in Notes XLVI and XLVII.

THEOREM B. *The conclusions of Theorem A are also true if the value of the summable integral is defined as the limit of*

$$\frac{1}{2\pi i} \int_{a-i\infty}^{a+i\infty} e^{\eta s^2} f(s) x^{-s} ds \quad (\eta > 0)$$

when η tends to zero.

We have

$$\frac{1}{2\pi i} \int_{a-i\infty}^{a+i\infty} e^{\eta s^2} f(s) x^{-s} ds = \frac{1}{2\pi i} \int_{a-i\infty}^{a+i\infty} e^{\eta s^2} ds \int_0^\infty \phi(xy) y^{s-1} dy$$

$$= \frac{1}{2\pi i} \int_0^\infty \phi(xy) \frac{dy}{y} \int_{a-i\infty}^{a+i\infty} e^{\eta s^2} y^s ds,$$

if the inversion is permissible. Now this would certainly be the case, in virtue of the uniform convergence of the integral with respect to s, if the limits of integration with respect to y were δ and Δ. It is therefore sufficient to show that

$$\frac{1}{2\pi i}\int_{a-i\infty}^{a+i\infty} e^{\eta s^2} ds \int_{\Delta}^{\infty} \phi(xy) y^{s-1} dy \ldots\ldots\ldots(3)$$

is convergent and tends to zero when $\Delta \to \infty$; and that a similar condition is satisfied for the interval $(0, \delta)$.

Choose c so that $a < a + c < \beta$; and write

$$\int_{y}^{\infty} \phi(xw) w^{a+c-1} dw = \Phi(y).$$

Then

$$\int_{\Delta}^{\infty} \phi(xy) y^{a+it-1} dy = -\int_{\Delta}^{\infty} \Phi'(y) y^{it-c} dy$$

$$= \Phi(\Delta) \Delta^{it-c} + (it-c)\int_{\Delta}^{\infty} \Phi(y) y^{it-c-1} dy.$$

The first term tends to zero when Δ tends to infinity, uniformly in t; the second is less in absolute value than a constant multiple of

$$\Delta^{-c} |t|.$$

It follows that (3) is convergent, and less in absolute value than a constant multiple of

$$\Delta^{-c} \int_{-\infty}^{\infty} e^{-\eta t^2} |t| dt,$$

which tends to zero. A similar argument may be applied to the interval $(0, \delta)$. Thus the inversion of the order of integration is justified. Now

$$\int_{a-i\infty}^{a+i\infty} e^{\eta s^2} y^s ds = i \int_{-\infty}^{\infty} e^{-\eta t^2} y^{it} dt = i\sqrt{\left(\frac{\pi}{\eta}\right)} e^{-(\log y)^2/4\eta};$$

so that

$$\frac{1}{2\pi i}\int_{a-i\infty}^{a+i\infty} e^{\eta s^2} f(s) x^{-s} ds = \frac{1}{2\sqrt{(\pi\eta)}}\int_{0}^{\infty} \phi(xy) e^{-(\log y)^2/4\eta} \frac{dy}{y}.$$

The limit of the right-hand side, when $\eta \to 0$, may be calculated by putting $y = e^u$ and applying the known results concerning Weierstrass's singular integral.* We thus complete the proof of Theorem B.

---

* See Note XLII.

4. Neither of the theorems establishes the *convergence* of (2), and this is an essentially more difficult question. We have

$$\frac{1}{2\pi i}\int_{a-iT}^{a+iT} x^{-s} f(s)\, ds = \frac{1}{2\pi}\int_{-T}^{T} dt \int_{0}^{\infty} y^{a+it-1}\phi(xy)\, dy$$

$$= \frac{1}{\pi}\int_{0}^{\infty} y^{a-1}\phi(xy)\,\frac{\sin(T\log y)}{\log y}\, dy.$$

On writing e^u for y, we obtain an integral of Dirichlet's type, and the conditions that it should tend to a limit are of a more special character. In the first place we must suppose that $\phi(x)$ satisfies (as in the theory of the convergence of Fourier series) some condition more stringent than that of (*e.g.*) mere continuity. Secondly we must assume more about the integral (1) than mere convergence.*

Suppose for example that

$$\phi(x) = \sin x,$$

$$f(s) = \int_{0}^{\infty} \sin x\, x^{s-1}\, dx = \sin\tfrac{1}{2}s\pi\, \Gamma(s) \quad (-1 < \sigma < 1).$$

Then $\phi(x)$ is continuous and of bounded variation in any finite interval. But

$$\frac{1}{2\pi i}\int_{a-i\infty}^{a+i\infty} \sin\tfrac{1}{2}s\pi\, \Gamma(s)\, x^{-s}\, ds$$

is only convergent if $a < \tfrac{1}{2}$.

It is easily shown that the integral is convergent, when defined as

$$\lim \int_{a-iT}^{a+iT} f(s)\, x^{-s}\, ds,$$

if $\phi(x)$ satisfies any of the ordinary conditions which occur in the theory of convergence of Fourier series and the integral (1) is *absolutely* convergent for $\alpha < \sigma < \beta$. In particular cases the convergence of the integral, in the ordinary sense, can usually be recognised directly.

5. I conclude by remarking that an elegant formula given recently by Mr. Ramanujan† may be deduced without difficulty from (1) and (2).

---

* See, *e.g.*, my paper 'Fourier's double integral and the theory of divergent integrals', *Trans. Camb. Phil. Soc.*, vol. xxi., pp. 427–451 (p. 434).

† S. Ramanujan, 'Some definite integrals', *Messenger of Mathematics*, vol xliv., 1915, pp. 10–18 [formula (2)].

Take $\phi(x) = x^p(1+x)^{-q}$ $(0 < p < q)$

so that $f(s) = \dfrac{\Gamma(s+p)\Gamma(q-p-s)}{\Gamma(q)}$

if $-p < \sigma < q-p$. We obtain

$$\frac{1}{2\pi i}\int_{a-i\infty}^{a+i\infty} y^{-s}\Gamma(s+p)\Gamma(q-p-s)\,ds = \Gamma(q)y^p(1+y)^{-q}$$

if $-p < a < q-p$.* If in particular we take $q = 2p$, $a = 0$, we find

$$\int_0^\infty \cos(t\log y)\,|\Gamma(p+it)|^2\,dt = \pi\Gamma(2p)y^p(1+y)^{-2p};$$

or, writing $y = e^{2m}$ and using the multiplication formula for the Gamma-function,

$$\int_0^\infty \cos(2mt)\,|\Gamma(p+it)|^2\,dt = \tfrac{1}{2}\sqrt{\pi}\,\Gamma(p)\Gamma(p+\tfrac{1}{2})\operatorname{sech}^{2p} m,$$

which is Mr. Ramanujan's formula.

In order to give an example which is not covered by Mellin's conditions, we may take

$$\phi(x) = (1-x)^{q-1}\quad (0 < x < 1),\quad \phi(x) = 0\quad (x > 1),$$

where q is positive. We find that

$$\frac{1}{2\pi i}\int_{a-i\infty}^{a+i\infty} \frac{\Gamma(s)}{\Gamma(s+q)}x^{-s}\,ds = \frac{1}{\Gamma(q)}(1-x)^{q-1}$$

if $a > 0$, $0 < x < 1$. If $x > 1$, the value of the integral is zero.

---

* This is formula (64) of Mellin's paper in the *Math. Annalen* quoted above.

CORRECTIONS

p. 179, *Theorem* A, *line* 3. Both integrals should be $\int_0^\infty$.

p. 184, *line* 8. 'multiplication' should be 'duplication'.

NOTES ON SOME POINTS IN THE INTEGRAL CALCULUS.

By *G. H. Hardy.*

LII.

On some definite integrals considered by Mellin.

1. In Note XLIX* I applied the ideas of the modern theory of integration to the pair of reciprocal integral relations

$$f(s) = \int_0^\infty \phi(x) x^{s-1} dx \quad \ldots\ldots\ldots\ldots (1),$$

$$\phi(x) = \frac{1}{2\pi i} \int_{a-i\infty}^{a+i\infty} f(s) x^{-s} ds \quad \ldots\ldots (2),$$

usually associated with the name of Mellin. Mellin has also considered† integrals of the type

$$\frac{1}{2\pi i} \int_{a-i\infty}^{a+i\infty} f_1(s) f_2(s) f_3(s) \ldots x^{-s} ds \quad \ldots\ldots (3),$$

where $f_1(s), f_2(s), \ldots$ are functions of the type (1); and it is interesting to extend the results of my earlier note in this direction. I consider here in particular the case of two functions

$$f(s) = \int_0^\infty \phi(x) x^{s-1} dx, \quad g(s) = \int_0^\infty \psi(x) x^{s-1} dx \ldots (4).$$

2. We shall use the following lemmas.

LEMMA 1. *If F, G, F^{1+p}, and $G^{1+1/p}$, where $p > 0$, are summable, then*

$$\int_a^b F(x+t) G(t) dt$$

is a continuous function of t.

This is a known theorem due to W. H. Young‡.

LEMMA 2. *If F, G, F^{1+p}, and $G^{1+1/p}$, where $p > 0$, are summable in any finite interval, and*

$$\int_{-\infty}^\infty |F|^{1+p} dx, \quad \int_{-\infty}^\infty |G|^{1+1/p} dx$$

---

* *Messenger,* vol. xlvii. (1918), pp. 178–184.
† See, *e.g.,* pp. 336–337 of his memoir in the *Math. Annalen* quoted in Note XLIX.
‡ 'On a class of parametric integrals and their application in the theory of Fourier series', *Proc. Roy. Soc.* (A), vol. lxxxv., 1911, pp. 401–414 (p. 407).

are convergent, then
$$J(T) = \frac{1}{2\pi T} \int_{-\infty}^{\infty} F(u) G(v) \left\{ \frac{\sin \tfrac{1}{2} T(u+v)}{\tfrac{1}{2}(u+v)} \right\}^2 du\, dv$$
$$\to \int_{-\infty}^{\infty} F(x) G(-x) dx$$
when $T \to \infty$.

The lemma rests on the following formal transformations:
$$u+v = 2x, \quad u-v = 2y, \quad du\, dv = 2 dx\, dy,$$
$$J(T) = \frac{1}{\pi T} \int_{-\infty}^{\infty} F(x+y) G(x-y) \left(\frac{\sin Tx}{x} \right)^2 dx\, dy$$
$$= \frac{1}{\pi T} \int_{-\infty}^{\infty} \left(\frac{\sin Tx}{x} \right)^2 H(x)\, dx \to H(0),$$
where
$$H(x) = \int_{-\infty}^{\infty} F(x+y) G(x-y) dy.$$

In order to justify these equations, we observe first that
$$\int_{-\infty}^{\infty} |F(x+y) G(x-y)| \left(\frac{\sin Tx}{x} \right)^2 dx\, dy$$
$$\leq \int_{-\infty}^{\infty} \operatorname{Min}\left(T^2, \frac{1}{x^2} \right) dx \int_{-\infty}^{\infty} |F(x+y) G(x-y)|\, dy$$
$$\leq \int_{-\infty}^{\infty} \operatorname{Min}\left(T^2, \frac{1}{x^2} \right) dx \left\{ \int_{-\infty}^{\infty} |F(x+y)|^{1+p} dy \right\}^{1/1+p}$$
$$\times \left\{ \int_{-\infty}^{\infty} |G(x-y)|^{1+1/p} dy \right\}^{p/1+p}$$
$$= \int_{-\infty}^{\infty} \operatorname{Min}\left(T^2, \frac{1}{x^2} \right) dx \left\{ \int_{-\infty}^{\infty} |F|^{1+p} dx \right\}^{1/1+p} \left\{ \int_{-\infty}^{\infty} |G|^{1+1/p} dx \right\}^{p/1+p}.$$

Thus all our integrals are absolutely convergent, and the transformation of $J(T)$ by substitution, and its reduction to a repeated integral, are legitimate. The integral involving $H(x)$ is of Fejér's type, and so all that remains to be proved is that $H(x)$ is continuous for $x=0$. Now
$$H(x) = \int_{-\infty}^{\infty} F(2x-t) G(t) dt = \int_{-\infty}^{a} + \int_{a}^{b} + \int_{b}^{\infty} = H_1 + H_2 + H_3.$$
Suppose that $-\delta \leq x \leq \delta$. Then
$$|H_1| < \left\{ \int_{-\infty}^{a} |F(2x-t)|^{1+p} dt \right\}^{1/1+p} \left\{ \int_{-\infty}^{a} |G(t)|^{1+1/p} dt \right\}^{p/1+p},$$
and may be made less than ϵ by choice of an a independent of x. The same argument applies to H_3; and H_2 is con-

tinuous by Lemma 1. Thus H is continuous, and Lemma 2 is proved.

3. THEOREM A. *Suppose that ϕ, ψ, ϕ^{1+p}, and $\psi^{1+1/p}$, where $p > 0$, are summable in any interval $0 < \xi_1 \leq x \leq \xi_2$. Further, suppose that constants K, α, β, γ, δ exist such that $\alpha < \gamma$, $\beta < \delta$, $\alpha < \delta$, $\beta < \gamma$,*
$$|\phi(x)| < Kx^{-\alpha}, \quad |\psi(x)| < Kx^{-\beta},$$
for all sufficiently small positive values of x, and
$$|\phi(x)| < Kx^{-\gamma}, \quad |\psi(x)| < Kx^{-\delta}$$
for all sufficiently large positive values of x; so that the functions $f(s)$ and $g(s)$ of $s = \sigma + it$, defined by (4), are regular for
$$\alpha < \sigma < \gamma, \quad \beta < \sigma < \delta$$
respectively. Then the integral
$$J(x) = \frac{1}{2\pi i} \int_{a-i\infty}^{a+i\infty} f(s) g(s) x^{-s} ds,$$
where $x > 0$, $\alpha < a < \gamma$, $\beta < a < \delta$, is summable $(C, 1)$; and its sum is
$$\int_0^\infty \phi(w) \psi\left(\frac{x}{w}\right) \frac{dw}{w}.$$

Suppose that x_1 and x_2 are any two positive numbers whose product is x. Then
$$\frac{1}{2\pi i T} \int_{a-iT}^{a+iT} (T-|t|) f(s) g(s) x^{-s} ds$$
$$= \frac{1}{2\pi T} \int_{-T}^T (T-|t|) dt \int_0^\infty y^{a+it-1} \phi(x_1 y) dy \int_0^\infty z^{a+it-1} \psi(x_2 z) dz.$$

The triply repeated integral just written is absolutely convergent, as may be seen at once by comparison with
$$\int_{-T}^T dt \int_0^\infty y^{a-1} |\phi(x_1 y)| dy \int_0^\infty z^{a-1} |\psi(x_2 z)| dz.$$

Hence it may be written in the form
$$\frac{1}{2\pi T} \int_0^\infty y^{a-1} \phi(x_1 y) dy \int_0^\infty z^{a-1} \psi(x_2 z) dz \int_{-T}^T (T-|t|)(yz)^{it} dt$$
$$= \frac{1}{2\pi T} \int_0^\infty (yz)^{a-1} \phi(x_1 y) \psi(x_2 z) \left\{\frac{\sin(\tfrac{1}{2} T \log yz)}{\tfrac{1}{2} \log yz}\right\}^2 dy\, dz$$
$$= \frac{1}{2\pi T} \int_{-\infty}^\infty e^{a(u+v)} \phi(x_1 e^u) \psi(x_2 e^v) \left\{\frac{\sin \tfrac{1}{2} T(u+v)}{\tfrac{1}{2}(u+v)}\right\}^2 du\, dv.$$

Let $e^{au} \phi(x_1 e^u) = \chi(u)$. Then, since ϕ^{1+p} is summable in any interval $0 < \xi_1 \leq x \leq \xi_2$, χ^{1+p} is summable in any finite interval. Also $|\chi(u)| = O\{e^{(a-\alpha)u}\}$ for large negative values of u, and $|\chi(u)| = O\{e^{(a-\beta)u}\}$ for large positive values; and $a - \alpha > 0$, $a - \beta < 0$. Hence

$$\int_{-\infty}^{\infty} |\chi|^{1+p} du$$

is convergent. Similarly, if $e^{av}\psi(x_2 e^v) = \omega(v)$,

$$\int_{-\infty}^{\infty} |\omega|^{1+1/p} dv$$

is convergent. Thus all the conditions of Lemma 2 are satisfied, and

$$\frac{1}{2\pi i T}\int_{a-iT}^{a+iT} (T-|t|) f(s) g(s) x^{-s} ds \to \int_{-\infty}^{\infty} \chi(\rho)\omega(-\rho) d\rho$$

$$= \int_{-\infty}^{\infty} \phi(x_1 e^\rho) \psi(x_2 e^{-\rho}) d\rho = \int_{0}^{\infty} \phi(x_1 w) \psi\left(\frac{x_2}{w}\right) \frac{dw}{w}$$

$$= \int_{0}^{\infty} \phi(w) \psi\left(\frac{x}{w}\right) \frac{dw}{w};$$

which proves the theorem.

4. THEOREM B. *The conclusions of Theorem A are also true if the value of the summable integral is defined as the limit of*

$$\frac{1}{2\pi i}\int_{a-i\infty}^{a+i\infty} e^{\eta s^2} f(s) g(s) x^{-s} ds \quad (\eta > 0)$$

when η tends to zero.

It is hardly necessary to give the proof in detail. The reader may compare § 3 of Note XLIX., and the corresponding developments in Notes XLVI. and XLVII.

THEOREM C. *If the conditions of Theorem A are satisfied, and the integral is convergent, then its value is that given in Theorem A.*

This is an immediate corollary of Theorem A. To establish the convergence of the integral is (compare § 4 of Note XLIX.) essentially more difficult. In applications it can usually be recognised directly.

Examples.

5. (i) Suppose

$$\phi(x) = \frac{x^{-\alpha}}{(1+x)^{\gamma-\alpha}}, \quad \psi(x) = \frac{x^{-\beta}}{(1+x)^{\delta-\beta}},$$

$$f(s) = \frac{\Gamma(s-\alpha)\Gamma(\gamma-s)}{\Gamma(\gamma-\alpha)}, \quad g(s) = \frac{\Gamma(s-\beta)\Gamma(\delta-s)}{\Gamma(\delta-\beta)},$$

α, β, γ, δ being real. Then

$$\frac{1}{2\pi i}\int_{a-i\infty}^{a+i\infty}\Gamma(s-\alpha)\Gamma(s-\beta)\Gamma(\gamma-s)\Gamma(\delta-s)x^{-s}ds$$
$$=\Gamma(\gamma-\alpha)\Gamma(\delta-\beta)x^{-\beta}\int_0^\infty\frac{w^{\delta-\alpha-1}dw}{(1+w)^{\gamma-\alpha}(w+x)^{\delta-\beta}}$$

if $\alpha < a < \gamma$, $\beta < a < \delta$. Suppose in particular that $x=1$, $a=0$, $\alpha+\gamma=\beta+\delta=0$. Then we obtain

$$\frac{1}{2\pi}\int_{-\infty}^\infty|\Gamma(\gamma+it)|^2|\Gamma(\delta+it)|^2 dt = \frac{\Gamma(2\gamma)\Gamma(2\delta)\{\Gamma(\gamma+\delta)\}^2}{\Gamma(2\gamma+2\delta)}$$
$$=\frac{1}{2\sqrt{\pi}}\frac{\Gamma(\gamma)\Gamma(\gamma+\tfrac{1}{2})\Gamma(\delta)\Gamma(\delta+\tfrac{1}{2})\Gamma(\gamma+\delta)}{\Gamma(\gamma+\delta+\tfrac{1}{2})},$$

a formula given by Mr. Ramanujan* and valid for all positive values of γ and δ.

(ii) Suppose

$$\phi(x) = x^{-\alpha}(1-x)^{\lambda-1} \ (0<x<1), \quad \phi(x)=0 \ (x>1),$$
$$\psi(x)=0 \ (0<x<1), \quad \psi(x)=x^{1-\delta-\mu}(x-1)^{\mu-1} \ (x>1).$$

In order that our conditions should be satisfied, it must be possible to choose a positive p such that

$$(1+p)(\lambda-1) > -1, \quad (1+1/p)(\mu-1) > -1;$$

and this will be so if λ and μ are positive and $\lambda+\mu > 1$. We have then

$$f(s) = \frac{\Gamma(\lambda)\Gamma(s-\alpha)}{\Gamma(\lambda+s-\alpha)}, \quad g(s) = \frac{\Gamma(\mu)\Gamma(\delta-s)}{\Gamma(\mu+\delta-s)}$$

and

$$\frac{1}{2\pi i}\int\frac{\Gamma(s-\alpha)\Gamma(\delta-s)}{\Gamma(\lambda+s-\alpha)\Gamma(\mu+\delta-s)}ds = \frac{1}{\Gamma(\lambda)\Gamma(\mu)}\int_0^1 w^{\delta-\alpha-1}(1-w)^{\lambda+\mu-2}dw$$
$$=\frac{\Gamma(\delta-\alpha)\Gamma(\lambda+\mu-1)}{\Gamma(\lambda)\Gamma(\mu)\Gamma(\delta-\alpha+\lambda+\mu-1)} \quad (\alpha<a<\delta).$$

Suppose in particular that

$$\alpha+\delta=0, \quad a=0, \quad \lambda=\mu=\epsilon-\delta.$$

Then

$$\frac{1}{2\pi}\int_{-\infty}^\infty\left|\frac{\Gamma(\delta+it)}{\Gamma(\epsilon+it)}\right|^2 dt = \frac{\Gamma(2\delta)\Gamma(2\epsilon-2\delta-1)}{\{\Gamma(\epsilon-\delta)\}^2\Gamma(2\epsilon-1)}$$
$$=\frac{1}{2\sqrt{\pi}}\frac{\Gamma(\delta)\Gamma(\delta+\tfrac{1}{2})\Gamma(\epsilon-\delta-\tfrac{1}{2})}{\Gamma(\epsilon-\tfrac{1}{2})\Gamma(\epsilon)\Gamma(\epsilon-\delta)},$$

---

* S. Ramanujan, 'Some definite integrals', *Messenger*, vol. xliv. (1915), pp. 10–18 (p. 15).

where $\epsilon > \delta + \tfrac{1}{2} > \tfrac{1}{2}$. This result also agrees with that given by Mr. Ramanujan*.

(iii) Suppose
$$\phi(x) = x^{-\alpha} e^{-x}, \qquad \psi(x) = x^{-\beta} e^{-x},$$
$$f(s) = \Gamma(1-\alpha+s), \qquad g(s) = \Gamma(1-\beta+s).$$

Then we obtain
$$\frac{1}{2\pi i}\int_{a-i\infty}^{a+i\infty} \Gamma(s-\alpha)\Gamma(s-\beta) x^{-s} ds = x^{-\beta}\int_0^\infty e^{-w-x/w} w^{\beta-\alpha-1} dw,$$
provided $\alpha < a$, $\beta < a$. In particular, if $\alpha = \beta = 0$, we obtain
$$\frac{1}{2\pi i}\int_{a-i\infty}^{a+i\infty} \{\Gamma(s)\}^2 x^{-s} ds = \int_0^\infty e^{-w-x/w}\frac{dw}{w},$$
where a and x are positive. This formula (which can of course be extended at once to complex values of x) is one of importance in the analytic theory of numbers†.

6. The following example is more novel. It is clear that, if we take $\psi(x) = x^{-u}\phi(1/x)$, we have
$$g(s) = \int_0^\infty x^{s-u-1}\phi(1/x) dx = \int_0^\infty x^{u-s-1}\phi(x) dx = f(u-s).$$

If $$\phi(x) = \sum_1^\infty e^{-n^2 x},$$
we have $$f(s) = \Gamma(s)\sum n^{-2s} = \Gamma(s)\zeta(2s),$$
and
$$\frac{1}{2\pi i}\int_{a-i\infty}^{a+i\infty}\Gamma(s)\Gamma(u-s)\zeta(2s)\zeta(2u-2s) ds = \int_0^\infty w^{u-1}\{\phi(w)\}^2 dw.$$

This formula is valid if $\tfrac{1}{2} < a < u - \tfrac{1}{2}$. Now
$$\{\phi(w)\}^2 = \tfrac{1}{4}\{\vartheta(x)-1\}^2,$$
where $$\vartheta(x) = 1 + 2e^{-x} + 2e^{-4x} + \ldots;$$
and so is equal to
$$\tfrac{1}{4}\{1 + \sum_1^\infty r(n)e^{-nx} - 2 - 4\sum_1^\infty e^{-n^2x} + 1\} = \tfrac{1}{4}\sum_1^\infty r(n)e^{-nx} - \sum_1^\infty e^{-n^2 x},$$

---

* *Loc. cit.*, p. 12.
† See G. Voronoï, 'Sur une fonction transcendante et ses applications à la sommation de quelques séries', *Annales scientifiques de l'École Normale Supérieure*, ser. 3, vol. xxi., 1904, pp. 207–268, 459–534; G. H. Hardy and J. E. Littlewood, 'Contributions to the theory of the Riemann Zeta-function and the theory of the distribution of primes', *Acta Mathematica*, vol. xli., 1917, pp. 119–196.

where $r(n)$ is the number of representations of n as the sum of two squares. Hence we obtain the formula

$$\frac{1}{2\pi i}\int_{a-i\infty}^{a+i\infty} \Gamma(s)\,\Gamma(u-s)\,\zeta(2s)\,\zeta(2u-2s)\,ds$$

$$=\tfrac{1}{4}\Gamma(u)\left(\sum_{1}^{\infty}\frac{r(n)}{n^u} - 4\sum_{1}^{\infty}\frac{1}{n^{2u}}\right) = \Gamma(u)\{\zeta(u)\eta(u) - \zeta(2u)\},$$

where $\qquad \eta(u) = 1^{-u} - 3^{-u} + 5^{-u} - \ldots$.

7. The corresponding formula for p functions
$$f_1(s),\ f_2(s),\ \ldots,\ f_p(s)$$
is

$$\frac{1}{2\pi i}\int_{a-i\infty}^{a+i\infty} f_1(s)f_2(s)\ldots f_p(s)\,x^{-s}\,ds$$

$$= \int_0^\infty \phi_1(w_1)\phi_2(w_2)\ldots\phi_{p-1}(w_{p-1})\,\phi_p\left(\frac{x}{w_1 w_2 \ldots w_{p-1}}\right)\frac{dw_1\,dw_2\ldots dw_{p-1}}{w_1 w_2 \ldots w_{p-1}}.$$

CORRECTIONS

p. 90, line 5. $\Gamma(1-\alpha+s)$ should be $\Gamma(s-\alpha)$, and $\Gamma(1-\beta+s)$ should be $\Gamma(s-\beta)$.

NOTES ON SOME POINTS IN THE INTEGRAL CALCULUS.

By *G. H. Hardy*.

LIV.

Further notes on Mellin's inversion formulæ.

1. The formulæ discussed in Notes XLIX. and LII. lend themselves to various transformations. It is worth while to add a few remarks concerning some of the most interesting formulæ thus obtained.

Suppose, in Theorem A of Note XLIX.*, that $\phi(x) = 0$ when $x > 1$, in which case the second convergence condition may naturally be dropped; and write

$$x = e^{-y}, \quad \phi(x) = \phi(e^{-y}) = \chi(y).$$

---

* There is a misstatement in the enunciation of this theorem. In lines 6, 7, instead of 'is uniformly convergent throughout any strip $\alpha < \alpha' \leq \sigma \leq \beta' < \beta$, and represents an analytic function $f(s)$ regular in the strip' read 'is uniformly convergent throughout any rectangle

$$\alpha < \alpha' \leq \sigma \leq \beta' < \beta, \quad -T \leq t \leq T,$$

and represents an analytic function $f(s)$ regular in the strip $\alpha < \sigma < \beta$'. The error is verbal only, as the second statement embodies all that is required or used in the proof.

We thus obtain

Theorem A. *Suppose that $\chi(y)$ is integrable (in the sense of Lebesgue) in every interval $0 \leq y \leq \Delta$, and that*

$$\int_0^\infty e^{-ay} \chi(y)\, dy,$$

where a is real, is convergent, so that

(1) $$\int_0^\infty e^{-sy} \chi(y)\, dy$$

is uniformly convergent throughout any region

$$\sigma \geq a' > a, \quad -T \leq t \leq T,$$

and represents an analytic function $f(s)$ regular for $\sigma > a$. Then the integral

(2) $$\frac{1}{2\pi i} \int_{a-i\infty}^{a+i\infty} e^{sy} f(s)\, ds \quad (a > a)$$

is summable $(C, 1)$, to sum $\chi(y)$, for almost all positive values of y. In particular this is true at all points of continuity of $\chi(y)$: more generally, the integral is summable to sum

(3) $$\tfrac{1}{2}\{\chi(y-0) + \chi(y+0)\}$$

whenever this expression has a meaning.

The formulæ

(4) $$f(s) = \int_0^\infty e^{sy} \chi(y)\, dy,$$

(5) $$\chi(y) = \frac{1}{2\pi i} \int_{a-i\infty}^{a+i\infty} e^{sy} f(s)\, ds$$

are of course very familiar, and have been studied in a variety of forms by different writers[*].

It is hardly necessary to point out that (5) embodies a proof of the uniqueness of any solution of the integral equation (4).

2. I add some examples to illustrate the use of the formulæ (4) and (5).

(a) Suppose that $0 \leq \lambda_1 < \lambda_2 < \lambda_3 < \ldots$, $\lambda_n \to \infty$, and

$$A(y) = a_1 + a_2 + \ldots + a_n \quad (\lambda_n < y \leq \lambda_{n+1}).$$

---

[*] See for example H. Bateman, 'Report on the history and present state of the Theory of Integral Equations', *Reports of the British Association* (Sheffield, 1910), p. 61.

Finally suppose that the series Σa_n is convergent. Then it may be verified at once that

(6) $$f(s) = \Sigma a_n e^{-\lambda_n s} = \int_0^\infty s e^{-sy} A(y)\, dy$$

if $\sigma > 0$. It follows that

(7) $$\frac{1}{2\pi i} \int_{a-i\infty}^{a+i\infty} e^{sy} \frac{f(s)}{s}\, ds = a_1 + a_2 + \ldots + a_n$$

if $a > 0$, $\lambda_n < y < \lambda_{n+1}$. If $y = \lambda_{n+1}$ the value of the integral is

$$a_1 + a_2 + \ldots + a_n + \tfrac{1}{2} a_{n+1}.$$

This is the well-known formula of Cahen, Hadamard, and Perron, which is so fundamental in the general theory of Dirichlet's series. It should be observed however that Theorem A shows only that the integral (5) is summable, and not that it is convergent.

It is easily shown that* *if the integral (1) is absolutely convergent for $\sigma > \alpha$, and $\chi(y)$ is of bounded variation in the neighbourhood of the value of y considered, then (2) is convergent in the sense that*

$$\lim_{T \to \infty} \int_{a-iT}^{a+iT} e^{sy} f(s)\, ds$$

(the principal value of Cauchy) *exists and has the value* (3). Applying this result to the present problem, we obtain Perron's theorem in a form much nearer to that in which it is usually stated†. There is still something lacking; for in fact

$$\lim_{T_1, T_2 \to \infty} \int_{a-iT_2}^{a+iT_2} e^{sy} \frac{f(s)}{s}\, ds$$

exists, the use of the principal value being only required when y has one of the particular values λ_n. This, however, is of no particular importance; and it is of some interest to have alternative methods for the proof of so fundamental a formula. It is plain that the more general formulæ for the Rieszian means‡

$$\Sigma a_n (\omega - \lambda_n)^\kappa, \quad \Sigma a_n (w - l_n)^\kappa,$$

where $l_n = e^{\lambda_n}$, $w = e^\omega$, may be arrived at in a similar manner.

---
* See Note XLIX., p. 181.
† See Dr. Riesz's and my tract, 'The general theory of Dirichlet's series', pp. 12–14.
‡ Hardy and Riesz, *l.c.*, pp. 50–51.

3. (b) It is easily verified that, if
$$\vartheta(y) = 1 + 2\sum_{1}^{\infty} e^{-n^2\pi^2 y},$$
then
$$f(s) = \int_0^\infty e^{-sy}\vartheta(y)\,dy = \frac{1}{\sqrt{s}\tanh\sqrt{s}} \quad (\sigma > 0).$$

The simplest method of proof is to substitute for $\vartheta(y)$ from the functional equation
$$\vartheta(y) = \frac{1}{\sqrt{(\pi y)}}\vartheta\left(\frac{1}{\pi^2 y}\right)$$
and integrate term by term.

Now the function $f(s)$ satisfies the Riccati's differential equation
$$f^2 + 2f' + \frac{f-1}{s} = 0.$$

Also
$$f^2 = \left(\int_0^\infty e^{-sy}\vartheta(y)\,dy\right)^2 = \int_0^\infty e^{-sy}\,dy\int_0^y \vartheta(u)\vartheta(y-u)\,du,$$
$$f' = -\int_0^\infty e^{-sy}y\vartheta(y)\,dy,$$
and
$$\frac{f}{s} = \frac{1}{s}\int_0^\infty e^{-sy}\vartheta(y)\,dy = \int_0^\infty e^{-sy}\,dy\int_0^y \vartheta(u)\,du.$$

Hence, if we write
$$\phi(y) = \int_0^y \vartheta(u)\vartheta(y-u)\,du - 2y\vartheta(y) + \int_0^y \vartheta(y)\,du - 1,$$
we have
$$\int_0^\infty e^{-sy}\phi(y)\,dy = f^2 + 2f' + \frac{f-1}{s} = 0,$$
so that
$$\phi(y) = 0.$$

The discovery of this very curious integral equation for $\vartheta(y)$ is due to F. Bernstein*. The proof originally given by Bernstein is quite different; but I understand that he had also found the one which I have sketched, and I therefore shall not enter further into its details.

4. (c) It is known that†

(8)
$$\int_0^\infty e^{-sy}y^\alpha J_\alpha(my)\,dy = \frac{\Gamma(\alpha+\tfrac{1}{2})}{\sqrt{\pi}}\frac{(2m)^\alpha}{(m^2+s^2)^{\alpha+\tfrac{1}{2}}},$$

---

* F. Bernstein, 'Die Integralgleichung der elliptische Thetanullfunktion', *Berliner Sitzungsberichte*, 21 Oct. 1920. [The proof has been published since this note was written in the *Proceedings of the Royal Society of Amsterdam*.]
† Nielsen, *Cylinderfunktionen*, p. 186.

if $\sigma > 0$, $m > 0$, $\alpha > -\tfrac{1}{2}$. Hence, if also $\beta > -\tfrac{1}{2}$, and

$$\phi(y) = \int_0^y u^\alpha (y-u)^\beta J_\alpha(mu) J_\beta(my-mu)\, du,$$

we have

$$\int_0^\infty e^{-sy} \phi(y)\, dy = \frac{\Gamma(\alpha + \tfrac{1}{2})\, \Gamma(\beta + \tfrac{1}{2})}{\pi} \frac{(2m)^{\alpha+\beta}}{(m^2+s^2)^{\alpha+\beta+1}}.$$

Hence we deduce the formula

$$(9) \quad \int_0^y u^\alpha (y-u)^\beta J_\alpha(mu) J_\beta(my-mu)\, du$$

$$= \frac{\Gamma(\alpha+\tfrac{1}{2})\,\Gamma(\beta+\tfrac{1}{2})}{\Gamma(\alpha+\beta+1)\sqrt{(2\pi m)}}\, y^{\alpha+\beta+\tfrac{1}{2}} J_{\alpha+\beta+\tfrac{1}{2}}(my).$$

This formula appears to be new: the particular case in which $\alpha = 0$, $\beta = 0$, viz.

$$(10) \quad \int_0^y J_0(mu) J_0(my-mu)\, du = \frac{\sin my}{m},$$

has been given before by Kapteyn and Bateman*.

We have also

$$(11) \quad \int_0^\infty e^{-sy} J_\alpha(my)\, \frac{dy}{y} = \frac{1}{\alpha}\left\{\frac{\sqrt{(m^2+s^2)}-s}{m}\right\}^\alpha$$

if $\alpha > 0$, and

$$(12) \quad \int_0^\infty e^{-sy} J_\alpha(my)\, dy = \frac{1}{\sqrt{(m^2+s^2)}}\left\{\frac{\sqrt{(m^2+s^2)}-s}{m}\right\}^\alpha$$

if $\alpha > -1$. Multiplying two equations of the type (11), or of the types (11) and (12), and pursuing the same line of argument as above, we are led to the formulæ

$$(13) \quad \int_0^y \frac{J_\alpha(mu) J_\beta(my-mu)}{u(y-u)}\, du = \frac{\alpha+\beta}{\alpha\beta}\, \frac{J_{\alpha+\beta}(my)}{y},$$

$$(14) \quad \int_0^y J_\alpha(mu) J_\beta(my-mu)\, \frac{du}{u} = \frac{1}{\alpha}\, J_{\alpha+\beta}(my).$$

---

* W. Kapteyn, 'On a series of Bessel functions', *Proc. Roy. Soc. Amsterdam*, vol. 8 (1904), pp. 494–500, and 'Recherches sur les fonctions cylindriques', *Mém. Soc. Roy. des Sc. de Liège* (3), vol. 6 (1906), no. 5, pp. 1–24; H. Bateman, 'A generalisation of the Legendre Polynomial', *Proc. London Math. Soc.* (2), vol. 3 (1905), pp. 111–123. I am indebted for these references to Prof. G. N. Watson.

In (13) $\alpha > 0$, $\beta > 0$; in (14) $\alpha > 0$, $\beta > -1$. The latter is given by both Kapteyn and Bateman; and the former is an obvious corollary. The integral

$$(15) \quad \phi(y) = \int_0^y J_\alpha(mu) J_\beta(my - mu) \, du,$$

where $\alpha > -1$, $\beta > -1$, is less simple. Multiplying two equations of the type (12), we obtain

$$\int_0^\infty e^{-sy} \phi(y) \, dy = \frac{1}{m^2 + s^2} \left\{ \frac{\sqrt{(m^2 + s^2)} - s}{m} \right\}^{\alpha+\beta}$$

$$= \frac{2}{m} (S^{\alpha+\beta+1} - S^{\alpha+\beta+3} + \ldots),$$

where
$$S = \frac{\sqrt{(m^2 + s^2)} - s}{m}.$$

And from this we deduce Bateman's formula

$$(16) \quad \int_0^y J_\alpha(mu) J_\beta(my - mu) \, du$$
$$= \frac{2}{m} \{ J_{\alpha+\beta+1}(my) - J_{\alpha+\beta+3}(my) + \ldots \},$$

which also reduces to (10) when $\alpha = 0$, $\beta = 0$. The series can be summed in finite terms when $\alpha + \beta$ is an integer, by means of the formulæ

$$\cos x = J_0(x) - 2J_2(x) + 2J_4(x) - \ldots, \quad \sin x = 2J_1(x) - 2J_3(x) + \ldots.$$

5. There are of course also a variety of formulæ which are, in substance, transformations of those of Note LII. Among these I may quote the following:

$$\left. \begin{array}{l} f(s) = \int_0^\infty x^{s-1} \phi(x) \, dx, \quad g(s) = \int_0^\infty x^{s-1} \psi(x) \, dx, \\ \int_0^\infty x^{2a-1} \phi(xe^\lambda) \psi(xe^{-\lambda}) \, dx = \frac{1}{2\pi} \int_{-\infty}^\infty e^{2\lambda it} f(a+it) g(a-it) \, dt \end{array} \right\},$$

$$\left. \begin{array}{l} f(s) = \int_0^\infty e^{-sy} \phi(y) \, dy, \quad g(s) = \int_0^\infty e^{-sy} \psi(y) \, dy, \\ \frac{1}{bc} \int_0^y \phi\left(\frac{u}{b}\right) \psi\left(\frac{y-c}{c}\right) du = \frac{1}{2\pi i} \int_{a-i\infty}^{a+i\infty} e^{sy} f(bs) g(cs) \, ds \end{array} \right\},$$

$$\left. \begin{array}{l} f(s) = \int_0^\infty e^{-sy} \phi(y) \, dy, \quad g(s) = \int_0^\infty e^{-sy} \psi(y) \, dy, \\ \int_\lambda^\infty e^{-2au} \phi(u-\lambda) \psi(u+\lambda) \, du = \frac{1}{2\pi} \int_{-\infty}^\infty e^{2\lambda it} f(a+it) g(a-it) \, dt \end{array} \right\}.$$

None of these are essentially different from the formulæ of Note LII., and the reader will be able to supply the conditions to be satisfied by the various functions and parameters which occur in them.

If in the last result we take
$$\phi(y) = \psi(y) = y^{\alpha-1},$$
we obtain, after some elementary transformations,
$$\int_0^\infty \frac{\cos 2\lambda t}{(t^2 + a^2)^\alpha} dt = \frac{\pi e^{-2a\lambda}(2\lambda)^{2\alpha-1}}{\{\Gamma(\alpha)^2\}} \int_0^\infty e^{-4a\lambda t}\{t(1+t)\}^{\alpha-1} dt.$$

Each of these integrals is in fact expressible in terms of Bessel's functions*.

CORRECTIONS

p. 166, *formula* (4). e^{sy} should be e^{-sy}.

p. 170, §5, *line* 5. $e^{2\lambda it}$ should be $e^{-2\lambda it}$.

——, §5, *line* 9. $\phi(u-\lambda)\psi(u+\lambda)$ should be $\phi(u+\lambda)\psi(u-\lambda)$.

---

* See G. F. Meyer's edition of Dirichlet's lectures on definite integrals, p. 289, and G. H. Hardy, 'Some multiple integrals', *Quarterly Journal*, vol. xxxix. (1908), pp. 357–375 (357–358).

NOTES ON SOME POINTS IN THE INTEGRAL CALCULUS.

By *G. H. Hardy*.

LVII.

On Fourier transforms.

1. The theory of 'Fourier transforms' is due primarily to Plancherel.* Plancherel proved that, if $f(x)$ is integrable† over any finite interval $(0, X)$, and its square is integrable over $(0, \infty)$, then a function $g(x)$ exists which has the same properties, and is connected with $f(x)$ by the reciprocal relations

$$(1.11) \quad f(x) = \sqrt{\left(\frac{2}{\pi}\right)} \frac{d}{dx} \int_0^\infty \frac{\sin xy}{y} g(y)\, dy,$$

$$(1.12) \quad g(x) = \sqrt{\left(\frac{2}{\pi}\right)} \frac{d}{dx} \int_0^\infty \frac{\sin xy}{y} f(y)\, dy,$$

---

* M. Plancherel, 'Contribution à l'étude de la représentation d'une fonction arbitraire par des intégrales définies', *Rend. Circ. Mat. Palermo*, 30 (1910), pp. 289–335.

† In the sense of Lebesgue.

for almost all values of x. Further

$$(1.13) \qquad \int_0^\infty \{f(x)\}^2 dx = \int_0^\infty \{g(x)\}^2 dx.$$

The formulae (1.1) reduce to the well-known formulae of Fourier when differentiation under the sign of integration is permissible, and assert, in a symmetrical form, the reciprocity between two functions implied in 'Fourier's integral theorem'.

Plancherel obtained the formulae (1.1) as corollaries of a very general theory, and his actual deduction of them is somewhat artificial. More recently Titchmarsh[*] has investigated them in a more direct and natural manner, and has established the corresponding formulae associated with Hankel's generalisation of Fourier's theorem.

In this note I give an alternative investigation, which differs from Titchmarsh's in that all relics of the general theory of 'orthogonal functions' have disappeared.

2. The integral

$$(2.1) \qquad \psi(x) = \int_0^\infty \frac{\sin xy}{y} f(y)\, dy$$

is absolutely convergent, since

$$\left(\int_0^\infty \left| \frac{\sin xy}{y} f(y) \right| dy \right)^2 \leq \int_0^\infty \left(\frac{\sin xy}{y} \right)^2 dy \int_0^\infty \{f(y)\}^2 dy.$$

If

$$(2.2) \quad \Delta(x, h) = \frac{\psi(x+h) - \psi(x-h)}{2h} = \int_0^\infty \frac{\sin hy}{hy} \cos xy\, f(y)\, dy,$$

we have

$$(2.3) \quad \Delta(x, h)\, \Delta(x, h')$$
$$= \int_0^\infty \int_0^\infty \frac{\sin hy}{hy} \frac{\sin h'y'}{h'y'} \cos xy \cos xy'\, f(y)\, f(y')\, dy\, dy',$$

the double integral being absolutely convergent, and uniformly convergent in x. Hence

$$(2.4) \quad \int_0^X \Delta(x, h)\, \Delta(x, h')(X-x)\, dx$$
$$= \int_0^\infty \int_0^\infty \frac{\sin hy}{hy} \frac{\sin h'y'}{h'y'} f(y)f(y')\, dy\, dy' \int_0^X (X-x) \cos xy \cos xy'\, dx$$
$$= \int_0^\infty \int_0^\infty \frac{\sin hy}{hy} \frac{\sin h'y'}{h'y'} \left\{ \frac{\sin^2 \tfrac{1}{2} X (y+y')}{(y+y')^2} + \frac{\sin^2 \tfrac{1}{2} X (y-y')}{(y-y')^2} \right\}$$
$$\times f(y) f(y')\, dy\, dy'.$$

---

[*] E. C Titchmarsh, 'Hankel transforms', *Proc. Camb. Phil. Soc.*, 21 (1923), pp. 463–473.

3. LEMMA A. *There is a constant K such that*

$$(3.1) \quad \int_0^\infty \int_0^\infty |f(y)||f(y')| \frac{\sin^2 m(y \pm y')}{(y \pm y')^2} \, dy\, dy' < Km$$

for all real values of m.

We define $f(y)$ for negative y so that it shall be even. Then

$$\int_{-\infty}^\infty \int_{-\infty}^\infty |f(y)||f(y')| \frac{\sin^2 m(y \pm y')}{(y \pm y')^2}\, dy\, dy'$$

$$= \int_{-\infty}^\infty |f(y)|\, dy \int_{-\infty}^\infty |f(y+u)| \left(\frac{\sin mu}{u}\right)^2 du$$

$$= \int_{-\infty}^\infty \left(\frac{\sin mu}{u}\right)^2 du \int_{-\infty}^\infty |f(y)||f(y+u)|\, dy$$

$$\leq \int_{-\infty}^\infty \left(\frac{\sin mu}{u}\right)^2 du \sqrt{\left\{\int_{-\infty}^\infty \{f(y)\}^2 dy \int_{-\infty}^\infty \{f(y+u)\}^2 dy\right\}}$$

$$= \pi m \int_{-\infty}^\infty \{f(y)\}^2 dy < Km.$$

4. It follows that the integral on the right of (2.4) is for any fixed value of X, uniformly convergent in a rectangle

$$-h_0 \leq h \leq h_0, \quad -h_0' \leq h' \leq h_0';$$

and therefore that

$$(4.1) \quad \lim_{h,h' \to 0} \int_0^X \Delta(x,h)\Delta(x,h')(X-x)\, dx$$

$$= \int_0^\infty \int_0^\infty \left\{\frac{\sin^2 \tfrac{1}{2}X(y+y')}{(y+y')^2} + \frac{\sin^2 \tfrac{1}{2}X(y-y')}{(y-y')^2}\right\} f(y)f(y')\, dy\, dy'.$$

If we write, for shortness, $\Delta(x,h) = \Delta$, $\Delta(x,h') = \Delta'$, the three integrals

$$\int_0^X \Delta^2 (X-x)\, dx, \quad \int_0^X \Delta\Delta'(X-x)\, dx, \quad \int_0^X \Delta'^2(X-x)\, dx$$

tend to the same limit when h and h' tend to zero, and therefore

$$(4.2) \quad \int_0^X (\Delta - \Delta')^2 (X-x) = \int_0^X (\Delta^2 - 2\Delta\Delta' + \Delta'^2)(X-x)dx \to 0.$$

It follows that $\Delta(x,h)\sqrt{(X-x)}$ 'converges in mean'[*] to a function $\chi(x)\sqrt{(X-x)}$ integrable, with its square, in $(0, X)$;

---

[*] See Plancherel, *l.c.*, p. 4. The idea of 'convergence en moyenne', which is fundamental in this theory, is due to E. Fischer ('Sur la convergence en moyenne', *Comptes Rendus*, 13 May, 1907).

and that
$$(4.3) \quad \int_0^X (\Delta - \chi)^2 (X-x)\, dx \to 0$$
and
$$(4.4) \quad \int_0^X \Delta^2 (X-x)\, dx \to \int_0^X \chi^2 (X-x)\, dx,$$
when $h \to 0$.

If $0 < \xi < X$,
$$(4.5) \quad \int_0^\xi (\Delta - \chi)^2\, dx \leq \frac{1}{X-\xi} \int_0^\xi (\Delta-\chi)^2 (X-x)\, dx \to 0.$$

Hence Δ converges in mean to χ in $(0, \xi)$, that is to say in any finite interval, and χ and χ^2 are integrable in any finite interval. Also
$$\left\{ \int_0^\xi (\Delta - \chi)\, dx \right\}^2 \leq \xi \int_0^\xi (\Delta - \chi)^2\, dx \to 0,$$
$$(4.6) \quad \lim_{h \to 0} \int_0^\xi \Delta(x, h)\, dx = \int_0^\xi \chi(x)\, dx.$$

Next, integrating (2.2) over $(0, \xi)$, we obtain
$$\int_0^\xi \Delta(x, h)\, dx = \int_0^\infty \frac{\sin hy}{hy} \frac{\sin \xi y}{y} f(y)\, dy,$$
and the last integral is uniformly convergent in h. Hence
$$\lim_{h \to 0} \int_0^\xi \Delta(x, h)\, dx = \int_0^\infty \frac{\sin \xi y}{y} f(y)\, dy$$
or
$$(4.7) \quad \int_0^\xi \chi(x)\, dx = \int_0^\infty \frac{\sin \xi y}{y} f(y)\, dy\,;$$
and
$$\chi(\xi) = \frac{d}{d\xi} \int_0^\infty \frac{\sin \xi y}{y} f(y)\, dy$$
for almost all ξ. If
$$(4.8) \quad \chi(\xi) = \sqrt{\left(\frac{\pi}{2}\right)} g(\xi),$$
this is (1.12).

5. We prove next that g^2 is integrable over $(0, \infty)$, and that
$$(5.1) \quad \int_0^\infty \{g(x)\}^2\, dx = \int_0^\infty \{f(x)\}^2\, dx.$$

We have, from (4.1),

$$(5.2) \qquad \int_0^X \chi^2(X-x)\,dx$$

$$= \int_0^\infty \int_0^\infty \left\{\frac{\sin^2 \tfrac{1}{2}X(y+y')}{(y+y')^2} + \frac{\sin^2 \tfrac{1}{2}X(y-y')}{(y-y')^2}\right\} f(y) f(y')\,dy\,dy'$$

or

$$(5.3) \qquad \frac{1}{X}\int_0^X \chi^2(X-x)\,dx = \int_0^\infty \{F_1(y,X) + F_2(y,X)\} f(y)\,dy,$$

where

$$(5.4) \quad \begin{cases} F_1(y,X) = \dfrac{1}{X}\displaystyle\int_0^\infty f(y') \dfrac{\sin^2 \tfrac{1}{2}X(y+y')}{(y+y')^2}\,dy', \\[6pt] F_2(y,X) = \dfrac{1}{X}\displaystyle\int_0^\infty f(y') \dfrac{\sin^2 \tfrac{1}{2}X(y-y')}{(y-y')^2}\,dy'. \end{cases}$$

When $X \to \infty$,

$$(5.5) \qquad F_1(y,X) \to 0, \quad F_2(y,X) \to \tfrac{1}{2}\pi f(y),$$

for almost all y, by Lebesgue's generalisation of Fejér's theorem. If then we may proceed to the limit under the integral sign, on the right-hand side of (5.3), we obtain

$$(5.6) \qquad \lim \frac{1}{X}\int_0^X \chi^2(X-x)\,dx = \tfrac{1}{2}\pi \int_0^\infty f^2\,dy,$$

or, since χ^2 is positive,

$$(5.7) \qquad \int_0^\infty \chi^2\,dx = \tfrac{1}{2}\pi \int_0^\infty f^2\,dy,$$

which is (5.1). It is only necessary, then, to justify the limiting process applied to (5.3). This process will certainly be legitimate if

$$(5.8) \qquad \int_0^\infty \{F_1(y,X)\}^2\,dy, \quad \int_0^\infty \{F_2(y,X)\}^2\,dy$$

are bounded functions of X*.

6. It is plainly sufficient to prove that, if we define $f(y)$ for negative y as in § 3, and write

$$(6.1) \quad F(y,X) = \frac{1}{X}\int_{-\infty}^\infty f(y') \frac{\sin^2 \tfrac{1}{2}X(y-y')}{(y-y')^2}\,dy',$$

---

* See, e.g., W. H. Young, 'The application of expansions to definite integrals', Proc. Lond. Math. Soc. (2), 9 (1910), pp. 463–485 (p. 469).

then
(6.2) $$\int_{-\infty}^{\infty} \{F(y, X)\}^2 dy$$
is bounded. But this integral is

(6.3) $$\frac{1}{X^2}\int_{-\infty}^{\infty} dy \int_{-\infty}^{\infty}\int_{-\infty}^{\infty} f(y')f(y'') \frac{\sin^2 \tfrac{1}{2}X(y-y')}{(y-y')^2}$$
$$\times \frac{\sin^2 \tfrac{1}{2}X(y-y'')}{(y-y'')^2} dy' dy''$$
$$= \frac{\pi}{2X^2}\int_{-\infty}^{\infty}\int_{-\infty}^{\infty} f(y')f(y'') \frac{X(y'-y'')-\sin X(y'-y'')}{(y'-y'')^3} dy' dy'',$$

since

(6.4) $$\int_{-\infty}^{\infty} \frac{\sin^2 m(x-a)}{(x-a)^2} \frac{\sin^2 m(x-b)}{(x-b)^2} dx$$
$$= \frac{\pi}{2} \frac{m(a-b)-\sin m(a-b)}{(a-b)^3}.$$

But the integral (6.3) is
$$\frac{\pi}{2X}\int_0^X \phi(\xi) d\xi,$$
where
$$\phi(\xi) = \int_{-\infty}^{\infty}\int_{-\infty}^{\infty} f(y')f(y'') \frac{1-\cos X(y'-y'')}{(y'-y'')^2} dy' dy'',$$

which is bounded, by Lemma A. Hence (6.3) is bounded, which completes the proof of (5.1).

7. We have now established the existence and integrability of g and g^2 and the equations (1.12) and (1.13). It remains to prove (1.11).

We have

(7.1) $$\int_0^{\infty} \Delta(x, h) \frac{\sin xu}{x} dx$$
$$= \int_0^{\infty} \frac{\sin xu}{x} dx \int_0^{\infty} \frac{\sin hy}{hy} \cos xy f(y) dy$$
$$= \int_0^{\infty} \frac{\sin hy}{hy} f(y) dy \int_0^{\infty} \frac{\sin xu}{x} \cos xy dx$$
$$= \tfrac{1}{2}\pi \int_0^u \frac{\sin hy}{hy} f(y) dy,$$

if the integrations may be inverted. Now
$$\int_0^\infty \frac{\sin hy}{hy} \cos xy f(y)\, dy$$
is uniformly convergent over any finite interval of values of x, so that

(7.2) $$\int_\alpha^\beta \Delta \frac{\sin xu}{x}\, dx = \int_0^\infty \frac{\sin hy}{hy} f(y)\, dy \int_\alpha^\beta \frac{\sin xu}{x} \cos xy\, dx$$

for $0 \leq \alpha < \beta$. As the inner integral on the right-hand side is bounded for all α, β, and y, we may write 0 for α and ∞ for β, which completes the proof of (7.1).

Making $h \to 0$ in (7.1), we obtain
$$\int_0^\infty \chi(x) \frac{\sin xu}{x}\, dx = \tfrac{1}{2}\pi \int_0^u f(y)\, dy,$$

and so $$f(u) = \frac{2}{\pi} \frac{d}{du} \int_0^\infty \chi(x) \frac{\sin xu}{x}\, dx,$$

for almost all u, which is (1.11). This process is legitimate if only

(7.3) $$\lim_{h \to 0} \int_0^\infty \Delta \frac{\sin xu}{x}\, dx = \int_0^\infty \chi \frac{\sin xu}{x}\, dx.$$

Now $$\lim_{h \to 0} \int_0^\xi \Delta \frac{\sin xu}{x}\, dx = \int_0^\xi \chi \frac{\sin xu}{x}\, dx$$

for every finite ξ, since Δ converges in mean to χ. It is therefore sufficient, in order to prove (7.3), to show that

(7.4) $$\int_0^\infty \Delta \frac{\sin xu}{x}\, dx$$

converges uniformly in h.

Suppose that α is large. Then, by (7.2),
$$\left| \int_\alpha^\beta \Delta \frac{\sin xu}{x}\, dx \right| \leqq \left(\int_0^{2u} + \int_{2u}^\infty \right) |f(y)|\, |\phi(\alpha, \beta, y)|\, dy = J_1 + J_2$$

say, where
$$\phi(\alpha, \beta, y) = \int_\alpha^\beta \frac{\sin xu}{x} \cos xy\, dx.$$

In J_2 we have
$$|\phi(\alpha, \beta, y)| = \left| \tfrac{1}{2} \int_{\alpha(y+u)}^{\beta(y+u)} \frac{\sin z}{z}\, dz - \tfrac{1}{2} \int_{\alpha(y-u)}^{\beta(y-u)} \frac{\sin z}{z}\, dz \right| < \frac{K}{\alpha y}$$

where K is a constant. Hence
$$J_2 < \frac{K}{\alpha} \int_{2u}^{\infty} \frac{|f(y)|}{y} dy \to 0,$$
when $\alpha \to \infty$. Also $\phi(\alpha, \beta, y)$ is bounded, and tends to zero, when $\alpha \to \infty$, for every y in $(0, 2u)$ save $y = u$. Hence $J_1 \to 0$. As J_1 and J_2 are independent of h, our conclusion follows.

8. The functions $f(x)$ and $g(x)$ are Fourier *cosine* transforms of one another. There is a similar theory of Fourier *sine* transforms, in which

(8.1) $\quad f(x) = \sqrt{\left(\frac{2}{\pi}\right)} \frac{d}{dx} \int_0^{\infty} \frac{1 - \cos xy}{y} g(y) dy,$

(8.2) $\quad g(x) = \sqrt{\left(\frac{2}{\pi}\right)} \frac{d}{dx} \int_0^{\infty} \frac{1 - \cos xy}{y} f(y) dy,$

and which may be developed in exactly the same manner.

CORRECTIONS

p. 137, *line* 8 *up*. $\Delta(x, h') = \Delta$ should be $\Delta(x, h') = \Delta'$.
p. 139, *formula* (5.3). } missing after $F_2(y, X)$.
p. 140, *line* 13. X should be ξ.

NOTES ON SOME POINTS IN THE INTEGRAL CALCULUS.

By *G. H. Hardy.*

LVIII.

On Hilbert transforms.

1. My first object in this and a succeeding note is to prove the following theorem:

THEOREM A. *Suppose that $f(x)$ is a function which is integrable in any finite interval and whose square is integrable in the infinite interval $(-\infty, \infty)$. Then there is a function $g(x)$ which possesses the same properties and which is connected with $f(x)$ by the formulæ*

$$(1.11) \qquad g(x) = \frac{1}{\pi} \frac{d}{dx} \int_{-\infty}^{\infty} f(t) \log \left| 1 - \frac{x}{t} \right| dt,$$

$$(1.12) \qquad f(x) = -\frac{1}{\pi} \frac{d}{dx} \int_{-\infty}^{\infty} g(t) \log \left| 1 - \frac{x}{t} \right| dt,$$

each valid for almost all values of x.

I call two functions thus related *Hilbert transforms* of one another, since functional reciprocities of this particular character seem to have been considered first by Hilbert in his lectures on integral equations.* The reciprocity may be stated in various forms, such as

$$(1.21) \quad g(x) - \int_0^1 g(t) dt = P \int_0^1 f(t) \cot \pi(x-t) dt,$$

$$(1.22) \quad f(x) - \int_0^1 f(t) dt = -P \int_0^1 g(t) \cot \pi(x-t) dt;$$

$$(1.31) \quad g(x) = \frac{1}{\pi} \int_{-\pi}^{\pi} f'(t) \log \{2 | \sin \tfrac{1}{2}(x-t) |\} dt,$$

---

* See O D. Kellogg, 'Unstetigkeiten in den linearen Integralgleichungen', *Math. Ann.*, lviii. (1904), 441–456, and 'Unstetigkeiten bei den linearen Integralgleichungen, mit Anwendung auf ein Problem von Riemann', *ibid.*, lx. (1905), 424–433; D. Hilbert, *Grundzüge einer allgemeinen Theorie der linearen Integralgleichungen*, (1912), 75–77.

(1.32) $\quad f(x) = -\dfrac{1}{\pi} \displaystyle\int_{-\pi}^{\pi} g'(t) \log\{2\,|\sin\tfrac{1}{2}(x-t)|\}\,dt$;

(1.41) $\quad\quad g(x) = \dfrac{1}{\pi} P \displaystyle\int_{-\infty}^{\infty} \dfrac{f(t)}{x-t}\,dt$,

(1.42) $\quad\quad f(x) = -\dfrac{1}{\pi} P \displaystyle\int_{-\infty}^{\infty} \dfrac{g(t)}{x-t}\,dt$.

Here P is the sign of Cauchy's principal value. The first form is the form in which Kellogg quotes the formulæ from Hilbert; the second that in which Hilbert gives them himself; and the third that in which I have considered them before.* It will be observed that the form which I give them now, and which seems to me on the whole the best, is different again.

No general and independent theory of the formulæ (1.11) and (1.12) appears to exist. In all the writings which I have quoted, very special assumptions are made as regards the nature of $f(x)$. Thus in my own discussion I assumed that (i) $f(x)$ possesses a continuous second derivative, (ii) $f(x)$ is bounded, and (iii) the integrals

$$\int^{\infty} \frac{f(x)\log x}{x}\,dx, \quad \int_{-\infty} \frac{f(x)\log(-x)}{x}\,dx$$

are (not necessarily absolutely) convergent. These conditions are general enough in so far as the behaviour of $f(x)$ at infinity is concerned; far more so, indeed, than those which I shall suppose satisfied here; but they are very restrictive in other respects, and the same may be said of those assumed by Hilbert and Kellogg.

The only authors, so far as I know, who have discussed the formulæ from the standpoint of modern theories of integration are Young, Titchmarsh, Plessner, and M. Riesz. Young, in

---

* G. H. Hardy, 'The theory of Cauchy's principal values (fourth paper)', *Proc. London Math. Soc.* (2), vii. (1908), 181–208. In the third paper of this series [*Proc. London Math. Soc.* (1), **xxxv**. (1902), 81–107 (107)] I stated, without detailed discussion, the formula

$$\int_a^A dx\, P\int_b^B f(x,y)\,dy = \int_b^B dy\, P\int_a^A f(x,y)\,dx + \Delta,$$

where $\quad\quad f(x,y) = \dfrac{\psi(x,y)}{\lambda(x,y)\,\mu(x,y)}$,

the curves $\lambda=0$ and $\mu=0$ intersect at a point (α, β) inside the rectangle of integration, and

$$\Delta = 2\pi^2 \psi(\alpha, \beta) \Big/ \frac{\delta(\lambda, \mu)}{\delta(\alpha, \beta)}$$

This formula is essentially equivalent to the formulæ of Hilbert.

a note published here in 1912,* noted that the formulæ

$$(1.51) \quad g(x) = \frac{1}{\pi}\int_0^\infty \frac{f(x+t)-f(x-t)}{t}\,dt,$$

$$(1.52) \quad f(x) = -\frac{1}{\pi}\int_0^\infty \frac{g(x+t)-g(x-t)}{t}\,dt,$$

arise naturally, in the special case in which $f(x)$ and $g(x)$ are periodic, in the theory of trigonometrical series, and indicated various interesting problems which they suggest, especially when the functions involved satisfy 'Lipschitz conditions'. These formulæ are equivalent to (1.41) and (1.42). Some of the problems suggested by Young have been solved by Titchmarsh in a paper as yet unpublished. Plessner,† in a recent dissertation, has deduced the formulæ (1.21) and (1.22), for functions of integrable square, from the theory of conjugate trigonometrical series; and the formulæ have also been considered from this point of view by Riesz, in a very beautiful memoir, whose main results I have seen, but which is at present unpublished.

2. *The theory of Hilbert transforms may be reduced to dependence on that of Fourier transforms.*‡ If f^2 is integrable in $(-\infty, \infty)$, and F is the Fourier cosine transform of f, then g, the Hilbert transform of f, is the Fourier sine transform of F. It is, however, desirable to develop the theory of the correspondence independently, and I propose to do so without making any use either of the theory of Fourier transforms or of the theory of trigonometrical series. I begin by proving a number of lemmas.

LEMMA 1. *Suppose that* (i) $f(x)$ *and* $\phi(x, h, h', \ldots)$ *are functions, of which the second depends on certain parameters* $h, h', \ldots,$ *integrable over any finite range of values of* x; (ii) *that*

$$(2.1) \qquad \phi(x, h, h', \ldots) \to \phi(x)$$

when $h, h', \ldots$ *tend to certain limiting values, for almost all values of* x; (iii) *that* f^2 *is integrable over* $(-\infty, \infty)$; *and* (iv) *that*

$$(2.2) \qquad \int_{-\infty}^\infty \{\phi(x, h, h', \ldots)\}^2\,dx < A,$$

---

* W. H. Young, 'Note on a certain functional reciprocity in the theory of Fourier series', *Messenger*, vol. xli. (1912), pp. 161–166.

† A. Plessner, 'Zur theorie der konjugierten trigonometrischen Reihen', Giessen, 1922.

‡ See Note LVII, *Messenger*, vol. liii. (1924), pp. 135–142, and the papers of Plancherel and Titchmarsh there referred to.

where A is a constant. Then $f(x)\phi(x)$ is integrable over $(-\infty, \infty)$, and

$$(2.3) \quad \int_{-\infty}^{\infty} f(x)\phi(x, h, h', \ldots)\,dx \to \int_{-\infty}^{\infty} f(x)\phi(x)\,dx.$$

This theorem is in substance due to Lebesgue, but the form here given to it is Young's.*

LEMMA 2. *Suppose that lx stands for $\log|x|$, and that a, a', b, b' are real. Then the value of the definite integral*

$$(2.4) \quad I(a, a', b, b') = \int_{-\infty}^{\infty} l\left(\frac{x-a}{x-a'}\right) l\left(\frac{x-b}{x-b'}\right) dx$$

is zero if the intervals (a, a') and (b, b') are separated from one another. The value of the integral is

if $\qquad\qquad\pi^2(a'-b)$
$\qquad\qquad\qquad a < b < a' < b'$,

and $\qquad\qquad\pi^2(b'-b)$

if $\qquad\qquad a < b < b' < a'$.

These results enable us to write down the value of the integral for all possible arrangements of a, a', b, b'.

The integral is most quickly evaluated either (a) by complex integration or (b) by differentiation with respect to the parameters $a, a', \ldots$. The second method introduces principal values,† and neither is quite appropriate here. I therefore proceed differently. I prove first that

$$(2.5) \quad I(c, d, c, d) = \int_{-\infty}^{\infty} \left\{l\left(\frac{x-c}{x-d}\right)\right\}^2 dx = \pi^2 |c-d|.$$

Suppose, for example, that $c > d$, and write u for $(x-c)/(x-d)$. We obtain

$$I(c, d, c, d) = (c-d)\int_{-\infty}^{\infty} \frac{(lu)^2}{(u-1)^2}\,du$$

$$= 2(c-d)\int_0^1 (lu)^2 \left\{\frac{1}{(1-u)^2} + \frac{1}{(1+u)^2}\right\} du = \pi^2(c-d)$$

---

* See W. H. Young, 'The application of expansions to definite integrals', *Proc. London Math. Soc.* (2), ix. (1910), 463–485, where the reference to Lebesgue is given.

† See the third of my series of papers on 'Cauchy's principal values' already referred to.

by expansion and integration term by term.

This proves (2.5). We have now

$$2l\left(\frac{x-a}{x-a'}\right) l\left(\frac{x-b}{x-b'}\right)$$
$$= \left(l\frac{x-a}{x-b'}\right)^2 + \left(l\frac{x-a'}{x-b}\right)^2 - \left(l\frac{x-a}{x-b}\right)^2 - \left(l\frac{x-a'}{x-b'}\right)^2,$$

(2.6) $\quad I(a, a', b, b')$
$$= \tfrac{1}{2}\pi^2 \{|a-b|+|a'-b|-|a-b|-|a'-b'|\},$$

which contains all the results of the lemma.

LEMMA 3. *If h and h' are positive, then*

(2.7) $\quad J(t, u, h, h') = \int_{-\infty}^{\infty} l\left(\frac{t-x-h}{t-x+h}\right) l\left(\frac{u-x-h'}{u-x+h'}\right) dx = 0$

if $|t-u| \geq h+h'$, and

(2.8) $\quad J(t, u, h, h') = \pi^2 \operatorname{Min}(h+h'-|t-u|, h+h'-|h-h'|)$

if $|t-u| \leq h+h'$.

We have only to write $t-h$, $t+h$, $u-h'$, and $u+h'$ for a, a', b, b', and use Lemma 2.

3. LEMMA 4. *If h and h' are positive,*

(3.11) $\quad \mu = \mu(t, w, h, h') = 0 \quad (|w| \geq h+h'),$

(3.12) $\quad \mu = \operatorname{Min}(h+h'-|w|, h+h'-|h-h'|)$
$$(|w| \leq h+h'),$$

and

(3.2) $\quad \chi(t, h, h') = \dfrac{1}{4hh'} \int_{-\infty}^{\infty} f(t+w)\, \mu(t, w, h, h')\, dw,$

then

(3.3) $\quad\quad\quad \chi(t, h, h') \to f(t)$

when h and h' tend to zero, for almost all values of t.

Suppose for example that $h \leq h'$, so that
$$h+h'-|h-h'| = 2h,$$
and write $\quad \psi(w) = f(t+w) + f(t-w).$

Then $\chi = \dfrac{1}{2h'} \displaystyle\int_0^{h'-h} \psi(w)\,dw + \dfrac{1}{4hh'} \displaystyle\int_{h'-h}^{h+h'} \psi(w)(h+h'-w)\,dw$.

If we substitute $\psi(0) = 2f(t)$ for $\psi(w)$, we obtain

$$\frac{h'-h}{h'} f(t) + \frac{h}{h'} f(t) = f(t).$$

Since $h + h' - w < 2h$ if $h' - h < w < h + h'$, the difference between χ and $f(t)$ is numerically less than

$$\frac{1}{2h'} \int_0^{h+h'} |\psi(w) - \psi(0)|\,dw \leq \frac{1}{2h'} \int_0^{2h'} |\psi(w) - \psi(0)|\,dw = 0 \quad (1)$$

for almost all values of t.

LEMMA 5. *There is a constant A such that*

(3.4) $$\int_{-\infty}^{\infty} \{\chi(t, h, h')\}^2\,dt < A.$$

We have

$$\chi = \frac{1}{4hh'} \int_{-h-h'}^{h+h'} f(t+w)\,\mu\,dw,$$

$$\chi^2 < \frac{1}{16h^2h'^2} \int_{-h-h'}^{h+h'} \{f(t+w)\}^2\,dw \int_{-h-h'}^{h+h'} \mu^2\,dw.$$

Suppose for example that $h \leq h'$. Then

$$\int_{-h-h'}^{h+h'} \mu^2\,dw = 2 \int_0^{h'-h} (2h)^2\,dw + 2 \int_{h'-h}^{h+h'} (h+h'-w)^2\,dw$$

$$= 8h^2h' + \tfrac{16}{3}h^3 < Ah^2h';$$

and so $$\chi^2 < \frac{A}{h'} \int_{-h-h'}^{h+h'} \{f(t+w)\}^2\,dw,$$

$$\int_{-\infty}^{\infty} \chi^2\,dt < \frac{A}{h'} \int_{-\infty}^{\infty} dt \int_{-h-h'}^{h+h'} \{f(t+w)\}^2\,dw$$

$$< \frac{A}{h'} \int_{-h-h'}^{h+h'} dw \int_{-\infty}^{\infty} \{f(t+w)\}^2\,dt < \frac{A(h+h')}{h'} < A.$$

4. I write now

(4.1) $$\phi(x) = \int_{-\infty}^{\infty} f(t)\,l\left(1 - \frac{x}{t}\right) dt.$$

The integral is plainly absolutely convergent. I suppose h and h' positive, and write

$$(4.2) \quad \Delta = \Delta(x, h) = \frac{\phi(x+h) - \phi(x-h)}{2h}$$

$$= \frac{1}{2h} \int_{-\infty}^{\infty} f(t) \, l\left(\frac{t-x-h}{t-x+h}\right) dt,$$

$\Delta' = \Delta(x, h')$, and

$$(4.3) \quad I(h, h') = \int_{-\infty}^{\infty} \Delta \Delta' \, dx.$$

We have

$$(4.4) \quad I(h, h')$$
$$= \frac{1}{4hh'} \int_{-\infty}^{\infty} dx \int_{-\infty}^{\infty} f(t) \, l\left(\frac{t-x-h}{t-x+h}\right) dt \int_{-\infty}^{\infty} f(u) \, l\left(\frac{u-x-h'}{u-x+h'}\right) du$$
$$= \frac{1}{4hh'} \int_{-\infty}^{\infty} f(t) \, dt \int_{-\infty}^{\infty} f(u) \, du \int_{-\infty}^{\infty} l\left(\frac{t-x-h}{t-x+h}\right) l\left(\frac{u-x-h'}{u-x+h'}\right) dx,$$

if inversion of the order of integration is legitimate. In order to prove this, we must establish the absolute convergence of the triple integral. We divide the domain of integration into four parts, defined respectively by

$$t > x, \ u > x; \quad t < x, \ u > x; \quad t > x, \ u < x; \quad t < x, \ u < x;$$

and it is sufficient to consider one of these partial domains, say the first.

If $t > x, u > x$, we have

$$l\left(\frac{t-x-h}{t-x+h}\right) < 0, \quad l\left(\frac{u-x-h'}{u-x+h'}\right) < 0;$$

and we have now to verify the convergence of

$$(4.5) \quad \int_x^{\infty} |f(t)| \, dt \int_x^{\infty} |f(u)| \, du \int_{-\infty}^{\infty} l\left(\frac{t-x+h}{t-x-h}\right) l\left(\frac{u-x+h'}{u-x-h'}\right) dx,$$

which is, by Lemma 3,

$$\pi^2 \iint |f(t)| |f(u)| \, \mathrm{Min}\,(h+h' - |t-u|, \ h+h' - |h-h'|) \, dt \, du,$$

the area of integration being now defined by $t > x$, $u > x$, $|t-u| < h+h'$. The third factor in the integrand is less than a constant. The part of the integral for which $t < u$ is therefore less than

$$A \int_x^\infty |f(t)| \, dt \int_t^{t+h+h'} |f(u)| \, du$$
$$= A \int_0^{h+h'} dw \int_x^\infty |f(t)| \, |f(t+w)| \, dt < A.$$

It follows that (4.5) is convergent and (4.4) true.

5. We have now

$$I(h, h') = \frac{1}{4hh'} \int_{-\infty}^\infty f(t) \, dt \int_{-\infty}^\infty f(u) \, J(t, u, h, h') \, du$$
$$= \pi^2 \int_{-\infty}^\infty f(t) \, dt \frac{1}{4hh'} \int_{-\infty}^\infty f(t+w) \, \mu(t, w, h, h') \, dw$$
$$= \pi^2 \int_{-\infty}^\infty f(t) \, \chi(t, h, h') \, dt,$$

by (2.7), (2.8), (3.11), (3.12), and (3.2). We may apply Lemma 1 to this integral, since χ, by Lemmas 4 and 5, satisfies the conditions there imposed on ϕ; and we conclude that

$$I(h, h') \to \pi^2 \int_{-\infty}^\infty \{f(t)\}^2 \, dt$$

when h and h' tend in any manner to zero. From this it follows that

$$\int_{-\infty}^\infty (\Delta - \Delta')^2 \, dx = I(h, h) - 2I(h, h') + I(h', h') \to 0.$$

Hence $\Delta(x, h)$ *converges in mean to a function* $q(x)$ *of integrable square.*

I call

$$g(x) = \frac{1}{\pi} q(x)$$

the Hilbert transform of $f(x)$. It remains to verify that it possesses the properties prescribed by Theorem A. This is not difficult when the existence of $g(x)$ is once established; but I postpone the proof to the next note of the series.

Prof. Hardy, On some points in the integral calculus.

CORRECTIONS

p. 21, *2nd line of footnote.* 1908 should be 1909.

p. 25, *line* 6. 0(1) should be $o(1)$.

———, *line* 6 *up.* = should be <.

———, *line* 3 *up.* The 1st < should be =.

p. 26, *line* 11 to *p.* 27, *line* 5. The passage 'In order to prove this ... (4.4) is true' is wrong. See 1932, 2, where papers 1925, 9 and 1925, 10 are rewritten and simplified.

NOTE ON SOME POINTS IN THE INTEGRAL CALCULUS.

By *G. H. Hardy*.

LIX.

On Hilbert transforms (continued).

1. In Note LVIII. I proved that $\Delta(x, h)$ tends in mean to a function $q(x) = \pi g(x)$ of integrable square. We must now prove that $\phi(x)$ *is the integral of* $q(x)$. We require

LEMMA 6. *The integral which defines* $\phi(x)$ *is uniformly convergent throughout any finite interval of values of* x.

This follows at once from the inequality

$$\left\{\int_{T'}^{T''} f(t)\, l\left(1 - \frac{x}{t}\right) dt\right\}^2 \leq \int_{T'}^{T''} \{f(t)\}^2\, dt \int_{-\infty}^{\infty} \left\{l\left(1 - \frac{x}{t}\right)\right\}^2 dt.$$

2. We have therefore

$$(2.1) \quad \int_a^\xi \Delta(x, h)\, dx = \int_a^\xi dx\, \frac{1}{2h} \int_{-\infty}^{\infty} f(t)\, l\left(\frac{t - x - h}{t - x + h}\right) dt$$

$$= \frac{1}{2h} \int_{-\infty}^{\infty} f(t)\, dt \int_a^\xi l\left(\frac{t - x - h}{t - x + h}\right) dx$$

$$= \int_{-\infty}^{\infty} f(t)\, L_1(t, h)\, dt + \int_{-\infty}^{\infty} f(t)\, L_2(t, h)\, dt = J_1 + J_2,$$

say, where

$$(2.21) \quad L_1(t, h) = \tfrac{1}{2} l\left(\frac{t - \xi + h}{t - a + h}\, \frac{t - \xi - h}{t - a - h}\right),$$

$$(2.22) \quad L_2(t, h) = \frac{t - \xi}{2h}\, l\left(\frac{t - \xi + h}{t - \xi - h}\right) - \frac{t - a}{2h}\, l\left(\frac{t - a + h}{t - a - h}\right).$$

I prove first that

$$(2.3) \qquad\qquad J_2 \to 0$$

when $h \to 0$. Write

$$(2.4) \quad J_2 = \int_{-\infty}^{-T} + \int_{-T}^{T} + \int_{T}^{\infty} = J_{2,1} + J_{2,2} + J_{2,3},$$

where $T > 2\,\text{Max}\,(|\xi|, |a|)$. In $J_{2,1}$ and $J_{2,3}$ we have
$$L_2 = \frac{h^2}{3\,(t-\xi)^2} + \ldots - \frac{h^2}{3\,(t-a)^2} - \ldots = O\left(\frac{h^2}{t^2}\right),$$
from which it follows at once that $J_{2,1}$ and $J_{2,3}$ tend to zero. Also
$$\int_{-T}^{T} \left(\frac{t-\xi}{2h}\, l\,\frac{t-\xi+h}{t-\xi-h}\right)^2 dt < \int_{-2T}^{2T} \frac{u^2}{h^2}\left(l\,\frac{u+h}{u-h}\right)^2 du$$
$$= h \int_{-2T/h}^{2T/h} w^2 \left(l\,\frac{w+1}{w-1}\right)^2 dw = h\, O\left(\frac{1}{h}\right) < A.$$

Hence, by Lemma 1,

(2.5) $\quad \displaystyle\int_{-T}^{T} f(t)\,\frac{t-\xi}{2h}\, l\left(\frac{t-\xi+h}{t-\xi-h}\right) dt \to \int_{-T}^{T} f(t)\,dt.$

Combining (2.5) with the corresponding relation in which ξ is replaced by a, we find that $J_{2,2} \to 0$, which completes the proof of (2.3).

On the other hand

(2.6) $\quad \displaystyle J_1 \to \int_{-\infty}^{\infty} f(t)\, l\left(\frac{t-\xi}{t-a}\right) dt = \phi(\xi) - \phi(a)$

by Lemma 6; and so

(2.7) $\quad \displaystyle\int_{a}^{\xi} \Delta(x, h)\,dx = J_1 + J_2 \to \phi(\xi) - \phi(a).$

It follows that
$$\int_{a}^{\xi} q(x)\,dx = \phi(\xi) - \phi(a),$$
which is equivalent to (1.11) of Note LVIII.

3. In order to complete the proof of Theorem A, we must establish the reciprocal formula (1.12).

LEMMA 7. *The integral*
$$\int_{-\infty}^{\infty} l\left(1 - \frac{x}{t}\right) dt \int_{-\infty}^{\infty} f(y)\, l\left(\frac{y-t-h}{y-t+h}\right) dy$$
is absolutely convergent.

We can prove at once* that the part of the integral in which $-T < t < T$ is absolutely convergent, and we may there-

---

* As in the proof of Lemma 6.

fore confine our attention to (say) the part in which $t > T$. We consider separately the regions in which (a) $t < \tfrac{1}{2}y$, (b) $t > \tfrac{3}{2}y$, and (c) $\tfrac{1}{2}y < t < \tfrac{3}{2}y$.

(a) This part of the integral may be compared with
$$\int_{2T}^{\infty}|f(y)|\,dy \int_{T}^{\frac{1}{2}y}\frac{dt}{yt} \leqq \int_{2T}^{\infty}|f(y)|\log y\,\frac{dy}{y},$$
which is convergent.

(b) If $\mu = \text{Max}\,(T, \tfrac{3}{2}y)$, this part may be compared with
$$\int_{-\infty}^{\infty}|f(y)|\,dy\int_{\mu}^{\infty}\frac{dt}{t(t-y)} = \int_{-\infty}^{\infty}|f(y)|\,\text{Min}\left(\log 3,\ \log\frac{T}{T-y}\right)\frac{dy}{y},$$
which is convergent.

(c) The third part may be compared with
$$\int_{\frac{2}{3}T}^{\infty}|f(y)|\,dy\int_{\frac{1}{2}y}^{\frac{3}{2}y}\left|l\left(\frac{y-t-h}{y-t+h}\right)\right|\frac{dt}{t}$$
$$\leqq 2\int_{\frac{2}{3}T}^{\infty}\frac{|f(y)|}{y}\,dy\int_{\frac{1}{2}y}^{\frac{3}{2}y}\left|l\left(\frac{y-t-h}{y-t+h}\right)\right|dt.$$

An elementary calculation shows that the inner integral here is
$$2\int_{0}^{\frac{1}{2}y}l\left(\frac{u+h}{u-h}\right)du = 2h\int_{0}^{y/2h}l\left(\frac{w+1}{w-1}\right)dw = O\left(h\log\frac{y}{2h}\right) = O(\log y);$$
so that this repeated integral is also convergent.

4. We have now

(4.1) $\quad\displaystyle\int_{-\infty}^{\infty}\Delta(t,h)\,l\left(1-\frac{x}{t}\right)dt$

$\displaystyle\qquad = \int_{-\infty}^{\infty}l\left(1-\frac{x}{t}\right)dt\,\frac{1}{2h}\int_{-\infty}^{\infty}f(y)\,l\left(\frac{y-t-h}{y-t+h}\right)dy$

$\displaystyle\qquad = \frac{1}{2h}\int_{-\infty}^{\infty}f(y)\,dy\int_{-\infty}^{\infty}l\left(1-\frac{x}{t}\right)l\left(\frac{y-t-h}{y-t+h}\right)dt,$

by Lemma 7. We may suppose $x > 0$ and $0 < 2h < x$; and the value of the inner integral is then, by Lemma 2,

$-\pi^2(x-y+h)\ (x-h \leqq y \leqq x+h),\quad -2\pi^2 h\ (h \leqq y \leqq x-h),$
$\qquad -\pi^2(y+h)\ (-h \leqq y \leqq h),$

and zero otherwise. Thus (4.1) takes the form

$$(4.2) \quad \int_{-\infty}^{\infty} \Delta(t, h) \, l\left(1 - \frac{x}{t}\right) dt$$
$$= -\frac{\pi^2}{2h} \int_{-h}^{h} (y+h) f(y) \, dy - \pi^2 \int_{h}^{x-h} f(y) \, dy$$
$$- \frac{\pi^2}{2h} \int_{x-h}^{x+h} (x-y+h) f(y) \, dy.$$

The first and third integrals tend to zero with h, for almost all values of x; and so

$$(4.3) \quad \int_{-\infty}^{\infty} \Delta(t, h) \, l\left(1 - \frac{x}{t}\right) dt \to -\pi^2 \int_{0}^{x} f(y) \, dy.$$

But

$$\int_{-\infty}^{\infty} \Delta(t, h) \, l\left(1 - \frac{x}{t}\right) dt \to \int_{-\infty}^{\infty} q(t) \, l\left(1 - \frac{x}{t}\right) dt,$$

since the square of the logarithm is integrable over the infinite range. It follows that

$$(4.4) \quad \int_{-\infty}^{\infty} g(t) \, l\left(1 - \frac{x}{t}\right) dt = -\pi \int_{0}^{x} f(y) \, dy,$$

which is (1.12).

5. THEOREM B. *If $f(x)$ and $F(x)$ are two functions of integrable square, and $g(x)$ and $G(x)$ are their Hilbert transforms, then*

$$\int_{-\infty}^{\infty} f(x) F(x) \, dx = \int_{-\infty}^{\infty} g(x) G(x) \, dx.$$

It is sufficient to prove the theorem when $f(x) = g(x)$, the general formula being an immediate corollary. The proof is substantially included in the analysis of Note LVIII., §4. We have in fact*

$$I(h, h) = \int_{-\infty}^{\infty} \{\Delta(x, h)\}^2 dx = \frac{1}{4h^2} \int_{-\infty}^{\infty} \int_{-\infty}^{\infty} f(t) f(u) J(t, u, h, h) \, dt \, du$$
$$= \frac{1}{4h^2} \iint f(t) f(u) (2h - |t-u|) \, dt \, du,$$

the integration being extended over the region $|t-u| < 2h$. The limit of this integral when $h \to 0$ is

$$\pi^2 \int_{-\infty}^{\infty} \{f(t)\}^2 \, dt,$$

---

* *l.c.* (4.3), (4.4), and (2.8).

and so
$$\int_{-\infty}^{\infty} \{g(t)\}^2 \, dt = \int_{-\infty}^{\infty} \{f(t)\}^2 \, dt.$$

6. I have not up to the present introduced the notion of the principal value. I shall now prove that, if we use principal values, we may, for almost all values of x, carry out the differentiations of Theorem A under the integral sign.

THEOREM C. *The formulæ* (1.11) *and* (1.12) *may be written in the form*

$$(6.11) \qquad g(x) = \frac{1}{\pi} P \int_{-\infty}^{\infty} \frac{f(t)}{x-t} \, dt,$$

$$(6.12) \qquad f(x) = -\frac{1}{\pi} P \int_{-\infty}^{\infty} \frac{g(t)}{x-t} \, dt,$$

for almost all values of x.

We require an additional lemma, viz.

LEMMA 8. *If* f^2 *is integrable then*

$$(6.2) \qquad \int_{-\infty}^{\infty} \frac{f(t)}{t-x} \, l\left(\frac{t-x+\epsilon}{t-x-\epsilon}\right) dt \to \pi^2 f(x),$$

when $\epsilon \to 0$, *for almost all values of* x.

We may suppose without loss of generality that $x=0$, $f(x)=0$, and

$$(6.3) \qquad \int_{-t}^{t} \{f(u)\}^2 \, du = o(t).^*$$

It is then plainly sufficient to discuss

$$(6.4) \quad J(\epsilon) = \int_0^{\infty} \frac{f(t)}{t} \, l\left(\frac{t+\epsilon}{t-\epsilon}\right) dt = \int_0^{2\epsilon} + \int_{2\epsilon}^{\infty} = J_1 + J_2,$$

say. Here

$$J_1 = O\left[\left(\int_0^{2\epsilon} \{f(t)\}^2 \, dt\right)^{\frac{1}{2}} \left(\int_{-\infty}^{\infty} \left\{l\left(\frac{t+\epsilon}{t-\epsilon}\right)\right\}^2 \frac{dt}{t^2}\right)^{\frac{1}{2}}\right].$$

The second integral here is a constant multiple of $1/\epsilon$, and the first is $o(\epsilon)$, by (6.3). Hence

$$(6.5) \qquad J_1(\epsilon) = o(1).$$

---

* If f^2 is integrable then
$$\int_{x-t}^{x+t} \{f(u) - f(x)\}^2 \, du = o(t)$$
for almost all values of x.

In J_2 we introduce the function

(6.6) $$\chi(t) = \int_0^t |f(u)| \, du,$$

which is $O(\sqrt{t})$ for large and $o(t)$ for small values of t. Then

$$|J_2(\epsilon)| \leq \int_{2\epsilon}^\infty \frac{\chi'(t)}{t} l\left(\frac{t+\epsilon}{t-\epsilon}\right) dt = -\frac{\chi(2\epsilon)}{2\epsilon} \log 3$$

$$+ \int_{2\epsilon}^\infty \frac{\chi(t)}{t^2} l\left(\frac{t+\epsilon}{t-\epsilon}\right) dt + 2\epsilon \int_{2\epsilon}^\infty \frac{\chi(t) \, dt}{t(t^2-\epsilon^2)} = j_1 + j_2 + j_3,$$

say. Here $j_1 = o(1)$, by (6.3). We choose δ so that $|\chi| < \eta t$ for $0 < t < \delta$, and divide j_2 and j_3 each into two parts, over the ranges $(2\epsilon, \delta)$ and (δ, ∞) respectively. The contributions of the second range, for fixed δ, plainly tend to zero with ϵ. Finally

$$\int_{2\epsilon}^\delta \frac{\chi(t)}{t^2} l\left(\frac{t+\epsilon}{t-\epsilon}\right) dt \leq \eta \int_{2\epsilon}^\delta l\left(\frac{t+\epsilon}{t-\epsilon}\right) \frac{dt}{t}$$

$$< \eta \int_0^\infty l\left(\frac{t+1}{t-1}\right) \frac{dt}{t} = \tfrac{1}{2} \pi^2 \eta;$$

and $$2\epsilon \int_{2\epsilon}^\delta \frac{\chi(t) \, dt}{t(t^2-\epsilon^2)} \leq 2\epsilon\eta \int_{2\epsilon}^\delta \frac{dt}{t^2-\epsilon^2} < \eta \log 3.$$

Hence j_2 and j_3 also tend to zero, and

(6.7) $$J_2(\epsilon) = o(1);$$

which completes the proof.

7. We are now in a position to prove Theorem C. If $g(x)$ is the Hilbert transform of $f(x)$, we have

(7.1) $$\int_\epsilon^\infty \frac{f(x+t) - f(x-t)}{t} \, dt$$

$$= -\frac{1}{\pi} \int_\epsilon^\infty \frac{dt}{t} \frac{d}{dt} \int_{-\infty}^\infty g(u) l\left\{1 - \frac{t^2}{(u-x)^2}\right\} du = -\frac{1}{\pi} \int_\epsilon^\infty W'(t) \frac{dt}{t},$$

say. If we assume provisionally that

(7.2) $$W(t) = o(t)$$

when $t \to \infty$, we may write this in the form

(7.3) $$\int_\epsilon^\infty \frac{f(x+t) - f(x-t)}{t} \, dt = \frac{W(\epsilon)}{\pi\epsilon} - \frac{1}{\pi} \int_\epsilon^\infty \frac{W(t)}{t^2} dt.$$

We substitute for $W(t)$ its expression as a definite integral, and make the further provisional assumption that we may invert the order of integration. We thus obtain

$$(7.4) \quad \int_\varepsilon^\infty \frac{W(t)}{t^2} dt = \int_\varepsilon^\infty \frac{dt}{t^2} \int_{-\infty}^\infty g(u)\, l\left\{1 - \frac{t^2}{(u-x)^2}\right\} du$$

$$= \int_{-\infty}^\infty g(u)\, du \int_\varepsilon^\infty l\left\{1 - \frac{t^2}{(u-x)^2}\right\} \frac{dt}{t^2}.$$

The inner integral here is

$$\frac{1}{\varepsilon} l\left\{1 - \frac{\varepsilon^2}{(u-x)^2}\right\} - 2P\int_\varepsilon^\infty \frac{dt}{(u-x)^2 - t^2}$$

$$= \frac{1}{\varepsilon} l\left\{1 - \frac{\varepsilon^2}{(u-x)^2}\right\} + \frac{1}{u-x} l\left(\frac{u-x+\varepsilon}{u-x-\varepsilon}\right).$$

Substituting this value in (7.4), and combining with (7.3), we obtain

$$(7.5) \quad \int_\varepsilon^\infty \frac{f(x+t) - f(x-t)}{t} dt$$

$$= -\frac{1}{\pi}\int_{-\infty}^\infty \frac{g(u)}{u-x} l\left(\frac{u-x+\varepsilon}{u-x-\varepsilon}\right) du.$$

And hence, by Lemma 8, we have

$$(7.6) \quad \int_0^\infty \frac{f(x+t) - f(x-t)}{t} dt = -\pi g(x)$$

for almost all values of x. This is equivalent to (6.11), and (6.12) may of course be proved in the same way. It is be observed that the symbol of the principal value is required only to deal with the infinity at $t = x$: the integrals are absolutely convergent at infinity.

8. It remains only that we should justify our two provisional assumptions in § 7.

(i) To prove (7.2), we have to show that

$$(8.1) \quad \int_{-\infty}^\infty g(u) \cdot \frac{1}{t} l\left\{1 - \frac{t^2}{(u-x)^2}\right\} \cdot du \to 0$$

when $t \to \infty$. The second factor in the integrand tends to zero, and

$$\int_{-\infty}^\infty \frac{1}{t^2}\left[l\left\{1 - \frac{t^2}{(u-x)^2}\right\}\right]^2 du = \frac{1}{t}\int_{-\infty}^\infty \left\{l\left(1 - \frac{1}{w^2}\right)\right\}^2 dw < A.$$

Hence (8.1) is a corollary of Lemma 1.

(ii) We have to show that
$$\int_\varepsilon^\infty \frac{dt}{t^2} \int_{-\infty}^\infty g(u)\, l\left\{1 - \frac{t^2}{(u-x)^2}\right\} du$$
$$= \int_{-\infty}^\infty g(u)\, du \int_\varepsilon^\infty l\left\{1 - \frac{t^2}{(u-x)^2}\right\} \frac{dt}{t^2}.$$

There is plainly no real loss of generality in supposing $x = 0$; and it is enough to establish the convergence of

(8.2) $$\int_{-\infty}^\infty |g(u)|\, du \int_\varepsilon^\infty \left| l\left(1 - \frac{t^2}{u^2}\right) \right| \frac{dt}{t^2}.$$

It is sufficient to consider the part of the integral for which u is positive. The inner integral is
$$-\int_\varepsilon^{u\sqrt{2}} l\left(1 - \frac{t^2}{u^2}\right) \frac{dt}{t^2} + \int_{u\sqrt{2}}^\infty l\left(1 - \frac{t^2}{u^2}\right) \frac{dt}{t^2}$$
if $u\sqrt{2} > \varepsilon$, and
$$\int_\varepsilon^\infty l\left(1 - \frac{t^2}{u^2}\right) \frac{dt}{t^2}$$
if $u\sqrt{2} < \varepsilon$. An elementary calculation gives the values
$$-\frac{1}{\varepsilon} l\left(1 - \frac{\varepsilon^2}{u^2}\right) - \frac{1}{u} l\left(\frac{u+\varepsilon}{u-\varepsilon}\right) + \frac{2}{u} l\left(\frac{\sqrt{2}+1}{\sqrt{2}-1}\right)$$
and
$$\frac{1}{\varepsilon} l\left(1 - \frac{\varepsilon^2}{u^2}\right) + \frac{1}{u} l\left(\frac{u+\varepsilon}{u-\varepsilon}\right)$$
in the two cases. Every term here gives an integral convergent over the range for which it occurs, and so (8.2) is convergent.

CORRECTIONS

p. 84, *line* 9 up to *p.* 85, *line* 1. The proof of Theorem B is wrong, as it depends on the mistake in 1925, 9, pp. 26–7. See 1932, 2 for a correct proof.

NOTES ON TWO THEOREMS OF NORBERT WIENER

S. Bochner *and* G. H. Hardy*.

1. The following notes concern Theorems II and III of Wiener's recent and important memoir in the *Mathematische Zeitschrift*†. His results, so far as we are concerned with them now, may be summarized as follows. He considers functions $f(x)$ integrable, with their squares, over any finite range. He calls f *nearly bounded* if for some (or, what is the same thing, for any) h

$$\int_\xi^{\xi+h} f^2\, dx$$

is bounded in ξ. His Theorem I, which is (as he shows) an easy deduction from the Plancherel-Titchmarsh theory of "Fourier transforms"‡, asserts that in these circumstances the functions

$$\gamma(x) = \frac{1}{\pi}\int_{-\infty}^{\infty} f(t)\,\frac{\sin xt}{t}\,dt, \quad \delta(x) = \frac{1}{\pi}\int_{-\infty}^{\infty} f(t)\,\frac{e^{-t^2}-\cos xt}{t}\,dt$$

(where the integrals are to be interpreted in the sense of "mean convergence") exist for all x.

Wiener then introduces another new idea, that of *convergence almost in the mean*. We say that $F(x, y)$ converges almost in the mean to $f(x)$ in (a, b), when $y \to \infty$, if

$$\lim_{\eta \to \infty} \int_\eta^{\eta+k} dy \int_a^b \{F(x, y) - f(x)\}^2 dx = 0$$

for some (or any) positive k. Wiener shows that a certain function formed from $\gamma(x)$ and $\delta(x)$, and of the general character of a Fourier integral, converges almost in the mean to $f(x)$ over any finite range. This is his Theorem II.

In his Theorem III he shows that if f is nearly bounded, and f^2 possesses a mean value

$$\mu = Mf^2 = \lim_{X \to \infty} \frac{1}{2X}\int_{-X}^{X} f^2\, dx,$$

---

* Received and read 29 April, 1926.

† N. Wiener, "On the representation of functions by trigonometrical integrals", *Math. Zeitschrift*, 24 (1925), 575-616. See also "The harmonic analysis of irregular motion", *Publications of the Massachusetts Institute of Technology*, 61 (1926), 99-121, 158-189.

‡ We assume acquaintance with the principal results of this theory. The necessary references will be found in E. C. Titchmarsh, "A contribution to the theory of Fourier transforms", *Proc. London Math. Soc.* (2), 23 (1925), 279-289.

then an integral composed from squared differences of γ and δ exists and has the limit μ when a parameter tends to zero. With his further results, and their applications to almost periodic functions, we are not at present concerned.

2. An examination of Wiener's argument shows that its kernel lies in the proofs of the two following propositions.

A. *If f is nearly bounded and*

$$F = F(x, y) = \frac{1}{\pi} \int_{-\infty}^{\infty} f(t) \frac{\sin y(t-x)}{t-x} \, dt$$

(*where the integral is taken in the sense of mean convergence*), *then F converges almost in the mean to f over any finite range (a, b).*

B. *If $f(x)$ is nearly bounded and has a mean value μ, then*

$$\lim_{h \to 0} \frac{1}{\pi h} \int_{-\infty}^{\infty} f^2 \frac{\sin^2 hx}{x^2} \, dx = \mu.$$

It will clear ideas to observe that, if f^2 were integrable over $(-\infty, \infty)$, then F would converge to f in the mean over $(-\infty, \infty)$.

The proofs which Wiener gives of these propositions are considerably larger and more difficult than is necessary, and our primary object here is to give simpler proofs which result, incidentally, in a considerable broadening of the conditions. We must add that we communicated our proofs to Dr. Wiener, and learnt from him that he had already proved Theorem I below by a method substantially the same as ours. We have, however, retained our proof here, since its omission would seriously detract from what value our note may have as a guide to Wiener's memoir. Dr. Wiener has also communicated to us an alternative proof of our Theorems 2 and 3, which we hope may also be published in the *Journal*.

3. We prove first

THEOREM 1. *It is a sufficient condition for the conclusion of Wiener's Theorem A that the integral*

$$\int_{-\infty}^{\infty} f^2(x) \operatorname{Min}\left(1, \frac{1}{x^2}\right) dx$$

should exist.

We may suppose, without real loss of generality, $k = 1$, $a = -1$, $b = 1$. We write $f = f_1 + f_2$, where $f_1 = 0$ if $|x| > c > 1$ and $f_2 = 0$ if $|x| \leq c$, and we denote by F_1, F_2 the functions formed from f_1, f_2 as F is from f. The existence of F_1 and F_2 follows immediately from the Plancherel-Titchmarsh theory. It is sufficient to prove that

(1) $$J_1 + J_2 = \int_\eta^{\eta+1} dy \int_{-1}^1 (F_1 - f_1)^2 dx + \int_\eta^{\eta+1} dy \int_{-1}^1 F_2^2 dx < \epsilon,$$

if $\eta > \eta_0(\epsilon)$.

(i) We may write

$$F_2 = \frac{1}{\pi} \left(\int_c^\infty + \int_{-\infty}^{-c} \right) f \frac{\sin y(t-x)}{t-x} dt = \phi + \psi.$$

Then
$$\phi = \frac{1}{\pi} \int_{c-x}^\infty \frac{f(u+x)}{u} \sin yu \, du,$$

and therefore, after Plancherel and Titchmarsh,

$$\int_0^\infty \phi^2 dy = \frac{1}{2\pi} \int_{c-x}^\infty \left(\frac{f(u+x)}{u} \right)^2 du \leq \frac{1}{2\pi} \left(\frac{c}{c-1} \right)^2 \int_c^\infty \left(\frac{f}{t} \right)^2 dt < \frac{\epsilon}{16},$$

if $|x| \leq 1$ and $c > c_0(\epsilon)$. Since ψ satisfies a similar inequality, we have

$$\int_0^\infty F_2^2 dy \leq 2 \int_0^\infty (\phi^2 + \psi^2) dy < \frac{\epsilon}{4},$$

(2) $$J_2 = \int_\eta^{\eta+1} dy \int_{-1}^1 F_2^2 dx \leq 2 \int_\eta^{\eta+1} F_2^2 dy < \frac{\epsilon}{2} \quad (c > c_0).$$

(ii) Having chosen c, we have only to observe that f_1, being of integrable square over $(-\infty, \infty)$, is subject to the Plancherel-Titchmarsh theory, so that

$$\lim_{y \to \infty} \int_{-\infty}^\infty (F_1 - f_1)^2 dx = 0$$

and

(3) $$J_1 = \int_\eta^{\eta+1} dy \int_{-1}^1 (F_1 - f_1)^2 dx \leq \int_\eta^{\eta+1} dy \int_{-\infty}^\infty (F_1 - f_1)^2 dx < \frac{\epsilon}{2},$$

if $\eta > \eta_0(c, \epsilon) = \eta_0(\epsilon)$. From (2) and (3) we deduce (1) and so the theorem.

We may observe that (as is easily substantiated by examples) the hypothesis does *not* involve either (a) that F converges *in the mean* to f, even over a finite range, or (b) that F converges almost in the mean to f over the infinite range.

4. Wiener's Theorem B is a particular case of

THEOREM 2. *If ϕ is any integrable function which has a mean value*

$$\mu = M\phi = \lim_{X\to\infty} \frac{1}{X} \int_0^X \phi\, dx,$$

and

$$\frac{1}{X} \int_0^X |\phi|\, dx$$

is bounded for $X > 1$, then

$$\lim_{h\to 0} \frac{2}{\pi h} \int_0^\infty \phi\, \frac{\sin^2 hx}{x^2}\, dx = \mu.$$

In fact, if we take $\phi = f^2(x) + f^2(-x)$,

we obtain Wiener's result, generalized by the omission of the condition that f is nearly bounded. In this case ϕ is positive, so that our second condition is included in the first.

We may suppose $\mu = 0$, so that

$$\Phi(x) = \int_0^x \phi\, dt = o(x)$$

for large x. Hence, integrating by parts,

$$\int_0^\infty \phi\, \frac{\sin^2 hx}{x^2}\, dx = -\int_0^\infty \Phi\, \frac{d}{dx}\left(\frac{\sin^2 hx}{x^2}\right) dx$$

$$= -\int_0^1 \Phi\, \frac{d}{dx}\left(\frac{\sin^2 hx}{x^2}\right) dx + 2\int_1^\infty \frac{\Phi}{x} \left(\frac{\sin hx}{x}\right)^2 dx - h\int_1^\infty \Phi\, \frac{\sin 2hx}{x^2}\, dx$$

$$= I_1 + I_2 + I_3,$$

say. It is plain that $I_1 = O(h^4) = o(h)$, and $I_2 = o(h)$, since it is an integral of Fejér's type*, so that everything is reduced to proving that $I_3 = o(h)$. That this is so follows from the following theorem, which, in fact, gives a little more.

5. THEOREM 3. *If $g(x) = o(x)$, and the variation of $g(x)$ in $(1, x)$ is $O(x)$, then*

$$\lim_{k\to 0} \int_1^\infty g(x)\, \frac{\sin kx}{x^2}\, dx = 0.$$

---

* The parameter tends to 0 instead of to ∞, and it is large values of x which are important; but the argument is essentially the same.

It should be observed that even the convergence of the integral is not obvious. We write

$$J = \int_1^\infty = \int_1^{c/k} + \sum_{n=0}^\infty \int_{2^n c/k}^{2^{n+1} c/k} = I + \sum_{n=0}^\infty I_n \quad (c > 0),$$

assuming momentarily the convergence of the series. When c is fixed

(4) $$I = \int_1^{c/k} o(t) \, O\!\left(\frac{kt}{t^2}\right) dt = o(1).$$

On the other hand

$$I_n = k \int_{2^n c}^{2^{n+1} c} g\!\left(\frac{t}{k}\right) \frac{\sin t}{t^2} dt.$$

Now we may write g in the form $g = g_1 - g_2$, where g_1 and g_2 are increasing and $O(x)$, and so express I_n as the difference of the two corresponding integrals $I_{n,1}$ and $I_{n,2}$. Then by a double application of the second mean value theorem, we obtain

$$I_{n,1} = k g_1\!\left(\frac{2^{n+1} c}{k}\right) \int_\tau^{2^{n+1} c} \frac{\sin t}{t^2} dt = k g_1\!\left(\frac{2^{n+1} c}{k}\right) \frac{1}{\tau^2} \int_\tau^{\tau'} \sin t \, dt,$$

where τ and τ' lie between $2^n c$ and $2^{n+1} c$; and this is $O(2^{-n} c^{-1})$, uniformly in k, n, c. It follows that $\Sigma I_{n,1}$, and so ΣI_n, is less than a constant multiple of $1/c$. From this and (4) our conclusion follows, by choice of first c and then k. In the particular case relevant to Theorem 2, g is Φ, and its variation is the integral of $|\phi|$.

NOTES ON SOME POINTS IN THE INTEGRAL CALCULUS.

By *G. H. Hardy.*

LXII.

A singular integral.

1. I was led in Note LIX to the singular integral

(1.1) $\quad J = J(\epsilon) = J(x, \epsilon) = \dfrac{1}{\pi^2} \displaystyle\int_{-\infty}^{\infty} \dfrac{f(t)}{t-x} \, l\left(\dfrac{t-x+\epsilon}{t-x-\epsilon}\right) dt,$ *

and proved that, if f^2 is integrable over $(-\infty, \infty)$, then

(1.2) $\qquad\qquad J(\epsilon) \to f(x),$

for almost all values of x, when $\epsilon \to 0$. The integral (1.1) is of considerable intrinsic interest, and I propose now to consider its properties in more detail.

THEOREM A. *If $p > 1$ and $|f|^p$ is integrable over $(-\infty, \infty)$, then (1.2) is true for almost all values of x, and in particular at all points of continuity of f.*

The proof is a simple generalisation of that which I gave before when $p = 2$. We require the following lemma.

2. LEMMA. *If $|f|^p$ is integrable, then*

(2.1) $\qquad \displaystyle\int_0^t |f(x+u) - f(x)|^p \, du = o(t),$

when $t \to 0$, for almost all values of x.

We may suppose without loss of generality that $f \geq 0$. The function

$$f(x) = \dfrac{|x-1|^p}{|x^p - 1|}$$

is less than 1 for all positive values of x. Hence

$$|f(x+u) - f(x)|^p \leq |f^p(x+u) - f^p(x)|.$$

---

* $lu = \log|u|$.

But it is known that
$$\int_0^t |f^p(x+u) - f^p(x)|\,du = o(t)$$
for almost all values of x, which proves the lemma. There is in fact, associated with any integrable function $F(x)$, a set of points, whose complementary set is of measure zero, in which $|F(x) - q|$ is the derivative of its integral for every q,* and in particular in which
$$\int_0^t |F(x+u) - F(x)|\,du = o(t).$$
We shall call this set the *Lebesgue set* of F. Here $F = f^p$.

3. We assume now that x has a value which satisfies (2.1). We may then simplify the problem, without real loss of generality, by supposing that $x = 0$, $f(x) = 0$, and by limiting ourselves to the part of the integral for which $t > 0$. We have then to show that

(3.1) $$K(\epsilon) = \int_0^\infty \frac{f(t)}{t}\, l\left(\frac{t+\epsilon}{t-\epsilon}\right) dt \to 0$$

when

(3.2) $$\int_0^t |f(u)|^p\,du = o(t).$$

We write

(3.3) $$K = \int_0^{2\epsilon} + \int_{2\epsilon}^\infty = K_1 + K_2.$$

As regards K_1, we have by the inequality of Hölder,
$$|K_1| \leq \left(\int_0^{2\epsilon} |f|^p\,dt\right)^{1/p} \left[\int_0^{2\epsilon} \left| l\left(\frac{t+\epsilon}{t-\epsilon}\right)\right|^q \frac{dt}{t^q}\right]^{1/q},$$
where $q = p/(p-1)$. The first factor is $o(\epsilon^{1/p})$, by (3.2). The integral in the second factor is less than that from 0 to ∞, which is a constant multiple of ϵ^{1-q}. Hence
$$K_1 = o\{\epsilon^{1/p}\,\epsilon^{(1-q)/q}\} = o(1).$$

As regards K_2, there is nothing to add to the argument of Note LIX.† In fact, no use is made there of the integrability of f^2, except in the assertion that the function there called

---

* H. Lebesgue, *Leçons sur les séries trigonométriques*, 12-13.
† p. 86.

$\chi(t)$ is $O(\sqrt{t})$ for large values of t. This $O(\sqrt{t})$ must now be replaced by $O(t^{1/q})$, which is equally effective for the argument. Thus $K_s = o(1)$, and Theorem A is proved.

4. In so far as the behaviour of $f(t)$ at infinity, the conditions of Theorem A may be greatly relaxed.

THEOREM B. *The conclusion of Theorem A holds whenever* (i) $|f|^p$, *where* $p > 1$, *is integrable over any finite range, and either* (ii)
$$|t|^{-r}|f|^r,$$
where $r > 1$, *is integrable at infinity, or* (ii') *the integrals*
$$\int^{\infty} \frac{f(t)}{t^2} dt, \quad \int_{-\infty} \frac{f(t)}{t^2} dt$$
are convergent (not necessarily absolutely).

It is in fact sufficient to verify that, if either of (ii) or (ii') is satisfied, the integral

(4.1) $$\int_A^{\infty} \frac{f(t)}{t} l\left(\frac{t+\epsilon}{t-\epsilon}\right) dt \quad (A > 0)$$

is uniformly convergent in an interval $0 < \epsilon \leq \epsilon_0$. If (ii) is satisfied, and $s = r/(r-1)$, we have

$$\left|\int_A^{A'} \frac{f(t)}{t} l\left(\frac{t+\epsilon}{t-\epsilon}\right) dt\right| \leq \left\{\int_A^{A'} \left(\frac{f}{t}\right)^r dt\right\}^{1/r} \left\{\int_{-\infty}^{\infty} \left|l\left(\frac{t+\epsilon}{t-\epsilon}\right)\right|^s dt\right\}^{1/s}.$$

The second factor is less than a constant multiple of $\epsilon^{1/s}$, and the conclusion follows. If (ii') is satisfied, we write

$$\int_A^{\infty} \frac{f(t)}{t} l\left(\frac{t+\epsilon}{t-\epsilon}\right) dt = \epsilon \int_A^{\infty} \frac{f(t)}{t^2} \frac{t}{\epsilon} l\left(\frac{t+\epsilon}{t-\epsilon}\right) dt.$$

The function
$$\frac{t}{\epsilon} l\left(\frac{t+\epsilon}{t-\epsilon}\right)$$

is monotomic when t/ϵ is large enough, and tends to the limit 2 when $t \to \infty$. It follows at once that the integral (4.1) is uniformly convergent.

The first set of conditions is satisfied, for example, if $f(t) = \sin at$, and the second if $f(t) = t \sin at$.

5. A more interesting question is whether Theorem A can be extended to the case $p=1$. The answer is negative. We begin however by proving

THEOREM C. *If f is integrable over $(-\infty, \infty)$, then (1.2) holds at every point of continuity of f.*

The argument of Note LIX still applies to the integral $K_2(\epsilon)$, $\chi(t)$ being now bounded for large values of t. Since f is continuous for $t=0$, we may suppose that $|f| < \delta$ in $(0, 2t)$, and then

$$|K_1(\epsilon)| \leq \delta \int_0^{2\epsilon} l\left(\frac{t+\epsilon}{t-\epsilon}\right) \frac{dt}{t} < \delta \int_0^{\infty} l\left(\frac{t+1}{t-1}\right) \frac{dt}{t}.$$

Hence $K_1 \to 0$, which proves the theorem.

THEOREM D. *It is almost always true, when $p = 1$, that*

$$J(\epsilon) \to f(x) \quad (C, 1),$$

i.e. that
$$\frac{1}{\epsilon} \int_0^{\epsilon} J(\eta)\, d\eta \to f(x).$$

We prove that the conclusion holds in the Lebesgue set of f. It is plainly only necessary to show that, in the notation of §3,

$$\frac{1}{\epsilon} \int_0^{\epsilon} K_1(\eta)\, d\eta \to 0.$$

Now
$$\int_0^{\epsilon} K_1(\eta)\, d\eta = \int_0^{\epsilon} d\eta \int_0^2 \frac{f(\eta u)}{u} l\left(\frac{u+1}{u-1}\right) du$$
$$= \int_0^2 l\left(\frac{u+1}{u-1}\right) \frac{du}{u} \int_0^{\epsilon} f(\eta u)\, d\eta = \epsilon \int_0^2 \frac{F(\epsilon u)}{u} l\left(\frac{u+1}{u-1}\right) du,$$

where
$$F(t) = \frac{1}{t} \int_0^t f(u)\, du.$$

Since F is continuous for $t=0$, we have

$$\int_0^{\epsilon} K_1(\eta)\, d\eta = o(\epsilon),$$

which proves the theorem.

6. THEOREM E. *The conclusion of Theorem A, viz. that $J(\epsilon) \to f(x)$ almost always, is false when $p = 1$.*

I prove this by constructing a function $f(t)$ such that (i) $f=0$ outside $(0, 1)$, (ii) f is integrable over $(0, 1)$, and (iii) the limit
$$\lim_{\epsilon \to 0} J(\epsilon)$$
does not exist for any x in $(0, 1)$, and in fact $J(\epsilon)$ is divergent for every such x and an infinity of values of ϵ whose limit is zero.

Suppose that $\alpha_1, \alpha_2, \ldots$ is an enumerable set of numbers all lying in $(0, 1)$ and dense in $(0, 1)$; and that $\delta_1, \delta_2, \ldots$ is a decreasing sequence of positive numbers less than $\frac{1}{2}$, whose limit is zero; and let

(6.1) $\qquad g_\nu(t) = \dfrac{1}{|t-\alpha_\nu|\{l(t-\alpha_\nu)\}^2} - \dfrac{1}{\delta_\nu (l\delta_\nu)^2}$

in that part of $(\alpha_\nu - \delta_\nu, \alpha_\nu + \delta_\nu)$ which lies in $(0, 1)$, and $g_\nu(t) = 0$ elsewhere. Then

(6.2) $\qquad \displaystyle\int_0^1 g_\nu(t)\, dt \leq \int_{-\delta_\nu}^{\delta_\nu} \left\{ \dfrac{1}{u(lu)^2} - \dfrac{1}{\delta_\nu(l\delta_\nu)^2} \right\} du < \dfrac{2}{|l\delta_\nu|}.$

If now we define $f(t)$ by

(6.3) $\qquad\qquad f(t) = \displaystyle\sum_\nu A_\nu g_\nu(t),$

where A_ν is positive, we have

(6.4) $\qquad\qquad \displaystyle\int_0^1 f(t)\, dt < 2 \sum \dfrac{A_\nu}{|l\delta_\nu|},$

so that f is integrable over $(0, 1)$ if the series on the right hand side is convergent, as for example if $A_\nu = 1/\nu$ and $\delta_\nu = e^{-\nu}$.

On the other hand it is plain that
$$J(\epsilon) = \dfrac{1}{\pi^2} \int_0^1 \dfrac{f(t)}{t-x}\, l\left(\dfrac{t-x+\epsilon}{t-x-\epsilon}\right) dt$$
is divergent if $x - \epsilon = \alpha_\nu$ or $\epsilon = x - \alpha_\nu$. Whatever value x may have, there are values of ϵ, as small as we please, for which the integral diverges, so that the limit (1.2) cannot exist.

7. I conclude by a few words concerning the connection of these results with the theory of Fourier series. If f is periodic and integrable, and

(7.1) $\quad f(x) \sim \frac{1}{2}a_0 + \displaystyle\sum_1^\infty (a_n \cos nx + b_n \sin nx) = \sum_0^\infty A_n,$

then

(7.2) $\quad \psi(t) = \tfrac{1}{2}\{f(x+t) - f(x-t)\} \sim \sum_{1}^{\infty} B_n \sin nt,$

where

(7.21) $\quad B_n = b_n \cos nx - a_n \sin nx.$

The series ΣB_n is the 'allied' series of the Fourier series ΣA_n.

The series (7.2) is the Fourier series of $\psi(t)$, and we may multiply it by $1/t$ and integrate term by term over (ϵ, ∞), ϵ being any positive number.* If we do this we obtain

(7.3) $\quad \displaystyle\int_{\epsilon}^{\infty} \frac{f(x+t) - f(x-t)}{t} dt = 2\int_{\epsilon}^{\infty} \frac{\psi(t)}{t} dt = 2\sum_{1}^{\infty} B_n\, Si(n\epsilon),$

where

(7.31) $\quad Si(w) = \displaystyle\int_{w}^{\infty} \frac{\sin u}{u} du.$

The allied series ΣB_n is not necessarily a Fourier series. Let us suppose however that it is one, and that of the function $g(x)$. Then

(7.4) $\quad g_\epsilon(x) = \dfrac{1}{\pi} \displaystyle\int_{-\infty}^{\infty} \dfrac{g(u)}{u-x} l\left(\dfrac{u-x+\epsilon}{u-x-\epsilon}\right) du$

exists for almost all values of x, and is integrable. We proceed to calculate its Fourier constants. We have

(7.5) $\quad \dfrac{1}{\pi} \displaystyle\int_{0}^{2\pi} g_\epsilon(x) \cos nx\, dx$

$= \dfrac{1}{\pi^2} \displaystyle\int_{0}^{2\pi} \cos nx\, dx \int_{-\infty}^{\infty} \dfrac{g(x+w)}{w} l\left(\dfrac{w+\epsilon}{w-\epsilon}\right) dw$

$= \dfrac{1}{\pi^2} \displaystyle\int_{-\infty}^{\infty} l\left(\dfrac{w+\epsilon}{w-\epsilon}\right) \dfrac{dw}{w} \int_{0}^{2\pi} g(x+w) \cos nx\, dx\dagger$

$= \dfrac{1}{\pi^2} \displaystyle\int_{-\infty}^{\infty} (b_n \cos nw - a_n \sin nw) \, l\left(\dfrac{w+\epsilon}{w-\epsilon}\right) \dfrac{dw}{w}.$

Here the coefficient of a_n is zero; and

$\displaystyle\int_{-\infty}^{\infty} \dfrac{\cos nw}{w} l\left(\dfrac{w+\epsilon}{w-\epsilon}\right) dw = 2\pi\, Si(n\epsilon).\ddagger$

---

* By a theorem of W. H. Young. See G. H. Hardy, "Notes on some points in the integral calculus", LV, *Messenger*, li. (1922), 186–192.

† The double integral is absolutely convergent.

‡ The simplest way to evaluate the integral is by an application of Cauchy's Theorem to the function

$$\dfrac{e^{niz}}{z} \log\left(\dfrac{z+\epsilon}{z-\epsilon}\right).$$

Hence

(7.6) $$\frac{1}{\pi} \int_0^{2\pi} g_\epsilon(x) \cos nx\, dx = 2b_n\, Si(n\epsilon);$$

and similarly we find

(7.7) $$\frac{1}{\pi} \int_0^{2\pi} g_\epsilon(x) \sin nx\, dx = -2a_n\, Si(n\epsilon).$$

It follows that the Fourier series of $g_\epsilon(x)$ is $2\Sigma B_n\, Si(n\epsilon)$, that is to say the series (7.3). And hence we have

(7.8) $$\int_\epsilon^\infty \frac{f(x+t)-f(x-t)}{t}\, dt = \frac{1}{\pi} \int_{-\infty}^\infty \frac{g(u)}{u-x}\, l\left(\frac{u-x+\epsilon}{u-x-\epsilon}\right) du$$

for almost all values of x.

8. It follows from (7.8) and Theorem D that

(8.1) $$\int_\epsilon^\infty \frac{f(x+t)-f(x-t)}{t}\, dt \to \pi g(x) \quad (C, 1)$$

for almost all x, and in particular in the Lebesgue set of g. And if also x is in the Lebesgue set of f, and therefore almost always, then the left hand side of (8.1), if it tends to a limit $(C, 1)$, must tend to a limit in the ordinary sense.* It follows that

(8.2) $$\int_0^\infty \frac{f(x+t)-f(x-t)}{t}\, dt = \pi g(x)$$

for almost all values of x. Finally, since the allied series of the allied series is the original series reversed in sign, the reciprocal formula

(8.3) $$\int_0^\infty \frac{g(x+t)-g(x-t)}{t}\, dt = -\pi f(x)$$

is also valid for almost all values of x.

---

* See G. H. Hardy and J. E. Littlewood, "The allied series of a Fourier series", *Proc. London Math. Soc.* (2), 24 (1925), 211–246 (245–246, lemma o).

CORRECTION

p. 15, *line* 2. } after $f(x-t)$ missing.

NOTE ON SOME POINTS IN THE INTEGRAL CALCULUS.

By *G. H. Hardy.*

LXIII.

Some further applications of Mellin's inversion formula.

1. In this note I shall apply Mellin's inversion formula, in the form in which I proved it in Note 49*, to the evaluation of a number of curious integrals containing Bessel functions or allied transcendents.

---

* *Messenger*, vol. xlvii (1918), pp. 178–184.

I follow the notation of Watson's treatise, except that I do not adopt his distinction between the general parameter ν and the necessarily integral n. In particular I denote by $K_n(x)$ Macdonald's function defined, for non-integral n, by

(1.1) $\qquad K_n(x) = \tfrac{1}{2}\pi \operatorname{cosec} n\pi \{I_{-n}(x) - I_n(x)\}.$

2. I consider first the integral

(2.1) $\qquad f(x) = \int_0^\infty \dfrac{e^{-t}}{t} K_n\left(\dfrac{x}{t}\right) dt,$

n and x being positive. Supposing in the first instance that s also is positive, and $s > n$, we have

(2.2) $\displaystyle\int_0^\infty x^{s-1} f(x)\, dx = \int_0^\infty \dfrac{e^{-t}}{t} dt \int_0^\infty x^{s-1} K_n\left(\dfrac{x}{t}\right) dx$

$\qquad\qquad = 2^{s-2} \Gamma\left(\dfrac{s+n}{2}\right) \Gamma\left(\dfrac{s-n}{2}\right) \int_0^\infty t^{s-1} e^{-t}\, dt^*$

$\qquad\qquad = 2^{s-2} \Gamma(s) \Gamma\left(\dfrac{s+n}{2}\right) \Gamma\left(\dfrac{s-n}{2}\right).$

There is no difficulty in the inversion, since everything is positive and the repeated integral convergent.

The formula (2.2) remains true for all values of s whose real part exceeds n, and Mellin's formula† is applicable. We have therefore

(2.3) $\quad f(x) = \dfrac{1}{8\pi i} \displaystyle\int_{c-i\infty}^{c+i\infty} (\tfrac{1}{2}x)^{-s} \Gamma(s) \Gamma\left(\dfrac{s+n}{2}\right) \Gamma\left(\dfrac{s-n}{2}\right) ds,$

if $c > n$.

Now it is easily verified that, if $\epsilon = e^{\frac{1}{4}\pi i}$ and $\eta = e^{-\frac{1}{4}\pi i}$ are the principal square roots of i and $-i$, y is positive, and the values of the K's are those derived by rectilinear continuation from the positive real axis, then

(2.4) $\quad 8 K_n(\epsilon y) K_n(\eta y)$

$\qquad = \dfrac{1}{2\pi i} \displaystyle\int_{c-i\infty}^{c+i\infty} (\tfrac{1}{2}y)^{-2s} \Gamma(s) \Gamma\left(\dfrac{s+n}{2}\right) \Gamma\left(\dfrac{s-n}{2}\right) ds.$

---

* Watson, p 388, formula (8).
† Mellin's formula (as I proved it in Note 49) shows only that the s integral is summable $(C, 1)$. Its convergence in the ordinary sense is an immediate consequence of Stirling's theorem.

We have in fact
$$K_n(x) K_n(-ix) = \left(\frac{\pi}{2\sin n\pi}\right)^2 (e^{-\frac{1}{2}n\pi i} I_n J_n - e^{\frac{1}{2}n\pi i} I_n J_{-n}$$
$$- e^{-\frac{1}{2}n\pi i} I_{-n} J_n + e^{\frac{1}{2}n\pi i} I_{-n} J_{-n}),$$

the arguments of the I's and J's being x. The expansions of the products on the right-hand side are given by Watson*, and are immediately expressible as integrals of Mellin's type: thus

$$I_n(x) J_n(x) = \sum_{p=0}^{\infty} \frac{(-1)^p (\tfrac{1}{2}x)^{2n+4p}}{p!\,\Gamma(n+p+1)\,\Gamma(n+2p+1)}$$
$$= \frac{1}{2\pi i} \int_{c-i\infty}^{c+i\infty} (\tfrac{1}{2}x)^{2n-4s} \frac{\Gamma(s)}{\Gamma(n-s+1)\,\Gamma(n-2s+1)}\, ds.$$

Combining four formulæ of this type, and writing ϵx for y, (2.4) follows by a simple calculation. Comparing (2.3) and (2.4), we obtain

(2.5) $\quad\displaystyle\int_0^\infty \frac{e^{-t}}{t} K_n\!\left(\frac{x}{t}\right) dt = 2 K_n\{\sqrt{(2xi)}\}\, K_n\{\sqrt{(-2xi)}\}.$

This formula has been proved on the assumption that n and x are positive, but its range of validity may be extended at once by analytical continuation. Thus it holds for $\mathfrak{R}(x) \geq 0$ and for all values of n.

From (2.5) we may derive a number of other formulæ. We may suppose x complex, and equate real and imaginary parts, or we may do this after a preliminary transformation $t = xu$. We thus obtain, for example, the formulæ

(2.6) $\quad\displaystyle\int_0^\infty \frac{e^{-t}}{t} J_n\!\left(\frac{2x^2}{t}\right) dt = 2 K_n(2x) J_n(2x),$

(2.7) $\quad\displaystyle\int_0^\infty \frac{e^{-t}}{t} Y_n\!\left(\frac{2x^2}{t}\right) dt = 2 K_n(2x) Y_n(2x),$

(2.8) $\quad\displaystyle\int_0^\infty \frac{\cos t}{t} K_n\!\left(\frac{2x^2}{t}\right) dt$
$\qquad = -\pi K_n(2x)\{\sin\tfrac{1}{2}n\pi\, J_n(2x) + \cos\tfrac{1}{2}n\pi\, Y_n(2x)\},$

(2.9) $\quad\displaystyle\int_0^\infty \frac{\sin t}{t} K_n\!\left(\frac{2x^2}{t}\right) dt$
$\qquad = \pi K_n(2x)\{\cos\tfrac{1}{2}n\pi\, J_n(2x) - \sin\tfrac{1}{2}n\pi\, Y_n(2x)\}.$

---

* p. 148.

3. A number of other integrals may be derived from those of § 2 by means of the formula

$$(3.1) \quad x^n \int_0^\infty t^{2\nu-\lambda-1} K_n(t) K_{n-\lambda}\left(\frac{x}{t}\right) dt$$

$$= x^\nu \int_0^\infty t^{2n-\lambda-1} K_\nu(t) K_{\nu-\lambda}\left(\frac{x}{t}\right) dt.$$

This formula may be proved by substituting integral representations for the K's and changing the order of integration; but it is more natural to prove it here by showing that the Mellin transform of the left hand side is a symmetric function of n and ν. Denoting the left hand side of (3.1) by $f(x)$, we have

$$\int_0^\infty x^{s-1} f(x) \, dx = \int_0^\infty t^{2\nu-\lambda-1} K_n(t) \, dt \int_0^\infty x^{s+n-1} K_{n-\lambda}\left(\frac{x}{t}\right) dx$$

$$= 2^{s+n-2} \Gamma\left(\frac{s+\lambda}{2}\right) \Gamma\left(\frac{s+2n-\lambda}{2}\right) \int_0^\infty t^{s+n+2\nu-\lambda-1} K_n(t) \, dt$$

$$= 2^{2s+2n+2\nu-\lambda-4} \Gamma\left(\frac{s+\lambda}{2}\right) \Gamma\left(\frac{s+2n-\lambda}{2}\right) \Gamma\left(\frac{s+2\nu-\lambda}{2}\right) \Gamma\left(\frac{s+2n+2\nu-\lambda}{2}\right);$$

which is, as required, symmetric.

If $\Re(x) > 0$, (3.1) is true for all values of n, ν and λ: if $\Re(x) = 0$, $2n$ and 2ν must each be greater than $\lambda - \frac{3}{2}$.

If in (3.1) we put $n = \frac{1}{2} - \mu$, $\lambda = \frac{1}{2} - 2\mu$, $\nu = \frac{1}{2}$, and then write n again for μ, we find

$$(3.2) \quad \int_0^\infty t^{2n-\frac{1}{2}} K_{\frac{1}{2}-n}(t) K_n\left(\frac{x}{t}\right) dt = x^n \sqrt{(\tfrac{1}{2}\pi)} \int_0^\infty \frac{e^{-t}}{t} K_{2n}\left(\frac{x}{t}\right) dt$$

$$= x^n \sqrt{(2\pi)} K_{2n}\{\sqrt{(2xi)}\} K_{2n}\{\sqrt{(-2xi)}\}.$$

From this we deduce

$$(3.3) \quad \int_0^\infty t^{2n-\frac{1}{2}} K_{\frac{1}{2}-n}(t) J_n\left(\frac{2x^2}{t}\right) dt = 2^n x^{2n} \sqrt{(2\pi)} K_{2n}(2x) J_{2n}(2x),$$

$$(3.4) \quad \int_0^\infty t^{2n-\frac{1}{2}} K_{\frac{1}{2}-n}(t) Y_n\left(\frac{2x^2}{t}\right) dt = 2^n x^{2n} \sqrt{(2\pi)} K_{2n}(2x) Y_{2n}(2x),$$

$$(3.5) \quad \int_0^\infty t^{2n-\frac{1}{2}} J_{\frac{1}{2}-n}(t) J_n\left(\frac{2x^2}{t}\right) dt$$

$$= \frac{2^n x^{2n}}{2i \sin 2n\pi} \{e^{2n\pi i} J_{-2n}(\epsilon x) J_{2n}(\eta x) - e^{-2n\pi i} J_{2n}(\epsilon x) J_{-2n}(\eta x)\}.$$

4. Another elegant system of formulæ may be derived from (3.1) by taking $\lambda = 0$, $\nu = \frac{1}{2}$. We then obtain

$$x^n \int_0^\infty K_n(t) K_n\left(\frac{x}{t}\right) dt = \tfrac{1}{2}\pi \int_0^\infty t^{n-1} \exp\left(-t - \frac{x}{t}\right) dt = \pi x^n K_{2n}(2\sqrt{x}),$$

(4.1) $\qquad \int_0^\infty K_n(t) K_n\left(\frac{x^2}{t}\right) dt = \pi K_{2n}(2x).$

This formula is valid for all values of n if $|\arg x| < \tfrac{1}{4}\pi$. From it we derive

(4.2) $\qquad \int_0^\infty K_n(t) J_n\left(\frac{x^2}{t}\right) dt = i\{e^{\frac{1}{2}n\pi i} K_{2n}(2\epsilon x) - e^{-\frac{1}{2}n\pi i} K_{2n}(2\eta x)\},$

(4.3) $\qquad \int_0^\infty K_n(t) Y_n\left(\frac{x^2}{t}\right) dt = -e^{\frac{1}{2}n\pi i} K_{2n}(2\epsilon x) - e^{-\frac{1}{2}n\pi i} K_{2n}(2\eta x),$

x being real and $-\tfrac{5}{2} < n < \tfrac{5}{2}$;

(4.4) $\qquad \int_0^\infty J_n(t) J_n\left(\frac{x^2}{t}\right) dt = J_{2n}(2x),$

(4.5) $\qquad \int_0^\infty J_n(t) Y_n\left(\frac{x^2}{t}\right) dt = Y_{2n}(2x) + \frac{2}{\pi} K_{2n}(x),$

(4.6) $\qquad \int_0^\infty Y_n(t) Y_n\left(\frac{x^2}{t}\right) dt = -J_{2n}(2x).$

The first of the last three formulæ is valid if $n > -\tfrac{1}{2}$, the second and third if $-\tfrac{1}{2} < n < \tfrac{1}{2}$. The formula (4.4) has been given before by Cailler and Bateman.* Another formula of the same type is

(4.7) $\qquad \int_0^\infty \mathfrak{H}_n(t) \mathfrak{H}_n\left(\frac{x^2}{t}\right) dt = -J_{2n}(2x),$

where $\mathfrak{H}_n$ is Struve's function.†

From these formulæ many others may be deduced by differentiation, or by the use of theorems of inversion of Hankel's type. Thus we have

$$\int_0^\infty J_n(t) J_{2n}\{2\sqrt{(xt)}\} dt = J_n(x),$$

$$\int_0^\infty Y_n(t) J_{2n}\{2\sqrt{(xt)}\} dt = -J_n(x),$$

$$\int_0^\infty \mathfrak{H}_n(t) J_{2n}\{2\sqrt{(xt)}\} dt = -J_n(x).$$

---

* See the references given by Watson, p. 437.
† Watson, p. 328.

The first of these formulæ is also due to Bateman. The conditions for validity, when x is positive and n real, are $n > -\tfrac{1}{2}$, $n > -\tfrac{1}{2}$, and $n > -1$.

5. There is some resemblance between the formulæ of §§ 2–3 and Macdonald's formula*

$$(5.1) \quad \int_0^\infty \exp\left(-\frac{t}{2} - \frac{x^2+y^2}{2t}\right) K_n\left(\frac{xy}{t}\right) \frac{dt}{t} = 2 K_n(x) K_n(y),$$

though this has a second exponential factor. This formula is an immediate consequence of

$$(5.2) \quad \int_0^\infty \exp\left(-\frac{t}{2} - \frac{x^2+y^2}{2t}\right) I_n\left(\frac{xy}{t}\right) \frac{dt}{t}$$
$$= \begin{array}{ll} 2 I_n(x) K_n(y) & (0 < x \leq y) \\ 2 K_n(x) I_n(y) & (0 < y \leq x) \end{array}.$$

This formula is easily proved as follows. We have†

$$\int_0^\infty e^{-\tfrac{1}{2}tu^2} J_n(xu) J_n(yu) u\, du = \frac{1}{t} \exp\left(-\frac{x^2+y^2}{2t}\right) I_n\left(\frac{xy}{t}\right);$$

and so

$$\int_0^\infty \exp\left(-\frac{t}{2} - \frac{x^2+y^2}{2t}\right) I_n\left(\frac{xy}{t}\right) \frac{dt}{t}$$
$$= \int_0^\infty J_n(xu) J_n(yu) u\, du \int_0^\infty e^{-\tfrac{1}{2}t(1+u^2)}\, dt$$
$$= 2 \int_0^\infty J_n(xu) J_n(yu) \frac{u\, du}{1+u^2} = 2 I_n(x) K_n(y),\ddagger$$

if $0 < x \leq y$.

I should add that the integral

$$\int_0^\infty J_m(x) J_n\left(\frac{x}{t}\right) t^{s-1} dt$$

has been calculated by C. V. Hanumanta Rao,§ for all values of the parameters for which it is convergent, as the sum of two generalised hypergeometric series. It is not altogether easy to derive the most interesting particular cases from the general formula.

---

* See Watson, p. 439. † Watson, p. 395.
‡ Watson, p. 429. § See Watson, p. 437.

6. A good many of the preceding formulæ may also be derived from Theorem A of Note 52,[*] which asserts that if

$$f(s) = \int_0^\infty \phi(t) t^{s-1} dt, \quad g(s) = \int_0^\infty \psi(t) t^{s-1} dt,$$

then, under certain conditions,

$$\int_0^\infty \phi(t) \psi\left(\frac{x}{t}\right) \frac{dt}{t} = \frac{1}{2\pi i} \int_{c-i\infty}^{c+i\infty} f(s) g(s) x^{-s} ds.$$

If for example we take

$$\phi(t) = t^{n+\frac{1}{2}} K_{\frac{1}{2}-n}(t), \quad \psi(t) = t^{-n} K_n(t),$$

we have

$$f(s) = 2^{s+n-\frac{3}{2}} \Gamma\left(\frac{s+1}{2}\right) \Gamma\left(\frac{s}{2}+n\right), \quad g(s) = 2^{s-n-2} \Gamma\left(\frac{s}{2}\right) \Gamma\left(\frac{s}{2}-n\right);$$

and we are led to

$$\int_0^\infty t^{2n-\frac{1}{2}} K_{\frac{1}{2}-n}(t) K_n\left(\frac{x}{t}\right) dt$$
$$= \frac{x^n \sqrt{(2\pi)}}{8} \frac{1}{2\pi i} \int_{c-i\infty}^{c+i\infty} \left(\frac{x}{2}\right)^{-s} \Gamma(s) \Gamma\left(\frac{s}{2}+n\right) \Gamma\left(\frac{s}{2}-n\right) ds;$$

and (3.2) follows from this and (2.4).

---

[*] *Messenger*, vol. xlix. (1920), pp. 85–91.

NOTES ON SOME POINTS IN THE INTEGRAL CALCULUS.

By G. H. Hardy.

LXVII.

On the repeated integral which occurs in the theory of conjugate functions.

1. It is known that in certain circumstances

(1.1) $$-\frac{1}{\pi^2}\int_{-\infty}^{\infty}\frac{dt}{t-x}\int_{-\infty}^{\infty}\frac{f(u)}{u-t}du = f(x),$$

the two integrals being each principal values in Cauchy's sense. Suppose, for example, that $f(x)$ is real and belongs to L^2 in $(-\infty, \infty)$.* Then there is a function of $g(x)$, the conjugate or 'Hilbert transform' of $f(x)$, which (i) belongs also to L^2 and (ii) satisfies the equations

(1.2) $$-\frac{1}{\pi}\int_{-\infty}^{\infty}\frac{f(t)}{t-x}dt = g(x), \quad \frac{1}{\pi}\int_{-\infty}^{\infty}\frac{g(t)}{t-x}dt = f(x),$$

each for almost all values of x.† If we substitute from the first of (1.2) into the second, we obtain (1.1), which therefore holds for almost all x.

There is however a gap in the theory. No one, so far as I know, has stated any simple and general criterion for the truth of (1.1) for a particular x. Such a criterion should involve $f(x)$ only, whereas those which result from the general theory involve both $f(x)$ and $g(x)$. My object here is to fill this gap.‡

I begin by proving

THEOREM 1. *It is sufficient for the truth of* (1.1) *that* $f(t)$ *should belong to* L^2 *and be continuous for* $t = x$.

In Theorem 2 I extend this result to functions of L^p, for any $p > 1$. I keep the special case and the generalisation apart, because the former can be deduced from the theorems proved in my earlier notes, while the latter depends upon the later and deeper work of Riesz and Titchmarsh.§

---

* f is measurable and f^2 integrable over $(-\infty, \infty)$.

† One possible method of developing the theory is set out in Notes 58 and 59 of this series, to which I shall refer freely in what follows.

‡ See also G. H. Hardy and J. E. Littlewood, 'A point in the theory of conjugate functions', *Journal London Math. Soc.*, 4 (1929). We discuss there the corresponding problem for the finite range $(-\pi, \pi)$, from the standpoint of the theory of trigonometrical series.

§ M. Riesz, 'Sur les fonctions conjuguées', *Math. Zeitschrift*, 27 (1927), 218–244 (see also a preliminary note in the *Comptes Rendus* for 28 April 1924); E. C. Titchmarsh, 'Reciprocal formulae for series and integrals', *Math. Zeitschrift*, 25 (1926), 321–347.

2. The proof of Theorem 1 depends upon the chain of formulae

$$(2.1) \quad \int_{-\infty}^{\infty} \frac{dt}{t-x} \int_{-\infty}^{\infty} \frac{f(u)}{u-t} du$$

$$= \lim_{\varepsilon \to 0} \left(\int_{x+\varepsilon}^{\infty} + \int_{-\infty}^{x-\varepsilon} \right) \frac{dt}{t-x} \int_{-\infty}^{\infty} \frac{f(u)}{u-t} du$$

$$= \lim_{\varepsilon \to 0} \int_{-\infty}^{\infty} f(u) du \left(\int_{x+\varepsilon}^{\infty} + \int_{-\infty}^{x-\varepsilon} \right) \frac{dt}{(t-x)(u-t)}$$

$$= \lim_{\varepsilon \to 0} \int_{-\infty}^{\infty} \frac{f(u)}{u-x} \log \left| \frac{u-x-\varepsilon}{u-x+\varepsilon} \right| du = -\pi^2 f(x).$$

Let us call the five expressions here J_1, J_2, J_3, J_4, J_5. Then $J_1 = J_2$ by the definition of J_1; $J_3 = J_4$ by an elementary calculation; and $J_4 = J_5$ by Lemma 8 of Note 59.* What we have to do is therefore to prove that $J_2 = J_3$, that is to say to justify the inversion of the order of integration. We may plainly take $x = 0$ without real loss of generality, and we have then to prove

$$(2.2) \quad \left(\int_{\varepsilon}^{\infty} + \int_{-\infty}^{-\varepsilon} \right) \frac{dt}{t} \int_{-\infty}^{\infty} \frac{f(u)}{u-t} du$$

$$= \int_{-\infty}^{\infty} f(u) du \left(\int_{\varepsilon}^{\infty} + \int_{-\infty}^{-\varepsilon} \right) \frac{dt}{t(u-t)}.$$

3. (i). We prove first that the order of integration in

$$(3.1) \quad \int_{\varepsilon}^{\infty} \frac{dt}{t} \left(\int_{t+\beta}^{\infty} + \int_{-\infty}^{t-\beta} \right) \frac{f(u)}{u-t} du$$

(and in a corresponding integral over $-\infty$, $-\varepsilon$) may be inverted for any positive β. This part of the proof is easy; we shall in fact prove that the corresponding double integrals are absolutely convergent whenever $f(u)$ is integrable over any finite interval and $f(u) \log |u|/u$ is integrable at infinity. These conditions are satisfied if $f(u)$ belongs to any L^p for which $p > 1$.

We consider separately the parts of (3.1) in which $t < u$ and $t > u$. When $t < u$, $u > \varepsilon + \beta$, and t goes from ε to $u-\beta$.† Replacing the integrand by its modulus, we obtain

$$\int_{\varepsilon+\beta}^{\infty} |f(u)| du \int_{\varepsilon}^{u-\beta} \frac{dt}{t(u-t)} = \int_{\varepsilon+\beta}^{\infty} \frac{|f(u)|}{u} \log \frac{(u-\beta)(u-\varepsilon)}{\varepsilon\beta} du,$$

which is convergent.

---

* The result of the lemma is stated in terms of 'almost all' x; the condition imposed upon $f(x)$ is obviously satisfied at any point of continuity.
† The reader should draw a figure.

When $t > u$, u may have any value and t goes from $u+\beta$ to ∞ or from ϵ to ∞, according as $u > \epsilon - \beta$ or $u < \epsilon - \beta$. In the first case we obtain

$$\int_{\epsilon-\beta}^{\infty} |f(u)|\, du \int_{u+\beta}^{\infty} \frac{dt}{t(t-u)} = \int_{\epsilon-\beta}^{\infty} \frac{|f(u)|}{u} \log\left(\frac{u+\beta}{\beta}\right) du,$$

and in the second

$$\int_{-\infty}^{\epsilon-\beta} |f(u)|\, du \int_{\epsilon}^{\infty} \frac{dt}{t(t-u)} = \int_{-\infty}^{\epsilon-\beta} \frac{|f(u)|}{u} \log\left(\frac{\epsilon}{\epsilon-u}\right) du;$$

and again each of these integrals is convergent.

It follows that we may invert the integrations in (3.1), and a similar proof applies to the corresponding integral over $(-\infty, -\epsilon)$.

4. (ii). We write

$$(4.1) \quad \psi(t, \beta) = \left(\int_{t+\beta}^{\infty} + \int_{-\infty}^{t-\beta}\right) \frac{f(u)}{u-t}\, du$$

$$= \int_{\beta}^{\infty} \frac{f(t+w) - f(t-w)}{w}\, dw,$$

so that

$$(4.2) \qquad \psi(t, \beta) \to \psi(t) = -\pi g(t),$$

when $\beta \to 0$, for almost all t. Then

$$(4.3) \quad \int_{\epsilon}^{\infty} \psi(t, \beta) \frac{dt}{t} = \int_{\epsilon}^{\infty} \frac{dt}{t} \left(\int_{t+\beta}^{\infty} + \int_{-\infty}^{t-\beta}\right) \frac{f(u)}{u-t}\, du$$

$$= \int_{\epsilon+\beta}^{\infty} f(u)\, du \left(\int_{\epsilon}^{u-\beta} + \int_{u+\beta}^{\infty}\right) \frac{dt}{t(u-t)} + \int_{\epsilon-\beta}^{\epsilon+\beta} f(u)\, du \int_{u+\beta}^{\infty} \frac{dt}{t(u-t)}$$

$$+ \int_{-\infty}^{\epsilon-\beta} f(u)\, du \int_{\epsilon}^{\infty} \frac{dt}{t(u-t)}$$

$$= \int_{\epsilon+\beta}^{\infty} \frac{f(u)}{u} \log\left(\frac{(u-\beta)(u-\epsilon)}{\epsilon(u+\beta)}\right) du + \int_{\epsilon-\beta}^{\epsilon+\beta} \frac{f(u)}{u} \log\left(\frac{\beta}{u+\beta}\right) du$$

$$+ \int_{-\infty}^{\epsilon-\beta} \frac{f(u)}{u} \log\left(\frac{\epsilon-u}{\epsilon}\right) du$$

$$= \left(\int_{\epsilon+\beta}^{\infty} + \int_{-\infty}^{\epsilon-\beta}\right) \frac{f(u)}{u} \log\left|\frac{u-\epsilon}{\epsilon}\right| du + \int_{\epsilon+\beta}^{\infty} \frac{f(u)}{u} \log\left(\frac{u-\beta}{u+\beta}\right) du$$

$$+ \int_{\epsilon-\beta}^{\epsilon+\beta} \frac{f(u)}{u} \log\left(\frac{\beta}{u+\beta}\right) du.$$

56 Prof. Hardy, *On some points in the integral calculus.*

When $\beta \to 0$, the second integral here is

$$O\left(\beta \int_\varepsilon^\infty \frac{|f(u)|}{u^2} du\right) = o(1)$$

and the third is

$$O\left(|\log \beta| \int_{\varepsilon-\beta}^{\varepsilon+\beta} |f(u)|\,du\right) = O\left(|\log \beta| \beta^{(p-1)/p}\right) = o(1),$$

if $f(u)$ belongs to L^p, and in particular if it belongs to L^2. It therefore follows from (4.3) that

$$(4.4) \quad \lim_{\beta \to 0} \int_\varepsilon^\infty \psi(t, \beta) \frac{dt}{t} = \int_{-\infty}^\infty \frac{f(u)}{u} \log\left|\frac{u-\varepsilon}{\varepsilon}\right| du.$$

A similar argument shows that

$$(4.5) \quad \lim_{\beta \to 0} \int_{-\infty}^{-\varepsilon} \psi(t, \beta) \frac{dt}{t} = \int_{-\infty}^\infty \frac{f(u)}{u} \log\left|\frac{\varepsilon}{u+\varepsilon}\right| du.$$

Hence

$$(4.6) \quad \lim_{\beta \to 0} \left(\int_\varepsilon^\infty + \int_{-\infty}^{-\varepsilon}\right) \psi(t, \beta) \frac{dt}{t} = \int_{-\infty}^\infty \frac{f(u)}{u} \log\left|\frac{u-\varepsilon}{u+\varepsilon}\right| du.$$

5. (iii). Suppose now that we can prove that

$$(5.1) \quad \lim_{\beta \to 0} \left(\int_\varepsilon^\infty + \int_{-\infty}^{-\varepsilon}\right) \psi(t, \beta) \frac{dt}{t} = \left(\int_\varepsilon^\infty + \int_{-\infty}^{-\varepsilon}\right) \psi(t) \frac{dt}{t}.$$

Comparing (4.6) and (5.1), and replacing $\psi(t)$ by the integral which defines it, we shall obtain (2.2). The proof of Theorem 1 is thus reduced to that of (5.1).

The proof of (5.1), which is the kernel of the proof of the theorem, depends upon the formula

$$(5.2) \quad \psi(t, \beta) = \frac{1}{\pi} \int_{-\infty}^\infty \frac{g(t+u)}{u} \log\left|\frac{u-\beta}{u+\beta}\right| du.$$

This is formula (7.5) of Note 59, with ε, x, t, u replaced by β, t, w, $t+u$. There is however a simpler proof of the formula than that given in Note 59. It may be verified by elementary calculation that the functions

$$l(u) = \frac{1}{u-t} \quad (|u-t| > \varepsilon), \quad l(u) = 0 \quad (|u-t| < \varepsilon)$$

and

$$m(u) = \frac{1}{u-t} \log\left|\frac{u-t-\varepsilon}{u-t+\varepsilon}\right|$$

are conjugate, so that

$$\int_{-\infty}^{\infty} f(u)\, l(u)\, du = \int_{-\infty}^{\infty} g(u)\, m(u)\, du,$$

by Theorem B of Note 59.[*] This is equivalent to (5.2).

From (5.2) we deduce

$$\int_{-\infty}^{\infty} \psi^2(t, \beta)\, dt = \frac{1}{\pi^2} \int_{-\infty}^{\infty} dt \left(\int_{-\infty}^{\infty} \frac{g(t+u)}{u} \log \left| \frac{u+\beta}{u-\beta} \right| du \right)^2$$

$$\leq \frac{1}{\pi^2} \int_{-\infty}^{\infty} dt \left(\int_{-\infty}^{\infty} \frac{g^2(t+u)}{u} \log \left| \frac{u+\beta}{u-\beta} \right| du \int_{-\infty}^{\infty} \frac{1}{u} \log \left| \frac{u+\beta}{u-\beta} \right| du \right)$$

$$= \int_{-\infty}^{\infty} \frac{1}{u} \log \left| \frac{u+\beta}{u-\beta} \right| du \int_{-\infty}^{\infty} g^2(t+u)\, dt = \pi^2 \int_{-\infty}^{\infty} g^2(t)\, dt.$$

Also $1/t$ belongs to L^2 in $(-\infty, -\epsilon)$ and (ϵ, ∞). It follows, by Lemma 1 of Note 58, that the passage to the limit in (5.1) is legitimate. This completes the proof of Theorem 1.

6. The argument of §§ 3, 4 applies as it stands to functions of any L^p ($p > 1$). The argument of § 5 will apply to such functions when we have generalised Lemma 1 of Note 58 to the case in which $f(x)$[†] belongs to $L^{p/(p-1)}$ and

$$\int_{-\infty}^{\infty} |\phi(x, h, h', \ldots)|^p\, dx < A,$$

and we have proved (a) that (5.2) is still true, (b) that $\psi(t, \beta)$ tends to $\psi(t)$ for almost all t, and (c) that

$$\int_{-\infty}^{\infty} |\psi(t, \beta)|^p\, dt$$

is bounded. The necessary generalisation of the lemma will be found in a paper by Young later than that referred to in Note 58[‡]; and (c) demands merely the use of the general Hölder inequality in the last paragraph of § 5. The truth of (a) and (b) results from the general theory of Riesz and Titchmarsh[§]; it is only here that an appeal to this theory is required.

---

[*] 'Parseval's Theorem' for conjugate functions.
[†] The $f(x)$ of the lemma, not of the argument here.
[‡] W. H. Young, 'Successions of integrals and Fourier series', *Proc. London Math. Soc.* (2), 11 (1913), 43–75.
[§] (b) is the fundamental theorem of the existence of the conjugate, while (a) follows from 'Parseval's Theorem', and is used by Riesz, equation (54), p. 240.

We have thus

THEOREM 2. *It is sufficient for the truth of* (1.1) *that* $f(t)$ *should belong to* L^p, *where* $p > 1$, *and be continuous for* $t = x$.

7. We may ask in conclusion whether it is not possible to replace the condition of continuity by a broader condition. This question is subsidiary here, since we know already that the formula holds for almost all x, and it is as a means to proving this that such generalisations are most interesting. The answer may however be derived from the results of Note 62. It was only at one stage that the continuity of $f(t)$ was relevant, namely in the evaluation of the final limit in (2.1). The results of Note 62 show that we can replace continuity by the condition

$$\int_0^t |f(x+u) + f(x-u) - 2f(x)|^p \, du = o(t),$$

but that it is *not* sufficient to suppose that x belongs to the 'Lebesgue set' of $f(t)$. We could employ the more general condition

$$\int_0^t \{f(x+u) + f(x-u) - 2f(x)\} \, du = o(t),$$

if we supposed that the limit when $\epsilon \to 0$ was taken in a Cesarò sense.*

CORRECTION

p. 58, *last line.* Cesarò should be Cesàro. Delete asterisk.

SELF-RECIPROCAL FUNCTIONS

By G. H. HARDY and E. C. TITCHMARSH

[Received 16 July 1930]

1. Introduction

1.1. THE Fourier reciprocity between two real even functions $f(x)$ and $g(x)$ is expressed by the formulae

$$g(x) = \sqrt{\left(\frac{2}{\pi}\right)} \int_0^\infty \cos xy\, f(y)\, dy, \qquad f(x) = \sqrt{\left(\frac{2}{\pi}\right)} \int_0^\infty \cos xy\, g(y)\, dy, \quad (1.11)$$

and two functions so connected are said to be *cosine transforms* of one another; they were described by Cauchy as *reciprocal* functions.†
The integrals may be integrals of the classical kind (as in Cauchy) or integrals in some generalized sense (as in the modern theory of transforms‡).

In special cases $f(x) = g(x)$, or

$$f(x) = \sqrt{\left(\frac{2}{\pi}\right)} \int_0^\infty \cos xy\, f(y)\, dy, \qquad (1.12)$$

a singular homogeneous integral equation of the second kind. We then describe $f(x)$ as a *self-reciprocal* function. There are familiar examples such as

$$x^{-\frac{1}{2}}, \qquad e^{-\frac{1}{2}x^2}, \qquad \operatorname{sech} x\sqrt{(\tfrac{1}{2}\pi)};$$

a more novel one§ is

$$f(x) = x^{\frac{1}{2}} J_{-\frac{1}{4}}(\tfrac{1}{2}x^2).$$

1.2. Our problem is *to determine all self-reciprocal functions*, or rather (since complete generality is hardly attainable) all such functions of certain classes, such as the class L^2 of functions of integrable square.

It may be well to observe at once that there is a sense in which

---

† 'Of the first kind', reciprocal functions of the second kind being sine transforms of each other. See Cauchy (**8**), Burkhardt (**5**).

‡ For the standard results of this theory, which we shall take for granted in what follows, see Plancherel (**21**, **22**), Titchmarsh (**25**, **26**), Hobson (**16**), §§ 481–8.

§ See § 3 below.

there is an immediate solution. Suppose, to be definite, that $\phi(x)$ belongs to L^2, i.e. that $\phi(x)$ is measurable and

$$\int_0^\infty \phi^2(x)\, dx$$

finite. Then $\phi+\phi^*$, where the star denotes the transform, is also a function of L^2, and is self-reciprocal, since

$$(\phi+\phi^*)^* = \phi^*+\phi.$$

Also any self-reciprocal f may be expressed as $\tfrac{1}{2}f+\tfrac{1}{2}f = \tfrac{1}{2}f+(\tfrac{1}{2}f)^*$, so that the solution is complete.

On the other hand it is obvious that none of the examples we have quoted have been obtained in this way, and the solution does not enable us to decide (unless by actual verification) whether a given $f(x)$ is self-reciprocal. To determine whether $f(x)$ is of the form $\phi+\phi^*$ is in fact to solve another integral equation, viz.

$$f(x) = \phi(x) + \sqrt{\left(\frac{2}{\pi}\right)} \int_0^\infty \cos xy\, \phi(y)\, dy$$

(a non-homogeneous singular equation).

1.3. We therefore look for a solution of a different type; but it is convenient first to enlarge the formal basis of our analysis. The same problem arises for the equation

$$f(x) = \sqrt{\left(\frac{2}{\pi}\right)} \int_0^\infty \sin xy\, f(y)\, dy, \tag{1.31}$$

and, more generally, for

$$f(x) = \int_0^\infty \sqrt{(xy)} J_\nu(xy) f(y)\, dy, \tag{1.32}$$

where $J_\nu(x)$ is a Bessel function of order $\nu \geqslant -\tfrac{1}{2}$.† The cosine and sine cases are the cases in which $\nu = -\tfrac{1}{2}$, and $\nu = \tfrac{1}{2}$.

The functions

$$x^{-\frac{1}{2}}, \qquad xe^{-\frac{1}{2}x^2}, \qquad \frac{1}{e^{x\sqrt{(2\pi)}}-1} - \frac{1}{x\sqrt{(2\pi)}}$$

are self-reciprocal in the sine transform, and the functions

$$x^{-\frac{1}{2}}, \qquad x^{\nu+\frac{1}{2}}e^{-\frac{1}{2}x^2}, \qquad x^{\frac{1}{2}}J_{\frac{1}{2}\nu}(\tfrac{1}{2}x^2)$$

in the transform of order ν.‡

† For the theory of 'Hankel transforms' of order $\nu \geqslant -\tfrac{1}{2}$ see Titchmarsh (**25, 27**), Plancherel (**23**). There does not seem to have been any discussion of the Hankel formulae for $\nu < -\tfrac{1}{2}$.

‡ The appropriate references are given in § 3.

1.4. We can obtain two formal solutions of our problem as follows. Here, and throughout the paper, we give details of the analysis for the equation (1.12) only. The general equation (1.32) provides us with a greater wealth of formal illustration, but presents no additional difficulty of principle, so that we are content with stating results.

(i) *First solution.* We multiply (1.12) by x^{s-1} and integrate from 0 to ∞ with respect to x. If we write

$$\phi(s) = \int_0^\infty x^{s-1} f(x)\, dx \tag{1.41}$$

and assume the convergence of all the integrals concerned, and the legitimacy of the inversion of the order of integration, we obtain

$$\phi(s) = \sqrt{\left(\frac{2}{\pi}\right)} \int_0^\infty x^{s-1}\, dx \int_0^\infty \cos xy\, f(y)\, dy$$

$$= \sqrt{\left(\frac{2}{\pi}\right)} \int_0^\infty f(y)\, dy \int_0^\infty x^{s-1} \cos xy\, dx = \sqrt{\left(\frac{2}{\pi}\right)} \Gamma(s) \cos \tfrac{1}{2} s\pi \int_0^\infty y^{-s} f(y)\, dy,$$

or

$$\phi(s) = \sqrt{\left(\frac{2}{\pi}\right)} \Gamma(s) \cos \tfrac{1}{2} s\pi\, \phi(1-s). \tag{1.42}$$

If now we write

$$\phi(s) = 2^{\tfrac{1}{2}s} \Gamma(\tfrac{1}{2}s) \psi(s), \tag{1.43}$$

(1.42) becomes

$$\psi(s) = \psi(1-s). \tag{1.44}$$

Finally, inverting (1.41) by Mellin's formula,† we obtain

$$f(x) = \frac{1}{2\pi i} \int_{c-i\infty}^{c+i\infty} \phi(s) x^{-s}\, ds \tag{1.45}$$

or

$$f(x) = \frac{1}{2\pi i} \int_{c-i\infty}^{c+i\infty} 2^{\tfrac{1}{2}s} \Gamma(\tfrac{1}{2}s) \psi(s) x^{-s}\, ds, \tag{1.46; A}$$

where $\psi(s)$ is an even function of $s-\tfrac{1}{2}$. This is our first solution.

The simplest example is

$$\psi(s) = 1, \qquad f(x) = 2e^{-\tfrac{1}{2}x^2}.$$

Formal analysis on the general lines of that just set out is naturally to be found in many places in the literature,‡ though we have never

---

† For discussions of the Mellin formulae see Mellin (**17**), Hardy (**11**), Pollard (**24**), Burkill (**6**).

‡ See, for example, citations of Ramanujan by Hardy and Littlewood (**15**), § 2.53; or that of Schwarzschild by Eddington (**9**), 208–10, where everything is stated in terms of the Fourier inversion formulae, of which Mellin's formulae are, from the strictly formal point of view, merely a variant.

seen it applied to this particular problem, and the formula (A) appears to be new.

1.5. (ii) *Second solution.* Let

$$\lambda(s) = \int_0^\infty e^{-\tfrac{1}{2}sx^2} f(x)\, dx. \qquad (1.51)$$

Then multiplying (1.12) by $e^{-\tfrac{1}{2}sx^2}$, integrating with respect to x, and again assuming the convergence of the integrals and the legitimacy of the inversion, we find

$$\lambda(s) = \int_0^\infty e^{-\tfrac{1}{2}sx^2} f(x)\, dx = \sqrt{\left(\tfrac{2}{\pi}\right)} \int_0^\infty e^{-\tfrac{1}{2}sx^2}\, dx \int_0^\infty \cos xy\, f(y)\, dy$$

$$= \sqrt{\left(\tfrac{2}{\pi}\right)} \int_0^\infty f(y)\, dy \int_0^\infty e^{-\tfrac{1}{2}sx^2} \cos xy\, dx = \frac{1}{\sqrt{s}} \int_0^\infty e^{-y^2/2s} f(y)\, dy$$

or

$$\lambda(s) = s^{-\tfrac{1}{2}} \lambda\!\left(\tfrac{1}{s}\right). \qquad (1.52)$$

Hence, if

$$\mu(s) = s^{\tfrac{1}{4}} \lambda(s), \qquad (1.53)$$

we have

$$\mu(s) = \mu\!\left(\tfrac{1}{s}\right). \qquad (1.54)$$

Again

$$\lambda(s) = \int_0^\infty e^{-su} \frac{f\{\sqrt{(2u)}\}}{\sqrt{(2u)}}\, du = \int_0^\infty e^{-su} g(u)\, du, \qquad (1.55)$$

say. There is again a well-known formula for the inversion of this equation, viz.†

$$g(u) = \frac{1}{2\pi i} \int_{c-i\infty}^{c+i\infty} e^{us} \lambda(s)\, ds. \qquad (1.56)$$

Substituting into (1.56) from (1.53), we obtain

$$f(x) = \frac{x}{2\pi i} \int_{c-i\infty}^{c+i\infty} e^{\tfrac{1}{2}x^2 s} s^{-\tfrac{1}{4}} \mu(s)\, ds, \qquad (1.57;\ B)$$

where $\mu(s)$ is a solution of (1.54). This is our second solution. The simplest example is

$$\mu(s) = 1, \qquad f(x) = \frac{2^{\tfrac{3}{4}}}{\Gamma(\tfrac{1}{4})} \frac{1}{\sqrt{x}}.$$

† Apparently due, in this form, to Pincherle (**20**); see Bateman (**3**), 61. The formulae are, of course, again variants of the Fourier formulae.

The formulae corresponding to (A) and (B) for general ν are

$$f(x) = \frac{1}{2\pi i} \int_{c-i\infty}^{c+i\infty} 2^{\frac{1}{2}s}\Gamma(\tfrac{1}{2}s+\tfrac{1}{2}\nu+\tfrac{1}{4})\psi(s)x^{-s}\,ds \qquad (1.58;\,A')$$

where $\psi(s)$ satisfies (1.44), and

$$f(x) = \frac{x^{\frac{1}{2}-\nu}}{2\pi i} \int_{c-i\infty}^{c+i\infty} e^{\frac{1}{2}x^2 s} s^{-\frac{1}{2}(\nu+1)}\mu(s)\,ds, \qquad (1.59;\,B')$$

where $\mu(s)$ satisfies (1.54).

1.6. In § 2 we discuss formula (A). Our main object is to find theorems which give a complete solution of the problem for stated classes of functions $f(x)$, that is to say theorems of the type 'in order that $f(x)$ should be a self-reciprocal function of a certain class P, it is necessary and sufficient that $f(x)$ should be given by the formula (A), $\psi(s)$ being a function of a certain class Q'. Theorems **1**, **3**, and **4** are of this type, the class P being in Theorem **1** the class L^2, in Theorem **3** an associated class of functions of a real variable, and in Theorem **4** a certain class (the most natural class for this problem) of analytic functions.

In § 3 we consider the known examples of self-reciprocal functions, and others of our own, in the light of the theorems we have proved. There are some of them which do not fall under any of the theorems, and we are thus led to Theorems **13** and **14**, in which the conditions are not of the 'necessary and sufficient' type, but which enable us to render an account of all our examples but one.

In § 6 we prove the theorem for our second solution (Theorem **16**) which corresponds to Theorem **4**. This theorem covers the example which evaded the first solution. Generally, the second solution is more powerful for analytic functions and the first for real functions. There is a theorem concerning the second solution analogous to Theorem **1**, but it does not seem to be so useful.

We also state, but without proof, a theorem (Theorem **17**) corresponding to Theorem **13**. Our general principle in selecting theorems has been to concentrate on 'necessary and sufficient' theorems, with the reservation that there should be at least one theorem to cover any well-known example.

2. The first solution
Functions of L^2

2.1. Theorem 1. *A necessary and sufficient condition that a function $f(x)$ of $L^2(0, \infty)$† should be its own cosine transform is that it should be of the form (1.45), where $c = \frac{1}{2}$, the integral is a mean-square integral, and $\phi(\frac{1}{2}+it) = \chi(t)$ belongs to $L^2(-\infty, \infty)$ and satisfies (1.42), i.e.*

$$\frac{\chi(t)}{2^{\frac{1}{4}+\frac{1}{2}it}\Gamma(\frac{1}{4}+\frac{1}{2}it)} \tag{2.11}$$

is an even function of t.

(1) *The condition is necessary.* Since

$$\int_0^\infty f^2(x)\, dx = \int_{-\infty}^\infty f^2(e^y)e^y\, dy,$$

the function
$$F(y) = f(e^y)e^{\frac{1}{2}y}$$
belongs to $L^2(-\infty, \infty)$. It follows from the theory of transforms that

$$\chi(t) = \int_{-\infty}^\infty F(y)e^{ity}\, dy = \int_0^\infty f(x)x^{-\frac{1}{2}+it}\, dx$$

exists as a mean-square integral for almost all t. Also $\chi(t)$ belongs to $L^2(-\infty, \infty)$, and

$$F(y) \equiv \frac{1}{2\pi}\int_{-\infty}^\infty \chi(t)e^{-iyt}\, dt,\ddagger$$

the integral being again a mean-square integral. This formula is equivalent to (1.45).

It remains to verify that the function (2.11) is even. The formal proof of this is obtained by putting $s = \frac{1}{2}+it$ in the analysis of § 1.4. A direct justification of the inversion of the order of integration is troublesome, and we therefore proceed differently.

We have to prove that $\chi(t)$ is equivalent to

$$\omega(t) = \sqrt{\left(\frac{2}{\pi}\right)}\Gamma(\frac{1}{2}+it)\cos(\tfrac{1}{4}\pi+\tfrac{1}{2}\pi it)\chi(-t),$$

and for this it is sufficient to prove that

$$\int_0^T \chi(t)\, dt = \int_0^T \omega(t)\, dt.$$

---

† By $L^p(a, b)$ we mean the class of functions which belong to L^p in (a, b).
‡ $f(x) \equiv g(x)$ means that f is equivalent to g, i.e. that the functions differ at most in a set of measure zero.

for all T. Now $\chi(t)$ is the mean-square limit of
$$\int_{\xi}^{X} f(x) x^{-\frac{1}{2}+it} \, dx$$
when $\xi \to 0$, $X \to \infty$; and so
$$\int_0^T \chi(t) \, dt = \lim_{\xi \to 0, X \to \infty} \int_0^T dt \int_{\xi}^X f(x) x^{-\frac{1}{2}+it} \, dx$$
$$= \lim \int_{\xi}^X f(x) \, dx \int_0^T x^{-\frac{1}{2}+it} \, dt = \int_0^{\infty} f(x) \, dx \int_0^T x^{-\frac{1}{2}+it} \, dt.$$

Similarly
$$\int_0^T \omega(t) \, dt = \int_0^{\infty} f(x) \, dx \int_0^T \sqrt{\left(\frac{2}{\pi}\right)} \Gamma(\tfrac{1}{2}+it) \cos(\tfrac{1}{4}\pi + \tfrac{1}{2}\pi it) x^{-\frac{1}{2}-it} \, dt;$$

and we have to prove that the last two repeated integrals are equal. Since f belongs to L^2, and is its own cosine transform, the result required is a case of Parseval's theorem for transforms, provided that the two functions
$$\int_0^T x^{-\frac{1}{2}+it} \, dt, \quad \int_0^T \sqrt{\left(\frac{2}{\pi}\right)} \Gamma(\tfrac{1}{2}+it) \cos(\tfrac{1}{4}\pi + \tfrac{1}{2}\pi it) x^{-\frac{1}{2}-it} \, dt$$
are cosine transforms of each other and belong to L^2.

The first function is
$$x^{-\frac{1}{2}} \frac{x^{iT} - 1}{i \log x},$$
and therefore belongs to L^2. Its cosine transform is
$$\sqrt{\left(\frac{2}{\pi}\right)} \int_0^{\infty} \cos xu \, du \int_0^T u^{-\frac{1}{2}+it} \, dt,$$
while the second function is the result of inverting this integral. All that remains, then, is to justify this inversion; and since
$$\int_0^U \cos xu \, du \int_0^T u^{-\frac{1}{2}+it} \, dt = \int_0^T dt \int_0^U u^{-\frac{1}{2}+it} \cos xu \, du$$
for any finite U, the result will follow if only
$$\lim_{U \to \infty} \int_0^T dt \int_U^{\infty} u^{-\frac{1}{2}+it} \cos xu \, du = 0. \tag{2.12}$$

But for fixed x

$$\int_U^\infty u^{-\frac{1}{2}+it}\cos xu\,du = \left[u^{-\frac{1}{2}+it}\frac{\sin xu}{x}\right]_U^\infty + \frac{\frac{1}{2}-it}{x}\int_U^\infty u^{-\frac{3}{2}+it}\sin xu\,du$$

$$= O(U^{-\frac{1}{2}}) + O\left(\int_U^\infty u^{-\frac{3}{2}}\,du\right) = O(U^{-\frac{1}{2}})$$

uniformly for $0 \leqslant t \leqslant T$, and this proves (2.12).

(2) *The condition is sufficient.* In the first place, if $\chi(t)$ belongs to $L^2(-\infty, \infty)$, so does $F(y)$, and therefore $f(x)$ belongs to $L^2(0, \infty)$.

We have to prove that $g(x)$, the cosine transform of $f(x)$, is equivalent to $f(x)$. Now

$$g_1(x) = \int_0^x g(u)\,du = \sqrt{\left(\frac{2}{\pi}\right)}\int_0^\infty f(y)\frac{\sin xy}{y}\,dy.$$

Since $(\sin xy)/y$ belongs to L^2, and $f(y)$ is the mean-square limit of

$$\frac{1}{2\pi}\int_{-T}^T \chi(t) y^{-\frac{1}{2}-it}\,dt,$$

it follows that

$$g_1(x) = \lim_{T\to\infty}\frac{1}{2\pi}\sqrt{\left(\frac{2}{\pi}\right)}\int_0^\infty \frac{\sin xy}{y}\,dy \int_{-T}^T \chi(t) y^{-\frac{1}{2}-it}\,dt$$

$$= \lim_{T\to\infty}\frac{1}{2\pi}\sqrt{\left(\frac{2}{\pi}\right)}\int_{-T}^T \chi(t)\,dt \int_0^\infty \frac{\sin xy}{y^{\frac{3}{2}+it}}\,dy,$$

since the last integral with respect to y is uniformly convergent in t. Hence

$$g_1(x) = \frac{1}{4}\sqrt{\left(\frac{2}{\pi}\right)}\int_{-\infty}^\infty \frac{\chi(t)}{\cos(\frac{1}{4}\pi + \frac{1}{2}\pi it)\Gamma(\frac{1}{2}+it)}\frac{x^{\frac{1}{2}+it}}{\frac{1}{2}+it}\,dt$$

$$= \frac{1}{2\pi}\int_{-\infty}^\infty \chi(t)\frac{x^{\frac{1}{2}-it}}{\frac{1}{2}-it}\,dt, \qquad (2.15)$$

on using the functional equation for $\chi(t)$ and then changing the sign of t.

On the other hand

$$f_1(x) = \int_0^x f(u)\, du = \lim_{T \to \infty} \frac{1}{2\pi} \int_0^x du \int_{-T}^T \chi(t) u^{-\frac{1}{2}-it}\, dt$$

$$= \lim_{T \to \infty} \frac{1}{2\pi} \int_{-T}^T \chi(t)\, dt \int_0^x u^{-\frac{1}{2}-it}\, du = \frac{1}{2\pi} \int_{-\infty}^\infty \chi(t) \frac{x^{\frac{1}{2}-it}}{\frac{1}{2}-it}\, dt. \qquad (2.16)$$

Comparing (2.15) and (2.16), we see that $g_1(x) = f_1(x)$, and so $g(x) \equiv f(x)$. This completes the proof of Theorem **1**.

Functions of L^p $(1 < p < 2)$

2.2. We suppose now that $1 < p < 2$, and denote by p' the index conjugate to p, that is to say the index defined by

$$p' = \frac{p}{p-1}, \qquad \frac{1}{p} + \frac{1}{p'} = 1.$$

THEOREM 2. *If a function $f(x)$ of $L^p(0, \infty)$, where $1 < p < 2$, is its own cosine transform, then $f(x)$ is of the form* (1.45); *where $\phi(s)$ is an analytic function which* (i) *is regular in the strip*

$$1/p' < \sigma < 1/p, \qquad (2.21)$$

(ii) *tends to zero uniformly when s tends to infinity inside any interior strip, and* (iii) *satisfies the equation* (1.42); *and the integral in* (1.45) *is a mean-square integral along any line of the strip* (2.21).

It will be observed that Theorem **2** is a one-sided theorem with conditions which are necessary only and not sufficient.

If $f(x)$ belongs to L^p, its cosine transform belongs to $L^{p'}$, so that $f(x)$ here belongs both to L^p and to $L^{p'}$, and therefore to all intermediate classes L^q. In particular it belongs to L^2, and so satisfies the conditions of Theorem **1**.

(i) The function $\phi(s)$ defined by (1.41) reduces to the $\chi(t)$ of Theorem **1** when $s = \frac{1}{2} + it$, but $\phi(s)$ is now analytic and regular in the strip (2.21). In fact

$$\int_0^\infty |f(x)| x^{\sigma-1}\, dx \leqslant \left(\int_0^1 |f|^{p'}\, dx\right)^{1/p'} \left(\int_0^1 x^{p(\sigma-1)}\, dx\right)^{1/p} +$$
$$+ \left(\int_1^\infty |f|^p\, dx\right)^{1/p} \left(\int_1^\infty x^{p'(\sigma-1)}\, dx\right)^{1/p'},$$

and these integrals converge for the values of σ stated. It follows in the usual manner that $\phi(s)$ is regular in the strip and bounded in any interior strip.

(ii) We can write $\phi(s)$ in the form

$$\phi(s) = \left(\int_0^\delta + \int_\delta^\Delta + \int_\Delta^\infty\right) f(x) x^{\sigma-1} x^{it}\, dx = \phi_1(s) + \phi_2(s) + \phi_3(s).$$

Suppose now that $\eta > 0$ and

$$\frac{1}{p'} + \eta \leqslant \sigma \leqslant \frac{1}{p} - \eta. \tag{2.23}$$

Then

$$|\phi_1(s)| \leqslant \left(\int_0^\delta |f|^{p'}\, dx\right)^{1/p'} \left(\int_0^\delta x^{p(\sigma-1)}\, dx\right)^{1/p} = O(\delta^\eta)$$

when $\delta \to 0$, and we can therefore choose δ so that $|\phi_1| < \epsilon$ for all s of (2.21). Similarly, we can choose Δ so that $|\phi_3| < \epsilon$. When δ and Δ are fixed, $\phi_2 \to 0$ uniformly when $s \to \infty$ in (2.21). It follows that $\phi \to 0$ uniformly in (2.21).

(iii) It follows from Theorem 1 that $\phi(s)$ satisfies (1.42) on the line $s = \tfrac{1}{2} + it$, and therefore throughout (2.21).

Thus $\phi(s)$ possesses the properties stated in the theorem, and it remains only to prove (1.45). By Theorem 1, (1.45) is true for $c = \tfrac{1}{2}$, so that it is sufficient to prove that the value of the integral is independent of c.

If c and c' are two values of σ in (2.21), and

$$f(x, c, T) = \frac{1}{2\pi i} \int_{c-iT}^{c+iT} \phi(s) x^{-s}\, ds,$$

then, by Cauchy's theorem,

$$f(x, c, T) - f(x, c', T) = \frac{1}{2\pi i} \int_{c-iT}^{c'-iT} \phi(s) x^{-s}\, ds - \frac{1}{2\pi i} \int_{c+iT}^{c'+iT} \phi(s) x^{-s}\, ds.$$

Since $\phi(s) \to 0$, this difference tends to zero when $T \to \infty$, and uniformly in any interval (a, b) of positive values of x. Hence

$$\lim_{T \to \infty} \int_a^b \{f(x, c, T) - f(x, c', T)\}^2\, dx = 0,$$

so that $f(x, c, T)$ and $f(x, c', T)$ converge in mean to the same limit $f(x)$ independent of c.

Functions of L_p^* $(1 < p < 2)$

2.3. Theorem **2** is a one-sided theorem, and it is perhaps unreasonable, in view of the asymmetry of the theory of transforms about the number 2, to expect in this case a theorem as satisfactory as Theorem **1**. There is, however, a very similar class of functions for which we can obtain a complete solution.

We shall say that $f(x)$ belongs to $L_p^*(0, \infty)$, where $1 < p < 2$, if $x^\alpha f(x)$ belongs to $L_2(0, \infty)$ for

$$-\alpha_0 = -\frac{1}{p} + \frac{1}{2} < \alpha < \frac{1}{p} - \frac{1}{2} = \alpha_0. \qquad (2.31)$$

It is plain that $f(x)$ then belongs to $L^q(0, 1)$ for $q \leqslant 2$. Suppose now that $p < q < 2$. Then we can choose $\alpha < \alpha_0$ so that $2q\alpha > 2-q$; and then

$$\int_1^\infty |f|^q \, dx \leqslant \left(\int_1^\infty x^{2\alpha} |f|^2 \, dx \right)^{q/2} \left(\int_1^\infty x^{-2q\alpha/(2-q)} \, dx \right)^{(2-q)/2} < \infty,$$

so that $f(x)$ belongs to L^q. If also $f(x)$ is its own cosine transform, it belongs to $L^{q'}$, so that a self-reciprocal $f(x)$ of L_p^* belongs to all Lebesgue classes between L^p and $L^{p'}$ (though not usually to either of these).

The class of self-reciprocal functions of L_p^* is thus in this respect a little wider than the class of those of L^p. In other respects it is narrower. Suppose, for example, that $h(x)$ is defined by

$$h(x) = 2^{-n} \ (n!-1 \leqslant x \leqslant n!+1, \ n = 2,3,\ldots)$$

and $h(x) = 0$ elsewhere. Then $h(x)$ belongs to L^r for every positive r, but to no L_p^*, since $\sum 2^{-rn}$ is convergent but $\sum (n!)^{2\alpha} 2^{-2n}$ divergent for every positive α. The cosine transform of $h(x)$ is

$$h^*(x) = \sqrt{\left(\frac{2}{\pi}\right)} \sum 2^{-n} \int_{n!-1}^{n!+1} \cos xy \, dy = 2 \sqrt{\left(\frac{2}{\pi}\right)} \frac{\sin x}{x} \sum 2^{-n} \cos n!x,$$

which is continuous and $O(x^{-1})$ at infinity, so that $h^*(x)$ belongs to L^r for $r > 1$ and to L_p^* for $1 < p < 2$. Thus $h(x) + h^*(x)$ is a self-reciprocal function which belongs to L^r for all $r > 1$, but to no L_p^*.

THEOREM 3. *A necessary and sufficient condition that a function $f(x)$ of $L_p^*(0, \infty)$ should be its own cosine transform is that it should be of the form* (1.45), *where $\phi(s)$ satisfies the conditions* (i), (ii), (iii) *of Theorem* **2** *and* (iv) *belongs to $L^2(-\infty, \infty)$, qua function of t, for all σ of* (2.21).

(1) *The condition is necessary.* Since $f(x)$ belongs to L^r for $p < r < p'$, we have only to show that $\phi(s)$ satisfies condition (iv). This results immediately from the theory of transforms, since

$$\phi(s) = \int_0^\infty x^{-\frac{1}{2}+it} \cdot x^{\sigma-\frac{1}{2}} f(x)\, dx,$$

and $x^{\sigma-\frac{1}{2}} f(x)$ belongs to L^2 if $|\sigma - \frac{1}{2}| < \alpha_0$, i.e. if $1/p' < \sigma < 1/p$.

(2) *The condition is sufficient.* Since $\phi(s)$ belongs to L^2 on the line $s = c + it$, the integral (1.45) exists as a mean-square integral for all c in question, and (as in § 2.2) its value is independent of c. It therefore defines a function $f(x)$ independent of c. Since

$$x^{c-\frac{1}{2}} f(x) = \frac{1}{2\pi} \int_{-\infty}^{\infty} \phi(c+it) x^{-\frac{1}{2}-it}\, dt$$

for $|c - \frac{1}{2}| < \alpha_0$, and the right-hand side belongs to L^2 for every such c, $f(x)$ belongs to L_p^*. Finally, by Theorem 1, $f(x)$ is self-reciprocal.

Analytic functions

2.4. We shall say that $f(x)$ belongs to $A(\omega, a)$, where $0 < \omega \leqslant \pi$,† $a < \frac{1}{2}$, if (i) it is an analytic function of $x = re^{i\theta}$ regular in the angle A defined by $r > 0$, $|\theta| < \omega$, and (ii) it is $O(|x|^{-a-\delta})$ for small x, and $O(|x|^{a-1+\delta})$ for large x, for every positive δ and uniformly in any angle $|\theta| \leqslant \omega - \eta < \omega$.

THEOREM 4. *A necessary and sufficient condition that a function $f(x)$ of $A(\omega, a)$ should be its own cosine transform is that it should be of the form (A); where $\psi(s)$ is regular, and satisfies (1.44), in the strip*

$$a < \sigma < 1-a; \tag{2.41}$$

$$\psi(s) = O\{e^{(\frac{1}{2}\pi - \omega + \eta)|t|}\} \tag{2.42}$$

for every positive η and uniformly in any strip interior to (2.41); and c is any value of σ in (2.41).

(1) *The condition is necessary.* The integral (1.41) is absolutely convergent for $a < \sigma < 1-a$, so that $\phi(s)$ is regular in (2.41). Also $f(x)$ belongs to L^2, and $\phi(\frac{1}{2}+it)$ is the $\chi(t)$ of Theorem 1. It follows from Theorem 1 that $\phi(s)$ satisfies (1.42) on $\sigma = \frac{1}{2}$ and therefore throughout (2.41), or (what is the same thing) that $\psi(s)$ satisfies (1.44).

We may deform the integral (1.41), by Cauchy's theorem, into one

---

† This restriction on ω is not essential, but when $\omega > \pi$ the angle is re-entrant, and $f(x)$ is not usually one-valued in it.

along any radius vector $\theta = \beta$, where $|\beta| < \omega$, and we may suppose that β has the sign of t. Then

$$|\phi(s)| = \left| e^{i\beta\sigma-\beta t} \int_0^\infty r^{\sigma+it-1} f(re^{i\beta})\, dr \right| \leqslant e^{-\beta t} \int_0^\infty r^{\sigma-1} |f(re^{i\beta})|\, dr = O(e^{-|\beta||t|}).$$

Since $|\beta|$ is any number less than ω, and

$$|\Gamma(\tfrac{1}{2}s)| \sim C e^{-\frac{1}{4}\pi|t|} |\tfrac{1}{2}t|^{\frac{1}{2}\sigma-\frac{1}{2}}, \tag{2.43}$$

where C is a constant, this implies (2.42).

The integral (A) is absolutely convergent, and uniformly convergent for any fixed x and all c of any interval interior to (2.41). It represents $f(x)$ when $c = \tfrac{1}{2}$, by Theorem **1**, and therefore generally.

(2) *The condition is sufficient.* The modulus of the integrand in (A) is, by (2.42) and (2.43), less than a constant multiple of

$$r^{-\sigma} e^{-|t|(\omega-\eta-|\theta|)} |t|^{\frac{1}{2}c-\frac{1}{2}}.$$

The integral is therefore uniformly convergent throughout any domain of x interior to the angle A; and $f(x)$ is regular in A, and $O(|x|^{-c})$ in any smaller angle. Since c may be taken as near to a or $1-a$ as we please, $f(x)$ belongs to $A(\omega, a)$. Finally, $f(x)$ belongs to L^2, and is therefore, by Theorem **1**, self-reciprocal.

2.5. All of Theorems **2–4** have been deduced from Theorem **1** (though Theorem **4** might naturally be proved independently), and their proofs depend upon the theory of transforms, of which we shall make no further use. This is therefore a convenient place to consider some of the most interesting special cases in the light of what we have proved. In order to have a greater variety of illustrations, we state first the theorems corresponding to Theorems **1–4** for functions reciprocal in the Hankel transform.

THEOREMS **5–8**. *Theorems* **1–4** *are true for Hankel transforms with the following formal changes. In Theorem* **1** *the function* (2.11) *is to be replaced by*

$$\frac{\chi(t)}{2^{\frac{1}{2}+\frac{1}{2}\nu+\frac{1}{2}it}\Gamma(\tfrac{1}{2}+\tfrac{1}{2}\nu+\tfrac{1}{2}it)}. \tag{2.51}$$

In Theorems **2**, **3**, *and* **4**, (1.43) *is to be replaced by*

$$\phi(s) = 2^{\frac{1}{2}s}\Gamma(\tfrac{1}{2}s+\tfrac{1}{2}\nu+\tfrac{1}{4})\psi(s) \tag{2.52}$$

and (A) *by* (A').†

† Since $\nu \geqslant -\tfrac{1}{2}$, $\Gamma(\tfrac{1}{2}s+\tfrac{1}{2}\nu+\tfrac{1}{4})$ has no poles for which $\sigma > 0$, so that conditions concerning the regularity of $\phi(s)$ and $\psi(s)$ in the strips considered are equivalent.

SELF-RECIPROCAL FUNCTIONS

There is a corresponding series of theorems for *skew-reciprocal* functions, i.e. functions for which $f(x) = -g(x)$, where $g(x)$ is the cosine or Hankel transform of $f(x)$.

THEOREMS 9–12. *The results of all the preceding theorems become true for skew-reciprocal functions if we change the sign of the right-hand side of the functional equations satisfied by $\chi(t)$, $\phi(s)$, and $\psi(s)$.*

3. Examples

(1) If we take $\psi(s) = 1$ in Theorems **4** and **8**, and evaluate the integrals (A) and (A') by the calculus of residues, we obtain
$$f(x) = 2e^{-\frac{1}{2}x^2}, \qquad f(x) = 2^{\frac{3}{2}-\frac{1}{2}\nu}x^{\nu+\frac{1}{2}}e^{-\frac{1}{2}x^2}$$
respectively. Conversely, we may assume these forms of $f(x)$, and calculate $\phi(s)$ and $\psi(s)$ by (1.41) and (1.43) or (2.52). The conditions of Theorems **4** or **8** (and *a fortiori* those of the less special theorems) are satisfied in these cases.

(2) If we take $\psi(s) = P(\frac{1}{2}-s)$, where $P(u)$ is an even polynomial, or an even integral function of order less than 1, we find that
$$f(x) = 2\sum_0^\infty \frac{(-\frac{1}{2}x^2)^n}{n!} P(2n+\tfrac{1}{2})$$
is its own cosine transform. If $P(u)$ is a polynomial, $f(x) = e^{-\frac{1}{2}x^2}Q(x^2)$, where $Q(u)$ is a polynomial. The simplest example, after $P(u) = 1$, is $P(u) = u^2$, when
$$f(x) = 2e^{-\frac{1}{2}x^2}(x^4 - 3x^2 + \tfrac{1}{4}).$$

If $P(u)$ is odd we obtain examples of Theorem **12**, such as $e^{-\frac{1}{2}x^2}(x^2-\frac{1}{2})$.

(3) If $f(x) = \operatorname{sech} x\sqrt{(\frac{1}{2}\pi)}$, we find that
$$\phi(s) = 2\left(\frac{2}{\pi}\right)^{\frac{1}{2}s}\Gamma(s)L(s) = 2\left(\frac{2}{\pi}\right)^{\frac{1}{2}s}\Gamma(s)(1^{-s}-3^{-s}+5^{-s}-\ldots),$$
which satisfies (1.42). This is an example of Theorem **4**.

If
$$f(x) = \frac{1}{e^{x\sqrt{(2\pi)}}-1} - \frac{1}{x\sqrt{(2\pi)}},$$
we find that $\phi(s) = (2\pi)^{-\frac{1}{2}s}\Gamma(s)\zeta(s)$. Taking $\nu = \frac{1}{2}$, we obtain
$$\psi(s) = \frac{\xi(s)}{s(s-1)},$$
where $\xi(s)$ is Riemann's ξ-function. This is an example of Theorem **8**, with $\nu = \frac{1}{2}$.

(4) If†
$$f(x) = x^{\frac{1}{2}-\nu}e^{\frac{1}{4}x^2}\int_x^\infty t^{2\nu-1}e^{-\frac{1}{2}t^2}\,dt,$$

$f(x)$ is $O(x^{\frac{1}{2}-|\nu|})$ for small x‡ and $O(x^{\nu-\frac{3}{2}})$ for large x. The function satisfies the conditions of Theorem 8 when $\nu < 1$, but not when $\nu \geq 1$. In this case we find
$$\psi(s) = 2^{\frac{1}{2}\nu-\frac{7}{4}}\frac{\Gamma(\tfrac{1}{4}-\tfrac{1}{2}\nu+\tfrac{1}{2}s)\Gamma(\tfrac{3}{4}-\tfrac{1}{2}\nu-\tfrac{1}{2}s)}{\Gamma(1-\nu)}.$$

This function satisfies the conditions of Theorem 8 when $\nu < 1$, but a pole comes on to the line $\sigma = \tfrac{1}{2}$ for $\nu = 1$. The integral (A') is convergent if $\nu < 1$ and $\nu - \tfrac{1}{2} < c < \tfrac{3}{2} - \nu$, but is never convergent if $\nu > 1$.

In the case $\nu = -\tfrac{1}{2}$ the function is
$$f(x) = 1 - xe^{\frac{1}{4}x^2}\int_x^\infty e^{-\frac{1}{4}t^2}\,dt = \int_0^\infty e^{-\frac{1}{2}w^2-xw}\,w\,dw.$$

(5) If §
$$f(x) = \int_0^\infty \frac{e^{-xw}}{\cosh bw}\,dw,$$

where $b = \sqrt{(\tfrac{1}{2}\pi)}$, and $\nu = \tfrac{1}{2}$, we find that
$$\psi(s) = \frac{2^{\frac{3}{2}-s}\pi^{\frac{1}{2}+\frac{1}{2}s}L(1-s)}{\sin s\pi\,\Gamma(\tfrac{1}{2}+s)},$$

where $L(s)$ is defined as under (3). This is an example of Theorem 8, with $\nu = \tfrac{1}{2}$. Here
$$f(x) = 2\left(\frac{1}{x+b} - \frac{1}{x+3b} + \frac{1}{x+5b} - \cdots\right)$$

is regular except for $x = -b, -3b, \ldots$, and is $O(1)$ for small and $O(1/x)$ for large x. Hence we may take $\omega = \pi$ and $a = 0$ in Theorem 8; and $\psi(s)$ is regular in $0 < \sigma < 1$, with poles at $s = 0$ and $s = 1$, and is $O\{e^{-(\frac{1}{2}\pi-\eta)|t|}\}$, for every positive η, for large t.

(6) Suppose that‖
$$f(x) = x^{\frac{1}{2}+\nu}\int_0^\infty w^\nu F(w)e^{-\frac{1}{2}x^2w^2}\,dw,$$

where $F(w) = F(1/w)$, and that
$$F(w) = O(w^\lambda) \quad (w \to 0), \qquad F(w) = O(w^{-\lambda}) \quad (w \to \infty).$$

† For the case $\nu = \tfrac{1}{2}$ see Hardy (**10**), Phillips (**19**). The general formula was given by Bailey (**1**).
‡ Except when $\nu = 0$, when there is an additional factor $\log x$.
§ Phillips (**19**).
‖ For this example see Bailey (**2**).

SELF-RECIPROCAL FUNCTIONS

where $0 < \lambda < 1+\nu$.† It may easily be verified that $f(x) = f(re^{i\theta})$ is regular inside the angle $|\theta| \leqslant \tfrac{1}{4}\pi$, and that

$$f(x) = O(r^{-\tfrac{1}{2}+\lambda}) \quad (r \to 0), \qquad f(x) = O(r^{-\tfrac{1}{2}-\lambda}) \quad (r \to \infty)$$

uniformly in any smaller angle. Thus the conditions of Theorem **8** are satisfied with $\omega = \tfrac{1}{4}\pi$, $a = \tfrac{1}{2}-\lambda$.

We find by straightforward calculation that

$$\psi(s) = 2^{\tfrac{1}{2}\nu-\tfrac{1}{2}} \int_0^\infty w^{-\tfrac{1}{2}-s} F(w)\, dw$$

if $\tfrac{1}{2}-\lambda < \sigma < \tfrac{1}{2}+\lambda$. It easily results that $\psi(s)$ is regular in this strip, bounded in any interior strip, and satisfies (1.44).

(7) A very interesting example of a more recondite character is

$$f(x) = x^{\tfrac{1}{2}-\nu}(x^2-b^2)^{\tfrac{1}{2}(\nu-1)} J_{\tfrac{1}{2}(\nu-1)}\{b\sqrt{(x^2-b^2)}\} \quad (x > b > 0)$$
$$f(x) = 0 \quad (0 < x < b).$$

That this function is its own Hankel transform results, by elementary substitutions and specializations, from a formula of Sonine.‡

Here $f(x)$ is $O\{(x-b)^{\tfrac{1}{2}\nu-\tfrac{1}{2}}\}$ near $x = b$, and $O(x^{-\tfrac{1}{2}\nu-\tfrac{1}{2}})$ at infinity, and it belongs to L^2 if $\nu > 0$, in which case only do our theorems apply. The conditions of Theorem **3** are satisfied if $\nu > 2\alpha_0$.

To calculate $\psi(s)$ we observe that

$$\phi(s) = \int_b^\infty x^{s-\tfrac{1}{2}-\nu}(x^2-b^2)^{\tfrac{1}{2}(\nu-1)} J_{\tfrac{1}{2}(\nu-1)}\{b\sqrt{(x^2-b^2)}\}\, dx$$

$$= b^{\tfrac{1}{2}s+\tfrac{1}{2}\nu} \int_0^\infty (1+u^2)^{\tfrac{1}{2}(s-\nu-\tfrac{3}{2})} u^{\tfrac{1}{2}(\nu+1)} J_{\tfrac{1}{2}(\nu-1)}(b^2 u)\, du.$$

This integral may be calculated by another formula of Sonine,§ and we find that

$$\psi(s) = b^{\nu-\tfrac{1}{2}} 2^{\tfrac{1}{2}-\tfrac{1}{2}\nu} \frac{K_{\tfrac{1}{2}(s-\tfrac{1}{2})}(b^2)}{\Gamma(\tfrac{3}{4}+\tfrac{1}{2}\nu-\tfrac{1}{2}s)\Gamma(\tfrac{1}{4}+\tfrac{1}{2}\nu+\tfrac{1}{2}s)}.$$

From this, and the fact that $K_\mu(x)$ is an even function of μ, it follows that $\psi(s)$ satisfies (1.44).

(8) Taking
$$f(x) = x^{\tfrac{1}{2}} J_{\tfrac{1}{4}\nu}(\tfrac{1}{2}x^2),$$
we find
$$\psi(s) = \frac{\pi^{\tfrac{1}{2}} 2^{-\tfrac{1}{2}\nu-\tfrac{1}{4}}}{\Gamma(\tfrac{1}{4}\nu+\tfrac{5}{8}+\tfrac{1}{4}s)\Gamma(\tfrac{1}{4}\nu+\tfrac{7}{8}-\tfrac{1}{4}s)},$$

† The analysis can be adapted by slight changes to the case $\lambda > 1+\nu$, but our theorems fail if $\lambda < 0$.
‡ Watson (**30**), § 13.47 (1).
§ See Watson (**30**), § 13.6 (2).

so that $\psi(s) = \psi(1-s)$. Since

$$\int_0^\infty t J_\nu(xt) J_{\frac{1}{2}\nu}(\tfrac{1}{2}t^2)\, dt = J_{\frac{1}{2}\nu}(\tfrac{1}{2}x^2),\dagger$$

$f(x)$ is its own Hankel transform. In this case, however, $f(x)$ behaves like $x^{\nu+\frac{1}{2}}$ for small x and like $x^{-\frac{1}{2}}$ for large x, and does not belong to L^2, or to any L^p for which $p \leqslant 2$, so that it is not an example of any of the theorems so far proved.

The integral (A') is convergent if $-\nu-\frac{1}{2} < \sigma < \frac{5}{2}$ (absolutely if $-\nu-\frac{1}{2} < \sigma < \frac{1}{2}$), and $\phi(s)$ satisfies the conditions of our theorems concerning regularity, but its order of magnitude for large t is that of $|t|^{\frac{1}{2}\sigma-\frac{3}{4}}$, and it does not belong to L^2 on $\sigma = \frac{1}{2}$.

(9) There are very interesting examples of self-reciprocal functions in the analytic theory of numbers. Suppose that $r(n)$ is the number of representations of n as a sum of two squares, and that

$$\overline{P}(x) = \sum_{0 \leqslant n \leqslant x}{}' r(n) - \pi x, \qquad f(x) = x^{-\frac{1}{2}}\left\{\overline{P}\!\left(\frac{x^2}{2\pi}\right)-1\right\}, \tag{3.1}$$

the dash implying the insertion of a factor $\frac{1}{2}$ in the last term of the sum when x is an integer. Then $f(x)$ *is its own Hankel transform when* $\nu = 2$. In fact the well-known identity

$$\overline{P}(x) = \sqrt{x}\sum_1^\infty \frac{r(n)}{\sqrt{n}} J_1\{2\pi\sqrt{(nx)}\} \tag{3.2}$$

is easily transformed by partial summation into the form

$$\int_0^\infty \frac{1}{y}\left\{\overline{P}\!\left(\frac{y^2}{2\pi}\right)-1\right\} J_2(xy)\, dy = \frac{1}{x^2}\left\{\overline{P}\!\left(\frac{x^2}{2\pi}\right)-1\right\}, \tag{3.3}$$

in which it expresses the property stated.‡

Here $f(x) = O(x^{\frac{5}{2}})$ for small x. If we use the known (but rather difficult) theorem that $\overline{P}(x) = O(x^{\frac{1}{3}})$ for large x, then $f(x) = O(x^{-\frac{2}{3}})$ for large x, and $f(x)$ satisfies the conditions of Theorem 7 for $\nu = 2$ and $\alpha_0 = \frac{1}{3}$.

Calculation shows that

$$\phi(s) = \tfrac{1}{2}(2\pi)^{\frac{1}{2}s-\frac{3}{4}} \frac{Z(\tfrac{3}{4}-\tfrac{1}{2}s)}{\tfrac{3}{4}-\tfrac{1}{2}s},$$

† This is a trivial transformation of a formula due to Bateman. See Bateman (4), 185, Hardy (13), and Watson (30), 437.

‡ See Hardy (14). It should have been mentioned in this paper that the formula had been found independently by A. L. Dixon, who communicated it to Hardy when the paper was completed but not in print.

where $Z(s)$ is the analytic function defined for $\sigma > 1$ by
$$Z(s) = \sum \frac{r(n)}{n^s} = \zeta(s)(1^{-s} - 3^{-s} + 5^{-s} - \ldots) = \zeta(s)L(s),$$
and that $\psi(s)$ satisfies (1.44).

It should be observed that we can hardly hope to *prove* the difficult identity (3.2) as a corollary of a general theorem of this character. What results from Theorem **7** is the existence of the integral (3.3) as a mean-square integral; what is required to prove (3.2) is its convergence in the ordinary sense.

It has been proved by Walfisz and Oppenheim that, if $r_p(n)$ is the number of representations of n as a sum of p squares, and $\overline{P}_p(x)$ is defined as in (3.1), then
$$\overline{P}_p(x) = x^{\frac{1}{2}p} \sum_1^\infty \frac{r_p(n)}{n^{\frac{1}{2}p}} J_{\frac{1}{2}p}\{2\pi\sqrt{(nx)}\},$$
the series being summable by Cesàro's means of sufficiently high order.† The corresponding result in our present language would be that
$$x^{-\frac{1}{2}-\frac{1}{2}p}\left\{\overline{P}_p\!\left(\frac{x^2}{2\pi}\right) - 1\right\}$$
is its own Hankel transform of order $1 + \frac{1}{2}p$. If we take $p = 3$, and use Walfisz's result $\overline{P}_3(x) = O(x^{\frac{19}{28}+\epsilon})$,‡ we find that $f(x)$ falls under Theorem **5**. This is not true for any larger p.

If we take $p = 1$, we find that
$$f(x) = \frac{1}{x}\left\{\frac{x}{\sqrt{(2\pi)}} - \left[\frac{x}{\sqrt{(2\pi)}}\right]\right\}$$
is its own Hankel transform of order $\frac{3}{2}$, as may be verified directly.

(10) The functions
$$f(x) = x^{-\frac{1}{2}}, \qquad f(x) = 2^{\frac{1}{2}a}\Gamma(\tfrac{1}{2}a)x^{-a} + 2^{\frac{1}{2}(1-a)}\Gamma(\tfrac{1}{2}-\tfrac{1}{2}a)x^{a-1}$$
are their own cosine transforms. Neither of them belongs to L^2 (or to any other Lebesgue class). The integral (1.41) is not convergent for any s. It is summable by an appropriate adaptation of Riesz's logarithmic means, i.e. as
$$\lim_{\lambda\to\infty} \frac{1}{\log\lambda} \int_1^\lambda \frac{dw}{w} \int_{1/w}^w f(x)x^{s-1}\,dx,\S$$

† For a precise statement of the final results see Oppenheim (**18**), Walfisz (**28**).

‡ Walfisz (**29**). § Compare Pollard (**24**), 463.

but its value is 0 except for certain special values of s† and then is ∞. It is clear that our theorems fail to give any account of these solutions.

4. Further theorems concerning the first solution

4.1. There are several of the preceding examples which are not covered by any of the theorems which we have proved. Our next theorem, which was suggested by a theorem of Pollard,‡ is a one-sided theorem, which asserts only that reciprocal functions of a certain class have necessarily a certain form; but the class in question is a good deal more general than those which we have considered so far.

THEOREM 13. *Suppose that*
(i) *the integrals*
$$\int_0^1 |f(x)|\, dx, \qquad \int_1^\infty \frac{|f(x)|}{x^2}\, dx \qquad (4.11)$$
are finite;
(ii) *the integral* (1.41), *defined as a Cauchy integral at both limits, is convergent for* $|\sigma-\tfrac{1}{2}| < \alpha \leqslant \tfrac{1}{2}$;
(iii) *the integral* (1.12), *defined as a Cauchy integral at infinity, is convergent, and equal to* $f(x)$, *for all positive* x (*so that* $f(x)$ *is finite for all positive* x *and is its own cosine transform*).

Then $f(x)$ *is of the form* (A), *where the integral is summable* $(C, 1)$ *for* $|c-\tfrac{1}{2}| < \alpha$, *and where* $\psi(s)$ *is regular, and satisfies* (1.44), *for* $|\sigma-\tfrac{1}{2}| < \alpha$.

The integral (1.41) is uniformly convergent in any strip interior to $|\sigma-\tfrac{1}{2}| < \alpha$, so that it represents an analytic function $\phi(s)$ regular for $|\sigma-\tfrac{1}{2}| < \alpha$. We have next to prove that $\phi(s)$ satisfies (1.42).

In what follows we suppose that s has a fixed value in the strip $|\sigma-\tfrac{1}{2}| < \alpha$. Let
$$f_1(x) = \int_0^x f(u)\, du, \qquad f_2(x) = \int_0^x f_1(u)\, du.$$

Then, by partial integration,
$$\int_\delta^X x^{s-1} f(x)\, dx = X^{s-1} f_1(X) - (s-1) X^{s-2} f_2(X) -$$
$$- \delta^{s-1} f_1(\delta) + (s-1)\delta^{s-2} f_2(\delta) + (s-1)(s-2)\int_\delta^X x^{s-3} f_2(x)\, dx. \qquad (4.12)$$

---
† $s = \tfrac{1}{2}$ in the first case, $s = a$ and $s = 1-a$ in the second.
‡ Pollard (**24**).

Let
$$g(x) = \int_0^x u^{s-1} f(u)\, du.$$
Then
$$f_1(\delta) = \int_0^\delta u^{1-s} \cdot u^{s-1} f(u)\, du = \delta^{1-s} g(\delta) - (1-s) \int_0^\delta u^{-s} g(u)\, du$$
$$= o(\delta^{1-\sigma}) + \int_0^\delta o(u^{-\sigma})\, du = o(\delta^{1-\sigma})$$

when $\delta \to 0$; and therefore also $f_2(\delta) = o(\delta^{2-\sigma})$. Hence the integrated terms in (4.2) which involve δ tend to zero with δ. Similarly, using the function
$$h(x) = \int_x^\infty u^{s-1} f(u)\, du$$
we can show that the integrated terms which involve X tend to zero when $X \to \infty$. It follows that, if $\phi(s)$ is defined by (1.41),
$$\phi(s) = (s-1)(s-2) \int_0^\infty x^{s-3} f_2(x)\, dx. \tag{4.13}$$

4.2. Let
$$F(x) = \sqrt{\left(\frac{2}{\pi}\right)} \int_0^\infty f(y) \frac{1-\cos xy}{y^2}\, dy. \tag{4.21}$$

The integral is absolutely and uniformly convergent, in virtue of condition (i), so that $F(x)$ is continuous. Also, if $x > 0$ and h is sufficiently small,
$$\frac{F(x+2h) + F(x-2h) - 2F(x)}{4h^2} = \sqrt{\left(\frac{2}{\pi}\right)} \int_0^\infty f(y) \cos xy \left(\frac{\sin hy}{hy}\right)^2 dy;$$
and the limit of this, when $h \to 0$, is the value of the integral (1.12),† i.e. $f(x)$. It follows that
$$f_2(x) = F(x). \tag{4.22}$$
Substituting from (4.21) and (4.22) in (4.13), we obtain
$$\phi(s) = (s-1)(s-2) \sqrt{\left(\frac{2}{\pi}\right)} \lim \int_\delta^X x^{s-3}\, dx \int_0^\infty f(y) \frac{1-\cos xy}{y^2}\, dy$$
$$= (s-1)(s-2) \sqrt{\left(\frac{2}{\pi}\right)} \lim \int_0^\infty \frac{f(y)}{y^2}\, dy \int_\delta^X x^{s-3}(1-\cos xy)\, dx, \tag{4.23}$$

† The argument which proves this is familiar, and the details are given by Pollard (**24**), 457–8.

the inversion being justified by the uniform convergence of the integral with respect to y. We wish to take the limit under the integral sign, and to justify this we must show that

$$J_1 = \int_0^\infty \int_X^\infty \to 0, \qquad J_2 = \int_0^\infty \int_0^\delta \to 0 \qquad (4.24)$$

when $X \to \infty$ and $\delta \to 0$ respectively.

(i) We write

$$J_1 = \int_0^1 \int_X^\infty + \int_1^\infty \int_X^\infty = J_3 + J_4.$$

Then $J_4 \to 0$ because the integral with respect to x converges uniformly for $y \geqslant 1$. Also

$$J_3 = \int_0^1 \frac{f(y)}{y^2} dy \int_X^\infty x^{s-3}(1-\cos xy)\, dx = \int_0^1 \frac{f(y)}{y^s} dy \int_{Xy}^\infty u^{s-3}(1-\cos u)\, du.$$

If now we write

$$G(y) = \int_0^y z^{-s} f(z)\, dz$$

and integrate by parts, we obtain

$$J_3 = G(1) \int_X^\infty u^{s-3}(1-\cos u)\, du + \int_0^1 G(y)(Xy)^{s-3}(1-\cos Xy)\, X\, dy.$$

The first term plainly tends to zero. To prove that the second does so also, suppose that $A = A(s)$ is the maximum of $|G(y)|$, and that $|G(y)| < \epsilon$ for $y \leqslant \eta$. Then the modulus of the second term does not exceed

$$\epsilon \int_0^\eta (Xy)^{\sigma-3}(1-\cos Xy)\, X\, dy + A \int_\eta^1 (Xy)^{\sigma-3}(1-\cos Xy)\, X\, dy$$

$$= \epsilon \int_0^{\eta X} u^{\sigma-3}(1-\cos u)\, du + A \int_{\eta X}^\infty u^{\sigma-3}(1-\cos u)\, du,$$

which may be made as small as we please by choice of ϵ, η, and X in succession.

(ii) We write

$$J_2 = \int_0^1 \int_0^\delta + \int_1^\infty \int_0^\delta = J_5 + J_6,$$

and $J_5 \to 0$ because $1 - \cos xy \leqslant \tfrac{1}{2} x^2 y^2$. If we write

$$H(y) = \int_y^\infty z^{-s} f(z)\, dz,$$

we have

$$J_6 = \int_1^\infty \frac{f(y)}{y^2}\, dy \int_0^\delta x^{s-3}(1-\cos xy)\, dx = \int_1^\infty \frac{f(y)}{y^s}\, dy \int_0^{\delta y} u^{s-3}(1-\cos u)\, du$$

$$= H(1) \int_0^\delta u^{s-3}(1-\cos u)\, du + \int_1^\infty H(y)(\delta y)^{s-3}(1-\cos \delta y)\, \delta dy,$$

and from this point the proof proceeds substantially as before. We may therefore take the limit under the integral sign in (4.23).

We have thus proved that

$$\phi(s) = (s-1)(s-2) \sqrt{\left(\frac{2}{\pi}\right)} \int_0^\infty \frac{f(y)}{y^2}\, dy \int_0^\infty x^{s-3}(1-\cos xy)\, dx,$$

and, inserting the value of the inner integral, we see that $\phi(s)$ satisfies (1.42).

Finally (A) requires no further proof, since it follows from a form of Mellin's inversion theorem already proved by Hardy.†

4.3. We may generalize Theorem 13 as follows:

THEOREM 14. *The conclusions of Theorem* 13 *are still true if condition* (iii) *is satisfied except for a finite number of values of x, provided that $f(x) \to 0$ when $x \to \infty$.*

The only part of the proof which requires reconsideration is the proof that $F(x) = f_2(x)$. This is still true when $F(x)$ ceases to have a generalized second derivative for a finite number of values of x, provided that
$$F(x+2h) + F(x-2h) - 2F(x) = o(h)$$
for the exceptional values of x, that is to say here if

$$\int_0^\infty f(y) \cos xy \, \frac{\sin^2 hy}{y^2}\, dy = o(h);$$

and the integral is

$$O\left\{ h^2 \int_0^{1/h} |f(y)|\, dy + \int_{1/h}^\infty \frac{|f(y)|}{y^2}\, dy \right\} = o(h),$$

since $f(y) \to 0$.

4.4. We may test Theorems 13 and 14 on those of the examples

† Hardy (11), Theorem A.

of § 3 which escaped our earlier theorems and in which $\nu = -\tfrac{1}{2}$. These are (7), (8), and (10).

Taking (8) first, so that $f(x) = x^{\frac{1}{2}} J_{-\frac{1}{4}}(\tfrac{1}{2}x^2)$, we have $f(x) = O(1)$ for small x and $f(x) = O(x^{-\frac{1}{2}})$ for large x, so that (i) is satisfied. The integral (1.41) is convergent in this case if $0 < \sigma < \tfrac{5}{2}$, so that (ii) is satisfied with $\alpha = \tfrac{1}{2}$. Finally, the integral in (iii) is convergent for all positive x. This example therefore falls under Theorem 13.

In (7), $$f(x) = x(x^2-b^2)^{-\frac{3}{8}} J_{-\frac{3}{4}}\{b\surd(x^2-b^2)\}$$
for $x > b$, and behaves at infinity like a multiple of $x^{-\frac{1}{4}} \cos(bx + \tfrac{1}{8}\pi)$. Condition (i) is satisfied, and condition (ii) is satisfied if $0 < \sigma < \tfrac{5}{4}$, so that we may take $\alpha = \tfrac{1}{2}$. Condition (iii) is satisfied except for $x = b$. Thus Theorem 13 fails, but the example falls under Theorem 14.

Finally, the examples under (10) still evade our theorems, since the integral (1.41) is not convergent for any s. We shall return to these examples later, considering first some extensions in another direction.

5. Stieltjes integrals

5.1. Extensions of the Fourier formula to Stieltjes integrals have been given by various writers, notably by Hahn, N. Wiener, and Burkill. The extension with which we are concerned now is Burkill's.† Burkill proves that *if $f_1(x)$ is of bounded variation in any finite interval* $(0, X)$; *if*
$$f_1(0) = 0, \qquad f_1(x) = \tfrac{1}{2}\{f_1(x+0) + f_1(x-0)\} \quad (x > 0); \qquad (5.11)$$

if
$$\int^\infty \frac{|df_1(x)|}{x}$$

is finite; and if
$$g_1(x) = \sqrt{\left(\frac{2}{\pi}\right)} \int_0^\infty \frac{\sin xt}{t} \, df_1(t);$$

then
$$f_1(x) = \sqrt{\left(\frac{2}{\pi}\right)} \int_0^\infty \frac{\sin xt}{t} \, dg_1(t).$$

The theorem may be illustrated by the example
$$f_1(x) = 0, \tfrac{1}{2}\lambda, \lambda, \tfrac{1}{2}\lambda, 0 \quad (x < a, x = a, a < x < b, x = b, x > b);$$
here
$$g_1(x) = \lambda\left(\frac{\sin ax}{a} - \frac{\sin bx}{b}\right).$$

It suggests that we should consider the circumstances in which $f_1(x)$ and $g_1(x)$ are identical.

† Burkill (6).

SELF-RECIPROCAL FUNCTIONS

5.2. THEOREM 15. *Suppose that $f_1(x)$ has bounded variation in any finite interval $(0, X)$, and satisfies (5.11).*

Suppose also that (i) *the integral*

$$\int_1^\infty \frac{|df_1(x)|}{x^2} \qquad (5.21)$$

is finite; (ii) *the integral*

$$\phi(s) = \int_0^\infty x^{s-1}\, df_1(x) = \lim_{\delta \to 0,\, X \to \infty} \int_\delta^X x^{s-1}\, df_1(x) \qquad (5.22)$$

is convergent for $|\sigma - \tfrac{1}{2}| < \alpha \leqslant \tfrac{1}{2}$; (iii) *the integral*

$$\sqrt{\left(\frac{2}{\pi}\right)} \int_0^\infty \frac{\sin xy}{y}\, df_1(y) = \sqrt{\left(\frac{2}{\pi}\right)} \lim_{Y \to \infty} \int_0^Y \frac{\sin xy}{y}\, df_1(y)$$

is convergent, and equal to $f_1(x)$, for all positive x.

Then

$$f_1(x) = \frac{1}{2\pi i} \int_{c-i\infty}^{c+i\infty} \phi(s) \frac{x^{1-s}}{1-s}\, ds = \frac{1}{2\pi i} \lim_{T \to \infty} \int_{c-iT}^{c+iT} \phi(s) \frac{x^{1-s}}{1-s}\, ds,$$

where $|c - \tfrac{1}{2}| < \alpha$ and $\phi(s)$ has the properties stated in Theorem **13**.

It is plain that $\phi(s)$ is still regular in $|\sigma - \tfrac{1}{2}| < \alpha$. The partial integrations of § 4.1 are still valid, and (4.13) is still true.

Now let

$$G(x) = \sqrt{\left(\frac{2}{\pi}\right)} \int_0^\infty \frac{xy - \sin xy}{y^3}\, df_1(y). \qquad (5.23)$$

Then $G(x)$ is continuous, and

$$\frac{G(x+2h) + G(x-2h) - 2G(x)}{4h^2} = \sqrt{\left(\frac{2}{\pi}\right)} \int_0^\infty \frac{\sin xy}{y}\left(\frac{\sin hy}{hy}\right)^2 df_1(y). \qquad (5.24)$$

If we put

$$H(y) = \int_y^\infty \frac{\sin xu}{u}\, df_1(u),$$

and integrate by parts, we see that the integral in (5.24) is equal to

$$\sqrt{\left(\frac{2}{\pi}\right)} H(0) + \sqrt{\left(\frac{2}{\pi}\right)} \int_0^\infty H(y) \frac{d}{dy}\left(\frac{\sin hy}{hy}\right)^2 dy$$

$$= f_1(x) + \sqrt{\left(\frac{2}{\pi}\right)} \int_0^\infty H\left(\frac{u}{h}\right) \frac{d}{du}\left(\frac{\sin u}{u}\right)^2 du. \qquad (5.25)$$

Now $H(y)$ is bounded, and tends to 0 when $y \to \infty$, and
$$\int_0^\infty \left| \frac{d}{du}\left(\frac{\sin u}{u}\right)^2 \right| du$$
is convergent. It follows that the integral in (5.25) tends to zero with h, and so, from (5.24), that $G(x)$ has a generalized second derivative equal to $f_1(x)$ for all positive x. Since $G(x)$ has also a continuous first derivative equal to
$$F(x) = \sqrt{\left(\frac{2}{\pi}\right)} \int_0^\infty \frac{1-\cos xy}{y^2} df_1(y),$$
it follows that $F(x) = f_2(x)$, as in (4.22).

It now follows from (4.13) that
$$\frac{\phi(s)}{(s-1)(s-2)} = \sqrt{\left(\frac{2}{\pi}\right)} \int_0^\infty x^{s-3} dx \int_0^\infty \frac{1-\cos xy}{y^2} df_1(y),$$
and we can deduce, substantially as in § 4, that $\phi(s)$ satisfies (1.42).

5.3. We cannot in this case complete the proof by an appeal to a known theorem. We have
$$\int_{c-iT}^{c+iT} \phi(s) \frac{x^{1-s}}{1-s} ds = \int_{c-iT}^{c+iT} \frac{x^{1-s}}{1-s} ds \int_0^\infty u^{s-1} df_1(u) = \int_0^\infty df_1(u) \int_{c-iT}^{c+iT} \left(\frac{x}{u}\right)^{1-s} \frac{ds}{1-s}.$$
(5.31)

the inversion being justified by the uniform convergence of the integral with respect to u. We divide this into three parts corresponding to the intervals $(0, \delta)$, (δ, Δ), and (Δ, ∞) of u; we suppose $\delta < x < \Delta$. The inner integral on the right-hand side of (5.31) is bounded for $\delta \leqslant u \leqslant \Delta$ and all T,† and its limit when $T \to \infty$ is $2\pi i$ for $u < x$ and 0 for $u > x$. Hence
$$\lim_{T \to \infty} \int_\delta^\Delta df_1(u) \int_{c-iT}^{c+iT} \left(\frac{x}{u}\right)^{1-s} \frac{ds}{1-s} = 2\pi i \int_\delta^x df_1(u).$$

Our proof will be completed if we can show that the corresponding integrals over $(0, \delta)$ and (Δ, ∞) tend to zero when $\delta \to 0$ and $\Delta \to \infty$, uniformly in T.

† This is a familiar result, and may be proved directly by applying Cauchy's theorem to a contour formed by the path of integration and a semicircle described on it as diameter.

SELF-RECIPROCAL FUNCTIONS

Since $\frac{1}{2}-\alpha < c$, we can choose β so that $\frac{1}{2}-\alpha < \beta < c$. Let

$$\chi(u) = \int_0^u v^{\beta-1} \, df_1(v).$$

Then

$$\int_0^\delta df_1(u) \int_{c-iT}^{c+iT} \left(\frac{x}{u}\right)^{1-s} \frac{ds}{1-s} = \int_0^\delta u^{\beta-1} f_1(u) \, du \int_{c-iT}^{c+iT} x^{1-s} u^{s-\beta} \frac{ds}{1-s} \quad (5.32)$$

$$= \chi(\delta) \int_{c-iT}^{c+iT} x^{1-s} \delta^{s-\beta} \frac{ds}{1-s} - \int_0^\delta \chi(u) \, du \int_{c-iT}^{c+iT} x^{1-s} u^{s-\beta-1} \frac{s-\beta}{1-s} \, ds.$$

We can estimate the integrals with respect to s by applying Cauchy's theorem to the line $(c-iT, c+iT)$ and the perpendiculars through its ends in the direction of positive infinity. We find thus that

$$\int_{c-iT}^{c+iT} x^{1-s} \delta^{s-\beta} \frac{ds}{1-s} = 2\pi i \delta^{1-\beta} + O\left(\delta^{c-\beta} \int_c^\infty x^{-\sigma} \, d\sigma\right) = O(\delta^{c-\beta}),$$

uniformly in T; and also that

$$\int_{c-iT}^{c+iT} x^{1-s} u^{s-\beta-1} \frac{s-\beta}{1-s} \, ds = 2\pi i (1-\beta) u^{-\beta} + O\left(\int_c^\infty x^{1-\sigma} u^{\sigma-\beta-1} \, d\sigma\right)$$

$$= O(u^{-\beta}) + O\left\{\frac{u^{c-\beta-1}}{\log(x/u)}\right\} = O(u^{c-\beta-1})$$

when $u \to 0$, uniformly in T. Hence (5.32) is

$$\chi(\delta) O(\delta^{c-\beta}) + \int_0^\delta \chi(u) O(u^{c-\beta-1}) \, du = o(\delta^{c-\beta})$$

and tends to zero with δ, uniformly in T.

The integral involving Δ may be treated similarly; and this completes the proof of the theorem.

5.4. To obtain an example of Theorem **15**, suppose that $a = \sqrt{(2\pi)}$ and that

$$f_1(x) = \frac{x}{a} - \left[\frac{x}{a}\right], \qquad f_1(x) = \tfrac{1}{2},$$

according as x is not or is a multiple of a. It may be verified at once that conditions (i) and (ii) of Theorem **15** are satisfied. Also

$$\int_0^\infty \frac{\sin xy}{y} \, df_1(y) = \frac{1}{a} \int_0^\infty \frac{\sin xy}{y} \, dy - \sum_1^\infty \frac{\sin nax}{na} = \sqrt{(\tfrac{1}{2}\pi)} f_1(x),$$

so that $g_1(x) = f_1(x)$.

In this case
$$\phi(s) = \lim_{N\to\infty} \int_0^{(N+\frac{1}{2})a} x^{s-1}\, df_1(x) = \lim_{N\to\infty} \left\{ \frac{1}{a} \int_0^{(N+\frac{1}{2})a} x^{s-1}\, dx - \sum_1^N (na)^{s-1} \right\}$$

$$= \lim_{N\to\infty} a^{s-1} \left\{ \frac{(N+\frac{1}{2})^s}{s} - \sum_1^N n^{s-1} \right\} = -(2\pi)^{\frac{1}{2}s-\frac{1}{2}}\zeta(1-s),$$

and so satisfies (1.42).

5.5. This example may be expressed more picturesquely in a slightly different manner. Let us agree to write $f(x) = f'_1(x)$ in any interval in which $f_1(x)$ is an integral, and
$$f(x) = \lambda \infty$$
at a discontinuity of $f_1(x)$ where the saltus is λ. Further, let us agree that $\int \phi f\, dx$ means $\int \phi\, df_1$, and that
$$\int_0^\infty \phi(y) \cos xy\, dy = \mu \infty$$
means that
$$\int_0^{\pi Y} \phi(y) \cos xy\, dy \sim \mu Y$$
when $Y \to \infty$.

In our example
$$f(x) = -1\infty \ (x = na), \qquad f(x) = 1/a \ (x \neq na),$$
where $a = \sqrt{(2\pi)}$. An easy calculation shows that
$$\int_0^{\pi Y} f(y) \cos xy\, dy = \frac{\sin \pi Y x}{ax} + \frac{1}{2} - \frac{\sin(\nu+\frac{1}{2})ax}{2\sin\frac{1}{2}ax},$$
where ν is the greatest integer such that $\nu a \leqslant \pi Y$, and the last two terms are to be replaced by $-\nu$ if x is a multiple of a. It follows that
$$\sqrt{\left(\frac{2}{\pi}\right)} \int_0^\infty f(y) \cos xy\, dy = \frac{1}{a} \qquad (C,1)$$
if $x \neq na$, and
$$\sqrt{\left(\frac{2}{\pi}\right)} \int_0^\infty f(y) \cos xy\, dy = -\sqrt{\left(\frac{2}{\pi}\right)}\sqrt{\left(\frac{\pi}{2}\right)} \infty = -1\infty$$
if $x = na$. In other words, $f(x)$ is its own cosine transform.

All this is in the main merely a translation into other language of what Theorem **15** asserts in this particular case. There is, however, nothing in Theorem **15** to justify any general assertion concerning the summability $(C, 1)$ of

$$\int_0^\infty \cos xy \, df_1(y).$$

5.6. The examples in § 3 (10), which do not fall under any of the theorems proved so far, may also be fitted into our scheme by appropriate conventions. Take, for example,

$$f(x) = x^{-\frac{1}{2}}, \qquad \phi(\tfrac{1}{2}+it) = \int_0^\infty x^{-1-it} \, dx.$$

This integral is never convergent or summable by a mean of Cesàro type, but it is summable by 'logarithmic means'†, except when $t = 0$, and has the value 0. It amounts to the same thing to say that, if we effect the formal transformation $x = e^u$, we obtain

$$\phi(\tfrac{1}{2}+it) = \int_{-\infty}^\infty e^{-itu} \, du,$$

and that the last integral is summable $(C, 1)$ to 0 except when $t = 0$, when it is infinite.

The integral is the formal derivative of the integral

$$\Phi(t) = i \int_{-\infty}^\infty \frac{e^{-itu}}{u} \, du = \pi \operatorname{sgn} t$$

and we may interpret the formula (1.45) as meaning

$$f(x) = \frac{1}{2\pi} \int_{-\infty}^\infty x^{-\frac{1}{2}-it} \, d\Phi(t) = \frac{1}{\sqrt{x}}.$$

The function $\phi(\tfrac{1}{2}+it)$, which is zero except for $t = 0$, and then infinite, is in a sense a solution of (1.42).

Similar interpretations may be found for the other solutions mentioned in § 3 (10). These functions are, however, interpreted much more naturally as examples of our second general solution, and we shall not attempt to fit our formalities into exact theorems.

6. The second solution

6.1. We shall say that $f(x)$ belongs to $A^*(\omega, a)$, where $0 < \omega \leqslant \tfrac{1}{4}\pi$, $0 < a < \tfrac{1}{2}$, if (i) it is an analytic function of $x = re^{i\theta}$ regular in the angle A^* defined by $r > 0$, $|\theta| < \omega$, and (ii) it is $O(|x|^{-a-\frac{1}{2}-\delta})$ for small

† For a formal definition see the end of § 3.

x, and $O(|x|^{a-\frac{1}{2}+\delta})$ for large x, for every positive δ and uniformly in any angle $|\theta| \leqslant \omega - \eta < \omega$.

It is to be observed that a function of $A^*(\omega, a)$, unlike the functions of § 2.4, does not usually belong to L^2, so that we cannot base our analysis on the theory of transforms.

6.2. THEOREM 16. *A necessary and sufficient condition that a function $f(x)$ of $A^*(\omega, a)$ should be its own cosine transform is that it should be of the form* (B), *where c is any positive number, the integral is the limit of an integral over $(c-iT, c+iT)$, and $\mu(s)$ has the properties*

(i) $\mu(s) = \mu(\rho e^{i\phi})$ *is an analytic function of s, regular in the angle $B(\omega, a)$ defined by $\rho > 0$, $|\phi| < \frac{1}{2}\pi + 2\omega$;*

(ii) $\mu(s)$ *is $O(|s|^{-\frac{1}{2}a-\delta})$ for small s, and $O(|s|^{\frac{1}{2}a+\delta})$ for large s, for every positive δ and uniformly in any angle $|\phi| \leqslant \frac{1}{2}\pi + 2\omega - \zeta < \frac{1}{2}\pi + 2\omega$;*

(iii) $\mu(s)$ *satisfies the equation* (1.54) *in $B(\omega, a)$.*

(i) *The condition is necessary.* The integrals

$$\int_0^1 |f|\, dx, \quad \int_1^\infty \frac{|f|}{x}\, dx$$

are finite, and (*ex hypothesi*) the integral (1.12) converges for every positive x to the integrable value $f(x)$. The conditions of Pollard's theorem quoted in § 4.2 are therefore satisfied, and (in the notation of § 4.2) $F(x) = f_2(x)$. We may differentiate this equation with respect to x, since the resulting integral is uniformly convergent. Hence

$$\sqrt{\left(\frac{2}{\pi}\right)} \int_0^\infty f(y) \frac{\sin xy}{y}\, dy = f_1(x). \tag{6.21}$$

We define $\lambda(s)$ and $\mu(s)$ as in § 1.5. It is plain first that these functions are regular for $\sigma > 0$. We may move the path of integration to the line $\theta = \theta_1$ of A^*†, provided that $\cos(\phi + 2\theta) > 0$ for all θ between 0 and θ_1. Then

$$\lambda(s) = e^{i\theta_1} \int_0^\infty \exp(-\tfrac{1}{2}sr^2 e^{2i\theta_1}) f(re^{i\theta_1})\, dr,$$

and this formula shows that $\lambda(s)$ is regular in any angle in which $\cos(\phi + 2\theta_1) > 0$. Varying θ_1 in A^*, we see that $\lambda(s)$ and $\mu(s)$ are regular in B.

† We use A^* both for the angle and for the class of functions associated with it.

Suppose now that s is in the angle of condition (ii). Then $\lambda(s)$ possesses a majorant of the type

$$\int_0^\infty e^{-C\rho r^2}|f(re^{i\theta})|\,dr,$$

where $C = C(\theta_1, \zeta)$. For small ρ this is

$$O\left(\int_0^\infty e^{-C\rho r^2} r^{a-\frac{1}{2}+\delta}\,dr\right) = O(\rho^{-\frac{1}{2}a-\frac{1}{4}-\frac{1}{2}\delta}),$$

and for large ρ it is

$$O\left(\int_0^\infty e^{-C\rho r^2} r^{-a-\frac{1}{2}-\delta}\,dr\right) = O(\rho^{\frac{1}{2}a-\frac{1}{4}+\frac{1}{2}\delta}).\dagger$$

Hence $\mu(s)$ satisfies condition (ii). It remains to verify (1.54) and (B).

We suppose now that $\sigma = c > 0$. Integrating by parts, and using (6.21), we find

$$\lambda(s) = s\int_0^\infty e^{-\frac{1}{2}sx^2} xf_1(x)\,dx = \sqrt{\left(\frac{2}{\pi}\right)}s\int_0^\infty e^{-\frac{1}{2}sx^2}x\,dx\int_1^\infty f(y)\frac{\sin xy}{y}\,dy.$$

The double integral is absolutely convergent, by comparison with

$$\int_0^\infty e^{-\frac{1}{2}cx^2}x^2\,dx\int_0^1 |f|\,dy + \int_0^\infty e^{-\frac{1}{2}cx^2}x\,dx\int_1^\infty \frac{|f|}{y}\,dy,$$

and we may invert the integrations. We thus obtain

$$\lambda(s) = \sqrt{\left(\frac{2}{\pi}\right)}s\int_0^\infty \frac{f(y)}{y}\,dy\int_0^\infty e^{-\frac{1}{2}sx^2}x\sin xy\,dx$$

$$= s^{-\frac{1}{2}}\int_0^\infty f(y)e^{-y^2/2s}\,dy = s^{-\frac{1}{2}}\lambda\left(\frac{1}{s}\right),$$

which is equivalent to (1.54).

Finally, putting $x = \sqrt{(2y)}$ and $f(x) = xg(y)$, we obtain

$$\lambda(s) = \int_0^\infty e^{-sy}\frac{f(x)}{x}\,dy = \int_0^\infty e^{-sy}g(y)\,dy,$$

$$\frac{1}{2\pi i}\int_{c-iT}^{c+iT} e^{ys}\lambda(s)\,ds = \frac{1}{2\pi i}\int_{c-iT}^{c+iT} e^{ys}\,ds\int_0^\infty e^{-su}g(u)\,du.$$

† Large r are important for small ρ, and conversely.

The inner integral is uniformly convergent, and so, inverting the order of integration, we obtain

$$\frac{1}{2\pi i}\int_{c-iT}^{c+iT} e^{ys}\lambda(s)\,ds = \frac{1}{\pi}\int_0^\infty e^{c(y-u)}\frac{\sin T(y-u)}{y-u}g(u)\,du.$$

Since $g(u)=O(u^{-\frac{1}{2}a-\frac{3}{4}-\frac{1}{2}\delta})$ for small u, the integral

$$\int_0^\infty e^{-cu}|g(u)|\,du$$

is finite. Also $g(u)$ is analytic for $u=y$. It therefore follows from the classical theory of Dirichlet's integral that

$$\frac{1}{2\pi i}\lim_{T\to\infty}\int_{c-iT}^{c+iT} e^{ys}\lambda(s)\,ds = g(y);$$

and this is equivalent to (B).

(ii) *The condition is sufficient.* Suppose $f(x)$ defined by (B). We begin by deforming the integral into one taken along the radii $|\phi|=\frac{1}{2}\pi+\lambda$, where $2|\theta|<\lambda<2\omega$. The integral then becomes absolutely convergent, and has a majorant of the type

$$O\left(r\int_0^1 e^{-C\rho r^2}\rho^{-\frac{1}{2}a-\frac{1}{4}-\delta}\,d\rho\right) + O\left(r\int_1^\infty e^{-C\rho r^2}\rho^{\frac{1}{2}a-\frac{1}{4}+\delta}\,d\rho\right).$$

It follows that $f(x)$ is regular in $A^*(\omega, a)$, and is

$$O\{r\cdot(r^2)^{-\frac{1}{2}a-\frac{3}{4}-\delta}\} = O(r^{-a-\frac{1}{2}-2\delta})$$

for small r, and $O(r^{a-\frac{1}{2}-2\delta})$ for large r. Hence $f(x)$ belongs to $A^*(\omega, a)$.

6.3. It is to be observed that the results just proved concerning the order of $f(x)$, and those assumed concerning that of $\mu(s)$, carry with them as corollaries corresponding results concerning the derivatives of these functions. In fact

$$f^{(n)}(x) = O(|x|^{-a-\frac{1}{2}-n-2\delta}), \qquad f^{(n)}(x) = O(|x|^{a-\frac{1}{2}-n+2\delta})$$

for small and large x respectively, uniformly in any angle interior to A^* (and in particular for real x); and similarly

$$\mu^{(n)}(s) = O(|s|^{-\frac{1}{2}a-n-\delta}), \qquad \mu^{(n)}(s) = O(|s|^{\frac{1}{2}a-n+\delta})$$

for small and large s, uniformly in any angle $|\phi|\leqslant\frac{1}{2}\pi+2\omega-\zeta$ (and in particular on the line $\sigma=c$).

Suppose, for example, that x is positive and small, and let C be a circle in the u-plane with its centre at $u = x$ and touching the lines $|\operatorname{am} u| = \omega - \delta$. Then

$$f^{(n)}(x) = \frac{n!}{2\pi i} \int_C \frac{f(u)}{(u-x)^{n+1}} du = O\left(\frac{x^{-a-\frac{1}{2}-2\delta} \cdot x}{x^{n+1}}\right) = O(x^{-a-\frac{1}{2}-n-2\delta}).$$

Similarly in the other cases.

Taking in particular $n = 1$, and integrating by parts, we see that the integral (1.12) is uniformly convergent in any finite interval of positive values of x; and similarly the integral (B) is uniformly convergent in any similar interval.

6.4. We now define $\lambda(s)$ by (1.51), and prove that it satisfies (1.53). Suppose that $\Re(u) = v > 0$, take the integral (B) along the contour Γ used at the end of § 6.2, and substitute into (1.51). We thus obtain

$$\lambda(u) = \frac{1}{2\pi i} \int_0^\infty x e^{-\frac{1}{2}ux^2} dx \int_\Gamma e^{\frac{1}{2}x^2 s} s^{-\frac{1}{2}} \chi(s) \, ds.$$

Here the integral is absolutely convergent. This is obvious so far as concerns the finite part of Γ, while on the distant parts there is a convergent majorant of the type

$$\int_0^\infty x e^{-\frac{1}{2}vx^2} dx \int_P^\infty e^{-\frac{1}{2}Cx^2\rho} \rho^{\frac{1}{2}a-\frac{1}{2}+\delta} d\rho = \int_P^\infty \frac{\rho^{\frac{1}{2}a-\frac{1}{2}+\delta} d\rho}{v + C\rho}.$$

We may therefore invert the integrations, and so obtain

$$\lambda(u) = \frac{1}{2\pi i} \int_\Gamma \frac{s^{-\frac{1}{2}} \mu(s)}{u - s} \, ds = u^{-\frac{1}{2}} \mu(u).$$

This proves (1.53). It follows that $\lambda(s)$ satisfies (1.52).

6.5. We start now from the formula

$$\frac{\sin xy}{y} = \frac{\sqrt{\pi}}{4\pi i} \int_{c-i\infty}^{c+i\infty} e^{x^2 s} s^{-\frac{3}{2}} e^{-y^2/4s} \, ds, \qquad (6.51)$$

valid for all positive x and y.† Hence

$$\frac{1 - \cos xy}{y^2} = \frac{\sqrt{\pi}}{4\pi i} \int_{c-i\infty}^{c+i\infty} e(x, s) s^{-\frac{3}{2}} e^{-y^2/4s} \, ds, \qquad (6.52)$$

† The formula is a special case of one of the classical integral representations of the Bessel functions: take $\nu = \frac{1}{2}$ in Watson (**30**), 177, formula (8).

where
$$e(x,s) = \int_0^x e^{w^2 s} \, dw. \qquad (6.53)$$

It is plain that $e(x, c+it)$ is bounded in any finite t interval; also
$$e(x, c+it) = e^{x^2 c} \int_{x'}^x e^{w^2 it} \, dt = O(|t|^{-\frac{1}{2}}) \qquad (6.54)$$
for large t.

Multiplying (6.52) by $f(y)$, and integrating,
$$\int_0^\infty \frac{1-\cos xy}{y^2} f(y) \, dy = \frac{\sqrt{\pi}}{4\pi i} \int_0^\infty f(y) \, dy \int_{c-i\infty}^{c+i\infty} e(x,s) s^{-\frac{3}{2}} e^{-y^2/4s} \, ds. \qquad (6.55)$$

The integral is absolutely convergent. To prove this, it is plainly sufficient to show that
$$\int_{-\infty}^\infty |e(x, c+it)|(1+|t|)^{-\frac{3}{2}} e^{-cy^2/4(c^2+t^2)} \, dt$$
is $O(1)$ for small y and $O(y^{-1})$ for large y. The first result follows at once from the boundedness of $e(x, c+it)$ and the convergence of $\int (1+|t|)^{-\frac{3}{2}} \, dt$. As regards the second, the integral is less than a constant multiple of
$$\int_0^\infty \frac{e^{-cy^2/4(c^2+t^2)}}{(1+t)^2} \, dt < \int_0^c e^{-y^2/8c} \, dt + \int_c^\infty \frac{e^{-cy^2/8t^2}}{t^2} \, dt$$
$$< ce^{-y^2/8c} + \frac{1}{y} \int_0^\infty e^{-\frac{1}{8}cu^2} \, du = O\left(\frac{1}{y}\right).$$

Inverting the integrations in (6.55), we obtain
$$\int_0^\infty \frac{1-\cos xy}{y^2} f(y) \, dy = \frac{\sqrt{\pi}}{4\pi i} \int_{c-i\infty}^{c+i\infty} e(x,s) s^{-\frac{3}{2}} \, ds \int_0^\infty f(y) e^{-y^2/4s} \, dy$$
$$= \frac{\sqrt{(2\pi)}}{4\pi i} \int_{c-i\infty}^{c+i\infty} \frac{e(x,s)}{s} \lambda(2s) \, ds,$$
by (1.52). If we differentiate twice with respect to x, we find
$$\int_0^\infty \cos xy \, f(y) \, dy = \frac{\sqrt{(\frac{1}{2}\pi)}x}{2\pi i} \int_{c-i\infty}^{c+i\infty} e^{\frac{1}{2}x^2 s} \lambda(s) \, ds = \sqrt{(\frac{1}{2}\pi)} f(x).$$

SELF-RECIPROCAL FUNCTIONS

The legitimacy of the differentiations results from the uniform convergence of the final integrals, proved in § 6.3; and this completes the proof of the theorem.

6.6. The corresponding formula for general ν is

$$f(x) = \frac{x^{\frac{1}{2}-\nu}}{2\pi i} \int_{c-i\infty}^{c+i\infty} e^{\frac{1}{2}sx^2} s^{-\frac{1}{2}(\nu+1)} \mu(s) \, ds.$$

As examples, besides the one mentioned in § 1.4, we have:

$$\mu(s) = 2^{-\frac{1}{2}} \Gamma(\tfrac{1}{2}a)\Gamma(\tfrac{1}{2}-\tfrac{1}{2}a)(s^{\frac{1}{2}a-\frac{1}{4}} + s^{\frac{1}{4}-\frac{1}{2}a}),$$

$$f(x) = 2^{\frac{1}{2}a}\Gamma(\tfrac{1}{2}a)x^{-a} + 2^{\frac{1}{2}(1-a)}\Gamma\{\tfrac{1}{2}(1-a)\}x^{-(1-a)},$$

$$\mu(s) = \sqrt{(\tfrac{1}{2}\pi)}\frac{1}{\sqrt{(s^{\frac{1}{2}}+s^{-\frac{1}{2}})}}, \qquad f(x) = e^{-\frac{1}{2}x^2},$$

$$\mu(s) = \frac{\Gamma(\tfrac{1}{4})}{\sqrt{(2\pi)}}\left(s+\frac{1}{s}\right)^{-\frac{1}{4}}, \qquad f(x) = x^{\frac{1}{2}}J_{-\frac{1}{4}}(\tfrac{1}{2}x^2)$$

(all for $\nu = -\tfrac{1}{2}$).

As an example of the general case, take

$$\mu(s) = \exp\left\{-\tfrac{1}{2}b^2\left(s+\frac{1}{s}\right)\right\}.$$

It will be found that we obtain the example (7) of § 3.

6.7. There is an 'L^2' theorem connected with the second solution, but it involves the convergence of

$$\int_0^\infty e^{-cx^2} f^2(x) \frac{dx}{x} \qquad (c > 0)$$

and is not very useful. The theorem corresponding to Theorem 13 is

THEOREM 17. *If the integrals*

$$\int_0^1 |f|\frac{dx}{x}, \qquad \int_1^\infty |f|\frac{dx}{x^2}$$

are finite, and (1.12) *holds for all positive* x, *the integral being defined as a Cauchy integral at infinity, then* $f(x)$ *is of the form* (B), $\mu(s)$ *being regular, and satisfying* (1.54), *for* $\sigma > 0$.

The proof is very similar to that of Theorem 13, but rather simpler, and there is no condition corresponding to (ii) of Theorem 13.

REFERENCES

1. W. N. Bailey, 'On a function which is its own reciprocal in the Fourier-Bessel integral transform': *Journal London Math. Soc.* 5 (1930), 92–5.
2. ——, 'Some classes of functions which are their own reciprocals in the Fourier-Bessel integral transform': ibid. 5 (1930), 258–65.
3. H. Bateman, 'Report on the history and present state of the theory of integral equations': *Reports of the British Association*, 1910, 345–424.
4. ——, 'The solution of linear differential equations by means of definite integrals': *Trans. Cambridge Phil. Soc.* 21 (1912), 171–96.
5. H. Burkhardt, 'Trigonometrische Reihen und Integrale': *Encykl. der Math. Wiss.* II A. 12.
6. J. C. Burkill, 'The expression in Stieltjes integrals of the inversion formulae of Fourier and Hankel': *Proc. London Math. Soc.* (2), 25 (1926), 513–24.
7. ——, 'On Mellin's inversion formula': *Proc. Camb. Phil. Soc.* 23 (1927), 356–60.
8. A. L. Cauchy, 'Sur les fonctions réciproques': *Œuvres*, Sér. 1, 1, 300–3.
9. A. S. Eddington, *Stellar Motions and the Structure of the Universe*, Macmillan, 1914.
10. G. H. Hardy, 'Note on the function $\int_x^\infty e^{\frac{1}{2}(x^2-t^2)} dt$': *Quarterly Journal*, 35 (1904), 193–207.
11. ——, 'On Mellin's inversion formula': *Messenger of Math.* 47 (1918), 178–84.
12. ——, 'Further notes on Mellin's inversion formula': ibid. 50 (1921), 165–71.
13. ——, 'Some further applications of Mellin's inversion formula': ibid. 56 (1926), 186–92.
14. ——, 'A discontinuous integral': ibid. 57 (1927), 113–20.
15. G. H. Hardy and J. E. Littlewood, 'Contributions to the theory of the Riemann zeta-function and the theory of the distribution of primes': *Acta Math.* 41 (1918), 119–96.
16. E. W. Hobson, *The Theory of Functions of a Real Variable*, vol. 2, ed. 2, Cambridge, 1926.
17. Hj. Mellin, 'Ueber die fundamentalen Wichtigkeit des Satzes von Cauchy für die Theorien der Gamma- und der hypergeometrischen Funktionen': *Acta Soc. Fennicae*, 21 (1896), No. 1, 1–115.
18. A. Oppenheim, 'Some identities in the theory of numbers': *Proc. London Math. Soc.* (2), 26 (1927), 295–350.
19. E. G. Phillips, 'Note on a problem of Ramanujan': *Journal London Math. Soc.* 4 (1925), 310–13.
20. S. Pincherle, 'Studi sopra alcune operazioni funzionali': *Mem. di Bologna* (4), 7 (1886), 393–442.
21. M. Plancherel, 'Contribution à l'étude de la représentation d'une fonction arbitraire par des intégrales définies': *Rend. di Palermo*, 30 (1910), 289–335.
22. ——, 'Sur la convergence et sur la sommation par les moyennes de Cesàro de $\lim_{z=\infty} \int_a^z f(x)\cos xy\, dx$': *Math. Annalen*, 76 (1915), 315–26.
23. ——, 'Sur les formules d'inversion de Fourier et de Hankel': *Proc. London Math. Soc.* (2), 24 (1925), 62–70.

24. S. Pollard, 'Identification of the coefficients in a trigonometrical integral': *Proc. London Math. Soc.* (2), 25 (1926), 451–68.
25. E. C. Titchmarsh, 'Hankel transforms': *Proc. Camb. Phil. Soc.* 21 (1922), 463–73.
26. ——, 'A contribution to the theory of Fourier transforms': *Proc. London Math. Soc.* (2), 23 (1924), 279–89.
27. ——, 'A note on Hankel transforms': *Journal London Math. Soc.* 1 (1926), 195–6.
28. A. Walfisz, 'Über die sommatorischen Funktionen einiger Dirichletscher Reihen': *Dissertation*, Göttingen, 1922.
29. ——, 'Über zwei Gitterpunktprobleme': *Math. Annalen*, 95 (1926), 69–83.
30. G. N. Watson, *Theory of Bessel Functions*, Cambridge, 1922.

CORRECTIONS

p. 211, *line 8 up*. $b^{\frac{1}{2}s+\frac{1}{2}\nu}$ should be $b^{s-\frac{1}{2}\nu}$.

——, *line 5 up*. $b^{\nu-\frac{1}{2}}$ should be $b^{\frac{1}{2}\nu-\frac{1}{2}}$.

p. 213, *line 2*. $\zeta(s)L(s)$ should be $4\zeta(s)L(s)$.

p. 221, *line 4*. $f_1(u)du$ should be $df_1(u)$.

p. 225, *line 3*. θ should be θ_1.

——, *line 11*. The last integral should be $\int_0^\infty$ (not $\int_1^\infty$).

p. 227, *line 13*. χ should be μ.

p. 228, *line 3*. dt should be dw.

A NOTE ON PARSEVAL'S THEOREM FOR FOURIER TRANSFORMS

G. H. HARDY *and* E. C. TITCHMARSH[†].

1. Parseval's theorem for Fourier cosine transforms asserts that, if f and F are connected by the Fourier reciprocity formulae

(1.11) $$F(x) = \sqrt{\left(\frac{2}{\pi}\right)} \int_0^\infty f(t) \cos xt \, dt,$$

(1.12) $$f(x) = \sqrt{\left(\frac{2}{\pi}\right)} \int_0^\infty F(t) \cos xt \, dt,$$

and g and G are similarly related, then

(1.2) $$J = \int_0^\infty F(x) G(x) \, dx = \int_0^\infty f(x) g(x) \, dx = j. \qquad \text{(P)}$$

The first satisfactory discussion of the formula was that of Plancherel[‡].

Plancherel's discussion depends on his theory of transforms, in which the functions belong to the Lebesgue class L^2; the formula is then always true. The theory and the range of validity of the formula were afterwards extended by Titchmarsh. There are, however, a number of cases which seem to have been overlooked, and it may be useful that we should give here a short but rather more systematic discussion. This will be abbreviated if we begin with a number of formal definitions.

We suppose generally that $p > 1$, and write q for $p/(p-1)$. We say that $f(x)$ belongs to L^p, or is L^p, in a certain interval, $(0, \infty)$ unless the contrary is stated, if f is measurable and $|f|^p$ integrable in that interval. We write L for L^1. It is sometimes convenient to use the interval $(-\infty, \infty)$; in that case it is to be understood that the functions considered are even.

The integrals (1.1) may exist in various senses. First, as Lebesgue, or L, integrals. Secondly, as Cauchy, or **C**, integrals, that is to say as limits of L integrals, of one or other of the types

$$\lim_{T \to \infty} \int_0^T f(t) \cos xt \, dt, \qquad \lim_{\tau \to 0, \, T \to \infty} \int_\tau^T f(t) \cos xt \, dt;$$

[†] Received 3 November, 1930; read 13 November, 1930.

[‡] Plancherel, **3**. It is not easy to say to whom the formula was originally due. It has been much used by physicists, and Bateman (**1**, 6) attributes it to Rayleigh (1889); but it is difficult to believe that it does not stand somewhere in Fourier, Cauchy, or Poisson.

we describe such integrals as **C** integrals at ∞, or at 0 and ∞. Thirdly, as Cesàro, or C integrals, that is to say as limits of the type

$$\lim_{T\to\infty} \frac{1}{T}\int_0^T (T-t)f(t)\cos xt\, dt.$$

Finally, as "mean p-th power", or M_p, integrals, the integral (1.11), for example, being said to exist as an M_p integral, and have the value $F(x)$, if

$$\lim_{T\to\infty}\int_0^\infty \left| F(x)-\sqrt{\left(\frac{2}{\pi}\right)}\int_0^T f(t)\cos xt\, dt\right|^p dx = 0.$$

If (1.11) is true, in any of these senses, we shall say that F is the transform of f, and write $F = f^*$.

The integrals J and j may also exist in any of the first three senses.

2. **Theorem 1.** *If f and g are L^2, then F and G exist as M_2 integrals for almost all x, and* (P) *is true, both integrals being L integrals.*

Theorem 2. *If f and G are L^2, then F and g exist as M_2 integrals for almost all x, and* (P) *is true, both integrals being L integrals.*

Both of these theorems are due to Plancherel: they are equivalent, but logically distinct.

Theorem 3. *If f and G are L^p, where $1 < p \leqslant 2$, then F and g exist as M_q integrals for almost all x, and* (P) *is true, both integrals being L integrals.*

This is Titchmarsh's generalisation of **2**†; there is no corresponding extension of **1**.

Theorem 4. *If f and G are L, then* (P) *is true, both integrals being L integrals.*

Here it is obvious that F and g exist as L integrals for all x, and are continuous. The theorem is trivial, since we may replace F in (1.2) by its expression as an integral in (1.11), and invert the repeated integral, which is absolutely convergent.

It is to be observed in all these theorems that the functions on which the hypotheses bear (*e.g.* f and G in **4**) are the *given* functions, and that the others are defined as their transforms (*e.g.* $F = f^*$, $g = G^*$). There is nothing in **4**, for example, to justify the assertion that if f is L, g is given, $G = g^*$ (in one sense or another) and G is L, then (P) is necessarily true;

---

† Titchmarsh, **6**.

for it is not obvious that $y = G^*$. We require further conditions, such as appear in the next theorem.

THEOREM 5. *If f is L, g is L in $(0, 1)$, and g/x^2 is L in $(1, \infty)$, and if G exists, for all x, as a Cauchy integral at ∞, and is L, then* (P) *is true, the integrals being L integrals. If $g \to 0$, the condition on G may be broken for a finite number of values of x.*

In particular (P) *is true whenever f, g, and G are all L.*

For in these circumstances, by theorems of Pollard†, $g = G^*$.

3. Before going further, we write down the formulae on which our argument depends. There are three sets, viz. (i)

$$(3.1) \quad \int_0^X F(x) G(x) dx = \frac{1}{\sqrt{(2\pi)}} \int_0^X G(x) dx \int_{-\infty}^{\infty} f(t) \cos xt \, dt$$

$$= \frac{1}{\sqrt{(2\pi)}} \int_{-\infty}^{\infty} f(t) dt \int_0^X G(x) \cos xt \, dx$$

$$= \frac{1}{2\pi} \int_{-\infty}^{\infty} f(t) dt \int_0^X \cos xt \, dx \int_{-\infty}^{\infty} g(u) \cos xu \, du$$

$$= \frac{1}{2\pi} \int_{-\infty}^{\infty} f(t) dt \int_{-\infty}^{\infty} g(u) du \int_0^X \cos xt \cos xu \, dx$$

$$= \frac{1}{2\pi} \int_{-\infty}^{\infty} f(t) dt \int_{-\infty}^{\infty} g(u) \frac{\sin X(u-t)}{u-t} du$$

$$= \frac{1}{2\pi} \int_{-\infty}^{\infty} f(t) \Phi(t, X) dt \to \tfrac{1}{2} \int_{-\infty}^{\infty} f(t) g(t) dt;$$

(ii) a corresponding set of formulae

$$(3.2) \quad \int_\xi^X F(x) G(x) dx$$

$$= \ldots = \frac{1}{2\pi} \int_{-\infty}^{\infty} f(t) \{\Phi(t, X) - \Phi(t, \xi)\} dt \to \tfrac{1}{2} \int_{-\infty}^{\infty} f(t) g(t) dt;$$

and (iii) a set

$$(3.3) \quad \frac{1}{X} \int_0^X (X-x) F(x) G(x) dx$$

$$= \ldots = \frac{1}{\pi} \int_{-\infty}^{\infty} f(t) dt \int_{-\infty}^{\infty} g(u) \frac{\sin^2 \tfrac{1}{2} X(u-t)}{X(u-t)^2} du$$

$$= \frac{1}{\pi} \int_{-\infty}^{\infty} f(t) \Psi(t, X) dt \to \tfrac{1}{2} \int_{-\infty}^{\infty} f(t) g(t) dt.$$

---

† Pollard, 4. The three clauses of the theorem depend (i) on his Theorem II, (ii) on this and his Theorem V, and (iii) on his (comparatively trivial) Theorem I.

In each case we have to justify two inversions and a final passage to the limit.

4. **Theorem 6.** *If f is L, g is of bounded variation in $(0, \infty)$, and $g(\infty) = 0$, then* (P) *is true, J being a* **C** *integral at 0 and ∞ and j an L integral.*

We use (3.2). The first inversion is justified by the uniform convergence of (1.11) for all x, the second by the uniform convergence of the corresponding integral for G for $\xi \leqslant x \leqslant X$. Finally, $\Phi(t, X)$ and $\Phi(t, \xi)$ are bounded, and tend to $g(t)$ and 0, when $X \to \infty$ and $\xi \to 0$, for almost all t, so that the final passage to the limit is legitimate†.

Theorem 7. *If f is L, and g is L^2 and bounded, then* (P) *is true, J being a C integral and j an L integral.*

In this case g is L^p for all $p \geqslant 2$. We use formulae (3.3), and the first inversion is justified as in the proof of **6**. For the second we appeal to the theorem that

$$\int \psi(x)\, dx \int \phi(x, t)\, dt = \int dt \int \psi(x)\, \phi(x, t)\, dx$$

when $\int \phi(x, t)\, dt$ exists as an M_q integral and $\psi(x)$ is L^p‡. Here q is 2 and ψ continuous. Finally, since g is bounded, $\Psi(t, X)$ is bounded, and the result follows as before.

In particular the result holds if f is L, and g is L and bounded, and then both inversions are trivial, the integrals for both F and G being uniformly convergent. This remark leads us to state the next theorem, which is of no particular interest, but which we happen to require elsewhere.

Theorem 8. *The conclusion of Theorem* **7** *holds whenever f is L, g is bounded, and the integral for G is uniformly convergent§ in any interval $0 < \xi \leqslant x \leqslant X$.*

---

† By Lebesgue's fundamental criterion. More precisely, the limits of Φ are
$$\tfrac{1}{2}\{g(t+0) + g(t-0)\}$$
and 0 for all t. The proofs of the assertions which precede depend on the second mean value theorem, and are of a familiar type.

‡ The proof follows almost immediately from the definitions; compare Hobson (2, 251).

§ Here, and in **10** below, the hypothesis of uniform convergence may be replaced by wider hypotheses; but we do not require these extensions.

5. Similar arguments may be applied to the case in which f is L^p, where $1 < p \leqslant 2$, and g is L^2 and L^q (and so L^r for $2 \leqslant r \leqslant q$); and lead to the same conclusion. In this case, however, there is a stronger theorem.

THEOREM 9. *If f is L^p, where $1 < p \leqslant 2$, and g is L^2 and L^q, then* (P) *is true, J being a* **C** *integral at ∞ and j an L integral.*

We use formulae (3.1). Both inversions are justified by the theorem appealed to in § 4. In the second, the argument is unchanged. In the first, (1.11) is an M_q integral, and G is L^2, so *a fortiori* L^p in the finite interval of x. Finally,

$$\frac{1}{\pi}\int_{-\infty}^{\infty} f(t)\, dt \int_{-\infty}^{\infty} g(u)\, \frac{\sin X(u-t)}{u-t}\, du \to \int_{-\infty}^{\infty} f(t)\, g(t)\, dt$$

whenever f is L^p and g is L^q, by a theorem of M. Riesz†.

It is to be observed that there is nothing in Riesz's theorem corresponding to our condition that g is L^2. His hypotheses do not ensure the existence of G, and it is for this reason that he does not state his theorem as a form of Parseval's theorem.

Finally we add the theorem which corresponds to **9** as **8** corresponds to **7**, viz.

THEOREM 10. *The conclusion of* **9** *holds whenever f is L^p, g is L^q, and the integral for G is uniformly convergent in any interval $0 < \xi \leqslant x \leqslant X$.*

References.

1. H. Bateman, "Report on the history and present state of the theory of integral equations", *British Assoc. Reports* (1910), 345-424.
2. E. W. Hobson, *The theory of functions of a real variable* (2, second ed., 1926).
3. M. Plancherel, "Contribution a l'étude de la représentation d'une fonction arbitraire par des intégrales définies", *Rend. di Palermo*, 30 (1910), 289-335.
4. S. Pollard, "Identification of the coefficients in a trigonometrical integral", *Proc. London Math. Soc* (2), 25 (1926), 451-468.
5. M. Riesz, "Sur les fonctions conjuguées", *Math. Zeitschrift*, 27 (1927), 218-244.
6. E. C. Titchmarsh, "A contribution to the theory of Fourier transforms", *Proc. London Math. Soc.* (2), 23 (1924), 279-289.

---

† M. Riesz, **5**, 244. Riesz does not give the details of the proof, but the argument is substantially the same as that of pp. 230-231, where he proves the corresponding theorem for series.

FORMULAE CONNECTING DIFFERENT CLASSES OF SELF-RECIPROCAL FUNCTIONS

By G. H. HARDY *and* E. C. TITCHMARSH.

[Received 3 November, 1930.—Read 13 November, 1930.]

1. This note originates from a remark of Ramanujan[*], that the function

$$f(x) = \frac{1}{x+}\frac{1}{x+}\frac{2}{x+}\frac{3}{x+}\cdots,$$

which may also be expressed in the forms

$$f(x) = \int_0^\infty \frac{e^{-xt}}{\cosh at}\,dt = 2\left(\frac{1}{x+a} - \frac{1}{x+3a} + \frac{1}{x+5a} - \cdots\right),$$

where $a = \sqrt{(\tfrac{1}{2}\pi)}$, is its own sine transform. Proofs have been given by Phillips and Watson[†]. An examination of the second proof shows that it rests on the fact that $\operatorname{sech} ax$ is its own cosine transform, and suggests the general rule

I. *If $f(x)$ is its own cosine (sine) transform, then*

$$h(x) = \int_0^\infty e^{-xt} f(t)\,dt$$

is its own sine (cosine) transform.

We shall use the following language. If $f(x)$ is its own Hankel transform of order μ, so that

$$f(x) = \int_0^\infty \sqrt{(xt)}\, J_\mu(xt)\, f(t)\,dt,$$

we shall say that $f(x)$ belongs to R_μ; and we shall write R_C for $R_{-\frac{1}{2}}$ and R_S for $R_{\frac{1}{2}}$. Our rule may then be stated in the form: *if $f(x)$ belongs to R_C (R_S), then $h(x)$ belongs to R_S (R_C).* We shall also express this by saying that *the kernel e^{-xt} transforms R_C (R_S) into R_S (R_C).*

---

[*] Ramanujan, **5**, 334.
[†] See Phillips, **4**.

The lines of the proof are as follows. Supposing, for example, that (x) belongs to R_C, we have

$$\sqrt{\left(\frac{2}{\pi}\right)} \int_0^\infty h(t) \sin xt\, dt = \sqrt{\left(\frac{2}{\pi}\right)} \int_0^\infty \sin xt\, dt \int_0^\infty e^{-ty} f(y)\, dy$$

$$= \sqrt{\left(\frac{2}{\pi}\right)} \int_0^\infty f(y)\, dy \int_0^\infty e^{-yt} \sin xt\, dt$$

$$= \sqrt{\left(\frac{2}{\pi}\right)} \int_0^\infty \frac{x}{x^2+y^2} f(y)\, dy.$$

Now $\sqrt{\left(\frac{2}{\pi}\right)} \frac{x}{x^2+y^2}$ is the cosine transform of e^{-xy}, and $f(y)$ is its own cosine transform. Hence Parseval's theorem for cosine transforms gives

(1.1) $$\sqrt{\left(\frac{2}{\pi}\right)} \int_0^\infty \frac{x}{x^2+y^2} f(y)\, dy = \int_0^\infty e^{-xy} f(y)\, dy = h(x),$$

which establishes the rule formally.

2. We shall justify this argument under appropriate conditions later. First, however, we obtain a more general formal rule of the same character.

II. *The kernel* $(xt)^{\frac{1}{2}(\mu+\nu+1)} K_{\frac{1}{2}(\nu-\mu)}(xt)$ *transforms* R_μ *into* R_ν.

The rule may be established (when its form has been found) by transformations similar to those of § 1. It is more instructive to proceed as follows. A general formula for functions of R_μ is*

(2.1) $$f(x) = \frac{1}{2\pi i} \int_{c-i\infty}^{c+i\infty} 2^{\frac{1}{2}s}\, \Gamma(\tfrac{1}{2}s+\tfrac{1}{2}\mu+\tfrac{1}{4})\, \psi(s)\, x^{-s}\, ds,$$

where $0 < c < 1$ and

(2.2) $$\psi(s) = \psi(1-s).$$

Hence we obtain

$$h(x) = \int_0^\infty (xt)^{\frac{1}{2}(\mu+\nu+1)} K_{\frac{1}{2}(\nu-\mu)}(xt) f(t)\, dt$$

$$= \frac{1}{2\pi i} \int_{c-i\infty}^{c+i\infty} 2^{\frac{1}{2}s}\, \Gamma(\tfrac{1}{2}s+\tfrac{1}{2}\mu+\tfrac{1}{4})\, \psi(s)\, ds \int_0^\infty (xt)^{\frac{1}{2}(\mu+\nu+1)} K_{\frac{1}{2}(\nu-\mu)}(xt)\, t^{-s}\, dt\,;$$

---

* Hardy and Titchmarsh, 1 (1.58).

and, on inserting the value of the inner integral*,

$$(2.3) \quad h(x) = \frac{1}{2\pi i} \int_{c-i\infty}^{c+i\infty} 2^{\frac{1}{2}(\mu+\nu-s-1)} \Gamma(\tfrac{1}{2}s+\tfrac{1}{2}\mu+\tfrac{1}{4}) \Gamma(\tfrac{1}{2}\mu+\tfrac{3}{4}-\tfrac{1}{2}s)$$
$$\times \Gamma(\tfrac{1}{2}\nu+\tfrac{3}{4}-\tfrac{1}{2}s) \psi(s) x^{s-1} ds$$
$$= \frac{1}{2\pi i} \int_{1-c-i\infty}^{1-c+i\infty} 2^{\frac{1}{2}w} \Gamma(\tfrac{1}{2}w+\tfrac{1}{2}\nu+\tfrac{1}{4}) \psi_1(w) x^{-w} dw,$$

where

$$\psi_1(w) = 2^{\frac{1}{2}\mu+\frac{1}{2}\nu-1} \Gamma(\tfrac{3}{4}+\tfrac{1}{2}\mu-\tfrac{1}{2}w) \Gamma(\tfrac{1}{4}+\tfrac{1}{2}\mu+\tfrac{1}{2}w) \psi(w).$$

It may be verified at once that $\psi_1(w) = \psi_1(1-w)$; and then, comparing (2.3) and (2.1), we see that $h(x)$ belongs to R_ν.

Taking $\mu = \nu$, we see that

III. *The kernel $(xt)^{\mu+\frac{1}{2}} K_0(xt)$ transforms R_μ into itself; in particular, $K_0(xt)$ transforms R_C into itself.*

3. Rule I is the case $\mu = -\tfrac{1}{2}$, $\nu = \tfrac{1}{2}$ of rule II (or the converse case). So far as rigorous analysis is concerned, we confine ourselves to rule I. There are two things in the analysis of § 1 which require justification, the inversion of the integrations and the appeal to Parseval's theorem.

It is easy to show, first, that the inversion is legitimate whenever the integrals

$$(3.1) \qquad \int_0^1 |f(y)| dy, \quad \int_1^\infty \frac{|f(y)|}{y^2} dy$$

are finite. It is obvious that we may invert the integrations over the ranges (τ, T) of t and $(0, \infty)$ of y; and it is therefore sufficient to prove that the integrals

$$I_1 = \int_0^\infty f(y) dy \int_0^\tau e^{-yt} \sin xt\, dt, \quad I_2 = \int_0^\infty f(y) dy \int_T^\infty e^{-yt} \sin xt\, dt$$

exist, and tend to zero when $\tau \to 0$ and $T \to \infty$. But

$$|I_1| \leqslant x \int_0^Y |f(y)| dy \int_0^\tau t\, dt + x \int_Y^\infty |f(y)| dy \int_0^\infty t e^{-yt} dt$$
$$= \tfrac{1}{2} x \tau^2 \int_0^Y |f(y)| dy + x \int_Y^\infty \frac{|f(y)|}{y^2} dy < \epsilon$$

---

* Watson, **6**, § 13.21 (8).

for $Y = Y(\epsilon)$, $\tau < \tau_0(Y, \epsilon) = \tau_0(\epsilon)$; and

$$|I_2| \leq \int_0^\infty |f(y)| \left| \frac{x \cos Tx + y \sin Tx}{x^2 + y^2} \right| e^{-Ty} dy \leq \frac{1}{x} \int_0^\infty |f(y)| e^{-Ty} dy$$

$$\leq \frac{1}{x} \int_0^\eta |f(y)| dy + \frac{e^{-\frac{1}{2}T\eta}}{x} \int_\eta^\infty |f(y)| e^{-\frac{1}{2}y} dy < \epsilon$$

for $\eta = \eta(\epsilon)$ and $T > T_0(\eta, \epsilon) = T_0(\epsilon)$. Hence I_1 and I_2 tend to zero.

4. It remains to consider the appeal to Parseval's theorem. For this, we may use the criteria enumerated in our note **2**. Here $g(y) = e^{-xy}$ and its transform $G(y)$ satisfy any of the conditions there imposed on either $f(x)$ or $g(x)$, so that the theorem is true whenever our present $f(y)$ satisfies any of these conditions. The situation is also simplified by the fact that $f(y)$ is its own transform. Thus our rule I is applicable whenever $f(x)$ belongs to L (in which case it is also bounded), or to any L^p for which $1 < p \leq 2$ (in which case it also belongs to $L^{p'}$ and to all intervening Lebesgue classes); or when it is of bounded variation in $(0, \infty)$, and tends to zero when $x \to \infty$; or when it is bounded, and its cosine integral is uniformly convergent over any finite range of the variable. It will be found, for example, that if we take $f(x)$ to be any of the special functions enumerated in our paper **1**, there is no difficulty in picking out applicable criteria. It will, however, sometimes be necessary to write $f(x)$ as the sum of two functions and use different criteria for them; and sometimes our criteria will show only that one of the integrals of the theorem exists in some generalised sense, and it will be necessary to appeal to the special form of $f(x)$ to see that this integral is in fact convergent.

We may illustrate these remarks by considering two particular examples.

(i) Suppose that $f(x) = x^{-\frac{1}{2}}$. We write $f = f_1 + f_2$, where $f_1 = 0$ in $(1, \infty)$ and $f_2 = 0$ in $(0, 1)$. Then criterion **4** shows that

$$\int_0^\infty e^{-xy} f_1(y) dy = \sqrt{\left(\frac{2}{\pi}\right)} \int_0^\infty \frac{x}{x^2 + y^2} F_1(y) dy,$$

where F_1 is the transform of f_1, and criterion **6** gives the corresponding equation for f_2. Adding the equations, and observing that $F_1 + F_2 = F = f$, we obtain (1.1).

(ii) Suppose that $f(x) = x^{\frac{1}{2}} J_{-\frac{1}{4}}(\frac{1}{2}x^2)$. Here $f(x)$ does not belong to any Lebesgue class, nor is it of bounded variation. But $f(x)$ is bounded,

and
$$\int_0^\infty t^{\frac{1}{2}} J_{-\frac{1}{4}}(\tfrac{1}{2}t^2)\cos xt\, dt$$

is uniformly convergent in any finite interval of x, so that we can apply criterion **8**. We observe finally that the integrals (1.1) are, in fact, absolutely convergent.

5. We conclude the paper by determining a number of further rules for transformation of the classes R_μ. Our point of view here is purely formal; the analysis of §§ 3–4 will be sufficient to show how the validity of a rule may be discussed when its form is known.

It is natural to ask: *what is the most general kernel $k(xt)$ which transforms R_C into R_S?* Let us suppose that $f(x)$ belongs to R_C, and write

(5.1) $$h(x) = \int_0^\infty k(xt) f(t)\, dt,$$

(5.2) $$K(s) = \int_0^\infty k(u) u^{s-1}\, du.$$

Taking $\mu = -\tfrac{1}{2}$ in (2.1), and substituting for $f(t)$ from (2.1) into (5.1), we obtain

$$\begin{aligned}
h(x) &= \int_0^\infty k(xt)\, dt\, \frac{1}{2\pi i}\int_{c-i\infty}^{c+i\infty} 2^{\frac{1}{2}s}\Gamma(\tfrac{1}{2}s)\,\psi(s)\, t^{-s}\, ds\\
&= \frac{1}{2\pi i}\int_{c-i\infty}^{c+i\infty} 2^{\frac{1}{2}s}\Gamma(\tfrac{1}{2}s)\,\psi(s)\, ds \int_0^\infty k(xt) t^{-s}\, dt\\
&= \frac{1}{2\pi i}\int_{c-i\infty}^{c+i\infty} 2^{\frac{1}{2}s}\Gamma(\tfrac{1}{2}s)\,\psi(s)\, K(1-s)\, x^{s-1}\, ds\\
&= \frac{1}{2\pi i}\int_{1-c-i\infty}^{1-c+i\infty} 2^{\frac{1}{2}w}\Gamma(\tfrac{1}{2}w+\tfrac{1}{2})\,\psi_1(w)\, x^{-w}\, dw,
\end{aligned}$$

where
$$\psi_1(w) = 2^{\frac{1}{2}-w}\frac{\Gamma(\tfrac{1}{2}-\tfrac{1}{2}w)}{\Gamma(\tfrac{1}{2}w+\tfrac{1}{2})} K(w)\,\psi(1-w).$$

The condition that $h(x)$ should belong to R_S is that $\psi_1(w) = \psi_1(1-w)$.

Inverting (5.2) by Mellin's formula, we obtain

(5.3) $$k(x) = \frac{1}{2\pi i}\int_{a-i\infty}^{a+i\infty} K(s) x^{-s}\, ds = \frac{1}{2\pi i}\int_{a-i\infty}^{a+i\infty} \Gamma(s)\chi(s) x^{-s}\, ds,$$

where
$$\chi(s) = 2^{s-\frac{1}{2}}\frac{\Gamma(\tfrac{1}{2}s+\tfrac{1}{2})}{\Gamma(s)\,\Gamma(\tfrac{1}{2}-\tfrac{1}{2}s)}\frac{\psi_1(s)}{\psi(1-s)} = \frac{\sqrt{(2\pi)}}{\Gamma(\tfrac{1}{2}s)\,\Gamma(\tfrac{1}{2}-\tfrac{1}{2}s)}\frac{\psi_1(s)}{\psi(1-s)},$$

so that

(5.4) $$\chi(s) = \chi(1-s).$$

Hence we derive the rule

IV. *The most general kernel $k(xt)$ which transforms R_C into R_S is that defined by* (5.3), *where $\chi(s)$ is a solution of* (5.4).

If we take $\chi(s) = 1$, we find that $k(x) = e^{-x}$. Taking

$$\chi(s) = \frac{\sqrt{\pi}}{\Gamma(\tfrac{1}{2}+\tfrac{1}{2}s)\,\Gamma(1-\tfrac{1}{2}s)}, \quad \frac{2\sqrt{\pi}}{\Gamma(\tfrac{1}{2}s)\,\Gamma(\tfrac{1}{2}-\tfrac{1}{2}s)}, \quad \frac{\sqrt{3}}{2\sqrt{\pi}}\Gamma(s-\tfrac{1}{3})\,\Gamma(\tfrac{2}{3}-s),$$

we obtain*
$$k(x) = J_0(x), \quad xJ_0(x), \quad x^{-\tfrac{1}{6}}e^{\tfrac{1}{2}x}K_{\tfrac{1}{6}}(\tfrac{1}{2}x).$$
Thus

V. *The kernels $J_0(xt)$, $xtJ_0(xt)$, $(xt)^{-\tfrac{1}{6}}e^{\tfrac{1}{2}xt}K_{\tfrac{1}{6}}(\tfrac{1}{2}xt)$ transform R_C into R_S.*

6. Further rules may be found by substituting for $f(t)$, in one of the rules given already, its expression as its own transform; or, again, by iterating a rule, or combining two different rules. Applying the first process to rule I, we find

$$h(x) = \sqrt{\left(\frac{2}{\pi}\right)} \int_0^\infty e^{-xt}\,dt \int_0^\infty f(t)\cos ty\,dy = \sqrt{\left(\frac{2}{\pi}\right)} \int_0^\infty \frac{x}{t^2+x^2} f(t)\,dt.$$

If we replace the e^{-xt} of rule I by the $J_0(xt)$ of rule V, we find

$$h(x) = \sqrt{\left(\frac{2}{\pi}\right)} \int_0^x \frac{f(t)}{\sqrt{(x^2-t^2)}}\,dt.$$

If we iterate rule I, we obtain

$$g(x) = \int_0^\infty e^{-xy} h(y)\,dy = \int_0^\infty e^{-xy}\,dy \int_0^\infty e^{-yt} f(t)\,dt = \int_0^\infty \frac{f(t)}{t+x}\,dt.$$

Here $f(t)$ belongs to R_C, $h(y)$ to R_S, and $g(x)$ to R_C (or conversely). If we replace one of the exponentials by a J_0, we obtain

$$g(x) = \int_0^\infty \frac{f(t)}{\sqrt{(t^2+x^2)}}\,dt.$$

---

* See Watson, **6**, 192–193.

Thus

VI. *The kernels*
$$\frac{x}{t^2+x^2}; \quad \frac{1}{\sqrt{(x^2-t^2)}} \quad (t<x), \quad 0 \quad (t>x)$$

transform R_C *into* R_S. *The kernels*
$$\frac{1}{t+x}, \quad \frac{1}{\sqrt{(t^2+x^2)}}$$

transform R_C *into itself.*

7. It is plain that rules of this character might be multiplied indefinitely, and our object is merely to indicate a few of the simplest. We may observe finally that it is interesting to apply them to a "singular" example of a function of R_C given in our paper **1**. This function was defined conventionally* by

(7.1) $$f(x) = 1\infty, \quad f(x) = -(2\pi)^{-\frac{1}{2}},$$

according as x is or is not a positive integral multiple of $\sqrt{(2\pi)}$; an integral $\int \phi(x) f(x) dx$ is to be interpreted as a Stieltjes integral $\int \phi(x) df_1(x)$, where

$$f_1(x) = \left[\frac{x}{\sqrt{(2\pi)}}\right] - \frac{x}{\sqrt{(2\pi)}}.$$

Applying rules I and V, we find that

$$h(x) = \int_0^\infty e^{-xt} df_1(t) = \sum_1^\infty e^{-nx\sqrt{(2\pi)}} - \frac{1}{\sqrt{(2\pi)}} \int_0^\infty e^{-xt} dt$$

$$= \frac{e^{-x\sqrt{(2\pi)}}}{1-e^{-x\sqrt{(2\pi)}}} - \frac{1}{x\sqrt{(2\pi)}},$$

and†

$$h(x) = \int_0^\infty J_0(xt) df_1(t) = \sum_1^\infty J_0\{nx\sqrt{(2\pi)}\} - \frac{1}{\sqrt{(2\pi)}} \int_0^\infty e^{-xt} dt$$

$$= \sqrt{\left(\frac{2}{\pi}\right)} \sum_{n<x/\sqrt{(2\pi)}} \frac{1}{\sqrt{(x^2-2n^2\pi)}} - \frac{1}{2} - \frac{1}{x\sqrt{(2\pi)}}$$

belong to R_S: the first of these results is well known. We cannot discuss such applications of our rules on the lines of §§ 3–4 without exten-

---

* See Hardy and Titchmarsh, **1**, § 5.5.
† Nielsen, **3**, 336.

sions of the Parseval formula to Stieltjes integrals. These would carry us beyond our present object; but they are, in any case, unnecessary here, because, when $f(x)$ is the function (7.1), the Parseval formula reduces to "Poisson's summation formula".

References.

1. G. H. Hardy and E. C. Titchmarsh, "Self-reciprocal functions", *Quarterly Journal* (Oxford series), **1** (1930), 196-231 [this paper contains a comprehensive list of references to the literature].
2. ———— "A note on Parseval's theorem for Fourier transforms", *Journal London Math. Soc.*, **6** (1930), 44-48.
3. N. Nielsen, *Handbuch der Theorie der Cylinderfunktionen* (Leipzig, 1904).
4. E. G. Phillips, "Note on a problem of Ramanujan", *Journal London Math. Soc.*, **4** (1929), 114-116.
5. S. Ramanujan, *Collected mathematical papers* (Cambridge, 1927).
6. G. N. Watson, *A treatise on the theory of Bessel functions* (Cambridge, 1922).

CORRECTIONS

p. 225, *lines* 1-6 (Correction by E. C. Titchmarsh). '1. This note ... where' should read:
1. This note originates from a remark of Ramanujan*, that if

$$g(x) = \frac{1}{x+} \frac{1^2}{x+} \frac{2^2}{x+} \frac{3^2}{x+} \cdots,$$

then $g(x/a)$, which may also be expressed in the forms

$$a \int_0^\infty \frac{e^{-xt}}{\cosh at} dt = 2a\left(\frac{1}{x+a} - \frac{1}{x+3a} + \frac{1}{x+5a} - \cdots\right),$$

where ....

p. 226, *line* 2 (x) should be $f(x)$.

p. 232, *reference* 4. The page numbers should be 310-313.

ON HILBERT TRANSFORMS

By G. H. HARDY (*Cambridge*)

[Received 14 January 1932]

1. Two functions $f(x)$ and $g(x)$ of the real variable x may be said to be *conjugate*, or to be *Hilbert transforms* of one another, if they are connected by the formulae

$$g(x) \equiv -\frac{1}{\pi}\frac{d}{dx}\int_{-\infty}^{\infty} f(t)\log\left|1-\frac{x}{t}\right|dt, \qquad (1.1.1)$$

$$f(x) \equiv \frac{1}{\pi}\frac{d}{dx}\int_{-\infty}^{\infty} g(t)\log\left|1-\frac{x}{t}\right|dt; \qquad (1.1.2)$$

or by

$$g(x) \equiv \frac{1}{\pi}\int_{-\infty}^{\infty} \frac{f(t)}{t-x}dt, \qquad (1.2.1)$$

$$f(x) \equiv -\frac{1}{\pi}\int_{-\infty}^{\infty} \frac{g(t)}{t-x}dt. \qquad (1.2.2)$$

Here '$\equiv$' denotes equivalence, i.e. equality for almost all x, and the integrals (1.2) are 'principal values' in the sense of Cauchy. The relation is 'real', but is a derivative of that between the conjugate components of an analytic function of the complex variable.

In two notes published in 1924 in the *Messenger of Mathematics*,† I set out to give a direct and self-contained theory of Hilbert transforms of the class L^2.‡ The theory may be based on the theory of analytic functions, on the theory of Fourier transforms, or on reciprocal relations between infinite series;§ but it is interesting to develop it independently, and Watson‖ has shown that the method which I used can be applied to more general problems.

There is, however, a mistake in my analysis, which was pointed out to me by Professor Watson.†† This is not difficult to rectify; but

---

† Hardy (**1, 2**). In these notes $g(x)$ is defined with the opposite sign. According to (1.1) and (1.2), the conjugate of $\cos mx$ is $-\sin mx$ if $m > 0$, $\sin mx$ if $m < 0$.

‡ The class of measurable functions whose squares are integrable over $(-\infty, \infty)$.

§ See M. Riesz (**5, 6**), Titchmarsh (**7, 8, 9**). Riesz and Titchmarsh deal also with the more difficult case in which f and g belong to a general Lebesgue class L^p with $p > 1$.   ‖ Watson (**10**).

†† In **1**, p. 26, § 4. From 'In order to prove this...' to the end of § 4 is nonsense.

while reconsidering the proof I have found a number of places where the analysis can be simplified and shortened materially. I therefore restate the whole argument in §§ 2–10 of the present note. The proofs of Lemmas 2–4 are substantially the same as before, but the rest of the argument has been entirely recast. In § 11, I indicate shortly the lines of an alternative, and in some ways still simpler, proof of Theorem 1, suggested to me by Professor Titchmarsh.

2. The theorems to be proved are as follows:

THEOREM 1. *If $f(x)$ is L^2, then there is a function $g(x)$ which has the properties*:

(a) $g(x)$ is L^2,

(b) $\int_{-\infty}^{\infty} g^2(x)\, dx = \int_{-\infty}^{\infty} f^2(x)\, dx,$ (2.1)

(c) $f(x)$ and $g(x)$ satisfy (1.1).

THEOREM 2. *The functions $f(x)$ and $g(x)$ also satisfy* (1.2), *the integrals being Lebesgue integrals at infinity, but principal values about $t = x$.*

It is to be observed that both the statement and the proof of Theorem 1 are independent of Theorem 2, although it is in the form (1.2) that the relation has usually been stated. Theorem 1 corresponds to Plancherel's fundamental theorem concerning Fourier transforms, in which the cosine transform of $f(x)$ is defined by

$$g(x) \equiv \sqrt{\left(\frac{2}{\pi}\right)} \frac{d}{dx} \int_0^{\infty} f(t) \frac{\sin xt}{t}\, dt.$$

Here we cannot usually differentiate under the integral sign. Theorem 2 asserts that, in the formulae for Hilbert transforms, we can; and this is one of the most interesting properties of the transformation.

Lemmas for Theorem 1

3. In what follows all numbers which occur are real, and

$$lx = \log|x|.$$

LEMMA 1. *Suppose that $\chi(t) = \chi(t, h, h',...)$ has the following properties*:

(a) *$\chi(t)$ belongs to L^2 for all values of $h, h',...$ in question;*

(b) $\int_{-\infty}^{\infty} \chi^2(t, h, h',...)\, dt < A,$

where A is independent of $h, h',...$;

(c) $$\chi(t, h, h', \ldots) \to \chi^*(t),$$

when $h, h', \ldots$ tend to certain limiting values, for almost all t. Then $f(t)\chi^*(t)$ belongs to L,† and

$$\int_{-\infty}^{\infty} \chi(t, h, h', \ldots) f(t) \, dt \to \int_{-\infty}^{\infty} \chi^*(t) f(t) \, dt,$$

for every $f(t)$ of L^2.‡

LEMMA 2. If
$$I(a, a', b, b') = \int_{-\infty}^{\infty} l\left(\frac{x-a}{x-a'}\right) l\left(\frac{x-b}{x-b'}\right) dx, \tag{3.1}$$

then
$$I(a, a', b, b') = \tfrac{1}{2}\pi^2(|a-b'|+|a'-b|-|a-b|-|a'-b'|). \tag{3.2}$$

In particular, if h and h' are positive, and
$$J(t, u, h, h') = I(t-h, t+h, u-h', u+h')$$
$$= \int_{-\infty}^{\infty} l\left(\frac{t-x+h}{t-x-h}\right) l\left(\frac{u-x+h'}{u-x-h'}\right) dx, \tag{3.3}$$

then
$$J(t, u, h, h') = \pi^2 \mu(t-u, h, h'), \tag{3.4}$$

where
$$\mu(w, h, h') = 0 \qquad (|w| \geqslant h+h'), \tag{3.5.1}$$

$$\mu(w, h, h') = \min(h+h'-|w|, h+h'-|h-h'|) \qquad (|w| \leqslant h+h'). \tag{3.5.2}$$

The proof of this lemma is a matter of elementary calculation. Since
$$2l\left(\frac{x-a}{x-a'}\right) l\left(\frac{x-b}{x-b'}\right)$$
$$= \left\{l\left(\frac{x-a}{x-b'}\right)\right\}^2 + \left\{l\left(\frac{x-a'}{x-b}\right)\right\}^2 - \left\{l\left(\frac{x-a}{x-b}\right)\right\}^2 - \left\{l\left(\frac{x-a'}{x-b'}\right)\right\}^2,$$

the general formula (3.2) will follow from the particular case
$$I(\alpha, \beta, \alpha, \beta) = \int_{-\infty}^{\infty} \left\{l\left(\frac{x-\alpha}{x-\beta}\right)\right\}^2 dx = \pi^2 |\alpha-\beta|.$$

And if we suppose, for example, that $\alpha > \beta$, then simple transformations give
$$I(\alpha, \beta, \alpha, \beta) = (\alpha-\beta) \int_{-\infty}^{\infty} \frac{(lu)^2}{(u-1)^2} du = 2(\alpha-\beta) \int_{-1}^{1} \frac{(lu)^2}{(1-u)^2} du$$

† Is integrable over $(-\infty, \infty)$.
‡ See Young (11), and the writings of Lebesgue there referred to.

$$= 2(\alpha-\beta) \int_{-1}^{1} (lu)^2(1+2u+3u^2+\ldots)\, du$$

$$= 8(\alpha-\beta)\left(\frac{1}{1^2}+\frac{1}{3^2}+\frac{1}{5^2}+\ldots\right) = \pi^2(\alpha-\beta).$$

This proves (3.2), and (3.4) is a special case. Geometrically, μ is an even non-negative function of w which has the constant value $2h$ or $2h'$ (whichever is smaller) from $w=0$ to $w=|h-h'|$, decreases linearly until it vanishes for $w = h+h'$, and then remains zero. If $h \leqslant h'$,

$$\frac{1}{4hh'}\int_{-\infty}^{\infty} \mu\, dw = \frac{1}{2hh'}\int_{0}^{h'-h} 2h\, dw + \frac{1}{2hh'}\int_{h'-h}^{h+h'} (h+h'-w)\, dw$$

$$= \frac{h'-h}{h'}+\frac{h}{h'} = 1; \tag{3.6}$$

and this is true by symmetry for all positive h, h'.

4. LEMMA 3. *If f is L^2, and*

$$\chi(t,h,h') = \frac{1}{4hh'}\int_{-\infty}^{\infty} f(t+w)\mu(w,h,h')\, dw,$$

then

$$\int_{-\infty}^{\infty} \chi^2(t,h,h')\, dt \leqslant 2\int_{-\infty}^{\infty} f^2(t)\, dt.$$

We may suppose $h \leqslant h'$. Then, using Schwarz's inequality, and observing that $\mu = 0$ if $|w| > h+h'$, we find

$$\chi^2 \leqslant \frac{1}{16h^2h'^2}\int_{-h-h'}^{h+h'} f^2(t+w)\, dw \int_{-\infty}^{\infty} \mu^2\, dw.$$

Since

$$\int_{-\infty}^{\infty} \mu^2\, dw \leqslant 2h\int_{-\infty}^{\infty} \mu\, dw = 8h^2h',$$

we deduce

$$\int_{-\infty}^{\infty} \chi^2\, dt \leqslant \frac{1}{2h'}\int_{-\infty}^{\infty} dt \int_{-h-h'}^{h+h'} f^2(t+w)\, dw$$

$$= \frac{1}{2h'}\int_{-h-h'}^{h+h'} dw \int_{-\infty}^{\infty} f^2(t+w)\, dw \leqslant 2\int_{-\infty}^{\infty} f^2\, dt.$$

LEMMA 4. *If h and h' tend to zero, then*
$$\chi(t,h,h') \to f(t)$$
for almost all t.

We may consider separately systems (h,h') for which $h \leqslant h'$ and $h' \leqslant h$. Suppose, then, that $h \leqslant h'$, and let
$$\psi(w) = \psi(w,t) = \tfrac{1}{2}\{f(t+w)+f(t-w)-2f(t)\}. \tag{4.1}$$
After (3.6), we have
$$\chi(t,h,h')-f(t) = \frac{1}{2hh'} \int_0^\infty \psi(w)\mu(w)\,dw$$
$$= \frac{1}{h'} \int_0^{h'-h} \psi(w)\,dw + \frac{1}{2hh'} \int_{h'-h}^{h+h'} \psi(w)(h+h'-w)\,dw,$$
$$|\chi(t,h,h')-f(t)| \leqslant \frac{1}{h'} \int_0^{2h'} |\psi(w)|\,dw,$$
which tends to zero for almost all t.

It follows from Lemmas 3 and 4 that $\chi(t)$ satisfies the conditions of Lemma 1, with $\chi^* = f$.

Proof of Theorem 1

5. We suppose first that $f(t) = 0$ for $|t| > T$.† We write
$$\phi(x) = -\frac{1}{\pi} \int_{-\infty}^{\infty} f(t)\, l\!\left(1-\frac{x}{t}\right) dt; \tag{5.1}$$
the integral (in which the infinite limits are at present apparent only) is convergent because f is L^2, and uniformly convergent in any finite interval of x. Further, we write
$$\Delta = \Delta(x,h) = \frac{\phi(x+h)-\phi(x-h)}{2h} = \frac{1}{2\pi h} \int_{-\infty}^{\infty} f(t)\, l\!\left(\frac{t-x+h}{t-x-h}\right) dt, \tag{5.2}$$
$$I(h,h') = \int_{-\infty}^{\infty} \Delta(x,h)\Delta(x,h')\,dx = \int_{-\infty}^{\infty} \Delta\Delta'\,dx. \tag{5.3}$$
Then
$$I(h,h') = \frac{1}{4\pi^2 hh'} \int_{-\infty}^{\infty} dx \int_{-\infty}^{\infty} f(t)\, l\!\left(\frac{t-x+h}{t-x-h}\right) dt \int_{-\infty}^{\infty} f(u)\, l\!\left(\frac{u-x+h'}{u-x-h'}\right) du.$$

† The mistake in (**1**) arose from an attempt to avoid this distinction.

The triple integral is absolutely convergent, since f vanishes for large t or u, and therefore

$$I(h,h') = \frac{1}{4\pi^2 hh'} \int_{-\infty}^{\infty} f(t)\, dt \int_{-\infty}^{\infty} f(u)\, du \int_{-\infty}^{\infty} l\!\left(\frac{t-x+h}{t-x-h}\right) l\!\left(\frac{u-x+h'}{u-x-h'}\right) dx$$

$$= \frac{1}{4hh'} \int_{-\infty}^{\infty} f(t)\, dt \int_{-\infty}^{\infty} f(u)\, \mu(t-u,h,h')\, du$$

$$= \frac{1}{4hh'} \int_{-\infty}^{\infty} f(t)\, dt \int_{-\infty}^{\infty} f(t+w)\, \mu(w,h,h')\, dw = \int_{-\infty}^{\infty} f(t)\, \chi(t,h,h')\, dt.$$

Since χ satisfies the conditions of Lemma 1, it follows that

$$I(h,h') \to \int_{-\infty}^{\infty} f^2(t)\, dt,$$

and hence that

$$\int_{-\infty}^{\infty} (\Delta - \Delta')^2\, dx = I(h,h) - 2I(h,h') + I(h',h') \to 0. \tag{5.4}$$

6. We can now apply the classical theory of mean-square convergence. It follows from (5.4) that $\Delta(x,h)$ converges in mean square to a function $g(x)$ of L^2, and that

$$\int_{-\infty}^{\infty} g^2(x)\, dx = \lim \int_{-\infty}^{\infty} \Delta^2\, dx = \int_{-\infty}^{\infty} f^2(t)\, dt.$$

This is (2.1), for an f which vanishes for $|t| > T$.

Also
$$\int_{-\infty}^{\infty} (\Delta - g) k\, dx \to 0,$$

for any k of L^2. If in particular we suppose that $k = 1$ when x lies between 0 and y, and $k = 0$ otherwise, we find

$$\int_0^y (\Delta - g)\, dx \to 0$$

and so

$$\int_0^y g\, dx = \lim \int_0^y \Delta\, dx = \lim \frac{1}{2h} \int_0^y \{\phi(x+h) - \phi(x-h)\}\, dx$$

$$= \lim \left\{ \frac{1}{2h} \int_{y-h}^{y+h} \phi(u)\, du - \frac{1}{2h} \int_{-h}^{h} \phi(u)\, du \right\} = \phi(y) - \phi(0),$$

since ϕ is continuous. That is to say, ϕ is the integral of g, and g is defined in terms of f by (1.1.1). We call g the *transform* of f.

We have thus proved Theorem 1, for f which vanish for large t, except for the reciprocity asserted by (1.1.2). This it is more convenient to defer for a moment.

7. We now suppose that f is any function of L^2, that f_T is f in $(-T, T)$ and 0 outside, and that

$$g_T \equiv -\frac{1}{\pi}\frac{d}{dx}\int_{-T}^{T} f(t)\, l\!\left(1-\frac{x}{t}\right) dt$$

is the transform of f_T.

If $-X \leqslant x \leqslant X$ and $T > X$, then

$$-\frac{d}{dx}\int_{T}^{\infty} f(t)\, l\!\left(1-\frac{x}{t}\right) dt = \int_{T}^{\infty} \frac{f(t)}{t-x}\, dt$$

exists and is continuous, since the second integral is uniformly convergent; and the same things are true of the integrals over $(-\infty, -T)$. It follows that $g(x)$, defined by (1.1.1), exists for almost all x of $(-X, X)$, and so for almost all x; that ϕ, defined by (5.1), is the integral of g; and that $g_T \to g$, when $T \to \infty$, uniformly in any finite interval of x.†

If $T' > T$, then $g_{T'} - g_T$ is the transform of the function which is f when $T \leqslant |t| \leqslant T'$ and zero otherwise; and so

$$\int_{-\infty}^{\infty} (g_{T'}-g_T)^2\, dx = \left(\int_{-T'}^{-T} + \int_{T}^{T'}\right) f^2\, dt \to 0$$

when T and T' tend to infinity. Hence g_T tends in mean square to a function g^* when $T \to \infty$, and

$$\int_{-\infty}^{\infty} g^{*2}\, dx = \lim \int_{-\infty}^{\infty} g_T^2\, dx = \lim \int_{-T}^{T} f^2\, dt = \int_{-\infty}^{\infty} f^2\, dt.$$

Finally, as in § 6,

$$\int_{0}^{y} (g_T - g^*)\, dx \to 0,$$

$$\int_{0}^{y} g^*\, dx = \lim \int_{0}^{y} g_T\, dx = \int_{0}^{y} g\, dx$$

---

† Strictly, g_T and g are equivalent to functions h_T and h, and $h_T \to h$ uniformly.

(since $g_T \to g$ uniformly in any finite interval), and so $g^* \equiv g$. This completes the proof of Theorem 1, except for (1.1.2).

8. If f and f_1 are two functions of L^2, g and g_1 their transforms, then, by (2.1),
$$\int_{-\infty}^{\infty} (f \pm f_1)^2 \, dt = \int_{-\infty}^{\infty} (g \pm g_1)^2 \, dt;$$
and so, by subtraction,
$$\int_{-\infty}^{\infty} f f_1 \, dt = \int_{-\infty}^{\infty} g g_1 \, dt. \tag{8.1}$$
If in particular f_1 is π in $(0, y)$ and 0 outside, then
$$g_1 = -\frac{d}{dx} \int_0^y l\left(1 - \frac{x}{t}\right) dt = -\frac{d}{dx} \int_{-x}^{y-x} lw \, dw = l\left(1 - \frac{y}{x}\right).$$
Taking this f_1 and g_1 in (8.1), we obtain
$$\pi \int_0^y f(t) \, dt = \int_{-\infty}^{\infty} g(t) l\left(1 - \frac{y}{t}\right) dt,$$
and (1.1.2) follows by differentiation. This completes the proof.

An auxiliary theorem

9. The proof of Theorem 2 depends on another theorem of some intrinsic interest.

THEOREM 3. *If f belongs to L^2, then*
$$L(x, \epsilon) = \frac{1}{\pi^2} \int_{-\infty}^{\infty} \frac{f(t)}{t-x} l\left(\frac{t-x+\epsilon}{t-x-\epsilon}\right) dt \to f(x), \tag{9.1}$$

when $\epsilon \to 0$, for almost all x and in particular at any point of continuity of f.

The assertion is equivalent to
$$M(\epsilon) = \int_0^{\infty} \frac{\psi(t)}{t} l\left(\frac{t+\epsilon}{t-\epsilon}\right) dt \to 0, \tag{9.2}$$
where $\psi(t) = \psi(t, x)$ is defined as in (4.1); and it is sufficient to prove that (9.2) is true when
$$\Psi(t) = \int_0^t \psi^2(u) \, du = o(t) \tag{9.3}$$
for small t (this being true for almost all x when f is L^2).

We write
$$M(\epsilon) = \int_0^{2\epsilon} + \int_{2\epsilon}^{\infty} = M_1(\epsilon) + M_2(\epsilon).$$
Then
$$M_1^2(\epsilon) \leqslant \int_0^{2\epsilon} \psi^2(t)\,dt \int_{-\infty}^{\infty} \left\{l\!\left(\frac{t+\epsilon}{t-\epsilon}\right)\right\}^2 \frac{dt}{t^2}$$
$$= \frac{1}{\epsilon}\int_0^{2\epsilon} \psi^2(t)\,dt \int_{-\infty}^{\infty} \left\{l\!\left(\frac{w+1}{w-1}\right)\right\}^2 \frac{dw}{w^2} \to 0,$$
by (9.3). Also
$$M_2^2(\epsilon) \leqslant \int_{2\epsilon}^{\infty} \frac{\psi^2(t)}{t^2}\,dt \int_{-\infty}^{\infty} \left\{l\!\left(\frac{t+\epsilon}{t-\epsilon}\right)\right\}^2 dt.$$
The second factor is $O(\epsilon)$, while the first is
$$\int_{2\epsilon}^{\infty} \frac{\Psi'(t)}{t^2}\,dt = -\frac{\Psi(2\epsilon)}{4\epsilon^2} + 2\int_{2\epsilon}^{\infty} \frac{\Psi(t)}{t^3}\,dt \leqslant 2\int_{2\epsilon}^{\infty} \frac{\Psi(t)}{t^3}\,dt$$
$$= 2\int_{2\epsilon}^{1} o\!\left(\frac{1}{t^2}\right)dt + 2\int_1^{\infty} O\!\left(\frac{1}{t^3}\right)dt = o\!\left(\frac{1}{\epsilon}\right).$$
Hence $M_2(\epsilon) \to 0$.†

Proof of Theorem 2

10. Let
$$f_1 = \frac{1}{t-x} \quad (|t-x| \geqslant \epsilon), \qquad f_1 = 0 \quad (|t-x| < \epsilon).$$
The transform g_1 of f_1 is
$$\frac{1}{\pi}\left(\int_{-\infty}^{x-\epsilon} + \int_{x+\epsilon}^{\infty}\right) \frac{du}{(u-t)(u-x)} = \frac{1}{\pi(t-x)} l\!\left(\frac{t-x+\epsilon}{t-x-\epsilon}\right).$$
Hence, by (8.1),
$$\left(\int_{-\infty}^{x-\epsilon} + \int_{x+\epsilon}^{\infty}\right) \frac{f(t)}{t-x}\,dt = \frac{1}{\pi}\int_{-\infty}^{\infty} \frac{g(t)}{t-x} l\!\left(\frac{t-x+\epsilon}{t-x-\epsilon}\right) dt.$$
Making $\epsilon \to 0$, and using Theorem 3, we find that
$$\int_{-\infty}^{\infty} \frac{f(t)}{t-x}\,dt = \pi g(x)$$
for almost all x. This is (1.2.1), and (1.2.2) is proved similarly.

† In this note I am concerned only with functions of L^2; but it is worth remarking that Theorem 3 is true for functions of L^p whenever $p > 1$, and that the proof generalizes immediately. The result is true at a point of continuity, but *not* necessarily almost always, when $p = 1$. See Hardy (3).

An alternative method

11. The alternative proof of Theorem 1 suggested by Professor Titchmarsh,† and referred to in § 1, proceeds as follows.

Suppose first that $f(t)$ has a continuous derivative and vanishes for $|t| > T$; we may call such a function a function of class $\mathbf{F}_T$. Then

$$\phi(x) = -\frac{1}{\pi}\int_{-\infty}^{\infty} f(t)\, l\!\left(1-\frac{x}{t}\right) dt = -\frac{x}{\pi}\int_{-\infty}^{\infty} f(xu)\, l\!\left(1-\frac{1}{u}\right) du,$$

and so

$$g(x) = \phi'(x) = -\frac{1}{\pi}\int_{-\infty}^{\infty} f(xu)\, l\!\left(1-\frac{1}{u}\right) du - \frac{x}{\pi}\int_{-\infty}^{\infty} uf'(xu)\, l\!\left(1-\frac{1}{u}\right) du$$

$$= -\frac{1}{x\pi}\int_{-\infty}^{\infty} F(t)\, l\!\left(1-\frac{x}{t}\right) dt,$$

where
$$F(t) = f(t) + tf'(t).$$

The transformations are trivial when f belongs to $\mathbf{F}_T$.

We now have

$$\int_{-\infty}^{\infty} g^2(x)\, dx = \frac{1}{\pi^2}\int_{-\infty}^{\infty}\frac{dx}{x^2}\int_{-\infty}^{\infty} F(t)\, l\!\left(1-\frac{x}{t}\right) dt \int_{-\infty}^{\infty} F(u)\, l\!\left(1-\frac{x}{u}\right) du$$

$$= \frac{1}{\pi^2}\int_{-\infty}^{\infty} F(t)\, dt \int_{-\infty}^{\infty} F(u)\, du \int_{-\infty}^{\infty} l\!\left(1-\frac{x}{t}\right) l\!\left(1-\frac{x}{u}\right) \frac{dx}{x^2} \quad (11.1)$$

(the triple integral being absolutely convergent for f of $\mathbf{F}_T$). If we write $1/x$ for x, the inner integral becomes a special case of the integral of Lemma 2, and its value is

$$\pi^2 \min\!\left(\frac{1}{|t|}, \frac{1}{|u|}\right)$$

if t and u have the same sign, 0 if t and u have opposite signs.

Suppose that $0 < u < t$. Then the corresponding part of (11.1) is

$$\int_0^\infty F(t)\, dt \int_0^t \frac{F(u)}{t}\, du = \int_0^\infty \frac{f(t)+tf'(t)}{t}\, dt \int_0^t \{f(u)+uf'(u)\}\, du$$

$$= \int_0^\infty (f^2 + tff')\, dt = \tfrac{1}{2}\int_0^\infty f^2\, dt,$$

† In a rather different setting suggested by Watson's work.

by partial integration. The part for which $0 < t < u$ gives the same contribution, and the parts for which t and u are both negative contribute corresponding integrals over $(-\infty, 0)$. Hence

$$\int_{-\infty}^{\infty} g^2(x)\, dx = \int_{-\infty}^{\infty} f^2(t)\, dt. \tag{11.2}$$

This is (2.1) for f of $\mathbf{F}_T$. But any f of L^2 is a mean-square limit of a sequence of functions f_n of $\mathbf{F}_T$, and (11.2) may therefore be extended to all f of L^2 in the ordinary manner. The formulae (1.1) may then be deduced from (2.1) as (1.1.2) was deduced in § 8.

[Mr. J. Cossar suggests to me another interesting alternative to the argument of § 5. The function

$$F(t, u) = F(t-u) = \int_{-\infty}^{\infty} f(t+x) f(u+x)\, dx$$

is continuous. If we write $\Delta(x, h)$ in the form

$$\Delta(x, h) = \frac{1}{2\pi h} \int_{-\infty}^{\infty} f(t+x)\, l\!\left(\frac{t+h}{t-h}\right) dt,$$

and evaluate $I(h, h')$ with the help of Lemma 2, we find
$I(h, h')$
$$= \frac{1}{8hh'} \int_{-\infty}^{\infty} F(t)\{|t+h+h'| + |t-h-h'| - |t+h-h'| - |t-h+h'|\}\, dt;$$

and, $F(t)$ being continuous, it is easy to show that $I(h, h') \to F(0)$.]

REFERENCES

1, 2, 3, 4. G. H. Hardy, 'Notes on some points in the integral calculus', 58, 59, 62, 67: *Messenger of Math.* 54 (1924), 20–7 and 81–8; 56 (1926), 10–16; 58 (1928), 53–8.
5. M. Riesz, 'Les fonctions conjuguées et les séries de Fourier': *Comptes Rendus*, 28 April 1924.
6. ——, 'Sur les fonctions conjuguées': *Math. Zeitschrift*, 27 (1927), 218–44.
7. E. C. Titchmarsh, 'Conjugate trigonometrical integrals': *Proc. London Math. Soc.* (2), 24 (1925), 109–30.
8. ——, 'Reciprocal formulae involving series and integrals': *Math. Zeitschrift*, 25 (1926), 321–47.
9. ——, 'A series inversion formula': *Proc. London Math. Soc.* (2), 26 (1927), 1–11.
10. G. N. Watson, 'General transforms' (unpublished).
11. W. H. Young, 'The application of expansions to definite integrals': *Proc. London Math. Soc.* (2), 9 (1910), 463–85.

CORRECTION

p. 105, *last line*. The second dw should be dt.

A CLASS OF FOURIER KERNELS

By G. H. HARDY and E. C. TITCHMARSH.

[Received and read 14 January, 1932.]†

Introduction and formalities.

1.1. We may say that $K(x)$ is a *Fourier kernel* if it gives rise to a "Fourier formula"

$$(1.1.1) \qquad F(x) = \int_0^\infty K(xu)\, du \int_0^\infty K(uy)\, F(y)\, dy$$

for an arbitrary function $F(x)$, or (what is, from a formal point of view, the same thing) to a "reciprocity"

$$(1.1.2) \qquad G(x) = \int_0^\infty K(xy)\, F(y)\, dy.$$

$$(1.1.3) \qquad F(x) = \int_0^\infty K(xy)\, G(y)\, dy,$$

The classical example is $K(x) = \sqrt{(2/\pi)} \cos x$.

Suppose that

$$(1.1.4) \qquad k(s) = \int_0^\infty x^{s-1} K(x)\, dx$$

(and similarly with other letters). Then, starting from (1.1.2), and integrating formally, we obtain

$$g(s) = \int_0^\infty x^{s-1}\, dx \int_0^\infty K(xy)\, F(y)\, dy = \int_0^\infty F(y)\, dy \int_0^\infty x^{s-1} K(yx)\, dx$$

$$= \int_0^\infty y^{-s} F(y)\, dy \int_0^\infty w^{s-1} K(w)\, dw = f(1-s)\, k(s).$$

Similarly (1.1.3) gives

$$f(s) = g(1-s)\, k(s);$$

---

† We are indebted to Mr. J. Cossar for reading the proofs and correcting a number of mistakes.

and, if we change s into $1-s$ in one of these equations, and multiply, we deduce that

(1.1.5) $$k(s)k(1-s) = 1.$$

Solving (1.1.4) by Mellin's inversion formula, we obtain

(1.1.6) $$K(x) = \frac{1}{2\pi i} \int_{c-i\infty}^{c+i\infty} x^{-s} k(s) \, ds.$$

We may therefore expect that any Fourier kernel must be expressible in this form, with a $k(s)$ subject to (1.1.5).

1.2. This condition may also be expected to be in some sense sufficient.

A characteristic property of a Fourier kernel $K(x)$ is that, **if**

(1.2.1) $$K_1(x) = \int_0^x K(u) \, du,$$

then

(1.2.2) $$\int_0^\infty K(yu) \frac{K_1(xu)}{u} du = \begin{cases} 1 & (0 < y < x) \\ 0 & (y > x). \end{cases}$$

In fact, (1.2.2) leads to

$$\int_0^x F(y) \, dy = \int_0^\infty F(y) \, dy \int_0^\infty K(yu) \frac{K_1(xu)}{u} du = \int_0^\infty \frac{K_1(xu)}{u} du \int_0^\infty K(uy) F(y) \, dy,$$

from which (1.1.1) follows by formal differentiation.

On the other hand, (1.1.6) gives

(1.2.3) $$K_1(x) = \frac{1}{2\pi i} \int_{c-i\infty}^{c+i\infty} \frac{x^{1-s}}{1-s} k(s) \, ds.$$

Taking $c = \tfrac{1}{2}$ in (1.1.6) and (1.2.3), we obtain

$$\int_0^\infty K(yu) \frac{K_1(xu)}{u} du = \lim_{\lambda \to \infty} \int_{1/\lambda}^{\lambda} K(yu) \frac{K_1(xu)}{u} du$$

$$= \lim_{\lambda \to \infty} \frac{1}{4\pi^2} \int_{1/\lambda}^{\lambda} \frac{du}{u} \int_{-\infty}^{\infty} (yu)^{-\frac{1}{2}-it} k(\tfrac{1}{2}+it) \, dt \int_{-\infty}^{\infty} \frac{(xu)^{\frac{1}{2}-it'}}{\tfrac{1}{2}-it'} k(\tfrac{1}{2}+it') \, dt'$$

$$= \lim_{\lambda \to \infty} \frac{1}{4\pi^2} \int_{-\infty}^{\infty} y^{-\frac{1}{2}-it} k(\tfrac{1}{2}+it) \, dt \int_{-\infty}^{\infty} \frac{x^{\frac{1}{2}-it'}}{\tfrac{1}{2}-it'} k(\tfrac{1}{2}+it') \, dt' \int_{1/\lambda}^{\lambda} u^{-1-it-it'} du$$

$$= \lim_{\lambda \to \infty} \frac{1}{2\pi^2} \int_{-\infty}^{\infty} y^{-\frac{1}{2}-it} k(\tfrac{1}{2}+it) \, dt \int_{-\infty}^{\infty} \frac{x^{\frac{1}{2}-it'}}{\tfrac{1}{2}-it'} k(\tfrac{1}{2}+it') \frac{\sin \mu(t'+t)}{t'+t} \, dt',$$

where

(1.2.4) $$\mu = \log \lambda.$$

If we pass to the limit formally in accordance with the ordinary rules, we obtain

$$\frac{1}{2\pi} \int_{-\infty}^{\infty} \left(\frac{x}{y}\right)^{\frac{1}{2}+it} k(\tfrac{1}{2}+it)\, k(\tfrac{1}{2}-it) \frac{dt}{\tfrac{1}{2}+it} = \frac{1}{2\pi} \int_{-\infty}^{\infty} \left(\frac{x}{y}\right)^{\frac{1}{2}+it} \frac{dt}{\tfrac{1}{2}+it},$$

by (1.1.5), and the value of this integral is 1 if $y < x$ and 0 if $y > x$.

1.3. We have stated our analysis in terms of the symmetrical formulae arising from a Fourier kernel. We may, however, apply similar analysis to the unsymmetrical equations

(1.3.1) $$F(x) = \int_0^\infty K(xu)\, du \int_0^\infty H(uy)\, F(y)\, dy,$$

(1.3.2) $$G(x) = \int_0^\infty H(xy)\, F(y)\, dy,$$

(1.3.3) $$F(x) = \int_0^\infty K(xy)\, G(y)\, dy.$$

The equation (1.1.5) is then replaced by

(1.3.4) $$k(s)\, h(1-s) = 1,$$

and the "solving kernel" of the equation (1.3.2) is defined by

(1.3.5) $$H(x) = \frac{1}{2\pi i} \int_{c-i\infty}^{c+i\infty} x^{-s} h(s)\, ds.$$

1.4. If $K(x)$ is a Fourier kernel, and

(1.4.1) $$F(x) = \lambda \int_0^\infty K(xy)\, F(y)\, dy,$$

then it follows from the Fourier formulae that $\lambda^2 = 1$. Thus a Fourier kernel has two *Eigenwerte* only, viz. ± 1. With each of these values is associated a mass of *Eigenfunktionen* (self-reciprocal functions) of very diverse types.

If $K(x)$ is not a Fourier kernel, and we operate on (1.4.1) as in §1.1, we obtain

$$\lambda^2 k(s)\, k(1-s) = 1,$$

which is in general false. A typical case is that in which

$$K(x) = e^{-x}, \quad k(s) = \Gamma(s).$$

We considered this case in our paper **11**. The inference from our present analysis is that there can be no solutions for which

$$\int_0^\infty f(x) x^{s-1} dx$$

exists. Actually, every real λ is an *Eigenwert*, and with each is associated (substantially) just one function.

1.5. Before proceeding further we illustrate our formalities by examples.

(1) If

(1.5.1) $$K(x) = x^{\frac{1}{2}} J_\nu(x),$$

then†

(1.5.2) $$k(s) = 2^{s-\frac{1}{2}} \frac{\Gamma(\frac{1}{2}\nu + \frac{1}{2}s + \frac{1}{4})}{\Gamma(\frac{1}{2}\nu - \frac{1}{2}s + \frac{3}{4})},$$

a solution of (1.1.5). The cases $\nu = -\frac{1}{2}$ and $\nu = \frac{1}{2}$ give

$$K(x) = \sqrt{\left(\frac{2}{\pi}\right)} \cos x, \quad k(s) = 2^{s-\frac{1}{2}} \frac{\Gamma(\frac{1}{2}s)}{\Gamma(\frac{1}{2} - \frac{1}{2}s)};$$

$$K(x) = \sqrt{\left(\frac{2}{\pi}\right)} \sin x, \quad k(s) = 2^{s-\frac{1}{2}} \frac{\Gamma(\frac{1}{2} + \frac{1}{2}s)}{\Gamma(1 - \frac{1}{2}s)}.$$

These functions give the Hankel and Fourier formulae.

(2) If $$K(x) = x^{\frac{1}{2}} Y_\nu(x),$$

then‡

$$k(s) = -2^{s-\frac{1}{2}} \pi^{-1} \Gamma(\tfrac{1}{2}s + \tfrac{1}{2}\nu + \tfrac{1}{4}) \Gamma(\tfrac{1}{2}s - \tfrac{1}{2}\nu + \tfrac{1}{4}) \cos(\tfrac{1}{2}s - \tfrac{1}{2}\nu + \tfrac{1}{4})\pi.$$

This is not a solution of (1.1.5), and $K(x)$ is not a Fourier kernel. Here (1.3.4) gives

$$h(s) = 2^{s-\frac{1}{2}} \frac{\Gamma(\tfrac{1}{2}s + \tfrac{1}{2}\nu + \tfrac{1}{4})}{\Gamma(\tfrac{1}{2}\nu - \tfrac{1}{2}s + \tfrac{3}{4})} \tan \tfrac{1}{2}(s + \nu + \tfrac{1}{2})\pi.$$

If we substitute this value of $h(s)$ in (1.3.5), with $0 < c < 1$, and evaluate the integral by calculation of the residues at poles on the left, we find that

$$H(x) = x^{\frac{1}{2}} \mathbf{H}_\nu(x),$$

---

† Watson (**18**), §13.24 (1).
‡ Watson (**18**), §13.24 (5).

where $\mathbf{H}_\nu(x)$ is Struve's function†. This transformation was found by Titchmarsh‡.

(3) Fox§ has discovered and discussed a number of very general transformations in which $K(x)$ and $H(x)$ are linear combinations of generalized hypergeometric functions. From our present point of view, these originate as follows.

Suppose that $a_1 > 0$, that ρ_1 and ρ_2 are any real numbers other than negative integers, and that

$$\phi = a_1 - \rho_1 - \rho_2 + \tfrac{1}{2},$$

and let
$$k(s) = 2^{s-1} \frac{\Gamma(a_1 + \tfrac{1}{2}\phi - \tfrac{1}{2}s)\,\Gamma(\tfrac{1}{2}s - \tfrac{1}{2}\phi)}{\Gamma(\rho_1 + \tfrac{1}{2}\phi - \tfrac{1}{2}s)\,\Gamma(\rho_2 + \tfrac{1}{2}\phi - \tfrac{1}{2}s)}.$$

Then $K(x)$ may be calculated from (1.1.6), with $0 < c < a_1$; we find

$$K(x) = (\tfrac{1}{2}x)^{-\phi} \sum_{n=0}^{\infty} \frac{\Gamma(a_1+n)}{\Gamma(\rho_1+n)\,\Gamma(\rho_2+n)} \frac{(-\tfrac{1}{4}x^2)^n}{n!} = (\tfrac{1}{2}x)^{-\phi}\,{}_1F_2(a_1, \rho_1, \rho_2;\, -\tfrac{1}{4}x^2),$$

in the notation of Barnes. If now we calculate $h(s)$ from (1.3.4), and $H(x)$ from (1.3.5), by summation of residues, we find

$$H(x) = H_1(x) + H_2(x),$$

where
$$H_1(x) = \frac{\sin(a_1 - \rho_1)\pi}{\sin(\rho_2 - \rho_1)\pi} (\tfrac{1}{2}x)^{2\rho_1 + \phi - 1}\,{}_1F_2(1 - a_1 + \rho_1,\, 1 - \rho_2 + \rho_1,\, \rho_1;\, -\tfrac{1}{4}x^2),$$

and $H_2(x)$ is derived from $H_1(x)$ by interchange of ρ_1 and ρ_2. The formulae thus obtained are those of Fox's Theorem 1, in the special case $p = 1$. In the general case $k(s)$ is a more complex product of gamma-functions of the same type.

The case $a_1 = 1$, $\rho_1 = \tfrac{3}{2}$, $\rho_2 = \nu + \tfrac{3}{2}$ gives example (2) above. The case $a_1 = 1$, $\rho_1 = a+1$, $\rho_2 = \nu + a + 1$ gives a more general transformation found by Hardy‖ and discussed by Cooke¶. The case $a_1 = \nu + \tfrac{3}{2}$, $\rho_1 = \nu + 1$, $\rho_2 = 2\nu + 1$ gives

$$K(x) = \tfrac{1}{2}\sqrt{\pi}\,\frac{d}{dx}\{xJ_\nu^2(\tfrac{1}{2}x)\}, \quad H(x) = -\sqrt{\pi}\,J_\nu(\tfrac{1}{2}x)\,Y_\nu(\tfrac{1}{2}x);$$

---

† Watson (18), §10.4, §13.24 (2).
‡ Titchmarsh (15).
§ Fox (4).
‖ Hardy (7).
¶ Cooke (3).

a transformation due to Bateman†. The case $a_1 = \nu+a+\tfrac{1}{2}$, $\rho_1 = \nu+a+1$, $\rho_2 = 2\nu+a+1$ gives a more general transformation due to Titchmarsh‡. Fuller details concerning these transformations will be found in §5.2 of Fox's paper.

If we take
$$k(s) = 2^{s-\tfrac{1}{2}} \frac{\Gamma(a_1+\tfrac{1}{2}s)\,\Gamma(a_2+\tfrac{1}{2}s)\,\Gamma(a_3-\tfrac{1}{2}s)}{\Gamma(b_1+\tfrac{1}{2}s)\,\Gamma(b_2-\tfrac{1}{2}s)\,\Gamma(b_3-\tfrac{1}{2}s)},$$
where
$$a_1+a_2+a_3+\tfrac{1}{2} = b_1+b_2+b_3,$$
we obtain examples of Fox's Theorem 2. For example, if
$$a_1 = \tfrac{1}{2}\mu+\tfrac{1}{2}\nu+\tfrac{1}{2}, \quad a_2 = \tfrac{1}{2}-\tfrac{1}{2}\mu-\tfrac{1}{2}\nu, \quad a_3 = \tfrac{1}{2},$$
$$b_1 = 1, \quad b_2 = \tfrac{1}{2}\nu-\tfrac{1}{2}\mu+\tfrac{1}{2}, \quad b_3 = \tfrac{1}{2}\mu-\tfrac{1}{2}\nu+\tfrac{1}{2},$$
$K(x)$ and $H(x)$ are each combinations of two hypergeometric functions, and can be reduced to the forms§

$$K(x) = \frac{\sqrt{\pi}}{2\sqrt{2}\sin\tfrac{1}{2}(\mu+\nu)\pi}\, x\{J_{-\mu}(\tfrac{1}{2}x)\,J_{-\nu}(\tfrac{1}{2}x) - J_{\mu}(\tfrac{1}{2}x)\,J_{\nu}(\tfrac{1}{2}x)\},$$

$$H(x) = \frac{\sqrt{\pi}}{2\sqrt{2}\cos\tfrac{1}{2}(\mu-\nu)\pi}\, \frac{d}{dx}\{x[J_{\mu}(\tfrac{1}{2}x)\,J_{-\nu}(\tfrac{1}{2}x) + J_{-\mu}(\tfrac{1}{2}x)\,J_{\nu}(\tfrac{1}{2}x)]\}.$$

(4) If
$$K(x) = x^{\tfrac{1}{2}}\left\{Y_{\nu}(x) + \frac{2}{\pi}\cos a\pi\, K_{\nu}(x)\right\},$$
where $K_{\nu}(x)$ is the function of Basset and Macdonald, then∥ $k(s)$ is the product of 2^{2s-1} by

$$\frac{\Gamma(\tfrac{1}{4}s+\tfrac{1}{4}\nu+\tfrac{1}{8})\,\Gamma(\tfrac{1}{4}s+\tfrac{1}{4}\nu+\tfrac{5}{8})\,\Gamma(\tfrac{1}{4}s-\tfrac{1}{4}\nu+\tfrac{1}{8})\,\Gamma(\tfrac{1}{4}s-\tfrac{1}{4}\nu+\tfrac{5}{8})}{\Gamma(\tfrac{1}{8}-\tfrac{1}{4}\nu+\tfrac{1}{8}+\tfrac{1}{2}a)\,\Gamma(\tfrac{7}{8}+\tfrac{1}{4}\nu-\tfrac{1}{8}s-\tfrac{1}{2}a)\,\Gamma(\tfrac{1}{8}s-\tfrac{1}{4}\nu+\tfrac{1}{8}-\tfrac{1}{2}a)\,\Gamma(\tfrac{7}{8}+\tfrac{1}{4}\nu-\tfrac{1}{8}s+\tfrac{1}{2}a)}.$$

In this case again $H(x)$ is the sum of two hypergeometric series. There are two interesting particular cases. If $\nu = 0$, $a = 1$, then
$$k(s) = -2^{2s-1}\left\{\frac{\Gamma(\tfrac{1}{4}s+\tfrac{1}{8})}{\Gamma(\tfrac{3}{8}-\tfrac{1}{4}s)}\right\}^2,$$
which satisfies (1.1.5), so that
$$K(x) = x^{\tfrac{1}{2}}\left\{Y_0(x) - \frac{2}{\pi}K_0(x)\right\}$$

---

† Bateman (1), 482, and (2), 75.
‡ Titchmarsh (17).
§ See Fox (4), §5.3.
∥ Watson (18), §3.7, §13.21 (8), §13.24 (5).

is a Fourier kernel. If $\nu = 2$, $a = 0$, then

$$k(s) = 2^{2s-1} \left\{ \frac{\Gamma(\tfrac{1}{4}s + \tfrac{9}{8})}{\Gamma(\tfrac{11}{8} - \tfrac{1}{4}s)} \right\}^2 \frac{\tfrac{1}{4}s - \tfrac{3}{8}}{\tfrac{1}{4}s + \tfrac{1}{8}},$$

and $k(s)k(1-s) = -1$, so that

$$K(x) = -H(x) = x^{\tfrac{1}{2}} \left\{ Y_2(x) + \frac{2}{\pi} K_2(x) \right\}.$$

The formulae in this case are due to A. L. Dixon and Hardy†. Much more general formulae of a similar character have been found by Steen‡ and by Kuttner, whose results are not yet published.

(5) If
$$k(s) = e^{-ai(s-\tfrac{1}{2})^2} \quad (a > 0),$$

then
$$h(s) = e^{ai(s-\tfrac{1}{2})^2}.$$

Taking $c = \tfrac{1}{2}$ in (1.1.6), we find that

$$K(x) = \frac{1}{2\pi \sqrt{x}} \int_{-\infty}^{\infty} e^{ait^2} \cos(t \log x)\, dt = \frac{e^{\tfrac{1}{4}\pi i}}{2\sqrt{(\pi a x)}} e^{-i(\log x)^2/4a},$$

while $H(x)$ is the conjugate function. The Fourier formula thus obtained may be reduced by a change of variable to the exponential form of the ordinary Fourier formula. It is

$$F(x) = \frac{1}{4\pi a} \int_0^{\infty} e^{-i\{\log(xu)\}^2/4a} \frac{du}{\sqrt{(xu)}} \int_0^{\infty} e^{i\{\log(uy)\}^2/4a} \frac{F(y)}{\sqrt{(uy)}}\, dy\,;$$

and, if we put $a = \tfrac{1}{2}$ and

$$x = e^{\xi}, \quad u = e^{\zeta}, \quad y = e^{\eta}, \quad f(\xi) = e^{\tfrac{1}{2}i\xi^2 + \tfrac{1}{2}\xi} F(e^{\xi}),$$

we obtain
$$f(\xi) = \frac{1}{2\pi} \int_{-\infty}^{\infty} e^{-i\xi\zeta}\, d\zeta \int_{-\infty}^{\infty} e^{i\zeta\eta} f(\eta)\, d\eta.$$

This formula is not included in our standard form, since the limits are $-\infty$ and ∞ instead of 0 and ∞.

(6) If
$$k(s) = e^{a(s-\tfrac{1}{2})^3} \quad (a > 0),$$

then $K(x)$ is a Fourier kernel. Taking $c = \tfrac{1}{2}$ in (1.1.6), we find

$$K(x) = \frac{1}{\pi \sqrt{x}} \int_0^{\infty} \cos(at^3 + t \log x)\, dt.$$

---

† Hardy (10); see also Hardy and Titchmarsh (12), 212.
‡ See Steen (14).

This is an "Airy" integral, and†

$$K(x) = \frac{1}{3\sqrt{(3ax)}} \left(\log \frac{1}{x}\right)^{\frac{1}{2}} \left[J_{\frac{1}{3}}\left\{ \frac{2}{3\sqrt{(3a)}} \left(\log \frac{1}{x}\right)^{\frac{3}{2}} \right\} \right.$$

$$\left. + J_{-\frac{1}{3}}\left\{ \frac{2}{3\sqrt{(3a)}} \left(\log \frac{1}{x}\right)^{\frac{3}{2}} \right\} \right] \quad (0 < x < 1),$$

$$K(x) = \frac{1}{3\pi\sqrt{(ax)}} (\log x)^{\frac{1}{2}} K_{\frac{1}{3}}\left\{ \frac{2}{3\sqrt{(3a)}} (\log x)^{\frac{3}{2}} \right\} \quad (x > 1).$$

(7) If
$$k(s) = 1,$$

then (1.1.5) is satisfied, but the integral (1.1.6) is not convergent. If, however, we regard (1.2.3), with $0 < c < 1$, as the definition of $K_1(x)$, we have

$$K_1(x) = \frac{1}{2\pi i} \int_{c-i\infty}^{c+i\infty} \frac{x^{1-s}}{1-s} ds = \begin{cases} 1 & (x > 1) \\ 0 & (0 < x < 1). \end{cases}$$

And if we replace (1.1.2) and (1.1.3) by

$$F(x) = \frac{1}{x} \int_0^\infty G(y) \, dK_1(xy), \quad G(x) = \frac{1}{x} \int_0^\infty F(y) \, dK_1(xy),$$

then our formulae become

$$F(x) = \frac{1}{x} G\left(\frac{1}{x}\right), \quad G(x) = \frac{1}{x} F\left(\frac{1}{x}\right),$$

which are plainly consequences of one another.

(8) In all these examples $k(s)$ is analytic. Suppose, however, that $c = \frac{1}{2}$ and
$$k(\tfrac{1}{2} + it) = i \operatorname{sgn} t.$$

Then h, defined by (1.3.4), is $-k$. The integral (1.1.6) is not convergent, but it is formally

$$\frac{i}{2\pi x^{\frac{1}{2}}} \left(\int_0^\infty x^{-it} dt - \int_{-\infty}^0 x^{-it} dt \right) = -\frac{i}{2\pi x^{\frac{1}{2}}} \int_0^\infty (x^{it} - x^{-it}) dt$$

$$= \frac{1}{\pi x^{\frac{1}{2}}} \int_0^\infty \sin(t \log x) \, dt = \frac{1}{\pi x^{\frac{1}{2}} \log x},$$

---

† Watson (18), § 6.4 (1), (2).

the integral being summable $(C, 1)$†. Our formulae become

$$F(x) = \frac{1}{\pi} \int_0^\infty \frac{G(y)\,dy}{(xy)^{\frac{1}{2}} \log(xy)}, \qquad G(x) = -\frac{1}{\pi} \int_0^\infty \frac{F(y)\,dy}{(xy)^{\frac{1}{2}} \log(xy)}.$$

If we replace x and y by e^x and e^y, and interpret the integrals as principal values, we obtain formulae equivalent to those of the theory of "Hilbert transforms".

(9) If $k(s) = \cot \tfrac{1}{2} s\pi$, then (1.1.5) is satisfied. The integral (1.1.6) is of the same type as in (8). A formal application of the theorem of residues gives

$$K(x) = \frac{2}{\pi} \frac{1}{1-x^2},$$

and we again obtain formulae of the "Hilbert transform" type.

(10) We obtain formulae of a somewhat different type by taking

$$k(\tfrac{1}{2}+it) = e^{i/t}.$$

Then (1.1.5) is satisfied, and the formula for $K(x)$ is

$$K(x) = \frac{1}{2\pi x^{\frac{1}{2}}} \int_{-\infty}^\infty e^{i/t} x^{-it} dt = \frac{1}{\pi x^{\frac{1}{2}}} \int_0^\infty \cos\left(\frac{1}{t} - t \log x\right) dt.$$

The integral is summable $(C, 1)$ if $x \neq 1$, and has the value‡

$$-\left(x \log \frac{1}{x}\right)^{-\frac{1}{2}} J_1\left\{2\left(\log \frac{1}{x}\right)^{\frac{1}{2}}\right\} \quad (0 < x < 1), \qquad 0 \quad (x > 1).$$

If $x = 1$, the integral for $K(x)$ diverges to infinity, and $K_1(x)$ has a discontinuity, as in example (7). The formula which results is therefore

$$G(x) = \frac{1}{x} F\left(\frac{1}{x}\right) - \int_0^{1/x} \frac{J_1\{2(-\log xy)^{\frac{1}{2}}\}}{(-xy \log xy)^{\frac{1}{2}}} F(y)\,dy.$$

If we put

$$x = e^{-\xi}, \quad y = e^{-\eta}, \quad e^{-\frac{1}{2}\xi} F(e^{-\xi}) = \phi(\xi), \quad e^{-\frac{1}{2}\xi} G(e^{-\xi}) = \psi(\xi),$$

we obtain

$$\psi(\xi) = \phi(-\xi) - \int_{-\xi}^\infty \frac{J_1\{2\sqrt{(\xi+\eta)}\}}{\sqrt{(\xi+\eta)}} \phi(\eta)\,d\eta.$$

---

† We could use $K_1(x)$ as in the last example, but this would involve a certain amount of calculation.

‡ See Watson (18), §6.23 (1), (2): we have to consider a limiting case.

The reciprocal formula is obtained by interchanging ϕ and ψ. The Fourier formula which results may be verified by using the integral†

$$\int_{-\lambda}^{\infty} \frac{J_1\{2\sqrt{(x+\lambda)}\} J_1\{2\sqrt{(x+\mu)}\}}{\sqrt{\{(x+\lambda)(x+\mu)\}}} dx = \frac{J_1\{2\sqrt{(\mu-\lambda)}\}}{\sqrt{(\mu-\lambda)}}.$$

1.6. There are two different directions in which we may look for a general theory of all these transformations.

(1) We may attempt to construct a theory of "transforms" similar to Plancherel's theory of Fourier transforms. Our functions will then, in the first instance at any rate, be functions of integrable square (of the Lebesgue class L^2); and our object will be to show that the formulae (1.1.2) and (1.1.3) are valid (at any rate in some generalized sense) for almost all x.

This is what has been done by Watson in the paper which follows. Watson has shown that, if only ‡

(1.6.1) $$k(\tfrac{1}{2}+it)\, k(\tfrac{1}{2}-it) = 1,$$

(1.6.2) $$\int_{-\infty}^{\infty} \left| \frac{k(\tfrac{1}{2}+it)}{\tfrac{1}{2}+it} \right|^2 dt < \infty,$$

then there is a satisfactory generalization of Plancherel's theory.

(2) We may look for an analogue of the older theory of the Fourier integral. This is a "convergence theory", in the widened sense in which "convergence" includes summability by Cesàro's or similar means; and the functions to which it applies are not necessarily of integrable square. We wish to impose conditions on $k(s)$ such that, when $K(x)$ is defined by (1.1.6), then (1.1.1) shall be true in the classical or in some more general sense. It is essential (from our present point of view) that our conditions shall bear directly on $k(s)$ and on $k(s)$ alone.

This is what we have tried to do in the present paper. Our results have none of the generality of Watson's; we have found it necessary to subject $k(s)$ to very much more stringent conditions. We do not suggest that our theorems are at all so general as they should be; but we are

---

† Watson (**18**), §13.47 (10).
‡ If $K(x)$ is real, $k(s)$ is real for real s, (1.6.1) is $|k(\tfrac{1}{2}+it)|^2 = 1$, and (1.6.2) is a consequence of (1.6.1).

disposed to think that our restrictions bear some relation to the facts, and that a convergence theory really requires conditions of a more drastic type.

We begin by supposing that $k(\tfrac{1}{2}+it)$, while not necessarily analytic, behaves much as

$$k_1(s) = \sqrt{\left(\frac{2}{\pi}\right)}\,\Gamma(s)\cos\tfrac{1}{2}s\pi$$

behaves on the line $\sigma = \tfrac{1}{2}$. This function is the $k(s)$ which corresponds to the standard case of the Fourier cosine transformation. In other words we assume that $k(\tfrac{1}{2}+it)$ behaves, for large t, very much like

$$e^{i(t\log t - t)}.$$

On this hypothesis we prove Theorem 1. In the repeated integral here the inner integral is a Lebesgue integral, but the outer integral is defined by a summability convention of Rieszian type. It is a peculiarity of the theorem, which we cannot remove without further restrictions on $k(s)$, that this convention is required even when $F(y)$ is of bounded variation, when the ordinary Fourier formula is true in the classical sense. For this reason, though our formulae include those of Fox and other writers, Theorem 1 does not include the theorems which they prove.

We have proved a corresponding theorem (Theorem 2) in which $k(\tfrac{1}{2}+it)$ behaves like

$$e^{it^c} \quad (c > 1),$$

but the proof is long and we do not reproduce the details. There are doubtless similar theorems involving $e^{i\chi(t)}$, where χ is a general function subject to appropriate restrictions, but it is probable that the proof of any such theorem would be a very tedious task.

In §6 we strengthen our hypotheses concerning $k(s)$, supposing now that $k(s)$ is analytic and behaves very much like $k_1(s)$ in a strip parallel to the imaginary axis. We are then able to prove a theorem (Theorem 3) in which the integral converges in the classical sense. This theorem includes (for example) the theorems of Fox.

In §7 we prove a theorem concerning the special case considered in example (6) above. It would, no doubt, be possible to deal with the more general cases in which

$$k(s) = e^{-ai(s-\tfrac{1}{2})^{2p}}$$

or

$$k(s) = e^{a(s-\tfrac{1}{2})^{2p+1}},$$

and p is any positive integer, or in which $k(s)$ mimics the behaviour of these functions sufficiently closely. The proofs would be long, and we confine ourselves to the simplest case†.

Our selection of cases for detailed discussion may seem somewhat arbitrary, but has been dictated by the desire (a) to keep the paper within reasonable bounds and (b) to include so far as possible the special transformations catalogued in §1. Actually our analysis is sufficient to cover all the examples (1) to (7).

We conclude, in §7, by proving a theorem which, unlike all those which precede, was suggested to us by Watson's work. The theorem covers only a special class of $F(x)$, and is of a different character in that we impose a condition directly on $K(x)$. If we do this, it is reasonable to expect results applicable to less specialized kernels.

Summability.

2.1. In the analysis which follows, $k(s) = k(\sigma+it)$ is defined for $\sigma = \frac{1}{2}$ only, and is a complex function of the real variable t. We retain the notation $k(\frac{1}{2}+it)$ for formal convenience.

THEOREM 1. *Suppose that $k(\frac{1}{2}+it)$ satisfies the following conditions*: (i) *it is bounded*; (ii)

(2.1.1) $$k(\tfrac{1}{2}+it) = e^{i(t\log t - t)}\{a+\beta(t)+\gamma(t)\},$$

for $t > 1$, where a is constant, β is a constant multiple of a function which tends steadily to 0, and γ is (bounded and) integrable in $(1, \infty)$; and

$$k(\tfrac{1}{2}+it) = e^{i(t\log|t|-t)}\{a'+\beta'(t)+\gamma'(t)\}$$

for $t < -1$, a', β', and γ' satisfying the same conditions as a, β, and γ; (iii)

(2.1.2) $$k(\tfrac{1}{2}+it)\,k(\tfrac{1}{2}-it) = 1.$$

Suppose also that $y^{-\frac{1}{2}} F(y)$ is integrable in $(0, 1)$, and $F(y)$ in $(1, \infty)$, that x is positive, and that $F(y)$ is of bounded variation near $y = x$. Then

(2.1.3) $$\lim_{\lambda\to\infty} \int_{1/\lambda}^{\lambda} \left(1 - \frac{|\log u|}{\log \lambda}\right) K(xu)\,du \int_0^\infty K(uy)\,F(y)\,dy$$
$$= \tfrac{1}{2}\{F(x+0)+F(x-0)\}.$$

We say that a function $k(\frac{1}{2}+it)$ which satisfies (i), (ii), (iii) belongs to class **k**; the class includes all the examples of §1 in which $k(s)$ is a

---

† Apart from the case $p = 1$, which, after what was said under (5) above, requires no special discussion.

combination of gamma-functions. Consider, for example, Hankel's formulae, in which $k(s)$ is given by (1.5.2). Stirling's formula gives

$$k(s) = e^{\frac{1}{2}\pi i \nu} e^{i(t\log t - t)} t^{\sigma - \frac{1}{2}} \left\{ 1 + \frac{b}{t} + O\left(\frac{1}{t^2}\right) \right\} \quad (t > 1),$$

where b is a function of σ; and this is of the form (2.1.1) when $\sigma = \frac{1}{2}$.

We base the proof of Theorem 1 on a series of lemmas. Lemmas 1–4 are concerned with integrals of the form

$$\int e^{i\psi(t)} dt,$$

and are of a type familiar in the work of van der Corput and other writers. Lemmas 5–7 assert properties of the $K(x)$ corresponding to a $k(s)$ of the class **k**; and Lemma 8 is a case of a well-known theorem of Hobson.

Lemmas concerning integrals $\int e^{i\psi(t)} dt$.

3.1. LEMMA 1. *If* $\psi'(t)$ *is monotonic, and* $\psi'(t) \geq r > 0$ *or* $\psi'(t) \leq -r < 0$, *then*

$$\left| \int_a^b e^{i\psi(t)} dt \right| \leq \frac{4}{r}.$$

Suppose, for example, that $\psi'(t)$ increases and is not less than r. Then (with an obvious notation)

$$\int_a^b e^{i\psi(t)} dt = \int_\alpha^\beta \frac{e^{iu}}{\psi'(t)} du = \int_\alpha^\beta \frac{\cos u}{\psi'(t)} du + i \int_\alpha^\beta \frac{\sin u}{\psi'(t)} du$$

$$= \frac{1}{\psi'(a)} \int_\alpha^\gamma \cos u \, du + \frac{i}{\psi'(a)} \int_\alpha^{\gamma'} \sin u \, du,$$

and the modulus of this does not exceed $4/r$.

3.2. LEMMA 2. *If* $\psi''(t) \geq r > 0$ *or* $\psi''(t) \leq -r < 0$, *then*

$$\left| \int_a^b e^{i\psi(t)} dt \right| \leq \frac{8}{\sqrt{r}}.$$

Since $\psi'(t)$ is monotonic, it can vanish at most once, say for $t = t_0$. We prove the lemma on the hypothesis that $a < t_0 < b$, leaving to the reader the simple modifications required in other cases.

We have

(3.2.1) $\quad I = \int_a^b e^{i\psi(t)} dt = \int_a^{t_0-\rho} + \int_{t_0-\rho}^{t_0+\rho} + \int_{t_0+\rho}^b = I_1 + I_2 + I_3;$

some parts of the integrals may be absent. Then $|I_2| \leqslant 2\rho$. Also

$$|\psi'(t_0+\rho)| = \left|\int_{t_0}^{t_0+\rho} \psi''(t)\,dt\right| \geqslant r\rho;$$

and so $|\psi'(t)| \geqslant r\rho$ for $t \geqslant t_0+\rho$. Hence, by Lemma 1, $|I_3| \leqslant 4/r\rho$.

A similar inequality holds for I_1, so that

$$|I| \leqslant 2\rho + \frac{8}{r\rho}.$$

Taking $\rho = 2r^{-\frac{1}{2}}$, we obtain the result.

3.3. LEMMA 3. *Suppose that $\psi'(t)$ is monotonic and that $|\psi''(t)|$ and $|\psi'''(t)|$ are decreasing; and that*

$$\psi'(t_0) = 0, \quad \psi''(t_0) \neq 0,$$
$$|\psi'''(t_0)| \leqslant B|\psi''(t_0)|^{\frac{3}{2}}.$$

Then
$$\left|\int_a^b e^{i\psi(t)}\,dt\right| \leqslant \frac{C}{|\psi''(t_0)|^{\frac{1}{2}}},$$

where $C = C(B)$ depends only on B.

It is not necessary that t_0 should fall in the range of integration, which may be any part of the range in which the conditions are satisfied; but we state the proof in the typical case in which $a < t_0 < b$. We divide I into I_1, I_2, I_3 as in (3.2.1), and write ψ_2 and ψ_3 for $|\psi''(t_0)|$ and $|\psi'''(t_0)|$. Then $|I_2| \leqslant 2\rho$ as before. Also

$$|\psi'(t_0-\rho)| = \left|\int_{t_0-\rho}^{t_0} \psi''(t)\,dt\right| \geqslant \rho\psi_2,$$

and

$$|\psi'(t_0+\rho)| = \left|\int_{t_0}^{t_0+\rho} \psi''(t)\,dt\right| \geqslant \rho|\psi''(t_0+\rho)| = \rho\left|\psi''(t_0) + \int_{t_0}^{t_0+\rho} \psi'''(t)\,dt\right|$$

$$\geqslant \rho\psi_2 - \rho^2\psi_3 = \rho\psi_2\left(1-\rho\psi_2^{\frac{1}{2}}\frac{\psi_3}{\psi_2^{\frac{3}{2}}}\right) \geqslant \rho\psi_2(1-B\rho\psi_2^{\frac{1}{2}}).$$

Hence, if we take
$$\rho = (2B)^{-1}\psi_2^{-\frac{1}{2}},$$

we obtain
$$|I| \leqslant 2\rho + \frac{4}{\rho\psi_2} + \frac{8}{\rho\psi_2} < C\psi_2^{-\frac{1}{2}}.$$

3.4. **Lemma 4.** *If $\chi(t)$ is a positive decreasing function, and*

$$\left|\int_a^b e^{i\psi(t)}\,dt\right| \leqslant C$$

for all b, then
$$\left|\int_a^b e^{i\psi(t)}\chi(t)\,dt\right| \leqslant 4C\chi(a).$$

This is an obvious corollary of the second mean value theorem.

Lemmas concerning $K(x)$.

4.1. **Lemma 5.** *If $k(\tfrac{1}{2}+it)$ belongs to $\mathbf{k}$, then the integral $(1.1.6)$, with $c = \tfrac{1}{2}$, is uniformly convergent in any finite interval $0 < a \leqslant x \leqslant b$, and $K(x)$ is continuous in such an interval. Also $K(x)$ is bounded as $x \to \infty$, and is $O(x^{-\frac{1}{2}})$ as $x \to 0$.*

By Lemma 1, with $\psi(t) = t\log t - t - t\log x$,

$$\left|\int_T^{T'} e^{i(t\log t - t)} x^{-it}\,dt\right| \leqslant \frac{4}{\log T - \log x},$$

for $x < T < T'$. Hence the part of $(1.1.6)$ arising from the a in $(2.1.1)$ is uniformly convergent. The uniform convergence of the part arising from β then follows from Lemma 4, and that of the part arising from γ is obvious. This proves the first clause of the lemma.

Next let

$$(4.1.1)\quad K(x) = \frac{1}{2\pi x^{\frac{1}{2}}}\left\{\int_{-\infty}^{-1} + \int_{-1}^{1} + \int_{1}^{\infty}\right\} k(\tfrac{1}{2}+it)\,x^{-it}\,dt = \frac{1}{2\pi x^{\frac{1}{2}}}(J_1 + J_2 + J_3).$$

Plainly J_2 is bounded for all x. Also

$$J_3 = a\int_1^\infty e^{i\psi(t)}\,dt + \int_1^\infty e^{i\psi(t)}\beta(t)\,dt + \int_1^\infty e^{i\psi(t)}\gamma(t)\,dt.$$

The first integral here satisfies the conditions of Lemma 3 for large x, since $t_0 = x$, $\psi'' = 1/t$, $\psi''' = -1/t^2$. Hence it is $O(x^{\frac{1}{2}})$ for large x. If $x < 1$,

$$\psi'(t) = \log t - \log x > \log\frac{1}{x},$$

and the integral is bounded, by Lemma 1. Lemma 4 then shows that the second integral has the same properties; and the third is obviously bounded. Hence J_3 is bounded for small x, and $O(x^{\frac{1}{2}})$ for large x. A similar argument applies to J_1. The second clause of the lemma then follows from $(4.1.1)$.

It should be observed that (1.2.3) now follows from (1.1.6).

4.2. LEMMA 6. *If $k(\tfrac{1}{2}+it)$ belongs to* **k**, *and*
$$\Phi(\lambda, x, y) = \int_{1/\lambda}^{\lambda} \left(1 - \frac{|\log u|}{\log \lambda}\right) K(xu) K(yu) \, du,$$
where $\lambda > 2$ and $x > 0$, then
$$|y^{\frac{1}{2}} \Phi| < B = B(x, \zeta)\dagger$$
for all positive y for which $|y-x| \geqslant \zeta$.

We have
$$(4.2.1) \quad \Phi(\lambda, x, y) = \frac{1}{4\pi^2} \int_{1/\lambda}^{\lambda} \left(1 - \frac{|\log u|}{\log \lambda}\right) du \int_{-\infty}^{\infty} k(\tfrac{1}{2}+it)(xu)^{-\frac{1}{2}-it} dt$$
$$\times \int_{-\infty}^{\infty} k(\tfrac{1}{2}+it')(yu)^{-\frac{1}{2}-it'} dt'.$$

Let
$$(4.2.2) \quad \Phi(\lambda, x, y, T) = \frac{1}{4\pi^2} \int_{1/\lambda}^{\lambda} \left(1 - \frac{|\log u|}{\log \lambda}\right) du \int_{-T}^{T} k(\tfrac{1}{2}+it)(xu)^{-\frac{1}{2}-it} dt$$
$$\times \int_{-T}^{T} k(\tfrac{1}{2}+it')(yu)^{-\frac{1}{2}-it'} dt'.$$

By Lemma 5, $\Phi(\lambda, x, y)$ is the limit of $\Phi(\lambda, x, y, T)$, and it is therefore enough to prove that
$$(4.2.3) \quad |y^{\frac{1}{2}} \Phi(\lambda, x, y, T)| < B(x, \zeta).$$

We may invert the order of integration in (4.2.2). Since
$$\int_{1/\lambda}^{\lambda} \left(1 - \frac{|\log u|}{\log \lambda}\right) u^{-1-it-it'} du = \frac{4 \sin^2 \tfrac{1}{2}\mu(t+t')}{\mu(t+t')^2},$$
where $\quad \mu = \log \lambda,$

we have
$$(4.2.4) \quad \Phi(\lambda, x, y, T)$$
$$= \frac{1}{\pi^2 (xy)^{\frac{1}{2}}} \int_{-T}^{T} k(\tfrac{1}{2}+it) x^{-it} dt \int_{-T}^{T} k(\tfrac{1}{2}+it') y^{-it'} \frac{\sin^2 \tfrac{1}{2}\mu(t+t')}{\mu(t+t')^2} dt'.$$

---

† Here, and in all our later analysis, x is "fixed". It is to be observed that B depends upon the function $k(\tfrac{1}{2}+it)$ as well as upon the parameters shown, any number occurring in a datum concerning k being treated as a constant. When we say, in the argument which follows, that a function is bounded, or is $O(1)$, we mean that it is bounded by a B.

We split up the integral into

$$\int_{-1}^{1}\int_{-T}^{T} + \int_{1}^{T}\int_{-1}^{1} + \int_{-T}^{-1}\int_{-1}^{1} + \int_{1}^{T}\int_{1}^{T} + \int_{1}^{T}\int_{-T}^{-1} + \int_{-T}^{-1}\int_{1}^{T} + \int_{-T}^{-1}\int_{-T}^{-1}$$

$$= J_1 + J_2 + J_3 + J_4 + J_5 + J_6 + J_7.$$

Since $k(\tfrac{1}{2}+it)$ is bounded,

$$J_1 = O\left\{\int_{-1}^{1} dt \int_{-T}^{T} \frac{\sin^2 \tfrac{1}{2}\mu(t+t')}{\mu(t+t')^2} dt'\right\} = O\left(\int_{-1}^{1} dt \int_{-\infty}^{\infty} \frac{\sin^2 \tfrac{1}{2}\mu z}{\mu z^2} dz\right) = O(1),$$

and similarly J_2 and J_3 are bounded. Since J_7 is of the same type as J_4, and J_6 as J_5, it is enough to prove J_4 and J_5 bounded.

The leading term in J_4 is

$$a^2 \int_1^T e^{i(t\log t - t)} x^{-it} dt \int_1^T e^{i(t'\log t' - t')} y^{-it'} \frac{\sin^2 \tfrac{1}{2}\mu(t+t')}{\mu(t+t')^2} dt'.$$

If we make the substitution

$$t = u+v, \quad t' = u-v,$$

we obtain

$$\tfrac{1}{2} \int_1^{\tfrac{1}{2}T+\tfrac{1}{2}} (xy)^{-iu} \frac{\sin^2 \mu u}{\mu u^2} du \int_{1-u}^{u-1} e^{i\psi(v)} dv + \int_{\tfrac{1}{2}T+\tfrac{1}{2}}^{T} \int_{u-T}^{T-u} = I_1 + I_2,$$

where

$$\psi(v) = (u+v)\log(u+v) + (u-v)\log(u-v) - 2u + v\log(y/x),$$

and the integrands are the same in I_2 as in I_1. The inner integrals are of the type considered in Lemma 2, with

$$\psi''(v) = \frac{1}{u+v} + \frac{1}{u-v} \geqslant \frac{2}{u}.$$

Hence each inner integral is $O(u^{\tfrac{1}{2}})$, and the whole is

$$O\left(\int_1^T \frac{du}{\mu u^{\tfrac{3}{2}}}\right) = O(1).$$

Similarly (using Lemma 4) the terms arising from $a\beta(t)$, $a\beta(t')$, and $\beta(t)\beta(t')$ are bounded; and those involving γ are of the type

$$O\left\{\int_1^T |\gamma(t)|dt \int_1^T \frac{dt'}{t'^2}\right\} = O(1).$$

Hence J_4 is bounded.

Passing to J_5, the leading term is

$$aa' \int_1^T e^{i(t\log t - t)} x^{-it} dt \int_1^T e^{-i(t'\log t' - t')} y^{it'} \frac{\sin^2 \tfrac{1}{2}\mu(t-t')}{\mu(t-t')^2} dt'.$$

Making the same substitution, we obtain

(4.2.5) $$aa' \int_0^{\tfrac{1}{2}T - \tfrac{1}{2}} (xy)^{-iv} \frac{\sin^2 \mu v}{\mu v^2} dv \int_{v+1}^{T-v} e^{i\chi(u)} du,$$

where

$$\chi(u) = (u+v)\log(u+v) - (u-v)\log(u-v) - 2v + u\log(y/x),$$

together with a similar integral over $-\tfrac{1}{2}T + \tfrac{1}{2} \leqslant v \leqslant 0$. We may confine our attention to (4.2.5).

We have

$$\chi'(u) = \log(u+v) - \log(u-v) + \log(y/x),$$

$$\chi''(u) = \frac{1}{u+v} - \frac{1}{u-v}.$$

Then $\chi'(u) = 0$ when $\quad u = u_0 = \dfrac{x+y}{x-y} v,$

and $\chi''(u) < 0$.

First suppose that $y \geqslant x + \zeta$. Then

$$\chi'(u) > \log\left(1 + \frac{\zeta}{x}\right) > B,$$

and, by Lemma 1,

(4.2.6) $$\left| \int_{v+1}^{T-v} e^{i\chi(u)} du \right| < \frac{4}{B} = O(1).$$

If $y \leqslant x - \zeta$, we must distinguish two cases. If $v \leqslant (x-y)/(4y)$, then

$$u_0 = \frac{x+y}{x-y} v \leqslant v + \tfrac{1}{2} < v+1,$$

$$\chi'(u) \leqslant \chi'(v+1) = \log\left\{(2v+1)\frac{y}{x}\right\} \leqslant \log\frac{x+y}{2x} < -B,$$

and the integral (4.2.6) is bounded. If $v \geqslant (x-y)/(4y)$, then, since

$$\chi''(u_0) = -\frac{(x-y)^2}{2xyv}, \quad \chi'''(u_0) = \frac{(x+y)(x-y)^3}{4x^2 y^2 v^2},$$

the conditions of Lemma 3 are satisfied†. Hence the integral (4.2.6) is

$$O\left\{\frac{1}{|\chi''(u_0)|^{\frac{1}{2}}}\right\} = O(v^{\frac{1}{2}}).$$

It follows that the inner integral in (4.2.5) is in any case

$$O(1+|v|^{\frac{1}{2}})$$

and so that (4.2.5) is bounded. This accounts for the dominant term in J_5, and the other terms may be accounted for like the corresponding terms in J_4‡. Hence J_5 is bounded; and this completes the proof.

4.3. LEMMA 7. *If* $k(\tfrac{1}{2}+it)$ *belongs to* **k** *and*

$$\Psi(\lambda, x, y) = \int_{1/\lambda}^{\lambda} \left(1 - \frac{|\log u|}{\log \lambda}\right) \frac{K(xu) K_1(yu)}{u} \, du,$$

where $\lambda > 2$ *and* $x > 0$, *then*

(4.3.1) $\qquad\qquad\qquad |\Psi| < B(x, \zeta)$

for $0 < x-\zeta \leqslant y \leqslant x+\zeta$; *and* $\Psi(\lambda, x, y)$ *converges (boundedly), when* $\lambda \to \infty$, *to the limit*

$$0 \ (y<x), \quad \tfrac{1}{2} \ (y=x), \quad 1 \ (y>x).$$

As in the proof of Lemma 6, we may replace $\Psi(\lambda, x, y)$ by

$$\Psi(\lambda, x, y, T) = \frac{y^{\frac{1}{2}}}{\pi^2 x^{\frac{1}{2}}} \int_{-T}^{T} k(\tfrac{1}{2}+it) x^{-it} dt \int_{-T}^{T} k(\tfrac{1}{2}+it') \frac{y^{-it'}}{\tfrac{1}{2}-it'} \frac{\sin^2 \tfrac{1}{2}\mu(t+t')}{\mu(t+t')^2} dt'.$$

We have first to prove this bounded; and, if we divide up the repeated integral as in §4.2, it appears at once, as in the proof of Lemma 6, that the only parts of the integral which require a detailed discussion are the parts J_4' and J_5' corresponding to the J_4 and J_5 of §4.2.

In the discussion of J_4 we made no use of the hypothesis $|x-y| \leqslant \zeta$, and J_4' differs from J_4 (apart from the outside factor) only in the occurrence of an additional factor $\tfrac{1}{2}-i(u-v)$ in the denominator. It is

---

† The main condition reduces to $(xyv)^{\frac{1}{2}} > B(x+y)$, which is true because $yv > B$.

‡ There is a term of the type

$$O\left\{\int_1^T |\gamma(t)| dt \int_1^T |\gamma(t')| \frac{\sin^2 \tfrac{1}{2}\mu(t-t')}{\mu(t-t')^2} dt'\right\}$$

which is bounded because $\gamma(t')$ is bounded and $\gamma(t)$ integrable over $(1, \infty)$.

plain then that the argument for J_4 will dispose of J_4' also, if we add an appeal to Lemma 4 at the appropriate places.

The part of J_5' corresponding to (4.2.5) (omitting the outside factor) is

$$(4.3.2) \qquad \int_0^{\frac{1}{2}T-\frac{1}{2}} (xy)^{-iv} \frac{\sin^2 \mu v}{\mu v^2} dv \int_{v+1}^{T-v} \frac{e^{i\chi(u)}}{\frac{1}{2}-i(u-v)} du.$$

We may replace the last denominator by $-iu$, since the error so introduced is

$$O\left\{\int_0^{\frac{1}{2}T-\frac{1}{2}} \frac{\sin^2 \mu v}{\mu v^2} dv \int_{v+1}^\infty \frac{v+1}{u(u-v)} du\right\} = O\left\{\int_0^{\frac{1}{2}T-\frac{1}{2}} \frac{\sin^2 \mu v}{\mu v^2} \frac{v+1}{v} \log(1+v) dv\right\}$$

$$= O\left\{\int_0^1 \frac{\sin^2 \mu v}{\mu v^2} dv + \int_1^\infty \frac{\log(v+1)}{\mu v^2} dv\right\} = O(1).$$

The inner integral in (4.3.2) then becomes

$$i \int_{v+1}^{T-v} \frac{e^{i\chi(u)}}{u} du = i \int_{u \leqslant 1/|\xi|} + i \int_{u \geqslant 1/|\xi|} = iI_1 + iI_2,$$

say, where $\xi = \log(x/y)$. Here $|\xi| < B$, and one of the integrals may disappear.

We consider I_2 first. If $y > x$, then $\chi'(u) > -\xi = |\xi|$ and

$$I_2 = |\xi| \int_{1/|\xi|}^{u_1} e^{i\chi(u)} du = O\left(|\xi| \frac{1}{|\xi|}\right) = O(1),$$

by Lemmas 4 and 1.

If $y = x$, there is no I_2.

If $y < x$, we argue much as at the end of §4.2. If $v \leqslant (x-y)/(4y)$, then

$$I_2 = O\left\{\frac{\xi}{|\chi'(v+1)|}\right\} = O\left(\frac{\xi}{|\log\{(x+y)/2x\}|}\right) = O(1),$$

since $\xi = O(x-y)$. If $v \geqslant (x-y)/(4y)$, then

$$I_2 = O\left\{\frac{\xi}{|\chi''(u_0)|^{\frac{1}{2}}}\right\} = O\left[\xi \left\{\frac{2xyv}{(x-y)^2}\right\}^{\frac{1}{2}}\right] = O(v^{\frac{1}{2}}).$$

Substituting these estimates in (the simplified form of) the integral (4.3.2), we obtain a bounded integral.

It remains to estimate the contribution of I_1 to (4.3.2), and here we cannot deal with (4.3.2) as it stands, but must associate with it the corresponding term arising from J_6. The leading term in J_6 is the conjugate of that in J_5, so that every term in our decomposition is accompanied by its conjugate. We associate the part of (4.3.2) arising from I_1 with its conjugate, and transform the latter by writing $-v$ for v. We thus obtain an inner integral involving

$$e^{i\{(u+v)\log(u+v)-(u-v)\log(u-v)-2v\}}\left(\frac{y}{x}\right)^{iu}\frac{1}{-iu}$$
$$-e^{-i\{(u-v)\log(u-v)-(u+v)\log(u+v)+2v\}}\left(\frac{y}{x}\right)^{-iu}\frac{1}{iu},$$

and a repeated integral

$$2\int_0^{\frac{1}{2}T+\frac{1}{2}}(xy)^{-iv}\frac{\sin^2\mu v}{\mu v^2}\,dv\int_{u\leqslant 1/|\xi|}e^{i\{(u+v)\log(u+v)-(u-v)\log(u-v)-2v\}}\sin\left(u\log\frac{x}{y}\right)\frac{du}{u}.$$

Since $\sin u\xi = O(u|\xi|)$, the inner integral is

$$O\left(|\xi|\int_{u\leqslant 1/|\xi|}du\right) = O(1),$$

and the repeated integral is bounded.

We have thus disposed of the leading terms (involving $\alpha\alpha'$) in J_5' (and J_6'). The discussion of these integrals may then be completed as in the proof of Lemma 6; we use Lemma 4 for the terms in $\alpha\beta'$, ..., while the boundedness of the terms involving a γ is almost evident. This completes the proof that Ψ is bounded.

We have finally to prove that, when x and y are fixed, Ψ tends to a certain limit. In the first place, the argument which precedes is easily adapted to show that

(4.3.3) $\qquad \Psi(\lambda, x, y, T_2) - \Psi(\lambda, x, y, T_1) \to 0,$

when T_1 and T_2 tend to infinity, uniformly with respect to λ. If $y \neq x$, then (since x and y are now fixed) the difficulties which arose when ξ is small are avoided. The discussion is then similar to that of Lemma 6, the additional term $\frac{1}{2}-it'$ in the denominator leading to convergence instead of boundedness. If $y = x$, we must associate conjugate terms as in the argument which precedes, but the term involving $\sin u\xi$ vanishes. In either case (4.3.3) follows without difficulty. It then follows from (4.3.3) that

(4.3.4) $\qquad \Psi(\lambda, x, y, T) \to \Psi(\lambda, x, y),$

when $T \to \infty$, uniformly in λ.

The theory of Fejér's integral shows that, when T is fixed and $\mu \to \infty$,

$$\int_{-T}^{T} k(\tfrac{1}{2}+it') \frac{y^{-it'}}{\tfrac{1}{2}-it'} \frac{\sin^2 \tfrac{1}{2}\mu(t+t')}{\mu(t+t')^2} dt' \to \tfrac{1}{2}\pi k(\tfrac{1}{2}-it) \frac{y^{it}}{\tfrac{1}{2}+it}$$

boundedly in t. Hence

$$\lim_{\lambda \to \infty} \Psi(\lambda, x, y, T) = \frac{y^{\frac{1}{2}}}{2\pi x^{\frac{1}{2}}} \int_{-T}^{T} k(\tfrac{1}{2}+it) k(\tfrac{1}{2}-it) \left(\frac{y}{x}\right)^{it} \frac{dt}{\tfrac{1}{2}+it}$$

$$= \frac{1}{2\pi} \int_{-T}^{T} \left(\frac{y}{x}\right)^{\frac{1}{2}+it} \frac{dt}{\tfrac{1}{2}+it} = \varpi(x, y) + O\left(\frac{1}{T}\right),$$

where $\varpi(x, y)$ is the limit-function of the lemma. Hence

$$\lim_{T \to \infty} \lim_{\lambda \to \infty} \Psi(\lambda, x, y, T) = \varpi;$$

and we can invert the order of the limits on account of the uniformity of (4.3.4). This completes the proof of the lemma.

Proof of Theorem 1.

5.1. We require one further lemma.

LEMMA 8. *Suppose that* $F(y)$ *is integrable in* (a, b), *that* $g(\lambda, y)$ *is bounded for* $a \leqslant y \leqslant b, \lambda > \lambda_0$, *and that*

$$\lim_{\lambda \to \infty} \int_{\alpha}^{\beta} g(\lambda, y) dy = 0$$

for every α, β *such that* $a \leqslant \alpha < \beta \leqslant b$. *Then*

$$\lim_{\lambda \to \infty} \int_{a}^{b} F(y) g(\lambda, y) dy = 0.$$

This is a special case of a well-known theorem of Hobson†.

5.2. By Lemma 5, the inner integral in (2.1.3) is uniformly convergent for $1/\lambda \leqslant u \leqslant \lambda$, so that we may invert the integrations. We thus obtain

$$\int_{1/\lambda}^{\lambda} \left(1 - \frac{|\log u|}{\log \lambda}\right) K(xu) du \int_{0}^{\infty} K(uy) F(y) dy = \int_{0}^{\infty} F(y) \Phi(\lambda, x, y) dy.$$

We write

$$\int_{0}^{\infty} F(y) \Phi dy = \int_{0}^{\delta} + \int_{\delta}^{x-\zeta} + \int_{x-\zeta}^{x} + \int_{x}^{x+\zeta} + \int_{x+\zeta}^{\Delta} + \int_{\Delta}^{\infty} = I_1 + I_2 + I_3 + I_4 + I_5 + I_6.$$

† Hobson (12), 422-3.

It follows from Lemma 6 that

$$(5.2.1) \quad |I_1| \leqslant B \int_0^\delta y^{-\frac{1}{2}} |F| \, dy < \epsilon, \quad |I_6| < B \int_\Delta^\infty |F| \, dy < \epsilon,$$

for $\delta = \delta(\epsilon)$, $\Delta = \Delta(\epsilon)$, $\lambda > 2$.

Next

$$\int_\alpha^\beta \Phi \, dy = \int_{1/\lambda}^\lambda \left(1 - \frac{|\log u|}{\log \lambda}\right) K(xu) \frac{K_1(\beta u) - K_1(\alpha u)}{u} \, du$$

$$= \Psi(\lambda, x, \beta) - \Psi(\lambda, x, \alpha).$$

If $\alpha < \beta < x$, or $x < \alpha < \beta$, this tends to 0, when $\lambda \to \infty$, by Lemma 7; and so, by Lemma 8,

$$(5.2.2) \quad \lim_{\lambda \to \infty} I_2 = 0, \quad \lim_{\lambda \to \infty} I_5 = 0,$$

when ζ, δ, and Δ are fixed.

We may suppose ζ small enough to ensure that $F(y)$ is of bounded variation in $(x-\zeta, x+\zeta)$, and then

$$F(y) - F(x-0) = F_1(y) - F_2(y),$$

where $F_1(y)$ and $F_2(y)$ are positive and decreasing and tend to 0 when $y \to x$ from below. Then

$$I_3 = F(x-0)\{\Psi(\lambda, x, x) - \Psi(\lambda, x, x-\zeta)\}$$

$$+ \int_{x-\zeta}^x F_1(y) \Phi(\lambda, x, y) \, dy - \int_{x-\zeta}^x F_2(y) \Phi(\lambda, x, y) \, dy.$$

The first term tends to $\frac{1}{2} F(x-0)$. The second is

$$F_1(x-\zeta) \int_{x-\zeta}^\eta \Phi \, dy = F_1(x-\zeta)\{\Psi(\lambda, x, \eta) - \Psi(\lambda, x, x-\zeta)\},$$

where $x-\zeta < \eta < x$, and (since Ψ is bounded) this is less than ϵ (for all λ in question) if ζ is sufficiently small. A similar argument applies to the third term. Hence

$$\left| \varlimsup_{\lambda \to \infty} I_3 - \tfrac{1}{2} F(x-0) \right| \leqslant 2\epsilon$$

if ζ is sufficiently small. There is a corresponding result for I_4, and these results, together with (5.2.1) and (5.2.2), prove the theorem.

5.3. **Theorem 2.** *The conclusion of Theorem 1 still holds if (2.1.1) is replaced by*

$$k(\tfrac{1}{2}+it) = e^{it^c}\{\alpha + \beta(t) + \gamma(t)\},$$

where $c > 1$ (the other conditions being unchanged).

We do not prove this theorem in detail, for the reasons stated in the introduction. The proof is quite similar to that of Theorem 1 when $c < \frac{3}{2}$, but there are additional complications when $c \geqslant \frac{3}{2}$. We shall merely indicate shortly how these affect the proof of the crucial Lemma 6.

The critical integral, corresponding to (4.2.5), is now a multiple of

$$(5.3.1) \qquad \int_0^{\frac{1}{2}(T-1)} (xy)^{-iv} \frac{\sin^2 \mu v}{\mu v^2} dv \int_{v+1}^{T-v} e^{i\chi(u)} du,$$

with
$$\chi(u) = (u+v)^c - (u-v)^c + u \log \frac{y}{x},$$

$$\chi'(u) = c\{(u+v)^{c-1} - (u-v)^{c-1} - \xi\},$$

where
$$\xi = \frac{1}{c} \log \frac{x}{y} > B.$$

If $c = 2$, $\chi'(u)$ is independent of u; we ignore this special case. In general, we have to consider the position of the value u_0 of u defined by $\chi'(u_0) = 0$.

Suppose first that $1 < c < 2$. Then it is easily verified that u_0 falls outside the range of integration when v is small, and that u_0 is of higher order than v when v is large. Thus it is large v that are important in (5.3.1) when $c < 2$. Similarly it is small v that are important when $c > 2$.

There is in any case a standard formula for an integral $\int e^{i\chi} du$ when u_0 falls within the range of integration. This approximation, furnished by the "principle of stationary phase", is

$$\int e^{i\chi} du \sim C \frac{e^{i\chi(u_0)}}{\sqrt{\{|\chi''(u_0)|\}}},$$

where C is a constant. Let us examine the results of replacing the inner integral in (5.3.1) by such an approximation. Solving for u_0 in terms of v and ξ, we obtain

$$(5.3.2) \qquad \int e^{i\chi} du \sim C \xi^{-\frac{3-c}{2(2-c)}} v^{\frac{1}{2(2-c)}} \exp\left(Di \, \xi^{-\frac{c-1}{2-c}} v^{\frac{1}{2-c}}\right),$$

where C and D are constants. If $c < \frac{3}{2}$, then

$$0 < \frac{1}{2(2-c)} < 1;$$

and then (5.3.1) has a majorant of the type

$$\int \frac{\sin^2 \mu v}{\mu v^2} v^b \, dv \quad (0 < b < 1);$$

when $c \to 1$, $b \to \frac{1}{2}$, in agreement with the analysis of §4.2. In this case the proof of Theorem 2 runs quite parallel to that of Theorem 1.

There is no such simple proof when $c > \frac{3}{2}$. In this case we are compelled to pay attention to the exact form of (5.3.2). Let us take, to fix our ideas, $c = 3$, so that (5.3.2) becomes

$$\int e^{ix} du \sim Cv^{-\frac{1}{2}} e^{Di\xi^2/v};$$

and let us further simplify the issue by confining our attention to a particular y, so that ξ may be regarded as constant. Then (5.3.1) is reduced to the form

(5.3.3) $$\int_0^\infty \frac{\sin^2 \mu v}{\mu v^2} v^{-\frac{1}{2}} e^{piv + qi/v} dv.$$

The integral (5.3.3) is substantially the Fejér integral of

(5.3.4) $$v^{-\frac{1}{2}} e^{qi/v}$$

(the factor e^{piv} being unimportant); and this integral is known to be bounded†, as may be proved directly by partial integration of the factor $v^{-2} e^{qi/v}$. The indication of our analysis is therefore that (5.3.1) is bounded when $c = 3$. The argument applies equally to general c, except that, when $c < 2$, the emphasis is on *large* values of v.

We have thus a basis for a proof of Lemma 6, modified to fit the class of $k(s)$ under consideration. To complete the proof we have (a) to investigate the errors involved in the "stationary phase" approximation, (b) to pay regard to the variability of y, and (c) to make the corresponding adaptation of the proof of Lemma 7. All this is naturally tedious, and the details may well be suppressed.

Convergence.

6.1. In this section we make additional assumptions concerning $k(s)$. The function

$$k_1(s) = 2^{s-\frac{1}{2}} \frac{\Gamma(\frac{1}{2}s)}{\Gamma(\frac{1}{2} - \frac{1}{2}s)} = \sqrt{\left(\frac{2}{\pi}\right)} \Gamma(s) \cos \tfrac{1}{2} s\pi$$

is regular in any strip $\sigma_1 < \sigma < \sigma_2$, except for a finite number of simple

---

† See Hardy (5). The Fourier series, at $v = 0$, of a function which behaves like (5.3.4) is, in fact, of the type

$$\Sigma n^{-\frac{1}{2}} e^{rin^{\frac{1}{2}}},$$

and oscillates finitely.

poles at points where $\sigma \leqslant 0$. If t is large and positive, then

$$k_1(\sigma+it) = At^{\sigma-\frac{1}{2}} e^{i(t\log t - t)} \left\{ 1 + \frac{a}{t} + O\left(\frac{1}{t^2}\right) \right\},$$

and $k_1(\sigma-it)$ satisfies the conjugate formula. Here A and a are complex, and a depends upon σ†.

Let
$$k_2(s) = 2^{s-\frac{1}{2}} \frac{\Gamma(\frac{1}{2}+\frac{1}{2}s)}{\Gamma(1-\frac{1}{2}s)} = \sqrt{\left(\frac{2}{\pi}\right)} \Gamma(s) \sin \tfrac{1}{2} s\pi,$$

$$k_3(s) = 2^{s-\frac{3}{2}} \frac{\Gamma(\frac{1}{2}s)}{\Gamma(\frac{3}{2}-\frac{1}{2}s)} = \sqrt{\left(\frac{2}{\pi}\right)} \Gamma(s) \frac{\cos \tfrac{1}{2} s\pi}{1-s},$$

$$k_4(s) = 2^{s-\frac{3}{2}} \frac{\Gamma(\frac{1}{2}+\frac{1}{2}s)}{\Gamma(2-\frac{1}{2}s)} = \sqrt{\left(\frac{2}{\pi}\right)} \Gamma(s) \frac{\sin \tfrac{1}{2} s\pi}{2-s}.$$

These three functions are of the forms

$$k_1(s) \left\{ i + O\left(\frac{1}{|s|^2}\right) \right\}, \quad k_1(s) \left\{ -\frac{1}{s} + O\left(\frac{1}{|s|^2}\right) \right\}, \quad k_1(s) \left\{ -\frac{i}{s} + O\left(\frac{1}{|s|^2}\right) \right\}$$

when t is large and positive, and of the forms

$$k_1(s) \left\{ -i + O\left(\frac{1}{|s|^2}\right) \right\}, \quad k_1(s) \left\{ -\frac{1}{s} + O\left(\frac{1}{|s|^2}\right) \right\}, \quad k_1(s) \left\{ \frac{i}{s} + O\left(\frac{1}{|s|^2}\right) \right\}$$

when t is large and negative.

We now consider functions $k(s)$ which satisfy the following conditions: (i) $k(s)$ is regular in a strip $\sigma_1 < \sigma < \sigma_2$, where $\sigma_1 < 0$, $\sigma_2 > 1$, except perhaps for a finite number of simple poles on the imaginary axis; and (ii)

$$k(s) = k_1(s) \left\{ \alpha + \frac{\beta}{s} + O\left(\frac{1}{|s|^2}\right) \right\}$$

for large positive t, and

$$k(s) = k_1(s) \left\{ \gamma + \frac{\delta}{s} + O\left(\frac{1}{|s|^2}\right) \right\}$$

for large negative t. We call such a function a function of class $\mathbf{k}'$.

It may be verified that, in the examples of (1.4), each $k(s)$ which is a combination of gamma-functions belongs to $\mathbf{k}'$. For example, the $k(s)$ of (1.5.2) belongs to $\mathbf{k}'$, and

$$\alpha = e^{(\frac{1}{2}\nu - \frac{1}{4})\pi i}, \quad \beta = -\tfrac{1}{2} i (\nu^2 - \tfrac{1}{4}) e^{(\frac{1}{2}\nu - \frac{1}{4})\pi i},$$

while γ and δ are the conjugates of α and β.

---

† Actually $\quad A = e^{-\frac{1}{4}\pi i}, \quad a = -\tfrac{1}{2} i (\sigma^2 - \sigma + \tfrac{1}{6}).$

The functions k_1, k_2, k_3, k_4 all belong to $\mathbf{k}'$. Further, if k is any function of $\mathbf{k}'$, we can find constants a_1, a_2, a_3, a_4 such that

(6.1.1) $$k = a_1 k_1 + a_2 k_2 + a_3 k_3 + a_4 k_4 + k^*,$$

where

(6.1.2) $$k^* = O\left(\left|\frac{k_1}{s^2}\right|\right)$$

for large s of the strip†.

The four K corresponding to the special k are

$$K_1(x) = \sqrt{\left(\frac{2}{\pi}\right)} \cos x‡, \quad K_2(x) = \sqrt{\left(\frac{2}{\pi}\right)} \sin x,$$

$$K_3(x) = \sqrt{\left(\frac{2}{\pi}\right)} \frac{\sin x}{x}, \quad K_4(x) = \sqrt{\left(\frac{2}{\pi}\right)} \frac{\sin x - x \cos x}{x^2}.$$

If k is any function of $\mathbf{k}'$, then

(6.1.3) $$K = a_1 K_1 + a_2 K_2 + a_3 K_3 + a_4 K_4 + K^*,$$

where K^* corresponds to the k^* of (6.1.1).

6.2. LEMMA 9. *If $k(s)$ belongs to $\mathbf{k}'$, then $K(x)$ is bounded for all positive x.*

This is true of the four special K, so that it is enough to prove

$$K^*(x) = \frac{1}{2\pi i} \int_{c-i\infty}^{c+i\infty} x^{-s} k^*(s) \, ds$$

bounded. If $x \geqslant 1$, we take $c = 1+\delta$, where $0 < \delta < \frac{1}{2}$ and $1+\delta < \sigma_2$. Since $k^*(s)$ is then $O(|s|^{\delta-\frac{3}{2}})$, $K^*(x)$ is bounded, and indeed $O(x^{-1-\delta})$. If $0 < x \leqslant 1$, we take $c = -\delta$, where $\sigma_1 < -\delta < 0$. Then

$$K^*(x) = \frac{1}{2\pi i} \int_{-\delta-i\infty}^{-\delta+i\infty} x^{-s} k^*(s) \, ds + R,$$

the latter term being the sum of the residues at any poles on the imaginary axis. It is plain that R is bounded, and the integral is bounded because $k^*(s) = O(|s|^{-\delta-\frac{3}{2}})$. Hence $K^*(x)$ is bounded for all positive x.

6.3. LEMMA 10. *If $k(s)$ belongs to $\mathbf{k}'$ and*

(6.3.1) $$\phi(\lambda, x, y) = \int_{1/\lambda}^{\lambda} K(xu) K(yu) \, du,$$

---

† We have $a_1 + i a_2 = \alpha$, $a_1 - i a_2 = \gamma$, $-a_3 - i a_4 = \beta$, $-a_3 + i a_4 = \delta$.

‡ This use of K_1 is temporary and will not lead to confusion with the K_1 of (1.2.1).

where $\lambda > 1$ and x is positive and fixed, then

(6.3.2) $$|\phi| < B(x, \zeta)$$

for all positive y for which $|y-x| \geqslant \zeta$.

It is plain, after Lemma 9, that we may replace ϕ by

$$\chi = \int_1^\lambda K(xu)\, K(yu)\, du.$$

We express k and K in the forms (6.1.1) and (6.1.3), but simplify them by writing

$$k^{(1)} = a_1 k_1 + a_2 k_2, \quad k^{(2)} = a_3 k_3 + a_4 k_4, \quad k^{(3)} = k^*$$

(and similarly for K). Then

$$\chi = \sum_{p,q=1}^{3} \int_1^\lambda K^{(p)}(xu)\, K^{(q)}(yu)\, du = \Sigma \chi_{p,q}.$$

We prove that each of the nine $\chi_{p,q}$ satisfies (6.3.2). We express this shortly by saying that each of them is bounded.

In the first place, this is obvious for $\chi_{1,1}$.

Next, $\chi_{1,2}$ splits up into four terms, a typical term being

(6.3.3) $$\int_1^\lambda \sin xu \, \frac{\sin yu - yu \cos yu}{y^2 u^2}\, du$$

$$-\int_1^{1/y} \sin xu \, \frac{\sin yu - yu \cos yu}{y^2 u^2}\, du + \int_{1/y}^\lambda \sin xu \, \frac{\sin yu}{y^2 u^2}\, du$$

$$-\int_{1/y}^\lambda \sin xu \, \frac{\cos yu}{yu}\, du.$$

Here we suppose that $1/\lambda < y < 1$; if y lies outside these limits the argument is simpler.

Since $(\sin x - x \cos x)/x^2$ is positive and increasing in $0 < x < 1$, the first term on the right of (6.3.3) is

$$(\sin 1 - \cos 1) \int_{u_1}^{1/y} \sin xu\, du,$$

where $0 < u_1 < 1/y$. The second and third are

$$\int_{1/y}^{u_2} \sin xu \sin yu\, du, \quad -\int_{1/y}^{u_3} \sin xu \cos yu\, du,$$

where $u_2 > 1/y$, $u_3 > 1/y$. All these are bounded, and the other terms of $\chi_{1,2}$ may be shown to be bounded in the same way. Hence $\chi_{1,2}$ is bounded.

A similar argument applies to $\chi_{2,1}$ and $\chi_{2,2}$. Thus a typical term of $\chi_{2,2}$ is
$$\int_1^\lambda \frac{\sin xu}{xu} \frac{\sin yu}{yu} du.$$

Here (supposing again that $1/\lambda < y < 1$) we have
$$\int_1^{1/y} \frac{\sin xu}{xu} \frac{\sin yu}{yu} du = \frac{\sin y}{y} \int_1^{u_1} \frac{\sin xu}{xu} du,$$
$$\int_{1/y}^\lambda \frac{\sin xu}{xu} \frac{\sin yu}{yu} du = \int_{1/y}^{u_2} \frac{\sin xu}{xu} \sin yu \, du,$$

and each of these is bounded.

6.4. A typical term in $\chi_{1,3}$ is

(6.4.1) $$\frac{1}{2\pi i} \int_1^\lambda \cos xu \, du \int_{c-i\infty}^{c+i\infty} (yu)^{-s} k^{(3)}(s) \, ds.$$

Suppose first that $k^{(3)}(s)$ has no pole on the imaginary axis. Then we may take $c = 0$. If we do this, and invert the order of integration, we obtain

(6.4.2) $$\frac{1}{2\pi} \int_{-\infty}^\infty y^{-it} k^{(3)}(it) \, dt \int_1^\lambda u^{-it} \cos xu \, du.$$

The inner integral is $O(|t|)$ for large t. Suppose, for example, that $t > 0$. Then
$$\left| \int_1^\lambda u^{-it} \cos xu \, du \right| < t$$
if $t > \lambda$, while if $t < \lambda$ it is

$$O(t) + \int_t^\lambda u^{-it} \cos xu \, du$$

$$= O(t) + \frac{1}{x} \left[\sin xu \, u^{-it} \right]_t^\lambda - \frac{it}{x^2} \left[u^{-it-1} \cos xu \right]_t^\lambda$$

$$- \frac{it(it+1)}{x^2} \int_t^\lambda u^{-it-2} \cos xu \, du = O(t).$$

Since $k^{(3)}(it) = O(|t|^{-\frac{3}{2}})$ for large t, (6.4.2) is bounded†.

---

† In fact, with a B independent of ζ.

The argument requires modification if, as may happen, $k^{(3)}(s)$ has poles on the imaginary axis. It is sufficient to consider the case of one pole, say at $s = i\tau$ and with residue C. We write

$$k^{(3)}(s) = C\Gamma(s-i\tau) + \kappa(s)$$

and
$$K^{(3)}(x) = Cx^{-i\tau}e^{-x} + \mathrm{K}(x),$$

K being related to κ as K to k. Then κ satisfies the conditions imposed above upon $k^{(3)}$. The integral (6.4.1) falls into two, one of which is bounded in virtue of the preceding argument. The other is

$$Cy^{-i\tau}\int_1^\lambda u^{-i\tau} e^{-yu} \cos xu \, du,$$

which is also bounded, since

$$\int_1^\lambda u^{-i\tau} \cos xu \, du$$

is bounded†.

Hence $\chi_{1,3}$ is bounded. Practically the same argument proves $\chi_{2,3}$ to be bounded. A typical term here is

$$\frac{1}{2\pi i}\int_1^\lambda \frac{\sin xu - xu \cos xu}{x^2 u^2} du \int_{c-i\infty}^{c+i\infty} (yu)^{-s} k^{(3)}(s) \, ds.$$

In place of (6.4.2) we have

$$\frac{1}{2\pi}\int_{-\infty}^{\infty} y^{-it} k^{(3)}(it) \, dt \int_1^\lambda u^{-it} \frac{\sin xu - xu \cos xu}{x^2 u^2} du,$$

and the inner integral here is $O(\log|t|)$ for large t.

The integrals $\chi_{3,1}$ and $\chi_{3,2}$ are simpler. Thus a typical term in $\chi_{3,1}$ is

$$\frac{1}{2\pi i}\int_1^\lambda \cos yu \, du \int_{c-i\infty}^{c+i\infty} (xu)^{-s} k^{(3)}(s) \, ds.$$

If we take $c = 1+\delta$, where $0 < \delta < \frac{1}{2}$ and $1+\delta < \sigma_2$, then this is

$$\int_1^\lambda O(u^{-1-\delta}) \, du \int_{-\infty}^{\infty} O(|s|)^{\delta - \frac{3}{2}} dt,$$

which is plainly bounded. The same argument applies to $\chi_{3,2}$.

---

† By the argument which proved the inner integral in (6.4.2) to be $O(|t|)$.

6.5. It remains only to prove $\chi_{3,3}$ bounded. We have

$$\chi_{3,3} = -\frac{1}{4\pi^2} \int_1^\lambda du \int_{c-i\infty}^{c+i\infty} (xu)^{-s} k^{(3)}(s)\, ds \int_{c'-i\infty}^{c'+i\infty} (yu)^{-s'} k^{(3)}(s')\, ds'.$$

Suppose first, as in §6.4, that $k^{(3)}(s)$ has no poles on the imaginary axis, and take $c = 1+\delta$ (with the δ of the last paragraph) and $c' = 0$. The integral is absolutely convergent, and we may invert the order of integration. We thus obtain

$$-\frac{1}{4\pi^2} \int_{1+\delta-i\infty}^{1+\delta+i\infty} x^{-s} k^{(3)}(s)\, ds \int_{-i\infty}^{i\infty} y^{-s'} k^3(s') \frac{\lambda^{1-s-s'} - 1}{1-s-s'}\, ds'.$$

This is bounded because $\Re(s') = 0$, $\Re(s+s') > 1$, and $k^{(3)}(s)$ and $k^3(s')$ are $O(|s|^{\delta-\frac{3}{2}})$ and $O(|s'|^{-\frac{3}{2}})$ respectively.

If there is a pole at $s = i\tau$, with residue C, then, as in §6.4, we have to consider an additional term, which is in this case

$$\frac{C}{2\pi i} \int_1^\lambda (yu)^{-i\tau} e^{-yu}\, du \int_{1+\delta-i\infty}^{1+\delta+i\infty} (xu)^{-s} k^{(3)}(s)\, ds.$$

That this is bounded follows as at the end of §6.4.

Hence $\chi_{3,3}$ is bounded; and this completes the proof of Lemma 10.

6.6. LEMMA 11. *If $k(s)$ belongs to* $\mathbf{k}'$, *and satisfies* (1.1.5), *and*

$$\psi(\lambda, x, y) = \int_{1/\lambda}^\lambda K(xu) \frac{K_1(yu)}{u}\, du,$$

where $\lambda > 1$ and x is positive, then

$$|\psi| < B(x, \zeta)$$

for $0 < x - \zeta < y < x + \zeta$; *and $\psi(\lambda, x, y)$ converges (boundedly), when $\lambda \to \infty$, to the limit*

$$0 \ (y < x), \quad \tfrac{1}{2} \ (y = x), \quad 1 \ (y > x).$$

In proving this lemma we use a simpler decomposition of $k(s)$, viz.:

$$k(s) = k^{(1)}(s) + k^{(2)}(s),$$

where $k^{(1)}(s)$ is the same as before, but $k^{(2)}(s)$ takes the place of $k^{(2)}(s) + k^{(3)}(s)$, so that

$$k^{(2)}(s) = O(|s|^{\sigma-\frac{3}{2}}).$$

We have

$$(6.6.1) \quad \psi(\lambda, x, y) = -\frac{1}{4\pi^2} \int_{1/\lambda}^{\lambda} \frac{du}{u} \int_{c-i\infty}^{c+i\infty} (xu)^{-s} k(s) \, ds \int_{c'-i\infty}^{c'+i\infty} \frac{(yu)^{1-s'}}{1-s'} k(s') \, ds'$$

$$= \psi_{1,1} + \psi_{1,2} + \psi_{2,1} + \psi_{2,2},$$

where $\psi_{p,q}$ is the integral derived from ψ by replacing $k(s)k(s')$ by $k^{(p)}(s) k^{(q)}(s')$. We begin by considering $\psi_{2,2}$. The product of the inner integrals here is

$$O\left\{\int_{-\infty}^{\infty} (1+|t|)^{c-\frac{3}{2}} u^{-c} \, dt \int_{-\infty}^{\infty} (1+|t'|)^{c'-\frac{5}{2}} u^{1-c'} \, dt'\right\} = O(u^{1-c-c'}),$$

provided that $c < \frac{1}{2}$, $c' < \frac{3}{2}$. We take $c+c' = \frac{1}{2}$ when $u \leqslant 1$ and $c+c' = \frac{3}{2}$ when $u > 1$, thus obtaining

$$\psi_{2,2} = \int_{1/\lambda}^{1} O(u^{-\frac{1}{2}}) \, du + \int_{1}^{\lambda} O(u^{-\frac{3}{2}}) \, du.$$

Hence $\psi_{2,2}$ converges (boundedly) when $\lambda \to \infty$.

Next

$$\psi_{1,2} = \frac{1}{\pi i \sqrt{(2\pi)}} \int_{1/\lambda}^{\lambda} \frac{a_1 \cos xu + a_2 \sin xu}{u} \, du \int_{c'-i\infty}^{c'+i\infty} \frac{(yu)^{1-s'}}{1-s'} k^{(2)}(s') \, ds'.$$

The inner integral is

$$O\left\{\int_{-\infty}^{\infty} (1+|t'|)^{c'-\frac{5}{2}} u^{1-c'} \, dt'\right\} = O(u^{1-c'}),$$

provided that $c' < \frac{3}{2}$. We take $c' = \frac{1}{2}$ if $u \leqslant 1$ and $c' = \frac{5}{4}$ if $u > 1$, and it follows that $\psi_{1,2}$ converges boundedly. A similar argument applies to $\psi_{2,1}$.

Finally

$$\psi_{1,1} = \frac{2}{\pi} \int_{1/\lambda}^{\lambda} (a_1 \cos xu + a_2 \sin xu) \frac{a_1 \sin yu + a_2(1-\cos yu)}{u} \, du$$

$$= \frac{2}{\pi} \left\{ a_1^2 \int_{1/\lambda}^{\lambda} \frac{\cos xu \sin yu}{u} \, du + a_1 a_2 \int_{1/\lambda}^{\lambda} \frac{\cos xu - \cos(x+y)u}{u} \, du \right.$$

$$\left. + a_2^2 \int_{1/\lambda}^{\lambda} \frac{\sin xu (1-\cos yu)}{u} \, du \right\},$$

and each term here converges boundedly.

This proves the lemma except as regards the value of the limit. To calculate this directly requires some further examination of the argument, which presents no great difficulty. We can, however, avoid this calculation by an appeal to Lemma 7. The conditions of that lemma are satisfied, and therefore $\Psi'(\lambda, x, y)$ tends to the required limit. But it follows from the "consistency theorem" for Rieszian means that, if $\psi(\lambda, x, y)$ tends to a limit, then $\Psi'(\lambda, x, y)$ tends to the same limit, and this gives the result required.

6.7. If now we repeat the argument of §5.2, but with ϕ and ψ in place of Φ and Ψ', we obtain the following theorem:

THEOREM 3. *If* (i) $k(s)$ *belongs to* **k**' *and satisfies* (1.1.5), (ii) $F(y)$ *is integrable in* $(0, \infty)$, (iii) x *is positive and* $F(y)$ *is of bounded variation near* $y = x$, *then*

$$(6.7.1) \quad \int_0^\infty K(xu)\,du \int_0^\infty K(uy)\,F(y)\,dy = \tfrac{1}{2}\{F(x+0)+F(x-0)\},$$

the inner integral being a Lebesgue integral and the outer integral a Cauchy integral at infinity.

It need only be observed that the outer integral is in the first instance defined as

$$\lim_{\lambda \to \infty} \int_{1/\lambda}^{\lambda} \ldots du\,;$$

but that, since K is bounded, and F integrable, the convention about the lower limit may be dropped.

6.8. We have supposed that $k(s)$ satisfies (1.1.5), so that $K(x)$ is a Fourier kernel. There is, however, no new difficulty in the unsymmetrical case. If $k(s)$ belongs to **k**', and $h(s)$ is defined by (1.3.4) and also belongs to **k**', then our analysis is still valid when we replace k and K by h and H in the appropriate places. We thus obtain

THEOREM 4. *If* (i) $k(s)$ *belongs to* **k**', (ii) $h(s)$ *is defined by* (1.3.4) *and belongs to* **k**', *and* (iii) $F(y)$ *satisfies the conditions of Theorem* 3, *then*

$$\int_0^\infty K(xu)\,du \int_0^\infty H(uy)\,F(y)\,dy = \tfrac{1}{2}\{F(x+0)+F(x-0)\}.$$

There are naturally theorems corresponding to Theorems 1 and 2 as Theorem 4 corresponds to Theorem 3.

6.9. We have not attempted to prove any general theorem corresponding to Theorem 2 as Theorem 3 corresponds to Theorem 1. The

1932.] A CLASS OF FOURIER KERNELS. 149

most interesting cases are those in which $k(s)$ is

$$e^{-ai(s-\frac{1}{2})^r} \quad (a>0, \; r \text{ even}), \qquad e^{a(s-\frac{1}{2})^r} \quad (a>0, \; r \text{ odd}),$$

or a function behaving like one of these at infinity. The case $r=2$ reduces by substitution [as was pointed out in §1.5(5)] to a case of Fourier's theorem. We discuss here the next case $r=3$. The argument would be similar in the general case; but there are material simplifications when $r=3$, because some of the integrals which occur in the analysis can be evaluated in finite form.

In this case

$$K(x) = \frac{1}{2\pi i} \int_{c-i\infty}^{c+i\infty} e^{a(s-\frac{1}{2})^3} x^{-s} ds.$$

If we take $c > \frac{1}{2}$, the integral is absolutely convergent. If follows that $K(x)$ is continuous except for $x=0$ and $x=\infty$; that it is $O(x^{-\Delta})$, for every Δ, at infinity; and $O(x^{-\frac{1}{2}-\delta})$, for every $\delta > 0$, at the origin. More precise results follow from the explicit form of $K(x)$ given in §1(6).

LEMMA 12. *If*

(6.9.1) $$k(s) = e^{a(s-\frac{1}{2})^3} \quad (a>0),$$

then $$|y^{\frac{1}{2}} \phi| < B(x, \zeta)$$

for $|y-x| \geqslant \zeta$.

We first take $c = \frac{1}{2} + \delta$, where $\delta > 0$. The integral for $K(x)$ is then absolutely convergent, and we may substitute for $K(xu)$ and $K(yu)$, and invert the integrations, in (6.3.1). We thus obtain

$$\phi(\lambda, x, y) = \frac{1}{2\pi^2 (xy)^{\frac{1}{2}+\delta}} \int_{-\infty}^{\infty} e^{a(\delta+it)^3} x^{-it} dt \int_{-\infty}^{\infty} e^{a(\delta+it')^3} y^{-it'} \frac{\sin \mu(t+t'-2i\delta)}{t+t'-2i\delta} dt',$$

where $\mu = \log \lambda$. Putting $t = u+v$, $t' = u-v$, $\xi = \log(x/y)$, we get

$$\phi = \frac{1}{\pi^2(xy)^{\frac{1}{2}+\delta}} \int_{-\infty}^{\infty} e^{2a(\delta+iu)^3}(xy)^{-iu} \frac{\sin 2\mu(u-i\delta)}{2(u-i\delta)} du \int_{-\infty}^{\infty} e^{-6a(\delta+iu)v^2} \left(\frac{y}{x}\right)^{iv} dv$$

$$= \frac{1}{\pi(6a\pi)^{\frac{1}{2}}(xy)^{\frac{1}{2}+\delta}} \int_{-\infty}^{\infty} e^{2a(\delta+iu)^3}(xy)^{-iu} \frac{e^{-\xi^2/24a(\delta+iu)}}{(\delta+iu)^{\frac{1}{2}}} \frac{\sin 2\mu(u-i\delta)}{2(u-i\delta)} du.$$

The integral converges uniformly in δ, the integrand being $O(u^{-\frac{1}{2}})$ for small and $O(u^{-\frac{3}{2}})$ for large u. Hence we may put $\delta = 0$, so obtaining

$$\phi = \frac{1}{\pi(6a\pi xy)^{\frac{1}{2}}} \int_{-\infty}^{\infty} e^{-2aiu^3}(xy)^{-iu} \frac{e^{i\xi^2/24au}}{(iu)^{\frac{1}{2}}} \frac{\sin 2\mu u}{2u} du,$$

where $(iu)^{\frac{1}{2}}$ is $e^{\frac{1}{4}\pi i}u^{\frac{1}{2}}$ for $u > 0$ and $e^{-\frac{1}{4}\pi i}(-u)^{\frac{1}{2}}$ for $u < 0$. It is plain that the part of this integral for which $|u| > 1$ is bounded. In the remaining part we may suppress the factor e^{-2aiu^3}, with error

$$\int_0^1 O(u^3)\, O(u^{-\frac{3}{2}})\, du = O(1).$$

We then separate the sine into its exponential components, and obtain multiples of

(6.9.2) $$\int_0^1 \exp i\left(2\mu u - u\log xy + \frac{\xi^2}{24au}\right) u^{-\frac{3}{2}}\, du$$

and of a similar integral with $-\mu$ in the place of μ.

In (6.9.2) we may restore the infinite limit, with bounded error; and when we do this we obtain

$$\int_0^\infty \exp i\left(pu + \frac{q}{u}\right) u^{-\frac{3}{2}}\, du = \frac{|p|^{\frac{1}{2}}}{q^{\frac{1}{2}}} \int_0^\infty \exp\left\{i|p|^{\frac{1}{2}} q^{\frac{1}{2}}\left(\eta u + \frac{1}{u}\right)\right\} u^{-\frac{3}{2}}\, du,$$

where $p = 2\mu - \log xy$, $q = \xi^2/24a$, $\eta = \operatorname{sgn} p$. This is†

$$\frac{|p|^{\frac{1}{2}}}{q^{\frac{1}{2}}} O(|p|^{-\frac{1}{2}} q^{-\frac{1}{4}}) = O(q^{-\frac{3}{4}}) = O(|\xi|^{-\frac{3}{2}}) < B,$$

and this proves the lemma.

6.10. LEMMA 13. *The result of Lemma 11 is still true when $k(s)$ is defined by* (6.9.1).

Proceeding as before, we find that $\psi(\lambda, x, y)$ is

$$\frac{y^{1-\delta}}{\pi^2 x^{\frac{1}{2}+\delta}} \int_{-\infty}^\infty e^{2a(\delta+iu)^3}(xy)^{-iu} \frac{\sin 2\mu(u-i\delta)}{2(u-i\delta)}\, du \int_{-\infty}^\infty e^{-6a(\delta+iu)v^2} \frac{e^{-i\xi v}}{\frac{1}{2}-\delta+iv-iu}\, dv,$$

where $0 < \delta < \frac{1}{2}$. The inner integral is a sum of integrals of the type

$$\int e^{i\theta(v)} \chi(v)\, dv,$$

where χ is monotonic and bounded and

$$\theta(v) = -6auv^2 - v\xi, \quad \theta''(v) = -12au.$$

It follows from Lemmas 2 and 4 that the integral is $O(|u|^{-\frac{1}{2}})$, uniformly for

---

† See Watson (**18**, §§ 6.2–6.22) for general formulae expressing integrals of this type in terms of Bessel functions.

$0 \leqslant \delta \leqslant \frac{1}{4}$. Hence, as in the proof of Lemma 12, we may put $\delta = 0$, when we obtain

(6.10.1) $\quad \psi = \dfrac{y^{\frac{1}{2}}}{\pi x^{\frac{1}{2}}} \displaystyle\int_{-\infty}^{\infty} e^{-2aiu^3}(xy)^{-iu} \dfrac{\sin 2\mu u}{2u} du \int_{-\infty}^{\infty} e^{-6aiuv^2} \dfrac{e^{-i\xi v}}{\frac{1}{2}+iv-iu} dv.$

The part of (6.10.1) with $|u| > 1$ is

$$\int_1^\infty \sin 2\mu u \, O(u^{-\frac{3}{2}}) \, du,$$

and therefore (by the Riemann-Lebesgue theorem) converges boundedly to 0. If $|u| \leqslant 1$,

$$e^{-2aiu^3}(xy)^{-iu} = 1 + O(u),$$

and the $O(u)$ gives an integral which converges boundedly to 0. Finally, if we replace $\frac{1}{2}+iv-iu$ by $\frac{1}{2}+iv$ in the inner integral, the error introduced into it is less than

$$|u|\int_{-\infty}^{\infty} \dfrac{dv}{|\frac{1}{2}+iv||\frac{1}{2}+iv-iu|} = O\left\{|u|\int_{-2}^{2} \dfrac{dv}{\sqrt{(\frac{1}{4}+v^2)}}\right\}$$

$$+ O\left\{|u|\int_{|v|\geqslant 2} \dfrac{dv}{|v|\sqrt{(\frac{1}{4}+v^2)}}\right\} = O|u|;$$

and this again gives an integral which converges boundedly to 0. We may therefore replace ψ by

(6.10.2) $\quad \dfrac{y^{\frac{1}{2}}}{\pi^2 x^{\frac{1}{2}}} \displaystyle\int_{-1}^{1} \dfrac{\sin 2\mu u}{2u} du \int_{-\infty}^{\infty} \dfrac{e^{-6aiuv^2 - i\xi v}}{\frac{1}{2}+iv} dv.$

We write the inner integral here in the form

$$\int_{-\infty}^{\infty} e^{-6aiuv^2} \dfrac{e^{-i\xi v}}{iv} dv - \tfrac{1}{2}\int_{-\infty}^{\infty} \dfrac{e^{-i\xi v}}{(\frac{1}{2}+iv)iv} dv + \tfrac{1}{2}\int_{-\infty}^{\infty} (1 - e^{-6aiuv^2}) \dfrac{e^{-i\xi v}}{(\frac{1}{2}+iv)iv} dv$$

$$= P(u) + Q(u) + R(u),$$

say. The integrals which define $P(u)$ and $Q(u)$ are principal values at $v = 0$. Since

$$R(u) = O\left\{\int_{|v|<|u|^{-\frac{1}{2}}} \dfrac{|u||v|^2 dv}{|v|(\frac{1}{4}+v^2)^{\frac{1}{2}}}\right\} + O\left\{\int_{|v|>|u|^{-\frac{1}{2}}} \dfrac{dv}{|v|(\frac{1}{4}+v^2)^{\frac{1}{2}}}\right\} = O(|u|^{\frac{1}{2}}),$$

the contribution of $R(u)$ tends boundedly to 0, and we need only consider those of $P(u)$ and $Q(u)$.

A simple application of the calculus of residues shows that

$$Q(u) = 2\pi e^{\frac{1}{2}\xi} - \pi = 2\pi x^{\frac{1}{2}} y^{-\frac{1}{2}} - \pi \quad (\xi < 0),$$

$$Q(u) = \pi \quad (\xi \geqslant 0).$$

Hence the contribution of $Q(u)$ converges boundedly to the limit

(6.10.3) $\qquad 1 - \tfrac{1}{2} y^{\frac{1}{2}} x^{-\frac{1}{2}} \ (y > x), \quad \tfrac{1}{2} y^{\frac{1}{2}} x^{-\frac{1}{2}} \ (y \leqslant x).$

Finally, $\qquad P(u) = -(1-i)(\tfrac{1}{2}\pi)^{\frac{1}{2}} \int_0^{\xi^2/(24au)} e^{it} t^{-\frac{1}{2}} dt$

if u and ξ are positive†; $P(-u)$ is the conjugate of $P(u)$; and $P(u)$ is odd in ξ. If ξ and u are positive, $P(u)$ is of the forms

$$C + O\left(\frac{u^{\frac{1}{2}}}{\xi}\right) \ (u < \xi^2), \quad O\left(\frac{\xi}{u^{\frac{1}{2}}}\right) \ (u > \xi^2);$$

and there are similar forms for negative u or ξ. The corresponding contributions to (6.10.2) are of the type

$$O(1) + \int_0^{\xi^2} O\left(\frac{1}{\xi u^{\frac{1}{2}}}\right) du + \int_{\xi^2}^{\infty} O\left(\frac{\xi}{u^{\frac{3}{2}}}\right) du = O(1).$$

It follows that the contribution of $P(u)$ to (6.10.2) is bounded.

If $y < x$, $\xi > 0$, and $\mu \to \infty$, then

$$\frac{y^{\frac{1}{2}}}{\pi^2 x^{\frac{1}{2}}} \int_{-1}^{1} \frac{\sin 2\mu u}{2u} P(u) du \to \frac{y^{\frac{1}{2}}}{\pi^2 x^{\frac{1}{2}}} \cdot \tfrac{1}{4}\pi \{P(+0) + P(-0)\}$$

$$= -\frac{y^{\frac{1}{2}}}{4\pi x^{\frac{1}{2}}} \left\{ (\tfrac{1}{2}\pi)^{\frac{1}{2}}(1-i) \int_0^{\infty} e^{it} t^{-\frac{1}{2}} dt + (\tfrac{1}{2}\pi)^{\frac{1}{2}}(1+i) \int_0^{\infty} e^{-it} t^{-\frac{1}{2}} dt \right\} = -\frac{y^{\frac{1}{2}}}{2x^{\frac{1}{2}}}.$$

If $y > x$ it tends to $\tfrac{1}{2} y^{\frac{1}{2}} x^{-\frac{1}{2}}$, while if $y = x$, $P(u)$ vanishes. Adding these values to (6.10.3), we obtain the result of the lemma.

We have now only to repeat previous arguments in order to prove

THEOREM 5. *If $k(s)$ is defined by* (6.9.1), *$y^{-\frac{1}{2}-\delta} F(y)$ is integrable in* $(0, 1)$ *for some $\delta > 0$, and $y^{-\frac{1}{2}} F(y)$ is integrable in $(1, \infty)$‡, and $F(y)$ is of bounded variation near $y = x$, then* (6.7.1) *is true in the same sense as in Theorem* 3.

---

† This is easily proved by differentiation with respect to ξ.
‡ We are content to state those conditions of integrability at 0 and at ∞ which follow naturally from our analysis.

A special convergence theorem.

7.1. We conclude by proving the theorem referred to at the end of § 1. Here we assume acquaintance with the transform theory developed by Watson.

THEOREM 6. *Suppose* (i) *that* $k(\tfrac{1}{2}+it)$ *satisfies* (1.6.1) *and* (1.6.2), *so that*† $x^{-1} K_1(x)$, *defined by* (1.2.3), *with* $c = \tfrac{1}{2}$, *belongs to* $L^2(0, \infty)$; (ii) *that* $K_1(x)$ *is the integral of* $K(x)$; (iii) *that* $x^{-\tfrac{1}{2}} K_1(x)$ *is bounded.*
Further suppose that

$$(7.1.1) \qquad F(x) = \frac{1}{x} \int_0^x \phi(y)\, dy,$$

where $\phi(y)$ *belongs to* $L^2(0, \infty)$. *Then*

$$(7.1.2) \qquad \int_0^\infty K(xu)\, du \int_0^\infty K(uy)\, F(y)\, dy = F(x),$$

for every positive x, *each integral being a Cauchy integral at each limit.*

It follows from (7.1.1) and Schwarz's inequality that $F(x) = o(x^{-\tfrac{1}{2}})$ both for small and for large x, and from a theorem of Hardy‡ that $F(x)$ and

$$F^*(x) = \frac{1}{x} \int_0^x |\phi(y)|\, dy$$

belong to $L^2(0, \infty)$.

Let $\psi(x)$ be the transform of $\phi(x)$. Then, by Watson's theory, $\psi(x)$ belongs to L^2 and

$$\int_0^x \phi(y)\, dy = \int_0^\infty \frac{K_1(xu)}{u}\, \psi(u)\, du.$$

Let

$$(7.1.3) \qquad G(u) = \int_u^\infty \frac{\psi(v)}{v}\, dv.$$

Then

$$\int_0^x \phi(y)\, dy = -\int_0^\infty K_1(xu)\, G'(u)\, du$$

$$= \lim_{\delta \to 0,\, \Delta \to \infty} \left\{ -\Big[K_1(xu)\, G(u) \Big]_\delta^\Delta + x \int_\delta^\Delta K(xu)\, G(u)\, du \right\}.$$

---

† By Watson's theory.
‡ Hardy (**6, 8, 9**).

Now it follows from (7.1.3) and Schwarz's inequality that $G(u) = o(u^{-\frac{1}{2}})$ when $u \to 0$ or $u \to \infty$; and (*ex hypothesi*) $K_1(xu) = O(u^{\frac{1}{2}})$. Hence the integrated terms vanish in the limit, and

$$(7.1.4) \qquad F(x) = \frac{1}{x}\int_0^x \phi(y)\,dy = \int_0^\infty K(xu)\,G(u)\,du,$$

the integral being a Cauchy integral at each limit.

Again, (7.1.3) may be written

$$G(u) = \int_0^\infty \psi(v)\,\mu(v)\,dv,$$

where $\mu(v) = 0 \ (v < u), \quad \mu(v) = 1/v \ (v > u).$

Hence, by Parseval's theorem†,

$$G(u) = \int_0^\infty \phi(v)\,\lambda(v)\,dv,$$

where $\lambda(v)$ is the transform of $\mu(v)$. Now

$$\lambda(v) = \frac{d}{dv}\left\{\int_u^\infty \frac{K_1(vt)}{t^2}\,dt\right\} = \frac{d}{dv}\left\{v\int_{uv}^\infty \frac{K_1(w)}{w^2}\,dw\right\} = \int_{uv}^\infty \frac{K_1(w)}{w^2}\,dw - \frac{K_1(uv)}{uv};$$

and so

$$(7.1.5) \quad G(u) = \int_0^\infty \phi(v)\,dv \int_{uv}^\infty \frac{K_1(w)}{w^2}\,dw - \int_0^\infty \phi(v)\,\frac{K_1(uv)}{uv}\,dv = G_1(u) - G_2(u),$$

say. Here

$$(7.1.6) \quad G_1(u) = \int_0^\infty \frac{K_1(w)}{w^2}\,dw \int_0^{w/u} \phi(v)\,dv = \int_0^\infty \frac{K_1(w)}{w^2}\,\frac{w}{u}\,F\!\left(\frac{w}{u}\right)dw$$

$$= \int_0^\infty \frac{K_1(uy)}{uy}\,F(y)\,dy,$$

the inversion being justified because $K_1(w)/w$ and $F^*(w/u)$ are L^2, and so

$$\int_0^\infty \frac{|K_1(w)|}{w^2}\,dw \int_0^{w/u} |\phi(v)|\,dv = \frac{1}{u}\int_0^\infty \frac{|K_1(w)|}{w}\,F^*\!\left(\frac{w}{u}\right)dw$$

is finite. Also

$$G_2(u) = \lim_{\delta \to 0,\, \Delta \to \infty}\left\{\frac{K_1(uv)}{uv}\cdot vF(v)\right]_\delta^\Delta$$

$$- \int_\delta^\Delta \frac{K(uv)}{v}\cdot vF(v)\,dv + \int_\delta^\Delta \frac{K_1(uv)}{uv^2}\cdot vF(v)\,dv\bigg\}.$$

† In Watson's theory.

The terms integrated out again vanish in the limit, and so

$$(7.1.7) \qquad G_2(u) = -\int_0^\infty K(uv)\, F(v)\, dv + G_1(u),$$

the integral being again a Cauchy integral at both limits. From (7.1.5), (7.1.6), and (7.1.7) we deduce

$$(7.1.8) \qquad G(u) = \int_0^\infty K(uv)\, F(v)\, dv;$$

and (7.1.4) and (7.1.8) prove the theorem.

References.

1. H. Bateman, "The inversion of a definite integral", *Proc. London Math. Soc.* (2), 4 (1906), 461–498.
2. ———, "Report on the history and present state of the theory of integral equations", *Reports of the British Association*, 1910, 354–424.
3. R. G. Cooke, "The inversion formulae of Hardy and Titchmarsh", *Proc. London Math. Soc.* (2), 24 (1925), 381–420.
4. C. Fox, "A generalization of the Fourier-Bessel transform", *Proc. London Math. Soc.* (2), 29 (1929), 401–452.
5. G. H. Hardy, "Oscillating Dirichlet's integrals", *Quart. J. of Math.*, 44 (1913), 1–40 and 242–263.
6. ———, "Note on a theorem of Hilbert", *Math. Zeitschrift*, 6 (1920), 314–317.
7. ———, "Some formulae in the theory of Bessel functions", *Proc. London Math. Soc.* (2), 23 (1925), lxi (*Records* for 12 June, 1924).
8. ———, "An inequality between integrals", *Messenger of Math.*, 54 (1925), 150–156.
9. ———, "Further inequalities between integrals", *Messenger of Math.*, 57 (1927), 12–16.
10. ———, "A discontinuous integral", *Messenger of Math.*, 57 (1928), 113–120.
11. ——— and E. C. Titchmarsh, "Solution of an integral equation", *Journal London Math. Soc.*, 4 (1929), 300–304.
12. ——— and E. C. Titchmarsh, "Self-reciprocal functions", *Quart. J. of Math.* (Oxford), 1 (1930), 196–231.
13. E. W. Hobson, *The theory of functions of a real variable*, 2 ed., 2 (Cambridge, 1926).
14. S. W. P. Steen, "Divisor functions", *Proc. London Math. Soc.* (2), 31 (1930), 47–80, and *ibid.*, 32 (1931), 356–368.
15. E. C. Titchmarsh, "A pair of inversion formulae", *Proc. London Math. Soc.* (2), 22 (1923), xxxiv–xxxv (*Records* for 17 May, 1923).
16. ———, "Extensions of Fourier's integral formula to formulae involving Bessel functions", *Proc. London Math. Soc.* (2), 23 (1924), xxii–xxiv (*Records* for 17 January, 1924).
17. ———, "An inversion formula involving Bessel functions", *Proc. London Math Soc.* (2), 24 (1925), vi–vii (*Records* for 13 November, 1924).
18. G. N. Watson, *Theory of Bessel functions* (Cambridge, 1922).

CORRECTION

p. 154, *line 2 up*. After lim insert $\{[$.

A THEOREM CONCERNING FOURIER TRANSFORMS

G. H. Hardy†.

1. This note originates from a remark of Prof. N. Wiener, to the effect that "a pair of transforms f and g cannot both be very small". A little more precisely, if

$$(1.1) \qquad g(x) = \frac{1}{\sqrt{(2\pi)}} \int_{-\infty}^{\infty} f(t) e^{-xit} dt$$

and if f and g both tend to zero with great rapidity when $x \to \infty$ through positive or negative values, then f and g must be nul. Suppose, for example (to take the extreme case), that f and g are 0 for $|x| > X$. Then

$$g(z) = \frac{1}{\sqrt{(2\pi)}} \int_{-X}^{X} f(t) e^{-zit} dt$$

is an integral function of $z = x+iy$, which is 0 for $z = x > X$ and therefore for all z.

The theorems which follow give the most precise interpretation possible of Wiener's remark. They are suggested by the fact that $e^{-\frac{1}{2}x^2}$ is its own transform. More generally, if $H_n(x)$ is Hermite's polynomial, and

$$\phi_n(x) = e^{-\frac{1}{2}x^2} H_n(x) = e^{\frac{1}{2}x^2} \left(\frac{d}{dx}\right)^n e^{-x^2}$$

is the Hermite orthogonal function‡, then the transform of $\phi_n(x)$ is $(-i)^n \phi_n(x)$.

Theorem 1. *If f and g are both $O(|x|^m e^{-\frac{1}{2}x^2})$ for large x and some m, then each is a finite linear combination of Hermite functions.*

Theorem 2. *In particular, if f and g are both $O(e^{-\frac{1}{2}x^2})$, then $f = g = Ae^{-\frac{1}{2}x^2}$, where A is a constant; and if one is $o(e^{-\frac{1}{2}x^2})$, then both are nul.*

I give two proofs, the first of which depends on a theorem of "Phragmén-Lindelöf" type, and the second on the ideas of a recent paper

---

† Received 15 May, 1933; read 18 May, 1933.

‡ Not normalised. For the elementary properties of the Hermite functions see Courant-Hilbert (1), 77–82, Wiener (5), 46–71. It was from Wiener's lectures that I learnt their importance in the theory of the Fourier integral.

by Titchmarsh and myself†. The argument has been greatly improved by suggestions made to me after a lecture. The first proof (§§ 2–3) is the result of a discussion with Miss M. L. Cartwright, and the second (§§ 4–5) has been much simplified by a suggestion of Mr. H. D. Ursell.

First proof.

2. We may reduce the problem, in the usual manner, to the corresponding problem for a pair of even functions connected by the Fourier cosine transformation, or odd functions connected by the sine transformation. In what follows I suppose that f and g are even and that

$$(2.1) \qquad g(x) = \sqrt{\left(\frac{2}{\pi}\right)} \int_0^\infty f(t) \cos xt \, dt.$$

We require the following lemma: *if* (i) *$f(z)$ is an integral function;* (ii)

$$f(z) = O(|z|^n e^{a|z|}),$$

where $a > 0$, for large $|z|$; and (iii)

$$f(x) = O(x^n e^{-ax})$$

for large positive x; then $f(z) = e^{-az} P(z)$, where P is a polynomial of degree n.

We deduce this from a theorem of Pólya and Szegö. Let

$$h(\theta) = \overline{\lim} \, \frac{1}{r} \log |f(re^{i\theta})|$$

be the usual Phragmén-Lindelöf function associated with $f(z)$. Then $h(\theta) \leqslant a$ for all θ, and $h(0) \leqslant -a$. If $0 < \delta < \pi$ and

$$H(\theta) = -a \cos \theta + a \tan \tfrac{1}{2}\delta \sin \theta,$$

then $\qquad h(0) \leqslant -a = H(0), \quad h(\pi-\delta) \leqslant a = H(\pi-\delta),$

and so $h(\theta) \leqslant H(\theta)$ for $0 < \theta < \pi - \delta$‡. In particular,

$$h(\tfrac{1}{2}\pi) \leqslant H(\tfrac{1}{2}\pi) = a \tan \tfrac{1}{2}\delta.$$

Hence, since δ is at our disposal, $h(\tfrac{1}{2}\pi) \leqslant 0$.

---

† Hardy and Titchmarsh (2).
‡ This is the most familiar theorem of this type. See, for example, Titchmarsh (4), 183.

Now let
$$F(z) = e^{az} f(z), \quad G(z) = \frac{F(z)}{(z+i)^n} = \frac{e^{az} f(z)}{(z+i)^n}.$$

Then $G(z)$ is regular in the half-plane $y \geqslant 0$, and $|G(z)| < M e^{N|z|}$, for appropriate M, N, throughout the half-plane. Also $G(z)$ is bounded on the real axis, and

$$\varlimsup \frac{1}{y} \log |G(iy)| = \varlimsup \frac{1}{y} \log |f(iy)| \leqslant 0.$$

It follows† that $G(z)$ is bounded, and that $F(z) = O(|z|^n)$ in the half-plane.

A similar argument shows that $F(z) = O(|z|^n)$ in the lower half-plane. Hence $F(z)$ is a polynomial.

3. The equation (2.1), with z for x, defines $g(z)$ as an integral function of z. Also, since

$$|\cos zt| \leqslant \cosh |z| t, \quad |f(t)| < C t^m e^{-\frac{1}{2} t^2} \quad (t > 1),$$

where C is a constant, we have

$$|g(z)| < \sqrt{\left(\frac{2}{\pi}\right)} e^{|z|} \int_0^1 |f(t)| \, dt + C \int_0^\infty t^m e^{-\frac{1}{2} t^2} \cosh |z| t \, dt.$$

If, as we may suppose, m is even, the last integral is

$$\left(\frac{d}{d|z|}\right)^m \int_0^\infty e^{-\frac{1}{2} t^2} \cosh |z| t \, dt = \sqrt{(\tfrac{1}{2}\pi)} \left(\frac{d}{d|z|}\right)^m e^{\frac{1}{2}|z|^2};$$

and so
$$g(z) = O(|z|^m e^{\frac{1}{2}|z|^2})$$

for large $|z|$.

We now write $z^2 = u$, $g(z) = G(u)$, $m = 2n$. Then $G(u)$ is an integral function of u, is $O(|u|^n e^{\frac{1}{2}|u|})$ for all large $|u|$, and $O(u^n e^{-\frac{1}{2} u})$ for large positive u. Hence, by the lemma of §2, $G(u)$ is the product of $e^{-\frac{1}{2} u}$ by a polynomial, so that $g(z)$ is a finite linear combination of Hermite functions (here of even order). This proves Theorem 1.

Second proof.

4. We suppose, as in the first proof, that f and g are connected by (2.1).

---

† Pólya-Szegö (**3**), 147 (Theorem 325).

If a star indicates the transform, so that $g = f^*$, then $f = g^*$ and

$$(f+g)^* = f+g, \quad (f-g)^* = -(f-g).$$

Hence, in order to prove Theorem 1, it is enough to show that, *if f is self-reciprocal ($f^* = f$) or skew-reciprocal ($f^* = -f$), and $O(x^m e^{-\frac{1}{2}x^2})$, then f is a finite linear combination of Hermite functions*. The proof is suggested by the formula

$$(4.1) \qquad \int_0^\infty e^{-\frac{1}{2}sx^2} \phi_{2n}(x)\, dx = (-1)^n \frac{2n!}{n!} \sqrt{(\tfrac{1}{2}\pi)} \frac{(s-1)^n}{(s+1)^{n+\frac{1}{2}}},$$

valid for $\sigma = \Re(s) > -1$.

If

$$(4.2) \qquad \lambda(s) = \int_0^\infty e^{-\frac{1}{2}sx^2} f(x)\, dx,$$

then $\lambda(s)$ is regular in the half-plane D defined by $\sigma > -1$. If $\sigma > 0$, we have

$$\lambda(s) = \sqrt{\left(\frac{2}{\pi}\right)} \int_0^\infty e^{-\frac{1}{2}sx^2} dx \int_0^\infty f(y) \cos xy\, dy$$

$$= \sqrt{\left(\frac{2}{\pi}\right)} \int_0^\infty f(y)\, dy \int_0^\infty e^{-\frac{1}{2}sx^2} \cos xy\, dx$$

$$= s^{-\frac{1}{2}} \int_0^\infty e^{-y^2/2s} f(y)\, dy = s^{-\frac{1}{2}} \lambda\left(\frac{1}{s}\right).$$

Hence

$$(4.3) \qquad \mu(s) = \sqrt{(s+1)}\,\lambda(s)$$

satisfies the equation

$$(4.4) \qquad \mu(s) = \mu\left(\frac{1}{s}\right).$$

The equation (4.4) is proved, first, for $\sigma > 0$. The left-hand side being regular in D, the right-hand side must be so also, and $\mu(s)$ must be regular in the region E which is the inverse of D with respect to the origin. This is the outside of the circle on $(-1, 0)$ as diameter, and the two regions together cover the whole plane except the one point $s = -1$. Hence the only singularities of $\mu(s)$ and $\lambda(s)$ are at -1 and at infinity.

Since $\mu(s)$ is regular at the origin, and satisfies (4.4), it is regular at infinity. It is therefore a one-valued function with one singular point only, at $s = -1$; and

(4.5) $$\mu(s) = \sum_0^\infty \frac{a_n}{(s+1)^n}, \quad \lambda(s) = \sum_0^\infty \frac{a_n}{(s+1)^{n+\frac{1}{2}}}.$$

5. We have now to prove that the series (4.5) are finite, so that $\mu(s)$ is a rational function. It is enough to prove that

$$\lambda(s) = O(|s+1|^{-p}).$$

for some p, near $s = -1$.

Let $\sigma + 1 = \tau$. On the unit circle†

$$|s| = 1, \quad \tau = \tfrac{1}{2}|s+1|^2.$$

Hence, if s lies within or on the circle, we have $\tau \geqslant \tfrac{1}{2}|s+1|^2$ and

(5.1) $$\lambda(s) = \int_0^\infty e^{-\frac{1}{2}sx^2} f(x)\,dx = O\left(\int_0^\infty e^{-\frac{1}{2}\tau x^2} x^m\,dx\right)$$

$$= O\{\tau^{-\frac{1}{2}(m+1)}\} = O(|s+1|^{-m-1}).$$

If, on the other hand, s lies outside the circle, $1/s$ lies inside it, and

$$\lambda(s) = s^{-\frac{1}{2}} \lambda\!\left(\frac{1}{s}\right) = O\left|\lambda\!\left(\frac{1}{s}\right)\right| = O(|s+1|^{-m-1});$$

so that (5.1) is true in any case. Hence the series (4.5) are finite.

Finally, since

$$\lambda(s) = \sum_0^m \frac{a_n}{(s+1)^{n+\frac{1}{2}}},$$

we have

$$\int_0^\infty e^{-\frac{1}{2}sx^2}(f-\Phi)\,dx = 0,$$

where Φ is an appropriate combination of Hermite functions, for $s > 0$. It follows, by Lerch's theorem, that $f = \Phi$.

References.

1. R. Courant und D. Hilbert, *Methoden der math. Physik*, ed. 2 (Springer, 1931).
2. G. H. Hardy and E. C. Titchmarsh, "Self-reciprocal functions", *Quarterly Journal* (Oxford), 1 (1930), 196–231.
3. G. Pólya und G. Szegö, *Aufgaben und Lehrsätze aus der Analysis*, I (Springer, 1925).
4. E. C. Titchmarsh, *The theory of functions* (Oxford, 1932).
5. N. Wiener, *The Fourier integral and certain of its applications* (Cambridge, 1933).

† The simple argument which follows was suggested by Mr. Ursell.

CORRECTION

p. 227, *lines* 4–5 *up*. Read $e^{-\frac{1}{2}x^2}$.

THE RESULTANT OF TWO FOURIER KERNELS

By Prof. G. H. HARDY, Trinity College

[*Received* 20 October, *read* 26 November 1934]

1. A "Fourier kernel" means here a function $K(x)$ which gives rise to a formula

$$f(x) = \int_0^\infty K(xu)\, du \int_0^\infty K(ut) f(t)\, dt \qquad (1\cdot1)$$

of the Fourier type. Thus

$$\sqrt{\left(\frac{2}{\pi}\right)} \cos x, \quad \sqrt{\left(\frac{2}{\pi}\right)} \sin x, \quad x^{\frac{1}{2}} J_\nu(x), \quad \frac{2}{\pi} \frac{1}{1-x^2}, \quad \ldots$$

are Fourier kernels.* If $K(x)$ is a Fourier kernel, λ is real, and a positive, then

$$\frac{1}{x} K\left(\frac{1}{x}\right), \quad \lambda x^{\frac{1}{2}(\lambda-1)} K(x^\lambda), \quad a^{\frac{1}{2}} K(ax)$$

are Fourier kernels.

The *resultant*, or *Faltung*, $M(x)$ of $K(x)$ and $L(x)$ is defined by

$$M(x) = \int_0^\infty K(xt) L(t)\, dt. \qquad (1\cdot2)$$

If $M(x)$ is the resultant of $K(x)$ and $L(x)$, then

$$\frac{1}{x} M\left(\frac{1}{x}\right)$$

is the resultant of $L(x)$ and $K(x)$.

There are various formal reasons which suggest that *the resultant of two Fourier kernels is a Fourier kernel*. For example, we may argue as follows. Replacing K by M in the integral on the right of (1·1), and substituting from (1·2), we obtain

$$\iint M(xu) M(ut) f(t)\, du\, dt = \iiiint K(xuy) K(utz) L(y) L(z) f(t)\, du\, dt\, dy\, dz;$$

and the substitution $t = v/z$, $y = zw$ gives

$$\iint L(z) L(zw)\, dz\, dw \iint K(xzwu) K(uv) f\left(\frac{v}{z}\right) du\, dv = \iint L(z) L(zw) f(xw)\, dz\, dw = f(x).$$

The argument is naturally of a purely formal type, the multiple integrals being

* Further examples are given by Hardy and Titchmarsh (**2**) and Watson (**6**).

divergent, and the inversions and substitutions impossible to justify, even in the simplest standard cases.*

We can also appeal to the idea which underlies the recent work of Watson and of Titchmarsh and myself. If
$$k(s) = \int_0^\infty x^{s-1} K(x)\, dx$$
is the Mellin transform of a Fourier kernel $K(x)$, then
$$k(s)\, k(1-s) = 1;$$
and this is also a sufficient condition that $K(x)$ should be a Fourier kernel. Now the Mellin transform of $M(x)$ is
$$m(s) = \int_0^\infty x^{s-1}\, dx \int_0^\infty K(xt)\, L(t)\, dt = \int_0^\infty L(t)\, dt \int_0^\infty x^{s-1} K(xt)\, dx$$
$$= \int_0^\infty t^{-s} L(t)\, dt \int_0^\infty u^{s-1} K(u)\, du = k(s)\, l(1-s);$$
so that
$$m(s)\, m(1-s) = k(s)\, k(1-s)\, l(s)\, l(1-s) = 1.$$
This argument also is formal, but the transformations are a little nearer to reality than those of the first.

It is plain in any case that we must be prepared for a very liberal interpretation of (1·1) and (1·2). Thus
$$\frac{2}{\pi} \int_0^\infty \cos xt \cos t\, dt$$
is generally summable $(C, 1)$ to 0, but diverges to infinity when $x = 1$. The integral is never convergent. Similarly
$$\frac{2}{\pi} \int_0^\infty \cos xt \sin t\, dt = \frac{2}{\pi} \frac{1}{1-x^2} \quad (C, 1), \tag{1·3}$$
except for $x = 1$, when the value is $1/2\pi$. On the other hand
$$\frac{4}{\pi^2} \int_0^\infty \frac{dt}{(1-x^2 t^2)(1-t^2)}$$
converges to 0 in general (as a Cauchy principal value), but diverges to infinity when $x = 1$. And a similar freedom of interpretation is necessary in (1·1).

2. It is easy to reduce all this to order by means of Watson's theory.† We start from a function $K_1(x)$ with the properties (i) that $x^{-1} K_1(x)$ is L^2 in $(0, \infty)$, and (ii) that
$$\int_0^\infty \frac{K_1(ax)\, K_1(bx)}{x^2}\, dx = \mathrm{Min}\,(a, b) \tag{2·1}$$

---

* I have been familiar with these formal ideas for a good many years, but cannot say whence I derived them. Possibly from Ramanujan; but I can refer to nothing in his published work, and it is likely enough that the ideas are much older.

† Watson (6). Considerable simplifications in the theory have been made by Plancherel (3) and Titchmarsh (4).

if a and b are positive. In these circumstances, if $f(x)$ is L^2, and $g(x)$ is defined by

$$\int_0^x g(y)\,dy = \int_0^\infty \frac{K_1(xt)}{t} f(t)\,dt, \qquad (2\cdot 2)$$

then $g(x)$ is also L^2 and the relationship is reciprocal. We call $f(x)$ and $g(x)$ "K-transforms" of one another. If $F(x)$ and $G(x)$ are also K-transforms of one another, then

$$\int_0^\infty f(x)\,F(x)\,dx = \int_0^\infty g(x)\,G(x)\,dx. \qquad (2\cdot 3)$$

This is "Parseval's Theorem". In all this there is no direct reference to a function $K(x)$, but, if $K_1(x)$ is the integral of $K(x)$, then the transformation is that envisaged formally in §1.

Let us now suppose that $K_1(x)$ and $L_1(x)$ satisfy Watson's conditions, and define $M_1(x)$ by

$$\int_0^x M_1\!\left(\frac{1}{y}\right) dy = \int_0^\infty \frac{K_1(t)}{t}\frac{L_1(xt)}{t}\,dt. \qquad (2\cdot 4)$$

If K_1, L_1, M_1 are the integrals of K, L, M, then two differentiations reduce $(2\cdot 4)$ formally to $(1\cdot 2)$.

Since $M_1(1/x)$ is the L-transform of $x^{-1}K_1(x)$,

$$\int_0^\infty \frac{M_1^2(x)}{x^2}\,dx = \int_0^\infty M_1^2\!\left(\frac{1}{x}\right) dx < \infty.$$

Also $M_1(a/x)$, $M_1(b/x)$ are the L-transforms of $x^{-1}K_1(ax)$, $x^{-1}K_1(bx)$; and hence, by Parseval's Theorem,

$$\int_0^\infty \frac{M_1(ax)\,M_1(bx)}{x^2}\,dx = \int_0^\infty M_1\!\left(\frac{a}{x}\right) M_1\!\left(\frac{b}{x}\right) dx = \int_0^\infty \frac{K_1(ax)\,K_1(bx)}{x^2}\,dx = \text{Min}\,(a,b).$$

Hence M_1 satisfies the same conditions as K_1 and L_1, and there are formulae in M_1 similar to $(2\cdot 2)$ and its reciprocal. When K_1, L_1, M_1 are integrals, then K, L, M are Fourier kernels; and it is natural to call the M-transformation M the resultant of the K- and L-transformations K and L.

If $S_1(x)$ is 0 for $x < 1$, and 1 for $x \geq 1$, and $K_1(x) = S_1(x)$, then the transformation is

$$g(x) = \frac{1}{x} f\!\left(\frac{1}{x}\right), \quad f(x) = \frac{1}{x} g\!\left(\frac{1}{x}\right).$$

We call this transformation S. If $K_1 = L_1$ then

$$\int_0^x M_1\!\left(\frac{1}{y}\right) dy = \int_0^\infty \frac{K_1(t)\,K_1(xt)\,dt}{t^2} = \text{Min}\,(1,x)$$

and $M_1 \equiv S_1$. If $L_1 = S_1$, then

$$\int_0^x M_1\!\left(\frac{1}{y}\right) dy = \int_{1/x}^\infty \frac{K_1(t)}{t^2}\,dt = \int_0^x K_1\!\left(\frac{1}{t}\right) dt,$$

and $M_1 \equiv K_1$. Thus the resultant of K and K is S, and the resultant of K and S is K.

G. H. Hardy

Examples

3. The interest of the examples which follow is mainly formal, and I allow myself, as in §1, a certain latitude of expression, speaking in terms of K, L, M when precise expression demands a return to K_1, L_1, M_1.

(1) The equation (1·3) indicates that the resultant of the cosine and sine transformations is that defined by the kernel

$$\frac{2}{\pi} \frac{1}{1-x^2}.$$

Here
$$M_1(x) = \frac{1}{\pi} \log \left| \frac{x-1}{x+1} \right|$$

and is not (in the strict sense) an integral. The M transformation is

$$g(x) = \frac{2}{\pi} \int_0^\infty \frac{f(t)}{1-x^2 t^2} dt.$$

If we suppose $f(x)$ even, and make some trivial transformations, we obtain

$$\frac{1}{x} g\left(\frac{1}{x}\right) = \frac{1}{\pi} \int_{-\infty}^\infty \frac{f(t)}{x-t} dt,$$

the conjugate or "Hilbert transform" of $f(x)$.

If we call this transformation C then the resultant of K and C is defined by

$$M(x) = \frac{2}{\pi} \int_0^\infty \frac{K(xt)}{1-t^2} dt;$$

or, regarding $K(x)$ as even, by

$$M(x) = \frac{1}{\pi} \int_{-\infty}^\infty \frac{K(t)}{x-t} dt.$$

Thus *the conjugate of a Fourier kernel is a Fourier kernel.*

(2) The function
$$L_1(x) = x \quad (x < 1), \quad L_1(x) = 0 \quad (x \geqslant 1)$$

satisfies Watson's conditions.* We conclude that, if $K(x)$ is a Fourier kernel, then

$$M(x) = \int_0^\infty K(xt) \, dL_1(t) = \int_0^1 K(xt) \, dt - K(x) = \frac{1}{x} \int_0^x K(u) \, du - K(x)$$

is a Fourier kernel. Or again, taking another of Watson's examples, viz.

$$L_1(x) = 0 \quad (x < 1), \quad L_1(x) = \log x - 1 \quad (x \geqq 1),$$

we find that
$$\int_x^\infty \frac{K(u)}{u} du - K(x)$$
is a Fourier kernel.

* Watson (**6**, p. 197).

(3) Since*
$$\int_0^\infty J_\nu(xt) J_{\nu-1}\left(\frac{1}{t}\right)\frac{dt}{t} = x^{-\frac{1}{2}} J_{2\nu-1}(2x^{\frac{1}{2}}),$$
the resultant of $t^{\frac{1}{2}} J_\nu(t)$ and $t^{-\frac{3}{2}} J_\nu(1/t)$ is $J_{2\nu-1}(2t^{\frac{1}{2}})$.

(4) Since
$$\frac{2}{\pi}\int_0^\infty \frac{\cos}{\sin} xt \frac{\cos}{\sin}\left(\frac{1}{t}\right)\frac{dt}{t} = \frac{2}{\pi} K_0(2x^{\frac{1}{2}}) \mp Y_0(2x^{\frac{1}{2}}),\dagger$$
the functions just written are the resultants of
$$\sqrt{\left(\frac{2}{\pi}\right)}\frac{\cos}{\sin} x, \quad \sqrt{\left(\frac{2}{\pi}\right)}\frac{1}{x}\frac{\cos}{\sin}\left(\frac{1}{x}\right)$$
(the two cosines or the two sines going together). We conclude that the functions
$$x^{\frac{1}{2}}\left\{Y_0(x) \mp \frac{2}{\pi}K_0(x)\right\}$$
are Fourier kernels. The first of them is the kernel which occurs in the theory of Dirichlet's divisor problem. The functions may be generated differently. Thus
$$\frac{2}{\pi}\int_0^\infty \frac{J_0\{2(xt)^{\frac{1}{2}}\}}{1-t^2}dt = \frac{4}{\pi}\int_0^\infty \frac{uJ_0(2x^{\frac{1}{2}}u)}{1-u^4}du = Y_0(2x^{\frac{1}{2}}) + \frac{2}{\pi}K_0(2x^{\frac{1}{2}}),$$
so that this last kernel is the conjugate of $J_0(2x^{\frac{1}{2}})$.

(5) The resultant of $J_0(2x^{\frac{1}{2}})$ and $\cos x$ is $-\sin x$, and that of $J_0(2x^{\frac{1}{2}})$ and $\sin x$ is $\cos x$.

(6) It is easily proved that
$$x^{\frac{1}{2}}\int_0^\infty tJ_\mu(xt) J_{-\mu}(t)dt = -\frac{2\sin\mu\pi}{\pi}\frac{x^{\mu+\frac{1}{2}}}{1-x^2} \quad (C,1),$$
provided that $x \neq 1$, while when $x=1$ the integral diverges like
$$\frac{\cos\mu\pi}{\pi}\int^\infty dt.$$
This divergence indicates that, when we form the resultant of $x^{\frac{1}{2}} J_\mu(x)$ and $x^{\frac{1}{2}} J_{-\mu}(x)$, there will be a discontinuity in $M_1(x)$ at $x=1$. In fact, in this case,
$$M_1(x) = -\frac{2\sin\mu\pi}{\pi}\int_0^x \frac{t^{\mu+\frac{1}{2}}dt}{1-t^2} \quad (x<1), \quad M_1(x) = -\frac{2\sin\mu\pi}{\pi}\int_0^x \frac{t^{\mu+\frac{1}{2}}dt}{1-t^2} + \cos\mu\pi \quad (x \geq 1).$$
The inversion formulae are
$$g(x) = -\frac{2\sin\mu\pi}{\pi}\int_0^\infty \frac{(xt)^{\mu+\frac{1}{2}}}{1-x^2t^2}f(t)dt + \cos\mu\pi\frac{1}{x}f\left(\frac{1}{x}\right)$$
and the reciprocal formula. The transformation is a generalization of C, to which it reduces when $\mu = -\frac{1}{2}$, the extra term then disappearing.

---

* The formula is easily deducible from one due to Bateman. See Hardy (**1**).
† Here, and in (7), K_ν is used as in Watson (**5**).

(7) If we form the resultant $M(x)$ of
$$\sqrt{\left(\frac{2}{\pi}\right)}\cos x, \quad J_{\frac{1}{4}}(2x^{\frac{1}{2}}) = \pi^{-\frac{1}{2}}x^{-\frac{1}{4}}\sin 2x^{\frac{1}{2}},$$
and then replace it by
$$\frac{2^{-\frac{1}{4}}}{x}M\left(\frac{1}{2x}\right),$$
we obtain the Fourier kernel
$$(2x)^{\frac{1}{2}}\{\cos(x-\tfrac{1}{8}\pi)J_{\frac{1}{4}}(x) + \sin(x-\tfrac{1}{8}\pi)J_{-\frac{1}{4}}(x)\}.$$
The analysis involves the calculation of the integrals
$$\int_0^\infty e^{-x^4-4\alpha x^2}dx = \tfrac{1}{2}\alpha^{\frac{1}{2}}e^{2\alpha^2}K_{-\frac{1}{4}}(2\alpha^2),$$
$$\int_0^\infty e^{-x^4}\cos 4\alpha x^2 dx = 2^{-\frac{3}{2}}\pi\alpha^{\frac{1}{2}}e^{-2\alpha^2}I_{-\frac{1}{4}}(2\alpha^2),$$
$$\int_0^\infty \cos x^4 \cos 4\alpha x^2 dx = 2^{-\frac{3}{2}}\pi\alpha^{\frac{1}{2}}\cos(2\alpha^2-\tfrac{1}{8}\pi)J_{-\frac{1}{4}}(2\alpha^2).$$
In all of these α is positive.

REFERENCES

(1) G. H. HARDY, "Notes on some points in the integral calculus (LXIII)", *Messenger of Math.* 56 (1927), 186–92.

(2) G. H. HARDY and E. C. TITCHMARSH, "A class of Fourier kernels", *Proc. London Math. Soc.* (2), 35 (1933), 116–55.

(3) M. PLANCHEREL, "Sur les formules de réciprocité du type de Fourier", *Journal London Math. Soc.* 8 (1933), 220–26.

(4) E. C. TITCHMARSH, "A proof of a theorem of Watson", *Journal London Math. Soc.* 8 (1933), 217–20.

(5) G. N. WATSON, *Theory of Bessel functions*, Cambridge, 1922.

(6) G. N. WATSON, "General transforms", *Proc. London Math. Soc.* (2), 35 (1933), 156–99.

RAMANUJAN AND THE THEORY OF FOURIER TRANSFORMS

By G. H. HARDY (*Cambridge*)

[Received 9 August 1937]

1. DURING 1913 Ramanujan held a research studentship in the University of Madras, and submitted to the University three quarterly reports on the progress of his researches.* He was occupied mainly with definite integrals, and in particular with the formula

$$\int_0^\infty x^{s-1}\{\phi(0)-x\phi(1)+x^2\phi(2)-\ldots\}\,dx = \frac{\pi}{\sin s\pi}\phi(-s)$$

and its developments and corollaries.† He had no real proofs either of this or of any other of his formulae. All of them are valid under appropriate conditions; but it is sometimes not at all obvious what these conditions are, and nearly all of the formulae are worth a careful analysis.

The two which I examine here are

$$\int_0^\infty \left\{\phi(0)-\frac{\phi(1)}{1!}t+\frac{\phi(2)}{2!}t^2-\ldots\right\}\cos xt\,dt$$
$$= \phi(-1)-\phi(-3)x^2+\phi(-5)x^4-\ldots, \quad (1.1)$$

$$\int_0^\infty \left\{\phi(0)-\frac{\phi(1)}{1!}t+\frac{\phi(2)}{2!}t^2-\ldots\right\}\sin xt\,dt$$
$$= \phi(-2)x-\phi(-4)x^3+\phi(-6)x^5-\ldots. \quad (1.2)$$

2. Ramanujan had no doubt learnt Fourier's integral theorem from Carr,‡ and he had found many of the formal developments of this and other reciprocities. For example, as I stated in a recent lecture,§ 'he had most of the formal ideas which underlie the recent work of Watson, and of Titchmarsh and myself,‖ on "Fourier kernels" and "reciprocal functions"'. Thus he knew that, if

$$G(x) = \sqrt{\left(\frac{2}{\pi}\right)} \int_0^\infty F(t)\cos xt\,dt$$

---

* I owe my knowledge of the contents of these reports to Prof. Watson, who included them in his manuscript copy of Ramanujan's note-books.

† See Hardy (**2**) for a proof of this formula under the most obvious set of conditions.   ‡ Carr (**1**), vol. 2, 399–403.   § Hardy (**3**).

‖ Hardy and Titchmarsh (**4, 5**), Watson (**7**). See also Titchmarsh (**6**), Chs. VIII and IX.

is the Fourier cosine transform of $F(x)$, and

$$f(s) = \frac{1}{\sqrt{(2\pi)}} \int_0^\infty F(x) x^{s-1} \, dx, \quad g(s) = \frac{1}{\sqrt{(2\pi)}} \int_0^\infty G(x) x^{s-1} \, dx$$

are the 'Mellin transforms' of $F(x)$ and $G(x)$, then

$$g(s) = \sqrt{\left(\frac{2}{\pi}\right)} \Gamma(s) \cos \tfrac{1}{2} s\pi f(1-s)$$

or

$$\frac{g(s)}{2^{\frac{1}{2}s} \Gamma(\tfrac{1}{2}s)} = \frac{f(1-s)}{2^{\frac{1}{2}(1-s)} \Gamma\{\tfrac{1}{2}(1-s)\}};$$

and he was very much interested in other 'integral translations' of the Fourier reciprocity.

In particular, Ramanujan used the formulae (1.1) and (1.2) as a basis for a heuristic theory of Fourier transforms (a theory, naturally, valid only under very narrow conditions).

Suppose, for example, that $\phi(u)$ is an integral function, and that

$$\phi(1) = \phi(3) = \phi(5) = \ldots = 0, \tag{2.1}$$

so that

$$\sum_0^\infty \frac{(-1)^n \phi(n)}{n!} x^n = \sum_0^\infty \frac{\phi(2m)}{2m!} x^{2m}$$

is even; and write

$$\phi_1(u) = \sqrt{\left(\frac{2}{\pi}\right)} \Gamma(1+u) \cos \tfrac{1}{2} u\pi \, \phi(-u-1). \tag{2.2}$$

Then $\phi_1(u)$ is also integral.* Also

$$\phi_1(1) = \phi_1(3) = \phi_1(5) = \ldots = 0,$$

$$\phi_1(2n) = (-1)^n \sqrt{\left(\frac{2}{\pi}\right)} 2n! \, \phi(-2n-1),$$

$$\phi_1(-2n-1) = \sqrt{\left(\frac{2}{\pi}\right)} \phi(2n) \lim_{\epsilon \to 0} \{\Gamma(-2n+\epsilon) \cos \tfrac{1}{2}(2n+1-\epsilon)\pi\}$$

$$= (-1)^n \sqrt{\left(\frac{\pi}{2}\right)} \frac{\phi(2n)}{2n!}.$$

Hence, if we write

$$L(x) = \sum_0^\infty \frac{\phi(2n)}{2n!} x^{2n}, \quad M(x) = \sqrt{\left(\frac{2}{\pi}\right)} \sum_0^\infty (-1)^n \phi(-2n-1) x^{2n},$$

$$\tag{2.3}$$

* The poles of $\Gamma(1+u)$ at $u = -1, -3, -5, \ldots$ are cancelled by zeros of $\cos \tfrac{1}{2} u\pi$, and those at $u = -2, -4, -6, \ldots$ by zeros of $\phi(-u-1)$.

and denote by L_1 and M_1 the functions derived similarly from ϕ_1, then
$$L_1(x) = M(x), \qquad M_1(x) = L(x);$$
and (1.1) and the corresponding formula with ϕ_1 become
$$\sqrt{\left(\frac{2}{\pi}\right)} \int_0^\infty L(t)\cos xt\, dt = M(x), \qquad \sqrt{\left(\frac{2}{\pi}\right)} \int_0^\infty M(t)\cos xt\, dt = L(x),$$
in accordance with Fourier theory.

For example, if
$$\phi(u) = 2^{-1-\frac{1}{2}u}\frac{\Gamma(-\frac{1}{2}u)}{\Gamma(-u)} = \frac{2^{\frac{1}{2}u}\sqrt{\pi}}{\Gamma(\frac{1}{2}-\frac{1}{2}u)},$$
all the conditions are satisfied, and
$$L(x) = \sum_0^\infty \frac{(-1)^n}{n!}(\tfrac{1}{2}x^2)^n = e^{-\frac{1}{2}x^2}.$$
Also
$$\phi_1(u) = \sqrt{\left(\frac{2}{\pi}\right)}\Gamma(1+u)\cos \tfrac{1}{2}u\pi \cdot \frac{2^{\frac{1}{2}u-\frac{1}{2}}\Gamma(\tfrac{1}{2}+\tfrac{1}{2}u)}{\Gamma(1+u)} = \frac{2^{\frac{1}{2}u}\sqrt{\pi}}{\Gamma(\tfrac{1}{2}-\tfrac{1}{2}u)} = \phi(u),$$
and $M(x) = L(x)$; so that (1.1) expresses the 'self-reciprocal' property of $e^{-\frac{1}{2}x^2}$.

On the other hand, there are many familiar formulae which are not to be accounted for as cases of (1.1). Thus
$$\int_0^\infty e^{-t}\cos xt\, dt = \frac{1}{1+x^2}, \qquad \int_0^\infty \frac{\cos xt}{1+t^2}\, dt = \tfrac{1}{2}\pi e^{-|x|}.$$
The first of these is a case of (1.1) when $-1 < x < 1$, with $\phi(u) = 1$. But then
$$\phi_1(u) = \sqrt{\left(\frac{2}{\pi}\right)}\Gamma(1+u)\cos \tfrac{1}{2}u\pi$$
is not an integral function, and we cannot account in this way for the second formula (as is obvious because $e^{-|x|}$ is not expansible as a power-series in x).

3. I shall now obtain sufficient conditions for the validity of Ramanujan's formula. The results could be deduced from more general theories (and, in particular, from that of Titchmarsh and myself); but it is interesting to prove them directly by the methods of 'classical' analysis.

THEOREM 1. *Suppose that $\delta > 0$, and that $\phi(u)$ is an analytic function of*
$$u = v + iw$$
which satisfies the following conditions:

(1) $\phi(u)$ *is an integral function*;

(2) $|\phi(u)| < C|\Gamma(1+u)|e^{Pv+A|w|}$,

where $A < \pi$, for $v \geq -\delta$;

(3) $|\phi(u)| < De^{-Qv+B|w|}$,

where $B < \frac{1}{2}\pi$, for $v \leq -\delta$. Then

$$\sum_{0}^{\infty} \frac{(-1)^n \phi(n)}{n!} t^n \qquad (3.1)$$

is convergent for $-e^{-P} < t < e^P$, and represents an analytic function $L(t)$ regular for all positive t; and

$$\sqrt{\left(\frac{2}{\pi}\right)} \int_{0}^{\infty} L(t) e^{-ixt} \, dt = M(x), \qquad (3.2)$$

where

$$M(x) = \sqrt{\left(\frac{2}{\pi}\right)} \sum_{0}^{\infty} \phi(-n-1)(-ix)^n, \qquad (3.3)$$

for $-e^{-Q} < x < e^{-Q}$; so that Ramanujan's formulae (1.1) and (1.2) are true for $-e^{-Q} < x < e^{-Q}$.

If
$$\psi(u) = \frac{\phi(u)}{\Gamma(1+u)}, \qquad (3.4)$$
then $\psi(u)$ is integral,
$$\psi(-1) = \psi(-2) = \ldots = 0,$$
and
$$|\psi(u)| < Ce^{Pv+A|w|} \qquad (3.5)$$
for $v \geq \delta$.*

Suppose now that
$$-1 < -\delta < \kappa < 0, \qquad 0 < t < e^{-P}. \qquad (3.6)$$

Then a simple application of Cauchy's Theorem shows that

$$L(t) = -\frac{1}{2\pi i} \int_{\kappa - i\infty}^{\kappa + i\infty} \frac{\pi}{\sin u\pi} \psi(u) t^u \, du, \qquad (3.7)$$

---

* It would be sufficient to suppose that
$$|\psi(u)| < He^{K|u|}$$
for $v \geq -\delta$ and some H and K, and that (3.5) is true for $u = v > 0$ and for $v = -\delta + iw$. It then follows, by a simple argument of 'Phragmén-Lindelöf' type, that (3.5) is true as stated. A corresponding change may be made in conditions (2) and (3) of the theorem.

where t^u has its principal value. This is true, first, for $0 < t < e^{-P}$, but the integral is uniformly convergent in any interval
$$0 < \tau \leqslant t \leqslant T < \infty$$
and gives the analytic continuation of $L(t)$ for all positive t.

It follows that, if x is real and not 0,
$$\int_\tau^T L(t)e^{-ixt}\,dt = -\frac{1}{2\pi i}\int_{\kappa-i\infty}^{\kappa+i\infty}\frac{\pi}{\sin u\pi}\psi(u)\,du\int_\tau^T e^{-ixt}t^u\,dt,$$
for $0 < \tau < T < \infty$. Also
$$\left|\int_0^\tau e^{-ixt}t^u\,dt\right| \leqslant \int_0^\tau t^\kappa\,dt = \frac{\tau^{1+\kappa}}{1+\kappa};$$
and
$$\int_T^\infty e^{-ixt}t^u\,dt = \frac{e^{-ixT}}{ix}T^u + \frac{u}{ix}\int_T^\infty e^{-ixt}t^{u-1}\,dt,$$
so that
$$\left|\int_T^\infty e^{-ixt}t^u\,dt\right| \leqslant \frac{T^\kappa}{|x|} + \frac{|u|}{|x|}\int_T^\infty t^{\kappa-1}\,dt = \left(1+\frac{|u|}{|\kappa|}\right)\frac{T^\kappa}{|x|}.$$

Hence
$$\int_{\kappa-i\infty}^{\kappa+i\infty}\frac{\pi}{\sin u\pi}\psi(u)\,du\int_0^\tau e^{-ixt}t^u\,dt, \qquad \int_{\kappa-i\infty}^{\kappa+i\infty}\frac{\pi}{\sin u\pi}\psi(u)\,du\int_T^\infty e^{-ixt}t^u\,dt$$
are convergent and tend to 0 when $\tau \to 0$ and $T \to \infty$; and
$$\int_0^\infty L(t)e^{-ixt}\,dt = -\frac{1}{2\pi i}\int_{\kappa-i\infty}^{\kappa+i\infty}\frac{\pi}{\sin u\pi}\psi(u)\,du\int_0^\infty e^{-ixt}t^u\,dt$$
$$= -\frac{1}{2\pi i}\int_{\kappa-i\infty}^{\kappa+i\infty}\frac{\pi}{\sin u\pi}\phi(u)(ix)^{-u-1}\,du. \qquad (3.8)$$

Here
$$(ix)^{-u-1} = \exp\{-(u+1)(\log x + \tfrac{1}{2}\pi i)\}$$
if $x > 0$, and
$$(ix)^{-u-1} = \exp\{-(u+1)(\log|x| - \tfrac{1}{2}\pi i)\}$$
if $x < 0$, the logarithms having their real values.

Now
$$|\phi(u)| = O(e^{-Qv+B|w|})$$
and
$$(ix)^{-u-1} = O(e^{-v\log|x|+\frac{1}{2}\pi|w|})$$
for $v \leqslant -\delta$; and $B + \tfrac{1}{2}\pi < \pi$. Hence, if $Q + \log|x| < 0$, i.e. if

$-e^{-Q} < x < e^{-Q}$, we can evaluate the last integral in (3.8) by another application of Cauchy's Theorem; and

$$\int_0^\infty e^{-ixt} L(t)\, dt = \sum_1^\infty \phi(-n)(-ix)^{n-1} = \sum_0^\infty \phi(-n-1)(-ix)^n = \sqrt{\left(\frac{\pi}{2}\right)} M(x),$$

which is (3.2).

4. We now suppose that $\phi(u)$ satisfies (2.1), so that $L(x)$ is even, and define ϕ_1, L_1, M_1 as in § 2. Then

$$|\psi_1(u)| = \left|\frac{\phi_1(u)}{\Gamma(1+u)}\right| = \sqrt{\left(\frac{2}{\pi}\right)} |\cos \tfrac{1}{2} u \pi \, \phi(-u-1)| = O(e^{P_1 v + A_1 |w|}),$$

where $\qquad A_1 = B + \tfrac{1}{2}\pi < \pi, \qquad P_1 = Q,$ \hfill (4.1)

for $v \geqslant -\delta$; and

$$|\phi_1(u)| = \sqrt{\left(\frac{2}{\pi}\right)} |\Gamma(1+u) \cos \tfrac{1}{2} u \pi \, \phi(-u-1)|$$

$$= \sqrt{\left(\frac{\pi}{2}\right)} \left|\frac{\psi(-u-1)}{\sin \tfrac{1}{2} u \pi}\right| = O(e^{-Q_1 v + B_1 |w|}),$$

where $\qquad B_1 = A - \tfrac{1}{2}\pi < \tfrac{1}{2}\pi, \qquad Q_1 = P,$ \hfill (4.2)

for $v \leqslant -\delta$. Hence ϕ_1 satisfies conditions like those imposed upon ϕ, with A_1, B_1, P_1, Q_1 in the place of A, B, P, Q. Also

$$A = B_1 + \tfrac{1}{2}\pi, \qquad P = Q_1, \qquad B = A_1 - \tfrac{1}{2}\pi, \qquad P_1 = Q,$$

so that the relationship is fully symmetrical.

We thus obtain

THEOREM 2. *Suppose that $\phi(u)$ satisfies the conditions of Theorem 1, and that* $\qquad \phi(1) = \phi(3) = \phi(5) = \ldots = 0.$

Then the series

$$L(x) = \sum_0^\infty \frac{\phi(2n)}{2n!} x^{2n}, \qquad M(x) = \sqrt{\left(\frac{2}{\pi}\right)} \sum_0^\infty (-1)^n \phi(-2n-1) x^{2n}$$

are convergent for $-e^{-P} < x < e^{-P}$ *and* $-e^{-Q} < x < e^{-Q}$ *respectively. The functions $L(x)$ and $M(x)$ are regular for positive x; and*

$$\sqrt{\left(\frac{2}{\pi}\right)} \int_0^\infty L(t) \cos xt\, dt = M(x), \qquad \sqrt{\left(\frac{2}{\pi}\right)} \int_0^\infty M(t) \cos xt\, dt = L(x),$$

so that $L(x)$ and $M(x)$ are a pair of Fourier transforms.

There is, of course, a similar theorem for sine transforms. In this

$$\phi(0) = \phi(2) = \phi(4) = \ldots = 0,$$

$$\phi_1(u) = \sqrt{\left(\frac{2}{\pi}\right)}\Gamma(1+u)\sin\tfrac{1}{2}u\pi\,\phi(-u-1),$$

$$L(x) = \sum_0^\infty \frac{\phi(2n+1)}{(2n+1)!} x^{2n+1},$$

$$M(x) = \sqrt{\left(\frac{2}{\pi}\right)}\sum_0^\infty (-1)^{n+1}\phi(-2n-2)x^{2n+1}.$$

The inequality conditions on $\phi(u)$ are the same.

5. The self-reciprocal case. The condition that $L(x)$ should be its own reciprocal is
$$\phi_1(u) = \phi(u)$$

or
$$\phi(u) = \sqrt{\left(\frac{2}{\pi}\right)}\Gamma(1+u)\cos\tfrac{1}{2}u\pi\,\phi(-u-1). \tag{5.1}$$

If we write
$$\chi(u) = 2^{1-\frac{1}{2}u}\frac{\Gamma(u)}{\Gamma(\frac{1}{2}u)}\phi(-u) \tag{5.2}$$

then (5.1) reduces to
$$\chi(u) = \chi(1-u), \tag{5.3}$$

so that $\chi(u)$ is an even function of $u-\tfrac{1}{2}$. In this case

$$L(x) = M(x) = \sqrt{\left(\frac{2}{\pi}\right)}\sum_0^\infty (-1)^n 2^{n-\frac{1}{2}} \frac{\Gamma(n+\tfrac{1}{2})}{2n!}\chi(2n+1)x^{2n}$$

$$= \sum_0^\infty \frac{\chi(2n+1)}{n!}(-\tfrac{1}{2}x^2)^n.$$

When $\chi(u) = 1$, $L(x) = M(x) = e^{-\frac{1}{2}x^2}$.

The integral formula for $L(x)$ is

$$L(x) = -\frac{1}{2\pi i}\int_{\kappa-i\infty}^{\kappa+i\infty} \frac{\pi}{\sin u\pi}\frac{\phi(u)}{\Gamma(1+u)}x^u\,du = \frac{1}{2\pi i}\int_{\kappa-i\infty}^{\kappa+i\infty}\Gamma(-u)\phi(u)x^u\,du$$

$$= \frac{1}{2\pi i}\int_{c-i\infty}^{c+i\infty}\Gamma(u)\phi(-u)x^{-u}\,du = \frac{1}{2\pi i}\int_{c-i\infty}^{c+i\infty} 2^{\frac{1}{2}u-1}\Gamma(\tfrac{1}{2}u)\chi(u)x^u\,du,$$

where $0 < c < 1$. This is (apart from a factor 2) the formula of Hardy and Titchmarsh.* The conditions imposed upon $\chi(u)$ by our analysis are that (i) $\chi(u)$ is an integral function of u, and (ii) that
$$|\chi(u)| = O\{|\Gamma(\tfrac{1}{2}+\tfrac{1}{2}u)|e^{Kv+H|w|}\},$$

* Hardy and Titchmarsh (4), formula (1.46).

where $H < \tfrac{1}{2}\pi$ and $K = Q + \tfrac{1}{2}\log 2$, for $v \geqslant -\delta$. The series for $L(x)$ is convergent if $x^2 < e^{-2Q}$.

6. Examples.

(i) If we take
$$\phi(u) = 2^{-1-\tfrac{1}{2}u}\frac{\Gamma(-\tfrac{1}{2}u)}{\Gamma(-u)} = 2^{-\tfrac{1}{2}u}\cos\tfrac{1}{2}u\pi\,\frac{\Gamma(1+u)}{\Gamma(1+\tfrac{1}{2}u)},$$
then the conditions of Theorem 2 are satisfied, with $A = \tfrac{3}{4}\pi$, $B = \tfrac{1}{4}\pi$, and any values of P and Q. This is the case $L(x) = M(x) = e^{-\tfrac{1}{2}x^2}$.

(ii) The function $\eta(u)$, defined by the series
$$\eta(u) = 1^{-u} - 3^{-u} + 5^{-u} - \ldots$$
and its analytic continuations, satisfies the equation
$$\eta(1-u) = 2^u \pi^{-u} \sin\tfrac{1}{2}u\pi\,\Gamma(u)\eta(u).$$
If we take
$$\phi(u) = 2(\tfrac{1}{2}\pi)^{\tfrac{1}{2}u}\eta(-u),$$
it will be found that all the conditions of Theorem 2 are satisfied, with
$$A = \tfrac{1}{2}\pi, \quad P = -\tfrac{1}{2}\log\tfrac{1}{2}\pi, \quad B = 0, \quad Q = -\tfrac{1}{2}\log\tfrac{1}{2}\pi.$$
Here also $\phi_1(u) = \phi(u)$, and
$$L(x) = \operatorname{sech} x\sqrt{(\tfrac{1}{2}\pi)}$$
is self-reciprocal.

(iii) If
$$\phi(u) = \frac{\cos\tfrac{1}{2}u\pi}{u+1}$$
then $\phi(u)$ is integral and $\phi(1) = \phi(3) = \ldots = 0$. The conditions of Theorem 1 are not satisfied because*
$$\phi(iw) \sim \tfrac{1}{2}|w|^{-1}e^{\tfrac{1}{2}\pi|w|}.$$
The result of the theorem is, however, still true, the formulae (1.1) and (1.2) being
$$\int_0^\infty \frac{\sin t}{t}\cos xt\,dt = \tfrac{1}{2}\pi, \qquad \int_0^\infty \frac{\sin t}{t}\sin xt\,dt = \tfrac{1}{2}\log\frac{1+x}{1-x}.$$
These formulae hold for $-1 < x < 1$; but the values of the integrals are
$$0, \qquad \tfrac{1}{2}\log\frac{x+1}{x-1}$$
when $x > 1$, and $M(x)$ is not regular for all positive x.

* The values of A, P, B, Q in conditions (2) and (3) would be
$$A = \pi, \quad P = -\infty, \quad B = \tfrac{1}{2}\pi, \quad Q = 0,$$
$P = -\infty$ meaning that (2) is satisfied for any P.

ON THE THEORY OF FOURIER TRANSFORMS 253

We could, of course, generalize the conditions of Theorem 1 so as to make it cover this and similar cases.

7. General transforms. There are analogues of Ramanujan's formulae in which $\cos x$ or $\sin x$ are replaced by a general 'Fourier kernel' $K(x)$. I confine myself to formal indications.

The Mellin transform $k(s)$ of $K(x)$ satisfies
$$k(s)k(1-s) = 1,$$
and we may write
$$k(s) = \frac{p(s)}{p(1-s)}.$$

I suppose that $p(s)$ is the reciprocal of an integral function. We start from the formula
$$L(x) = \sum_0^\infty (-1)^n \frac{\chi(n)}{p(n+1)} x^n = -\frac{1}{2\pi i} \int_{\kappa-i\infty}^{\kappa+i\infty} \frac{\pi}{\sin u\pi} \frac{\chi(u)}{p(u+1)} x^u \, du,$$
where $-1 < \kappa < 0$, and argue as in § 3. We find that
$$\int_0^\infty K(xt)L(t)\,dt = M(x), \qquad \int_0^\infty K(xt)M(t)\,dt = L(x), \qquad (7.1)$$
where
$$L(x) = \sum_0^\infty (-1)^n \frac{\chi(n)}{p(n+1)} x^n, \qquad M(x) = \sum_0^\infty (-1)^n \frac{\chi(-n-1)}{p(n+1)} x^n. \quad (7.2)$$

The condition for a self-reciprocal $L(x)$ is
$$\chi(u) = \chi(-u-1).$$

For example, when $K(u) = u^{\frac{1}{2}} J_\nu(u),$

we have
$$k(s) = 2^{s-\frac{1}{2}} \frac{\Gamma(\tfrac{1}{2}s+\tfrac{1}{2}\nu+\tfrac{1}{4})}{\Gamma(\tfrac{1}{2}\nu+\tfrac{3}{4}-\tfrac{1}{2}s)};$$
and we may take $\quad p(s) = 2^{\frac{1}{2}s}\Gamma(\tfrac{1}{2}s+\tfrac{1}{2}\nu+\tfrac{1}{4}).$

The formulae (7.2) become
$$L(x) = \sum_0^\infty (-1)^n \frac{\chi(n)}{\Gamma(\tfrac{1}{2}n+\tfrac{1}{2}\nu+\tfrac{3}{4})} x^n,$$
$$M(x) = \sum_0^\infty (-1)^n \frac{\chi(-n-1)}{\Gamma(\tfrac{1}{2}n+\tfrac{1}{2}\nu+\tfrac{3}{4})} x^n.$$

When $\quad K(x) = \dfrac{2}{\pi}\dfrac{1}{1-x^2}$

we have $\quad k(s) = \cot \tfrac{1}{2}s\pi,$
and we may take $\quad p(s) = \operatorname{cosec} \tfrac{1}{2}s\pi.$

The formulae are then
$$L(x) = \sum_0^\infty (-1)^n \chi(2n) x^{2n}, \qquad M(x) = \sum_0^\infty (-1)^n \chi(-2n-1) x^{2n}.$$

Thus $\chi(u) = 1$ gives the self-reciprocal function
$$\frac{1}{1+x^2}.$$

REFERENCES

1. G. S. Carr, *A synopsis of elementary results in pure and applied mathematics*, 2 vols. (London, 1880, 1886).
2. G. H. Hardy, 'On two theorems of F. Carlson and S. Wigert', *Acta Math.* **42** (1920), 327–39.
3. ——, 'The Indian mathematician Ramanujan', *American Math. Monthly*, **44** (1937), 137–55.
4. G. H. Hardy and E. C. Titchmarsh, 'Self-reciprocal functions', *Quart. J. of Math.* (Oxford), **1** (1930), 196–231.
5. ——, 'A class of Fourier kernels', *Proc. London Math. Soc.* (2), **35** (1933), 116–55.
6. E. C. Titchmarsh, *Introduction to the theory of Fourier integrals* (Oxford, 1937).
7. G. N. Watson, 'General transforms', *Proc. London Math. Soc.* (2), **35** (1933), 156–99.

CORRECTION

p. 248, *line after* (3.1). e^p should be e^{-p}.

ON A THEOREM OF PALEY AND WIENER

By Prof. G. H. HARDY, Trinity College

[*Received* 7 November, *read* 23 November 1936]

1. The conjugate g of a periodic and integrable function f is not necessarily integrable, even when f is monotone inside its fundamental interval*. Paley and Wiener†, however, proved that g is integrable if f is monotone and odd. A simpler proof was given later by Zygmund‡.

Here I prove a rather more complete theorem, referring in the first instance to the infinite interval $(-\infty, \infty)$. The method of proof is substantially the same as Zygmund's.

2. **Theorem A.** *Suppose that*

$$\int_{-\infty}^{\infty} |f|\, dx < \infty, \tag{2.1}$$

$$\int_{-\infty}^{\infty} |x df| < \infty. \tag{2.2}$$

Then
$$g(x) = -\frac{1}{\pi x}\int_{-x}^{x} f(t)\,dt + h(x), \tag{2.3}$$

with
$$\int_{-\infty}^{\infty} |h(x)|\, dx < \infty. \tag{2.4}$$

It is sufficient to prove the theorem for odd and for even functions; i.e. to prove (1) that g is $L(-\infty, \infty)$§ when f is odd, and (2) that

$$g = -\frac{2}{\pi |x|}\int_{0}^{x} f\,dt + h, \tag{2.5}$$

where h is $L(-\infty, \infty)$, when f is even.

---

* Suppose for example that the interval is $(-\pi, \pi)$, and that
$$f(x) = 0 \quad (-\pi < x < a), \qquad f(x) = \frac{1}{\pi - x}\left(\log \frac{1}{\pi - x}\right)^{-2} \quad (a \leq x < \pi),$$
with $\pi - a$ small.

† *Trans. Amer. Math. Soc.* 35 (1933), 348–56.

‡ *Ibid.* 36 (1934), 615–16.

§ Lebesgue integrable over $(-\infty, \infty)$.

The conjugate is defined by
$$g(x) = \frac{1}{\pi}\int_{-\infty}^{\infty} \frac{f(t)}{t-x}dt = \frac{1}{\pi}\int_0^{\infty}\left(\frac{1}{t-x}+\frac{1}{t+x}\right)f(t)\,dt \tag{2.6}$$
if f is odd, and by
$$g(x) = \frac{1}{\pi}\int_{-\infty}^{\infty} \frac{f(t)}{t-x}dt = \frac{1}{\pi}\int_0^{\infty}\left(\frac{1}{t-x}-\frac{1}{t+x}\right)f(t)\,dt \tag{2.7}$$
if f is even, and is even in the first case and odd in the second. The integrals converge, as Cauchy principal values, for almost all x.

We suppose, in either case, that $x > 0$, and write
$$g = \frac{1}{\pi}\int_{\frac{3}{2}x}^{\infty} + \frac{1}{\pi}\int_{\frac{1}{2}x}^{\frac{3}{2}x} + \frac{1}{\pi}\int_0^{\frac{1}{2}x} = g_1 + g_2 + g_3.$$

If
$$\phi(x) = \int_x^{\infty} \frac{|f(t)|}{t}\,dt,$$
then
$$\int_0^{\infty}\phi(x)\,dx = \int_0^{\infty}dx\int_x^{\infty}\frac{|f(t)|}{t}\,dt = \int_0^{\infty}\frac{|f(t)|}{t}\,dt\int_0^t dx = \int_0^{\infty}|f(t)|\,dt < \infty.$$

Also
$$\frac{1}{t-x}+\frac{1}{t+x} \leq \frac{18}{5t}$$
for $t \geq \frac{3}{2}x$. Hence $\pi|g_1(x)| \leq \frac{18}{5}\phi(\frac{3}{2}x)$, and g_1 is $L(0,\infty)$.

Next, g_2 is $L(0,\infty)$. For
$$g_2 = \frac{1}{\pi}\int_{\frac{1}{2}x}^{\frac{3}{2}x}\frac{f(t)}{t-x}dt \pm \frac{1}{\pi}\int_{\frac{1}{2}x}^{\frac{3}{2}x}\frac{f(t)}{t+x}dt = g_4 \pm g_5,$$
the plus sign for odd and the minus for even f. Since $\pi|g_5| \leq \phi(\frac{1}{2}x)$, it is enough to prove g_4 integrable.

Now
$$g_4 = \frac{1}{\pi}\int_0^{\frac{1}{2}x}\frac{f(x+u)-f(x-u)}{u}du = \frac{1}{\pi}\int_0^{\frac{1}{2}x}\frac{du}{u}\int_{x-u}^{x+u}df(w),$$

$$|g_4| \leq \frac{1}{\pi}\int_0^{\frac{1}{2}x}\frac{du}{u}\int_{x-u}^{x+u}|df(w)|$$

$$= \frac{1}{\pi}\int_x^{\frac{3}{2}x}|df(w)|\int_{w-x}^{\frac{1}{2}x}\frac{du}{u}+\frac{1}{\pi}\int_{\frac{1}{2}x}^{x}|df(w)|\int_{x-w}^{\frac{1}{2}x}\frac{du}{u}$$

$$= \frac{1}{\pi}\int_{\frac{1}{2}x}^{\frac{3}{2}x}\log\frac{x}{2|w-x|}|df(w)|,$$

$$\int_0^{\infty}|g_4|\,dx \leq \frac{1}{\pi}\int_0^{\infty}dx\int_{\frac{1}{2}x}^{\frac{3}{2}x}\log\frac{x}{2|w-x|}|df(w)|$$

$$= \frac{1}{\pi}\int_0^{\infty}|df(w)|\int_{\frac{2}{3}w}^{2w}\log\frac{x}{2|w-x|}dx = \frac{1}{\pi}\int_{\frac{2}{3}}^{2}\log\frac{1}{2|1-t|}dt\int_0^{\infty}w|df(w)|.$$

Hence g_4, and therefore g_2, is integrable, and g differs from g_3 by an integrable function.

Thirdly, $$g_3 = \frac{1}{\pi}\int_0^{\frac{1}{2}x} \frac{f(t)}{t-x}\,dt \pm \frac{1}{\pi}\int_0^{\frac{1}{2}x}\frac{f(t)}{t+x}\,dt = g_6 \pm g_7.$$

Here $$g_6 = -\frac{1}{\pi x}\int_0^{\frac{1}{2}x} f(t)\,dt + \frac{1}{\pi}\int_0^{\frac{1}{2}x}\left(\frac{1}{x}+\frac{1}{t-x}\right)f(t)\,dt = g_8 + g_9,$$

and $$|g_9| \leqslant \frac{1}{\pi x}\int_0^{\frac{1}{2}x} \frac{t}{x-t}|f(t)|\,dt \leqslant \frac{2}{\pi x^2}\int_0^{\frac{1}{2}x} t|f(t)|\,dt,$$

$$\int_0^\infty |g_9|\,dx \leqslant \frac{2}{\pi}\int_0^\infty \frac{dx}{x^2}\int_0^{\frac{1}{2}x} t|f(t)|\,dt = \frac{2}{\pi}\int_0^\infty t|f(t)|\,dt \int_{2t}^\infty \frac{dx}{x^2} = \frac{1}{\pi}\int_0^\infty |f(t)|\,dt.$$

Hence g_3 differs from $$g_8 = -\frac{1}{\pi x}\int_0^{\frac{1}{2}x} f(t)\,dt$$

by an integrable function. Similarly g_7 differs from $-g_8$ by an integrable function. Hence g_3 is integrable when f is odd, and differs from an integrable function by $2g_8$ when f is even. Finally, since

$$\left|\frac{1}{x}\int_{\frac{1}{2}x}^x f(t)\,dt\right| \leqslant \phi(\tfrac{1}{2}x),$$

g_3, and therefore g, differs from an integrable function, when f is even and x positive, by

$$-\frac{2}{\pi x}\int_0^x f(t)\,dt;$$

and this completes the proof of the theorem.

3. We may notice two special cases in which the condition (2·2) is satisfied. I suppose $f(x)$ odd or even, so that we need only consider the interval $(0, \infty)$.

(1) Let $$f(x) = p(x)q(x),$$

where $p(x)$ is monotone and integrable, and

$$q(x) = O(1), \quad q'(x) = O\left(\frac{1}{x}\right),$$

both at 0 and ∞. Then

$$\int_0^\infty x|df| \leqslant \int_0^\infty x O(1)|dp| + \int_0^\infty x|p|O\left(\frac{1}{x}\right)dx.$$

The second integral is obviously finite, and

$$\int_0^\infty x|dp| = \left|\int_0^\infty x\,dp\right| = \left|\int_0^\infty p\,dx\right| < \infty$$

(since $xp \to 0$ both at 0 and at ∞).

For example, the condition is satisfied if $q = \cos(a\log x)$ and p is monotone and integrable.

(2) Suppose that $f(x)$ is positive and $x^{-a}f(x)$ decreasing for some value of a (which we may suppose positive). In this case, if $x^{-a}f = g$, we have
$$x\,df = af\,dx + x^{a+1}\,dg,$$
$$\int_0^\infty |x\,df| \leqslant a\int_0^\infty f\,dx - \int_0^\infty x^{a+1}\,dg$$
$$= a\int_0^\infty f\,dx + (a+1)\int_0^\infty x^a g\,dx = (2a+1)\int_0^\infty f\,dx$$
(since $x^{a+1}g = xf \to 0$ both at 0 and at ∞*).

When f is (odd or even and) absolutely continuous, (2·2) is equivalent to
$$\int_0^\infty x\,|f'(x)|\,dx < \infty.$$
This form of the theorem may be generalized by the use of fractional derivatives. We may suppose that
$$f(x) = \frac{1}{\Gamma(\alpha)}\int_x^\infty (t-x)^{\alpha-1} f^{(\alpha)}(t)\,dt \quad (0 < \alpha < 1),$$
so that $f^{(\alpha)}(x)$ is effectively the αth derivative of $f(x)$. It is then sufficient that
$$\int_0^\infty x^\alpha\,|f^{(\alpha)}(x)|\,dx < \infty.$$
In particular, this condition is satisfied if
$$f^{(\alpha)}(x) \geqslant 0.$$

4. In the Paley-Wiener case $f(x)$ is odd and has the period 2π, and
$$g(x) = \frac{1}{2\pi}\int_{-\pi}^{\pi} f(t)\cot\tfrac{1}{2}(t-x)\,dt.$$
If we suppose $f(x)$ odd, and put
$$\tan\tfrac{1}{2}t = \tau, \quad \tan\tfrac{1}{2}x = \xi, \quad \frac{f(t)}{1+\tau^2} = F(\tau), \quad \frac{g(x)}{1+\xi^2} = G(\xi),$$
then
$$G(\xi) = \frac{2}{\pi}\int_0^\infty \frac{\tau F(\tau)}{\tau^2 - \xi^2}\,d\tau,$$
which is of the same form as (2·6); and f, g are $L(-\pi, \pi)$ when F, G are $L(0, \infty)$. Hence g is integrable if
$$\int_0^\pi |f(x)|\,dx < \infty, \quad \int_0^\pi \tan\tfrac{1}{2}x\,|d(\cos^2\tfrac{1}{2}xf(x))| < \infty.$$

* For example,
$$xf(2x) = x(2x)^a g(2x) \leqslant 2^a \int_x^{2x} t^a g(t)\,dt - 2^a\int_x^{2x} f(t)\,dt \to 0.$$

Suppose for example that, as in the Paley-Wiener theorem, f is monotone, say increasing, throughout $(-\pi, \pi)$. Then $f \geqslant 0$ for positive x, and the last integral does not exceed

$$\int_0^\pi \sin^2 \tfrac{1}{2}x f(x)\, dx + \int_0^\pi \cos \tfrac{1}{2}x \sin \tfrac{1}{2}x\, df(x) \leqslant \int_0^\pi f(x)\, dx + \tfrac{1}{2}\int_0^\pi \sin x\, df(x)$$

$$= \int_0^\pi f(x)\, dx - \tfrac{1}{2}\int_0^\pi \cos x f(x)\, dx \leqslant \tfrac{3}{2}\int_0^\pi f(x)\, dx,$$

since $f(x)\sin x \to 0$ when $x \to \pi$. Hence Theorem 1 includes the Paley-Wiener theorem*.

Another interesting case is that in which $f(x)$ is positive and decreasing in $(0, \pi)$, with an infinite peak at the origin.

If f is odd, g is integrable†. If f is even, then

$$g(x) + \frac{1}{\pi} \cot \tfrac{1}{2}x \int_0^x f(t)\, dt = g(x) + \frac{1}{\pi} \cot \tfrac{1}{2}x f_1(x)$$

is integrable. In order that $\tfrac{1}{2} \cot \tfrac{1}{2}x f_1(x)$ or $x^{-1}f_1(x)$ should be integrable, it is necessary and sufficient that either of

$$f(x) \log \frac{1}{x}, \quad f(x) \log^+ f(x)$$

should be integrable‡. The last result, however, is only a special case of a known (though more difficult) theorem, since the integrability of $f \log^+ f$ is a necessary and sufficient condition for that of g whenever f is positive§.

* There is no special case of Theorem 1 corresponding to it exactly, since $f(t)$ cannot be odd, increasing for all t, and integrable.
† This is not exactly a case of the Paley-Wiener theorem, but becomes one if we consider $(0, 2\pi)$ instead of $(-\pi, \pi)$.
‡ See G. H. Hardy and J. E. Littlewood, *Acta Math.* 54 (1930), 99–102; G. H. Hardy, J. E. Littlewood and G. Pólya, *Inequalities*, 169, Theorems 240, 241.
§ By theorems of Zygmund and M. Riesz. See Zygmund, *Trigonometrical Series*, 150–151.

CORRECTION

p. 2, *line* 3 *up*. In the 2nd repeated integral $\dfrac{1}{2|1-t|}$ should be $\dfrac{t}{2|1-t|}$.

A DOUBLE INTEGRAL

G. H. HARDY*.

1. I shall say that $K(x)$ is a "Fourier kernel" if (i) it is real, (ii) it is integrable in any finite interval $(0, X)$ of positive x, and (iii) its integral

$$(1.1) \qquad K_1(x) = \int_0^x K(t)\, dt$$

satisfies the integral equation

$$(1.2) \qquad \int_0^\infty \frac{K_1(ax)\, K_1(bx)}{x^2}\, dx = \min(a, b)$$

for all positive (a, b). The equation (1.2) shows, in particular, that $x^{-1} K_1(x)$ belongs to $L^2(0, \infty)$. It was proved by Watson that in these circumstances any $f(x)$ of $L^2(0, \infty)$ possesses a transform $F(x)$, also of $L^2(0, \infty)$, which satisfies the reciprocal relations

$$(1.3) \qquad \int_0^x F(t)\, dt = \int_0^\infty \frac{K_1(at)}{t} f(t)\, dt, \qquad \int_0^x f(t)\, dt = \int_0^\infty \frac{K_1(at)}{t} F(t)\, dt.$$

I shall assume a knowledge of Watson's theory, as set out in Ch. 8 of Titchmarsh's *Fourier integrals*.

2. THEOREM A. *If* (i) $K(x)$ *is a Fourier kernel*;

(ii) $K_1(x) = O(x^{\frac{1}{2}})$ *for all* x;

(iii) $H(x, y)$ *is homogeneous of degree* -1, *and has derivatives* H_x *and* H_y *continuous except at the origin*;

then

$$(2.1) \qquad \int_0^\infty \int_0^\infty K(ax)\, K(by)\, H(x, y)\, dx\, dy = H(b, a)$$

for all positive a, b. *Here the integral is to be interpreted as a repeated Cauchy integral, i.e., as one of the limits*

$$(2.2) \qquad \begin{cases} \lim_{\xi \to 0,\, X \to \infty} \int_\xi^X \ldots dx \left(\lim_{\eta \to 0,\, Y \to \infty} \int_\eta^Y \ldots dy \right), \\ \lim_{\eta \to 0,\, Y \to \infty} \int_\eta^Y \ldots dy \left(\lim_{\xi \to 0,\, X \to \infty} \int_\xi^X \ldots dx \right). \end{cases}$$

---

* Received 10 February, 1941; read 29 May, 1941.

If also

(ii *a*) $K_1(x) = o(x^{\frac{1}{2}})$ *at* 0 *and* ∞;

(iii *a*) $H(x, y)$ *has second derivatives continuous except at the origin;*

then (2.1) *holds when the integral is interpreted as*

$$(2.3) \qquad \lim_{\xi \to 0,\, X \to \infty,\, \eta \to 0,\, Y \to \infty} \int_\xi^X \int_\eta^Y \ldots dx\, dy.$$

The first part of the theorem is a corollary of a theorem of Titchmarsh and myself which is proved in §8.10 of Titchmarsh's book. This theorem says that, if $K(x)$ satisfies (i) and (ii), and

$$(2.4) \qquad f(x) = \frac{1}{x} \int_0^x \phi(t)\, dt,$$

where ϕ is $L^2(0, \infty)$, then

$$(2.5) \qquad \int_0^\infty K(xt)\, dt \int_0^\infty K(tu) f(u)\, du = f(x),$$

both integrals being Cauchy integrals. Now

$$\int_0^\infty K(by) H(x, y)\, dy = \int_0^\infty K(bxw) H(1, w)\, dw = G(bx),$$

where $G(x)$ is the transform of $H(1, w)$; and

$$\int_0^\infty K(ax) G(bx)\, dx = \frac{1}{b} \int_0^\infty K\!\left(\frac{aw}{b}\right) G(w)\, dw = \frac{1}{b} H\!\left(1, \frac{a}{b}\right) = H(b, a).$$

Thus (2.1) is a case of (2.5)* and, in order to prove (2.1), it is only necessary to show that $f(w) = H(1, w)$ is of the form (2.4).

But $H(1, w) = O(1)$ for small w, and

$$H(1, w) = \frac{1}{w} H\!\left(\frac{1}{w}, 1\right) = O\!\left(\frac{1}{w}\right)$$

for large w, so that $H(1, w)$ is L^2; and $H_w(1, w)$ is $O(1)$ for small and $O(w^{-2})$ for large w, so that $wH_w(1, w)$ is also L^2. Hence

$$\phi(w) = \frac{d}{dw}\{wH(1, w)\} = H(1, w) + wH_w(1, w)$$

---

* That is to say, if (2.5) is true with some interpretation of the integrals, then (2.1) is true with the same interpretation.

is L^2, and continuous for all positive w; and

$$H(1, w) = \frac{1}{w} \int_0^w \phi(t)\, dt.$$

3. To prove the second part of the theorem, we need three lemmas.

LEMMA 1. *If* (i) f, F *and* g, G *are two pairs of K-transforms of L^2*;

(ii) $P(x, y)$ *is homogeneous of degree* -1, *and*

(3.1) $$\int_0^\infty \frac{|P(1, y)|}{\sqrt{y}}\, dy < \infty,$$

then

(3.2) $$\int_0^\infty \int_0^\infty f(x)\, g(y)\, P(x, y)\, dx\, dy = \int_0^\infty \int_0^\infty F(x)\, G(y)\, P(y, x)\, dx\, dy,$$

the integrals being absolutely convergent.

By Theorem 319 of *Inequalities**, the integral

$$\int_0^\infty g(y)\, P(x, y)\, dy = q(x)$$

is absolutely convergent, and $q(x)$ is L^2. It follows, by Parseval's theorem for K-transforms, that

$$\int_0^\infty f(x)\, q(x)\, dx = \int_0^\infty F(x)\, Q(x)\, dx,$$

where $Q(x)$ is the transform of $q(x)$. If we can prove that

(3.3) $$Q(x) \equiv R(x) = \int_0^\infty G(y)\, P(y, x)\, dy,$$

then the lemma will follow.

Now

$$\int_0^x Q(t)\, dt = \int_0^\infty \frac{K_1(xt)}{t}\, q(t)\, dt = \int_0^\infty \frac{K_1(xt)}{t}\, dt \int_0^\infty g(y)\, P(t, y)\, dy$$

$$= \int_0^\infty \frac{K_1(xt)}{t}\, dt \int_0^\infty g(tu)\, P(1, u)\, du$$

(by the homogeneity of P). This integral is absolutely convergent (again by Theorem 319 of *Inequalities*), and so

$$\int_0^x Q(t)\, dt = \int_0^\infty P(1, u)\, du \int_0^\infty \frac{K_1(xt)}{t}\, g(tu)\, dt = \int_0^\infty P(1, u)\, du \int_0^\infty K_1\!\left(\frac{xw}{u}\right) g(w)\, \frac{dw}{w}.$$

---

* Hardy, Littlewood and Pólya, *Inequalities*, 229.

But this is

$$\int_0^\infty P(1, u)\, du \int_0^{x/u} G(t)\, dt = \int_0^\infty P\left(1, \frac{1}{v}\right) \frac{dv}{v^2} \int_0^{xv} G(t)\, dt$$

$$= \int_0^\infty P(v, 1) \frac{dv}{v} \int_0^{xv} G(t)\, dt = \int_0^\infty P(v, 1)\, dv \int_0^x G(vw)\, dw$$

$$= \int_0^x dw \int_0^\infty P(v, 1)\, G(vw)\, dv$$

$$= \int_0^x \frac{dw}{w} \int_0^\infty P\left(\frac{y}{w}, 1\right) G(y)\, dy$$

$$= \int_0^x dw \int_0^\infty P(y, w)\, G(y)\, dy = \int_0^x R(w)\, dw.$$

Hence $Q(x) \equiv R(x)$.

LEMMA 2.

$$\int_0^\infty \int_0^\infty \frac{K_1(ax)}{x} \frac{K_1(by)}{y} P(x, y)\, dx\, dy = \int_0^a \int_0^b P(y, x)\, dx\, dy.$$

For the transform of $x^{-1} K_1(ax)$ is 1 for $x < a$ and 0 for $x > a$.

LEMMA 3. *If $H(x, y)$ satisfies conditions* (iii) *and* (iii a) *of Theorem A, then*

$$L(x, y) = xy \frac{\partial^2 H(x, y)}{\partial x\, \partial y}$$

satisfies condition (ii) *of Lemma* 1; *and*

$$\int_0^a \int_0^b L(x, y)\, dx\, dy = ab H(a, b).$$

First, $L(x, y)$ is homogeneous of degree -1, and $L(1, y)$ is $O(1)$ for small and $O(y^{-1})$ for large y. Hence $L(x, y)$ satisfies (3.1).

Next

$$\frac{\partial^2}{\partial x\, \partial y} \{xy\, H(x, y)\} = H + x H_x + y H_y + xy H_{xy} = xy H_{xy},$$

by Euler's theorem on homogeneous functions. Hence

$$\int_\epsilon^a \int_\eta^b xy H_{xy}\, dx\, dy = ab H(a, b) - a\eta H(a, \eta) - b\epsilon H(\epsilon, b) + \epsilon\eta H(\epsilon, \eta)$$

$$\to ab H(a, b).$$

when ϵ and η tend to 0. The integral is in fact absolutely convergent, since

$$L(x, y) = O\left\{\min\left(\frac{1}{x}, \frac{1}{y}\right)\right\},*$$

$$\int_0^c \int_0^c |L(x, y)| \, dx \, dy \leqslant A \int_0^c \frac{dx}{x} \int_0^x dy + A \int_0^c \frac{dy}{y} \int_0^y dx = 2Ac,$$

for a constant A.

4. If we integrate by parts with respect to each of x and y, and write

(4.1) $$M(x, y) = K_1(ax) K_1(by) H(x, y),$$

(4.2) $$R(\xi, X, y) = K_1(by) \int_\xi^X K_1(ax) \frac{\partial H(x, y)}{\partial x} \, dx,$$

(4.3) $$S(x, \eta, Y) = K_1(ax) \int_\eta^Y K_1(by) \frac{\partial H(x, y)}{\partial y} \, dy,$$

we obtain

(4.4) $$\int_\xi^X \int_\eta^Y K(ax) K(by) H(x, y) \, dx \, dy$$

$$= \frac{1}{ab} \left\{ M(X, Y) - M(X, \eta) - M(\xi, Y) + M(\xi, \eta) \right.$$

$$- R(\xi, X, Y) + R(\xi, X, \eta) - S(X, \eta, Y) + S(\xi, \eta, Y)$$

$$\left. + \int_\xi^X \int_\eta^Y K_1(ax) K_1(by) \frac{\partial^2 H(x, y)}{\partial x \, \partial y} \, dx \, dy \right\}.$$

The last term tends to the absolutely convergent integral

$$\frac{1}{ab} \int_0^\infty \int_0^\infty \frac{K_1(ax)}{x} \frac{K_1(by)}{y} L(x, y) \, dx \, dy = H(b, a),$$

by Lemmas 2 and 3. It remains to prove that the remaining terms on the right-hand side of (4.4) tend to zero.

Now, first

$$H(X, Y) = \frac{1}{X} H\left(1, \frac{Y}{X}\right) = O\left(\frac{1}{X}\right), \quad H(X, Y) = \frac{1}{Y} H\left(\frac{X}{Y}, 1\right) = O\left(\frac{1}{Y}\right),$$

---

* See the argument concerning $H(X, Y)$ in § 4.

according as $X > Y$ or $Y > X$, so that in any case
$$H(X, Y) = O(X^{-\frac{1}{2}} Y^{-\frac{1}{2}}),$$
$$M(X, Y) = o(X^{\frac{1}{2}}) o(Y^{\frac{1}{2}}) O(X^{-\frac{1}{2}} Y^{-\frac{1}{2}}) = o(1).$$

Similarly the other terms of type M tend to zero, the terms $M(X, \eta)$ and $M(\xi, Y)$ with powers of the variables to spare.

Next
$$\left\{ \int_\xi^X K_1(ax) \frac{\partial H(x, Y)}{\partial x} dx \right\}^2 \leqslant \int_0^\infty \left\{ \frac{K_1(ax)}{x} \right\}^2 dx \int_0^\infty \left\{ x \frac{\partial H(x, Y)}{\partial x} \right\}^2 dx,$$

and the last integral is
$$\frac{1}{Y} \int_0^\infty \left\{ u \frac{\partial H(u, 1)}{\partial u} \right\}^2 du = O\left(\frac{1}{Y}\right).$$

Hence
$$R(\xi, X, Y) = o(Y^{\frac{1}{2}}) O(Y^{-\frac{1}{2}}) = o(1).$$

Similarly the other terms of type R or S tend to zero; and this completes the proof of the theorem.

5. It is easy to see, by splitting up the range (ξ, X) or (η, Y) appropriately into two parts, that the terms of types R and S tend to zero whenever $K_1(x)$ satisfies condition (ii); and this is true also of $M(X, \eta)$ and $M(\xi, Y)$. But if, for example, $K_1(x)$ is actually of order $x^{\frac{1}{2}}$ for a sequence of large x, then $M(X, Y)$ will not necessarily tend to zero; and then the limit (2.3) will not necessarily exist. Thus it is essential, in the second part of the theorem, to keep the stronger condition (ii a). The condition (iii a) might be relaxed, but there is no particular interest in this.

We could naturally prove the first part of the theorem by an argument similar to that of § 4.

6. Typical examples are
$$\int_0^\infty \int_0^\infty \frac{\cos ax \cos by}{(\alpha x^2 + 2\beta xy + \gamma y^2)^{\frac{1}{2}}} dx\, dy = \frac{\pi}{2(\alpha b^2 + 2\beta ba + \gamma a^2)^{\frac{1}{2}}},$$
$$\int_0^\infty \int_0^\infty \frac{(xy)^{\frac{1}{2}} J_\nu(ax) J_\nu(by)}{x+y} dx\, dy = \frac{1}{a+b}.$$

Here $a > 0$, $\alpha\gamma - \beta^2 > 0$, $\nu > -1$.

Trinity College,
    Cambridge.

A DOUBLE INTEGRAL

G. H. HARDY[*].

1. My object here is to evaluate the integral

(1.1) $$J = \iint e^{\alpha i x + \beta i y} \frac{f(x) - f(y)}{x - y} \, dx \, dy,$$

where f is a function of the class $L^2(-\infty, \infty)$, and the integrations (as always in the sequel when no limits are shown) are over $(-\infty, \infty)$, in

---

[*] Received 24 June, 1947; read 16 October, 1947.

terms of the Fourier transform of f. We are to interpret J as

$$\int e^{\alpha i x} dx \int e^{\beta i y} \frac{f(x)-f(y)}{x-y} dy,$$

the inner integral being a Cauchy integral

$$\lim_{Y \to \infty} \int_{-Y}^{Y} \ldots dy,$$

and the outer integration having any one of the senses usual in the theory of Fourier transforms, viz.

$$\underset{X \to \infty}{\text{l.i.m.}} \int_{-X}^{X} \ldots dx, \quad \lim_{X \to \infty} \int_{-X}^{X} \left(1-\frac{|x|}{X}\right) \ldots dx, \quad \lim_{\delta \to 0} \int e^{-\delta|x|} \ldots dx.$$

We call these the M, (C, 1), and A senses respectively.

The conclusion is that

(1.2) $$J = -\pi i (\operatorname{sgn} \alpha + \operatorname{sgn} \beta) \int e^{i(\alpha+\beta)x} f(x) \, dx$$

for almost all α and β. It is easy enough to find this formula by formal calculations; but these depend upon inversions whose justification requires considerable care, since the senses in which J converges are all rather sophisticated. It can be shown, for example, that (1.1) is never absolutely convergent unless f is constant for almost all x.

I use the following notations:

(1.3) $$T^+f = T^+f(x) = \frac{1}{\sqrt{(2\pi)}} \int e^{xit} f(t) \, dt, \quad T^-f = \frac{1}{\sqrt{(2\pi)}} \int e^{-xit} f(t) \, dt,$$

(1.4) $$\tilde{f} = Hf = Hf(x) = \frac{1}{\pi} \int \frac{f(t)}{t-x} \, dt,$$

the first two integrals being M integrals and the third a principal value at $t=x$. It is familiar that these functions exist for almost all x and are L^2, and that*

(1.5) $$T^-T^+f = T^+T^-f = f, \quad HHf = -f.$$

2. I write

(2.1) $$\int e^{\alpha i x} dx \left\{ f(x) \int \frac{e^{\beta i y}}{x-y} dy - \int \frac{e^{\beta i y} f(y)}{x-y} dy \right\}$$

$$= \int e^{\alpha i x} \{L(x) - M(x)\} dx = P - Q:$$

---

* See for example Titchmarsh, *Fourier integrals*, chs. 3 and 5.

both $L(x)$ and $M(x)$ are principal values at $y=x$, while $L(x)$ is a Cauchy integral, $M(x)$ absolutely convergent, at infinity. Here, first,

(2.2) $\quad L(x) = -\pi i \operatorname{sgn} \beta\, e^{\beta i x} f(x), \quad P = -\pi i \operatorname{sgn} \beta \int e^{(\alpha+\beta) i x} f(x)\, dx.$

If we could invert the order of the integrations in Q, it would follow similarly that

(2.3) $\quad Q = \int e^{\beta i y} f(y)\, dy \int \dfrac{e^{\alpha i x}}{x-y}\, dx = \pi i \operatorname{sgn} \alpha \int e^{(\alpha+\beta) i y} f(y)\, dy,$

and this would lead at once to (1.2). Thus the whole difficulty lies in the proof of (2.3).

We use the theorem* that, if

(2.4) $\quad \phi(x) = T^- \psi(x) = \dfrac{1}{\sqrt{(2\pi)}} \int e^{-xit} \psi(t)\, dt,$

then

(2.5) $\quad H\phi(x) = T^-\{-i\psi(x)\operatorname{sgn} x\} = -\dfrac{i}{\sqrt{(2\pi)}} \int e^{-xit} \psi(t) \operatorname{sgn} t\, dt.$

Now

(2.6) $\quad M(x) = \int \dfrac{e^{\beta i y} f(y)}{x-y}\, dy = -\pi H\{f(x)\, e^{\beta i x}\}$

and

(2.7) $\quad T^+ f(x+\beta) = \dfrac{1}{\sqrt{(2\pi)}} \int e^{xit}\{e^{\beta i t} f(t)\}\, dt,$

so that, after (1.5),

(2.8) $\quad e^{\beta i x} f(x) = T^-\{T^+ f(x+\beta)\} = \dfrac{1}{\sqrt{(2\pi)}} \int e^{-xit}\{T^+ f(t+\beta)\}\, dt.$

It follows from (2.4)–(2.8) that

$$M(x) = \dfrac{\pi i}{\sqrt{(2\pi)}} \int e^{-xit}\, T^+\{f(t+\beta)\} \operatorname{sgn} t\, dt,$$

and so

$$\pi i T^+\{f(x+\beta)\} \operatorname{sgn} x = \dfrac{1}{\sqrt{(2\pi)}} \int e^{xit} M(t)\, dt.$$

---

* See Titchmarsh, ch. 5: actually (2.5) is equivalent to his (5.1.8).

Hence

$$Q = \int e^{ait} M(t)\,dt = \pi i \sqrt{(2\pi)}\,[T^+\{f(x+\beta)\}\,\operatorname{sgn} x]_{x=a}$$

$$= \pi i \operatorname{sgn} a \int e^{(a+\beta)ix} f(x)\,dx,$$

which is (2.3).

3. It is natural to ask for more special conditions in which the convergence of J is of a more elementary type. Two simple cases are

$$\text{(a)} \quad f(x) = \frac{1}{1+x^2}, \qquad \text{(b)} \quad f(x) = \frac{x}{1+x^2}.$$

It is easily verified that (1.2) is true for all a and β in case (a), both inner and outer integrals being Cauchy integrals, and that in case (b) it is true except when $a = \beta = 0$.* These special cases suggest the consideration of the following sets of conditions.

(1) Suppose that f satisfies a Lipschitz condition of positive order δ in every finite interval of x, and is L as well as L^2 in $(-\infty, \infty)$. The inner integral is then a Cauchy integral (absolutely convergent at $y = x$). If we write the formula

$$\int e^{aix}\,dx \int e^{\beta iy} \frac{f(x)-f(y)}{x-y}\,dy = -\pi i(\operatorname{sgn} a + \operatorname{sgn}\beta) \int e^{(a+\beta)ix} f(x)\,dx$$

as

$$\int e^{aix} \psi(x,\beta)\,dx = \chi(a,\beta),$$

then ψ is L^2. Hence $\int e^{aix} \psi(x,\beta)\,dx$ is summable (C, 1) for any a for which $\frac{1}{2}\{\chi(a+0,\beta)+\chi(a-0,\beta)\} = \chi(a,\beta)$; and so (since f is L) for all a. Thus (1.2) *is true for all a and β, the inner integral being a Cauchy and the outer a* (C, 1) *integral*.

(2) If
$$f = \frac{Ax}{1+x^2} + g,$$

where g satisfies the conditions of (1), then (1.2) is true, in the same sense, except for $a = \beta = 0$.

---

* In this case it reduces to $\pi^2 = 0$.

(3) We must impose stronger restrictions if we wish the outer integral also to be a Cauchy integral. I state only one special result which covers the two examples mentioned at the beginning of the section. If f' is continuous, and
$$f = \frac{A}{x} + O\left(\frac{1}{x^2}\right), \quad f' = O\left(\frac{1}{x^2}\right)$$
for large $|x|$, then the conditions of (2) are satisfied. It is also easy to verify that $\psi(x, \beta) = O(|x|^{-1})$ for large $|x|$. Hence the outer integral [being summable $(C, 1)$] also exists as a Cauchy integral, and the formula is true in the "obvious" sense, with both integrals Cauchy integrals, except when $\alpha = \beta = 0$, $A \neq 0$.

4. I add a few miscellaneous remarks.

(1) As already stated in §1,
$$\iint \left| \frac{f(x)-f(y)}{x-y} \right| dx\, dy < \infty$$
implies $f = K$ for almost all x.

(2) In order that
$$\iint \left| \frac{f(x)-f(y)}{x-y} \right|^2 dx\, dy < \infty$$
it is necessary and sufficient that $\int |x| |F(x)|^2 dx < \infty$, where F is the transform of f; and
$$\iint \left| \frac{f(x)-f(y)}{x-y} \right|^2 dx\, dy = 2\pi \int |x| |F(x)|^2 dx.$$
This is true in the first instance when f and F are L^2; but, with a proper interpretation of the Fourier transform relationship, it holds without reservation, *i.e.* whenever either side of the equation is finite.

(3) There is a "reciprocal" formula
$$\iint e^{-aix-biy} \phi(x, y)\, dx\, dy = 2\pi i \frac{f(b)-f(a)}{b-a},$$
where
$$\phi(x, y) = \tfrac{1}{2}(\operatorname{sgn} x + \operatorname{sgn} y) F(x+y).$$

Rather more generally,

$$\int_0^\infty \int_0^\infty e^{-aix-biy} F(x+y)\,dx\,dy = \pi i \left\{ \frac{f(b)-f(a)}{b-a} + i\frac{\tilde{f}(b)-\tilde{f}(a)}{b-a} \right\},$$

$$\int_{-\infty}^0 \int_{-\infty}^0 e^{-aix-biy} F(x+y)\,dx\,dy = \pi i \left\{ \frac{f(b)-f(a)}{b-a} - i\frac{\tilde{f}(b)-\tilde{f}(a)}{b-a} \right\}.$$

The double integrals here are most simply interpreted as "A" integrals

$$\lim_{\delta \to 0} \iint e^{-\delta(|x|+|y|)} \ldots dx\,dy.$$

Trinity College,
Cambridge.

(*b*) Integral Equations

INTRODUCTION TO THE PAPERS ON INTEGRAL EQUATIONS

Hardy wrote six papers on integral equations other than transform formulae. Of these, all but 1909, 3 were written in collaboration with Titchmarsh, who incorporated much of the work, with improvements and modifications, into his book on Fourier Integrals (8).

The papers 1909, 3, 1925, 8, 1930, 5, 1936, 3 are all concerned with integral equations of one of the forms

$$f(x) = \int_0^\infty \frac{\chi(y)}{y} \{f(x+y)+f(x-y)\}\, dy, \qquad (1)$$

$$f'(x) = \int_0^\infty \frac{\chi(y)}{y} \{f(x+y)-f(x-y)\}\, dy, \qquad (2)$$

where $\chi(y)$ is a known function. When

$$\chi(y) = \pi^{-1} \sin ny, \qquad (3)$$

(1) is known as Bateman's equation. When

$$\chi(y) = \tfrac{1}{2} J_1(y), \qquad (4)$$

(2) is Kapteyn's equation. And when

$$\chi(y) = \tfrac{1}{2} e^{-y}, \qquad (5)$$

(2) is Milne's equation.

Paper 1909, 3 is concerned with Bateman's equation only. It is treated by real variable methods, by contour integration, and as an eigenvalue problem. In 1925, 8, all three equations are treated (i) by the theory of Fourier transforms of the class L^2, (ii) by analytic function theory. The paper 1930, 5 corrects a mistake in 1925, 8 and improves some of the proofs. The last paper, 1936, 3, is on Milne's equation; it widens the conditions under which there is a non-trivial solution.

Kapteyn's equation arises as the equation to be satisfied by an odd function $f(x)$ if it can be expanded in a Neumann series of the form

$$\sum_{n=0}^{\infty} a_{2n+1} J_{2n+1}(x).$$

This expansion is the series part of an eigenfunction expansion related to the differential equation

$$\frac{d^2\phi}{dx^2} + (\lambda + e^{2x})\phi = 0 \quad (-\infty < x < \infty),$$

which has a partly discrete, partly continuous spectrum. The series expansion alone is obtained if the function $f(x)$ is such that the continuous spectrum makes no contribution to the expansion formula. The relationship of this condition to Kapteyn's equation has been investigated by Sears and Titchmarsh (6).

Milne's equation is of fundamental importance in the theory of the transfer of radiation through a stellar atmosphere and in the theory of the diffusion of neutrons. The homogeneous integral equation arising from the transfer of radiation through a semi-infinite, plane parallel, homogeneous atmosphere is

$$S(x) = \tfrac{1}{2}\omega \int_0^\infty S(t) E_1(|t-x|)\, dt \quad (x > 0,\, 0 < \omega \leq 1), \tag{6}$$

where

$$E_1(t) = \int_1^\infty e^{-tu}\, \frac{du}{u}.$$

Equation (2), with $\chi(y)$ given by (5), is

$$f'(x) = \tfrac{1}{2} \int_0^\infty e^{-y}\{f(x+y) - f(x-y)\}\, \frac{dy}{y}. \tag{7}$$

On writing

$$f(x) = \int_0^x S(t)\, dt$$

and (formally) inverting the order of the integrations on the right, (7) becomes

$$S(x) = \tfrac{1}{2} \int_{-\infty}^\infty S(t) E_1(|t-x|)\, dt. \tag{8}$$

This is the asymptotic form of (6) for large x and $\omega = 1$.

Equation (6), with $\omega = 1$, was first found by E. A. Milne in 1921 (see (5)) and the equations (6)–(8) were investigated at about the same time by a number of well-known mathematicians. References to all the papers are given by Smithies in (7). Hardy and Titchmarsh proved in 1925, 8 and 1930, 5 that (7) has no non-null solution in the class L^2 but that, if $f(x) = O(e^{A|x|})$ for large x ($0 < A < 1$), then $f(x)$ is a quadratic. In 1936, 3 this condition is relaxed to

$$\int_{-\infty}^{\infty} |g(x)|^2[1+\log^2|g(x)|]\,dx < \infty,$$

where $$g(x) = f(x)\operatorname{sech} x.$$

Smithies (7) proved that the only solution is a quadratic when $f(x)$ is absolutely continuous in every finite interval and

$$\int_{-\infty}^{\infty} \frac{e^{-|x|}}{1+|x|} |f(x)|\,dx < \infty.$$

This gives the best possible result for (8) in the sense that, if the equation has a meaning when the integration is taken in the sense of Lebesgue, then $S(x)$ is linear.

The existence of a non-null solution of (6) (with $\omega = 1$) was first proved by Hopf (4). Subsequently Wiener and Hopf (9) used Laplace transforms to obtain the solution in the form of a complex integral. Their method has been developed in various ways and applied to both homogeneous and non-homogeneous equations of the Milne type. See references (1)–(3), where further references and developments will be found.

The two remaining papers on integral equations bear no relation to the preceding ones. 1929, 5 is concerned with the equations

$$f(x) = \lambda \int_0^{\infty} e^{-xy} f(y)\,dy \qquad (9)$$

or, by iteration, with

$$f(x) = \lambda^2 \int_0^{\infty} \frac{f(y)}{x+y}\,dy. \qquad (10)$$

Assuming only that the equations are true for all real x, that f is integrable -L over any finite interval, and that the integrals have a meaning in the Cauchy sense, it is proved that any solution of (9) or (10) is necessarily of the form

$$f(x) = Ax^{-a} + Bx^{a-1} \quad (0 < \operatorname{re} a < 1)$$

provided $0 < \lambda^2 < 1/\pi$. The constants A, B, a satisfy

$$A^2\Gamma(1-a) = B^2\Gamma(a)$$
$$\sin a\pi = \lambda^2\pi.$$

When $\lambda^2 = 1/\pi$, any solution is necessarily of the form

$$f(x) = Ax^{-\frac{1}{2}} + Bx^{-\frac{1}{2}}\log x,$$

where $B = 0$ when $\lambda = \pi^{-\frac{1}{2}}$ and $2A = -B\pi^{-\frac{1}{2}}\Gamma'(\frac{1}{2})$ when $\lambda = -\pi^{-\frac{1}{2}}$. This is, of course, an eigenvalue problem. It is discussed by Smithies in (**7**).

The paper 1932, 7 is concerned with the equation

$$f(x) = \frac{\lambda}{\Gamma(x)} \int_x^\infty (y-x)^{\alpha-1} f(y)\, dy. \tag{11}$$

Under conditions similar to those quoted above for (9) and (10), it is shown that all solutions of (11) are finite sums of exponentials. This equation is also discussed by Smithies in (**7**).

REFERENCES

1. I. W. BUSBRIDGE, *The Mathematics of Radiative Transfer*, Cambridge Tracts, no. 50, 1960.
2. B. DAVISON (with J. B. SYKES), *Neutron Transport Theory*, Oxford, 1957.
3. E. HOPF, *Mathematical Problems of Radiative Equilibrium*, Cambridge Tracts, no. 31, 1934.
4. ——, Zum Problem des Strahlungsgleichgewichte in den äusseren Schichten der Sterne. Strenge Lösung der singulären Integralgleichung von Milne, *Z. Physik* 46, 374–82, 1928.
5. E. A. MILNE, Radiative equilibrium in the outer layers of a star: the temperature distribution and the law of darkening, *Monthly Notices of the Roy. Astr. Soc.* 81, 361–75, 1921.
6. D. B. SEARS and E. C. TITCHMARSH, Some eigenfunction formulae, *Quart. J. of Math.* (Oxford) (2), 1, 165–75, 1950.
7. F. SMITHIES, Singular integral equations, *Proc. London Math. Soc.* (2), 46, 409–66, 1940.
8. E. C. TITCHMARSH, *Theory of Fourier Integrals*, Oxford, 1937.
9. N. WIENER and E. HOPF. Über eine Klasse singulärer Integralgleichungen, *S.B. Preuss. Akad. Wiss.* 696–706, 1931.

I.W.B.

ON AN INTEGRAL EQUATION

By G. H. Hardy.

[Received February 20th, 1909.—Read March 11th, 1909.]

I.

Introductory.

1. In one of his papers on integral equations,* Mr. H. Bateman has stated and made use of the equation

(1) $$f(x) = \frac{1}{\pi} \int_{-\infty}^{\infty} \frac{\sin(t-x)}{t-x} f(t)\, dt,$$

a formula which is striking in itself and capable of interesting applications to various questions in the general theory of the representation of arbitrary functions by means of definite integrals and infinite series. The formula, as Mr. Bateman points out, "is evidently not satisfied by a perfectly arbitrary function," and it is desirable to "find a convenient description of a class of functions to which it is applicable." But so far as I am aware, neither Mr. Bateman himself nor any other writer has attempted to apply accurate methods to the determination of any such class of functions.†

---

* *Proc. London Math. Soc.*, 2nd Series, Vol. 4, pp. 483 *et seq.*
† Mr. Bateman observes that

$$\int_{-\infty}^{\infty} \frac{\sin(t-x)}{t-x} f(t)\, dt = \tfrac{1}{2} \int_{-\infty}^{\infty} f(t)\, dt \int_{-1}^{1} e^{i\mu(x-t)}\, d\mu;$$

and that therefore, if the function $f(x)$ satisfies the equation (1), and if we write

$$\psi(\mu) = \frac{1}{2\pi} \int_{-\infty}^{\infty} e^{-i\mu t} f(t)\, dt, \qquad (a)$$

we shall have (assuming the convergence of all the integrals concerned, and the legitimacy of the inversion of the order of integration)

$$f(x) = \int_{-1}^{1} e^{ix\mu} \psi(\mu)\, d\mu. \qquad (\beta)$$

But this, although a very interesting transformation, affords at most the lines of a proof that a function satisfying (1) possesses the inversion property expressed by (a) and (β), and not a proof that a function of the form given by (β) necessarily satisfies (1).

2. In the present communication I propose to discuss, *ab initio* and by rigorous methods, a slightly generalised form of the equation (1), viz.,

$$(2) \qquad f(x) = \frac{1}{\pi} \int_{-\infty}^{\infty} \frac{\sin m(t-x)}{t-x} f(t)\, dt,$$

where x, $f(x)$, and m are real and m positive. A function which satisfies this equation I shall call *an m-function*. I shall show how to define extensive classes of m-functions, using first a direct method depending on the evaluation of certain definite integrals, and second the method of contour integration, which is less direct and general, but, at any rate in the most simple cases, more obvious in its application. I shall show that, if a function is an m-function for $m = m_0$, it is (subject to certain conditions) an m-function for $m \geqslant m_0$; and I shall discuss more cursorily some other interesting properties which such functions possess.

II.

Some general considerations relating to the equation (2).

3. Before I proceed to any accurate discussion of the equation (2), I wish to point out certain formal relations in which it stands to various questions of analysis.

(i.) Fourier's "double integral theorem" may be expressed by the equation

$$(3) \qquad f(x) = \lim_{m \to \infty} \frac{1}{\pi} \int_{-\infty}^{\infty} \frac{\sin m(t-x)}{t-x} f(t)\, dt.$$

This equation is satisfied by what, from the point of view of the theory of functions of real variables, may reasonably be called a "perfectly arbitrary" function: that is to say, by *a function subject only to conditions of inequality depending only on the values of* $f(x)$ *for real values of* x. It is certainly satisfied, for example, if $f(t)$ is a function whose total fluctuation in any finite interval is limited, and if

$$\int_{-\infty}^{\infty} f(t)\, dt$$

is convergent.*

Now, if $f(x)$ is an m-function for $m = m_0$, and if we assume the result mentioned above, that it is then (subject to certain conditions, purely of

---

* Hobson, *Theory of Functions of a Real Variable*, pp. 758 et seq.; Carslaw, *Fourier's Series and Integrals*, pp. 178 et seq.; Bromwich, *Infinite Series*, pp. 471, 472.

inequality) an m-function for $m \geqslant m_0$, we see that the equation (2) is then satisfied by *all values of m from m_0 onwards, and not merely in the limit* as expressed by (3) Thus, as we shall see, $\cos \mu x$ ($\mu > 0$) is an m-function for $m > \mu$, though not for $m = \mu$; and $(\sin \mu x)/x$ is an m-function for $m \geqslant \mu$, whereas $e^{-|x|}$, e^{-x^2} are not m-functions for any value of m, though they of course satisfy (3).

But, as we shall see, m-functions cannot be defined purely by "conditions of inequality," but only by *conditions of functional form*, such as, e.g., that $f(x)$ is defined by a definite integral of a particular type, or that $f(x)$ is an integral function subject to inequalities involving complex values of x.

(ii.) If $f(x)$ is an m-function, we can at once obtain a formal expansion of $f(x)$ in a series of the type

$$\sum \frac{a_n}{n!} \left(\frac{d}{dx}\right)^n \frac{\sin mx}{x},$$

where
$$a_n = \frac{1}{\pi} \int_{-\infty}^{\infty} (-t)^n f(t)\, dt.$$

The integrals which define the coefficients may be convergent or "summable," and so may the series; in any case the expansion is at present a purely formal one. I mention it because such expansions are of interest on account of their connection with expansions of the form

$$\sum b_n J_{n+\frac{1}{2}}(mx);$$

the relation between the two types of series is indeed, in virtue of Lord Rayleigh's formula

$$\sqrt{\left(\frac{\pi}{2x}\right)} J_{n+\frac{1}{2}}(x) = i^{-n} P_n \left(\frac{d}{i\, dx}\right)\left(\frac{\sin x}{x}\right),$$

very similar to that between the series

$$\sum \frac{a_n x^n}{n!}, \quad \sum b_n P_n(x).$$

(iii.) Two formal proofs of the equation (2) are also worth mentioning: each of them might be made the basis of an accurate proof. The first is that by which (as he informs me) Mr. Bateman was originally led to the equation (1). If we start from the known expansion[*]

$$\frac{\sin m(t-x)}{t-x} = \pi \sum_{0}^{\infty} (2n+1) \frac{J_{n+\frac{1}{2}}(mt)}{\sqrt{t}} \frac{J_{n+\frac{1}{2}}(mx)}{\sqrt{x}};$$

---

[*] Hobson, *Proc. London Math. Soc.*, Vol. xxv,, pp. 60 *et seq.*; Nielsen, *Handbuch der Cylinderfunktionen*, pp. 278 *et seq.*

assume that $f(t)$ is expansible in a series of the form

$$\Sigma a_p \frac{J_{p+\frac{1}{2}}(mt)}{\sqrt{t}},$$

and that the two series may be multiplied together and integrated over the interval $(-\infty, \infty)$; and finally use the formula

$$\int_{-\infty}^{\infty} \frac{J_{n+\frac{1}{2}}(mx)}{\sqrt{x}} \frac{J_{p+\frac{1}{2}}(mx)}{\sqrt{x}} dx = 0 \ (n \neq p), \quad = \frac{1}{2n+1} \ (n = p),$$

we are at once led to the equation (2).

Or, again, (2) may be obtained by the use of the theory of "divergent integrals" that I have explained in two recent papers on that subject.* For if we put $t = u+x$, we obtain

$$\int_{-\infty}^{\infty} \frac{\sin mu}{u} f(u+x) \, du = \int_{0}^{\infty} \frac{\sin mu}{u} \{f(u+x)+f(u-x)\} \, du$$

$$= 2 \sum_{\nu=0}^{\infty} \frac{f^{2\nu}(x)}{2\nu!} G \int_{0}^{\infty} \sin mu \, u^{2\nu-1} du$$

$$= \pi f(x).$$

This formal process will, as I showed in the second of the two papers referred to above, certainly lead to a correct result, if (i.) $f(x)$ is an integral function, and (ii.) the series

$$\Sigma \frac{f^{(2\nu)}(x)}{2\nu!} G \int_{0}^{\infty} e^{-miu} u^{2\nu-1} du = \Sigma f^{2\nu}(x)(mi)^{-2\nu}$$

is convergent.

Although, as I stated above, each of these two processes may be made the foundation of an accurate proof that functions of a certain class are really m-functions, I shall not so use them, as other methods seem to be simpler and better.

III.

Direct discussion of the equation (2).

4. I shall now prove that, if $\phi(w)$ is subject to certain "conditions of inequality," *the function*

$$f(x) = \int_{a}^{A} \frac{\sin \mu (w-x)}{w-x} \phi(w) \, dw,$$

---

* *Quarterly Journal*, Vol. XXXV., pp. 22 *et seq.*; *Trans. Camb. Phil. Soc.*, Vol. XXI., pp. 1 *et seq.*

where μ is positive, and a and A have any real values, $-\infty$ and ∞ included, is an m-function for $m \geqslant \mu$.

The proof of this result depends on the formula

(4) $$\int_{-\infty}^{\infty} \frac{\sin m(t-x)}{t-x} \frac{\sin \mu(t-w)}{t-w} dt = \pi \frac{\sin \mu(w-x)}{w-x} \quad (0 < \mu \leqslant m).$$

In fact, if we denote the integral by J, we have

$$J = \int_{-\infty}^{\infty} \frac{\sin \mu(t-w)}{t-w} dt \int_{0}^{m} \cos \xi(t-x) d\xi$$

$$= \int_{0}^{m} d\xi \int_{-\infty}^{\infty} \frac{\sin \mu(t-w)}{t-w} \cos \xi(t-x) d\xi$$

$$= \int_{0}^{m} d\xi \int_{-\infty}^{\infty} \frac{\sin \mu u}{u} \cos \xi(u-x+w) du$$

$$= 2 \int_{0}^{m} \cos \xi(w-x) d\xi \int_{0}^{\infty} \frac{\sin \mu u}{u} \cos \xi u \, du$$

$$= \pi \int_{0}^{\mu} \cos \xi(w-x) d\xi = \pi \frac{\sin(w-x)}{w-x};$$

since $\int_{0}^{\infty} \frac{\sin \mu u}{u} \cos \xi u \, du = \tfrac{1}{2}\pi \ (0 < \xi < \mu), \ = 0 \ (\mu < \xi).$

The only step in this transformation which requires justification is the inversion of the order of integration. The integral*

(5) $$\int_{-\infty}^{\infty} \frac{\sin \mu(t-w)}{t-w} \cos \xi t \, dt$$

$$= \tfrac{1}{2} \int_{-\infty}^{\infty} \frac{[\sin\{(\mu+\xi)t-\mu w\} + \sin\{(\mu-\xi)t-\mu w\}]}{t-w} dt$$

is uniformly convergent in the intervals $0 \leqslant \xi \leqslant \mu - \delta$, $\mu + \delta' \leqslant \xi \leqslant m$, however small δ and δ' may be; but it is not (as the discontinuity in its value clearly shows) uniformly convergent for $\mu - \delta \leqslant \xi \leqslant \mu + \delta'$. We have, however, the equations

$$\int_{0}^{\mu-\delta} \int_{-\infty}^{\infty} = \int_{-\infty}^{\infty} \int_{0}^{\mu-\delta}, \quad \int_{\mu+\delta'}^{m} \int_{-\infty}^{\infty} = \int_{-\infty}^{\infty} \int_{\mu+\delta'}^{m};$$

---

* In this discussion we have taken $x = 0$; as the general form of (4) may be deduced at once, by the transformation $t = x + u$, from the particular form in which $x = 0$, there is no real loss of generality in supposing $x = 0$.

and it remains only to prove that

(5a) $$\int_{-\infty}^{\infty}\int_{\mu-\delta}^{\mu+\delta'} = \int_{-\infty}^{\infty} \frac{\sin\mu(t-w)}{t-w} \frac{\sin(\mu+\delta')t - \sin(\mu-\delta)t}{t} dt$$

has the limit zero when δ and δ' converge in any manner to zero.

It is enough to prove this on the hypothesis that $\delta = \delta'$ (which simplifies the proof appreciably). For suppose this proved. Then

$$P\int_0^m \int_{-\infty}^{\infty} = \lim \left(\int_0^{\mu-\delta} + \int_{\mu+\delta}^m \right) \int_{-\infty}^{\infty}$$

exists, and has a value equal to $\int_{-\infty}^{\infty}\int_0^m$; but, as we know, by the actual calculations, that the first repeated integral is convergent even when the sign of the principal value is removed, we are justified in simply omitting this sign.

In order to effect the proof, we observe first that

$$\int_a^\beta \sin\mu(t-w)\{\sin(\mu+\delta)t - \sin(\mu-\delta)t\} dt$$
$$= \tfrac{1}{2}\int_a^\beta [\cos(\delta t + \mu w) - \cos(\delta t - \mu w)$$
$$- \cos\{(2\mu+\delta)t - \mu w\} + \cos\{(2\mu-\delta)t - \mu w\}] dt$$

is, for all values of a and β, numerically less than

$$\frac{2}{\delta} + \frac{1}{2\mu+\delta} + \frac{1}{2\mu-\delta}.$$

If now we divide the integral (5a) (in which now $\delta = \delta'$) into three parts

$$\int_{-\infty}^{-p} + \int_{-p}^{p} + \int_{p}^{\infty} = i_1 + i_2 + i_3,$$

it is clear that $\quad |i_2| < Kp\delta.$

Also, since $\dfrac{1}{(t-w)t}$ is monotonic, we can apply the second mean-value theorem to i_3; we thus obtain

$$|i_3| < \frac{K}{p^2}\left(\frac{2}{\delta} + \frac{1}{2\mu+\delta} + \frac{1}{2\mu-\delta} \right) < \frac{K}{p^2\delta}.$$

If now we observe that i_1 satisfies the same inequality, and suppose p such a function of δ that

$$p\delta \to 0, \quad p^2\delta \to \infty,$$

we see that
$$|i_1| + |i_2| + |i_3| \to 0;$$
and our proof is accordingly completed.*

5. Now consider the integral
$$f(t) = \int_a^A \frac{\sin \mu (w-t)}{w-t} \phi(w) \, dw,$$
where
$$0 < \mu \leqslant m.$$

I suppose first that a and A are finite. Multiplying by
$$\frac{\sin m(t-x)}{t-x},$$
and integrating from $-\infty$ to ∞, we obtain

$$\int_{-\infty}^{\infty} \frac{\sin m(t-x)}{t-x} f(t) \, dt = \int_{-\infty}^{\infty} \frac{\sin m(t-x)}{t-x} dt \int_a^A \frac{\sin \mu (t-w)}{t-w} \phi(w) \, dw$$

$$= \int_a^A \phi(w) \, dw \int_{-\infty}^{\infty} \frac{\sin m(t-x)}{t-x} \frac{\sin \mu(t-w)}{t-w} dt$$

$$= \pi \int_a^A \frac{\sin \mu(w-x)}{w-x} \phi(w) \, dw = \pi f(x).$$

This transformation will certainly be valid if

(i.) $\phi(w)$ is continuous for $a \leqslant w \leqslant A$, with the possible exception of a finite number of points;†

(ii.) $\int_a^A |\phi(w)| \, dw$ is convergent.

For the only point which requires proof is that
$$\int_{-\infty}^{\infty} \frac{\sin m(t-x)}{t-x} \frac{\sin \mu(t-w)}{t-w} dt$$

---

* It is, I confess, surprising to me that a rigid proof of the equation (4) should be, comparatively speaking, so difficult to obtain.

[*Sept.*, 1909.—Mr. Berry and Dr. Bromwich have each pointed out to me that a simpler proof may be obtained as follows. We have
$$\int_{-\infty}^{\infty} \frac{\sin m(t-x)}{t-x} \frac{\sin \mu(t-w)}{t-w} dt = \frac{1}{w-x} \int_{-\infty}^{\infty} \left(\frac{1}{t-w} - \frac{1}{t-x} \right) \sin m(t-x) \sin \mu(t-w) \, dt.$$
Divide the integral into two parts, and put $t = u + w$ in the first and $t = u + x$ in the second. We then obtain
$$\frac{1}{\lambda} \int_{-\infty}^{\infty} \frac{\sin \mu u}{u} \sin m(u+\lambda) \, du - \frac{1}{\lambda} \int_{-\infty}^{\infty} \frac{\sin mu}{u} \sin \mu(u-\lambda) \, du,$$
where $\lambda = w - x$; and the result then follows easily from familiar formulæ. The integral may thus be evaluated without any inversion of limit operations.]

† This supposition could easily be generalised, but is general enough as it stands for our present purposes.

is uniformly convergent for $a \leqslant w \leqslant A$; and this is almost obvious, since
$$\left| \int_T^\infty \frac{\sin m(t-x)}{t-x} \frac{\sin \mu(t-w)}{t-w} dt \right| < K \int_T^\infty \frac{dt}{t^2} < \frac{K}{T}.$$

6. There remains the more difficult case in which a or A or both are infinite. The formal argument is then precisely the same, but the inversion of the integrations is a good deal harder to justify.

Let us suppose that $A = \infty$, and that
$$\int^\infty \frac{|\phi(w)|}{w} dw$$
is convergent. If this is so it is easy to see that

(6) $$\int_a^\beta \frac{\sin m(t-x)}{t-x} dt \int_a^\infty \frac{\sin \mu(w-t)}{w-t} \phi(w) dw$$
$$= \int_a^\infty \phi(w) dw \int_a^\beta \frac{\sin m(t-x)}{t-x} \frac{\sin \mu(w-t)}{w-t} dt$$

for any finite values of a and β. For, if we replace ∞ by A, we obtain an equation already established; and so all that remains is to show that
$$\int_a^\beta \frac{\sin m(t-x)}{t-x} dt \int_A^\infty \frac{\sin \mu(w-t)}{w-t} \phi(w) dw$$
is convergent and tends to zero as $A \to \infty$. Now, if K is any constant greater than unity, we can choose A so that
$$\left| \frac{w}{w-t} \right| < K \quad (A \leqslant w,\ a \leqslant t \leqslant \beta);$$
and it follows that the inner integral above is absolutely and uniformly convergent for $a \leqslant t \leqslant \beta$, and so a continuous function of t; and moreover that
$$\left| \int_A^\infty \frac{\sin \mu(w-t)}{w-t} \phi(w) dw \right| < K \int_A^\infty \frac{|\phi(w)|}{w} dw,$$
and so tends to zero as $A \to \infty$, uniformly for $a \leqslant t \leqslant \beta$. Hence our conclusion is established.

7. In order to pass from the equation (6) to the equation

(7) $$\int_{-\infty}^\infty \frac{\sin m(t-x)}{t-x} dt \int_a^\infty \frac{\sin \mu(w-t)}{w-t} \phi(w) dw$$
$$= \int_a^\infty \phi(w) dw \int_{-\infty}^\infty \frac{\sin m(t-x)}{t-x} \frac{\sin \mu(w-t)}{w-t} dt,$$

all that is necessary is to prove that
$$\int_a^\infty \phi(w) dw \int_T^\infty \frac{\sin m(t-x)}{t-x} \frac{\sin \mu(w-t)}{w-t} dt$$

is convergent, and has the limit zero as $T \to \infty$, together with a corresponding result involving $-\infty$ and $-T$ as the limits of integration with respect to t. We can write this repeated integral in the form

$$(8) \qquad \int_a^\infty \phi(w) F(x-T, w-T) \, dw,$$

where

$$(9) \qquad F(y, z) = \int_0^\infty \frac{\sin m(\xi-y)}{\xi-y} \frac{\sin \mu(\xi-z)}{\xi-z} \, d\xi.$$

I shall now establish certain inequalities satisfied by the function $F(y, z)$: the most convenient method for doing this is based upon a preliminary transformation of $F(y, z)$ by means of Cauchy's theorem.*

8. Consider the complex integral

$$\int \frac{e^{i[(m-\mu)x - my + \mu z]} - e^{i[(m+\mu)x - my - \mu z]}}{(x-y)(x-z)} \, dx$$

taken round the contour formed by the positive halves of the real and imaginary axes, and an "infinite quadrant of a circle" whose centre is the origin.†

Taking the real parts only of each term in the resulting equation, we obtain

$$(10) \quad F(y, z) = -\tfrac{1}{2} R \int_0^\infty \left\{ e^{-(m-\mu)\eta - (my - \mu z)i} - e^{-(m+\mu)\eta - (my + \mu z)i} \right\} \frac{i \, d\eta}{(i\eta - y)(i\eta - z)}.$$

Suppose first that $m > \mu$. Then

$$(11) \qquad F(y, z) = -\frac{1}{2(y-z)} R(\chi_1 + \chi_2 + \chi_3 + \chi_4),$$

where

$$(12) \quad \begin{cases} \chi_1(y, z) = e^{-(my - \mu z)i} \int_0^\infty \dfrac{e^{-(m-\mu)\eta} \, i \, d\eta}{i\eta - y}, \\[6pt] \chi_2(y, z) = -e^{-(my + \mu z)i} \int_0^\infty \dfrac{e^{-(m+\mu)\eta} \, i \, d\eta}{i\eta - y}, \\[6pt] \chi_3(y, z) = -e^{-(my - \mu z)i} \int_0^\infty \dfrac{e^{-(m-\mu)\eta} \, i \, d\eta}{i\eta - z}, \\[6pt] \chi_4(y, z) = -e^{-(my + \mu z)i} \int_0^\infty \dfrac{e^{-(m+\mu)\eta} \, i \, d\eta}{i\eta - z}. \end{cases}$$

---

* The results of this transformation can be found directly, but then much less simply.

† The subject of integration is regular for $x = z$, but has a pole for $x = y$; but, in the applications which we have in view, y is large and negative, so that it is unnecessary to consider the point $x = y$.

It is clear that

(13) $$|\chi_1| < \int_0^\infty \frac{e^{-(m-\mu)\eta}}{\sqrt{(\eta^2+y^2)}} d\eta < \frac{1}{(m-\mu)|y|} < \frac{K}{|y|},$$

and, similarly,

(14) $$|\chi_2| < \frac{K}{|y|}, \quad |\chi_3| < \frac{K}{|z|}, \quad |\chi_4| < \frac{K}{|z|}.$$

We shall be considering only cases in which y is large and negative, and these inequalities will be sufficient so far as χ_1 and χ_2 are concerned. But z will be capable of assuming all real values, so that different inequalities will be needed for χ_3 and χ_4, when $|z|$ is comparatively small.

9. Now, if ρ is positive,

$$\int_0^\infty \frac{e^{-\rho\eta}d\eta}{\sqrt{(\eta^2+z^2)}} = \int_0^\infty \frac{e^{-\rho|z|u}}{\sqrt{(u^2+1)}} du < \int_0^{1/|z|} \frac{du}{\sqrt{(u^2+1)}} + \int_{1/|z|}^\infty \frac{e^{-\rho|z|u}}{u} du$$

$$= \log\left\{\frac{1}{|z|} + \sqrt{\left(1+\frac{1}{|z|^2}\right)}\right\} + \int_1^\infty \frac{e^{-\rho u}}{u} du < K(1+|\log|z||).$$

Hence

(15) $$|\chi_3|, \quad |\chi_4| < \frac{K}{|z|}, \quad K(1+|\log|z||),$$

where either function may be chosen on either side of the inequality. From (11), (13), (14), and (15), we see that

(16) $$|F(y,z)| < F_1 + F_2,$$

where

(17) $$\begin{cases} 0 < F_1 < \dfrac{K}{|y-z||y|}, \\ 0 < F_2 < \dfrac{K}{|y-z||z|}, \quad \dfrac{K}{|y-z|}(1+|\log|z||). \end{cases}$$

From these inequalities it follows at once that the integral (8) is convergent. Moreover

$$\left|\int_a^\infty \phi(w) F(-T, w-T) dw\right| < \int_a^\infty |\phi(w)| F_1 dw + \int_a^\infty |\phi(w)| F_2 dw.^*$$

And $$\int_a^\infty |\phi(w)| F_1 dw < \frac{K}{T}\int_a^\infty \frac{|\phi(w)|}{w} dw,$$

---

* In what follows I suppose $x=0$ and $a>0$; it is obvious that this involves no loss of generality.

which tends to zero as $T \to \infty$. Also

$$\int_a^\infty |\phi(w)| F_2 dw = \left(\int_a^{T-\tau} + \int_{T-\tau}^{T+\tau} + \int_{T+\tau}^\infty\right) |\phi(w)| F_2 dw = j_1 + j_2 + j_3,$$

say, τ being at our disposal. Then

$$j_1 < \frac{K}{\tau} \int_a^\infty \frac{|\phi(w)|}{w} dw, \quad j_3 < \frac{K}{\tau} \int_a^\infty \frac{|\phi(w)|}{w} dw,$$

and $j_2 < K \int_{-\tau}^\tau \frac{|\phi(T+u)|}{T+u}(1+|\log|u||) du < \frac{K}{T}\left(\tau + \int_{-\tau}^\tau |\log u^2| du\right);$

and if we choose τ to be such a function of T that

$$\tau \to \infty, \quad \frac{\tau \log \tau}{T} \to 0,$$

we see that j_1, j_2, j_3 all tend to zero as $T \to \infty$.

We have thus established the equation (7) on the hypotheses that (i.) $\phi(w)$ is continuous save for at most a finite number of values of w, (ii.) $\int |\phi(w)| dw$ is convergent over any finite interval, (iii.) $\int^\infty \frac{|\phi(w)|}{w} dw$ is convergent, and (iv.) $m > \mu$. Thus a function $f(x)$ defined as at the beginning of § 4 is certainly an m-function for $m > \mu$, even when $A = \infty$, provided $\phi(w)$ satisfies these conditions. A similar conclusion, of course, is valid when $a = -\infty$.

10. We have, however, still to complete the discussion in the case in which $m = \mu$. In this case the integrals χ_1 and χ_3 are not convergent. We have, however,

(18) $$F(y, z) = -\tfrac{1}{2}R\left(\frac{\chi_2+\chi_4}{y-z}+\chi\right),$$

where

(19) $$\chi(y, z) = e^{-m(y-z)i} \int_0^\infty \frac{i\, d\eta}{(i\eta - y)(i\eta - z)}.$$

The value of the last integral is

$$\frac{1}{y-z}\left\{\log\frac{|z|}{|y|} + \tfrac{1}{2}\pi i(1+\operatorname{sgn} z)\right\}.$$

Hence (i.) $\chi \to 0$ as $T \to \infty$, uniformly throughout any finite interval of values of w, and

(ii.) $|\chi| < \dfrac{K}{w}\left\{1+\left|\log\left(1-\dfrac{w}{T}\right)^2\right|\right\}.$

From this inequality, coupled with those already established for χ_2 and χ_4, it follows at once that the integral (8) is convergent. And all that remains is to prove that

$$\int_a^\infty |\phi(w)| |\chi| dw$$

has the limit zero as $T \to \infty$.

Now we may write

$$\int_a^\infty |\phi(w)| |\chi| dw = \left(\int_a^H + \int_H^\infty\right) |\phi(w)| |\chi| dw$$

$$\leqslant \int_a^H |\phi(w)| |\chi| dw + \int_H^\infty |\phi(w)| \chi_1 dw + \int_H^\infty |\phi(w)| \chi_2 dw$$

$$= k_1 + k_2 + k_3,$$

say; where

$$0 < \chi_1 < \frac{K}{w}, \quad 0 < \chi_2 < \frac{K}{w} \left|\log\left(1-\frac{w}{T}\right)^2\right|.$$

We shall be obliged to subject $\phi(w)$ to a slightly more stringent condition than was imposed upon it in § 6, viz., that

(20) $$\int^\infty \frac{|\phi(w)| \log w}{w} dw$$

is convergent. I have proved elsewhere* that if this condition is satisfied then

$$\int_H^\infty \frac{|\phi(w)|}{w} \left|\log\left(1-\frac{w}{T}\right)^2\right| dw$$

has the limit zero as $T \to \infty$.

Now $$k_2 < K \int_H^\infty \frac{|\phi(w)|}{w} dw < K \int_H^\infty \frac{|\phi(w)| \log w}{w} dw.$$

We can therefore choose H so that $k_2 < \sigma$. When H is fixed we can choose T_0 so that $k_1 < \sigma$ and $k_3 < \sigma$ for $T \geqslant T_0$, and then

$$\int_a^\infty |\phi(w)| |\chi| dw < 3\sigma \quad (T \geqslant T_0).$$

Our proof is accordingly completed, for the part of $F(y, z)$ which involves χ_2 and χ_4 may be treated as before.

The conclusion at which we arrived at the end of § 9 may therefore be extended to the case in which $m = \mu$, provided we replace (iii.) by the

---

* See *Proc. London Math. Soc.*, Ser. 2, Vol. 7, pp. 201 *et seq.*

slightly more stringent condition involved in the convergence of the integral (20).

11. We may sum up the conclusions of §§ 4–10 as follows :—

Any function $f(x)$ defined by an integral of the type

$$(21) \qquad f(x) = \int_a^A \frac{\sin \mu (w-x)}{w-x} \phi(w)\, dw,$$

where $\phi(w)$ is subject to the conditions—

(i.) *$\phi(w)$ is continuous save perhaps for a finite number of values of w,*

(ii.) $\int |\phi(w)|\, dw$ *is convergent over any finite interval,*

is an m-function for $m \geqslant \mu$.

This conclusion holds for $A = \infty$ or $a = -\infty$, provided that $m > \mu$, and that (iii.) *the integrals*

$$\int^\infty \frac{|\phi(w)|}{w}\, dw, \quad \int_{-\infty} \frac{|\phi(w)|}{w}\, dw$$

are convergent.

It holds for $A = \infty$ or $a = -\infty$, even when $m = \mu$, provided that (iii'.) *the integrals*

$$\int^\infty \frac{|\phi(w)| \log w}{w}\, dw, \quad \int_{-\infty} \frac{|\phi(w)| \log(-w)}{w}\, dw$$

are convergent.

Further, if $f(x)$ is an m-function for $m = \mu$, and satisfies the conditions (i.), (ii.), (iii.), *it is an m-function for $m \geqslant \mu$.*

Finally, if $f(x)$ is subject to the conditions (i.), (ii.), (iii'.), *then the necessary and sufficient condition that it should be an m-function is that it should be expressible in the form* (21), *with either finite or infinite limits.* For that this condition is *sufficient* has just been proved, and that it is *necessary* is obvious, an expression of the form required being obtained by taking $a = -\infty$, $A = \infty$, $\mu = m$, $\phi(w) \equiv f(w)$.

IV.

Generalisations and examples.

12. The results of the preceding sections are capable of various generalisations.

Suppose that $F(x, w)$ is an m-function for $a \leqslant w \leqslant A$, and that

(22) $$f(x) = \int_a^A \phi(w)\, F(x, w)\, dw.$$

Then
$$\int_{-\infty}^{\infty} \frac{\sin m(t-x)}{t-x} f(t)\, dt = \int_{-\infty}^{\infty} \frac{\sin m(t-x)}{t-x}\, dt \int_a^A \phi(w)\, F(t, w)\, dw$$
$$= \int_a^A \phi(w)\, dw \int_{-\infty}^{\infty} \frac{\sin m(t-x)}{t-x} F(t, w)\, dt$$
$$= \pi \int_a^A \phi(w)\, F(x, w)\, dw = \pi f(x);$$

so that $f(x)$ is an m-function. If a and A are finite, the conditions (i.) that $F(x, w)$ is continuous, (ii.) that $\phi(w)$ is continuous save for a finite number of values of w, (iii.) that $\int_a^A |\phi(w)|\, dw$ is convergent, and (iv.) that

$$\int_{-\infty}^{\infty} \frac{\sin m(t-x)}{t-x} F(t, w)\, dt$$

is uniformly convergent for $a \leqslant w \leqslant A$, will certainly be sufficient to ensure the validity of the transformation. In particular the last condition will certainly be satisfied, if

$$|F(t, w)| < K|t|^{-s} \quad (s > 0),$$

for $a \leqslant w \leqslant A$, and all large values of $|t|$. When a or A is infinite the argument (as will be clear from the preceding discussion of a particular case) will involve much greater difficulties.

13. The most interesting cases of this general result are obtained by taking
$$F(x, w) = F(w-x), \quad F(xw).$$

Let us consider in particular the case in which
$$F(x, w) = \begin{matrix}\cos\\ \sin\end{matrix}\, wx.$$

We have to justify the equation

$$\int_{-\infty}^{\infty} \frac{\sin m(t-x)}{t-x}\, dt \int_a^A \phi(w) \begin{matrix}\cos\\ \sin\end{matrix}\, wt\, dw = \int_a^A \phi(w)\, dw \int_{-\infty}^{\infty} \frac{\sin m(t-x)}{t-x} \begin{matrix}\cos\\ \sin\end{matrix}\, wt\, dt.$$

If $-m < a < A < m$, the proof of this presents no difficulty, since the inner integral on the right-hand side is uniformly convergent for $a \leqslant w \leqslant A$ (cf. § 4). But, if $a = -m$ or $A = m$ (say the latter) the convergence ceases to be uniform for $w = m$. We have then to prove in addition that

$$(23) \qquad \int_{-\infty}^{\infty} \frac{\sin m(t-x)}{t-x} dt \int_{m-\delta}^{m} \phi(w) \frac{\cos}{\sin} wt\, dw$$

is convergent and has the limit zero as $\delta \to 0$.

Let us, for example, take the cosine in the inner integral, and let us suppose that
$$\phi(w) = (m-w)^{-a} \psi(w),$$
where $0 \leqslant a < 1$ and $\psi(w)$ is continuous and monotonic in $(m-\delta, m)$. Suppose, e.g., that $\psi(w)$ is positive and decreases as $w \to m$. Then

$$\int_{m-\delta}^{m} \frac{\psi(w)}{(m-w)^a} \cos wt\, dw$$

$$= \int_0^\delta \frac{\psi(m-u)}{u^a} \cos t(m-u)\, du$$

$$= \cos mt \int_0^\delta \frac{\psi(m-u)}{u^a} \cos tu\, du + \sin mt \int_0^\delta \frac{\psi(m-u)}{u^a} \sin tu\, du.$$

But
$$\int_0^\delta \frac{\psi(m-u)}{u^a} \cos tu\, du = \psi(m-\delta) \int_{\delta'}^{\delta} \frac{\cos tu}{u^a} du \quad (0 < \delta' < \delta)$$

$$= \psi(m-\delta)\, t^{-(1-a)} \int_{\delta' t}^{\delta t} \frac{\cos u}{u^a} du,$$

and is therefore in absolute value less than a constant multiple of $t^{-(1-a)}$, and the same argument may be applied to the second integral. And from this it follows at once that (23) is absolutely and uniformly convergent in an interval of values of δ including $\delta = 0$, and so has the limit zero as $\delta \to 0$.

Hence *the integrals*

$$f(x) = \int_a^A \phi(w) \frac{\cos}{\sin} wx\, dw$$

represent m-functions, if $-m < a < A < m$, *provided only that*

$$\int_a^A |\phi(w)|\, dw$$

is convergent. They represent m-functions when $A = m$, *if*

$$\phi(w) = (m-w)^{-a} \psi(w),$$

where $0 \leqslant a < 1$ *and* $\psi(w)$ *is continuous and monotonic near* $w = m$.

14. Suppose, for example, that $a = 0$, $A = 1$ and $\phi(w) = (1-w^2)^{\beta-\frac{1}{2}}$, where $\beta > -\frac{1}{2}$. Then

$$f(x) = \int_0^1 (1-w^2)^{\beta-\frac{1}{2}} \cos wx\, dw = 2^{\beta-1}\sqrt{\pi}\,\Gamma(\beta+\tfrac{1}{2})\frac{J_\beta(x)}{x^\beta}.$$

Hence $x^{-\beta}J_\beta(x)$ is an m-function if $m \geqslant 1$ and $\beta > -\frac{1}{2}$. If $m > 1$, the conclusion can be extended to all cases in which $\beta > -\frac{3}{2}$.

It may be worth while to set down here a few further examples of m-functions defined by integrals such as have been considered here.

(i.) If $\phi(w)$ is subject to the restrictions (i.), (ii.), and (iii.') of § 11, we know that

$$f(x) = \int_{-\infty}^{\infty} \frac{\sin \mu(w-x)}{w-x} \phi(w)\, dw,$$

is an m-function for $m \geqslant \mu$. If we assume that $\phi(w)$ is *even*, and that the order of integration may be changed, we find

$$f(x) = \int_0^\mu d\xi \int_{-\infty}^\infty \cos\xi(w-x)\,\phi(w)\,dw = 2\int_0^\mu \cos x\xi\,d\xi \int_0^\infty \cos\xi w\,\phi(w)\,dw.$$

The assumptions

$$\phi(w) = \frac{1}{a^2+w^2}\ (a>0),\quad |w|^{-s}\ (0<s<1),\quad e^{-a|w|},\quad e^{-w^2/4a}\ (a>0)$$

lead to the following examples of m-functions (for $m \geqslant \mu$):

$$\frac{1}{\pi}\int_{-\infty}^\infty \frac{\sin\mu(w-x)}{w-x}\frac{dw}{a^2+w^2} = \int_0^\mu e^{-a\xi}\cos x\xi\,d\xi$$

$$= \frac{a(1-e^{-a\mu}\cos\mu x)+xe^{-a\mu}\sin\mu x}{a^2+x^2},$$

$$\frac{\cos\tfrac{1}{2}s\pi\,\Gamma(s)}{\pi}\int_{-\infty}^\infty \frac{\sin\mu(w-x)}{w-x}\frac{dw}{|w|^s} = \int_0^\mu \xi^{s-1}\cos x\xi\,d\xi,$$

$$\frac{1}{2a}\int_{-\infty}^\infty \frac{\sin\mu(w-x)}{w-x} e^{-a|w|}\,dw = \int_0^\mu \frac{\cos x\xi\,d\xi}{a^2+\xi^2},$$

$$\frac{1}{2\sqrt{(a\pi)}}\int_{-\infty}^\infty \frac{\sin\mu(w-x)}{w-x} e^{-w^2/4a}\,dw = \int_0^\mu e^{-a\xi^2}\cos x\xi\,d\xi.$$

(ii.) The function
$$\frac{J_\beta\{\sqrt{(x^2-2ax\cos\theta+a^2)}\}}{(x^2-2ax\cos\theta+a^2)^{\frac{1}{2}\beta}}$$

is an m-function for $m \geqslant 1$, if $\beta > -\frac{1}{2}$. This follows from the fact that it can be expressed in the form*

$$\frac{(a\sin\theta)^{\frac{1}{2}-\beta}}{\sqrt{(2\pi)}} \int_0^\pi \cos(x\cos\phi - a\cos\theta\cos\phi) J^{\beta-\frac{1}{2}}(a\sin\theta\sin\phi)(\sin\phi)^{\beta+\frac{1}{2}} d\phi,$$

if we put $\cos\phi = w$, and use the results of § 13. If $m > 1$, the conclusion can be extended to all cases in which $\beta > -\frac{3}{2}$.

V.

Treatment of the equation (2) by contour integration.

15. I shall now consider the equation (2) from a different point of view, which, however, is not entirely unconnected with that adopted in the preceding sections. In order to make the connection clear I shall consider the function

(23′) $$f(x) = \int_a^A \cos xw \, \phi(w) \, dw,$$

where a and A are finite and $\int_a^A |\phi(w)| \, dw$ convergent. We saw in § 14 that this function is certainly an m-function, if $-m < a < A < m$.

The integral (23′) is convergent for all values of x real or complex. The integral

$$\int_a^A (-w \sin xw) \, \phi(w) \, dw,$$

obtained by differentiation under the integral sign with respect to x, is also convergent, and uniformly so with respect to all values of x in any finite region of the x-plane. Hence, *the integral* (23′) *represents an integral function of* x.

Again, if $x = u + iv$, where $v > 0$, we have

$$|\cos xw| < Ke^{v|w|},$$

and so $$|f(x)| < \int_a^A e^{v|w|} |\phi(w)| \, dw < Ke^{\alpha v},$$

where α is the greater of $|a|$ and $|A|$, so that $\alpha < m$.

Similar conclusions might be established (though in some cases with

---

* Nielsen, *Cylinderfunktionen*, p. 182.

much greater difficulty) about the other types of definite integrals that we have considered. They suggest the possibility of defining large classes of m-functions by a reference to their properties when considered as analytic functions of a complex variable; and this I shall now proceed to do.

16. Let us suppose that $t = u+iv$ is a complex variable, and that $f(t)$ is an integral function of t, real when t is real. Further, suppose that, when $v > 0$, $f(t)$ satisfies an inequality

(24) $$|f(t)| < \lambda_1(|t|) e^{mv} + |t| \lambda_2(|t|) e^{\mu v},$$

where $0 < \mu < m$, and λ_1 and λ_2 are positive functions of $|t|$ which tend to zero as $|t| \to \infty$.

Let us consider the integral

$$\int \frac{e^{mi(t-x)}}{t-x} f(t) \, dt$$

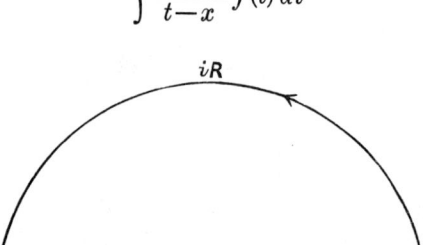

taken round the contour shown in the figure. It is clear that when the radius of the small semi-circle about the point $t = x$ is diminished indefinitely, we obtain

$$P \int_{-R}^{R} \frac{e^{mi(t-x)}}{t-x} f(t) \, dt = \pi i f(x) - J(R),$$

where $J(R)$ is the value of the integral round the large semi-circle.

Now, if on this semi-circle, we put $t = Re^{\phi i}$, we have

$$|e^{mi(t-x)}| = e^{-mR \sin \phi} = e^{-mv}, \qquad \left|\frac{1}{t-x}\right| < \frac{K}{R},$$

where K is a constant which, if R be large enough, may be taken as near to unity as we please. Hence

$$|J(R)| < j_1 + j_2,$$

where
$$0 < j_1 < K\lambda_1(R) \int_0^\pi d\phi = \pi K \lambda_1(R),$$

$$0 < j_2 < KR\lambda_2(R) \int_0^\pi e^{-(m-\mu)R\sin\phi} d\phi$$

$$< 2KR\lambda_2(R) \int_0^{\frac{1}{2}\pi} e^{-2(m-\mu)R\phi/\pi} d\phi$$

$$< \pi K \lambda_2(R)/(m-\mu).$$

Hence $J(R) \to 0$ as $R \to \infty$: and so, separating the real and imaginary parts, we obtain

(25)
$$\begin{cases} P\displaystyle\int_{-\infty}^\infty \frac{\cos m(t-x)}{t-x} f(t)\, dt = 0, \\ P\displaystyle\int_{-\infty}^\infty \frac{\sin m(t-x)}{t-x} f(t)\, dt = \pi f(x). \end{cases}$$

In these equations, it should be observed, the symbol P refers, not only to the infinity of the subject of integration of the first integral for $t = x$, but also to the fact that each integral is defined as a limit

$$\lim_{R\to\infty} \int_{-R}^R,$$

and not as a limit
$$\lim_{R,\,R'\to\infty} \int_{-R'}^R.$$

It is easy, however, to see that they remain true with the more general interpretation of the infinite limits in all cases in which *either* (a) the first term on the right-hand side of (24) may be omitted, or (b) the integral

$$\int^\infty \frac{\lambda_1(R)}{R} dR$$

is convergent. One or other of these circumstances will occur in all cases with which we shall be concerned, and then the symbol of the principal value in the second formula (25) may be omitted.

17. The following results are immediate applications of the theorem of § 16.

(i.) $\cos \mu x$ and $\sin \mu x$ are m-functions for $m > \mu$. The formulæ

$$P\int_{-\infty}^{\infty} \frac{\sin m(t-x)}{t-x} \genfrac{}{}{0pt}{}{\cos}{\sin} mt\, dt = \tfrac{1}{2}\pi \genfrac{}{}{0pt}{}{\cos}{\sin} mx,$$

show that they are *not* m-functions for $m = \mu$. On the other hand, $\sin\{\mu(x-a)\}/(x-a)$ is an m-function for $m \geqslant \mu$.

(ii.) The functions

$$\frac{J_\beta\{\mu(x-a)\}}{(x-a)^\beta}, \quad \frac{J_\beta\{\mu\sqrt{(x^2-2ax\cos\theta+a^2)}\}}{(x^2-2ax\cos\theta+a^2)^{\frac{1}{2}\beta}}$$

are m-functions, if $0 < \mu < m$ and $\beta > -\tfrac{3}{2}$, or if $m = \mu$ and $\beta > -\tfrac{1}{2}$.

(iii.) The function

$$\frac{J_\beta\{\mu(x-a)\}}{(x-a)^\beta} \genfrac{}{}{0pt}{}{\cos}{\sin} \mu'(x-a'),$$

is an m-function, if $0 < \mu < \mu+\mu' < m$ and $\beta > -\tfrac{3}{2}$, or if $\mu+\mu' = m$ and $\beta > -\tfrac{1}{2}$. And such examples may, of course, be multiplied indefinitely; some of them have already been obtained by other methods.

VI.

The function $\dfrac{\sin m(t-x)}{t-x}$ *as the kernel of an integral equation.*

18. In the preceding sections I have used only the classical methods of analysis, and have made no reference to the ideas of the theory of integral equations. It is, however, of some interest to consider briefly the properties of

(26) $$K(t, x) = \frac{\sin m(t-x)}{t-x}$$

as a "kernel" of an integral equation. It is convenient to divide the kernel by the constant factor π: it then possesses the property

(27) $$\int_{-\infty}^{\infty} K(t, x)\, K(t, y)\, dt = K(x, y),$$

and its properties are shared by any symmetrical kernel which possesses this particular property.

We have to consider the equations

(28) $$f(x) = \int_{-\infty}^{\infty} K(t,x)\, u(t)\, dt,$$

(29) $$u(x) = F(x) + \lambda \int_{-\infty}^{\infty} K(t,x)\, u(t)\, dt,$$

where $f(x)$ is a given function.

If we multiply (28) by $K(x, y)$, integrate from $-\infty$ to $+\infty$, and assume that the order of integration can be changed, we obtain

$$\int_{-\infty}^{\infty} K(x,y) f(x)\, dx = \int_{-\infty}^{\infty} u(t)\, dt \int_{-\infty}^{\infty} K(t,x) K(x,y)\, dx$$

$$= \int_{-\infty}^{\infty} K(t,y)\, u(t)\, dt = f(y).$$

Hence, *the equation* (28) *has no solution unless*

(28a) $$f(x) = \int_{-\infty}^{\infty} K(t,x) f(t)\, dt,$$

in which case it has the solution $u(x) = f(x)$. A more general solution is $u(x) = f(x) + F(x)$, where $F(x)$ is any solution of

(28b) $$\int_{-\infty}^{\infty} K(t,x) F(t)\, dt = 0.$$

Now apply the same process to (29). We find

$$\int_{-\infty}^{\infty} K(x,y) u(x)\, dx = \int_{-\infty}^{\infty} K(x,y) F(x)\, dx + \lambda \int_{-\infty}^{\infty} K(t,y) u(t)\, dt.$$

From this it follows that, *unless* $\lambda = 1$, *the equation* (29) *has the one solution*

$$u(x) = F(x) + \frac{\lambda}{1-\lambda} \int_{-\infty}^{\infty} K(t,x) F(t)\, dt.$$

On the other hand, if $\lambda = 1$, the equation has no solution unless $F(x)$ is a solution of (28b), in which case $u(x) = F(x)$ is a solution. A more general solution is $u(x) = F(x) + f(x)$, where $f(x)$ is any solution of (28a).

The kernel has the one *Eigenwert* 1, to which correspond an infinity of *Eigenfunktionen*. When the kernel is

$$\frac{1}{\pi} \frac{\sin m(t-x)}{t-x},$$

these latter are, in fact, our m-functions.

I have been content to give formal proofs of these assertions, as my only object is to indicate generally the nature of this particular kernel. It is well known that, when the interval of integration is infinite, the fundamental results of the ordinary theory of integral equations lose their validity. For example, Weyl* has discussed in detail a class of kernels (*Fourierkerne*) which have two *Eigenwerte* (−1 and +1).

VII.

Inversion formulæ satisfied by m-functions.

20. Suppose that we can express

$$\frac{\sin m(t-x)}{t-x}$$

in the form

(30) $$\frac{\sin m(t-x)}{t-x} = \int_a^\beta \phi(\lambda, x)\, \psi(\lambda, t)\, d\lambda.$$

If we multiply this equation by $f(t)$, an m-function, and integrate from $t = -\infty$ to $+\infty$, inverting the order of integration on the right-hand side, we are led formally to the equations

(31) $$\pi f(x) = \int_a^\beta \phi(\lambda, x)\, d\lambda \int_{-\infty}^\infty \psi(\lambda, t) f(t)\, dt,$$

(32) $$\begin{cases} F(\lambda) = \displaystyle\int_{-\infty}^\infty \psi(\lambda, t) f(t)\, dt, \\ \pi f(x) = \displaystyle\int_a^\beta \phi(\lambda, x) F(\lambda)\, d\lambda. \end{cases}$$

In each case, of course, the validity of our transformations requires a special discussion.

Mr. Bateman, in the paper already referred to, gave two examples of

---

* *Inaug. Diss.*, Göttingen, 1908, and *Math. Annalen*, Bd. LXVI., s. 273. See also Bôcher, *Introduction to Integral Equations* (Cambridge Tracts, No. 10, p. 70).

such inversion formulæ, based on the equations

$$(33) \quad \frac{\sin m(t-x)}{t-x} = \tfrac{1}{2}m \int_{-1}^{1} e^{-m(t-x)\lambda i} d\lambda,$$

$$(34) \quad \frac{\sin m(t-x)}{t-x} = \tfrac{1}{2}m\pi \int_{x}^{\infty} J_0\{m(\lambda-x)\} \frac{J_1\{m(\lambda-t)\}}{\lambda-t} d\lambda.$$

I do not propose to discuss the second of these two equations here, but I shall conclude this paper by (1) stating sufficient conditions for the truth of Mr. Bateman's first inversion formula, and (2) investigating another formula of a slightly different character.

21. (1) If

$$(35) \quad F(\lambda) = \int_{-\infty}^{\infty} e^{-m\lambda t i} f(t)\, dt,$$

$f(t)$ being an m-function such that the integral is convergent, and if we multiply by $e^{mx\lambda i}$, and integrate from $\lambda = -1$ to $\lambda = 1$, inverting the order of integration on the right-hand side, we obtain

$$\int_{-1}^{1} e^{mx\lambda i} F(\lambda)\, d\lambda = \int_{-1}^{1} e^{mx\lambda i} d\lambda \int_{-\infty}^{\infty} e^{-m\lambda t i} f(t)\, dt$$

$$= \int_{-\infty}^{\infty} f(t)\, dt \int_{-1}^{1} e^{-m(t-x)\lambda i} d\lambda$$

$$= \frac{2}{m} \int_{-\infty}^{\infty} \frac{\sin m(t-x)}{t-x} f(t)\, dt,$$

or

$$(35') \quad f(x) = \frac{m}{2\pi} \int_{-1}^{1} e^{mx\lambda i} F(\lambda)\, d\lambda.$$

The transformation is certainly justifiable, if (i.) $f(t)$ is continuous except for a finite number of values of t, (ii.) the integral $\int |f(t)|\, dt$ is convergent over any finite interval, and (iii.) the integral (35) is uniformly convergent for $-1 \leqslant \lambda \leqslant 1$.* But this last condition is unnecessarily restrictive

---

* Cf. Bateman, *l.c.*, p. 483.

and would prevent the application of the result to some of the simplest and most natural cases. For example, suppose that

$$f(t) = \frac{\sin \mu t}{t} \quad (0 < \mu < m).$$

Then
$$F(\lambda) = 2 \int_0^\infty \frac{\sin \mu t}{t} \cos m\lambda t \, dt = \pi \quad (-\mu < m\lambda < \mu),$$
$$= 0 \quad (|m\lambda| > \mu),$$

and
$$\frac{m}{2\pi} \int_{-1}^{1} e^{mx\lambda i} F(\lambda) \, d\lambda = \tfrac{1}{2} m \int_{-\mu/m}^{\mu/m} e^{mx\lambda i} d\lambda = \frac{\sin \mu x}{x}.$$

But the integral (35) is non-uniformly convergent near $\lambda = \pm \mu/m$.

I shall suppose that the integral (35) is uniformly convergent except for a finite number of values of λ, a number which we may without loss of generality suppose to be one; say for $\lambda = \lambda_0$. Then

$$\int_{-\infty}^{\infty} f(t) \, dt \left(\int_{-1}^{\lambda_0-\delta} + \int_{\lambda_0+\delta'}^{1} \right) e^{-m(t-x)\lambda i} \, d\lambda$$
$$= \left(\int_{-1}^{\lambda_0-\delta} + \int_{\lambda_0+\delta'}^{1} \right) e^{mx\lambda i} d\lambda \int_{-\infty}^{\infty} e^{-m\lambda t i} f(t) \, dt.$$

The inversion of the integrations will be justified if

$$\int_{-\infty}^{\infty} f(t) \, dt \int_{\lambda_0-\delta}^{\lambda_0+\delta'} e^{-m(t-x)\lambda i} d\lambda = \int_{-\infty}^{\infty} \frac{e^{-m(t-x)(\lambda_0-\delta)i} - e^{-m(t-x)(\lambda_0+\delta')i}}{mi(t-x)} f(t) \, dt$$

is convergent, and has the limit zero as δ and δ' tend to zero.

Write the last integral in the form

$$\int_{-\infty}^{-T} + \int_{-T}^{T} + \int_{T}^{\infty} = j_1 + j_2 + j_3,$$

say. Then j_2 is certainly convergent, and tends to zero with δ and δ'. A sufficient condition that j_3 should do so is that

$$\int_T^\infty \frac{e^{-m(t-x)\lambda i}}{t-x} f(t) \, dt$$

should be convergent and represent a continuous function of λ: and this will certainly be the case, if

$$\int^\infty \frac{|f(t)|}{t} dt$$

is convergent. A similar argument may be applied to j_1, and so we deduce the following result:—

Mr. Bateman's inversion formula, expressed by the equations (35), (35') *will certainly be true, if* (i.) $f(t)$ *is an m-function, with at most a finite number of discontinuities,* (ii.) $\int |f(t)| dt$ *is continuous over any finite interval,* (iii.) *the integral* (35) *is uniformly convergent save for a finite number of values of* λ, *and* (iv.) *the integrals*

$$\int^\infty \frac{|f(t)|}{t} dt, \quad \int_{-\infty} \frac{|f(t)|}{t} dt$$

are convergent.

22. In conclusion I shall investigate a new inversion formula, satisfied by m-functions. I shall begin by determining the value of the integral

(36) $$I_{\mu, \nu} = \int_{-\infty}^{\infty} \frac{J_\mu\{m(x-a)\}}{(x-a)^\mu} \frac{J_\nu\{m(x-b)\}}{(x-b)^\nu} dx,$$

which is convergent, if $\mu+\nu > 0$, a and b being supposed to be real, and m positive. I shall suppose first that

$$\mu > \tfrac{1}{2}, \quad \nu > -\tfrac{1}{2},$$

and that the integral has been reduced, by the substitution $x-b=\xi$, to the particular case in which $b=0$. We have then

$$I_{\mu,\nu} = \frac{2^{1-\nu} m^\nu}{\sqrt{\pi}\, \Gamma(\nu+\tfrac{1}{2})} \int_{-\infty}^{\infty} \frac{J_\mu\{m(x-a)\}}{(x-a)^\mu} dx \int_0^{\tfrac{1}{2}\pi} \cos(mx \sin\phi)(\cos\phi)^{2\nu} d\phi$$

$$= \frac{2^{1-\nu} m^\nu}{\sqrt{\pi}\, \Gamma(\nu+\tfrac{1}{2})} \int_0^{\tfrac{1}{2}\pi} (\cos\phi)^{2\nu} d\phi \int_{-\infty}^{\infty} \frac{J_\mu\{m(x-a)\}}{(x-a)^\mu} \cos(mx \sin\phi) dx$$

$$= \frac{2^{2-\nu} m^\nu}{\sqrt{\pi}\, \Gamma(\nu+\tfrac{1}{2})} \int_0^{\tfrac{1}{2}\pi} \cos(ma \sin\phi)(\cos\phi)^{2\nu} d\phi \int_0^\infty \frac{J_\mu(mx)}{x^\mu} \cos(mx \sin\phi) dx.$$

Now, if $\mu > -\tfrac{1}{2}$,*

$$\int_0^\infty \frac{J_\mu(mx)}{x^\mu} \cos(mx \sin\phi) dx = \frac{2^{-\mu} m^{\mu-1} \sqrt{\pi}}{\Gamma(\mu+\tfrac{1}{2})} (\cos\phi)^{2\mu-1}.$$

---

* Nielsen, *Cylinderfunktionen*, p. 198, where we must put $\rho = -\tfrac{1}{2}$.

Hence we obtain

$$\frac{2^{2-\mu-\nu} m^{\mu+\nu-1}}{\Gamma(\mu+\tfrac{1}{2})\,\Gamma(\nu+\tfrac{1}{2})} \int_0^{\tfrac{1}{2}\pi} \cos(ma\sin\phi)(\cos\phi)^{2\mu+2\nu-1}\,d\phi$$
$$= \frac{\Gamma(\mu+\nu)\sqrt{(2\pi/m)}}{\Gamma(\mu+\tfrac{1}{2})\,\Gamma(\nu+\tfrac{1}{2})} \frac{J_{\mu+\nu-\tfrac{1}{2}}(ma)}{a^{\mu+\nu-\tfrac{1}{2}}}.$$

Accordingly,

$$(37)\quad \int_{-\infty}^{\infty} \frac{J_\mu\{m(x-a)\}}{(x-a)^\mu}\,\frac{J_\nu\{m(x-b)\}}{(x-b)^\nu}\,dx$$
$$= \frac{\Gamma(\mu+\nu)\sqrt{(2\pi/m)}}{\Gamma(\mu+\tfrac{1}{2})\,\Gamma(\nu+\tfrac{1}{2})} \frac{J_{\mu+\nu-\tfrac{1}{2}}\{m(a-b)\}}{(a-b)^{\mu+\nu-\tfrac{1}{2}}}.$$

It is easy to verify that, when $\mu = \nu = \tfrac{1}{2}$, this formula reduces to (4).

If, as we supposed, $\mu > \tfrac{1}{2}$, the integral

$$\int_{-\infty}^{\infty} \frac{J_\mu\{m(x-a)\}}{(x-a)^\mu}\cos(mx\sin\phi)\,dx$$

is absolutely and uniformly convergent for $0 \leqslant \phi \leqslant \tfrac{1}{2}\pi$, and the inversion of the order of integration needs no special justification. The formula (37) holds, however, for a wider range of values of μ and ν, in fact for all values of μ and ν such that $\mu+\nu > 0$, and neither μ nor ν is a negative integer. It is, in fact, easy to see that both sides of (37) represent analytic functions of μ and ν regular throughout any region bounded by inequalities of the type

$$0 < \mu+\nu,\quad |\mu|+|\nu| < K,\quad |\mu+n| > \delta > 0,\quad |\nu+n| > \delta > 0,$$

for $n = 1, 2, \ldots$.

The most interesting case is that in which $\mu+\nu = 1$. We then obtain

$$(38)\quad \int_{-\infty}^{\infty} \frac{J_\mu\{m(x-a)\}}{(x-a)^\mu}\,\frac{J_{1-\mu}\{m(x-b)\}}{(x-b)^{1-\mu}}\,dx = \frac{4\cos\mu\pi}{(1-2\mu)\,m\pi}\,\frac{\sin\{m(a-b)\}}{a-b},$$

which holds for all values of μ other than $-1, -2, \ldots, 2, 3, \ldots$ [except for $\mu = \tfrac{1}{2}$, when the formula reduces to (4)].

23. From (38) we deduce, by the formal argument of § 20, the inversion formulæ

$$(39)\quad \frac{4\cos\mu\pi}{(1-2\mu)\,m}\,f(x) = \int_{-\infty}^{\infty} \frac{J_{1-\mu}\{m(\lambda-x)\}}{(\lambda-x)^{1-\mu}}\,d\lambda \int_{-\infty}^{\infty} \frac{J_\mu\{m(t-\lambda)\}}{(t-\lambda)^\mu}\,f(t)\,dt,$$

$$(40)\quad \begin{cases} F(\lambda) = \displaystyle\int_{-\infty}^{\infty} \frac{J_\mu\{m(t-\lambda)\}}{(t-\lambda)^\mu}\,f(t)\,dt, \\[1em] \dfrac{4\cos\mu\pi}{(1-2\mu)\,m}\,f(x) = \displaystyle\int_{-\infty}^{\infty} \frac{J_{1-\mu}\{m(\lambda-x)\}}{(\lambda-x)^{1-\mu}}\,F(\lambda)\,d\lambda, \end{cases}$$

where, of course, $f(t)$ is an m-function. The validity of the formula rests on the inversion of integrations expressed by

$$\int_{-\infty}^{\infty} \frac{J_{1-\mu}\{m(\lambda-x)\}}{(\lambda-x)^{1-\mu}} d\lambda \int_{-\infty}^{\infty} \frac{J_\mu\{m(t-\lambda)\}}{(t-\lambda)^\mu} f(t) dt$$
$$= \int_{-\infty}^{\infty} f(t) dt \int_{-\infty}^{\infty} \frac{J_{1-\mu}\{m(\lambda-x)\}}{(\lambda-x)^{1-\mu}} \frac{J_\mu\{m(t-\lambda)\}}{(t-\lambda)^\mu} d\lambda.$$

As all the limits are infinite it is not a very easy matter to investigate reasonably general conditions for the legitimacy of this inversion. And, in view of the length to which this paper has already extended, I shall not attempt to treat the question in any generality. I shall confine myself to taking the very simplest hypotheses possible, so as to show at any rate that there are *some* functions for which a rigorous proof is possible. I shall suppose that $-\frac{1}{2} < \mu < \frac{1}{2}$, and that the integral

$$\int_{-\infty}^{\infty} |f(t)| dt$$

is convergent. Then

$$\int^{\infty} \frac{|f(t)|}{t^{\mu+\frac{1}{2}}} dt, \quad \int_{-\infty} \frac{|f(t)|}{(-t)^{\mu+\frac{1}{2}}} dt$$

are convergent, and the inner integral on the left-hand side is absolutely and uniformly convergent throughout any finite interval of values of λ. Hence

$$\int_{-\lambda_1}^{\lambda_2} \int_{-\infty}^{\infty} = \int_{-\infty}^{\infty} \int_{-\lambda_1}^{\lambda_2},$$

for any positive values of λ_1 and λ_2, and it remains only to show that

$$\int_{-\infty}^{\infty} f(t) dt \int_{\lambda_2}^{\infty} \frac{J_{1-\mu}\{m(\lambda-x)\}}{(\lambda-x)^{1-\mu}} \frac{J_\mu\{m(t-\lambda)\}}{(t-\lambda)^\mu} d\lambda$$

is convergent, and has the limit zero when $\lambda_2 \to \infty$, together with a corresponding result relating to $-\lambda_1$. This will certainly be the case if the same is true of

$$\int_{-\infty}^{\infty} |f(t)| dt \int_{\lambda_2}^{\infty} \left| \frac{J_\mu\{m(t-\lambda)\}}{(t-\lambda)^\mu} \right| \frac{d\lambda}{(\lambda-x)^{\frac{1}{2}-\mu}}.$$

But

$$\left| \frac{J_\mu\{m(t-\lambda)\}}{(t-\lambda)^\mu} \right| < K,$$

$$K \int_{\lambda_2}^{\infty} \frac{d\lambda}{(\lambda-x)^{\frac{3}{2}-\mu}} < \frac{K}{\lambda_2^{\frac{1}{2}-\mu}}.$$

Hence the repeated integral last written is in absolute value less than

$$K\lambda_2^{-(\frac{1}{2}-\mu)},$$

and tends to zero as $\lambda_2 \to \infty$.

These conditions would be satisfied, *e.g.*, if

$$f(t) = t^{-\rho} J_\rho(\lambda t),$$

where $0 < \lambda \leqslant m$ and $\rho > \frac{1}{2}$. They are, of course, far from being as general as they might be made: but a more detailed investigation would carry us too far to be undertaken at this stage, and they suffice at any rate to show that the formulæ (39), (40) are amenable to accurate methods.

CORRECTIONS

p. 447, *bottom line.* π should be $\frac{1}{2}\pi$.

p. 448, *line* 5. $\dfrac{1}{2n+1}$ should be $\dfrac{2}{2n+1}$.

——, *line* 10. $f(u-x)$ should be $f(x-u)$.

p. 449, *line* 7. The second $d\xi$ should be dt.

——, *line* 10. $\sin(w-x)$ should be $\sin\mu(w-x)$.

p. 469, *line* 5. 'continuous' should be 'convergent'.

SOLUTIONS OF SOME INTEGRAL EQUATIONS CONSIDERED BY BATEMAN, KAPTEYN, LITTLEWOOD, AND MILNE

By G. H. Hardy and E. C. Titchmarsh.

[Received February 26th, 1923 —Read March 8th, 1923.]

1. *Introduction and Summary.*

1.1. The integral equations discussed in this paper are all very similar in form, but fall sharply into two classes as regards the character of their solutions. The three with which we are concerned primarily are

(1.11) (B)
$$f(x) = \frac{1}{\pi} \int_{-\infty}^{\infty} \frac{\sin(t-x)}{t-x} f(t)\, dt = \frac{1}{\pi} \int_{0}^{\infty} \frac{\sin t}{t} \{f(x+t)+f(x-t)\}\, dt,$$

(1.12) (K)
$$f'(x) = \tfrac{1}{2}\int_{0}^{\infty} \frac{J_1(t)}{t} \{f(x+t)-f(x-t)\}\, dt,$$

(1.13) (M)
$$f'(x) = \tfrac{1}{2}\int_{0}^{\infty} \frac{e^{-t}}{t} \{f(x+t)-f(x-t)\}\, dt.$$

These equations have all been considered before. Thus (B) occurs in a paper of Bateman,[*] and was discussed later in considerable detail by Hardy.[†] The second equation, in the slightly different form

(1.141) (K')
$$f'(x) = \tfrac{1}{2}\int_{0}^{\infty} \frac{J_1(t)}{t} \{f(t+x)+f(t-x)\}\, dt,$$

---

[*] H. Bateman, "The inversion of a definite integral", *Proc. London Math. Soc.* (2), 4 (1907), 461–498 (483).

[†] G. H. Hardy, "On an integral equation", *Proc. London Math. Soc.* (2), 7 (1909), 445–472.

was given by Kapteyn,* as the condition that an odd function $f(x)$ should be expansible in a " Neumann " series

$$(1.142) \qquad \Sigma a_{2n+1} J_{2n+1}(x)$$

in which the coefficients are determined by the formula

$$(1.143) \qquad a_{2n+1} = (4n+2) \int_0^\infty f(t) J_{2n+1}(t) \frac{dt}{t}.$$

This method of determining the coefficients had been suggested previously by Webb,† but his conclusions as regards the generality of the expansion were, as Kapteyn showed, erroneous.

It is obvious that all solutions of (K') are odd, and that, if $f(x)$ is odd, (K') and (K) are identical. A solution of (K), however, is not necessarily odd, so that (K) presents a slightly more general problem. Kapteyn's equation was discussed later by Bateman,‡ who was, however, unable to reach very definite conclusions as to the class of functions which satisfy the equation.

The equation (M) originates with Milne,§ who was led to it in an investigation of the distribution of temperature in the interior of a star. It was discussed later from the pure mathematical standpoint by Littlewood.∥ It is formally equivalent to the linear differential equation

$$(1.15) \qquad \tfrac{1}{3} f^{(3)}(x) + \tfrac{1}{5} f^{(5)}(x) + \ldots = 0,$$

with constant coefficients and of infinite order. Such equations have been discussed in an important memoir by Schürer, but the conclusions concerning the equation (M), which can be deduced from Schürer's theorems, are much less general than those which we propose to prove.

---

* W. Kapteyn, "On an expansion of an arbitrary function in a series of Bessel functions", *Messenger of Math.*, 35 (1906), 122-125. See G. N. Watson, *Theory of Bessel functions*, 533-535.

† H. A. Webb, "The expansion of an arbitrary function in a series of Bessel functions", *Messenger of Math.*, 33 (1904), 55-58.

‡ H. Bateman, "On the expansion of an arbitrary function in a series of Bessel functions", *Messenger of Math.*, 36 (1907), 31-37.

§ E. A. Milne, "Radiative equilibrium in the outer layers of a star", *Monthly Notices of the R.A.S.*, 81 (1921), 361-375.

∥ J. E. Littlewood, "On an integral equation", *Proc. Camb. Phil. Soc.*, 21 (1922), 205-215.

¶ F. Schürer, "Eine gemeinsame Methode zur Behandlung gewisser Funktionalgleichungsprobleme", *Leipziger Berichte*, 70 (1918), 185-246. Extensions of Schürer's work will be found in memoirs by Hilb in volumes 82 and 84 of the *Math. Annalen*.

There are three other equations, which form, with (B), (K), and (M), a symmetrical set of six. These equations are

$$(1.16) \quad (B_1) \qquad f'(x) = \frac{1}{\pi} \int_0^\infty \frac{\sin t}{t^2} \{f(x+t) - f(x-t)\} \, dt,$$

$$(1.17) \quad (K_1) \qquad f(x) = \tfrac{1}{2} \int_0^\infty J_1(t) \{f(x+t) + f(x-t)\} \, dt,$$

$$(1.18) \quad (M_1) \qquad f'(x) = \frac{1}{\pi} \int_0^\infty K_1(t) \{f(x+t) - f(x-t)\} \, dt,$$

where $K_1(t)$ is the function commonly so designated in mathematical physics.* The theory of these equations is substantially the same as that of the first three,† and we shall refer to them only incidentally.

We assume throughout that the formulae are significant and true for every real x. Hence, if $f'(x)$ occurs in an equation, any solution has necessarily a finite derivative for every x. It is of course not immediately obvious (though it may be true in fact) that this derivative is necessarily bounded in the neighbourhood of any x. The integrals are always Lebesgue integrals. Our notation, so far as Bessel functions are concerned, is that of Watson's treatise.

1.2. It appears at once that the theory of (M) and (M_1) is quite different from that of the other equations. Each has the solution

$$(1.21) \qquad f(x) = ax^2 + 2bx + c,$$

and there are no other obvious solutions. The simplest solution of each of the other equations is

$$(1.22) \qquad f(x) = A \cos ax + B \sin ax,$$

where $0 \leqslant a \leqslant 1$; this function is not a solution if $a > 1$. But these equations have much more general solutions involving arbitrary functions; thus any function of the form

$$(1.23) \qquad f(x) = \int_0^1 \{\lambda(u) \cos xu + \mu(u) \sin xu\} \, du,$$

---

* Watson, *loc. cit.*, 78.
† The theory of (B_1) is practically the same as that of (K), and that of (M_1) the same as that of (M). The equation (K_1) is rather more troublesome than (B); in particular, the results of §2 do not hold for (K_1) without reservation.

where λ and μ are subject only to very general conditions of inequality, is a solution. The reason underlying this difference will appear in § 2.

In § 2 we discuss the equations by the methods of the theory of functions of a real variable, and in particular by the use of the theory of "Fourier transforms" due to Plancherel,[*] and its generalisation to "Hankel transforms".[†] The use of this theory involves a restriction on the function $f(x)$, namely, that its square is integrable over the range $(-\infty, \infty)$; and it is plain that this assumption rules out some of the most obvious solutions, such as (1.21) and (1.22). But, within its range, the solution of the problem afforded by the theory of transforms is complete. Thus we prove:—

THEOREM 1.—*Suppose that the square of $f(x)$ is integrable over $(-\infty, \infty)$. Then, in order that $f(x)$ should be a solution of* (B), (B$_1$), *or* (K), *it is necessary and sufficient that $f(x)$ should be of the form* (1.23), λ *and μ being functions whose squares are integrable over* (0, 1).

We cannot prove so comprehensive a result for the equation (K$_1$), owing to the comparative slowness with which the kernel $J_1(t)$ tends to zero; and since additional hypotheses would be required, we have not worked out the results in detail.

THEOREM 2.—*There are no solutions of* (M) *or* (M$_1$) *of integrable square, other than* $f(x) = 0$.

Our methods give something even when the hypothesis that the square of $f(x)$ is integrable is abandoned; thus we can prove

THEOREM 3.—*Any function of the form* (1.23) *is a solution of* (B), (B$_1$), *or* (K); *and if $f(x)$ is of the form*

(1.24) $$f(x) = \int_0^\infty \{\lambda(u) \cos xu + \mu(u) \sin xu\} \, du,$$

then λ and μ are zero for $u > 1$, so that $f(x)$ is of the form (1.23).

The example (1.22), however, shows that a solution is not necessarily of this form. This remark leads us, in §§ 2.6–7, to discuss alternative forms of solution, and in particular one suggested by Bateman.

1.3. Our analysis brings out clearly the reason which underlies the difference between the theory of (B), (K), (B$_1$), or (K$_1$) on the one hand and

---

[*] M. Plancherel, "Contribution à l'étude de la représentation d'une fonction arbitraire par des intégrales définies", *Rend. di Palermo*, 30 (1910), 289–335.

[†] E. C. Titchmarsh, "Hankel Transforms", *Proc. Camb. Phil. Soc.*, 21 (1923), 463–473.

that of (M) or (M_1) on the other. If we consider generally the equation

(1.31) $$f'(x) = \tfrac{1}{2} \int_0^\infty \frac{\chi(t)}{t} \{f(x+t) - f(x-t)\} dt,$$

it will not generally be true that

(1.32) $$\int_0^\infty \frac{\chi(t)}{t} \sin tu \, dt = u,$$

except possibly for isolated values of u. In this case there will be no solutions of the equation, or a small number of exceptional solutions. But if it should happen that (1.32) holds over whole intervals of values of u, as it does for $0 < u < 1$ when $\chi(t)$ is $J_1(t)$ or $2 \sin t/\pi t$, then there is a multitude of solutions defined by definite integrals taken over these intervals.

The methods of § 2 may be applied to many other equations, such as

(1.33) $$f''(x) = \int_0^\infty \frac{\chi(t)}{t^2} \{f(x+t) + f(x-t) - 2f(x)\} dt.$$

If $\chi(t)$ is $J_1(t)$ or $2 \sin t/\pi t$, and f is of integrable square, the solution of this equation is the same as that of (B) or (K); if $\chi(t)$ is e^{-t}, there are no non-zero solutions of integrable square. In the last case the only obvious solution is a cubic polynomial.

They may also be applied to such equations as

(1.34) $$f'(x) = g(x) + \tfrac{1}{2} \int_0^\infty \frac{J_1(t)}{t} \{f(x+t) - f(x-t)\} dt,$$

(1.35) $$f'(x) = g(x) + \tfrac{1}{2} \int_0^\infty \frac{e^{-t}}{t} \{f(x+t) - f(x-t)\} dt,$$

and here again there is a complete difference between the natures of the solutions. If $f(x)$ and $g(x)$ are of integrable square and $x^{-3}g(x)$ is integrable at the origin, the equation (1.35) has a unique solution; while (1.34) has no solution unless $g(x)$ is a function of special form, and then an infinity, any two of which differ by a solution of (K). We have not included formal proofs of these results.

1.4. The theory of transforms gives, up to a certain point, very simple and satisfactory results, but it has serious limitations, inasmuch as it will not give an account of such obvious solutions as (1.21) and (1.22). In § 3 we consider the theory of (K) and (M) by more powerful methods depending on the theory of analytic functions. Our main results are:—

THEOREM 6.—*Suppose that $f(x)$ is real and*

(1.41) $$f(x) = O(|x|^{\frac{1}{2}-\delta})$$

where $\delta > 0$, for large real values of x. Then, in order that $f(x)$ should be a solution of (K), *it is necessary and sufficient that $f(z)$ should be an integral function of $z = x+iy$, and*

(1.42) $$|f(z)| < A |z|^{\frac{1}{2}-\delta} e^{|y|},$$

where A is a constant, for all sufficiently large values of $|z|$.

THEOREM 7.—*Suppose that $f(x)$ is real and*

(1.43) $$f(x) = O(e^{A|x|}),$$

where $A < 1$, for real values of x, and that $f(x)$ is a solution of (M). *Then $f(x)$ is a quadratic.*

These theorems give a reasonably, though not entirely, complete account of the equations, since the orders of magnitude prescribed by (1.41) and (1.43) are very near those demanded by the convergence of the integrals in the equations. The most that has been proved before is Littlewood's result that, roughly,* (M) *has no solution, other than a quadratic, whose second derivative is bounded.* This he proves by entirely elementary reasoning.

We conclude by proving a theorem, of a much more special character than Theorem 7, but interesting both for the method of proof and because, in one way, it goes further, encroaching, so to say, on the case $A = 1$.

THEOREM 8.—*Suppose that a solution of* (M) *is of the form*

$$f(x) = \Sigma a_n \frac{x^n}{n!},$$

and that $\quad a_0+a_2+\ldots+a_{2n}+\ldots, \quad a_1+a_3+\ldots+a_{2n+1}+\ldots$

are convergent. Then $f(x)$ is a quadratic.

This includes the case of Theorem 7 in which it is assumed that $f(z) = O(e^{A|z|})$, with $A < 1$, for all values of z; and some cases which are not covered by Theorem 7.

---

* Littlewood gives theorems more general than this in other directions, but not in regard to the order of magnitude of the function. We quote only the simplest theorem.

2. *Solutions by the methods of the theory of functions of a real variable.*

2.1. The theorems in the theory of Fourier transforms which will be made use of are as follows:—

(2.11) If the square of a function $f(x)$ is integrable over the range $(0, \infty)$, then the function

$$F_a(x) = \sqrt{\left(\frac{2}{\pi}\right)} \int_0^a \cos xt\, f(t)\, dt$$

converges in mean* to a function $F(x)$, whose square is also integrable over the range $(0, \infty)$. In these circumstances $F(x)$ is called the Fourier cosine transform of $f(x)$.

(2.12) $$\int_0^x F(t)\, dt = \sqrt{\left(\frac{2}{\pi}\right)} \int_0^\infty \frac{\sin xt}{t} f(t)\, dt.$$

(2.13) Reciprocally, $f(x)$ is the transform of $F(x)$.

(2.14) If the square of $g(x)$ is integrable over $(0, \infty)$, then

$$\lim_{a \to \infty} \int_0^\infty g(x)\{F(x) - F_a(x)\}\, dx = 0.$$

(2.15) Similar results hold if $\sin xt$ or $\sqrt{(xt)}\, J_0(xt)$ be substituted for $\cos xt$.

2.2. *Proofs of Theorems 1 and 2.*—The equations (K), (B$_1$), (M), and (M$_1$) can be conveniently considered together as particular cases of (1.31), the function $\chi(t)$ having the following forms:—

(2.21)   (K) $J_1(t)$,   (B$_1$) $\dfrac{2}{\pi}\dfrac{\sin t}{t}$,   (M) e^{-t},   (M$_1$) $\dfrac{2}{\pi} tK_1(t)$.

If $f(x)$ is a solution, so is each of the functions $\tfrac{1}{2}\{f(x) \pm f(-x)\}$, so that any solution is the sum of an odd and an even solution. Suppose first that $f(x)$ is an even solution whose square is integrable over $(-\infty, \infty)$. Then, for almost all x,

(2.22) $$f(x) = \frac{d}{dx} \int_0^\infty \frac{\sin xu}{u} \lambda(u)\, du,$$

---

* That is to say

$$\int_0^\infty \{F_a(x) - F(x)\}^2\, dx \to 0,$$

when $a \to \infty$.

where λ^2 is integrable over $(0, \infty)$. Hence

$$f(x+t) - f(x-t) = 2\frac{d}{dx}\int_0^\infty \frac{\cos xu \sin tu}{u} \lambda(u)\, du,$$

and the equation takes the form

$$f'(x) = \int_0^\infty \frac{\chi(t)}{t}\left\{\frac{d}{dx}\int_0^\infty \frac{\cos xu \sin tu}{u} \lambda(u)\, du\right\} dt.$$

Now in each case it is easily seen* that the integral with respect to t converges uniformly over a finite range of values of x, so that the order of integration with respect to t and differentiation with respect to x may be inverted. Thus

$$f'(x) = \frac{d}{dx}\int_0^\infty \frac{\chi(t)}{t}\, dt \int_0^\infty \frac{\cos xu \sin tu}{u} \lambda(u)\, du$$

$$= \frac{d}{dx}\int_0^\infty \frac{\cos xu}{u} \lambda(u)\, du \int_0^\infty \frac{\sin tu}{t} \chi(t)\, dt,$$

on inverting the order of integration.†

The integral
$$\int_0^\infty \frac{\sin tu}{t} \chi(t)\, dt$$

has the following values in the four cases under consideration:

(K) $\qquad u \quad (u<1), \qquad \{u+\sqrt{(u^2-1)}\}^{-1} \quad (u>1);$

(B$_1$) $\qquad u \quad (u<1), \qquad 1 \qquad (u>1);$

$\qquad\qquad$ (M) $\quad \arctan u;\qquad$ (M$_1$) $\quad u(u^2+1)^{-\frac{1}{2}}.$

The difference between (K), (B$_1$) on the one hand, and (M), (M$_1$) on the other, becomes clear at this point. Take (K) as typical of the former class. This equation is now reduced to

$$f'(x) = \frac{d}{dx}\left\{\int_0^1 \cos xu\, \lambda(u)\, du + \int_1^\infty \frac{\cos xu\, \lambda(u)}{u\{u+\sqrt{(u^2-1)}\}}\, du\right\},$$

and, integrating,

$$f(x) = \int_0^1 \cos xu\, \lambda(u)\, du + \int_1^\infty \frac{\cos xu\, \lambda(u)}{u\{u+\sqrt{(u^2-1)}\}}\, du.$$

---

* From the original form (1.31).
† The associated double integral is absolutely convergent, since
$$|\sin tu| \leqslant t^a u^a \quad (0 < a < \tfrac{1}{2}).$$

The constant of integration is zero; for both terms on the right tend to zero when x tends to infinity, and $f(x)$ cannot tend to a limit other than zero. But also, by (2.22),

$$f(x) = \int_0^1 \cos xu \, \lambda(u) \, du + \frac{d}{dx} \int_1^\infty \frac{\sin xu}{u} \lambda(u) \, du;$$

so that

(2.23) $\quad \dfrac{d}{dx} \displaystyle\int_1^\infty \dfrac{\sin xu}{u} \left\{ 1 - \dfrac{1}{u\{u + \sqrt{(u^2 - 1)}\}} \right\} \lambda(u) \, du = 0,$

Hence if

$$\phi(u) = \left\{ 1 - \frac{1}{u\{u + \sqrt{(u^2 - 1)}\}} \right\} \lambda(u) \quad (u > 1), \quad 0 \quad (u < 1),$$

then the transform of $\phi(u)$ is a null function. Thus $\phi(u)$ is a null function, and so $\lambda(u)$ is a null function for $u > 1$, since the other factor does not vanish. Hence

$$f(x) = \int_0^1 \cos xu \, \lambda(u) \, du.$$

Thus any even solution of (K), of integrable square, is of the form prescribed by the theorem. The proof for (B_1) is substantially the same; and it is plain that our argument shows the sufficiency of the condition as well as its necessity. The equation (M), which is typical of the second class, has been reduced to

$$f'(x) = \frac{d}{dx} \int_0^\infty \frac{\cos xu}{u} \arctan u \, \lambda(u) \, du,$$

and it follows as before that

$$\frac{d}{dx} \int_0^\infty \frac{\sin xu}{u} \left(1 - \frac{\arctan u}{u} \right) \lambda(u) \, du = 0.$$

Hence in this case $\lambda(u)$ is a null function for $u > 0$, i.e. $f(x)$ vanishes. The same conclusion evidently holds for (M_1).

If $f(x)$ is an odd solution of (1.31), whose square is integrable over $(0, \infty)$, then

$$f(x) = \frac{d}{dx} \int_0^\infty \frac{1 - \cos xu}{u} \mu(u) \, du$$

where μ^2 is integrable over $(0, \infty)$, for almost all x. It follows as before that for (K) and (B_1)

$$f(x) = \int_0^1 \sin xu \, \mu(u) \, du,$$

while for (M) and (M$_1$) $f(x)$ vanishes. This completes the proof of Theorems 1 and 2, except as regards equation (B).

The discussion of (B) proceeds on the same lines, except that the inversion of the order of integration of the repeated integral* cannot be so easily justified. In this case the integral is (for an even solution)

$$\int_0^\infty \frac{\sin t}{t} dt \int_0^\infty \frac{\sin xu \cos tu}{u} \lambda(u)\, du.$$

The integral with respect to u converges uniformly over any finite range of values of t, so that

$$\int_0^A \frac{\sin t}{t} dt \int_0^\infty \frac{\sin xu \cos tu}{u} \lambda(u)\, du = \int_0^\infty \frac{\sin xu}{u} \lambda(u)\, du \int_0^A \frac{\sin t \cos tu}{t}\, dt.$$

Hence it is sufficient to prove that

(2.24) $$\lim_{A \to \infty} \int_0^\infty \frac{\sin xu}{u} \lambda(u)\, du \int_A^\infty \frac{\sin t \cos tu}{t}\, dt = 0.$$

But here the integral with respect to t is bounded for all values of A and u, and its limit is zero for all values of u. It follows from a classical theorem of Lebesgue† that (2.24) is true, and so the inversion is justified.‡

2.4. *Proof of Theorem 3.*—Here λ and μ are integrable instead of λ^2 and μ^2. The proof given under the previous hypotheses remains valid, except that, having obtained the formula (2.23), we must integrate again and use Fourier's theorem directly.

2.5. *The equation* (K') *and the theory of Neumann series.*—Before we proceed to a discussion of (K'), with a less stringent condition for the convergence of the integral than that of Theorem 1, we must say something about the connection of this equation with the theory of Neumann series. It is convenient to refer to Watson's account of this subject,§ in which it is proved that, in order that $f(x)$ should satisfy (K'), it is necessary and sufficient that the expansion

(2.51) $$f(x) = \sum_{n=0}^\infty (4n+2) J_{2n+1}(x) \int_0^\infty \frac{J_{2n+1}(t)}{t} f(t)\, dt$$

---

* See p. 8, f.n.†
† See, for example, de la Vallée Poussin, *Cours d'analyse* (ed. 3, 1, 1914), 264.
‡ We shall refer to this argument as "inversion by bounded convergence".
§ Watson, *loc. cit.*, 533-535.

should be valid for all real values of x. Watson assumes that $f'(x)$ is continuous, and that $f(x)$ is absolutely integrable over the range $(0, \infty)$.

If the integrals are taken in the sense of Lebesgue, it is unnecessary to assume more about $f'(x)$ than that it exists and is finite everywhere. It is clear, in fact, from Watson's proof that all that is necessary is that $f(x)$ should be the integral of $f'(x)$. Now if it is assumed that the integral

$$(2.52) \qquad \int_0^\infty f(t)\, t^{-\frac{3}{2}}\, dt$$

is absolutely convergent, then it follows from (K') that

$$|f'(x)| \leqslant \tfrac{1}{2} \int_0^\infty \frac{|J_1(t)|}{t} |f(t+x)+f(t-x)|\, dt.$$

The integral

$$\int_0^\infty \frac{|J_1(t)|}{t}\, dt \int_{x_1}^{x_2} |f(t+x)+f(t-x)|\, dx$$

is plainly convergent, so that $f'(x)$ is integrable over any finite interval (x_1, x_2); and so $f(x)$ is the integral of $f'(x)$.*

Further, *the theorem stated by Watson is true, provided only that the integral* (2.52) *is absolutely convergent*. To prove this, it is necessary to sum the series (2.51) by changing the order of summation and integration. This is justified by uniformity of convergence if the integrals are only taken over a finite range; and it will be justified when the range is infinite if the integral

$$(2.53) \qquad \int_k^\infty |f(t)| \sum_{n=0}^\infty (4n+2) \left| J_{2n+1}(x) \frac{J_{2n+1}(t)}{t} \right| dt \quad (k > 0)$$

is convergent. Let

$$(2.54) \qquad \sum_{n=0}^\infty (4n+2)|J_{2n+1}(x) J_{2n+1}(t)| = \sum_{n<\theta t} + \sum_{n>\theta t} = \Sigma_1 + \Sigma_2,$$

where θ is a constant less than unity. In Σ_2

$$|J_{2n+1}(t)| < 1, \qquad |J_{2n+1}(x)| < A\delta^n \quad (\delta = \delta(\theta) < 1).$$

Hence

$$(2.55) \qquad \Sigma_2 < A \sum_{n>\theta t} n\delta^n < A t \delta^{\theta t} = O(t^{-\frac{1}{2}}).$$

In Σ_1 we use the asymptotic formula.†

$$J_n(n \sec \beta) \sim \frac{\cos\{n(\tan \beta - \beta) - \tfrac{1}{4}\pi\}}{\sqrt{(\tfrac{1}{2} n\pi \tan \beta)}} \quad (\beta > 0),$$

---

* See E. W. Hobson, *Theory of Functions of a Real Variable* (ed. 2, 1, 1921), 554.
† Watson, *loc. cit.*, 234, 244.

which shows that $J_n(t) = O(t^{-\frac{1}{2}})$ uniformly for $n < \theta t$. It follows that

(2.56) $\qquad \Sigma_1 = O(t^{-\frac{1}{2}}) \Sigma(4n+2)|J_{2n+1}(x)| = O(t^{-\frac{1}{2}})$,

and from (2.54), (2.55), and (2.56), we deduce

$$\sum_{n=0}^{\infty}(4n+2)|J_{2n+1}(x)J_{2n+1}(t)| = O(t^{-\frac{1}{2}}).$$

Hence (2.53) is convergent whenever (2.52) is absolutely convergent. The remainder of Watson's discussion requires no modification.

2.6. *An alternative solution of* (K').—We now prove

THEOREM 4.—*Suppose that the square of* $x^{-\frac{1}{2}}f(x)$ *is integrable over* $(0, \infty)$. *Then in order that* $f(x)$ *should be a solution of* (K'), *it is necessary and sufficient that it should be of the form*

(2.61) $\qquad f(x) = x \int_0^1 u^{\frac{1}{2}} J_0(ux) \lambda(u) \, du,$

where λ^2 *is integrable over* (0, 1).*

Since $\{f(x)\}^2/x$ is integrable over $(0, \infty)$, the integral (2.52) is absolutely convergent. Hence $f(x)$, if it is a solution of (K'), is expansible in a Neumann series in accordance with the formulæ (1.142) and (1.143); and conversely.

To verify the sufficiency of the condition, we can use Bateman's expansion †

$$xJ_0(ux) = \sum_{n=0}^{\infty}(4n+2) P_n(1-2u^2) J_{2n+1}(x),$$

which is valid and uniformly convergent for $0 \leqslant u \leqslant 1$. Thus if $f(x)$ is of the form (2.61),

$$f(x) = x \int_0^1 u^{\frac{1}{2}} J_0(ux) \lambda(u) \, du = \sum_{n=0}^{\infty}(4n+2) J_{2n+1}(x) \int_0^1 P_n(1-2u^2) u^{\frac{1}{2}} \lambda(u) \, du.$$

---

* This is the form of solution suggested by Bateman.

† See Bateman, *loc. cit.*, 32; and "Certain definite integrals and expansions connected with the Legendre and Bessel functions", *Messenger of Math.*, 38 (1904), 182–188 (this reference is given incorrectly in Bateman's later paper).

Hence $f(x)$ is expansible in a Neumann series convergent for all real values of x. Also

$$\int_0^\infty J_{2n+1}(x) f(x) \frac{dx}{x} = \int_0^\infty J_{2n+1}(x)\, dx \int_0^1 u^{\frac{1}{2}} J_0(ux) \lambda(u)\, du$$

$$= \int_0^1 u^{\frac{1}{2}} \lambda(u)\, du \int_0^\infty J_{2n+1}(x) J_0(ux) dx *$$

$$= \int_0^1 P_n(1-2u^2)\, u^{\frac{1}{2}} \lambda(u)\, du,$$

since† $\int_0^\infty J_{2n+1}(x) J_0(ux)\, dx = P_n(1-2u^2) \quad (u<1) \quad = 0 \quad (u>1).$

Hence the coefficients are of the Webb-Kapteyn form; and so $f(x)$ is a solution of (K'). The sufficiency of the condition can also be verified without difficulty by direct substitution.

We have next to prove the necessity of the condition. Suppose then that $f(x)$ satisfies (K'). It follows from the theory of Hankel transforms‡ that there is a function $\lambda(u)$, whose square is integrable over $(0, \infty)$, such that the function

$$x^{-\frac{1}{2}} f_a(x) = \int_0^a x^{\frac{1}{2}} u^{\frac{1}{2}} J_0(xu) \lambda(u)\, du$$

converges in mean to $x^{-\frac{1}{2}} f(x)$, when $a \to \infty$; and that

$$\lim_{a\to\infty} \int_0^\infty x^{-\frac{1}{2}} J_{2n+1}(x) \{x^{-\frac{1}{2}} f(x) - x^{-\frac{1}{2}} f_a(x)\}\, dx = 0, §$$

or

$$\int_0^\infty J_{2n+1}(x) f(x) \frac{dx}{x} = \lim_{a\to\infty} \int_0^\infty J_{2n+1}(x) f_a(x) \frac{dx}{x}.$$

Hence

$$a_{2n+1} = (4n+2) \lim_{a\to\infty} \int_0^\infty J_{2n+1}(x)\, dx \int_0^a u^{\frac{1}{2}} J_0(ux) \lambda(u)\, du$$

$$= (4n+2) \lim_{a\to\infty} \int_0^a u^{\frac{1}{2}} \lambda(u)\, du \int_0^\infty J_{2n+1}(x) J_0(ux)\, dx$$

$$= (4n+2) \int_0^1 P_n(1-2u^2) u^{\frac{1}{2}} \lambda(u)\, du.$$

---

* The inversion may be justified by bounded convergence.

† The proof of these formulae is included implicitly in Bateman's analysis. See also Watson, *loc. cit.*, 408–410.

‡ Titchmarsh, *loc. cit.*, 470.

§ Since the square of $x^{-\frac{1}{2}} J_{2n+1}(x)$ is integrable.

It follows that the coefficients in the Neumann series, and so also $f(x)$, are independent of the values of $\lambda(u)$ for $u > 1$. Let $\phi(u)$ be equal to $\lambda(u)$ when $u < 1$ and to zero otherwise. Then* $\phi(u)$ possesses a Hankel transform $x^{-\frac{1}{2}} g(x)$, where

$$g(x) = \frac{d}{dx}\left\{ x \int_0^\infty u^{-\frac{1}{2}} J_1(ux) \phi(u)\, du \right\} = x \int_0^1 u^{\frac{1}{2}} J_0(ux) \lambda(u)\, du.$$

This function is, as we have seen, a solution of (K'): and the coefficients in its Neumann expansion are the same as those of $f(x)$, so that it is identical with $f(x)$. Hence $\lambda(u)$ and $\phi(u)$ have the same Hankel transform, and so $\lambda(u) = \phi(u)$ almost everywhere. That is to say, $f(x)$ is of the form (2.61).

A result analogous to Theorem 3 can be obtained in a similar way, viz. that *if*

$$f(x) = x \int_0^\infty u^{\frac{1}{2}} J_0(ux) \lambda(u)\, du,$$

then it is necessary and sufficient that $\lambda(u) = 0$ *for* $u > 1$.

2.7. In Theorems 1 and 4 two forms for the solution of (K') have been obtained, viz.

$$(\alpha)\ \int_0^1 \sin xu\, \lambda(u)\, du, \qquad (\beta)\ x \int_0^1 J_0(xu) u^{\frac{1}{2}} \lambda(u)\, du,$$

$\lambda(u)$ being a function whose square is integrable. It is clear that these cannot be precisely equivalent, for a function of type (α) has its square integrable, while a function of type (β) is the product of $\sqrt{x}$ by a function of integrable square; thus (α) is more restricted at infinity, (β) more restricted at $x = 0$. The relation between the two types is as follows:—

If $f(x)$ is of type (α), then, in order that it should be of type (β), it is necessary and sufficient that the square of $x^{-\frac{1}{2}} f(x)$ should be integrable over $(0, \infty)$.

If $f(x)$ is of type (β), then, in order that it should be of type (α), it is necessary and sufficient that the square of $f(x)$ should be integrable over $(0, \infty)$.

In each case the necessity of the condition is obvious, from the theory of transforms. To prove the sufficiency of the condition in the first pro-

---

* Titchmarsh, *loc. cit.*, 471.

position, suppose $f(x)$ of the form (a). Since $\{f(x)\}^2/x$ is integrable, we have

$$f(x) = \frac{d}{dx} \int_0^\infty x u^{-\frac{1}{2}} J_1(ux) \mu(u)\, du,$$

where μ^2 is integrable; and

$$x^{\frac{1}{2}} \mu(x) = \frac{d}{dx} \int_0^\infty \frac{x J_1(ux)}{u} f(u)\, du = \frac{d}{dx} \left\{ x \int_0^\infty \frac{J_1(ux)}{u} du \int_0^1 \sin ut\, \lambda(t)\, dt \right\}$$

$$= \frac{d}{dx} \left\{ x \int_0^1 \lambda(t)\, dt \int_0^\infty \frac{J_1(ux)}{u} \sin ut\, du \right\} = 0,$$

if $x > 1$, since
$$\int_0^\infty \frac{J_1(ux)}{u} \sin ut\, du = \frac{t}{x} \quad (x > t).$$

Hence $\quad f(x) = \dfrac{d}{dx} \int_0^1 x u^{-\frac{1}{2}} J_1(ux) \mu(u)\, du = x \int_0^1 u^{\frac{1}{2}} J_0(ux) \mu(u)\, du,$

which is the result stated. The second part of the theorem may be proved in a similar way.

2.8. *An alternative solution of* (B_1).—We can also prove

THEOREM 5.—*If the square of $f(x)/x$ is integrable over $(-\infty, \infty)$, then, in order that $f(x)$ should be a solution of (B_1), it is necessary and sufficient that $f(x)/x$ should be of the form* (1.23), λ *and* μ *being functions whose squares are integrable over* (0, 1).

The discussion proceeds on the lines of § 2.2. It is rather long, and will be omitted.

2.9. It is plain that some of the simplest solutions of our equations do not belong to the classes hitherto considered. They do not include, for example, the obvious quadratic solutions of (M) or (M_1). Again, $\sin ax$ $(a \leqslant 1)$ is a solution of (B), (K'), and (B_1). It is a case of Theorem 5, but not of Theorems 1 or 4. On the other hand

$$\sin ax = x \int_0^a J_0(xt)(a^2 - t^2)^{-\frac{1}{2}} t\, dt,$$

so that it is of the form mentioned at the end of § 2.6, when $\lambda(u)$, but not its square, is integrable. If the integrals are taken to be non-absolutely convergent integrals, or integrals generalised in the "Cesàro" or other senses, new solutions appear; thus any polynomial, or any polynomial multiplied by $\sin ax$ $(a \leqslant 1)$, is in this sense a solution of (K).

3. Solutions by the methods of the theory of functions of a complex variable.

3.1. Proof of Theorem 6.

3.11. To prove the sufficiency of the condition (1.42), consider the integral

$$\int \frac{H_1^{(1)}(z-a)}{z-a} f(z)\, dz,$$

where a is real and the contour consists of the real axis, with an indentation above the point $z = a$, and a large semi-circle above it; and $H_1^{(1)}$ is the Bessel function of the third kind. On the large semi-circle

$$H_1^{(1)}(z-a) = O(|z-a|^{-\frac{1}{2}} e^{-y}), \quad f(z) = O(|z|^{\frac{1}{2}-\delta} e^y),$$

so that this part of the integral tends to zero when the radius tends to infinity. The integral round the indentation at $z = a$ tends to $-2f'(a)$, since

$$H_1^{(1)}(z-a) = -\frac{2i}{\pi(z-a)} + O(z-a).$$

Hence
$$\int_{-\infty}^{\infty} \frac{H_1^{(1)}(z-a)}{z-a} f(z)\, dz = 2f'(a).$$

Taking the real parts of this equation, we see that $f(z)$ is a solution of (K). The proof that the condition is necessary is more elaborate.

3.12. LEMMA a.—*Suppose that $f(z)$ is an integral function which satisfies the following conditions:*—

(3.121) $\qquad |f(x)| < C_1|x|^a \quad (a > 0),\ |x| > x_0;$

(3.122) $\qquad |f(iy)| < C_2(\epsilon)\, e^{|y|(1+\epsilon)}$

for all positive values of ϵ;

(3.123) $\qquad |f(z)| < C_3\, e^{K|z|}$

for some value of K and all values of z. Then

(3.124) $\qquad |f(z)| \leqslant C|z|^a\, e^{|y|}$

for all values of z of sufficiently large modulus.

Here K and the C's are absolute constants, except C_2, which may depend upon ϵ. The lemma is contained in substance in well known theorems due to Phragmén and Lindelöf.* Let

(3.125) $$g(z) = (z+ik)^{-a} e^{i(1+\delta)z} f(z) \quad (k > 0, \ \delta > \epsilon).$$

Then $|g(z)| < C_1$ on the real axis. Also $g(iy) \to 0$ when $y \to \infty$, so that $|g(iy)|$ has a maximum C_4, and $|g(z)| \leqslant C_4$ on the positive imaginary axis. Finally $|g(z)| < C_5 e^{K|z|}$ for all values of z. Hence

$$|g(z)| \leqslant C_6 = \text{Max}\,(C_1, C_4),$$

for $y \geqslant 0$. This would not in itself suffice to prove the lemma, since C_4 depends *prima facie* on δ. But it is easily proved that $C_4 \leqslant C_1$.† For suppose $C_4 > C_1$, and consider the angle bounded by the lines $\theta = 0$, $\theta = \frac{1}{2}\pi + \beta$, where $\theta = \arg z$ and $0 < \beta < \frac{1}{2}\pi$. Then $|g(z)| \leqslant C_4$ on the boundary of the angle, and attains the value C_4 at a point inside it; from which it follows that $g(z)$ is constant, and that $|g(z)| \leqslant C_1$ in any case. Thus we may take $C_6 = C_1$.

It follows then from (3.125) that

$$|f(z)| \leqslant C |z|^a e^{(1+\delta)y} \quad (y > 0, \ |z| > A)$$

for all positive values of δ, and so

$$|f(z)| \leqslant C |z|^a e^y.$$

A similar result can be obtained for $y < 0$. Thus the lemma is completely proved.

3.13. Suppose now that $f(x)$ is a solution of (K) which satisfies the condition (1.41). Then it follows from § 2.5 that $f(x)$ is expansible in a Neumann series (1.142) with Webb-Kapteyn coefficients. The series is convergent for all real values of x, and so by the general theory of Neumann series, for all complex values; hence $f(z)$ is an integral function.

It remains to prove that condition (1.42) is satisfied, *i.e.* that (3.124) is satisfied with $a = \frac{1}{2} - \delta$. Since (1.41) is equivalent to (3.121) with $a = \frac{1}{2} - \delta$, it is sufficient to prove that (3.122) and (3.123) are satisfied.

The proof of the last assertion is immediate; for‡

(3.131) $$|J_{2n+1}(z)| \leqslant \frac{|\tfrac{1}{2}z|^{2n+1}}{(2n+1)!} e^{|y|},$$

---

* E. Phragmén and Ernst Lindelöf, "Sur une extension d'un principe classique de l'analyse", *Acta Math.*, 31 (1908), 381-406 (385).
† Cf. Phragmén and Lindelöf, *loc. cit.*, 387.
‡ Watson, *loc. cit.*, 49, § 3.31 (1).

and it is evident that

(3.132) $$a_{2n+1} = O(n).$$

Hence

(3.133) $$\Sigma |a_{2n+1} J_{2n+1}(z)| < C|z|^C e^{2|z|} < Ce^{3|z|}.$$

To prove that (3.122) is satisfied, we may use the inequality*

(3.134) $$|J_m(iy)| < \frac{y^m e^{\sqrt{(m^2+y^2)}}}{\{m+\sqrt{(m^2+y^2)}\}^m} \quad (y > 0).$$

It may be verified that, for a given value of y, the expression on the right decreases as m increases; and for $m = 0$ it reduces to e^y. Let

$$\sum_{n=0}^{\infty} a_{2n+1} J_{2n+1}(iy) = \sum_{2n+1 \leqslant 3y} + \sum_{2n+1 > 3y} = \Sigma_1 + \Sigma_2.$$

Then

$$|\Sigma_1| < Ce^y \sum_{n < Cy} n < Cy^2 e^y.$$

If $m > 3y$, (3.134) gives

$$|J_m(iy)| < \left(\frac{1}{3+\sqrt{10}}\right)^m e^{\frac{1}{2}m\sqrt{10}},$$

which is the general term of a convergent geometrical progression. Thus $|\Sigma_2| < C$ and so

$$\sum_{n=0}^{\infty} a_{2n+1} J_{2n+1}(iy) = O(y^2 e^y) = O(e^{(1+\epsilon)y}) \quad (\epsilon > 0),$$

which completes the proof.

3.2. *Proof of Theorem* 7.

3.21. It is convenient to assume at first that $f(x)$ has derivatives of all orders for all real values of x, a restriction which is to be removed later.

LEMMA β.—*If $f(x)$ is an indefinitely differentiable solution of* (M), *and*

(3.211) $$|f(x)| < Ce^{A|x|} \quad (A < 1)$$

---

* Watson, *loc. cit.*, 268, § 8.7 (1).

for all real values of x, then

(3.212) $$|f^{(p)}(x)| < C\left(\frac{8e^{\frac{1}{2}A}}{1-A}\right)^p e^{A|x|}$$

for all values of p and all real values of x.

It is sufficient to prove that (3.212) holds for $p = 1$; for then the integral obtained by differentiating under the integral sign is uniformly convergent, so that $f'(x)$ is also a solution of the equation, and the argument can be repeated. Since $f'(x)$ is continuous, we have

$$f'(x) = \tfrac{1}{2}\int_0^\infty \frac{e^{-t}}{t}\{f(x+t)-f(x-t)\}\, dt$$

$$= \int_0^\delta e^{-t} f'(x+\theta t)\, dt + \tfrac{1}{2}\int_\delta^\infty \frac{e^{-t}}{t}\{f(x+t)-f(x-t)\}\, dt,$$

where $|\theta| < 1$; and

$$|f'(x)| \leqslant \int_0^\delta e^{-t}|f'(x+\theta t)|\, dt + \frac{1}{\delta}\int_\delta^\infty e^{-t}\cdot Ce^{A|x|+A|t|}\, dt$$

$$\leqslant \mu\delta + \frac{Ce^{A|x|}}{\delta(1-A)},$$

where
$$\mu = \underset{|\xi-x|\leqslant\delta}{\text{Max}}|f'(\xi)|.$$

But $|f'(\xi)|$, being continuous, attains its upper bound for some value x_1 of ξ such that $|x_1-x| \leqslant \delta$. Hence

$$|f'(x)| \leqslant \delta |f'(x_1)| + \frac{Ce^{A|x|}}{\delta(1-A)}.$$

In exactly the same way a chain of inequalities can be formed, the $(n+1)$-th being

(3.213) $$|f'(x_n)| \leqslant \delta_n |f'(x_{n+1})| + \frac{Ce^{A|x_n|}}{\delta_n(1-A)}$$

$$(\delta_0 = \delta, \quad x_0 = x, \quad |x_{n+1}-x_n| \leqslant \delta_n, \quad n = 0, 1, 2, \ldots).$$

Let $\delta_n = 2^{-n-2}$. Then

$$|x_n| \leqslant |x| + 2^{-2} + 2^{-3} + \ldots + 2^{-n-1} < |x| + \tfrac{1}{2},$$

and (3.213) gives, for all values of n,

(3.214) $$|f'(x_{n+1})| > 2^{n+2}|f'(x_n)| - \frac{2^{2n+4}Ce^{A|x|+\frac{1}{2}A}}{1-A}.$$

If the lemma is false, for $p = 1$, there is a value of x for which

(3.215) $$|f'(x)| > \frac{8Ce^{A|x|+\frac{1}{2}A}}{1-A}.$$

But it follows from (3.214) that if, for any value of n,

$$|f'(x_n)| > \frac{2^{n+3}Ce^{A|x|+\frac{1}{2}A}}{1-A},$$

then this is also true with $n+1$ instead of n. Hence, if it is true for $n = 0$, as in (3.215), it is true for all values of n; and it follows that $f'(x)$ is not bounded in the interval $(x-\frac{1}{2}, x+\frac{1}{2})$, contrary to the hypothesis that $f'(x)$ is continuous. This proves Lemma β.

3.22. LEMMA γ.—*If $f(z)$ is an integral function such that*

$$|f(x)| < Ce^{A|x|}, \quad |f(z)| < Ce^{K|z|} \quad (K > A),$$

then

(3.221) $$|f(z)| < Ce^{A|x|+K|y|}.$$

This is another case of the theorems of Phragmén and Lindelöf already referred to. In fact, if

$$g(z) = e^{-(A-iK)z}f(z),$$

then $|g(z)| < C$ when z lies on the positive halves of the real and imaginary axes, and $|g(z)| < Ce^{K|z|}$ for all values of z in the first quadrant. Hence $|g(z)| < C$ for all values of z in the first quadrant, and (3.221) follows. The other quadrants may be treated in a similar way.

3.23. LEMMA δ.—*Under the hypotheses of Lemma α, $f(z)$ is an integral function, and satisfies* (3.221).

For $|f^{(p)}(0)| < CK^p$, by Lemma β. Hence the series

$$\sum_{p=0}^{\infty} \frac{z^p}{p!} f^{(p)}(0)$$

is convergent for all values of z, i.e. $f(z)$ is an integral function; and

$$|f(z)| < C \sum_{p=0}^{\infty} \frac{|z|^p}{p!} K^p = C e^{K|z|}.$$

Lemma δ now follows from Lemma γ.

3.24. LEMMA ϵ.—*The function $f(z)$ can be expressed in the form*

(3.241) $$f(z) = \frac{1}{2\pi i} \int_C e^{zu} \chi(u)\, du,$$

where $\chi(u)$ is regular at all points (including infinity) outside the rectangle whose vertices are $(\pm A, \pm K)$, and

(3.242) $$\chi(u) = \frac{b_1}{u} + \frac{b_2}{u^2} + \cdots$$

for sufficiently large values of u. The contour C is a closed curve surrounding the rectangle, but lying between the lines $\Re(u) = \pm 1$.

Let $$f(z) = \sum_{n=0}^{\infty} a_n z^n, \quad \phi(z) = \sum_{n=0}^{\infty} n!\, a_n z^n.$$

The function $\phi(z)$ is regular for sufficiently small values of z, since

$$|n!\, a_n| = |f^{(n)}(0)| < CK^n.$$

We proceed to prove it regular in a more extended domain. In the first place,

$$\phi(z) = \int_0^{\infty} e^{-u} f(zu)\, du$$

for sufficiently small values of z. The integral is convergent if

$$A|x| + K|y| < 1;$$

so that $\phi(z)$ is regular inside the rhombus **R** whose vertices are $(\pm A^{-1}, 0), (0, \pm K^{-1})$. The domain can be extended still further by using the formula

$$\phi(z) = e^{i\lambda} \int_0^{\infty} e^{-u e^{i\lambda}} f(z u e^{i\lambda})\, du,$$

which is valid if $-\tfrac{1}{2}\pi < \lambda < \tfrac{1}{2}\pi$ and

$$A|x \cos\lambda - y \sin\lambda| + K|x \sin\lambda + y \cos\lambda| < \cos\lambda.$$

This region also is a rhombus; and $\phi(u)$ is regular inside the region **D** formed by all points inside one rhombus or another. The region **D** is bounded by four circles, described on the lines joining the origin to the corners of **R**; and is the inverse of the rectangle of the lemma with respect to the origin. Hence $\phi(1/u)$ is regular outside this rectangle. If now C is the contour of the lemma, we have

$$\frac{1}{2\pi i} \int_C e^{zu} \phi\left(\frac{1}{u}\right) \frac{du}{u} = \frac{1}{2\pi i} \int_C e^{zu} \left(\sum_{n=0}^{\infty} \frac{n!\, a_n}{u^n}\right) \frac{du}{u} = \sum_{n=0}^{\infty} a_n z^n = f(z);$$

which proves (3.241), with

$$\chi(u) = \frac{1}{u} \phi\left(\frac{1}{u}\right).$$

3.25. *Proof of the main theorem.*—We have, by (3.241),

$$f(x+t) - f(x-t) = \frac{1}{2\pi i} \int_C e^{xu}(e^{tu} - e^{-tu}) \chi(u)\, du,$$

and hence

$$\tfrac{1}{2}\int_0^{\infty} \frac{e^{-t}}{t} \{f(x+t) - f(x-t)\}\, dt = \frac{1}{2\pi i} \int_C e^{xu} \chi(u)\, du \int_0^{\infty} \frac{e^{-t}}{2t}(e^{tu} - e^{-tu})\, dt$$

$$= \frac{1}{2\pi i} \int_C e^{xu} \chi(u)\, \tfrac{1}{2} \log \frac{1+u}{1-u}\, du,$$

that value of the logarithm being taken which vanishes for $u = 0$. Also

$$f'(x) = \frac{1}{2\pi i} \int_C e^{xu} u \chi(u)\, du\, ;$$

so that equation (M) takes the form

$$\int_C e^{xu} \chi(u) \psi(u)\, du = 0,$$

where

(3.251) $$\psi(u) = u - \tfrac{1}{2} \log \frac{1+u}{1-u}.$$

Let $\mu < 1$ be the maximum of $\Re(u)$ on C; multiply by $e^{-\xi x}$, where $\mu < \xi < 1$, and integrate over $(0, \infty)$. We obtain

(3.252) $$\int_C \chi(u) \psi(u) \frac{du}{\xi - u} = 0.$$

We may deform C into the contour $\Gamma = \Gamma_1 + \Gamma_2 + \Gamma_3$ shown in the figure.* The radii of the small circles are all equal to ρ, and Γ_3 comprises the whole of Γ except the large circle Γ_2 and the small circle Γ_1 round ξ. We have

$$\int_C = \int_\Gamma = \int_{\Gamma_1} + \int_{\Gamma_2} + \int_{\Gamma_3} = 0.$$

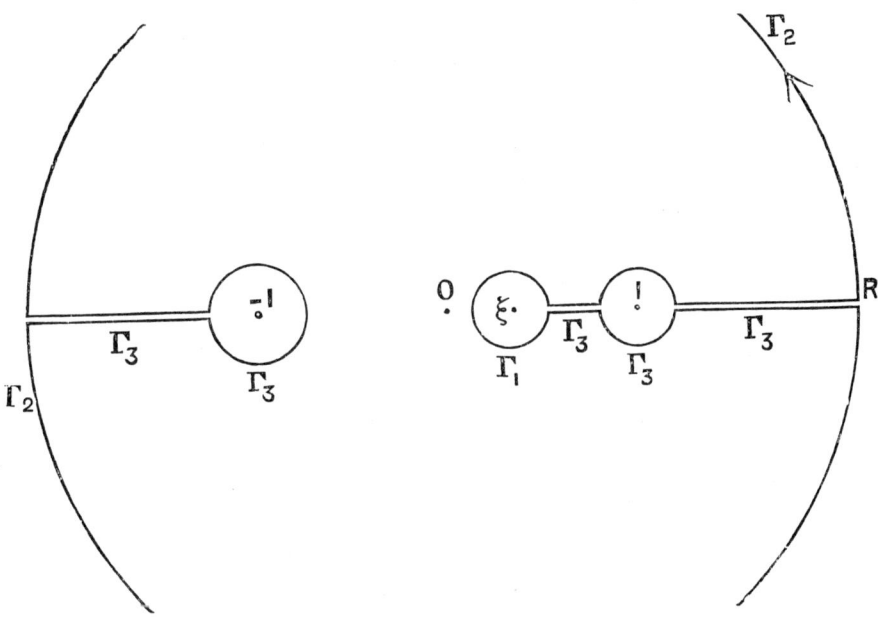

It is easily proved that

$$\lim_{\rho \to 0} \frac{1}{2\pi i} \int_{\Gamma_1} = \chi(\xi)\psi(\xi), \quad \lim_{R \to \infty} \frac{1}{2\pi i} \int_{\Gamma_2} = -b_1,$$

$$\lim \frac{1}{2\pi i} \int_{\Gamma_3} = -\int_1^\infty \frac{\chi(u)}{\xi - u} du - \int_{-\infty}^{-1} \frac{\chi(u)}{\xi - u} du.$$

Hence $\quad \chi(\xi)\psi(\xi) = b_1 + \int_1^\infty \frac{\chi(u)}{\xi - u} du + \int_{-\infty}^{-1} \frac{\chi(u)}{\xi - u} du.$

The expression on the right is an analytic function of ξ, regular for all values of ξ except real values such that $\xi \geqslant 1$ or $\xi \leqslant -1$; and so the same is true for $\chi(\xi)\psi(\xi)$.

---

* Most of the large circle Γ_2 is omitted in the figure. The point $u = \mu$ lies between O and the point nearest to O of the circle Γ_1, and $u = A$ lies between O and μ, so that the rectangle of Lemma ϵ lies entirely inside Γ.

We know already that $\chi(\xi)$ is regular outside the rectangle of Lemma ϵ; and it is easily verified that $\psi(\xi)$ has no zeros inside the rectangle, except its obvious zero, of the third order, for $\xi = 0$. Hence the only possible singularity of $\chi(u)$ is a pole of the third order at $u = 0$. Since $\chi(u)$ is regular at infinity, (3.242) reduces to

$$\chi(u) = \frac{b_1}{u} + \frac{b_2}{u^2} + \frac{b_3}{u^3};$$

and
$$f(x) = \frac{1}{2\pi i} \int_C e^{xu} \chi(u)\, du = b_1 + b_2 x + \tfrac{1}{2} b_3 x^2.$$

3.26. We have still to prove the theorem without assuming the existence of all the derivatives of $f(x)$. Suppose now simply that $f(x)$ satisfies (1.43), and let

$$\phi(x) = \int_{-\infty}^{\infty} f(y)\operatorname{sech}(y-x)\, dy = \int_{-\infty}^{\infty} f(x+u)\operatorname{sech} u\, du.$$

Then if $f(x)$ satisfies (M), so does $\phi(x)$; for

$$\tfrac{1}{2}\int_0^{\infty} \frac{e^{-t}}{t}\{\phi(x+t) - \phi(x-t)\}\, dt$$

$$= \tfrac{1}{2}\int_0^{\infty} \frac{e^{-t}}{t}\, dt \int_{-\infty}^{\infty} \{f(x+t+u) - f(x-t+u)\}\operatorname{sech} u\, du$$

$$= \int_{-\infty}^{\infty} f'(x+u)\operatorname{sech} u\, du = \phi'(x).$$

Also
$$\phi(x) = O\left(\int_{-\infty}^{\infty} e^{A|x|+A|u|}\operatorname{sech} u\, du\right) = O(e^{A|x|}),$$

and plainly all derivatives of $\phi(x)$ exist; so that by what has been proved already

$$\int_{-\infty}^{\infty} f(y)\operatorname{sech}(y-x)\, dy = \phi(x) = a + bx + cx^2.$$

We can determine constants α, β, γ, such that

$$\int_{-\infty}^{\infty} (\alpha + \beta y + \gamma y^2)\operatorname{sech}(y-x)\, dy = a + bx + cx^2.$$

Hence
$$\int_{-\infty}^{\infty} F(y)\operatorname{sech}(y-x)\, dy = 0$$

if
$$F(y) = f(y) - \alpha - \beta y - \gamma y^2.$$

Since $F(y) = O(e^{A|y|})$, we can differentiate repeatedly with respect to x, and then put $x = 0$; thus

$$(3.261) \quad \int_{-\infty}^{\infty} \frac{F(y)}{(\cosh y)^{2p+1}} dy = 0, \quad \int_{-\infty}^{\infty} \frac{\sinh y \, F(y)}{(\cosh y)^{2p}} dy = 0,$$

for all positive integral values of p. The first of these equations can be written

$$\int_0^{\infty} \frac{F(y)+F(-y)}{\cosh y} (\operatorname{sech}^2 y)^p \, dy = 0,$$

or, putting $\operatorname{sech}^2 y = e^{-u}$,

$$\int_0^{\infty} G(u) e^{-pu} \, du = 0,$$

where $$G(u) = \tfrac{1}{2} \operatorname{cosech} y \left(F(y) + F(-y) \right).$$

Hence* $G(u) = 0$, i.e. $F(y)$ is an odd function of y. It follows in the same way from the second equation (3.261) that $F(y)$ is an even function of y. Hence

$$F(y) = 0, \quad f(y) = \alpha + \beta y + \gamma y^2,$$

which completes the proof.

3.3. *Proof of Theorem* 8.

It follows without difficulty from our hypotheses that

$$f(x) = O(|x|^{-\frac{1}{2}} e^{|x|})$$

for real x, so that the integral (M) is absolutely convergent. But we shall not use this fact explicitly.

We have, for all values of x and t,

$$\frac{1}{2t} \{f(x+t) - f(x-t)\} = f'(x) + \frac{t^2}{3!} f^{(3)}(x) + \frac{t^4}{5!} f^{(5)}(x) + \ldots.$$

Multiplying by e^{-t}, and integrating term by term,

$$(3.31) \quad \tfrac{1}{2} \int_0^{\infty} \frac{e^{-t}}{t} \{f(x+t) - f(x-t)\} \, dt = f'(x) + \tfrac{1}{3} f^{(3)}(x) + \tfrac{1}{5} f^{(5)}(x) + \ldots.$$

---

* By the well-known theorem of Lerch: see M. Lerch, "Sur un point de la théorie des fonctions génératrices d'Abel", *Acta Math.*, 27 (1903), 339-351. For references to the many later proofs of the theorem, see G. H. Hardy, "Notes on some points in the integral calculus (42)", *Messenger of Math.*, 46 (1917), 43-48.

This step is justified provided the integrated series is convergent.* Now

$$\sum_{p=1}^{n} f^{(2p-1)}(x) = \sum_{p=1}^{n} a_{2p-1} + x \sum_{p=1}^{n} a_{2p} + \frac{x^2}{2!} \sum_{p=1}^{n} a_{2p+1} + \ldots$$

and hence

$$\left| \sum_{p=1}^{n} f^{(2p-1)}(x) \right| < 2M + |x| 2M + \frac{|x|^2}{2!} 2M + \ldots < 2M e^{|x|},$$

M being the upper bound of the moduli of the sums of alternate coefficients. Hence the series (3.31) is convergent for all values of x, and indeed uniformly for $|x| < K$. The term by term integration is thus justified; and (M) and (3.31) give

(3.32) $\qquad \frac{1}{3} f^{(3)}(x) + \frac{1}{5} f^{(3)}(x) + \ldots = 0.$

Since this series is uniformly convergent throughout any finite domain of values of x, we may multiply by x^{-p} and integrate term by term round a contour enclosing the origin. Thus

(3.33) $\qquad \frac{1}{3} a_p + \frac{1}{5} a_{p+2} + \frac{1}{7} a_{p+4} + \ldots = 0 \quad (p = 3, 4, \ldots),$

and so

(3.34) $\qquad \frac{1}{3} a_p = (\frac{1}{3} - \frac{1}{5}) a_{p+2} + (\frac{1}{5} - \frac{1}{7}) a_{p+4} + \ldots.$

Let $\mu_p = \text{Max}(|a_p|, |a_{p+2}|, \ldots)$. Then $\mu_{p+2} \leq \epsilon$ for $p \geq p_0(\epsilon)$, since $a_p \to 0$. It follows from (3.34) that $|a_p| \leq \mu_{p+2} \leq \epsilon$, and so

$$\mu_p = \text{Max}(|a_p|, \mu_{p+2}) = \mu_{p+2} \leq \epsilon.$$

Repeating the argument, we find that $|\mu_3| \leq \epsilon$, $|\mu_4| \leq \epsilon$, which is plainly only possible if $\mu_3 = 0$, $\mu_4 = 0$, i.e. if $a_3 = a_4 = \ldots = 0$. This proves the theorem.

---

* See G. H. Hardy, "Further researches in the theory of divergent series and integrals", *Trans. Camb. Phil. Soc.*, 21 (1908), 1–48 (5).

CORRECTIONS

p. 2, lines 9–11. The statement that 'all solutions of (K′) are odd' is incorrect. The derivative $f'(x)$ is even, but $f(x)$ is odd only if $f(0) = 0$, and this cannot be deduced from (K′). An odd solution of (K′) is a solution of (K), and an odd solution of (K) is a solution of (K′).

p. 8, lines 5–6. See *Proceedings of the London Mathematical Society* (2), 24, xxxi–iii (following this paper) for the correction of this statement.

p. 13, line 2 up. The second integral should be $\int_0^\infty$.

p. 16, §3.11. This section is incorrect as it stands. The integral round the indentation does not converge as the radius tends to zero and the final integral (line 12 of §3.11) diverges at the point a. The section can be amended as follows.

Consider the integral
$$\int \frac{H_1^{(1)}(z-a)}{z-a} \{f(z)-f(a)\}\, dz$$

round the contour of §3.11. The integral round the larger semicircle still tends to zero when the radius tends to infinity. Near to $z = a$
$$H_1^{(1)}(z-a) = -\frac{2i}{\pi(z-a)} + O[(z-a)\log(z-a)];$$

hence
$$\frac{H_1^{(1)}(z-a)}{z-a}\{f(z)-f(a)\} = -\frac{2if'(a)}{\pi(z-a)} + O(1),$$

and the integral round the indentation tends to $-2f'(a)$ as the radius tends to zero. Hence
$$\int_{-\infty}^{\infty} \frac{H_1^{(1)}(y)}{y}\{f(a+y)-f(a)\}\, dy = 2f'(a).$$

Taking real parts and using the fact that
$$H_1^{(1)}(\pm y) = J_1(y) \quad (y > 0),$$

we find that $f(x)$ is a solution of (K).

p. 17, §3.13 (*proof of the necessity of the conditions of Theorem* 6). This is based on §2.5, which is concerned with odd solutions of (K′), and hence of (K). Thus the proof only shows that the conditions of Theorem 6 are necessary when $f(x)$ is an odd function. Are they, in fact, necessary for any function to be a solution of (K)? The following proof that they are indeed necessary is due to Dr J. B. McLeod.

By direct substitution it is seen that, if $f(x)$ is a solution of (K), then so is $f(-x)$. So also therefore are $[f(x) \pm f(-x)]$. It follows that, by breaking $f(x)$ into its odd and even parts, it is sufficient to prove the necessity of the conditions when $f(x)$ is even. But when $f(x)$ is even, $f'(x)$ is odd, and $f'(x)$ also satisfies (K) (provided that differentiation under the integral sign in (K) is justified). Thus it should be possible to deduce the necessity of the conditions for $f(x)$ from those for $f'(x)$.

From (K) and the fact that
$$f(x) = O\{|x|^{\frac{1}{2}-\delta}\} \quad (\delta > 0)$$

for large real values of x, it is a routine estimation to verify that
$$f'(x) = O\{|x|^{\frac{1}{2}-\delta}\}$$
for large x. From this it follows that
$$\int_0^\infty \frac{J_1(t)}{t} \{f'(x+t) - f'(x-t)\}\, dt$$
converges uniformly on any finite interval, so that differentiation under the integral sign in (K) is justified. By Theorem 6 for odd $f(x)$, $f'(z)$ is an integral function and
$$|f'(z)| < A|z|^{\frac{1}{2}-\delta} e^{|y|}$$
for large $|z|$. Hence $f(z)$ is an integral function and we have only to prove that, for large $|z|$,
$$|f(z)| < A|z|^{\frac{1}{2}-\delta} e^{|y|}.$$
This comes from Lemma α (§3.12). From what has been proved above, $f'(z)$ satisfies the three hypotheses of the lemma. By integration, $f(z)$ satisfies the second and third, and it is given that it satisfies the first. Hence the inequality for $f(z)$ follows from the lemma.

p. 19, last line. This should be [see 1930, 5 (in this Vol.), footnote to p. 99]
$$|x-x_n| \leqq 2^{-2} + 2^{-3} + \cdots + 2^{-n-1} < \tfrac{1}{2}.$$

p. 20, line 4 up (statement of Lemma δ). α should be β.

p. 21, line 2 up. $\phi(u)$ should be $\phi(z)$.

p. 24, §3.26, lines 7–8. For the justification of the inversion of the order of integration, see 1930, 5 (in this Vol.), §3.2.

Correction of an error in a paper on certain integral equations

G. H. HARDY *and* E. C. TITCHMARSH.

We take this opportunity of correcting a mistake in our paper published in Vol. 23 of the *Proceedings*.† The mistake occurs on page 8, line 5, where we say: "in each case it is easily seen that the integral with respect to t is uniformly convergent over a finite range of values of x." We assert, in fact, that

$$\int_0^\infty \frac{\chi(t)}{t} \{f(x+t) - f(x-t)\}\, dt$$

---

† G. H. Hardy and E. C. Titchmarsh, "Solutions of some integral equations considered by Bateman, Kapteyn, Littlewood, and Milne", *Proc. London Math. Soc.* (2), 23 (1924), 1-26.

is uniformly convergent over any such interval. This is obvious enough (f being of integrable square) when $\chi(t)$ is $J_1(t)$, as in Kapteyn's equation, but anything but obvious when $\chi(t)$ is e^{-t}, as in Milne's equation. Our argument was written out originally with Kapteyn's equation in mind, and it is no doubt this which accounts for our having overlooked an obvious difficulty. But even in this case there is one point where our argument requires modification.

It seems necessary, if we are to correct the argument without introducing additional hypotheses concerning $f(x)$, to make use of the theory of "conjugate functions." It is known* that, if $f(x)$ is of integrable square, the function

$$g(x) = \frac{1}{\pi}\int_0^\infty \frac{f(x+t)-f(x-t)}{t}\,dt$$

exists (though not generally as a Lebesgue integral) for almost all x, and is itself of integrable square. Further, if $f(x)$ is (as, apart from a constant factor, it is in our argument) the Fourier cosine transform of $\lambda(x)$, then $-g(x)$ is the sine transform of the same function.

Suppose then that $\chi(t) = e^{-t}$, so that

$$f'(x) = \tfrac{1}{2}\int_0^\infty e^{-t}\frac{f(x+t)-f(x-t)}{t}\,dt,$$

the existence of this integral for all values of x being one of our data. Then

$$\tfrac{1}{2}\pi g(x) - f'(x) = \tfrac{1}{2}\int_0^\infty \psi(t)\frac{f(x+t)-f(x-t)}{t}\,dt,$$

where $\psi(t) = 1-e^{-t}$; and this integral is uniformly convergent over any finite interval in x. Hence we have

$$f(x+t)-f(x-t) = -2\frac{d}{dx}\int_0^\infty \frac{(1-\cos xu)\sin tu}{u}\lambda(u)\,du,$$

(1) $\quad f'(x) - \tfrac{1}{2}\pi g(x) = \int_0^\infty \frac{\psi(t)}{t}\left\{\frac{d}{dx}\int_0^\infty \frac{(1-\cos xu)\sin tu}{u}\lambda(u)\,du\right\}dt$

$$= \frac{d}{dx}\int_0^\infty \frac{\psi(t)}{t}\,dt \int_0^\infty \frac{(1-\cos xu)\sin tu}{u}\lambda(u)\,du.$$

---

* See M. Plessner, "Zur theorie der konjugierten trigonometrischen Reihen", Giessen, 1922; G. H. Hardy, "Notes on some points in the integral calculus", *Messenger of Math.*, 54 (1924), 20–27, 81–88; and other memoirs there referred to.

Now $f(x)$ and $-g(x)$ are, apart from the same constant factor in each case, the cosine and sine transform of $\lambda(u)$; so that

$$(2) \qquad -g(x) = \frac{d}{dx}\int_0^\infty \frac{1-\cos xu}{u} \lambda(u)\, du$$

$$= \frac{2}{\pi}\frac{d}{dx}\int_0^\infty \frac{1-\cos xu}{u}\lambda(u)\,du \int_0^\infty \frac{\sin tu}{t}\,dt$$

$$= \frac{2}{\pi}\frac{d}{dx}\int_0^\infty \frac{dt}{t}\int_0^\infty \frac{(1-\cos xu)\sin tu}{u}\lambda(u)\,du,$$

if we may invert the order of integration. The u-integral is uniformly (and absolutely) convergent over any finite interval $(0, T)$ in t, and*

$$\lim_{T\to\infty}\int_0^\infty \frac{1-\cos xu}{u}\lambda(u)\,du \int_T^\infty \frac{\sin tu}{t}\,dt = 0.$$

Hence the inversion is legitimate, and (2) is proved. From (1) and (2) we derive

$$(3) \qquad f'(x) = -\frac{d}{dx}\int_0^\infty \frac{e^{-t}}{t}\,dt \int_0^\infty \frac{(1-\cos xu)\sin tu}{u}\lambda(u)\,du.$$

Finally, we have to prove that this integral is absolutely convergent, so that the order of integration may be inverted. For this, we use the inequalities

$$1-\cos xu \leqslant 2, \qquad 1-\cos xu \leqslant \tfrac{1}{2}(xu)^2,$$

$$|\sin tu| \leqslant (tu)^a \quad (0 < a < \tfrac{1}{2}).$$

Our double integral may then be compared with the sum of the integrals

$$\tfrac{1}{2}x^2 \int_0^\infty t^{a-1}e^{-t}dt \int_0^1 u^{a+1}|\lambda(u)|\,du,$$

$$2\int_0^\infty t^{a-1}e^{-t}dt \int_1^\infty u^{a-1}|\lambda(u)|\,du,$$

each of which is convergent; and from this point the discussion proceeds as before.

This last argument applies equally when $\chi(t)$ is $J_1(t)$, or any of our other kernels. But, even when dealing with Kapteyn's equation, we must use $1-\cos xu$ in place of the $-\cos xu$ of our original argument.†

---

* By "bounded convergence" (*l.c.*, p. 10).
† So that, foot-note †, p. 8, is inaccurate.

SOLUTION OF AN INTEGRAL EQUATION

G. H. Hardy *and* E. C. Titchmarsh*.

1. The integral equation

(1.1) $$f(x) = \lambda \int_0^\infty e^{-xy} f(y)\, dy,$$

where λ is real, appears to have been considered first by Weyl†. It was discussed later (explicitly or implicitly) by Carleman‡ and Hyslop§ as an example of various points in the theory of singular integral equations. It is satisfied by

(1.2) $$f(x) = Ax^{-a} + Bx^{a-1},$$

provided that $A^2 \Gamma(1-a) = B^2 \Gamma(a)$, that

(1.3) $$\sin a\pi = \lambda^2 \pi,$$

and (since the integral must be convergent) that $0 < \Re(a) < 1$. These restrictions determine a as a function of λ. If $0 < \lambda^2 < 1/\pi$, a has two real values, of which we may select the one between 0 and $\tfrac{1}{2}$. If $\lambda^2 > 1/\pi$, a is complex and has the real part $\tfrac{1}{2}$. If $\lambda = \pi^{-\frac{1}{2}}$, the solution reduces to $Ax^{-\frac{1}{2}}$, with arbitrary A, and if $\lambda = -\pi^{-\frac{1}{2}}$ to $Ax^{-\frac{1}{2}} + Bx^{-\frac{1}{2}} \log x$, where $2A = -B\pi^{-\frac{1}{2}} \Gamma'(\tfrac{1}{2})$. There are also solutions for complex values of λ for which $\Re(\lambda^2) > 0$.

These solutions were given by the authors referred to, but none of them seems to have proved that they are the *only* solutions. Our object here is to supply this proof. Our first attempt at a proof depended on our method for the solution of Milne's astrophysical equation and others of similar types‖. Prof. Bateman then suggested to us that we might simplify the proof by arranging it as a reduction of the integral equa-

---

* Received 27 May, 1929; read 13 June, 1929.

† H. Weyl, "Singuläre Integralgleichungen", *Dissertation* (Göttingen, 1908). A revised version appeared in *Math. Annalen*, 66 (1909), 273-324, but without the passage referring to (1.1).

‡ T. Carleman, "Sur les équations intégrales singulières à noyau réel et symétrique" (Upsala, 1923). Carleman concerns himself explicitly only with the later equations (1.4) and (1.5).

§ J. Hyslop, "The integral expansions of arbitrary functions connected with integral equations", *Proc. Camb. Phil. Soc.*, 22 (1925), 169-185.

‖ G. H. Hardy and E. C. Titchmarsh "Solutions of certain integral equations considered by Bateman, Kapteyn, Littlewood, and Milne", *Proc. London Math. Soc.* (2), 23 (1923), 1-26.

tion to a difference equation. This we have been able to do, and at the same time to eliminate all restriction on the generality of the result.

We shall suppose that the integral in (1.1) exists as a "Cauchy integral", that is to say that it exists as a Lebesgue integral over any finite interval $0 < a \leqslant y \leqslant b$, and as

$$\lim_{\epsilon \to 0,\, Y \to \infty} \int_\epsilon^Y$$

over $(0, \infty)$. With this as our sole assumption, we shall prove that $f(x)$ is necessarily one of the functions indicated by Weyl.

There is another equation which we must consider at the same time as (1.1). If we multiply (1.1) by e^{-tx}, integrate over $(0, \infty)$, and assume that the order of the two integrations may be inverted, we find that $f(x)$ also satisfies

(1.4) $$f(x) = \lambda^2 \int_0^\infty \frac{f(y)}{y+x}\, dy.$$

This also has solutions of the Weyl type, but now with A and B independent. Finally if, with Carleman, we write $x = e^s$, $y = e^t$ and $e^{\frac{1}{2}s}f(e^s) = F(s)$, (1.4) takes the form

(1.5) $$F(s) = \lambda^2 \int_{-\infty}^\infty \frac{F(t)}{\cosh \frac{1}{2}(t-s)}\, dt,$$

and the "Weyl" solutions are now of the forms $Ae^{as} + Be^{-as}$ or $A + Bs$.

2. We begin by proving the following lemma: *if* (1.1) *is true, then* (1.4) *is true, the integrals in each case being Cauchy integrals, for all positive values of* x.

If $f(x)$ is complex, its real and imaginary parts are solutions of (1.1). We may therefore argue as if $f(x)$ were real. Since the integral in (1.1) is convergent for all positive x, it is uniformly convergent over any finite interval $0 < \xi \leqslant x < X$*.

Next, $f(x) \to 0$ when $x \to \infty$. For, if $x > 1$, we have, by the second mean value theorem,

$$\int_0^\delta e^{-xy} f(y)\, dy = \int_0^\epsilon f(y)\, dy, \quad \int_\delta^\infty e^{-xy} f(y)\, dy = e^{-(x-1)\delta} \int_\epsilon^Y e^{-y} f(y)\, dy,$$

where $0 < \epsilon < \delta < Y$. We can choose δ so as to make the first integral

---

* By familiar arguments based on the second mean value theorem.

arbitrarily small for all x, and the second tends to zero when δ is fixed and $x \to \infty$.

We have now, by uniform convergence,

$$\int_\xi^X f(x) e^{-tx} dx = \lambda \int_0^\infty f(y) dy \int_\xi^X e^{-(y+t)x} dx$$

$$= \lambda \int_0^\infty \frac{f(y)}{y+t} e^{-\xi(y+t)} dy - \lambda \int_0^\infty \frac{f(y)}{y+t} e^{-X(y+t)} dy.$$

The integral involving X plainly tends to zero when $X \to \infty$, so that

$$\int_\xi^\infty f(x) e^{-tx} dx = \lambda e^{-\xi t} \int_0^\infty \frac{f(y)}{y+t} e^{-\xi y} dy.$$

When $\xi \to 0$, the left-hand side tends to $f(t)/\lambda$, so that the right-hand side also tends to a limit. Since $f(y) = o(1)$ when $y \to \infty$, the integrand is $o(1/y)$. Hence we may apply the integral form of Tauber's theorem; and we obtain

$$f(t) = \lambda^2 \lim_{\xi \to 0} \int_0^\infty \frac{f(y)}{y+t} e^{-\xi y} dy = \lambda^2 \int_0^\infty \frac{f(y)}{y+t} dy.$$

3. This proves the lemma, and we may therefore confine our attention to (1.4). Any solution $f(x)$ of (1.4), whether arising from a solution of (1.1) or not, satisfies

(3.1) $\qquad\qquad\qquad f(x) = o(1) \quad (x \to \infty).$

For we may suppose as before that $f(x)$ is real, when the second mean value theorem gives

$$\int_0^{\sqrt{x}} \frac{f(y)}{y+x} dy = \frac{\sqrt{x}+1}{\sqrt{x}+x} \int_\xi^{\sqrt{x}} \frac{f(y)}{y+1} dy, \quad \int_{\sqrt{x}}^\infty \frac{f(y)}{y+x} dy = \int_{\xi'}^\infty \frac{f(y)}{y+1} dy,$$

where $0 < \xi < \sqrt{x} < \xi'$, and these equations lead at once to (3.1). We can also verify at once, by putting $y = 1/z$, that $x^{-1} f(x^{-1})$ is also a solution of (1.4). It then follows from (3.1) that

(3.2) $\qquad\qquad\qquad f(x) = o(x^{-1}) \quad (x \to 0).$

4. We observe next that the integral in (1.4) is uniformly convergent in any bounded domain of x which does not cross the negative real axis. This is easily seen, for example, by writing

$$\int_0^\infty \frac{f(y)}{y+x} dy = \frac{x-1}{x} \int_0^k \frac{y f(y)}{(y+x)(y+1)} dy + \frac{1}{x} \int_0^k \frac{f(y)}{y+1} dy$$

$$- (x-1) \int_k^\infty \frac{f(y)}{(y+x)(y+1)} dy + \int_k^\infty \frac{f(y)}{y+1} dy,$$

and using (3.1) and (3.2). It follows that $f(x)$ *is an analytic function of x regular except perhaps for* $-\infty < x \leqslant 0$.

Supposing now that $x = re^{i\theta}$, where $-\tfrac{1}{2}\pi \leqslant \theta \leqslant \tfrac{1}{2}\pi$, so that
$$|y+x| \geqslant \mathrm{Max}\,(y, r),$$
and putting $k = r$, we see that

(4.1) $\quad f(re^{i\theta}) = O\left\{\dfrac{1}{r}\displaystyle\int_0^r \dfrac{y\,|f(y)|}{y+1}\,dy + r\displaystyle\int_r^\infty \dfrac{|f(y)|}{y^2}\,dy\right\} + o(1) = o(1),$

when $r \to \infty$. Hence also

(4.2) $\qquad\qquad\qquad f(re^{i\theta}) = o(r^{-1}) \quad (r \to 0),$

both (4.1) and (4.2) holding uniformly for $|\theta| \leqslant \tfrac{1}{2}\pi$. These equations show that, so long as x does not lie in the third quadrant, we may rotate the line of integration to the imaginary axis, and obtain

(4.3) $\qquad\qquad\qquad f(x) = \lambda^2 \displaystyle\int_0^\infty \dfrac{if(i\eta)}{i\eta+x}\,d\eta.$

In particular, taking $x = i\xi$, where $\xi > 0$, we have

(4.4) $\qquad\qquad\qquad f(i\xi) = \lambda^2 \displaystyle\int_0^\infty \dfrac{f(i\eta)}{\eta+\xi}\,d\eta.$

In other words, $f(ix)$ *is also a solution of* (1.5).

The argument may now be repeated, and we see that $f(x)$ may be continued any number of times in either direction round the origin. Thus $x = 0$ *is the only finite singularity of $f(x)$, and* (4.1) *and* (4.2) *hold uniformly throughout any finite range of* θ.

5. If we put $x = re^{i\theta}$, where $\tfrac{1}{2}\pi < \theta < \pi$, in (4.3), we obtain

(5.1) $\qquad\qquad\qquad f(re^{i\theta}) = \lambda^2 \displaystyle\int_0^\infty \dfrac{if(i\eta)}{i\eta+re^{i\theta}}\,d\eta.$

If $x = re^{-i\theta}$, x lies in the third quadrant. We have then to allow for the fact that the axis of integration has been rotated across the pole at $y = -x$, so that (5.1) must be replaced by

(5.2) $\qquad f(re^{-i\theta}) = 2\pi i\lambda^2 f(re^{i\pi-i\theta}) + \lambda^2 \displaystyle\int_0^\infty \dfrac{if(i\eta)}{i\eta+re^{-i\theta}}\,d\eta.$

Comparing (5.1) and (5.2), and making $\theta \to \pi$, we obtain

(5.3) $\qquad\qquad f(re^{i\pi}) + 2\pi i\lambda^2 f(r) - f(re^{-i\pi}) = 0.$

If we replace r by $re^{i\theta}$, (5.3) is a difference equation in θ, valid for every r. We can then complete the solution by appealing to the general theory of difference equations. More directly, we may argue as follows. We can write (5.3) in the form

(5.4) $$f(re^{i\pi})-(p+q)f(r)+pqf(re^{-i\pi})=0,$$

where p and q are $-\pi i \lambda^2 \pm \sqrt{(1-\pi^2\lambda^4)}$. If we write

$$F(z)=f(z)-pf(ze^{-i\pi}), \quad q=e^{-i\pi a}, \quad \phi(z)=z^a F(z),$$

we obtain
$$\phi(re^{i\pi})=\phi(r),$$

for all r. Hence $\phi(z)$ is a one-valued (and in fact even) function of z, and

(5.51) $$f(z)-pf(ze^{i\pi})=z^{-a}\phi(z).$$

Similarly

(5.52) $$f(z)-qf(ze^{i\pi})=z^{-b}\psi(z),$$

where $p=e^{-\pi ib}$ and $\psi(z)$ is also one-valued (and even). Hence

(5.6) $$(q-p)f(z)=qz^{-a}\phi(z)-pz^{-b}\psi(z).$$

The equation (5.6) is, of course, significant only if $q\neq p$, i.e., if $\lambda^2\neq 1/\pi$. If $\lambda^2=1/\pi$, we write $f(z)=z^{-\frac{1}{2}}g(z)$, when (5.3) gives

$$g(re^{i\pi})-g(r)=g(r)-g(re^{-i\pi}).$$

It follows that
$$g(ze^{i\pi})-g(z)=i\pi\chi(z)$$

is one-valued (and even). If

$$\omega(z)=g(z)-\chi(z)\log z,$$

$$\omega(ze^{i\pi})-\omega(z)=i\pi\chi(z)-\chi(z)(\log z+i\pi)+\chi(z)\log z=0,$$

and $\omega(z)$ is one-valued (and even). Hence finally

(5.7) $$f(z)=z^{-\frac{1}{2}}\{\chi(z)\log z+\omega(z)\},$$

where $\chi(z)$ and $\omega(z)$ are one-valued.

To complete the solution we must show that the functions ϕ, ψ, χ, ω in (5.6) and (5.7) are, in fact, *constants*. It follows from (4.1), (4.2), and (5.51) that $\phi(z)=o(|z|^{\Re(a)})$ when $|z|\to\infty$ and $\phi(z)=o(|z|^{\Re(a)-1})$ when $|z|\to 0$. Since $0<\Re(a)<1$, and $\phi(z)$ is one-valued, $\phi(z)$ is constant. Similarly the other functions are constant.

To sum up, we have proved the following theorem :—*The only solutions of the equations* (1.1), (1.4), *and* (1.5) *are the "Weyl" solutions. Our only assumption is that the integrals have a meaning in the Cauchy sense and that the equations are true for all real* x.

CORRECTION

p. 301, equation (1.5). λ^2 should be $\tfrac{1}{2}\lambda^2$.

ADDITIONAL NOTE ON CERTAIN INTEGRAL EQUATIONS

By G. H. Hardy and E. C. Titchmarsh.

[Received 5 August, 1928.—Read 14 June, 1928.]

1. In a paper published in the *Proceedings* in 1924* we discussed the solution of the equations

(B) $$f(x) = \frac{1}{\pi} \int_0^\infty \frac{\sin t}{t} \{f(x+t) + f(x-t)\} \, dt,$$

(K) $$f'(x) = \tfrac{1}{2} \int_0^\infty \frac{J_1(t)}{t} \{f(x+t) - f(x-t)\} \, dt,$$

(M) $$f'(x) = \tfrac{1}{2} \int_0^\infty \frac{e^{-t}}{t} \{f(x+t) - f(x-t)\} \, dt,$$

of Bateman, Kapteyn, and Milne, using two different methods, a "real" method applicable only when $f(x)$ belongs to the class L^2†, and a "complex" method of wider scope. There was a mistake in our treatment of (M) by the first method, which we corrected in a later note‡.

We return to the subject here for two reasons. In the first place, we can simplify our "real" method by adopting a suggestion of Dr. Norbert Wiener, and we take the opportunity of simplifying the proof and incorporating our own correction simultaneously. This part of the paper (§ 2) is then complete in itself. Secondly, there is a passage later in our first paper, in the proof of Theorem 7, to which we have not referred before, where the same difficulty which necessitated our correction recurs again in a different form. We deal with this point in two different ways.

---

* G. H. Hardy and E. C. Titchmarsh, "Solutions of some integral equations considered by Bateman, Kapteyn, Littlewood, and Milne", *Proc. London Math. Soc.* (2), 23 (1924), 1-26. We refer to this paper as BKLM.

† f is measurable and f^2 integrable over $(-\infty, \infty)$.

‡ G. H. Hardy and E. C. Titchmarsh, "Correction of an error in a paper on certain integral equations", *Proc. London Math. Soc.* (2), 24 (1925), xxxi-xxxiii (*Records* for February, 1925).

In § 3 we show how to complete our original proof. Secondly, in § 4, we give a new, and in some respects simpler, method of dealing with this part of the proof of Theorem 7. This section is alternative to §§ 3.21 and 3.26 of BKLM and § 3 of the present paper. The "complex" argument of BKLM remains unchanged.

The case where $f(x)$ belongs to L^2.

2.1. THEOREM 1. *The necessary and sufficient condition that a function $f(x)$ of L^2 should satisfy* (B) *or* (K) *is that it should be of the form*
$$f(x) = \int_0^1 \{\lambda(u) \cos xu + \mu(u) \sin xu\} \, du,$$
where $\lambda(u)$ and $\mu(u)$ belong to L^2.

THEOREM 2. *The only solution of* (M) *belonging to L^2 is $f(x) = 0$.*

If $f(x)$ is a solution, so is $\tfrac{1}{2}\{f(x) \pm f(-x)\}$, so that any solution is the sum of an even and an odd solution. The argument is substantially the same in these two cases, and we may suppose $f(x)$ to be *even*.

2.2. *Solution of* (B). It follows from the theory of Fourier transforms* that there is a function $\lambda(u)$ of L^2 such that

(2.21) $$\int_0^a \lambda(u) \cos xu \, du \to f(x)$$

in mean square when $a \to \infty$ †. Since $t^{-1} \sin t$ belongs to L^2, we have
$$f(x) = \frac{1}{\pi} \int_0^\infty \frac{\sin t}{t} \{f(x+t) + f(x-t)\} \, dt$$
$$= \lim_{a \to \infty} \frac{2}{\pi} \int_0^\infty \frac{\sin t}{t} \, dt \int_0^a \lambda(u) \cos xu \cos tu \, du$$
$$= \lim_{a \to \infty} \frac{2}{\pi} \int_0^a \lambda(u) \cos xu \, du \int_0^\infty \frac{\sin t \cos ut}{t} \, dt,$$
the inversion of the order of integration being justified by "bounded

---

* See, for example, E. W. Hobson, *Theory of functions of a real variable*, 2 (ed. 2, 1926), 742–752.

† In BKLM we used the representation
$$f(x) = \frac{d}{dx} \int_0^\infty \lambda(u) \frac{\sin xu}{u} \, du.$$
Wiener's suggestion was that we should use (2.21) instead.

convergence"*. The inner integral is $\tfrac{1}{2}\pi$ if $u<1$ and 0 if $u>1$, so that
$$f(x) = \int_0^1 \lambda(u) \cos xu \, du.$$

2.3. *Solution of* (K). Using (2.21) again, we obtain
$$f'(x) = \tfrac{1}{2} \int_0^\infty \frac{J_1(t)}{t} \{f(x+t) - f(x-t)\} \, dt$$
$$= -\lim_{a \to \infty} \int_0^\infty \frac{J_1(t)}{t} dt \int_0^a \lambda(u) \sin xu \sin tu \, du$$
$$= -\lim_{a \to \infty} \int_0^a \lambda(u) \sin xu \, du \int_0^\infty \frac{J_1(t)}{t} \sin ut \, dt,$$
the inversion here being justified by absolute convergence. This gives
$$f'(x) = -\int_0^1 u \lambda(u) \sin xu \, du - \int_1^\infty \frac{\lambda(u) \sin xu}{u + \sqrt{(u^2 - 1)}} \, du\dagger.$$

Now $f(t)$ and $t^{-1} J_1(t)$ belong to L^2, and so the integrals
$$\int_0^\infty \frac{J_1(t)}{t} f(x \pm t) \, dt$$
are continuous‡. Hence f' is continuous, and its integral is f. Integrating, we obtain
$$f(x) = \int_0^1 \lambda(u) \cos xu \, du + \int_1^\infty \frac{\lambda(u) \cos xu}{u \{u + \sqrt{(u^2 - 1)}\}} \, du$$

(apart from a constant which must obviously be 0). Since the Fourier transform of $f(x)$ is $\lambda(u)$, $\lambda(u)$ must be null for $u > 1$, and we obtain the result.

2.4. *Solution of* (M). Since $f(x)$ belongs to L^2,
$$g(x) = \frac{1}{\pi} \int_0^\infty \{f(x+t) - f(x-t)\} \frac{dt}{t}$$
exists for almost all x and belong to L^2§.

---

* See BKLM, 10.

† See G. N. Watson, *Theory of Bessel functions*, 405.

‡ See E. W. Hobson, *loc. cit.*, 1 (ed. 3, 1927), 639.

§ f and $-g$ are "conjugate". For the theory of conjugate functions, see M. Riesz: (1) "Les fonctions conjuguées et les séries de Fourier", *Comptes rendus*, 178, 1464 (28 April 1924); (2) "Sur les fonctions conjuguées, *Math. Zeitschrift*, 27 (1927), 218–244; (3) "Sur les maxima des formes bilinéaires et sur les fonctionelles linéaires", *Acta Math.*, 49 (1927), 465–497: E. C. Titchmarsh, "Reciprocal formulae involving series and integrals", *Math. Zeitschrift*, 25 (1926), 321–347: G. H. Hardy, "On Hilbert transforms", *Messenger of Math.*, 54 (1925), 20–27 and 81-18; and other memoirs there referred to.

Hence
$$f'(x) - \tfrac{1}{2}\pi g(x) = \tfrac{1}{2}\int_0^\infty \frac{e^{-t}-1}{t}\{f(x+t)-f(x-t)\}\,dt$$
$$= \lim_{a\to\infty}\int_0^\infty \frac{1-e^{-t}}{t}\,dt\int_0^a \lambda(u)\sin xu \sin tu\,du$$
$$= \lim_{a\to\infty}\int_0^a \lambda(u)\sin xu\,du\int_0^\infty \frac{1-e^{-t}}{t}\sin ut\,dt$$
$$= \int_0^\infty \lambda(u)(\tfrac{1}{2}\pi - \arctan u)\sin xu\,du,$$

the inversion being justified by bounded convergence. But
$$\int_0^a \lambda(u)\sin xu\,du \to -g(x)$$
in mean, and so
$$-\int_0^a \lambda(u)\arctan u \sin xu\,du \to f'(x)$$
in mean. Now $f' - \tfrac{1}{2}\pi g$ is continuous (as in § 2.3), and g belongs to L^2, so that f' belongs to L^2. Also f' is finite everywhere, by hypothesis. It follows* that f is the integral of f'. Integrating, we obtain
$$\int_0^\infty \lambda(u)\arctan u\,\frac{1-\cos xu}{u}\,du = f(0)-f(x).$$

From the last equation, and the fact that $f(x)$ is the transform of $\lambda(u)$, it follows that
$$\int_0^\infty \lambda(u)\left(1-\frac{\arctan u}{u}\right)\cos xu\,du = f(0) - \int_0^\infty \lambda(u)\frac{\arctan u}{u}\,du,$$

where the integral on the left-hand side is a mean square integral and that on the right an ordinary integral. Since the left-hand side belongs to L^2, and the right-hand side is independent of x, both must vanish and
$$\lambda(u)\left(1-\frac{\arctan u}{u}\right)$$

must be a null function. It follows that $\lambda(u)$ is null and $f(x)$ identically zero.

---

* See E. W. Hobson, *loc. cit.*, 1 (ed. 3, 1927), 600.

2.5. We may add that the solutions which precede may be extended to the case in which $f(x)$ belongs to any Lebesgue class L^p, where $1 < p \leqslant 2$. The arguments are the same, but the theorems concerning transforms and conjugate functions on which they rest are naturally more difficult*. In this case the $\lambda(u)$ and $\mu(u)$ of the conclusion belong to $L^{p'}$, where $p' = p/(p-1)$.

The general case of Milne's equation.

3.1. We have to prove

THEOREM 7†. *If $f(x)$ satisfies* (M), *and*

(3.11) $$f(x) = O(e^{A|x|}),$$

where $0 < A < 1$, then $f(x)$ is a quadratic.

We proved this first on the assumption that $f(x)$ is indefinitely differentiable, and so far our proof requires no change‡. We then removed this restriction by an artifice, which demands a justification of the inversion of the order of integration in the repeated integral

$$\int_0^\infty \frac{e^{-t}}{t} dt \int_{-\infty}^\infty \{f(x+u+t) - f(x+u-t)\} \operatorname{sech} u \, du.$$

This inversion is not at all trivial; the point is similar to that which arose before, but there are additional complications.

3.2. There is no real loss of generality in taking x here to be 0, in which case what we have to prove is that

(3.21) $$\int_0^\infty \frac{e^{-t}}{t} dt \int_{-\infty}^\infty \{f(u+t) - f(u-t)\} \operatorname{sech} u \, du$$
$$= \int_{-\infty}^\infty \operatorname{sech} u \, du \int_0^\infty \frac{e^{-t}}{t} \{f(u+t) - f(u-t)\} dt$$

or

(3.22) $$\int_{-\infty}^\infty \frac{e^{-|t|}}{t} dt \int_{-\infty}^\infty \frac{f(u+t)}{\cosh u} du = \int_{-\infty}^\infty \frac{du}{\cosh u} \int_{-\infty}^\infty \frac{e^{-|t|}}{t} f(u+t) dt,$$

---

* See E. C. Titchmarsh, "A contribution to the theory of Fourier transforms", *Proc. London Math. Soc.* (2), 23 (1924), 279-289, and the works by Riesz, Titchmarsh, and Hobson already referred to.

† We follow the numeration of BKLM.

‡ We correct two misprints: in the enunciation of Lemma δ, a should be β, and the last line on p. 19 should be
$$|x - x_n| \leqslant 2^{-2} + 2^{-3} + \ldots + 2^{-n-1} < \tfrac{1}{2}.$$

where the integrals in t are principal values both at the origin and at infinity, *i.e.* are defined as
$$\lim_{\epsilon \to 0,\, T \to \infty} \left(\int_{-T}^{-\epsilon} + \int_{\epsilon}^{T} \right).$$

We write $$f(x) = e^{a|x|} g(x),$$
where $0 < a < 1$ and $g(x)$ belongs to L^2; this is possible after (3.11). If we do this we obtain

(3.23) $\displaystyle \int_{-\infty}^{\infty} \frac{dt}{t} \int_{-\infty}^{\infty} e^{-|t|+a|u+t|-a|u|} \frac{e^{a|u|}}{\cosh u} g(u+t)\, du$

$\displaystyle = \int_{-\infty}^{\infty} \frac{e^{a|u|}}{\cosh u} du \int_{-\infty}^{\infty} e^{-|t|+a|u+t|-a|u|} g(u+t)\, \frac{dt}{t}.$

Now it is easily verified that
$$0 \leqslant w(t, u) = 1 - e^{-|t|+a|u+t|-a|u|} \leqslant 1 - e^{-(1+a)|t|}.$$

It follows that the integral
$$\iint \frac{w(t, u)}{t} \frac{e^{a|u|}}{\cosh u} g(u+t)\, dt\, du$$
is absolutely convergent; for we may compare it with
$$\int_{-\infty}^{\infty} \frac{e^{a|u|}}{\cosh u} \int_{-\infty}^{\infty} \frac{1-e^{-(1+a)|t|}}{t} |g(u+t)|\, dt,$$
and the integral in t converges uniformly in u. Hence

(3.24) $\displaystyle \int_{-\infty}^{\infty} \frac{dt}{t} \int_{-\infty}^{\infty} w(t, u) \frac{e^{a|u|}}{\cosh u} g(u+t)\, du$

$\displaystyle = \int_{-\infty}^{\infty} \frac{e^{a|u|}}{\cosh u} du \int_{-\infty}^{\infty} w(t, u) g(u+t)\, \frac{dt}{t}.$

Comparing (3.23) and (3.24), we see that (3.23) will follow from

(3.25) $\displaystyle \int_{-\infty}^{\infty} \frac{dt}{t} \int_{-\infty}^{\infty} \frac{e^{a|u|}}{\cosh u} g(u+t)\, du = \int_{-\infty}^{\infty} \frac{e^{a|u|}}{\cosh u} du \int_{-\infty}^{\infty} g(u+t)\, \frac{dt}{t}.$

3.3. The proof of (3.25) depends upon the theory of conjugate functions. We state the result required in a more general form than we really need; but it has a certain interest of its own.

LEMMA a. *Suppose that $g(u)$ belongs to L^2 and that $-h(u)$ is its conjugate, and that $k(u)$ also belongs to L^2. Then the functions*

(3.31) $\displaystyle G(t) = \int_{-\infty}^{\infty} k(u) g(u+t)\, du, \quad H(t) = \int_{-\infty}^{\infty} k(u) h(u+t)\, du,$

are continuous, and

(3.32) $$\frac{1}{\pi}\int_{-\infty}^{\infty}\frac{G(t)}{t-x}dt = H(x)$$

for every x (so that G and $-H$ are also conjugate).

The continuity of G and H is a known result, which has been used already in § 2. Since all values of x are on the same footing, we may take $x = 0$. We have then to prove that

(3.33) $$\left(\int_{-T}^{-\epsilon}+\int_{\epsilon}^{T}\right)\frac{G(t)}{t}dt \to \pi\int_{-\infty}^{\infty}k(u)\,h(u)\,du,$$

when $\epsilon \to 0$ and $T \to \infty$.

Substituting from (3.31) for $G(t)$, and inverting the order of integration (by absolute convergence), we see that the left-hand side of (3.33) is equal to

$$\int_{-\infty}^{\infty}k(u)\,du\left(\int_{-T}^{-\epsilon}+\int_{\epsilon}^{T}\right)\frac{g(u+t)}{t}dt = \pi\int_{-\infty}^{\infty}k(u)\{h_\epsilon(u)-h_T(u)\}\,du,$$

where

(3.34) $$\pi h_\epsilon(u) = \left(\int_{-\infty}^{-\epsilon}+\int_{\epsilon}^{\infty}\right)\frac{g(u+t)}{t}dt.$$

The result of the lemma will follow if we can show that $h_\epsilon(u)$ tends to $h(u)$ in mean square, and $h_T(u)$ to zero, *i.e.* that

(3.35) $$\int_{-\infty}^{\infty}\{h_\epsilon(x)-h(x)\}^2\,dx \to 0$$

when $\epsilon \to 0$, and

(3.36) $$\int_{-\infty}^{\infty}\{h_T(x)\}^2\,dx \to 0$$

when $T \to \infty$.

3.4. To prove this, we use the formula*

(3.41) $$h_\epsilon(x) = \frac{1}{\pi^2}\int_{-\infty}^{\infty}h(y+x)\log\left|\frac{y+\epsilon}{y-\epsilon}\right|\frac{dy}{y}.$$

---

* This formula, which is very important in the theory of conjugate functions, is a corollary of "Parseval's theorem for conjugate functions", the theorem that, if g and $-h$, G and $-H$, are conjugate (and all belong to L^2), then

(P) $$\int_{-\infty}^{\infty}g(y)\,G(y)\,dy = \int_{-\infty}^{\infty}h(y)\,H(y)\,dy.$$

If we take $G(y)$ to be $(y-x)^{-1}$ when $|y-x| > \epsilon$ and zero otherwise, we find by an elementary calculation that

$$H(y) = \frac{1}{\pi(y-x)}\log\left|\frac{y-x+\epsilon}{y-x-\epsilon}\right|.$$

These values of G and H, substituted in (P), give (3.37).

For (P) see, for example, Titchmarsh, *loc. cit.*, 322, formula (5b); Hardy, *loc. cit.*, 84, Theorem B. Riesz does not state the formula explicitly, though it is naturally implied in his work.

The value of the right-hand side of (3.41) when $h(x) = 1$ is 1. Hence

$$h_\epsilon(x) - h(x) = \frac{1}{\pi^2} \int_{-\infty}^{\infty} \{h(x+y) - h(x)\} \log\left|\frac{y+\epsilon}{y-\epsilon}\right| \frac{dy}{y},$$

$$\int_{-\infty}^{\infty} \{h_\epsilon(x) - h(x)\}^2 dx$$

$$\leqslant \int_{-\infty}^{\infty} dx \left[\frac{1}{\pi^2} \int_{-\infty}^{\infty} \{h(x+y)-h(x)\}^2 \log\left|\frac{y+\epsilon}{y-\epsilon}\right| \frac{dy}{y} \cdot \frac{1}{\pi^2} \int_{-\infty}^{\infty} \log\left|\frac{y+\epsilon}{y-\epsilon}\right| \frac{dy}{y}\right]$$

$$= \frac{1}{\pi^2} \int_{-\infty}^{\infty} \log\left|\frac{y+\epsilon}{y-\epsilon}\right| \frac{dy}{y} \int_{-\infty}^{\infty} \{h(x+y) - h(x)\}^2 dx.$$

The x integral is bounded in y and tends to zero with y, and (3.35) follows.

To prove (3.36), let $h(x) = \phi(x) + \psi(x)$, where $\phi(x)$ is equal to $h(x)$ in $(-X, X)$ and to zero outside, and $\psi(x)$ conversely; and let ϕ_T, ψ_T be derived from ϕ, ψ as h_T is from h. Then

$$\int_{-\infty}^{\infty} h_T^2 dx \leqslant 2 \int_{-\infty}^{\infty} \phi_T^2 dx + 2 \int_{-\infty}^{\infty} \psi_T^2 dx.$$

The argument used above shows that

$$\int_{-\infty}^{\infty} \psi_T^2 dx \leqslant \frac{1}{\pi^2} \int_{-\infty}^{\infty} \log\left|\frac{y+T}{y-T}\right| \frac{dy}{y} \int_{-\infty}^{\infty} \{\psi(x+y)\}^2 dx$$

$$= \int_{-\infty}^{\infty} \{\psi(x)\}^2 dx = \left(\int_{-\infty}^{-X} + \int_{X}^{\infty}\right) \{h(x)\}^2 dx < \eta,$$

for any $\eta > 0$, if X is sufficiently large. Also

$$\int_{-\infty}^{\infty} \phi_T^2 dx = \int_{-\infty}^{\infty} dx \left(\frac{1}{\pi^2} \int_{-\infty}^{\infty} \phi(u) \log\left|\frac{u-x+T}{u-x-T}\right| \frac{du}{u-x}\right)^2$$

$$\leqslant A \int_{-\infty}^{\infty} dx \int_{-X}^{X} \left(\frac{1}{u-x} \log\left|\frac{u-x+T}{u-x-T}\right|\right)^2 du$$

$$= A \int_{-X}^{X} du \int_{-\infty}^{\infty} \left(\frac{1}{u-x} \log\left|\frac{u-x+T}{u-x-T}\right|\right)^2 du$$

$$= \frac{A}{T} \int_{-X}^{X} du \int_{-\infty}^{\infty} \left(\frac{1}{w} \log\left|\frac{w+1}{w-1}\right|\right)^2 dw < \frac{AX}{T},$$

where the A's are independent of T and X. We can make this also less than η, when X has been fixed, by choice of T. This completes the proof of (3.36) and so of the lemma.

The lemma may be extended to show that if g (and so h) belong to L^{1+p}, and k to L^{1+q}, where $p > 0$, $q > 0$, $pq \leqslant 1$, so that (in accordance with a theorem of Young*) G and H belong to L^ϖ, where

$$\varpi = \frac{(1+p)(1+q)}{1-pq}$$

if $pq < 1$, and are continuous if $pq = 1$, then G and H are conjugate, the formula expressing either in terms of the other holding almost always when $pq < 1$ and always when $pq = 1$.

3.4. To deduce (3.25) from the lemma, we take the $g(u)$ of the lemma to be the $g(u)$ of (3.25), and $k(u) = e^{a|u|}\operatorname{sech} u$. Then the left-hand side of (3.25) is

$$\int_{-\infty}^{\infty} \frac{dt}{t} \int_{-\infty}^{\infty} k(u)\, g(u+t)\, du = \int_{-\infty}^{\infty} \frac{G(t)}{t}\, dt,$$

and the right-hand side is

$$\pi \int_{-\infty}^{\infty} k(u)\, h(u)\, du = \pi H(0),$$

so that (3.25) becomes (3.32) with $x = 0$. This completes our original proof of Theorem 7.

4. *New proof of Theorem 7.* This also depends on the theory of conjugate functions. We use the following result, which does not seem to have been stated anywhere exactly in the form we require.

4.1. LEMMA β. *If $f(x)$ belongs to L^2, and*

$$g(x) = \int_{x-1}^{x+1} \frac{f(t)}{t-x}\, dt,$$

then there is an absolute constant K such that

(4.11) $$\int_{\xi-1}^{\xi+1} \{g(x)\}^2\, dx \leqslant K^2 \int_{\xi-2}^{\xi+2} \{f(t)\}^2\, dt.$$

Let
$$g_1(x) = \int_{\xi-2}^{\xi+2} \frac{f(t)}{t-x}\, dt.$$

Then

(4.12) $$\int_{\xi-1}^{\xi+1} \{g_1(x)\}^2\, dx \leqslant \int_{-\infty}^{\infty} \{g_1(x)\}^2\, dx = \pi^2 \int_{\xi-2}^{\xi+2} \{f(t)\}^2\, dt.$$

---

* W. H. Young, "On a class of parametric integrals", *Proc. Royal Soc.* (A), 85 (1911), 401–414, and "Sur la généralisation du théorème de Parseval", *Comptes rendus*, 155, 30 (1 July 1912).

Also for $\xi-1 < x < \xi+1$

$$|g_1(x)-g(x)| \leq \left|\int_{x+1}^{\xi+2} \frac{f(t)}{t-x} dt\right| + \left|\int_{\xi-2}^{x-1} \frac{f(t)}{t-x} dt\right|$$

$$\leq \left\{\int_{x+1}^{\xi+2} \frac{dt}{(t-x)^2} \int_{x+1}^{\xi+2} [f(t)]^2 dt\right\}^{\frac{1}{2}} + \left\{\int_{\xi-2}^{x-1} \frac{dt}{(t-x)^2} \int_{\xi-2}^{x-1} [f(t)]^2 dt\right\}^{\frac{1}{2}}$$

$$\leq 2 \left\{\int_{\xi-2}^{\xi+2} [f(t)]^2 dt\right\}^{\frac{1}{2}}.$$

Hence

(4.13) $$\int_{\xi-1}^{\xi+1} [g_1(x)-g(x)]^2 dx \leq 4 \int_{\xi-2}^{\xi+2} [f(t)]^2 dt,$$

and the result follows from (4.12) and (4.13).

4.2. We now begin the proof of Theorem 7 again at the beginning. The equation (M) gives

$$f'(x) = \tfrac{1}{2} \int_0^1 \frac{f(x+t)-f(x-t)}{t} dt$$

$$+ \tfrac{1}{2} \left[\int_0^1 \frac{e^{-t}-1}{t} \{f(x+t)-f(x-t)\} dt + \int_1^\infty \frac{e^{-t}}{t} \{f(x+t)-f(x-t)\} dt\right]$$

$$= \phi(x)+\psi(x),$$

say. Now $\psi(x)$ is bounded in a finite interval; more precisely

(4.21) $$|\psi(x)| \leq \tfrac{1}{2} \int_0^1 \{|f(x+t)|+|f(x-t)|\} dt$$

$$+ \tfrac{1}{2} \int_1^\infty e^{-t} \{|f(x+t)|+|f(x-t)|\} dt$$

$$< Ce^{A(|x|+1)} + \frac{Ce^{A|x|}}{1-A} < \frac{2Ce^{A(|x|+1)}}{1-A}.$$

Hence also

(4.22) $$\int_{x-1}^{x+1} \{\psi(u)\}^2 du < \frac{8C^2 e^{2A(|x|+2)}}{(1-A)^2}.$$

Also $f(x)$ belongs to L^2 over any finite interval, and so, by Lemma β,

(4.23) $$\int_{x-1}^{x+1} \{\phi(u)\}^2 du \leq \tfrac{1}{4} K^2 \int_{x-2}^{x+2} \{f(t)\}^2 dt < K^2 C^2 e^{2A(|x|+2)}.$$

We conclude that $f'(x)$ belongs to L^2 over any finite interval, and therefore that $f(x)$ satisfies a Lipschitz condition of order $\tfrac{1}{2}$; more precisely, for $|t|<1$,

$$|f(x+t)-f(x-t)| = \left|\int_{x-t}^{x+t} f'(u)\, du\right| \leq \left[\int_{x-t}^{x+t} du \int_{x-1}^{x+1} \{f'(u)\}^2\, du\right]^{\frac{1}{2}}$$

$$\leq \left[4t\int_{x-1}^{x+1} \{\phi(u)\}^2\, du\right]^{\frac{1}{2}} + \left[4t\int_{x-1}^{x+1} \{\psi(u)\}^2\, du\right]^{\frac{1}{2}}$$

(4.24)
$$< A_1 C e^{A|x|} \sqrt{t},$$

by (4.22) and (4.23), the constant A_1 depending on A only.

It now follows that $f'(x)$ is bounded in a finite interval; more precisely, by (4.24),

$$|\phi(x)| < \tfrac{1}{2}\int_0^1 A_1 C e^{A|x|} t^{-\frac{1}{2}}\, dt = A_1 C e^{A|x|},$$

and this and (4.21) give

(4.25)
$$|f'(x)| \leq |\phi(x)| + |\psi(x)| < A_2 C e^{A|x|},$$

say. Thus $f'(x)$ satisfies an inequality similar to that satisfied by $f(x)$.

4.3. We prove secondly that $f'(x)$ also is a solution of (M), *i.e.*

(4.31)
$$f''(x) = \tfrac{1}{2}\int_0^\infty \frac{e^{-t}}{t}\{f'(x+t)-f'(x-t)\}\, dt.$$

This can be obtained by differentiating (M), provided that the integral on the right of (4.31) is uniformly convergent. The uniform convergence at infinity follows from (4.25). Also

(4.32) $f'(x+t)-f'(x-t)$
$$= \tfrac{1}{2}\int_0^\infty \frac{e^{-u}}{u}\{f(x+t+u)-f(x+t-u)-f(x-t+u)+f(x-t-u)\}\, du$$

$$= \tfrac{1}{2}\int_0^\infty \frac{e^{-u}}{u} F(x, t, u)\, du,$$

say. Now

$$|f(x+t+u)-f(x+t-u)| = \left|\int_{x+t-u}^{x+t+u} f'(v)\, dv\right| \leq 2u A_2 C e^{A(|x|+|t|+|u|)},$$

and a similar result holds for the other difference. Hence

$$|F(x, t, u)| \leq 4u A_2 C e^{A(|x|+|t|+|u|)}.$$

Since $F(x, t, u)$ is symmetrical in t and u, we may replace the factor u on the right by t. Multiplying these two inequalities and taking the square root

$$|F(x, t, u)| \leq 4\sqrt{(ut)} A_2 C e^{A(|x|+|t|+|u|)}.$$

Inserting this in (4.32), we see that $f'(x)$ satisfies a Lipschitz condition of order $\frac{1}{2}$, and so that (4.31) is uniformly convergent at $t = 0$.

4.4. Our argument can now be repeated. We see that

$$|f''(x)| < A_2^2 C e^{A|x|},$$

and that $f''(x)$ satisfies M, and so generally. Then, as in Lemma δ of BKLM, $f(x)$ is an integral function, and from § 3.22 of BKLM onwards our proof proceeds as before.

4.5. We add, in conclusion, that the more general equation

$$f'(x) = \lambda \int_0^\infty \frac{e^{-t}}{t} \{f(x+t) - f(x-t)\} \, dt$$

has been investigated by E. Hopf*. The theory is not much changed by the generalization of λ, but the form of the solutions is different. If we assume that $f(x) = O(e^{A|x|})$, where $A < 1$, and that λ is real, then there are no solutions when $\lambda < 0$, and otherwise only the solutions

$$f(x) = a + be^{ax} + ce^{-ax},$$

where a and $-a$ are the two roots of

$$a = \lambda \log \frac{1+a}{1-a}$$

(other than $a = 0$) whose real part lies between -1 and 1. These roots are real when $\lambda < \frac{1}{2}$ and pure imaginary when $\lambda > \frac{1}{2}$. When $\lambda = \frac{1}{2}$ there is a treble root at the origin and the solution degenerates.

---

* E. Hopf, "Über eine Klasse singulärer Integralgleichungen: Bemerkungen zur Methode von Hardy und Titchmarsh", *Journal London Math. Soc.*, 4 (1929), 23-27.

CORRECTION

p. 97, *last footnote, last line.* 81-18 should be 81-88.

An integral equation. By Professor G. H. HARDY and E. C. TITCHMARSH.

[*Received* 11 January, *read* 8 February, 1932.]

1. The integral equation

(1·1) $$f(x) = \frac{\lambda}{\Gamma(\alpha)} \int_x^\infty (y-x)^{\alpha-1} f(y)\, dy,$$

where x, $f(x)$, and λ are real and α positive, may be regarded as a differential equation of order α. Suppose for example that α is a positive integer p, that $f(x)$ tends to 0, when $x \to \infty$, with sufficient rapidity, and that

$$f_1(x) = \int_x^\infty f(y)\, dy, \quad f_2(x) = \int_x^\infty f_1(y)\, dy, \ldots.$$

Then, if we integrate repeatedly by parts, and write z for $f_p(x)$, (1·1) becomes

$$\frac{d^p z}{dx^p} = (-1)^p \lambda z.$$

The only solutions are finite combinations of exponentials.

The general equation (1·1) also has exponential solutions. The conditions that $y = e^{-ax}$ should be a solution are that $\Re(a) > 0$ and $\lambda = a^\alpha$, where a^α means $e^{\alpha \log a}$ and $\log a$ has its principal value. If $\lambda > 0$, a may have any of the values

$$\lambda^{1/\alpha} e^{2r\pi i/\alpha} \qquad (r = 0, \pm 1, \ldots)$$

for which $|2r\pi/\alpha| < \tfrac{1}{2}\pi$. Appropriate linear combinations of these solutions will give real solutions, and it is natural to ask if they are the only solutions. If $\alpha \leqslant 4$, and in particular if $\alpha < 1$, the only admissible value of a is $\lambda^{1/\alpha}$. The question is then whether a solution of (1·1) is necessarily a multiple of e^{-ax}.

We take this last question as typical, and answer it by proving the following theorem.

THEOREM A. *If* (i) $\lambda > 0$, $0 < \alpha < 1$;

(ii) $f(x)$ *is integrable, in the sense of Lebesgue, in any finite interval of positive values of* x;

(iii) *the integral in* (1·1) *exists, as*

$$\lim_{X \to \infty,\, \delta \to 0} \int_{x+\delta}^X,$$

for all positive x;

(iv) (1·1) *is true for all positive x; then*

(1·2)
$$f(x) = Ce^{-ax},$$

where $a = \lambda^{1/\alpha}$ *and C is constant.*

In other words, the exponential solutions are the only solutions for which (1·1) is significant, in the most natural sense, and true for all positive x.

2. There are two natural methods for the solution of (1·1).

(i) Suppose that α is a rational number p/q, and that
$$\int_0^\infty x^n |f(x)| \, dx < \infty$$
for all n; and write (1·1) in the form
$$f = \lambda J^\alpha f.$$
The operator J^α satisfies
$$J^\alpha J^\beta f = J^{\alpha+\beta} f,$$
and so, if we iterate (1·1) q times, we obtain
$$f = \lambda^q J^p f,$$
which has only exponential solutions. Apart from the restriction on α, the method demands rather heavy restrictions on f (though not so stringent as the condition actually stated).

(ii) We may introduce the Mellin transform

(2·1)
$$\phi(s) = \int_0^\infty x^{s-1} f(x) \, dx$$

of $f(x)$, and translate (1·1) into a functional equation for $\phi(s)$. We have formally

(2·2)
$$\phi(s) = \frac{\lambda}{\Gamma(\alpha)} \int_0^\infty x^{s-1} dx \int_x^\infty (y-x)^{\alpha-1} f(y) \, dy$$
$$= \frac{\lambda}{\Gamma(\alpha)} \int_0^\infty f(y) \, dy \int_0^y x^{s-1} (y-x)^{\alpha-1} dx$$
$$= \frac{\lambda \Gamma(s)}{\Gamma(s+\alpha)} \phi(s+\alpha).$$

If

(2·3)
$$\chi(s) = \lambda^{s/\alpha} \frac{\phi(s)}{\Gamma(s)},$$

then (2·2) becomes

(2·4)
$$\chi(s+\alpha) = \chi(s),$$

so that $\chi(s)$ is periodic. If we can show that $\chi(s)$ also satisfies appropriate conditions of regularity and order, we shall be able to deduce that
$$\chi(s) = \Sigma a_n e^{2n\pi i s/a}, \qquad (2.5)$$
a finite sum of exponentials, and our conclusion will then follow from (2·5), (2·3), and Mellin's inversion formula.

The method is familiar in principle, and we have used it repeatedly ourselves. The manner of its application in the present instance is however not quite obvious, since the conditions of Theorem A are by no means sufficient to justify the transformations as they stand.

3. We begin by proving that $f(x)$ *is an integral function of exponential type*†.

Suppose that $c > 1$ and $0 < x \leqslant \tfrac{1}{2} c$, and write
$$f(x) = \frac{\lambda}{\Gamma(\alpha)} \int_x^c f(y)(y-x)^{\alpha-1} dy + \frac{\lambda}{\Gamma(\alpha)} \int_c^\infty f(y)(y-x)^{\alpha-1} dy$$
$$= \lambda f_\alpha^*(x) + \lambda g(x),$$
say. Then
$$f_\alpha^*(x) = \frac{\lambda}{\Gamma(\alpha)} \int_x^c \{f_\alpha^*(y) + g(y)\}(y-x)^{\alpha-1} dy = \lambda f_{2\alpha}^*(x) + \lambda g_\alpha^*(x),$$
since
$$(f_\alpha^*)_\beta^* = f_{\alpha+\beta}^* \,\ddagger.$$
Hence
$$f(x) = \lambda g(x) + \lambda^2 g_\alpha^*(x) + \lambda^2 f_{2\alpha}^*(x).$$
Repeating the argument, we obtain
$$f(x) = \lambda g(x) + \lambda^2 g_\alpha^*(x) + \ldots + \lambda^n g_{(n-1)\alpha}^*(x) + \lambda^n f_{n\alpha}^*(x).$$
But
$$\lambda^n f_{n\alpha}^*(x) = \frac{\lambda^n}{\Gamma(n\alpha)} \int_x^c f(y)(y-x)^{n\alpha-1} dy \to 0,$$
when $n \to \infty$, and so
$$f(x) = \sum_0^\infty \lambda^{n-1} g_{n\alpha}^*(x) \qquad (0 < x \leqslant \tfrac{1}{2} c). \qquad (3.1)$$

4. So far x is real and positive: we now use the series (3·1) to define $f(x)$ for complex x.

---

† $f(z)$ is 'of exponential type' if $|f(z)| < Ke^{K|z|}$ for some K.
‡ This is
$$\frac{1}{\Gamma(\beta)} \int_x^c f_\alpha^*(y)(y-x)^{\beta-1} dy = \frac{1}{\Gamma(\alpha+\beta)} \int_x^c f(y)(y-x)^{\alpha+\beta-1} dy,$$
and is familiar when all the integrals involved are absolutely convergent. The formula is proved, under conditions much more general than those required here, by Bosanquet (2).

We write

(4·1) $$F(x) = \int_x^\infty y^{a-1} f(y)\, dy$$

(at present for real positive x). By condition (iii) of Theorem A, $F(x)$ is bounded for $x > 1$, and tends to zero when $x \to \infty$. We denote the upper bound of $|F(x)|$ by M, and use $K = K(a)$ to denote a positive number depending on a only.

It is plain that $g(x)$ is an analytic function of x regular for $|x| < c$. Also

(4·2) $$g(x) = -\int_c^\infty \left(\frac{y-x}{y}\right)^{a-1} F'(y)\, dy$$
$$= \left(\frac{c-x}{c}\right)^{a-1} F(c) + \int_c^\infty F(y) \frac{d}{dy}\left(\frac{y-x}{y}\right)^{a-1} dy.$$

From (4·2) it follows (i) that $|g(x)|$ has an upper bound of the type KM for $|x| \leq \tfrac{1}{2}c$, and (ii) that

$$|g(x)| \leq 2M \left(\frac{c-x}{c}\right)^{a-1}$$

for real x between $\tfrac{1}{2}c$ and c.

We define $g_{na}^*(x)$, where $n > 0$, x is complex, and $|x| \leq \tfrac{1}{2}c$, by

$$g_{na}^*(x) = \frac{1}{\Gamma(na)} \int_x^{\frac{1}{2}c} (y-x)^{na-1} g(y)\, dy + \frac{1}{\Gamma(na)} \int_{\frac{1}{2}c}^c (y-x)^{na-1} g(y)\, dy,$$

each integration being rectilinear. The second integral is majorised by

$$\frac{2Mc^{1-a}}{\Gamma(na)} \int_{\frac{1}{2}c}^c |y-x|^{na-1} (c-y)^{a-1} dy < \frac{KMc^{na}}{\Gamma(na)}.$$

It is therefore regular for $|x| \leq \tfrac{1}{2}c$; and the first integral is plainly regular and has a majorant of the same form. Hence $g_{na}^*(x)$ is regular and has a similar majorant.

It follows that the series (3·1) is uniformly convergent for $|x| \leq \tfrac{1}{2}c$. Hence $f(x)$ is an analytic function regular for $|x| \leq \tfrac{1}{2}c$, and therefore an integral function. Further,

$$|f(x)| \leq KM + KM \sum_1^\infty \frac{\lambda^{n-1} c^{na}}{\Gamma(na)} < KM e^{K\lambda c}$$

for $|x| \leq \tfrac{1}{2}c$, so that $f(x)$ is of exponential type.

5. We prove next that

(5·1) $$f(x) = o(x^{1-a}),$$

as $x \to \infty$, uniformly in any finite half-strip parallel to the real axis. For this we consider $F(x)$ for complex x, defining it by (4·1) and taking the integral in the first instance from x to a point in the

real axis and then along the real axis to infinity. $F(x)$ is regular except at the origin, is of exponential type, and tends to zero along the positive real axis. Hence* it tends to zero uniformly in the half-strip, and therefore
$$F'(x) = -x^{a-1}f(x)$$
does the same. This proves (5·1).

It now follows from Cauchy's theorem that (1·1) is true for complex x, and that the integration may be taken along any path (of sufficient regularity) from x to infinity inside the half-strip, and in particular along a line parallel to the real axis.

6. We now introduce the function $\phi(s)$ defined, when
$$0 < \sigma = \Re(s) < \alpha,$$
by (2·1). It is essential for our purpose to use a more extensive representation of $\phi(s)$, and we select the representation

(6·1) $$\phi(s) = \frac{i}{2 \sin s\pi} \int_C (-x)^{s-1} f(x)\, dx,$$

where $$(-x)^{s-1} = e^{(s-1)\log(-x)},$$

the logarithm being real for $x < 0$, and C is a loop surrounding the positive real axis in the counter-clockwise direction. It follows from (6·1), in the classical manner, that $\phi(s)$ is regular for $\sigma < \alpha$, except perhaps for simple poles at $s = 0, -1, -2, \ldots$, and that
$$\psi(s) = \frac{\phi(s)}{\Gamma(s)}$$
is regular for $\sigma < \alpha$.

We have formally
$$\phi(s) = \frac{i}{2 \sin s\pi} \int_C (-x)^{s-1} dx \, \frac{\lambda}{\Gamma(\alpha)} \int_x^\infty (y-x)^{\alpha-1} f(y)\, dy$$
$$= \frac{i}{2 \sin s\pi} \int_C (-x)^{s-1} dx \, \frac{\lambda}{\Gamma(\alpha)} \int_0^\infty u^{\alpha-1} f(u+x)\, du$$
$$= \frac{\lambda}{\Gamma(\alpha)} \frac{i}{2 \sin s\pi} \int_0^\infty u^{\alpha-1} du \int_C (-x)^{s-1} f(u+x)\, dx.$$

If here we write w for $u+x$, and denote by C' a loop like C, but passing round $w = u$ instead of $w = 0$, we obtain
$$\phi(s) = \frac{\lambda}{\Gamma(\alpha)} \frac{i}{2 \sin s\pi} \int_0^\infty u^{\alpha-1} du \int_{C'} (u-w)^{s-1} f(w)\, dw.$$

* See, for example, Hardy and Titchmarsh (4), Lemma δ. We may prove first that $F(x)$ is bounded, and then use Montel's theorem. Theorems of this type are not affected by a possible singularity at the origin.

Here we may replace C' by C; and if we do this, and again invert the order of integration, we obtain

$$\phi(s) = \frac{\lambda}{\Gamma(\alpha)} \frac{i}{2 \sin s\pi} \int_C f(w)\, dw \int_0^\infty u^{\alpha-1}(u-w)^{s-1}\, du.$$

Finally, since

$$\int_0^\infty u^{\alpha-1}(u-w)^{s-1}\, du = (-w)^{\alpha+s-1} \frac{\Gamma(\alpha)\,\Gamma(1-s-\alpha)}{\Gamma(1-s)},$$

we deduce

$$\psi(s) = \frac{\phi(s)}{\Gamma(s)} = \lambda \frac{i}{2\Gamma(s+\alpha)\sin(s+\alpha)\pi} \int_C (-w)^{\alpha+s-1} f(w)\, dw$$
$$= \lambda \psi(s+\alpha),$$

which is (2·4). Our deduction is so far formal. We proceed to justify our formal analysis for values of s whose real part σ is negative and sufficiently large. It is sufficient to consider real values of s.

7. Suppose that $s < -B < 0$. An examination of the transformations of § 6, in the light of what was proved in § 5, shows that there are two steps only which require justification, viz. the two inversions of the order of integration.

(1) We have to show that

$$\int_C (-x)^{s-1}\, dx \int_0^\infty u^{\alpha-1} f(u+x)\, du = \int_0^\infty u^{\alpha-1}\, du \int_C (-x)^{s-1} f(u+x)\, dx.$$

It is plain from considerations of uniform convergence that the inversion would be valid if the limit ∞ were replaced by a finite U; and it is therefore sufficient to prove that

$$\int_C (-x)^{s-1}\, dx \int_U^\infty u^{\alpha-1} f(u+x)\, du$$

exists and tends to zero when $U \to \infty$. This will certainly be so if

$$\left| \int_U^\infty u^{\alpha-1} f(u+x)\, du \right| < k(|x|), \quad \int_C |(-x)^{s-1}|\, k(|x|)\, |dx| < \infty.$$

In particular, since B is at our disposal, it will be sufficient to prove that we may take for $k(|x|)$ a function of the type $K|x|^K$. Further, we need only consider the distant part of C, on which $\Re(x)$ is positive and large.

Now $\displaystyle J = \int_U^\infty u^{\alpha-1} f(u+x)\, du = \int_{x+U}^\infty (w-x)^{\alpha-1} f(w)\, dw,$

the w-path being parallel to the real axis. If $U < |x|$, then
$$J = \int_{x+U}^{x+|x|} + \int_{x+|x|}^{\infty} = J_1 + J_2,$$
say. It is plain from (5·1) that J_1 has a majorant of the type required, and
$$J_2 = -\int_{x+|x|}^{\infty} \left(\frac{w-x}{w}\right)^{a-1} F'(w)\, dw$$
$$= \left(\frac{|x|}{x+|x|}\right)^{a-1} F(x+|x|) + \int_{x+|x|}^{\infty} F(w) \frac{d}{dw}\left(\frac{w-x}{w}\right)^{a-1} dw$$
is bounded. If $U \geqslant |x|$ the argument is simpler, no decomposition of the integral being required. In any case J has a majorant $K|x|^K$, and the inversion is justified. It is in fact sufficient that $B \geqslant 1$.

8. (2) The second inversion is
$$\int_0^{\infty} u^{a-1} du \int_C (u-w)^{s-1} f(w)\, dw = \int_C f(w)\, dw \int_0^{\infty} u^{a-1}(u-w)^{s-1} du.$$
It is sufficient to show that
$$\int_C f(w)\, dw \int_U^{\infty} u^{a-1}(u-w)^{s-1} du$$
exists and tends to zero when $U \to \infty$. Further, this would plainly be true if C were replaced by any finite part of itself. It is therefore sufficient to prove that, if C_ϖ is the part of C on which $\mathfrak{R}(w) > \varpi$, then
$$(8 \cdot 1) \qquad |J| = \left| \int_{C_\varpi} f(w)\, dw \int_U^{\infty} u^{a-1}(u-w)^{s-1} du \right| < \epsilon$$
for $\varpi \geqslant \varpi_0(\epsilon)$ and all U in question. We need only consider the upper half of C_ϖ, which we may suppose defined by
$$w = i + W, \quad W > \varpi.$$

9. Integrating twice by parts, we have
$$\int_U^{\infty} u^{a-1}(u-w)^{s-1} du = -U^{a-1}\frac{(U-w)^s}{s} + (a-1)U^{a-2}\frac{(U-w)^{s+1}}{s(s+1)}$$
$$+ \frac{(a-1)(a-2)}{s(s+1)} \int_U^{\infty} u^{a-3}(u-w)^{s+1} du = j_1 + j_2 + j_3,$$
say. We denote the contributions of these three terms to J by J_1, J_2, and J_3 respectively.

(i) We estimate j_3 on the hypothesis that $U < \tfrac{1}{2}W$. The result is the same, and the argument simpler, in the contrary case. We suppose $B \geqslant 3$, so that $s < -3$.

Then

$$|j_3| < K \int_U^\infty u^{a-3} \{(u-W)^2+1\}^{\frac{1}{2}(s+1)} du = K\left(\int_U^{\frac{1}{2}W} + \int_{\frac{1}{2}W}^{\frac{3}{2}W} + \int_{\frac{3}{2}W}^\infty\right)$$

$$< K\left(\int_U^{\frac{1}{2}W} u^{a-3} W^{s+1} du + \int_{\frac{1}{2}W}^{\frac{3}{2}W} W^{a-3} \{(u-W)^2+1\}^{\frac{1}{2}(s+1)} du\right.$$
$$\left. + \int_{\frac{3}{2}W}^\infty u^{a-3} W^{s+1} du\right).$$

The first and last terms are $O(W^{s+1})$ and the second is $O(W^{a-3})$. Since $f(w) = o(W^{1-a})$, we deduce

$$J_3 = o\left(\int_\varpi^\infty W^{2-a+s} dW + \int_\varpi^\infty W^{-2} dW\right) = o(1).$$

It follows that (8·1) is true if J is replaced by J_3.

(ii) In estimating J_1 and J_2 we suppose that $\varpi < \frac{1}{2}U$. The argument is again simpler in the contrary case.

Since $f(w) = o(W^{1-a})$, we have

$$J_1 = o\left[U^{a-1} \int_\varpi^\infty W^{1-a} \{(U-W)^2+1\}^{\frac{1}{2}s} dW\right]$$

$$= o\left\{U^{a-1}\left(\int_\varpi^{\frac{1}{2}U} + \int_{\frac{1}{2}U}^{\frac{3}{2}U} + \int_{\frac{3}{2}U}^\infty\right)\right\}$$

$$= o\left\{U^{a-1} \int_\varpi^{\frac{1}{2}U} W^{1-a}(U-W)^s dW\right\}$$

$$+ o\left[U^{a-1} \int_{\frac{1}{2}U}^{\frac{3}{2}U} U^{1-a} \{(U-W)^2+1\}^{\frac{1}{2}s} dW\right]$$

$$+ o\left\{U^{a-1} \int_{\frac{3}{2}U}^\omega W^{1-a}(W-U)^s dW\right\}$$

$$= o(U^{s+1}) + o(1) + o(U^{s+1}) = o(1).$$

Here it is only necessary to suppose $B \geq 2$. The corresponding argument with J_2 demands that $B \geq 3$, and leads to

$$J_2 = o(U^{s+1}) + o(U^{-1}) + o(U^{s+1}) = o(1).$$

From these results (and the corresponding results when $\varpi > \frac{1}{2}U$) it follows that (8·1) is true if J is replaced by J_1 or J_2; and this completes the justification of the inversion.

It follows that (2·4) is true for real $s < -3$, and so wherever $\chi(s)$ is regular.

10. It is now easy to complete the proof of Theorem A. $\chi(s)$ is regular for $\sigma < a$, and has period a, so that it is an integral function. Also, since the argument of $-x$ on C does not exceed π,

$$\chi(s) = \lambda^{s/a} \frac{\phi(s)}{\Gamma(s)} = \lambda^{s/a} \frac{i\Gamma(1-s)}{2\pi} \int (-x)^{s-1} f(x)\, dx$$

is $O(e^{\frac{1}{2}\pi|t|})$ on any parallel to the imaginary axis. It follows from classical theorems that $\chi(s)$ is a constant*.

Hence $$\phi(s) = A\lambda^{-s/a}\Gamma(s),$$

where A is constant. In particular this is true in the strip $0 < \sigma < a$ in which $\phi(s)$ is defined by (2·1). Inverting by Mellin's formula†, we find that $f(x)$ is a multiple of e^{-ax}, where $a^a = \lambda$.

The proof would of course been very much simpler and shorter if we had imposed more restrictions on $f(x)$, but such restrictions would have destroyed the interest of the theorem, which lies almost entirely in its generality.

We may observe in conclusion that our equation is included in the general type

(10·1) $$f(x) = \lambda \int_{-\infty}^{\infty} K(y-x) f(y)\, dy,$$

with $K(u) = u^{a-1}$ for positive and $K(u) = 0$ for negative u. It has been proved by Bochner that, in certain circumstances, (10·1) has only exponential solutions; but his conditions are much too stringent for application to the problem considered here‡.

---

* A function $F(s)$ which is integral, has period a, and is $O(e^{A|t|})$ for some A, is necessarily a finite sum
$$\Sigma a_n e^{2n\pi i s/a}.$$
Here $A = \frac{1}{2}\pi < 2\pi/a$, so that $n = 0$ is the only admissible value.

† Appropriate conditions will be found in Hardy (3).

‡ Bochner (1). Bochner supposes that $K(u)$ tends to zero, when $u \to \infty$, more rapidly than any power of u, and that $|f(x)| < a(1+|x|)^b$ for some constant a, b.

REFERENCES.

1. S. BOCHNER, 'Über eine Klasse singulärer Integralgleichungen', *Berliner Sitzungsberichte*, 22 (1930), 403–411.

2. L. S. BOSANQUET, 'On Abel's integral equation and fractional integrals', *Proc. London Math. Soc.* (2), 31 (1930), 134–143.

3. G. H. HARDY, 'On Mellin's inversion formula', *Messenger of Math.*, 47 (1918), 178–184.

4. G. H. HARDY and E. C. TITCHMARSH, 'Solution of certain integral equations considered by Bateman, Kapteyn, Littlewood and Milne', *Proc. London Math. Soc.* (2), 23 (1924), 1–26. See also a supplementary paper, *ibid.*, 30 (1930), 95–106.

NEW SOLUTION OF AN INTEGRAL EQUATION

By G. H. Hardy *and* E. C. Titchmarsh.

[Received 15 June, 1935.—Read 20 June, 1935.]

1. *Statement of the theorem.*

1.1. The equation

(1.1.1:M) $$f'(x) = \tfrac{1}{2}\int_0^\infty \frac{e^{-u}}{u}\{f(x+u)-f(x-u)\}\,du$$

originates with Milne[†]. It is satisfied by any quadratic $f(x)$, and has no other obvious solutions. The problem is to prove that all solutions are quadratic[‡], or, failing this, that every solution belonging to some specified class of functions is quadratic.

We have already given several solutions of the problem[§], but that which we give here is a good deal more general than any obtained before. In our previous solutions we assumed that

$$f(x) = O(e^{A|x|}),$$

where $A < 1$, for all real x, and this is a great deal more than is needed for the convergence of the integral. A natural and much more general

---

[†] Milne (7).
[‡] That is to say, that every $f(x)$ such that (i) the integral exists, in some specified sense, for every x, (ii) the equation is true for every x, is quadratic. The precise meaning of the assertion would depend on the particular sense chosen in (i).
[§] Hardy and Titchmarsh (3, **4, 5**).

assumption would be that $f(x) = e^{|x|} g(x)$, or $f(x) = \cosh x\, g(x)$, where $g(x)$ is L^2 in $(-\infty, \infty)$†. We are still unable to prove that this assumption is sufficient, but the assumption which we prove sufficient is "all but" the same.

THEOREM. *Any solution of* (M) *of the form*

(1.1.2) $$f(x) = \cosh x\, g(x),$$

where

(1.1.3) $$\int_{-\infty}^{\infty} g^2 \{1 + (\log|g|)^2\}\, dx < \infty,$$

is a quadratic.

Our proof of this theorem is quite elaborate, and depends upon a long series of lemmas, many of which are more interesting than the final result. Some of our arguments are capable of considerable generalization, while others depend much more on the particular idiosyncrasies of the equation. We have not attempted to generalise the result, since our analysis must be justified by its intrinsic interest rather than by any importance in its conclusion.

Our new formal method consists in putting $f(x) = \cosh x\, g(x)$, and deducing from the given equation one satisfied by $G(x)$, the Fourier transform of $g(x)$. This is found to be

$$G(x-i) + G(x+i) = 0.$$

The analysis also shows that $G(x)$ is an analytic function, periodic and regular except possibly for a certain sequence of triple poles. These conditions determine $G(x)$, apart from the values of certain constants, and $G(x)$ in turn determines $g(x)$ and $f(x)$. The main argument begins in § 3.1, and the reader can begin there also, and refer to the lemmas as they are required. There is nothing particularly surprising in the results of the lemmas, however intricate the proofs may be.

2. *Preliminary lemmas.*

2.1. In this section we collect a number of lemmas (some of which are very well known theorems) required for the proof.

---

† The two forms of the hypothesis are plainly equivalent.

LEMMA α. *If $g(x)$ is L^2 in $(-\infty, \infty)$, then*

$$(2.1.1) \qquad h(x) = \frac{1}{\pi} \int_0^\infty \frac{g(x+u) - g(x-u)}{u} \, du$$

exists for almost all x, and is L^2.

This is a well-known theorem in the theory of conjugate functions†. The integral is a Lebesgue integral at ∞, a Cauchy integral at 0.

Lemmas concerning inequalities.

2.2. LEMMA β. *If $p(x)$ belongs to L, and $q(x)$ to L^2, in $(-\infty, \infty)$, then*

$$r(x) = \int_{-\infty}^\infty p(u) q(x+u) \, du$$

belongs to L^2.

For, if

$$\int_{-\infty}^\infty |p(u)| \, du = P, \quad \int_{-\infty}^\infty q^2(u) \, du = Q^2,$$

we have

$$\{r(x)\}^2 \leqslant P \int_{-\infty}^\infty |p(u)| q^2(x+u) \, du,$$

$$\int_{-\infty}^\infty \{r(x)\}^2 \, dx \leqslant P \int_{-\infty}^\infty |p(u)| \, du \int_{-\infty}^\infty q^2(x+u) \, dx = P^2 Q^2.$$

The lemma is a special case of "Young's inequality" ‡.

2.3. LEMMA γ. *If*

$$(2.3.1) \qquad \int_0^\infty \{g(x)\}^2 \{1 + (\log x)^2\} \, dx < \infty$$

and

$$g_1(x) = \int_0^x g(t) \, dt,$$

then

$$(2.3.2) \qquad \int_0^\infty \left\{\frac{g_1(x)}{x}\right\}^2 \{1 + (\log x)^2\} \, dx < \infty$$

and

$$(2.3.3) \qquad \int_0^\infty \frac{|g(x) g_1(x)|}{x} \{1 + (\log x)^2\} \, dx < \infty.$$

---

† See, for example, Titchmarsh (10).
‡ See Hardy, Littlewood, and Pólya (2), Theorem 280. The theorem stated there for $\lambda > 0$, $\mu > 0$ is still true when $\lambda = 0$ or $\mu = 0$, and it is one of these cases ($\lambda = 0$, $\mu = \frac{1}{2}$) which we need here. We therefore insert a proof.

Since $g(x) \log x$ is L^2,

$$\int_0^x g(t) \log t \, dt = o(x^{\frac{1}{2}})\dagger,$$

and
$$g_1(x) = o\left(\frac{x^{\frac{1}{2}}}{|\log x|}\right)$$

both at 0 and ∞. Also, if

$$1 + (\log x)^2 = \varpi(x),$$

then
$$\chi(x) = \int_x^\infty \frac{\varpi(t)}{t^2} dt < A \frac{\varpi(x)}{x},$$

where A is a constant, and $g_1^2 \chi = o(1)$ at 0 and ∞. Hence

$$\int_0^\infty \frac{g_1^2 \varpi}{x^2} dx = -\int_0^\infty g_1^2 \chi' \, dx = 2 \int_0^\infty g_1 g \chi \, dx$$

$$\leqslant 2A \int_0^\infty \left|\frac{g_1}{x}\right| |g| \varpi \, dx \leqslant 2A \left(\int_0^\infty \frac{g_1^2 \varpi}{x^2} dx \int_0^\infty g^2 \varpi \, dx\right)^{\frac{1}{2}},$$

$$\int_0^\infty \frac{g_1^2 \varpi}{x^2} dx \leqslant 4A^2 \int_0^\infty g^2 \varpi \, dx,$$

and
$$\int_0^\infty \frac{|gg_1| \varpi}{x} dx \leqslant 2A \int_0^\infty g^2 \varpi \, dx.$$

2.4. In the lemmas which follow we use g^* in the senses defined (with a slightly different notation) by Hardy, Littlewood, and Pólya. There are two such senses. In the first place, when the interval considered is $(0, \infty)$, g^* is the decreasing rearrangement of $|g|$. Secondly, when the interval is $(-\infty, \infty)$, g^* is the "symmetrically decreasing" rearrangement of $|g|$, the even function equimeasurable with $|g|$ and decreasing both ways from the origin‡.

LEMMA 8. *If g satisfies* (1.1.3), *then g^* satisfies* (2.3.1).

The hypothesis (1.1.3) is unaltered by the substitution of g^* for g. It is therefore sufficient to prove that (1.1.3) implies (2.3.1) when g is a positive symmetrically decreasing function.

---

† Hardy, Littlewood, and Pólya (2), 164, Theorems 222, 223.
‡ Hardy, Littlewood, and Pólya (2), ch. X. See in particular 276–278, where, however, the fundamental intervals are (0, 1) and $(-\infty, \infty)$, and what we should call g^* in the first case is denoted by $\bar{g}$.

We have then
$$\tfrac{1}{2}xg^2(x) \leqslant \int_{\frac{1}{2}x}^{x} g^2(t)\,dt < A$$

and so
$$g < Ax^{-\frac{1}{2}}, \quad (\log x)^2 < 4(\log g - \log A)^2$$

for $x > 1$, with A independent of x. Hence, first,

(2.4.1) $$\int_1^\infty g^2(\log x)^2\,dx \leqslant 8\int_1^\infty g^2\{(\log g)^2 + (\log A)^2\}\,dx < \infty.$$

On the other hand,
$$uv \leqslant u\log u + e^{v-1}$$

for $u > 0$ and all real v†. Taking $u = g$, $v = \tfrac{1}{4}\log(1/x)$, $x < 1$, we obtain
$$g\log\frac{1}{x} = 4\left(g \cdot \tfrac{1}{4}\log\frac{1}{x}\right) \leqslant 4(g\log g + e^{-1}x^{-\frac{1}{4}}),$$

(2.4.2) $$\int_0^1 g^2(\log x)^2\,dx \leqslant A\int_0^1 g^2(\log g)^2\,dx + A\int_0^1 x^{-\frac{1}{2}}\,dx < \infty.$$

The conclusion follows from (2.4.1) and (2.4.2).

LEMMA ϵ‡. *If $h(x)$ is symmetrically decreasing, then*
$$\int_{-\infty}^\infty \int_{-\infty}^\infty g(x)g(y)h(x-y)\,dx\,dy \leqslant \int_{-\infty}^\infty \int_{-\infty}^\infty g^*(x)g^*(y)h(x-y)\,dx\,dy.$$

2.5. LEMMA ζ. *If $g(x)$ satisfies (1.1.3), then*
$$h(x) = \int_1^\infty \frac{g(x+u)}{u}\,du$$

is L^2 in $(-\infty, \infty)$.

We may suppose g positive. The integral exists for all x, by Schwarz's inequality, and it is enough to show that
$$\int_{-\infty}^\infty h^2(x)\,dx = \int_{-\infty}^\infty dx \left(\int_{x+1}^\infty \frac{g(u)}{u-x}\,du\right)^2 < \infty.$$

This is
$$\int_{-\infty}^\infty dx \int_{x+1}^\infty \frac{g(u)}{u-x}\,du \int_{x+1}^\infty \frac{g(v)}{v-x}\,dv = \int_{-\infty}^\infty g(u)\,du \int_{-\infty}^\infty g(v)\,dv \int_{-\infty}^w \frac{dx}{(u-x)(v-x)},$$

---

† "Young's inequality": see Hardy, Littlewood, and Pólya (2), Theorem 63.
‡ See Hardy, Littlewood, and Pólya (2), Theorem 380.

where $w = \min(u-1, v-1)$. Hence

$$\int_{-\infty}^{\infty} h^2(x)\, dx = \int_{-\infty}^{\infty}\int_{-\infty}^{\infty} g(u)\, g(v)\, \frac{\log(1+|u-v|)}{|u-v|}\, du\, dv$$

$$\leq \int_{-\infty}^{\infty}\int_{-\infty}^{\infty} g^*(u)\, g^*(v)\, \frac{\log(1+|u-v|)}{|u-v|}\, du\, dv,$$

since
$$\frac{\log(1+|x|)}{|x|}$$

is symmetrically decreasing, so that we may use Lemma ϵ.

It is therefore sufficient to prove the last integral finite, and we may confine ourselves to the octant $0 \leq v \leq u$. But

$$\int_0^u g^*(v)\, \frac{\log(1+u-v)}{u-v}\, dv \leq \frac{1}{u}\int_0^u g^*(v)\, dv \int_0^u \frac{\log(1+u-v)}{u-v}\, dv,$$

by Tchebychef's inequality†; and the second factor is

$$\int_0^u \frac{\log(1+v)}{v}\, dv < A\{1 + (\log u)^2\}.$$

It is therefore sufficient that

$$\int_0^\infty \frac{g^*(u)\, g_1^*(u)}{u}\{1 + (\log u)^2\}\, du < \infty,$$

and this follows from Lemmas δ and γ.

2.6. LEMMA θ. *If $g(x)$ satisfies* (1.1.3), *then*

(2.6.1)
$$\phi(x) = \int_0^\infty \frac{g(x+u) - e^{-2u}g(x-u)}{u}\, du,$$

$$\psi(x) = \int_0^\infty \frac{e^{-2u}g(x+u) - g(x-u)}{u}\, du$$

are L^2.

It is sufficient to consider $\psi(x)$, which we write in the form

$$\psi(x) = \int_0^\infty \frac{g(x+u) - g(x-u)}{u}\, du - \int_0^1 \frac{1-e^{-2u}}{u}\, g(x+u)\, du$$

$$+ \int_1^\infty \frac{e^{-2u}}{u}\, g(x+u)\, du - \int_1^\infty \frac{g(x+u)}{u}\, du.$$

The four integrals here are L^2, the first by Lemma α, the second and third by Lemma β, and the last by Lemma ζ.

---

† Hardy, Littlewood, and Pólya (2), Theorems 43 and 236.

Lemmas concerning Fourier transforms.

2.7. We use the formulae

$$(2.7.1) \quad F(x) = \frac{1}{\sqrt{(2\pi)}} \int_{-\infty}^{\infty} f(t) e^{ixt} dt, \quad f(x) = \frac{1}{\sqrt{(2\pi)}} \int_{-\infty}^{\infty} F(t) e^{-ixt} dt,$$

the functions f and F being L^2 and the integrals "mean square integrals" in the sense of Plancherel†. The possible ambiguity in the meaning of "transform", in respect to the factors e^{ixt} and e^{-ixt}, may be avoided by standardising the distinction between small and capital letters as in the formulae (2.7.1).

LEMMA η. *If $f(x)$ and $f'(x)$ are L^2, and $F(x)$ and $H(x)$ are their transforms, then $xF(x)$ is L^2 and*

$$H(x) = -ixF(x).$$

Since ff' is L, we have

$$\left| \{f(X)\}^2 - \{f(x)\}^2 \right| = 2 \left| \int_x^X f(t) f'(t) dt \right| < \epsilon$$

for large x and X. Hence $f(x)$ tends to a limit when $x \to \infty$, and the limit is 0 since f is L^2. Also

$$\int_{-a}^{a} f'(t) e^{ixt} dt + ix \int_{-a}^{a} f(t) e^{ixt} dt = f(a) e^{iax} - f(-a) e^{-iax}.$$

The left-hand side has the mean square limit

$$\sqrt{(2\pi)} \{H(x) + ix F(x)\},$$

when $a \to \infty$, in any finite interval (x_1, x_2), and the right-hand side tends to 0, uniformly in this interval. Hence $H + ixF = 0$.

LEMMA κ. *If f is L^2 and F is its transform; and*

$$f(x) = \frac{1}{\sqrt{(2\pi)}} \int_{-\infty}^{\infty} \chi(t) e^{-ixt} dt,$$

where χ is L; then $\chi \equiv F$ (so that χ is L^2).

For

$$\lim_{T \to \infty} \frac{1}{\sqrt{(2\pi)}} \int_{-T}^{T} \left(1 - \frac{|t|}{T}\right) f(t) e^{ixt} dt$$

is equal, almost everywhere, to both $\chi(x)$ and $F(x)$‡.

---

† See Hobson (6), Paley and Wiener (8), or Titchmarsh (9).
‡ See Hobson (6), 740 and 748 respectively.

LEMMA λ. *If the transform of $g(x)$ is $G(x)$, then the transform of $h(x)$, defined by (2.1.1), i.e. the transform of the conjugate of $g(x)$, is*

$$-i\,\operatorname{sgn} x\, G(x).$$

If $g(x) = l(x)+m(x)$, where l is even and m odd, then

$$G(x) = L_c(x)+i\,M_s(x),$$

*where L_c is the cosine transform of l and M_s the sine transform of m. Also†
$h(x)$ is the sum of the sine transform of $-L_c(x)$ and the cosine transform of $M_s(x)$, and hence the transform of $h(x)$ is*

$$-i\,\operatorname{sgn} x\, L_c(x)+\operatorname{sgn} x\, M_s(x) = -i\,\operatorname{sgn} x\, G(x).$$

Lemmas concerning analytic functions.

2.8. LEMMA μ. *Suppose that (a, b) is a finite stretch of the real axis, and that $G_1(z)$ and $G_2(z)$ are analytic functions satisfying the following conditions:*

(i) *$G_1(z)$ is regular, except perhaps for a finite number of poles, not on the real axis, in the rectangle $a \leqslant x \leqslant b,\ -2 < y < 0$;*

(ii) *$G_2(z)$ has the same properties in the rectangle $a \leqslant x \leqslant b,\ 0 < y < 2$;*

(iii) *$G_1(z)$ and $G_2(z)$ have the same mean square limit $G(x)$, over the interval (a, b), when $y \to -0$ and $y \to +0$ respectively.*

Then $G_1(z)$ and $G_2(z)$ are elements of an analytic function $G(z)$ regular, except for the poles mentioned, in the region

$$a \leqslant x \leqslant b,\ -2 < y < 2.$$

This is a simple extension of a classical theorem‡.

Suppose, first, that $G_1(z)$ and $G_2(z)$ have no poles. Since

$$\int_a^b |G_2(x+iy)-G(x)|^2\,dx \to 0$$

when $y \to +0$,

$$\int_a^b |G_2(x+iy)|^2\,dx$$

is bounded for $0 < y < 2$, and

$$\int_a^b dx \int_0^1 |G_2(x+iy)|^2\,dy = \int_0^1 dy \int_a^b |G_2(x+iy)|^2\,dx < \infty.$$

---

† See Titchmarsh (10), §2.5–3.2. The sign of $g(x)$ on p. 115 should be changed and p. 119, line 5 should read " minus the sine transform ...".

‡ For which see, for example, Titchmarsh (9), 155–157.

Hence

(2.8.1) $$\int_0^1 |G_2(x+iy)|^2 \, dy < \infty$$

for almost all x. Similarly,

(2.8.2) $$\int_{-1}^0 |G_1(x+iy)|^2 \, dy < \infty$$

for almost all x. We suppose that α and β are two numbers of (a, b) for each of which both these conditions are satisfied.

Suppose that δ is small and positive and that z_0 lies in the upper of the two regions in question, and apply Cauchy's theorem to

$$\frac{G_2(z)}{z-z_0}$$

and the rectangle $\alpha+i\delta$, $\beta+i\delta$, $\beta+i$, $\alpha+i$. We obtain

$$2\pi i\, G_2(z_0) = \left(\int_{\alpha+i\delta}^{\beta+i\delta} + \int_{\beta+i\delta}^{\beta+i} + \int_{\beta+i}^{\alpha+i} + \int_{\alpha+i}^{\alpha+i\delta} \right) \frac{G_2(z)}{z-z_0}\, dz = J_1 + J_2 + J_3 + J_4,$$

say. When $\delta \to 0$,

$$J_1 = \int_{\alpha+i\delta}^{\beta+i\delta} \frac{G_2(z)}{z-z_0}\, dz \to \int_\alpha^\beta \frac{G(x)}{x-z_0}\, dx,$$

since $G(x)$ is the mean-square limit of $G(x+i\delta)$;

$$J_2 = \int_{\beta+i\delta}^{\beta+i} \frac{G_2(z)}{z-z_0}\, dz \to \int_\beta^{\beta+i} \frac{G_2(z)}{z-z_0}\, dz$$

on account of (2.8.1); and J_4 tends to a limit similarly. Hence

(2.8.3) $$2\pi i\, G_2(z_0) = \int_\alpha^\beta \frac{G(x)}{x-z_0}\, dx + \left(\int_\beta^{\beta+i} + \int_{\beta+i}^{\alpha+i} + \int_{\alpha+i}^\alpha \right) \frac{G_2(z)\, dz}{z-z_0}.$$

If we apply the same argument, with the same z_0, to a rectangle below the real axis, we obtain

(2.8.4) $$0 = \left(\int_\alpha^{\alpha-i} + \int_{\alpha-i}^{\beta-i} + \int_{\beta-i}^\beta \right) \frac{G_2(z)\, dz}{z-z_0} - \int_\alpha^\beta \frac{G(x)}{x-z_0}\, dx.$$

Combining (2.8.3) and (2.8.4), we have

$$G_2(z_0) = \frac{1}{2\pi i} \int \frac{G_2(z)\, dz}{z-z_0},$$

the contour being now the rectangle $(\alpha-i, \beta-i, \beta+i, \alpha+i)$. Since the integral is an analytic function $G(z_0)$, regular in this rectangle, $G_2(z_0)$ is an element of such a function. Similarly, starting with a z_0 below the axis, we show that $G_1(z_0)$ also is an element of $G(z_0)$.

The complication introduced by poles of $G_1(z)$ and $G_2(z)$ is trivial. There is a rational function $R(z)$ such that $G_1^* = G_1 - R$ and $G_2^* = G_2 - R$ are regular in the regions of the enunciation, and G_1^*, G_2^*, and so G_1, G_2, are analytic continuations of one another.

2.9. LEMMA v. *If $G(z)$ is regular for*

$$a \leqslant y \leqslant b, \quad x \geqslant 1,$$

where $b-a > 2$, and

$$\int_1^\infty |G(x+iy)|^2 dx$$

is bounded for $a \leqslant y \leqslant b$, then $G(z)$ is bounded for

(2.9.1) $\qquad a+1 \leqslant y \leqslant b-1, \quad x \geqslant 2$†.

For if x and y satisfy (2.9.1), and $\rho \leqslant 1$, we have

$$G(z) = \frac{1}{2\pi} \int_0^{2\pi} G(z+\rho e^{i\phi}) d\phi,$$

$$\tfrac{1}{2} G(z) = \frac{1}{2\pi} \int_0^1 \rho d\rho \int_0^{2\pi} G(z+\rho e^{i\phi}) d\phi,$$

$$\tfrac{1}{4}|G(z)|^2 \leqslant \frac{1}{4\pi^2} \int_0^1 \int_0^{2\pi} |G|^2 \rho d\rho d\phi \int_0^1 \int_0^{2\pi} \rho d\rho d\phi$$

$$\leqslant A \int_{y-1}^{y+1} dv \int_{x-1}^{x+1} |G(u+iv)|^2 du \leqslant A \int_{y-1}^{y+1} dv \int_1^\infty |G(u+iv)|^2 du.$$

Proof of the theorem.

3.1. When we substitute from (1.1.2) into (1.1.1), we obtain

(3.1.1) $\quad g'(x) \cosh x + g(x) \sinh x$

$$= \tfrac{1}{2} \int_0^\infty \frac{e^{-u}}{u} \{g(x+u) \cosh(x+u) - g(x-u) \cosh(x-u)\} du$$

$$= \tfrac{1}{4} e^x \phi(x) + \tfrac{1}{4} e^{-x} \psi(x),$$

where $\phi(x)$ and $\psi(x)$ are defined as in (2.6.1). The integrals exist for almost

---

† For more general theorems of the same character, see Hardy, Ingham, and Pólya (1), especially Theorem 11 (561).

all x, and $\phi(x)$ and $\psi(x)$ are L^2, by Lemma θ. We write (3.1.1) in the form

(3.1.2) $$e^x p(x) = e^{-x} q(x),$$

where

(3.1.3) $\quad p(x) = g'(x) + g(x) - \tfrac{1}{2}\phi(x), \quad q(x) = -g'(x) + g(x) + \tfrac{1}{2}\psi(x).$

Since

$$g'(x) = -\frac{e^x - e^{-x}}{e^x + e^{-x}} g(x) + \tfrac{1}{2} \frac{e^x}{e^x + e^{-x}} \phi(x) + \tfrac{1}{2} \frac{e^{-x}}{e^x + e^{-x}} \psi(x),$$

and ϕ and ψ are L^2, g' is L^2; and its transform is $-ixG(x)$, where $G(x)$ is the transform of $g(x)$, by Lemma η. Finally (3.1.3) shows that $p(x)$ and $q(x)$ are L^2; we proceed to calculate their transforms, $P(x)$ and $Q(x)$.

We write $p(x)$ in the form

$$p(x) = g'(x) + g(x) - \tfrac{1}{2} \int_0^\infty \frac{g(x+u) - g(x-u)}{u} du - \tfrac{1}{2} \int_0^\infty \frac{1 - e^{-2u}}{u} g(x-u) du$$

$$= g'(x) + g(x) - \tfrac{1}{2} \pi h(x) - \tfrac{1}{2} \phi_1(x).$$

The transforms of the first three terms are $-ix G(x)$, $G(x)$, and $\tfrac{1}{2}\pi i \operatorname{sgn} x \, G(x)$, the last by Lemma λ. To find that of the last term we make use of Parseval's theorem, in the form

$$\int_{-\infty}^\infty a(u) \overline{b}(u) \, du = \int_{-\infty}^\infty A(u) B(u) \, du,$$

taking

$$a(u) = g(x-u)$$

and

$$b(u) = \frac{1 - e^{-2u}}{u} \quad (u > 0), \quad b(u) = 0 \quad (u \leqslant 0).$$

Then $A(u) = e^{ixu} G(-u)$ and

$$B(u) = \frac{1}{\sqrt{(2\pi)}} \int_0^\infty \frac{1 - e^{-2t}}{t} e^{iut} dt = \frac{1}{\sqrt{(2\pi)}} \log \frac{2 - iu}{-iu},$$

where the logarithm has an imaginary part between $-\tfrac{1}{2}\pi$ and $\tfrac{1}{2}\pi$. Hence

$$\phi_1(x) = \frac{1}{\sqrt{(2\pi)}} \int_{-\infty}^\infty G(-u) \log \frac{2 + iu}{iu} e^{ixu} du$$

$$= \frac{1}{\sqrt{(2\pi)}} \int_{-\infty}^\infty G(u) \log \frac{2 - iu}{-iu} e^{-ixu} du.$$

The integral is absolutely convergent, since

$$G(u), \quad \log \frac{2-iu}{-iu}$$

are both L^2; and the transform of $\phi_1(x)$ is the product of these two functions, by Lemma κ.

Hence the transform of $p(x)$ is

$$P(x) = G(x)\left(1-ix+\tfrac{1}{2}\pi i \operatorname{sgn} x - \tfrac{1}{2}\log \frac{2-ix}{-ix}\right) = G(x)\left(1-ix-\tfrac{1}{2}\log \frac{2-ix}{ix}\right),$$

the imaginary part of the logarithm now lying between $-\pi$ and π. Similarly the transform of $q(x)$ is

$$Q(x) = G(x)\left(1+ix-\tfrac{1}{2}\log \frac{2+ix}{-ix}\right).$$

3.2. If

$$P(z) = P(x+iy) = \frac{1}{\sqrt{(2\pi)}} \int_{-\infty}^{\infty} p(u) e^{izu} du,$$

then, by (3.1.2),

$$P(z) = \frac{1}{\sqrt{(2\pi)}} \int_{-\infty}^{0} p(u) e^{ixu-yu} du + \frac{1}{\sqrt{(2\pi)}} \int_{0}^{\infty} q(u) e^{ixu-(y+2)u} du.$$

Each of these integrals is uniformly convergent in any strip

$$-2 < -2+\delta \leqslant y \leqslant -\delta < 0,$$

so that $P(z)$ is regular in the open strip $-2 < y < 0$. Also

$$P(x+iy) - P(x)$$

is the transform of

$$(e^{-yx} - 1) p(x),$$

so that

$$\int_{-\infty}^{\infty} |P(x+iy) - P(x)|^2 dx = \int_{-\infty}^{\infty} (e^{-yx}-1)^2 |p(x)|^2 dx,$$

which tends to zero with y. Hence $P(x)$ is the mean-square limit of $P(x+iy)$.

Similarly there is an analytic function $Q(x+iy)$, regular for $0 < y < 2$, whose mean square limit, when $y \to 0$, is $Q(x)$.

We define $G_1(z)$ and $G_2(z)$ by

$$G_1(z)\left(1-iz-\tfrac{1}{2}\log \frac{2-iz}{iz}\right) = P(z), \quad G_2(z)\left(1+iz-\tfrac{1}{2}\log \frac{2+iz}{-iz}\right) = Q(z).$$

Then $G_1(z)$ is regular for $-2 < y < 0$, except perhaps for a triple pole at $z = -i$, and $G_2(z)$ is regular for $0 < y < 2$, except perhaps for a triple pole at $z = i$. Also

$$\underset{y \to -0}{\text{l.i.m.}}\, P(x+iy) = P(x), \quad \underset{y \to +0}{\text{l.i.m.}}\, Q(x+iy) = Q(x)$$

over the whole interval $(-\infty, \infty)$ of x, and *a fortiori* over any finite interval (a, b); and so

$$\underset{y \to -0}{\text{l.i.m.}}\, G_1(x+iy) = G(x), \quad \underset{y \to +0}{\text{l.i.m.}}\, G_2(x+iy) = G(x)$$

over any such finite interval.

It now follows from Lemma μ that $G_1(z)$ and $G_2(z)$ are elements of an analytic function $G(z)$. This function is regular in the strip

$$-2 < y < 2,$$

except perhaps for triple poles at $z = \pm i$. In particular, it is regular at the origin.

3.3. Next,

$$P(x-i) = \frac{1}{\sqrt{(2\pi)}} \int_{-\infty}^{\infty} p(u)\, e^u \cdot e^{ixu}\, du,$$

and so

$$p(u)\, e^u = \frac{1}{\sqrt{(2\pi)}} \int_{-\infty}^{\infty} P(x-i)\, e^{-iux}\, dx;$$

and similarly

$$q(u)\, e^{-u} = \frac{1}{\sqrt{(2\pi)}} \int_{-\infty}^{\infty} Q(x+i)\, e^{-iux}\, dx.$$

All these integrals are mean square integrals. Using (3.1.2), we obtain

$$\int_{-\infty}^{\infty} \{P(x-i) - Q(x+i)\}\, e^{-iux}\, du = 0,$$

and therefore
$$P(x-i) = Q(x+i)$$

for almost all real x. This is

$$G(x-i)\left(-ix - \tfrac{1}{2} \log \frac{1-ix}{1+ix}\right) = G(x+i)\left(ix + \tfrac{1}{2} \log \frac{1-ix}{1+ix}\right);$$

and it follows that

(3.3.1) $$G(x-i) + G(x+i) = 0$$

for almost all, and therefore for all, real x.

We may now use (3.3.1) to effect the continuation of $G(z)$ all over the plane. The function $G(z)$ so defined has the period $4i$, and is regular, except perhaps for triple poles at

(3.3.2) $$z = \ldots, \quad -3i, \quad -i, \quad i, \quad 3i, \quad \ldots.$$

Further, the integrals
$$\int_1^\infty |G(x+iy)|^2 dx, \quad \int_{-\infty}^{-1} |G(x+iy)|^2 dx$$
are bounded, and therefore, by Lemma v, $G(z)$ is bounded for $|x| \geqslant 2$.

Finally, we write
$$z = \frac{2}{\pi} \log \zeta, \quad \zeta = e^{\frac{1}{2}\pi z}, \quad G(z) = G\left(\frac{2}{\pi}\log \zeta\right) = H(\zeta).$$

The points (3.3.2) correspond to $\zeta = -i$ and $\zeta = i$. Thus $H(\zeta)$ is odd, is regular except perhaps for triple poles at these two points, and is bounded for large ζ. Hence, for some A, B, C,
$$H(\zeta) - \left\{ \frac{A}{\zeta-i} + \frac{B}{(\zeta-i)^2} + \frac{C}{(\zeta-i)^3} + \frac{A}{\zeta+i} - \frac{B}{(\zeta+i)^2} + \frac{C}{(\zeta+i)^3} \right\}$$
is a constant, which must be 0 because $H(\zeta)$ is odd.

It follows that
$$G(z) = \alpha \frac{1}{\cosh \frac{1}{2}\pi z} + \beta \frac{d}{dz}\left(\frac{1}{\cosh \frac{1}{2}\pi z}\right) + \gamma \frac{d^2}{dz^2}\left(\frac{1}{\cosh \frac{1}{2}\pi z}\right),$$
for some α, β, γ, and therefore that
$$g(x) = \frac{a + bx + cx^2}{\cosh x}$$
for some a, b, c. Hence $f(x)$ is a quadratic.

3.4. We may add in conclusion that, if we assume a little more about $f(x)$, the "inequality" part of the proof may be much simplified without affecting the remainder. If $g(x)$ belongs to L^2 and also to some L^p with $1 < p < 2$, then
$$\int_1^\infty \frac{g(x+u)}{u} du$$
belongs to L^2 by Young's inequality†. The proof then proceeds as before.

---

† Hardy, Littlewood, and Pólya (2), Theorem 280: take
$$\lambda = \frac{1}{p} - \tfrac{1}{2}, \quad \mu = 1 - \frac{1}{p}, \quad f(-t) = \frac{1}{t} \ (t \geqslant 1), \quad f(-t) = 0 \ (t < 1).$$

References.

1. G. H. Hardy, A. E. Ingham, G. Pólya, " Theorems concerning mean values of analytic functions ", *Proc. Royal Soc.* (A), 113 (1927), 542–569.
2. G. H. Hardy, J. E. Littlewood, G. Pólya, *Inequalities* (Cambridge, 1934).
3. G. H. Hardy and E. C. Titchmarsh, " Solutions of some integral equations considered by Bateman, Kapteyn, Littlewood, and Milne ", *Proc. London Math. Soc.* (2), 23 (1923), 1–26.
4. ————, " Correction of an error in a paper on certain integral equations ", *Proc. London Math. Soc.* (2), 24 (1925), xxxi–xxxiii (*Records* for 12 February, 1925).
5. ————, " Additional note on certain integral equations ", *Proc. London Math. Soc.* (2), 30 (1928), 95–106.
6. E. W. Hobson, *The theory of functions of a real variable and the theory of Fourier series* (vol. 2, ed. 2, Cambridge, 1926).
7. E. A. Milne, " Radiative equilibrium in the outer layers of a star ", *Monthly notices of the R.A.S.*, 81 (1921), 361–375.
8. R. E. A. C. Paley and N. Wiener, *Fourier transforms in the complex domain* (American Math. Soc. Colloquium publications, 19, 1934).
9. E. C. Titchmarsh, *The theory of functions* (Oxford, 1932).
10. ————, " Conjugate trigonometrical integrals ", *Proc. London Math. Soc.* (2), 24 (1924), 109–130.

Trinity College,
    Cambridge.
New College,
    Oxford.

2. MISCELLANEOUS PAPERS

INTRODUCTION TO THE MISCELLANEOUS PAPERS

(a) *Set theory.*

Hardy's direct contribution to set theory was not great. In the paper on infinite cardinals [1904, 5] he gives a proof that the cardinal number of the continuum is greater than or equal to $\aleph_1$. This proof is more complicated than the earlier well-known diagonalization proof given by Cantor, and so has not received much attention in the literature. This was a period, of course, when Cantor's work was still a matter of heated controversy among professional mathematicians. It is clear from these papers that Hardy fully accepted Cantor's theories as a valid contribution to mathematics.

The paper in which Hardy rebuts Dr Hobson's criticisms [1907, 1] is interesting in that it contains one of the earliest discussions of the axiom of choice (referred to in the paper as the 'postulate of the existence of the multiplicative class'). The case put forward by Hardy is very much in line with Russell's views at that time.†

Two letters written on 11 December 1903 and 4 August 1904 to P. E. B. Jourdain (1879–1919) confirm Hardy's considerable interest at this time in the foundations of set theory; these can be found in the collection of Jourdain's correspondence at the Mittag-Leffler Institute in Djursholm, Sweden.

(b) *Differential equations.*

These two papers, the second of which proves a conjecture by Goldstein, constitute Hardy's sole contributions to the subject and are straightforward in character.

(c) *The Hardy–Weinberg law.*

Although the mathematical content of 1908, 10 is completely elementary, this seminal paper has proved to be of considerable importance in genetics and has been several times reproduced; see, for example,

M. C. Gabriel and S. Fogel (editors), *Great experiments in biology* (Prentice-Hall, 1955), pp. 295–7.

T. A. Peters (editor), *Classic papers in genetics* (Prentice-Hall, 1959), pp. 60–2.

G. A. Bousseau, Jr (editor), *Evolution* (William C. Brown Company, 1967), pp. 48–50.

---

† I am indebted to Dr M. T. Partis for these and other comments.

Hardy's communication to *Science* is dated 5 April, 1908 and was published on 10 July, 1908. It arose, as he states, from remarks made by the statistician G. Udny Yule in a discussion on 28 February, 1908 following a paper on 'Mendelism in relation to disease' delivered by R. C. Punnett to the epidemiological section of the Royal Society of Medicine.

A similar conclusion, although in a slightly simpler case where it is assumed at the outset that (in Hardy's notation) $q^2 = pr$, was earlier adduced by the German physician Wilhelm Weinberg (1862–1937) in a lecture given in Stuttgart on 13 January, 1908; this was published later that year (August–September?) under the title 'Über den Nachweis der Vererbung beim Menschen' in *Jahresheften des Vereins für vaterländische Naturkunde in Württemberg* 64 (1908), 369–82. The 'mathematics of the multiplication-table type' mentioned by Hardy is carried out in detail (with one minor error) on p. 379. See

C. Stern, 'The Hardy–Weinberg law', *Science*, 97 (1943), 137–8.

Reference may also be made to Weinberg's subsequent paper 'Über Vererbungsgesetze beim Menschen' in *Zeitschrift für induktive Abstammungs- und Vererbungslehre*, 1 (1908–09), 377–92, 440–60; 2 (1909), 276–330 (in particular, p. 283). Hardy's paper is listed in the list of references, but does not appear to be mentioned in the text. However, in a brief review of the paper on p. 395 of volume 1 of the same journal, Weinberg writes: 'Hardy weist nun darauf hin, dass Panmixie bei alternitiver Vererbung zu stabiler Bevölkerung führen müsse, was für einen speziellen Fall bereits 1904 Pearson und zu Anfang 1908 unabhäng von ihm und in einfacherer Weise Referent nachgewiesen hat. Siehe auch diese Zeitschrift S.377 ff.'

The special case considered by Karl Pearson is referred to in the postscript to Hardy's communication. The full reference is as follows:

Karl Pearson, 'III. Mathematical contributions to the theory of evolution.—XII. On a generalised theory of alternative inheritance, with special reference to Mendel's laws', *Philosophical Transactions of the Royal Society*, A 203 (1904), 53–86.

(d) *Elementary and expository notes*.

(e) *Addresses, invited lectures, etc.*

No general comments appear to be necessary on these publications.

R.A.R.

(a) Set Theory

A GENERAL THEOREM CONCERNING ABSOLUTELY CONVERGENT SERIES

By G. H. Hardy.

1. My object in this paper is to prove a theorem which contains as special cases a number of well known theorems concerning absolutely convergent series. Among these are, for instance, the following theorems:—

(i.) the terms of an absolutely convergent simple series may be permuted in any manner without altering the sum of the series;

(ii.) an absolutely convergent double series may be rearranged in any manner as a simple series;

(iii.) an absolutely convergent double series may be summed indifferently by rows or by columns (Cauchy's double-series theorem).

It occurred to me recently that all these theorems are very particular cases of a general theorem concerning the rearrangement of series. This theorem not only includes all the known theorems, but is, as I shall show, the most general possible theorem of its kind; and, in my opinion, its introduction lends to the theory a simplicity and generality which is otherwise lacking.

Series of Type β.*

2. The ordinary simply infinite series is *of type ω*, that is to say, its "ordinal type" is that of the class of numbers $< \omega$. The sum of two infinite series $a_0 + a_1 + \ldots + b_0 + b_1 + \ldots$ is of type $\omega \cdot 2$. A doubly infinite series, summed by rows or columns, is of type ω^2; summed by diagonals, it becomes of type ω. As the class of numbers $< \beta$ is enumerable, it is clear that we can arrange any enumerably infinite set of terms in type β.

---

* In what follows I shall denote by Greek letters $a, \beta, \gamma, \ldots$ numbers of Cantor's *first and second* classes $0, 1, 2, \ldots, \omega, \omega+1, \ldots, \omega \cdot 2, \ldots, \omega^2, \ldots, \omega^\omega, \ldots$. When it is necessary to distinguish specially the finite numbers (numbers of the *first* class) I shall use $m, n, p, \ldots$.

Suppose then that we have a series of type β, and, for the present, that its terms $u_0, u_1, \ldots, u_\omega, \ldots, u_{\omega.2}, \ldots, u_{\omega^2}, \ldots, u_{\omega^\omega}, \ldots$ are all $\geqslant 0$. There may or may not be a last term.

Sum of a Series of Type β.

3. Since all the u's are positive,

$$s_n = u_0 + u_1 + \ldots + u_{n-1} = \sum_{\gamma < n} u_\gamma$$

tends to a limit for $n = \infty$. We call this s_ω or $\sum_{\gamma < \omega} u_\gamma$. Of course it may be that $s_\omega = \infty$. If s_ω is finite, $s_{\omega+n} = s_\omega + u_\omega + \ldots + u_{\omega+n-1}$ has a limit for $n = \infty$, which we call $s_{\omega.2}$. This again may $= \infty$, and, if $s_\omega = \infty$, we regard $s_{\omega.2}$ as also $= \infty$. Thus we may define $s_{\omega.n}$ for any value of n, and it is clear that $s_\omega \leqslant s_{\omega.2} \leqslant s_{\omega.3} \ldots$; so that $\lim\limits_{n=\infty} s_{\omega.n}$ is determinate. We call this s_{ω^2}. It is clear that we may proceed thus and define s_α for all values of $\alpha \leqslant \beta$. We call s_β the sum of the series. If any $s_\alpha = \infty$, we agree that $s_{\alpha'} = \infty$ $(\alpha \leqslant \alpha' \leqslant \beta)$. In this case we say that the series is *divergent*. If s_β is finite, we say that the series is *convergent in type β*, and write $s_\beta = \sum\limits_{\gamma < \beta} u_\gamma$.

The object of this paper is to prove the following

THEOREM.—*If a series is absolutely convergent in type β, it remains absolutely convergent when its terms are rearranged in another type β', and its sums in the two types are the same.*

I may observe that the theorem covers the case of any rearrangement of the series in the *same* type.

Elementary Properties of Series of Type β.

4. The series thus defined possess some of the characteristic properties of ordinary series. Thus it is easy to prove that

$$(1) \quad \sum_{\gamma < \beta} (u_\gamma + v_\gamma) = \sum_{\gamma < \beta} u_\gamma + \sum_{\gamma < \beta} v_\gamma$$

if Σu_γ, Σv_γ are convergent, and

$$(2) \quad \sum_{\gamma < \beta} \kappa u_\gamma = \kappa \sum_{\gamma < \beta} u_\gamma.$$

These propositions follow readily from the equations

$$(a) \quad \sum_{\gamma < a+1} u_\gamma = \sum_{\gamma < a} u_\gamma + u_a$$

and

$$(b) \quad \sum_{\gamma < a} u_\gamma = \lim_{n = \infty} \sum_{\gamma < a_n} u_\gamma \quad (a = \lim a_n)$$

by induction.*

Again,

(3) *If* $\sum_{\gamma < \beta} u_\gamma$ *is convergent, and* $v_\gamma \leqslant u_\gamma$ *for all values of* γ, *then* Σv_γ *is convergent, and its sum* $\leqslant s_\beta$.

The proofs of these theorems are so simple that I need hardly write them out at length. In each case we show by means of the equations (*a*) and (*b*) that (i.), if they hold for series of type a, they hold for series of type $a+1$; and (ii.) that, if they hold for series of types $a_1, a_2, \ldots$ ($a_1 < a_2 < \ldots$, $\lim a_n = a$), they hold for series of type a. As they evidently hold for series containing only one term, they hold generally.

Proof of the Theorem.

5. In order to prove the theorem of § 3 we must establish a series of lemmas.

LEMMA 1.—*If a convergent series of type β is arranged in type ω, the new series is convergent and its sum* $\leqslant s_\beta$.

For the sum of any finite number of terms of the series $\leqslant s_\beta$. This follows from (3), § 4, if we replace all the other terms of the series by zeros. The lemma follows immediately.

LEMMA 2.—*If a series is convergent in type β, we can find a finite number of terms whose sum* $> s_\beta - \delta$, *where δ is an arbitrarily small positive quantity; and, if it is divergent, we can find a finite number of terms whose sum* $> G$, *an arbitrarily large quantity.*

For suppose the first part true for series of type a. Then it is true for series of type $a+1$. For, if $u_{\gamma_1} + u_{\gamma_2} + \ldots + u_{\gamma_N} > s_a - \delta$,

$$u_{\gamma_1} + u_{\gamma_2} + \ldots + u_{\gamma_N} + u_{a+1} > s_{a+1} - \delta.$$

Again, suppose the theorem true for series of the types $a_1, a_2, \ldots$, and $a = \lim_{n = \infty} a_n$. Then, since $s_a = \lim_{n = \infty} s_{a_n}$, we can choose n so that

$$s_{a_n} > s_a - \tfrac{1}{2}\delta.$$

---

* One fundamental property of an ordinary series which is obviously not possessed by the generalized series is that in an ordinary series we can assign a rank after which the terms are as small as we please.

But we can choose N numbers $\gamma_1, \ldots, \gamma_N < a_n$ so that
$$u_{\gamma_1} + \ldots + u_{\gamma_N} > s_{a_n} - \tfrac{1}{2}\delta > s_{a_n} - \delta.$$

As the first part of the lemma is true for series of one term, it is true generally. A similar proof applies to the second part.

LEMMA 3.—*If a series is convergent in type ω and its sum is s_ω, it will be convergent when arranged in type β and its sum will be $\leqslant s_\omega$.*

This follows at once by a *reductio ad absurdum* from Lemma 2.

LEMMA 4.—*If a series is convergent in type β, it will be convergent when arranged in type ω, and the sums will be the same, and conversely.*

Suppose it convergent in type β. By Lemma 1 it is convergent in type ω, and its sum $\leqslant s_\beta$. But, if its sum $< s_\beta$, it follows from Lemma 3 that its sum in type $\beta < s_\beta$, which is absurd. The converse is proved similarly. From Lemma 4 the theorem follows at once.

A well known particular case of the theorem is the theorem concerning the summation of a doubly infinite series by rows, columns, or diagonals. In this case we have to deal with two arrangements in type ω^2 and one in type ω. Similar theorems for multiple series of any order follow at once.

[It is perhaps worth while at this point to emphasize the distinction between a *multiple* and a *repeated* series, as it is only recently that the notion of a multiple series has been made at all precise. According to Jordan the double series $\sum_{m,n=1}^{\infty} u_{m,n}$ is convergent only if *all* possible simply infinite series formed out of it are convergent. According to Pringsheim and Stolz it is convergent if *a certain large class* of them are. The point of importance for my present purpose is that in any case to assert the convergence of a double or multiple series is not really to assert the convergence of *one* series of a type different from that of the ordinary simple series, but to assert the convergence of a *whole class* of series of the ordinary type. But to assert the convergence of a repeated series *is* to assert the convergence of one series of a type in any case $> \omega$. The theorem applies to either kind of series.]

6. The theorem is easily extended to any absolutely convergent series, real or complex.

An absolutely convergent series of type β is a series of type β which is convergent when every term is replaced by its modulus. In the first place it is to be observed that the definitions and explanations of § 3 were given only for series of positive and zero terms. They apply, however, to series of arbitrary terms, except that when the terms are not all positive we cannot be sure that a definite limit (finite or infinite) will be found at each stage of the process. However, we can prove that *an absolutely convergent series of type β is convergent*; that is to say, a definite limit s_β results if we carry out the process of § 3.

For suppose first that the terms are all real, and define two new series Σv_γ, Σw_γ by the equations

$$v_\gamma = u_\gamma \ (u_\gamma > 0) = 0 \ (u_\gamma < 0), \quad w_\gamma = 0 \ (u_\gamma > 0) = -u_\gamma \ (u_\gamma < 0);$$

then $u_\gamma = v_\gamma - w_\gamma$. Also, since $|v_\gamma| \leqslant |u_\gamma|$, $w_\gamma \leqslant |u_\gamma|$, Σv_γ, Σw_γ are convergent. Moreover, from (1) of § 4, $\sum_{\gamma < a} u_\gamma$ is convergent and

$$\sum_{\gamma < a} u_\gamma = \sum_{\gamma < a} v_\gamma - \sum_{\gamma < a} w_\gamma.$$

Now apply the theorem of the last paragraph to each of these series separately. We find that $\Sigma u_\gamma = \Sigma' v_\gamma - \Sigma' w_\gamma$, Σ' denoting a series of the new type. And the right-hand is $\Sigma' u_\gamma$.

Next suppose the terms u_γ complex, and let $u_\gamma = v_\gamma + i w_\gamma$. Then, since $|v_\gamma| \leqslant |u_\gamma|$, $|w_\gamma| \leqslant |u_\gamma|$, the series Σv_γ, Σw_γ are absolutely convergent; and the theorem follows as before. The theorem is therefore proved for all absolutely convergent series.

7. It only remains for me to justify the assertion that this theorem is the most general possible theorem of its kind. This is evident if we consider that *any* rearrangement of an ordinary series of type ω must lead to an arrangement in *some* type β. The possibility of a series with a non-enumerable infinity of positive terms is excluded by Cantor's theorem that any set of intervals on a straight line must be enumerable.*

8. In conclusion, I give an example, suggested by a passage in Schoenflies's *Mengenlehre*, of an arrangement of the series

$$\frac{1}{1^2} + \frac{1}{2^2} + \frac{1}{3^2} + \cdots$$

---

* See *e.g.*, W. H. Young, "On Sets of Intervals," *Proc. London Math. Soc.*, Vol. **xxxv.**, p. 248. Another question is whether it is possible to define the sum of a series with a non-enumerable number of terms which are *not* all positive. It seems to me that the answer is probably *no*; but, as the question involves considerations of some logical difficulty, I shall not at present attempt to justify this answer.

in type β, where β is at any rate $> \omega^\omega$. It is

$$\frac{1}{1^2} + \frac{1}{2^2} + \frac{1}{3^2} + \frac{1}{5^2} + \frac{1}{7^2} + \frac{1}{11^2} + \cdots$$

$$+ \frac{1}{2^2 \cdot 3^2} + \frac{1}{2^2 \cdot 5^2} + \frac{1}{2^2 \cdot 7^2} + \cdots$$

$$+ \frac{1}{3^2 \cdot 5^2} + \frac{1}{3^2 \cdot 7^2} + \cdots$$

$$+ \frac{1}{5^2 \cdot 7^2} + \cdots$$

$$+ \cdots$$

$$+ \frac{1}{2^2 \cdot 3^2 \cdot 5^2} + \frac{1}{2^2 \cdot 3^2 \cdot 7^2} + \frac{1}{2^2 \cdot 3^2 \cdot 11^2} + \cdots$$

$$+ \frac{1}{2^2 \cdot 5^2 \cdot 7^2} + \frac{1}{2^2 \cdot 5^2 \cdot 11^2} + \cdots$$

$$+ \frac{1}{2^2 \cdot 7^2 \cdot 11^2} + \cdots$$

$$+ \cdots$$

$$+ \frac{1}{3^2 \cdot 5^2 \cdot 7^2} + \cdots$$

$$+ \cdots.$$

It is easy to see that, if we carry out this scheme systematically, $\sum_{\gamma < \omega^2}$ contains every $1/n^2$ for which n is a product of two primes (including 1), $\sum_{\gamma < \omega^3}$ includes every $1/n^2$ for which n is a product of three primes, and so on. Hence $\sum_{\gamma < \omega^\omega}$ includes only those terms for which n is a product of different primes. The complete series (however we go on with it) is therefore of type $\beta > \omega^\omega$.

A THEOREM CONCERNING THE INFINITE CARDINAL NUMBERS.

By G. H. Hardy, Trinity College, Cambridge.

§ 1. THE principal object of this paper is to prove rigorously that the cardinal number of the continuum is *greater than or equal to* the cardinal number of Cantor's second number class; in symbols, that

$$2^{\alpha_0} \geq \alpha_1,$$

and, more generally, that

$$2^{\alpha_\beta} \geq \alpha_{\beta+1},$$

where β is any ordinal, and α_β the βth cardinal contained in the well ordered series of cardinals constructed by Cantor.* I do not claim any great degree of originality for this theorem, for, as I shall show in § 2, it follows at once from Cantor's theorem that

$$2^\alpha > \alpha,$$

where α is any cardinal (whether contained in the above mentioned well ordered series or not), and some general consideration of the nature of the cardinals contained in the series. But the theorem has not, as far as I know, ever been stated explicitly. Indeed, in so recent a work as Mr. Russell's *Principles of Mathematics*, it is expressly stated that 'it may be that 2^{α_0} is neither greater nor less than α_1 and α_2 and their successors.'† And it seems to me that the investigation of § 3, in which I construct a set of points of cardinal α_1 in the linear continuum, may be of some interest as throwing some light (though of course a very partial one) on one of the most fundamentally important and apparently hopeless questions in the whole range of pure mathematics.

---

* Schoenflies, *Mengenlehre*, p. 49. I use α_0, α_1, ... for Cantor's Aleph-Zero, Aleph-Eins, ... Mr. Whitehead and Mr. Russell use a_0, a_1, ...

† *l. c.*, p. 323.

§ 2. It is, however (as I have already remarked), possible to prove that $2^{\alpha_0} \geq \alpha_1$ without using the construction of § 3. For, although we are not entitled to assume that the cardinals of all aggregates occur in the series

$$\alpha_0, \alpha_1, \ldots, \alpha_\omega, \ldots, \alpha_\beta, \ldots,$$

it does seem to me clear that any cardinal must either occur in the series or be *greater than all* the terms of the series.

This can, I think, be shown by an extension of Cantor's argument proving that every infinite cardinal is greater than or equal to α_0. For, given any aggregate whose cardinal $> \alpha_0$, we can choose from it successively individuals

$$u_1, u_2, \ldots, u_\omega, \ldots, u_\beta, \ldots$$

corresponding to all the numbers of the first and second classes; if the process came to an end, the cardinal of the aggregate would be α_0. Its cardinal therefore $\geq \alpha_1$; and if $> \alpha_1$, $\geq \alpha_2$, and so on. And if $> \alpha_n$, for all finite values of n, it must be $\geq \alpha_\omega$; for we can choose individuals from the aggregate corresponding to all the numbers of the first, second, third, ..., n^{th}, ... classes. And by a repetition of these two arguments, we can show that if there is no α_β equal to the cardinal of the aggregate, it must be at least equal to the cardinal of the aggregate of all α_β's, and so greater than any α_β.

These considerations must, I imagine, have been familiar to Cantor. But he confines himself to showing that there is no cardinal *between* α_0 and α_1, *i.e.* $> \alpha_0$, but $< \alpha_1$, and never explicitly rejects the possibility contemplated by Mr. Russell.

§ 3. I come now to the actual construction of a set of points of cardinal α_1. I am not aware that such a set has been constructed before, unless indeed $2^{\alpha_0} = \alpha_1$; for all known sets have α_0 or 2^{α_0} as their cardinal.

Starting from the sequence of integral numbers,

$$(1) \quad 1, 2, 3, 4, 5, \ldots$$

infinite cardinal numbers.

we form a new sequence,

$$(2) \quad 2, 3, 4, 5, 6, \ldots$$

by omitting the first term; and by continuing the process we form

$$(3) \quad 3, 4, 5, 6, 7, \ldots$$
$$(4) \quad 4, 5, 6, 7, 8, \ldots$$
$$(5) \quad 5, 6, 7, 8, 9, \ldots$$
$$\ldots\ldots\ldots\ldots\ldots\ldots$$

We now form a new sequence

$$(\omega) \quad 1, 3, 5, 7, 9, \ldots$$

by traversing the above infinite array of sequences diagonally. Then we form

$$(\omega+1) \quad 3, 5, 7, 9, 11, \ldots$$
$$(\omega+2) \quad 5, 7, 9, 11, 13, \ldots$$
$$(\omega+3) \quad 7, 9, 11, 13, 15, \ldots$$
$$(\omega+4) \quad 9, 11, 13, 15, 17, \ldots$$
$$\ldots\ldots\ldots\ldots\ldots\ldots$$
$$(\omega.2) \quad 1, 5, 9, 13, 17, \ldots$$
$$(\omega.2+1) \quad 5, 9, 13, 17, 21, \ldots$$
$$(\omega.2+2) \quad 9, 13, 17, 21, 25, \ldots$$
$$(\omega.2+3) \quad 13, 17, 21, 25, 29, \ldots$$
$$\ldots\ldots\ldots\ldots\ldots\ldots$$
$$(\omega.3) \quad 1, 9, 17, 25, 33, \ldots$$
$$\ldots\ldots\ldots\ldots\ldots\ldots$$

Thus we form sequences corresponding to all the numbers

$$\omega.\mu + \nu,$$

where μ and ν are finite.*

---

* The method employed is similar in principle to that used by Borel to form a non-enumerable sequence of 'croissances'; v. *Leçons*, 1, p. 114.

To form the sequence corresponding to ω^2 we take the array of sequences

(ω) 1, 3, 5, 7, 9, ...
$(\omega.2)$ 1, 5, 9, 13, 17, ...
$(\omega.3)$ 1, 9, 17, 25, 33, ...
$(\omega.4)$ 1, 17, 33, 49, 65, ...
$(\omega.5)$ 1, 33, 65, 97, 129, ...

and traverse it diagonally; so that we obtain

(ω^2) 1, 5, 17, 49, 129, ...

Generally, if
$$b_1, b_2, b_3, b_4, ...$$
corresponds to β,
$$b_2, b_3, b_4, b_5, ...$$
corresponds to $\beta + 1$; while to obtain a set corresponding to a number γ which has no predecessor, we take the array of sequences corresponding to any ascending set of numbers
$$\beta_1, \beta_2, ...$$
whose limit is γ, and traverse it diagonally, as has been shown in the particular cases of the numbers
$$\omega, \omega.2, ..., \omega^2.$$

We might equally well have considered ω as the limit of
$$1, 3, 5, 7, ...,$$
or ω^2 as the limit of
$$\omega + 1, \omega.2 + 2, \omega.3 + 3, ...;$$
in fact we have an infinite freedom of choice whenever we wish to define the sequence corresponding to any number which has no immediate predecessor. This freedom of choice is important, as we are able, by exercising it suitably, to make it clear that all the sets we obtain are distinct.

It is clear that as we can always define a sequence corresponding to $\beta + 1$ if we know one corresponding to β, and a sequence corresponding to $\gamma = \lim. \beta_m$ if we know those corresponding to $\beta_1, \beta_2, ..., \beta_m, ...$, we can certainly find sequences corresponding to all the numbers of the second class. Hence, if we prove that all the sequences are distinct, their aggregate will have a_1 as its cardinal.

§ 4. I shall now establish this by proving that we can so construct our sequences
$$b_1, b_2, b_3, \ldots$$
that in every case $b_1 < b_2 < b_3 \ldots$, and that, if $b_1, b_2, b_3, \ldots$ and $b_1', b_2', b_3', \ldots$ correspond to β and β', and $\beta < \beta'$, there exists a number N such that
$$b_n' > b_n \quad (n \geq N).$$

Let us assume that we have constructed sequences corresponding to all the numbers $< \gamma$ in such a way that they satisfy this condition. Then there are two cases to consider, that in which γ has an immediate predecessor γ', and that in which it has not.

In the first place let $\gamma = \gamma' + 1$. Then if $\beta < \gamma'$ there is a number N such that
$$a_n' > b_n \quad (n \geq N).$$
But
$$a_n = a'_{n+1} > a_n' > b_n \quad (n \geq N).$$

Hence, if the construction is posible for all numbers $< \gamma$, it is possible for all numbers $\leq \gamma$.

Next suppose that γ has no immediate predecessor, and that
$$\gamma = \lim. \beta_m \quad (\beta_1 < \beta_2 < \beta_3 \ldots),$$
then
$$\gamma = \lim. (\beta_m + \nu_m),$$
where the ν's are any finite numbers. Now there is a number N_1 such that
$$b_{2,n} > b_{1,n}, \quad (n \geq N_1),$$
$b_{m,n}$ being the nth number in the sequence corresponding to β_m. A fortiori, if $\gamma_m = \beta_m + \nu_m$,
$$c_{2,n} = b_{2,n+\nu_2} > b_{2,n} > b_{1,n} \quad (n \geq N_1).$$

But if we take $\nu_2 > b_{1, N_1 - 1}$
$$c_{2,n} = b_{2, n+\nu_2} \geq n + \nu_2 > b_{1, N_1 - 1} > b_{1,n} \quad (n < N_1),$$

Hence $c_{2,n} > b_{1,n}$ for *all* values of n. Similarly we can choose ν_3 so that $\gamma_3 > \gamma_2$, and $c_{3,n} > c_{2,n}$ for all values of n; and so on generally.

For the sake of uniformity I write γ_1 for β_1, $c_{1,n}$ for $b_{1,n}$. Then we have a doubly infinite array

$$c_{1,1},\ c_{1,2},\ c_{1,3},\ \ldots,$$
$$c_{2,1},\ c_{2,2},\ c_{2,3},\ \ldots,$$
$$c_{3,1},\ c_{3,2},\ c_{3,3},\ \ldots,$$
$$\ldots\ldots\ldots\ldots\ldots,$$

and we define the sequence corresponding to γ by traversing it diagonally, so that

$$c_n = c_{n,n}.$$

If then $\beta < \gamma$, we can find m so that

$$\beta < \gamma_m.$$

Then there is a number K such that

$$c_{m,n} > b_n \quad (n \geq K).$$

But if $n > m$,

$$c_n = c_{n,n} > c_{m,n}.$$

If therefore n is greater than the greater of m, K

$$c_n > b_n.$$

And therefore, if the construction is possible for all numbers $< \gamma$, it is possible for all numbers $\leq \gamma$, whether γ has a predecessor or not. But it is evidently possible for small values of γ. Thus it is possible to carry out the construction so as to obtain $\mathfrak{a}_1$ distinct sequences.

In the case of the comparatively early numbers of the second class it is generally evident that a γ which has no predecessor is most naturally regarded as the limit of one particular set $\beta_1, \beta_2, \ldots$. Thus it is natural to regard ω^ω as the limit of

$$1,\ \omega,\ \omega^2,\ \omega^3,\ \ldots,$$

ω^{ω^ω} as the limit of

$$\omega,\ \omega^\omega,\ \omega^{\omega^2},\ \omega^{\omega^3},\ \ldots,$$

and ϵ_1, the first of Cantor's ϵ-numbers as the limit of

$$\omega,\ \omega^\omega,\ \omega^{\omega^\omega},\ \ldots.$$

And it is unnecessary to introduce the numbers ν, it being evident that the sequences which we obtain are all distinct. Thus I find

$$
\begin{array}{ll}
(\omega^2 + 1) & 5,\ 17,\ 49,\ 129,\ \ldots \\
(\omega^2 + \omega) & 1,\ 17,\ 129,\ \ldots \\
(\omega^2 + \omega.2) & 1,\ 129,\ \ldots \\
(\omega^2.2) & 1,\ 17,\ \ldots \\
(\omega^3) & 1,\ 17,\ \ldots \\
(\omega^\omega) & 1,\ 5,\ \ldots \\
(\epsilon_1) & 1,\ 5,\ \ldots .
\end{array}
$$

It will be seen that (ω^ω) and (ϵ_1) have the same two first terms, and we should find with higher numbers sequences which, though distinct, had more than any fixed finite number of terms in common. This will be clear when we have correlated the sequences with points in the linear continuum. The simplest way of effecting this correlation is perhaps to correlate

$$b_1,\ b_2,\ b_3,\ \ldots$$

with the binary decimal in which the b_1^{th}, b_2^{th}, ... figures are 1's and the remaining figures 0's. As the set of points thus obtained is non-enumerable, it must contain $\mathfrak{a}_1$ of its own limit points. Thus there are $\mathfrak{a}_1$ sequences such that, however great be n, we can find other sequences agreeing with them in their first n terms.

I may remark that we can divide the numbers 1, 2, ... into two classes, those which do not appear in any of the sequences after a certain one, and those which appear in a non-enumerable infinity of sequences. The latter we may call *persistent* numbers. If $\mu_1, \mu_2, \ldots$ are the non-persistent numbers, we can find numbers of the second class

$$\gamma_1,\ \gamma_2,\ \ldots,$$

such that μ_n does not appear in any sequences after the γ_n^{th}. If γ is the first number greater than any of $\gamma_1, \gamma_2, \ldots$, all the sequences after the γth are composed entirely of persistent numbers.

§5. A similar method may be employed to show that generally

$$2^{\mathfrak{a}_\beta} \geq \mathfrak{a}_{\beta+1}.$$

For instance, if $\beta = 1$, we start with the complete sequence

(1) $1, 2, 3, \ldots, \omega, \omega+1, \ldots, \omega.2, \ldots, \omega^2, \ldots$

of the numbers of the first and second classes, and form successively the sequences

(2) $2, 3, 4, \ldots, \omega+1, \omega+2, \ldots, \omega.2+1, \ldots, \omega^2+1, \ldots$
(3) $3, 4, 5, \ldots, \omega+2, \omega+3, \ldots, \omega.2+2, \ldots, \omega^2+2, \ldots$
(4)   ....

by increasing each term by $1, 2, \ldots$. We may obtain the sequence corresponding to ω by considering (1) as made up of α_1 sequences of the type

$$1, 2, 3, \ldots,$$

an aggregate of sequences which is ordinally similar to (1). We then get (ω) from (1) by moving these sequences in the same way in which we moved the single numbers when we obtained (2) from (1). Thus we find

(ω) $\omega, \omega+1, \omega+2, \ldots, \omega.2, \ldots, \omega^2+\omega, \ldots$.

Similarly we find

$(\omega.2)$ $\omega.2, \omega.2+1, \ldots, \omega.3, \ldots, \omega^2+\omega.2, \ldots$.

To find the sequence corresponding to ω^2, we consider (1) as made up of α_1 sequences of the type ω^2; this aggregate of sequences is again ordinally similar to (1).

By following this method we can find sequences for all the numbers of the second class. Then, by traversing the whole array of α_1 sequences we obtain a sequence for Ω, the first number of the third class. And then there is no difficulty in finding sequences for all the numbers of the third class. And these sequences may be correlated with a part of the aggregate of all possible sequences of α_1 figures each of which is either 0 or 1, an aggregate whose cardinal is clearly 2^{α_1}.* But I shall not attempt for this (still less for the general case) an investigation similar to that of §4, but shall content myself with the general line of proof indicated in §2, and the detailed construction given in §§3, 4 for the most interesting case.

---

* Mr. Whitehead worked out some of the most interesting properties of such aggregates in his lectures on the application of symbolic logic to the theory of aggregates, delivered during the winter of 1902-3.

THE CARDINAL NUMBER OF A CLOSED SET OF POINTS.

By *G. H. Hardy*, Trinity College, Cambridge.

§ 1. CANTOR proved that the cardinal number of a closed set of points contained in the linear continuum $(0, 1)$ is a_0 or 2^{a_0}. His proof* depends on two propositions, (i) that a closed set may be decomposed into the sum of an enumerable series of enumerable sets, and a set which is perfect or zero,

---

* v. Schoenflies, *Bericht über die Mengenlehre*, pp. 65—69.

(ii) that the cardinal number of a perfect set is 2^{α_0}. My object in this note is to give a proof of this theorem which seems to me simpler and more direct than Cantor's, in that (i) it does not involve any reference to the ordinal numbers of the second and higher classes, and the derived sets corresponding to them, and (ii) it avoids the necessity of dividing the theorem into several parts.

§ 2. Suppose that when any number x of the continuum (0, 1) is expressed as a binary decimal the a_1^{th}, a_2^{th}, ... figures are 1's, and the rest 0's. We represent x by the sequence

$$a_1, a_2, a_3, \ldots.$$

If now $x^1, x^2, x^3, \ldots$ is a sequence of numbers whose limit is x, it is clear that we can find a number M corresponding to any given number N, such that

$$a_n^m = a_n \ (n \leq N)$$

for every value of $m \geq M$.

And, conversely, if this condition is satisfied

$$\lim_{m = \infty} x^m = x.$$

§ 3. Now suppose we have a set of numbers x whose cardinal $> \alpha_0$. It may be that a_1 is the same for every x or for all but an enumerable number of x's. In this case we neglect this enumerable number, and we have left a non-enumerable set of x's all beginning, say, with b_1. It may then be that a_2 is the same (say $= b_2$) for all these x's, or all but an enumerable number of them; in this case we neglect these, and remain with a non-enumerable set of x's all beginning with b_1, b_2. This process may be repeated. It is, however, clear that we must come to a term a_{n_1} which is *not* the same for all but an enumerable set of x's.

For otherwise, after neglecting an enumerable number of enumerable sets of x's, we should be left with a single sequence

$$b_1, b_2, b_3, \ldots,$$

and so the whole set of x's would be enumerable. There are therefore two numbers $b^1_{n_1}$, $b^0_{n_1}$ such that a non-enumerable set of x's begin in each of the ways

(1) $\qquad b_1, b_2, \ldots, b_{n_1-1}, b^1_{n_1},$

(2) $\qquad b_1, b_2, \ldots, b_{n_1-1}, b^0_{n_1}.$

We neglect all the other x's.

We now repeat this argument with each of the sets of x's which begin with (1) or (2). We thus obtain *four* non-enumerable sets beginning respectively with

$$b_1, b_2, \ldots, b_{n_1-1}, \begin{cases} b^1{}_{n_1^1}, b^1{}_{n_1^1+1}, \ldots, b^1{}_{n_2^1-1} \begin{cases} b^{11}{}_{n_2^{11}}\ldots, \\ b^{10}{}_{n_2^{10}}\ldots, \end{cases} \\ b^0{}_{n_1^0}, b^0{}_{n_1^0+1}, \ldots, b^0{}_{n_2^0-1} \begin{cases} b^{01}{}_{n_2^{01}}\ldots, \\ b^{00}{}_{n_2^{00}}\ldots, \end{cases} \end{cases}$$

where $\quad n_1^1 = n_1^0 = n_1, \; n_2^{11} = n_2^{10} = n_2^1, \; n_2^{01} = n_2^{00} = n_2^0$,

but of course n_2^1 is not necessarily $= n_2^0$. And this process may be repeated indefinitely.

After n steps we have 2^n classes of sets characterised by indices which are the 2^n binary decimals of n figures.

Now consider any binary decimal, say

$$\cdot 010110\ldots.$$

The first figure 0 is the first index of all the b's in the above scheme which fall in the lower half after the first point of division. The second figure 1 is the second index of all these b's which fall in the upper half after the second point of division. The first n figures are the first n indices of all the b's which fall in a certain one of the 2^n sets formed by n divisions. The complete decimal thus guides us through the whole scheme and defines a certain sequence of b's. This sequence may or may not be one of the x's. But it must be a limit point of the set. Consider, for instance, the set

$$b_1, b_2, \ldots, b_{n_1-1}, b^1{}_{n_1^1}, b^1{}_{n_1^1+1}, \ldots, b^1{}_{n_2^1-1}, b^{11}{}_{n_2^{11}}, \ldots,$$

which is obtained by always keeping along the top of the scheme; *i.e.* the set corresponding to $\cdot 11111\ldots$. It is evident that the set of x's contains a non-enumerable number which agrees with this set as far as the Nth term, N being any integer. Moreover it is clear that two different binary decimals define different limit points.

Hence the set has $2^{\aleph_0}$ limit points; and if it is closed, its own cardinal is $2^{\aleph_0}$. In particular, the cardinal of a perfect set is $2^{\aleph_0}$.

THE CONTINUUM AND THE SECOND NUMBER CLASS

By G. H. Hardy.

[Received August 3rd, 1905.—Read November 9th, 1905.]

1. In a recent number of these *Proceedings** Dr. Hobson criticises (among other things) a construction which I gave in 1903† for a set of points of cardinal number $\aleph_1$ contained in the linear continuum (0, 1). This criticism is merely incidental to a much more comprehensive attack on the whole theory of Cantor's transfinite numbers, as it has been generally accepted by mathematicians, and in particular to the theory of cardinals elaborated by Mr. Whitehead and Mr. Russell, and expounded in the latter's *Principles of Mathematics*. It is, I believe, Mr. Russell's intention to reply to Dr. Hobson, and I should not wish to discuss the general question in the present communication, even if I felt competent to do so. My present object is a much more modest one. Besides the large question with which we are all concerned there is a smaller one which concerns only Dr. Hobson and myself. Each of us is of opinion that the other has made a mathematical mistake. It is with this smaller question that I propose to deal now, and I shall only refer to the larger issue in so far as is necessary if I am to make clear what the difference between us really is. If Dr. Hobson's views concerning cardinal numbers in general were correct, my construction would acquire a fundamental importance which I am not myself at all disposed to attach to it: I need therefore make no apology for considering in detail this particular part of Dr. Hobson's paper.

2. Before I proceed to discuss Dr. Hobson's objections to my construction, it will be convenient if I indicate a slight simplification‡ which can be made in it. Whether this alteration be made or not in no way affects the force of Dr. Hobson's arguments.

---

* *Proc. London Math. Soc.*, New Series, Vol. 3, p. 170.
† *Quarterly Journal*, Vol. xxxv., p. 88.
‡ Suggested by a passage in Baire's *Leçons sur les Fonctions discontinues*, p. 25.

The principle of my construction was to associate with every number a of the second class an ascending sequence

$$(a) \qquad a_1,\ a_2,\ a_3,\ \ldots$$

of positive integers by means of the three following rules:—

(i.) To the number 1 is to correspond the sequence

$$(1) \qquad 1,\ 2,\ 3,\ \ldots.$$

(ii.) The sequence for $a+1$ is to be formed by omitting the first term of the sequence for a.

(iii.) If a is a number of the second kind (one with no immediate predecessor), we are to select a fundamental sequence (a_ν) of which a is the limit, and we are to traverse the array

$$(a_1) \qquad a_{1,1},\ a_{1,2},\ a_{1,3},\ \ldots,$$
$$(a_2) \qquad a_{2,1},\ a_{2,2},\ a_{2,3},\ \ldots,$$
$$\ldots \qquad \ldots \qquad \ldots$$

diagonally, so that we obtain

$$(a) \qquad a_{1,1},\ a_{2,2},\ a_{3,3},\ \ldots.$$

I found then that in order to assure ourselves that the sequences thus generated are all distinct it may be necessary to substitute for the fundamental sequence (a_ν) another sequence $(a_\nu + m_\nu)$, where the m_ν's are finite numbers formed successively according to a definite rule which I gave in my former paper.* I now wish to point out that this slight complication is quite unnecessary if, instead of defining the sequence for a by the simple equation
$$a_n = a_{n,n},$$
we take a_n to be *the greatest of the integers* $a_{1,n},\ a_{2,n},\ \ldots,\ a_{n,n}$. It is then easy to prove, without the introduction of the numbers m_ν, that, if a, a' ($a < a'$) are any two numbers of the second class, there is a definite n from and after which $a'_n > a_n$; so that no two sequences can be the same. For, if this is true for all numbers $\leqslant a'$ (say), it is obviously true for all numbers $\leqslant a = a' + 1$. We have therefore only to show that, if it is true for all numbers $\leqslant a_\nu$, where a_ν is an arbitrary member of a fundamental sequence (a_ν), it is true for all numbers $\leqslant a$, the number which immediately follows this fundamental sequence. According to the construction, a_n is the greatest of
$$a_{1,n},\ a_{2,n},\ \ldots,\ a_{n,n}.$$

---

* *Loc. cit.*, p. 91.

Then, if $\beta < \alpha$, we can find m so that $\beta < a_m$, and we can find n_0 so that $n_0 > m$ and $a_{m,n} > b_n$ for $n \geqslant n_0$. Thus

$$a_n \geqslant a_{m,n} > b_n,$$

and so the sequences for β and for α are distinct.

It follows that, if all the sequences which correspond to numbers $< \alpha$ are distinct, all those which correspond to numbers $\leqslant \alpha$ are distinct, whether α has an immediate predecessor or not; and therefore that *all* the sequences are distinct.

3. Dr. Hobson argues (p. 187) that this construction must be faulty for the following reason. Given any integer m, a least number a_m, he contends, can be found, such that for all numbers $\alpha > a_m$ the second term of the corresponding sequence is greater than m. Taking a sequence of numbers

$$m_1 < m_2 < \ldots,$$

and forming the number α which is the limit of the fundamental sequence (a_{m_ν}), he deduces that the second term in the sequence for α is greater than m_ν, for all values of ν, *i.e.* is greater than any assignable integer, and therefore that no such sequence exists. And, in fact, he concludes that, if the sequences for the early numbers of the second class are formed as I formed them in my earlier paper, no sequence can be constructed to correspond to the number

$$\epsilon_\omega = \lim \epsilon_\nu,$$

where

$$\epsilon_1 = \lim \omega, \omega^\omega, \omega^{\omega^\omega}, \ldots,$$

$$\epsilon_2 = \lim \epsilon_1, \epsilon_1^{\epsilon_1}, \epsilon_1^{\epsilon_1^{\epsilon_1}}, \ldots,$$

$$\ldots \quad \ldots \quad \ldots$$

$$\epsilon_{\nu+1} = \lim \epsilon_\nu, \epsilon_\nu^{\epsilon_\nu}, \epsilon_\nu^{\epsilon_\nu^{\epsilon_\nu}}, \ldots .*$$

I hardly think that Dr. Hobson can have realised how paradoxical his conclusion is. For it follows from the definition of the construction that, if sequences have been assigned by it for all numbers $< \alpha$, a sequence is assigned by it for α. Therefore, if there is no sequence for α, we can find a number $< \alpha$, say $\alpha^{(1)}$, for which there is no sequence, a number $\alpha^{(2)} < \alpha^{(1)}$ for which there is no sequence, and so on. But, as it is impossible to find an infinite descending sequence $\alpha, \alpha^{(1)}, \alpha^{(2)}, \ldots$, we shall find ultimately that there are no sequences at all; and even Dr. Hobson is not so sceptical as this.

---

* It is not difficult to prove that the numbers thus defined are the same as the first few of Cantor's ϵ-numbers.

My answer to Dr. Hobson's argument consists simply of a denial of his major premiss. It is not true that, given *any* integer m, a number a_m can be found such that, for *every* $a > a_m$, $a_2 > m$. The second terms of our sequences may go on increasing for a while, but sooner or later a sudden jump downwards will occur. This is most obvious, perhaps, if we consider a fundamental sequence of the type $(a+\nu)$.

If for a the second term is a_2, for $a+\nu$ the second term is $a_2+\nu$, and we can find a value of ν for which this number is greater than any assignable number. But it by no means follows that the second term in the sequence for $a+\omega$ is greater than any assignable number; on the contrary, it is a_3, the second term in the sequence for $a+1$, and is *less* than the second terms in the sequences for $a+2, a+3, \ldots$. And, more generally, if b_n is the n-th term in the sequence for $a+\omega$, we can find a value of μ such that the n-th terms in the sequences for $a+\nu$ $(\nu \geqslant \mu)$ are all $> b_n$. It is equally true, as I showed in § 2, that, *given* ν, we can find n_0 so that for $n \geqslant n_0$ the figures in the sequence for $a+\omega$ are greater than those in the sequence for $a+\nu$. That these two propositions should simultaneously be true may seem paradoxical for a moment, but we have only to consider the sequences

$$
\begin{array}{ll}
(1) & 1, \ 2, \ 3, \ 4, \ \ldots, \\
(2) & 2, \ 3, \ 4, \ 5, \ \ldots, \\
& \ldots \quad \ldots \quad \ldots \quad \ldots \\
(\nu) & \nu, \ \nu+1, \ \nu+2, \ \nu+3, \ \ldots, \\
& \ldots \quad \ldots \quad \ldots \quad \ldots \\
(\omega) & 1, \ 3, \ 5, \ 7, \ \ldots,
\end{array}
$$

to see that they are in reality perfectly consistent with one another, just as the propositions—

(i.) given m, we can find x_0 so that, for $x \geqslant x_0$, $e^x > x^m$;

(ii.) given $x_0 (>1)$, we can find m_0 so that, for $m \geqslant m_0$, $x_0^m > e^{x_0}$

—are consistent with one another.

It may, no doubt, be the case that for a *particular* number m (*e.g.*. 1,000,000) all sequences later than an assignable sequence have second figures $> m$. Thus Dr. Hobson asserts that, for every $a \geqslant \omega^\omega$, $a_2 \geqslant 5$, and this may perfectly well be true. But, if it is true, it will depend on *two* facts:

(i.) that, for $a = \omega^\omega$, $a_2 \geqslant 5$;

(ii.) that *in constructing the sequences corresponding to higher numbers we never use a limit sequence* $a_1, a_2, \ldots$ *containing more than one term* $< \omega^\omega$.

Whether (ii.) is true or not depends entirely on the particular "norm" chosen for forming the sequences. If I had chosen $1, \omega, \omega^\omega, \omega^{\omega^\omega}, \ldots$ instead of $\omega, \omega^\omega, \omega^{\omega^\omega}, \ldots$ as the fundamental sequence for ϵ_1, the second term in the corresponding sequence would have been 3. And what Dr. Hobson's argument really proves is simply that, *however* the sequences are chosen, what may be true for some particular values of m cannot be true for *all* values of m, and that no such proposition as (ii.) above can be true for more than an enumerable sequence of values of a (such as ω^ω), which is otherwise obvious. And where his argument breaks down with regard to the particular sequence of numbers $(\epsilon_1, \epsilon_2, \ldots, \epsilon_\omega)$ which he considers is in his not having noticed that the second term in the sequence for ϵ_ω is less than the second term in the sequences for $\epsilon_3, \epsilon_4, \ldots$ (if, as is natural, we take $\epsilon_1, \epsilon_2, \ldots$ as the fundamental sequence for ϵ_ω and grant that, as he asserts, the second term in the sequence for ϵ_ν continually increases with ν).

4. The arguments which Dr. Hobson urges specifically against me appear to me therefore to be invalid. These arguments are (as I think Dr. Hobson and I agree) quite independent of those used by him in his general attack on the theory of transfinite cardinals; and I fully admit that the latter cannot be answered in so summary a manner. These arguments will be discussed in detail by Mr. Russell. I shall only refer to them now in order to make clear the point that Dr. Hobson has made against a great deal of generally accepted mathematical reasoning, of which my construction is an average specimen. The point is this, that a great deal of such reasoning really depends on the acceptance of a certain logical postulate of which no proof has yet been given, namely, *the postulate of the existence of the multiplicative class*. If we have a class of mutually exclusive classes k, no one of which is null, the *multiplicative class of the k's* is defined as the class of classes each member of which contains one and only one member of each of the k's.* The class can always be defined, but it has never been proved that it is *never null*, that is to say, that it always contains at least one member.

Among mathematical proofs in which the existence of the multiplicative

---

* A. N. Whitehead, "On Cardinal Numbers" (*American Journal*, Vol. XXIV., p. 383).

class is assumed, if not universally, at any rate in cases more extensive than those for which its existence has been proved, I may instance Bernstein's and König's theorems concerning the exponentiation of cardinals, Bernstein's and my own proofs that the cardinal number of the continuum is greater than or equal to $\aleph_1$, and Borel's construction for a function of arbitrary class,* among many others.

The last instance is peculiarly instructive, as it shows how assumptions equivalent to that of the existence of the multiplicative class find their way into the writings even of mathematicians who can recognise the assumption elsewhere.

M. Borel wishes to show that functions exist which cannot be represented as double series of polynomials.† If

$$P_{\alpha, \beta}(x) = \sum_{\gamma=0}^{\alpha+\beta} c_{\alpha, \beta, \gamma}\, x^\gamma$$

(he says), every double series of polynomials can be written in the form

(1) $$\sum_{\alpha=1}^{\infty} \sum_{\beta=1}^{\infty} P_{\alpha, \beta}(x).$$

If this series converges for $0 \leqslant x \leqslant 1$, it represents for those values of x a function of class 0, 1, or 2; and every function of class 0, 1, or 2 can be defined in this way by a suitable choice of the constants $c_{\alpha, \beta, \gamma}$. "*Chaque fonction est même définie d'une infinité de manières, mais cela n'a pas d'inconvénient pour ce qui suit.*"

M. Borel's subsequent reasoning depends entirely on *one* representation of every function of classes 0, 1, and 2 having been selected from among the infinity of representations which correspond to each function, *i.e.*, on the existence of the multiplicative class of the classes formed by all the representations of any given function. Yet M. Borel, criticising Zermelo's article in the *Annalen*,‡ uses language which might have been used by Dr. Hobson: "il me semble que les objections que l'on peut y opposer valent contre tout raisonnement où l'on suppose un *choix arbitraire* fait une infinité non dénombrable de fois; de tels raisonnements sont en dehors du domaine des mathématiques."§

---

* *Leçons sur les fonctions de variables réelles*, Note III., pp. 156–158.
† I am not implying any doubt of the correctness of the result.
‡ *Math. Ann.*, Bd. LX., p. 194.
§ The aggregate of functions considered above has the cardinal number of the continuum. The logical difficulty is the same whether the *choix arbitraire* has to be made an enumerable or a non-enumerable infinity of times; but, of course, it will generally be easier, in a particular case in which the infinity of times is enumerable, to get over the difficulty by substituting a "norm," *i.e.*, a set of rules for choosing, for the "arbitrary" acts of choice.

In the case of my construction the form in which the postulate is used is in the assumption of the existence of the multiplicative class of the class of classes formed by all the progressions whose limit is a given number a. To each a corresponds an infinity of such progressions: of these progressions we must select one for every a, and it certainly seems very paradoxical to suppose that the class whose members are defined to be all the various *aggregates of selected progressions* should be null, *i.e.*, possess no members. But, although we can define the class, we cannot (so far as can be seen at present) specify a single one among its members, and there seems to be no way of proving that there are members except by actually producing them.

The instance of the decimals referred to by Dr. Hobson is really not a parallel. For the class of decimals does exist: we can produce some at any rate of its members, for example ·0000..., ·0101.... Here the multiplicative class is that of an enumerable class of classes each of which contains the two members 0 and 1, and its existence can be proved.

The necessary axiom, if it is to be postulated, may be postulated in a variety of forms, that of the assumption of the universal existence of the multiplicative class, or its existence subject to restrictions,* or in either of Zermelo's two forms:

(*a*) That the product of any number of infinite cardinals cannot be zero.

(*b*) That a relation exists which correlates each class contained in a given class with *one* of its members.

It has been proved that (*b*) implies (*a*) and the universal existence of the multiplicative class, but whether or no the latter imply the former has not yet been decided. Mr. Russell has traced the consequences of the denial of the multiplicative axiom in the arithmetic of the transfinite numbers, and has shown that the question as to its truth or falsity has no bearing on the question of the Aleph-series and Burali-Forti's contradiction, which must be met in quite another way. There is therefore no reason for supposing that Zermelo's assumption is not valid except that it has not been proved. And that it has not been proved means simply that no general method has been given

---

* It is, of course, quite possible that the existence of the multiplicative class may not be universal, but may hold in extensive particular cases, *e.g.*, when the class of classes whose multiplicative class is required is well ordered. The multiplicative class certainly exists for any class of *well-ordered* classes, if a definite order is given with each class, but it has not been proved to exist for a *well-ordered* class of *any* classes.

for defining in finite terms, *i.e.*, by a finite number of repetitions of a finite number of symbols, *one* member of the multiplicative class whose existence is to be established. Even if we knew that it was *impossible* ever to define a single member of a class, it would not of course follow that members of the class did not exist, but there appears to be no way of proving the contrary, except by actually specifying a member or by showing that the hypothesis that there is no member leads to contradiction; and, awkward and paradoxical as the consequences of denying the multiplicative axiom are, it has yet to be shown that they are contradictory.

I am therefore, in default of proof, prepared to accept the multiplicative axiom* provisionally on the grounds

    (i.) that to deny it appears to be paradoxical;

    (ii.) that no *reason* has been given for denying it;

    (iii.) that to deny it reduces to a state of chaos a great deal of very interesting mathematics.

---

* I do not imply that I accept Zermelo's proof that every aggregate is well ordered. I agree with Dr. Hobson in thinking it open to objection on other grounds.

THE ORDINAL RELATIONS OF THE TERMS OF A CONVERGENT SEQUENCE

By G. H. HARDY.

[Received July 12th, 1909.—Read November 11th, 1909.]

1. Suppose that we are given a sequence

(1) $$(a_n) \quad (n = 1, 2, 3, \ldots)$$

and are able, when any two terms of the sequence, a_p and a_q say, are taken, to determine which of the relations

(2) $$a_p \gtreqless a_q$$

holds between a_p and a_q. We may express this by saying that we know the *ordinal relations* (as regards order of magnitude) which hold between the various terms of the sequence.

This note is devoted to a consideration of the following question, which seems to me of some logical interest: *How far is it possible to discriminate, on the ground* simply *of the ordinal relations that hold between their terms, between sequences which converge to a limit and sequences which do not?* And, in so far as this *is* possible, what is the simplest expression of the ordinal relations which characterise convergent sequences?

2. It is easy to recognise certain limitations to which the answers to these questions must be subject. In the first place, it is clearly hopeless to attempt to discriminate in this way between *convergent* and *properly divergent* sequences. This appears at once from a consideration of the simplest case of all, that of *monotonic* (say increasing) sequences. In this case
$$a_p \geqslant a_q,$$
if $p \geqslant q$. A classical theorem asserts that (a_n) is convergent or divergent to $+\infty$. But we cannot distinguish between the two cases: the sequences $(1-1/n)$ and (n) are characterised by exactly the same ordinal relations.

Thus we are obliged to regard (as, indeed, it is often convenient to regard) sequences which diverge to $+\infty$ or to $-\infty$ as a special case of convergent sequences, and to distinguish only between convergent and oscillatory series.

But even when this reservation is made, it is easy to see that the first of our two questions cannot be answered with a simple affirmative. Consider, for example, the sequence (a_n), where $a_n = (\sin \tfrac{1}{2} n\pi)/n$. Here the limit is zero. And now consider the sequence defined by the equations

$$b_n = 1 + \frac{1}{n} \sin \tfrac{1}{2} n\pi \quad (n = 4k+1),$$

$$b_n = \frac{1}{n} \sin \tfrac{1}{2} n\pi \quad (n = 4k,\ 4k+2),$$

$$b_n = -1 + \frac{1}{n} \sin \tfrac{1}{2} n\pi \quad (n = 4k+3).$$

Each of the inequalities $a_p \gtreqless a_q$ involves the corresponding one of the inequalities $b_p \gtreqless b_q$: the ordinal relations of the terms of the two sequences are precisely the same; but (a_n) is convergent, while (b_n) has the *three* limiting values $-1, 0, 1$.

We shall see later that this last example is of genuine theoretical interest as being as simple an example as possible of the one type of oscillatory sequence that cannot be distinguished from a convergent sequence simply by means of its ordinal relations. The answer to our questions is, in fact, that the ordinal relations of convergent sequences do possess an exceedingly simple characteristic which distinguishes them from all other sequences except one very special type of oscillatory sequence; but that this special type of oscillatory sequence cannot be distinguished from a convergent sequence by *any* marks of its ordinal relations.

The Classes P_n, Q_n, R_n.

3. We choose a definite term a_n of the sequence (a_n) and divide the terms of the sequence (including a_n itself) into three classes P_n, Q_n, R_n, which comprise the terms a_m for which

$$a_m < a_n, \quad a_m = a_n, \quad a_m > a_n$$

respectively. Thus in a monotonic (increasing) sequence, no two terms of which are equal, P_n includes $a_1, a_2, \ldots a_{n-1}$, Q_n includes a_n only, and R_n includes $a_{n+1}, a_{n+2}, \ldots$.

Quasi-monotonic Sequences.

4. A *quasi-monotonic (increasing) sequence* is a sequence such that the classes P_n, Q_n are always finite (or such that $P_n + Q_n$ is always finite).

THEOREM.—*The necessary and sufficient condition that a sequence should converge to a limit, greater than any of its terms, is that it should be quasi-monotonic (increasing).*

There are, of course, a corresponding definition and theorem for quasi-monotonic (decreasing) sequences.

Proof of the Theorem.—First, let the sequence converge to a limit greater than any a_n. If $P_n + Q_n$ were infinite, for any particular value of n, we could find values of m as large as we please and such that $a_m \leqslant a_n$. This would imply $\lim a_m \leqslant a_n$, contrary to our hypothesis.

Secondly, let $P_n + Q_n$ be always finite. Then, given n_1, we can choose n_2 so that
$$a_p > a_q \quad (p \geqslant n_2,\ q \leqslant n_1):$$
for the class formed by all the terms which belong to any of the classes $P_q + Q_p$ ($q \leqslant n_1$) is finite. Also we can suppose n_2 to be the *least* number satisfying this condition. Similarly we define n_3 as the least number such that
$$a_p > a_q \quad (p \geqslant n_3,\ q \leqslant n_2):$$
and so on. Thus we define a monotonic sequence (a_{n_ν}), which must tend to a limit a greater than any a_{n_ν} or to $+\infty$. If $a_{n_\nu} \to +\infty$, we have
$$a_{n_\nu} > G$$
by choice of ν, however large be G, and so
$$a_p > G \quad (p \geqslant n_{\nu+1}),$$
so that $a_n \to +\infty$.

On the other hand, if $a_{n_\nu} \to a$, a must be greater than any a_n. For, if $a_n \geqslant a$, and we choose ν so that $n_\nu > n$, we have $a_{n_{\nu+1}} > a$, which is impossible.

Finally, however small be ϵ, we can choose ν so that
$$a - a_{n_\nu} < \epsilon,$$
and so
$$a - a_n < a - a_{n_\nu} < \epsilon \quad (n \geqslant n_{\nu+1}).$$

Hence $a_n \to a$, and the theorem is established.

5. The proof may be shortened by the use of maximum and minimum limits of indetermination. If a_n does not converge to a limit, let G and H be its minimum and maximum limits: H may be $+\infty$, but neither can be $-\infty$, since only a finite number of terms are less than any a_n.

We can choose a subsequence (b_μ) whose limit is G, and a subsequence (c_ν) whose limit is H, and, as $G < H$, we can determine n_0 so that
$$b_\mu < c_\nu \quad (\mu, \nu \geqslant n_0).$$
Hence P_n is infinite for c_ν, in contradiction to our hypotheses.

6. The theorem of § 4 is verbally identical with one which I obtained some time ago,* but using a different definition. The definition of a quasi-monotonic (increasing) sequence that I adopted then is as follows: let ν be the largest number such that $a_1, a_2, \ldots, a_\nu$ are all less than a_n, so that ν is a function of n; then the sequence is quasi-monotonic (increasing) if $\nu \to \infty$ with n. It is easy to prove directly that the two definitions are equivalent. First, if the definition just stated is satisfied, then $P_n + Q_n$ is always finite. Otherwise we could, for some value of n, find an infinite number of terms $a_{m_1}, a_{m_2}, \ldots$, all less than or equal to a_n. But, by the definition, if ν is large enough,
$$a_m < a_{m_\nu} \quad (m \leqslant n),$$
which involves a contradiction.

Secondly, suppose that the definition just stated is not satisfied. Then we can find a number n_0 such that
$$a_{n_\nu} \geqslant a_{m_\nu},$$
where (m_ν) is an infinite ascending sequence, and $n_\nu \leqslant n_0$ for all values of ν. Since a_{n_ν} has at most n_0 possible values, there must be some one a_n $(n \leqslant n_0)$ which is not less than an infinity of the terms of the sequence (a_{m_ν}): and this shows that $P_n + Q_n$ is not always finite.

Quasi-monotonic Sequences in the wider sense.

7. A *quasi-monotonic (increasing) sequence in the wider sense* is a sequence such that P_n is always finite.

A quasi-monotonic increasing sequence in the wider sense has a limit not less than any term of the sequence. The proof is the same as that of the theorem of § 4: but there is one case in which the limit may be equal to terms of the sequence, viz., the trivial case in which all the terms of the sequence, from a certain rank onwards, are equal.

---

* The result is stated in substance, but without proof, as an example in Bromwich's *Infinite Series* (Ex. 14, p. 393).

The converse is not true. For let us take a quasi-monotonic (in the stricter sense) increasing sequence. It has a limit l greater than any term. In this sequence interpolate in any manner an infinity of terms equal to l. The sequence has still the limit l, not less than any of its terms: but it is no longer quasi-monotonic, even in the wider sense; for, if $a_n = l$, P_n is infinite.

Apart from this case of exception, it is easy to see that the theorem obtained from that of § 4 by changing *greater* to *not less* and *quasi-monotonic* to *quasi-monotonic in the wider sense* is true.

Convergent Sequences in General.

8 Now let us consider any convergent sequence (a_n). If its limit is $+\infty$ or $-\infty$, it is a quasi-monotonic sequence, increasing or decreasing.

Let us then suppose that it converges to a finite limit a. If Q_n is infinite for any value of n, then $a_n = a$, and the sequence contains an infinity of terms equal to a. The remaining terms may be divided into two classes, those less than and those greater than a. Hence *any convergent sequence is formed of* (i) *a sequence of equal terms*, (ii) *a quasi-monotonic (increasing) sequence, and* (iii) *a quasi-monotonic (decreasing) sequence*. Any of these classes, of course, may be finite or entirely absent. The terms of (ii) and (iii) are characterised by the facts that $P_n + Q_n$ and $Q_n + R_n$ are respectively finite: for the terms of (i) the classes P_n, Q_n, R_n may all be infinite.

Conversely, suppose that a sequence (a_n) possesses the properties (i) *that for at most one value of a are there infinitely many terms equal to a*, (ii) *that for any term of the sequence not equal to a either P_n or R_n is finite*.

For the latter terms Q_n is finite, and so either $P_n + Q_n$ or $Q_n + R_n$. Let (b_μ) be the sequence formed by the terms for which $P_n + Q_n$ is finite, and (c_ν) the sequence formed by the terms for which $Q_n + R_n$ is finite. If P'_μ, Q'_μ, R'_μ are the classes corresponding to b_μ, but formed from terms of the sequence (b_μ) only, $P'_\mu + Q'_\mu$ is, *a fortiori*, finite, and so (b_μ) is a quasi-monotonic (increasing) sequence, and converges to a limit β greater than any of its terms. Similarly (c_ν) is a quasi-monotonic (decreasing) sequence, and converges to a limit γ less than any of its terms. Also

$$\beta \leqslant a \leqslant \gamma.$$

For it is obvious that, if $\beta > a$ or $\beta > \gamma$, then P_n would be infinite for some terms of (b_μ).

If a, β, γ are equal, say all equal to a, the sequence converges to a limit, viz. a. But it is quite possible that $\beta < a < \gamma$: thus, in the case of the sequence considered in § 2, $\beta = -1$, $a = 0$, $\gamma = 1$.

If $\beta < a < \gamma$, the sequence has precisely the same ordinal relations as the sequence (b_n) defined by the equations

$$b_n = a_n \qquad (a_n = a),$$
$$b_n = a_n + (a - \beta) \quad (a_n < a),$$
$$b_n = a_n - (\gamma - a) \quad (a_n > a);$$

a sequence which plainly converges to the limit a.

Thus *the properties* (i) *and* (ii) *above characterise convergent sequences, in so far as these sequences can be characterised by their ordinal relations only. Every convergent sequence possesses them: and any sequence which possesses them is either convergent or is an oscillatory sequence of the special type formed by adding a constant to all the terms of a convergent sequence which are greater than the limit, and subtracting a constant from all those which are less than the limit. And this class of oscillatory sequences is indistinguishable from a convergent sequence by any test dependent on ordinal relations only.*

Oscillatory Sequences.

9. If (a_n) oscillates there must be terms for which both P_n and R_n are infinite (except in the special case discussed above). All of P_n, Q_n, R_n may always be infinite, as in the sequence

$$\tfrac{1}{2},\ \tfrac{1}{3},\ \tfrac{2}{3},\ \tfrac{1}{4},\ \tfrac{2}{4},\ \tfrac{3}{4},\ \tfrac{1}{5},\ \tfrac{2}{5},\ \tfrac{3}{5},\ \tfrac{4}{5},\ \tfrac{1}{6},\ \tfrac{2}{6},\ \tfrac{3}{6},\ \ldots\ ;$$

or, again, P_n and R_n may always be infinite, and Q_n consist of one term only, as in the sequence deduced from that above by rejecting all fractions which are not in their lowest terms.

(b) Differential Equations

SOME RESULTS CONCERNING THE BEHAVIOUR AT INFINITY OF A REAL AND CONTINUOUS SOLUTION OF AN ALGEBRAIC DIFFERENTIAL EQUATION OF THE FIRST ORDER

By G. H. HARDY.

[Received August 22nd, 1911.—Read December 8th, 1910.]

I.

1. The results obtained in this paper have reference to the algebraic differential equation

(1) $$f(x, y, y') \equiv \Sigma A x^m y^n y'^p = 0,$$

where m, n, p are positive integers. I suppose that this equation possesses a solution

(2) $$y = y(x),$$

which is real and possesses a continuous derivative for $x > x_0$.* The problem is to specify as completely as possible the various ways in which y may behave as $x \to \infty$.

This problem was first attacked by Borel, in his *Mémoire sur les Séries Divergentes*.† Borel proved that the equation (1) cannot have a solution y, such that

$$y > e^{e^x} = e_2(x)$$

for values of x surpassing all limit. He proved further that

(3) $$f(x, y, y', y'') = 0$$

cannot have a solution y such that

$$y > e_3(x)$$

for values of x surpassing all limit;‡ and there is no doubt of the truth of

---

* *I.e.*, for all values of x from some value onwards ("Orders of Infinity," *Camb. Math. Tracts*, No. 12, p. 6). We assume the existence of such a solution: it is not part of the problem to consider conditions for its existence.

† *Annales de l'École Normale*, t. 16, pp. 26 *et seq*.

‡ The proof is not complete, but its general lines are clearly indicated.

the corresponding general theorem, though, so far as I am aware, no strict proof has ever been given.

Borel also devoted a section of his memoir to the subject of *oscillating* solutions, but without obtaining any very definite results.

2. In a short memoir published in 1899, Lindelöf* returned to the questions raised by Borel, and proved the following much more precise result:

If the equation (1) *is of degree m in x, then there is a constant C, such that*

(4) $$y < e^{Cx^{m+1}}$$

for $x > x_0$.

Further, he proved that† either

(5) $$|y| < e^{x^\rho}, ‡$$

or

(6) $$e^{x^{\rho-\delta}} < |y| < e^{x^{\rho+\delta}} \quad (\rho > 0),$$

for $x > x_0$.

The solutions of the first class may oscillate, but those of the second are ultimately monotonic, together with all their derivatives.

3. The subject of the increase (*croissance*) of solutions of the equation (1) has also been considered by Boutroux.§

Boutroux confines himself to the equation

(7) $$y' = P(x, y)/Q(x, y),$$

where P and Q are polynomials; but he considers the whole subject from the point of view of the theory of functions of a complex variable. The distinction between the two classes of solutions (5) and (6) of course appears again, in a more precise form—there are solutions whose increase is less than that of some power of $|x|$, and solutions which, in certain angles, behave like exponentials.

---

* *Bulletin de la Société Mathématique de France*, t. 17, p. 205.

† Some of these results are contained in an additional note which is in part due to Borel.

‡ The notation is that explained in my tract cited above and my paper "Properties of Logarithmico-Exponential Functions," *Proc. London Math. Soc.*, Ser. 2, Vol. 10, p. 54.

§ *Leçons sur les fonctions définies par les équations différentielles du premier ordre*, Paris, Gauthier-Villars, 1908.

4. In this paper I consider first the equation (7); but, like Borel and Lindelöf, I consider it exclusively from the point of view of the real variable. I am thus able to obtain results very much more precise than those stated in the preceding sections. I show that *all* solutions of (7) are ultimately monotonic, and specify their possible modes of increase by simple asymptotic formulæ. I also show that substantially the same results hold for the equation

(8) $$y'^{\mu} = P(x, y)/Q(x, y),$$

where μ is *odd*.

I then return to the general equation (1). I find asymptotic formulæ, more precise than Lindelöf's, for the solutions which behave at infinity like exponentials, and I prove that any oscillating solution is ultimately less in absolute value than a power of x—in symbols,

$$y = O(x^{\Delta}).$$

In particular I show that, in the case of the equation (8), with μ *even*, every oscillating solution remains finite, *i.e.*,

$$y = O(1).$$

Finally, I discuss certain particular types of oscillating solutions.

Much of the argument is capable of extension, and results still more accurate may be obtained without the intervention of any fresh difficulty of principle. But, after a certain point, the work becomes too tedious to be justified by the interest of the results.

It would, however, be exceedingly interesting to see how far the methods used in the paper will go in proving the analogous results immediately suggested for equations of order higher than the first. Here I do not go beyond the first order, but I hope to return to the subject at a later opportunity.

II.

5. Let us consider the equation

(7) $$y' = P(x, y)/Q(x, y).$$

I shall prove first that it is impossible that y' should vanish for a series of values of x whose limit is infinity, except of course in the trivial case in which (7) has a solution $y = $ const. In other words, *every solution is ultimately monotonic*.

Suppose the contrary. Then the curves $y = y(x)$, $P = 0$ intersect at points corresponding to an infinity of values of x surpassing all limit.

But $P = 0$ consists of a finite number of branches, and so $y = y(x)$ must intersect at least one of these infinitely often.

Now the branches of $P = 0$, which extend to infinity in the direction of the axis of x consist of (i) a finite number of straight lines
$$y = c_s \quad (\gamma_s),$$
(ii) a finite number of branches
$$y = \phi_t(x) \quad (\delta_t),$$
along which y ultimately increases or decreases steadily.

In the first place, *$y = y(x)$ cannot cut any δ_t in an infinity of points.* Suppose, for example, that y ultimately increases along δ_t, and let P, Q be two successive points of intersection. Then $y = y(x)$ crosses δ_t at P and Q, in each case from above to below (Fig. 1), and a glance at the figure is

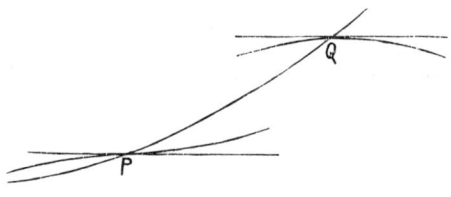

Fig. 1.

enough to show that this is impossible.*

We have now to consider the possible intersections of $y = y(x)$ and γ_s. These fall under the four types represented in Fig. 2.

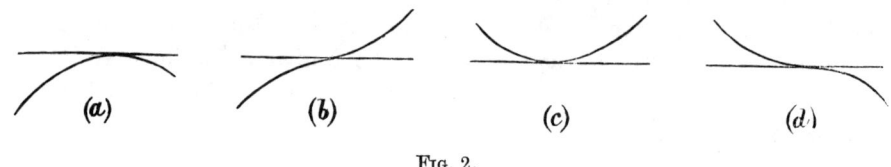

Fig. 2.

Of these we can at once rule out (*a*) and (*c*), since at such points y' would change its sign, and P would not. For a similar reason we can rule out (*b*) and (*d*), *unless the factor $y - c_s$ occurs an even number of times in P.* If this is so, and an intersection of (*e.g.*) type (*b*) occurs, it can

---

* We can suppose x large enough to ensure that P and Q cannot vanish simultaneously. Then it is easy to see that y is regular for a value of x which makes $P = 0$, and hence that there cannot be an infinity of intersections for values of x in the neighbourhood of any finite value. Hence there must be *successive* intersections. We need not elaborate this kind of point in future.

occur *once only*, so far as γ_s is concerned; for when y has once passed above γ_s it can plainly only come back to γ_s after an intersection of type (a) with some other γ_s.

Hence y is ultimately monotonic.

6. We can go further and say that every derivative $y^{(r)}$ is ultimately monotonic. For, by differentiation and substitution, we find

$$y^{(r)} = P_r(x, y)/Q_r(x, y),$$

where P_r, Q_r are polynomials, and in fact $Q_r = Q^{2r+1}$. Our assertion will therefore follow as a corollary from the following general theorem:—

THEOREM.—*Any rational function*

$$H(x, y) = K(x, y)/L(x, y)$$

is ultimately monotonic along the curve $y = y(x)$—unless $L = 0$ is a solution of the equation (7).

This theorem I shall now proceed to prove.

7. We have

(8) $$\frac{dH}{dx} = \frac{\partial H}{\partial x} + R\frac{\partial H}{\partial y} = T = \frac{U}{W},$$

where U and W are polynomials, and d/dx implies differentiation along the curve (2). If dH/dx is not ultimately of constant sign on the curve (2), it must vanish or become infinite infinitely often on (2). In the first case (2) must have an infinity of intersections with at least one of the finite number of branches of

(9) $$U = 0.$$

Now this branch may, for sufficiently large values of x, be represented in the form

(9') $$y = A_0 x^{a_0} + A_1 x^{a_1} + \ldots,$$

a convergent series of (not generally integral) descending powers of x. If $\delta/\delta x$ refers to differentiation along (9),

(10) $$\frac{\delta y}{\delta x} = A_0 a_0 x^{a_1-1} + A_1 a_1 x^{a_1-1} + \ldots.$$

Again, along (9), $R(x, y)$ is an algebraic function of x, which may, for sufficiently large values of x, be expressed in the form

(11) $$R = B_0 x^{\beta_0} + B_1 x^{\beta_1} + \ldots,$$

another series of descending powers. And, unless the series (10), (11) are identical, we shall have, at all points of (9) from some definite point onwards,

$$\frac{\delta y}{\delta x} > R \quad or \quad \frac{\delta y}{\delta x} < R.$$

From this it follows that, at the points of intersection, (2) always crosses (9) from one and the same side to the other and the same side; which is plainly impossible.

On the other hand, if the series (10) and (11) are identical, we have

$$\frac{\delta y}{\delta x} = R,$$

and $U = 0$ is a solution of (7). In other words, H is constant along (2).

There remains only the possibility that

$$\frac{dH}{dx} = \left(L \frac{dK}{dx} - K \frac{dL}{dx} \right) \bigg/ L^2$$

should become infinite infinitely often, as we describe (2). This cannot be true owing to K or L or

$$\frac{dK}{dx} = \frac{\partial K}{\partial x} + \frac{\partial K}{\partial y} \frac{dy}{dx}$$

or dL/dx becoming infinite, and so can only occur if L vanishes infinitely often. But then we can show as above that $L = 0$ is a solution of the equation (7).

Thus the proof of the theorem is completed.

COROLLARY.—*Any rational function*

$$H(x, y, y')$$

is ultimately monotonic, unless its denominator vanishes identically in virtue of (7).

The same is true of $\quad H(x, y, y', y'', ...)$.

8. We can obtain much more accurate information concerning the increase of the solutions of

(7) $$Qy' = P.$$

The ratio of any two terms is of one of the forms

$$A x^m y^n, \quad A x^m y^n y';$$

and is consequently ultimately monotonic, and so, between any two terms X_i, X_j, there subsists one of the relations

$$X_i \succ X_j, \quad X_i \asymp X_j, \quad X_i \prec X_j.$$

It follows that there must be one pair of terms at any rate such that

$$X_i \asymp X_j.$$

If these two terms come from the same side of (7), we obtain at once

(12) $$y \sim A x^s,$$

where s is *rational*. If they come from opposite sides, we obtain a relation of the form

(13) $$y^m y' \sim A x^n.$$

Here four cases present themselves. If $m \neq -1$, $n \neq -1$, we obtain a relation of the type (12). If $m \neq -1$, $n = -1$, we obtain a relation

(14) $$y \sim A (\log x)^{1/p},$$

where p is an integer. If $m = -1$, $n \neq -1$, we obtain a relation

$$\log y \sim A x^p,$$

(15) $$y = e^{A x^p (1+\epsilon)}.$$

Here p may be supposed a positive integer, as if p is negative $y \sim 1$.*
Finally, if $m = -1$, $n = -1$, we obtain

$$\log y \sim A \log x,$$

(16) $$y = x^{A+\epsilon}.$$

9. The relations (15), (16) are less precise than (12) and (14). We shall now proceed to examine them more closely.

Let us consider first the exponential solutions (15). We have

$$y' = \frac{P_0 y^r + P_1 y^{r-1} + \cdots}{Q_0 y^s + Q_1 y^{s-1} + \cdots},$$

where $P_0, \ldots, Q_0, \ldots$ are polynomials in x. It is clear that $s = r-1$, and that, for sufficiently large values of x, we have

$$y' = R_0 y + R_1 + O(x^{-\Delta}),$$

$$y'/y = R_0 + O(x^{-\Delta}),$$

---

* p is clearly at most equal to $r+1$, where r is the degree of (1) in x—this, of course, agrees with Lindelöf's result quoted in §2.

where R_0, R_1 are rational functions of x. Hence

$$y'/y = \Pi(x) + \frac{A}{x} + O\left(\frac{1}{x^2}\right),$$

where Π is a polynomial and A a constant (not necessarily rational). Hence, integrating, we deduce

(17) $$y \sim A x^\alpha e^{\Pi(x)}.^*$$

It is clear that this form includes the form (12).

10. We have now to consider the last case of § 8, which is rather more difficult. There are two terms

(18) $$\lambda x^s y^t y', \quad \mu x^{s-1} y^{t+1}$$

of equal order :† obviously we may suppose that no other term is of greater order. We may go further, and suppose that no other term is of equal order, since the contrary assumption leads at once to a relation of the type (12). We have also

$$y = x^{A+\epsilon}, \quad A = \mu/\lambda.$$

If follows from the theorem of §§ 6, 7 that, if X_i is any third term in the differential equation, the quotient

$$(\lambda x^s y^t y' - \mu x^{s-1} y^{t+1})/X_i$$

tends to a limit as $x \to \infty$. In other words, the difference of the two principal terms is definitely of order greater than, equal to, or less than that of any third term. We can now distinguish two possibilities.

(a) *There is a third term whose order is equal to that of the difference of the principal terms.*

In this case we have a relation of one of the forms

(19) $$\lambda x^s y^t y' - \mu x^{s-1} y^{t+1} \sim M x^\sigma y^\tau,$$

(20) $$\lambda x^s y^t y' - \mu x^{s-1} y^{t+1} \sim M x^\sigma y^\tau y'.$$

First, suppose (19) holds. Putting

$$y = x^A u = x^{\mu/\lambda} u,$$

---

* Not, of course, with the *same* A : cf. *Proc. London Math. Soc.*, Ser. 2, Vol. 10, p. 54.

† We say, of course, that X_i is of order greater than, equal to, or less than that of X_j, according as $X \succ X_j$, $X_i \succeq X_j$, or $X_i \prec X$.

and substituting, we obtain
$$u^{t-\tau}u' \sim Nx^{\sigma-s+(\tau-t-1)A}.$$
But, as $u = x^e$, this is only possible if
$$\sigma - s + (\tau - t - 1)A = -1,$$
which shows that A is *rational*, its denominator being $\tau - t - 1$. Also, integrating,
$$u^{t-\tau+1} \sim N \log x.$$
Thus

(21) $$y \sim A(x^p \log x)^{1/q},$$

where p and q are integers.

Next, suppose (20) holds. Making the same substitution, we obtain
$$\{\lambda x^{s+(t+1)A} u^t - M x^{\sigma+(\tau+1)A} u^\tau\} u' \sim MAx^{\sigma-1+(\tau+1)A} u^{\tau+1}.$$
But
$$x^s y^t \succ x^\sigma y^\tau, \quad x^{s+tA} u^t \succ x^{\sigma+\tau A} u^\tau,$$
and so
$$u^{t-\tau-1}u' \sim N x^{\sigma-s-1+(\tau-t)A};$$
and the argument may now be completed as before.

(b) *There is no third term whose order is equal to that of the difference of the principal terms.*

Let us denote the principal terms by $X_1, \bar{X}_1$. Then there must be at least one term X_2, such that
$$X_2 \succ X_1 - \bar{X}_1;$$
and therefore another term $\bar{X}_2$, such that
$$X_2 \sim \bar{X}_2:$$
and we may suppose, as in the case of $X_1, \bar{X}_1$, that these terms come from opposite sides of the equation. We may also suppose that $X_2, \bar{X}_2$ are of higher order than any other terms other than $X_1, \bar{X}_1$. Further, we may suppose them to be of the form
$$\lambda_2 x^{s_2} y^{t_2} y', \quad \mu_2 x^{s_2-1} y^{t_2+1},*$$
where
$$\mu_2/\lambda_2 = \mu/\lambda = A.$$
Putting $y = x^A u$, we obtain
$$X_1 - \bar{X}_1 = x^{s+(t+1)A} u^t u', \quad X_2 - \bar{X}_2 = x^{s_2+(t_2+1)A} u^{t_2} u';$$

---

* If they were not thus related, the increase of y could be determined at once as in §§ 8, 9.

from which it at once follows that
$$X_1 - \bar{X}_1 \succ X_2 - \bar{X}_2.$$

But in this case there must be a fifth term X_3, whose order is greater than that of $X_1 - \bar{X}_1$, and a sixth term $\bar{X}_3$, such that
$$X_3 \sim \bar{X}_3,$$
and we can prove that
$$X_1 \succ X_2 \succ X_3, \quad X_1 - \bar{X}_1 \succ X_2 - \bar{X}_2 \succ X_3 - \bar{X}_3.$$

And as this argument may be repeated indefinitely, and the number of terms is finite, we must find sooner or later that the supposition (b) leads either to the conclusion we desire or to a contradiction.

11. We have thus proved the following theorem :—

Any solution y^* of the equation
$$\frac{dy}{dx} = \frac{P(x, y)}{Q(x, y)}$$
is ultimately monotonic, together with all its derivatives, and satisfies one or other of the relations
$$y \sim A x^a e^{\Pi(x)}, \quad y \sim A (x^p \log x)^{1/q},$$
where $\Pi(x)$ is a polynomial, and p, q are integers.

These rates of increase are naturally included among the standard asymptotic forms for logarithmico-exponential functions of order 1,† of which they are quite special cases.

12. It is natural to attempt to extend our results to the more general equation

(8) $$y'^\mu = P(x, y)/Q(x, y).$$

---

* We are, of course, confining ourselves to continuous solutions : see § 1.

   Examples.—The solution of $x^2 y' = (x+1)^2 y$ is
   $$y = A x^2 e^{x - (1/x)} \sim A x^2 e^x;$$
   the solution of $2x(x+1) yy' = xy^2 + (x+1)^2$ is
   $$y = \sqrt{\{(x+1)(\log x + A)\}} \sim \sqrt{(x \log x)}.$$

† *Proc. London Math. Soc.*, Ser. 2, Vol. 10, p. 76.

If μ is odd this offers no new difficulties: all our arguments apply, with appropriate modification of detail.*

But if μ is even our results are obviously no longer true. Thus
$$y'^2 = 1 - y^2$$
possesses the oscillating solution
$$y = \sin x.$$

I shall now proceed to consider the general equation (1), and the particular equation (8), with μ even, with the especial idea of discovering to what limitations the existence of oscillating solutions is subject.

III.

13. I return now to the general equation (1). We can distinguish various possibilities.

(a) It may be possible to find a positive ρ such that

(22) $$y > e^{x^\rho},$$

for an infinity of values of x surpassing all limit. In this case Lindelöf has shown that this inequality holds for *all* sufficiently large values of x, and that y and all its derivatives are ultimately monotonic.†

In this case any rational function
$$H(x, y, y')$$
is ultimately monotonic. For, if we eliminate y' between
$$H = H(x, y, y'), \quad f(x, y, y') = 0,$$

---

* The standard forms of increase are
$$y \sim Ax^a e^{Bx^{p/\mu} + Cx^{(p/\mu)-1} + \cdots}, \quad y \sim A\,(x^p lx)^{1/q}, \quad y \sim A\,(lx)^{\mu/q}.$$
In the first of these x^a can occur only if p/μ is integral. The form
$$y \sim A\,(x^p lx)^{\mu/q},$$
can only occur if $\mu = 1$ or $p = 0$.

† Lindelöf (*l.c.*) shows that if $y = y(x)$ cuts $y = e^{x^\rho}$ at points whose abscissæ surpass all limit, we can find values of x surpassing all limit for which
$$y' = \rho x^{\rho-1} y, \quad y > e^{x^\rho}.$$
Substituting in (1), we obtain $\quad f_1(x, x^{\rho-1}, y) = 0,$
where f_1 is a polynomial; and it is impossible that f_1 should vanish for an infinity of pairs of values
$$(\xi_i, \eta_i), \quad \eta_i > e^{\xi_i^\rho},$$
unless it vanishes identically.

we obtain an algebraic relation
$$F(x, y, H) = 0 ;$$
and so the points at which $\dfrac{dH}{dx} = 0$

lie on an algebraic curve, which plainly contradicts (22).

We can now argue as in § 8. The equation (1) must contain two terms of equal order, and so we deduce
$$A_1 x^{m_1} y^{n_1} y'^{p_1} \sim A_2 x^{m_2} y^{n_2} y'^{p_2},$$
$$y^\mu y' \sim A x^\nu.$$

Here μ and ν are rational, and μ must plainly be -1. Hence

(23) $$y = e^{A x^s (1+\epsilon)}.$$

It is clear that s can be at most greater by unity than the degree of (1) in x.*

(b) It may be possible to find a number K such that
$$y = O(x^K).$$

14. It is obvious that (a) and (b) do not exhaust the *a priori* possibilities. It is our object now to prove that no other case is really possible.

If we are not in Case (b), it is possible, however large be Δ, to find values of x such that
$$y > x^\Delta.$$

We can therefore choose an increasing sequence (Δ_ν), whose limit is infinity, and a corresponding sequence (x_ν), such that
$$y(x) > x^{\Delta_\nu} \quad (x = x_\nu).$$

We shall now construct a curve

(24) $$y = x^{\Delta(x)} = e^{\Delta(x) \log x} = e^{\phi(x)},$$

passing through the points $(x_\nu, x_\nu^{\Delta_\nu})$ and satisfying certain conditions.

In the first place, we can suppose $\Delta'(x)$, and *a fortiori* $\phi'(x)$, positive and continuous. And we may suppose $\phi \prec x^\delta$, and *a fortiori* $\Delta \prec x^\delta$, since otherwise we should find ourselves again in Case (a).

---

* We can treat similarly the case in which y is ultimately negative.

1910.] ALGEBRAIC DIFFERENTIAL EQUATIONS OF THE FIRST ORDER. 463

Further, since we are at liberty to suppose the increase of the sequence (Δ_ν) as slow as we like, we may suppose that
$$x^{1-\delta}\Delta'(x) \to 0$$
for any positive δ.*

Now
$$\phi'(x) = \Delta'(x)\log x + \frac{\Delta(x)}{x}.$$

Hence

(25) $$x\phi' \to \infty, \quad x^{1-\delta}\phi' \to 0.$$

15. We have

(26) $$y(x) > e^{\phi(x)},$$

for an infinity of values of x surpassing all limit. We shall now show, by a modification of Lindelöf's argument, that this inequality must hold for all sufficiently large values of x.

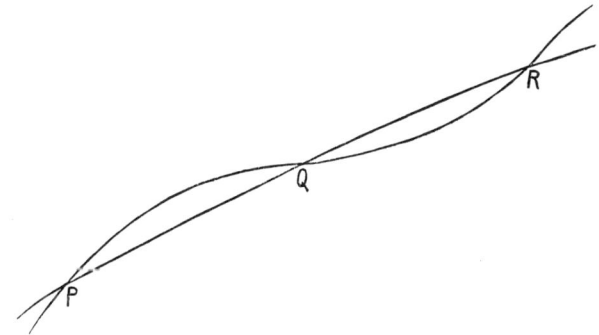

Fig. 3.

If this is not so the curves (2) and (24) must intersect in an infinity of points such as P, Q, R.†

At P (Fig. 3), we have
$$y = e^\phi, \quad y' \geqslant \phi'e^\phi = \phi'y,$$
and at Q we have
$$y = e^\phi, \quad y' \leqslant \phi'e^\phi = \phi'y.$$

As $y' - \phi'y$ is continuous, there must be a point between P and Q where
$$y' = \phi'y, \quad y \geqslant e^\phi,$$

---

* A supposition equivalent, in ordinary cases, to $\Delta(x) \prec x$.

† The argument is not affected if some of these points are points of *contact*.

and so there must be values of x surpassing all limit for which these relations hold. For these values of x,

$$f(x, y, \phi'y) = \Sigma A x^m y^{n+p} \phi'^p = 0.$$

All the terms in this equation (except those for which $n = 0$, $p = 0$) are large compared with any power of x,* and it is clear that, for any value of x for which the equation holds, there must be two terms such that

$$H x^m y^{n+p} \phi'^p < x^{m'} y^{n'+p'} \phi'^{p'} < K x^m y^{n+p} \phi'^p,$$

where H and K are numbers depending only on the form of the equation. Further, it is clear that

$$n+p = n'+p',$$

and so we have

$$H_1 < x^{m'-m} \phi'^{p'-p} < K_1,$$

say. But this plainly contradicts the relations (25), unless $m = m'$, $p = p'$, which is impossible.

16. The inequality (26) therefore holds for all sufficiently large values of x. But we can now prove, as in § 13, that any rational function $H(x, y, y')$ is ultimately monotonic, and thus arrive at the equation (23). We have thus proved the following theorem :†—

If y is any solution of the equation

$$f(x, y, y') = 0,$$

we have either

$$y = O(x^\Delta)$$

or

$$y = e^{Ax^s(1+\epsilon)},$$

where s is rational. All solutions of the latter class are monotonic, together with all their derivatives.

IV.

17. I shall now resume the consideration of the special equation

(8) $$y'^\mu = P/Q,$$

where μ is *even*—if μ is odd, we have already seen that there can be no oscillating solutions.

---

* Since $y \geqslant x^{\Delta(x)}$ and $\Delta(x) \to \infty$.

† It is hardly necessary to point out again that we are considering only continuous solutions.

We have seen that any oscillating solution of the general equation (1)*
must satisfy
$$y = O(x^K).$$

When the equation has the special form (8) we can go much further, and assert that any oscillating solution satisfies
$$y = O(1),$$

i.e., oscillates *finitely*.

In fact if, as in § 5, we denote by γ_s, δ_t the branches of $P = 0$ which stretch to infinity in the direction of the axis of x, we can still show, by the argument used there, that $y = y(x)$ cannot cut any δ_t infinitely often. It follows that y cannot (for sufficiently large values of x) increase beyond the greatest of the numbers c_s. For if it did so it would necessarily continue to increase until $y = y(x)$ met one of the branches δ_t. Hence y can oscillate at most finitely.

We can go further, and assert that, along any branch which does not remain finite, $H(x, y)$, any rational function of x and y—and so also any $H(x, y, y')$—is ultimately monotonic. For
$$\frac{dH}{dx} = \frac{\partial H}{\partial x} + R^{1/\mu}\frac{\partial H}{\partial y}.$$

Let
$$S = \left(\frac{\partial H}{\partial x}\right)^\mu - R\left(\frac{\partial H}{\partial y}\right)^\mu,$$

and suppose, if possible, that $y = y(x)$ meets a branch of $S = 0$ infinitely often.

Along such a branch we have, as in § 7,
$$y = A_0 x^{a_0} + A_1 x^{a_1} + \ldots,$$

(26)
$$\frac{\delta y}{\delta x} = A_0 a_0 x^{a_0-1} + A_1 a_1 x^{a_1-1} + \ldots.\dagger$$

Also, along this branch $R = B_0 x^{\beta_0} + B_1 x^{\beta_1} + \ldots$,

(27)
$$\frac{dy}{dx} = R^{1/\mu} = \pm(C_0 x^{\gamma_0} + C_1 x^{\gamma_1} + \ldots),$$

as $R^{1/\mu}$ has two real values, equal and opposite.

We can now prove without difficulty that the assumption of an infinity of intersections leads to a contradiction. Let $P, Q, R, \ldots$ be successive

---

* Such as $y = x \sin x$, which is a solution of
$$(xy' - y)^2 = x^2(x^2 - y^2).$$

† As in § 7, $\delta y/\delta x$ refers to $S = 0$, and dy/dx to $y = y(x)$.

intersections. These must correspond alternately to the two signs in (27). For, if, *e.g.*, P and Q corresponded to the same sign, $y = y(x)$ would cross $S = 0$ in the same sense at P and Q (Fig. 4*a*), which is manifestly impossible. On the contrary hypothesis (Fig. 4*b*) it is clear that we could find a system of values x_ν, tending to infinity, and such that

$$y(x_\nu) \to \infty, \quad y'(x_\nu) = 0;$$

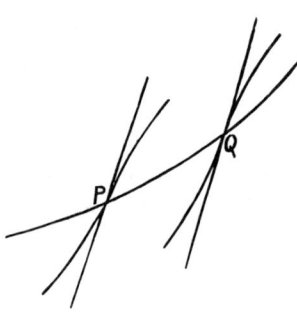
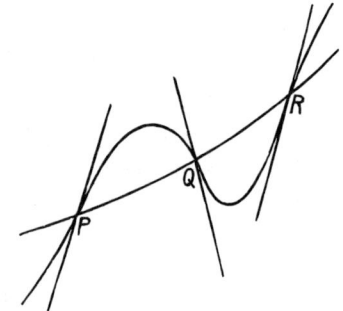

Fig. 4*a*.   Fig. 4*b*.

and this possibility has already been excluded. Thus it has been shown that, unless y remains finite, $H(x, y)$, $H(x, y, y')$, ... are ultimately monotonic.

We can now show, as in §§ 8, 13, that *any solution of* (8), *which does not remain finite, is determined asymptotically by one or other of the formulæ*

$$y = e^{Ax^s(1+\epsilon)}, \ldots$$

obtained in §§ 11, 12.

18. I shall conclude this paper by considering a few cases in which it is possible to obtain more precise information concerning the oscillating solutions.

First, let us suppose, in the equation (8), that P has no factors

$$(y - c_s)^{k_s},$$

in which k_s is even. Then $y = y(x)$ cannot *cross* a line $y = c_s$, since this would involve a change of sign on the part of P. Thus y remains continually between two adjacent lines $y = c_s$, attaining in succession maxima on the upper line and minima on the lower. In Borel's terminology, *the oscillation of y is of a simple and regular sinusoidal type*.

Suppose, in particular, that $\mu = 2$. Then it can be proved that, if y attains the value c_s, $y - c_s$ can occur in P as a simple factor only. For, if

$y = c_s$ for $x = \xi$, we have, near $x = \xi$,
$$y - c_s = A(x-\xi)^p + \ldots,$$
$$y'^2 = B(x-\xi)^{2p-2} + \ldots,$$

and $2p - 2 = pk_s$, which is only possible if $p = 2$, $k_s = 1$.*

We may suppose, without loss of generality, that the lines between which y oscillates are $y - 1 = 0$ and $y + 1 = 0$. We have then
$$y'^2 = (1-y^2) S(x, y),$$

where $S > 0$. Further,
$$S = \frac{P_0 x^m + P_1 x^{m-1} + \ldots}{Q_0 x^n + Q_1 x^{n-1} + \ldots},$$

where $P_0, P_1, \ldots, Q_0, \ldots$ are polynomials in y. Suppose, to avoid complications of detail, that $P_0 > 0$, $Q_0 > 0$ for $-1 \leqslant y \leqslant 1$.† Then
$$S = R_0 x^{m-n} + O(x^{m-n-1}),$$

where R_0 is a rational function of y. Putting $y = \sin \theta$, we obtain
$$\theta'^2 = R_0 (\sin \theta) x^{\frac{1}{2}(m-n)} (1+\epsilon),$$

or
$$\int \frac{d\theta}{\sqrt{R_0}} \sim A x^s,$$

say. This involves a relation of the type
$$\theta \sim B x^s.\ddagger$$

Thus y behaves, to put it roughly, like
$$\sin(B x^s).\S$$

19. When $\mu > 2$, we can, of course, obtain more complicated types of oscillating solutions.

---

* Consider, for example, the equation
$$y'^2 = (1-y^2)^2 x.$$
We find as the general solution $\quad y = \pm \tanh\left(\tfrac{2}{3} x^{\frac{3}{2}} + C\right),$
and y never attains the values ± 1. This is an example of a finite non-oscillating solution

† P_0/Q_0 cannot change its sign, but P_0 or Q_0 might vanish, when we should have to take account of the other terms.

‡ If $\theta = 2\nu\pi + \phi$, where $0 < \phi < 2\pi$,
$$\int \frac{d\theta}{\sqrt{R_0}} = \nu \int_0^{2\pi} \frac{d\theta}{\sqrt{R_0}} + O(1).$$

§ A simple example of such a solution is provided by the trochoidal curve
$$x^m = \theta - a \cos \theta, \quad y = \sin \theta \quad (|a| < 1),$$
which satisfies the equation $\quad y'^2 = \dfrac{m^2 x^{2m-2}(1-y^2)}{(1+ay)^2}.$

It is easily verified that $y = \sin^3 x$ satisfies
$$y'^6 + 27 y^2 y'^4 + 243 y^4 y'^2 = 729 y^4 (1-y^2).$$

This suggests that the equation
$$y'^6 = A y^4 (1-y^2)$$
has an oscillating solution of the type shown in Fig. 5, and it is easy to

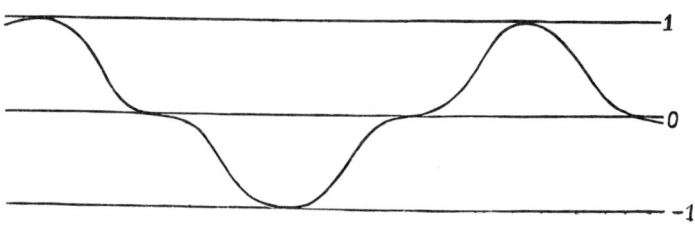

Fig. 5.

verify that this is the case. If we take the more general equation
$$y'^\mu = A y^a (1-y^2)^b, \quad (a, b > 0),$$
we find, as the conditions for the existence of continuous oscillating solutions
$$\frac{a}{\mu} < 1, \quad \frac{b}{\mu} < 1, \quad \frac{a}{\mu} = \frac{2p}{2q+1}, \quad 1 - \frac{2b}{\mu} = \frac{2r}{2s+1},$$
where p, q, r, s are integers. These cannot be satisfied if μ is odd (as is of course to be expected). For even values of μ we find, as possible cases, $\mu = 2$, $a = 0$, $b = 1$; $\mu = 6$, $a = 0$ or 4, $b = 1$, 3, or 5; and so on. The case mentioned above corresponds to $\mu = 6$, $a = 4$, $b = 1$.

[It is only since writing this paper that I have become acquainted with an important series of memoirs by Kneser and Horn, which deal with a variety of questions concerning the asymptotic behaviour of functions defined by differential equations. These memoirs are for the most part developments of the work of Poincaré on linear differential equations. The point of view adopted is very different from that of Borel and Lindelöf, and so far as I know none of the preceding results are contained in any of them. But the bibliographical indications of §§ 1–4 would be incomplete if I did not refer to them, and I accordingly add the following references, without professing that the list is complete:—

A. Kneser.—*Math. Annalen*, 42, p. 409. *Crelle's Journal*, 116, p. 178; 117, p. 72; 120, p. 267.
J. Horn.—*Math. Annalen*, 49, p. 453; 50, p. 525; 51, p. 346 and p. 360; 52, p. 271 and p. 340. *Crelle's Journal*, 116, p. 265; 117, p. 104 and p. 254; 118, p. 257; 119, p. 196 and p. 267; 120, p. 1.

Added January, 1912.]

A NOTE ON A DIFFERENTIAL EQUATION

By G. H. HARDY

Received 27 July 1939

1. In a note published recently in the *Proceedings** Goldstein suggests the truth of the following theorem.

Theorem A. *If the equation*
$$y''' - yy'' + 2(y'^2 - 1) = 0$$
has a solution y regular for large positive x, and
$$y' \to 1$$
when $x \to \infty$, then
$$y = x + b,$$
where b is a constant.

My object here is to supply a proof. It will be obvious that the argument could be generalized, but I confine myself to Goldstein's equation.

If (following Goldstein) we write
$$x + b = u, \quad y = x + b + z,$$
and then replace u and z again by x and y, the theorem becomes

Theorem B. *If*
$$y''' - (x+y)y'' + 4y' + 2y'^2 = 0 \tag{1}$$
and
$$y' \to 0,$$
then y is constant.

2. I require two lemmas, of which the second is familiar.

Lemma 1. *There is no solution of* (1) *for which y' is of fixed sign and tends to* 0.

Suppose, for example, that $y' > 0$, and write $y' = Y$. Since $y' \to 0$, $y = o(x)$. Hence
$$Y'' - x(1+\epsilon)Y' + (4+\epsilon)Y = 0,$$
where each ϵ is a function of x which tends to 0 when $x \to \infty$; and $Y > 0$ for large x.

If $Y' = 0$ for a large x, then $Y'' < 0$, so that all turning values of Y, for large x, are maxima, which is impossible. Hence Y' cannot vanish for large x, and Y is ultimately monotonic. Since it is positive, and tends to 0, it is decreasing, so that $Y' < 0$ and
$$Y'' = x(1+\epsilon)Y' - (4+\epsilon)Y < 0$$
for large x. Hence Y' tends to a negative limit, and Y does not tend to 0, a contradiction.

* S. Goldstein, *Proc. Cambridge Phil. Soc.* 35 (1939), 338–340.

LEMMA 2. *If* $$Z'' = QZ,$$
and $$Q = Q(x) \to l > 0$$
when $x \to \infty$, then either $Z = 0$ or Z is ultimately of fixed sign.

This (and much more) is well known*.

3. If we write Y for y' and Y_1 for y in (1), it becomes
$$Y'' - (x+Y_1)Y' + 4Y + 2Y^2 = 0;$$
and the substitution $$x^2 = u$$
gives $$\ddot{Y} - P\dot{Y} + \frac{Y + \tfrac{1}{2}Y^2}{u} = 0, \qquad (2)$$
where $$P = \frac{1}{2}\left(1 + \frac{Y_1}{u^{\frac{1}{2}}} - \frac{1}{u}\right)$$
and dots imply differentiations with respect to u.

Finally, if we put $$Y = Ze^{\frac{1}{2}\int P\,du},$$
we obtain $$\ddot{Z} = QZ,$$
where $$Q = \tfrac{1}{4}P^2 - \tfrac{1}{2}\dot{P} - \frac{1 + \tfrac{1}{2}Y}{u}.$$
But $P \to \tfrac{1}{2}$, since $$Y_1 = y = o(x) = o(u^{\frac{1}{2}}),$$
and $$\dot{P} = \frac{1}{2}\left(\frac{Y}{u^{\frac{1}{2}}} - \frac{Y_1}{2u^{\frac{3}{2}}} + \frac{1}{u^2}\right) \to 0.$$
Hence $$Q \to \tfrac{1}{16} > 0.$$

It now follows from Lemma 2 that Z, if not always zero, is ultimately of fixed sign, and so that $Y = y'$ is ultimately of fixed sign, in contradiction to Lemma 1. Hence $Z = 0$, $Y = 0$, and y is constant.

* See, for example, Bôcher, *Leçons sur les méthodes de Sturm*, 52–3.

TRINITY COLLEGE
    CAMBRIDGE

(c) The Hardy–Weinberg Law

DISCUSSION AND CORRESPONDENCE

MENDELIAN PROPORTIONS IN A MIXED POPULATION

To THE EDITOR OF SCIENCE: I am reluctant to intrude in a discussion concerning matters of which I have no expert knowledge, and I should have expected the very simple point which I wish to make to have been familiar to biologists. However, some remarks of Mr. Udny Yule, to which Mr. R. C. Punnett has called my attention, suggest that it may still be worth making.

In the *Proceedings of the Royal Society of Medicine* (Vol. I., p. 165) Mr. Yule is reported to have suggested, as a criticism of the Mendelian position, that if brachydactyly is dominant "in the course of time one would expect, in the absence of counteracting factors, to get three brachydactylous persons to one normal."

It is not difficult to prove, however, that such an expectation would be quite groundless. Suppose that Aa is a pair of Mendelian characters, A being dominant, and that in any given generation the numbers of pure dominants (AA), heterozygotes (Aa), and pure recessives (aa) are as $p:2q:r$. Finally, suppose that the numbers are fairly large, so that the mating may be regarded as random, that the sexes are evenly distributed among the three varieties, and that all are equally fertile. A little mathematics of the multiplication-table type is enough to show that in the next generation the numbers will be as

$$(p+q)^2 : 2(p+q)(q+r) : (q+r)^2,$$

or as $p_1 : 2q_1 : r_1$, say.

The interesting question is—in what circumstances will this distribution be the same as that in the generation before? It is easy to see that the condition for this is $q^2 = pr$. And since $q_1^2 = p_1 r_1$, whatever the values of p, q and r may be, the distribution will in any case continue unchanged after the second generation.

Suppose, to take a definite instance, that A is brachydactyly, and that we start from a population of pure brachydactylous and pure normal persons, say in the ratio of $1:10,000$. Then $p=1$, $q=0$, $r=10,000$ and $p_1=1$, $q_1=10,000$, $r_1=100,000,000$. If brachydactyly is dominant, the proportion of brachydactylous persons in the second generation is $20,001:100,020,001$, or practically $2:10,000$, twice that in the first generation; and this proportion will afterwards have no tendency whatever to increase. If, on the other hand, brachydactyly were recessive, the proportion in the second generation would be $1:100,020,001$, or practically $1:100,000,000$, and this proportion would afterwards have no tendency to decrease.

In a word, there is not the slightest foundation for the idea that a dominant character should show a tendency to spread over a whole population, or that a recessive should tend to die out.

I ought perhaps to add a few words on the effect of the small deviations from the theoretical proportions which will, of course, occur in every generation. Such a distribution as $p_1 : 2q_1 : r_1$, which satisfies the condition $q_1^2 = p_1 r_1$, we may call a *stable* distribution. In actual fact we shall obtain in the second generation not $p_1 : 2q_1 : r_1$ but a slightly different distribution $p_1' : 2q_1' : r_1'$, which is not "stable." This should, according to theory, give us in the third generation a "stable" distribution $p_2 : 2q_2 : r_2$, also differing slightly from $p_1 : 2q_1 : r_1$; and so on. The sense in which the distribution $p_1 : 2q_1 : r_1$ is "stable" is this, that if we allow for the effect of casual deviations in any subsequent generation, we should, according to theory, obtain at the next generation a new "stable" distribution differing but slightly from the original distribution.

I have, of course, considered only the very simplest hypotheses possible. Hypotheses other that that of purely random mating will give different results, and, of course, if, as appears to be the case sometimes, the character is not independent of that of sex, or has an influence on fertility, the whole question may be greatly complicated. But such complications seem to be irrelevant to the simple issue raised by Mr. Yule's remarks.

G. H. HARDY

TRINITY COLLEGE, CAMBRIDGE,
April 5, 1908

P. S. I understand from Mr. Punnett that he has submitted the substance of what I have said above to Mr. Yule, and that the latter would accept it as a satisfactory answer to the difficulty that he raised. The "stability" of the particular ratio $1:2:1$ is recognized by Professor Karl Pearson (*Phil. Trans. Roy. Soc.* (A), vol. 203, p. 60).

(*d*) Elementary and Expository Notes

223. [**D. 6. b.**] *Higher Trigonometry.*

Since I wrote my rather fragmentary 'Notes' on Higher Trigonometry I have had occasion to work out the theory of the elementary transcendental functions in a rather more systematic way. This attempt has led me to modify my views in some respects. The net result is that I disagree with Mr. Picken more decidedly than I should have done if I could have seen his method of developing the theory six months ago.

I have no time to discuss the question at length. My chief difference with Mr. Picken is about the use to be made of the theorem

$$\lim_{n=\infty}\left(1+\frac{x}{n}\right)^n = \exp x.$$

He makes it fundamental: in fact he takes it as his definition of the exponential function. I adhere to my statement that this is 'logically quite wrong.' Of course I do not mean by this that it is impossible to base a rigorous theory of $\exp x$ and $\log x$ upon this theorem: many writers have done so. What I mean is that to do so is to disturb the proper perspective of the subject. Lewis Carroll based a theory of parallels on the proposition, 'In every Circle, the inscribed equilateral Tetragon is greater than any one of the Segments which lie outside it.' He would have been the first to admit that this was, 'although possible, logically quite wrong.'

Moreover, the result is not encouraging. It is 'not for the immature schoolboy mind.' I am sanguine enough to believe, on the other hand, that it is quite possible for a clever schoolboy to master a good deal of the theory of these functions. But we must look about for methods other than Mr. Picken's. On the whole I incline to the integral definition of $\log x$ as the best starting point. Mr. Picken, I notice, in one place seems to presuppose this definition. If so, why not define the exponential as the inverse of the logarithm? From the equations

$$y = \int_1^x \frac{dt}{t}, \quad x = \exp y$$

the greater part of the theory follows with perfect rigour and extreme simplicity. In particular the theorem

$$\lim_{x=\infty}\left(1+\frac{1}{x}\right)^x = e$$

(which Mr. Picken, to judge by his remarks at the top of p. 360, seems to find rather a stumbling block) follows in two or three lines.

There is, of course, really no difference between starting from these definitions and starting from

$$\frac{dx}{dy} = x, \quad x_{y=0} = 1$$

as the definition of $x = \exp y$. In this case the logarithm is defined as an inverse function.

Otherwise it seems to me, in spite of what Mr. Picken says, that we must start from the exponential series. The only serious difficulty is the proof of the functional equation by multiplication of series, or the proof of the equation $\frac{dx}{dy} = x$ by differentiation of an infinite series. It is not necessary to face both difficulties: either may be used to avoid the other. Mr. Picken's criticisms (pp. 332-3) I cannot altogether follow.

May I make two other remarks? (1) How does Mr. Picken's investigation of the factors of $\sin x$ affect my statement that the factor theorem is 'really difficult'? Apart from details his proof is that which I was taught years ago—due, I believe, to Tannery (if not much older). And does Mr. Picken mean to imply that his proof is not 'really difficult.'?

(2) Mr Picken seems to me to follow Prof. Chrystal in a certain vagueness as to the distinction between a value for $x=1$ and a limit for $x=1$. The function $\frac{x^2-1}{x-1}$ has *no* value for $x=1$; for $x=1$, $\frac{x^2-1}{x-1}$ is strictly and absolutely meaningless. The fact that its limit for $x=1$ is 2 is entirely irrelevant. The functions $\frac{x^2-1}{x-1}$ and $x+1$ are different functions. They are equal when x is *not* equal to 1. Similarly the function $y=\frac{x}{x}$ is $=1$ when $x \neq 0$ and undefined for $x=0$. To calculate $f(x)$ for $x=0$ we must put $x=0$ in the expression of $f(x)$ and perform the arithmetical operations which the form of the function prescribes, and this we cannot do in this case.

Whether Mr. Picken agrees with me here I cannot say. I lay stress on the point because his language is not quite clear. Thus he says (p. 330) that 'a function of x may have a value for a given value a of the argument, although the expression $f(x)$ fails to provide a value when a is substituted for it'—and I might quote other sentences which I cannot regard as entirely satisfactory.

G. H. HARDY.

COMMENTS

This is a sequel to 1906, 9 [Vol. V, pp. 405-9]. Picken's paper appeared in *Math. Gaz.* 3 (1906), 329-335†, 357-65. David Kennedy Picken (1879-1956) of Jesus College, Cambridge, was bracketed 6th Wrangler in 1902. After a period (1907-14) as Professor of Mathematics at Victoria College, Wellington, New Zealand, he became Master of Ormond College, in the University of Melbourne.

---

† Not 333 as quoted in Vol. V, p. 409.

MATHEMATICAL NOTES.

224. [M¹. 8. g.] *A curious imaginary curve.*

The curve
$$(x+iy)^2 = \lambda(x-iy)$$
is (i) a parabola, (ii) a rectangular hyperbola, and (iii) an equiangular spiral. The first two statements are evidently true. The polar equation is
$$r = \lambda e^{-3i\theta},$$
the equation of an equiangular spiral. The intrinsic equation is easily found to be $\rho = 3is$.

It is instructive (i) to show that the equation of any curve which is both a parabola and a rectangular hyperbola can be put in the form given above, or in the form
$$(x+iy)^2 = x \text{ (or } y\text{)},$$
and (ii) to determine the intrinsic equation directly from one of the latter forms of the Cartesian equation.   G. H. HARDY.

COMMENT

This note is reprinted in the *Mathematical Gazette* 55 (1971), 221, as part of the centenary number published in March 1971.

225. [L¹. 1. a.] *The line at infinity, etc.*

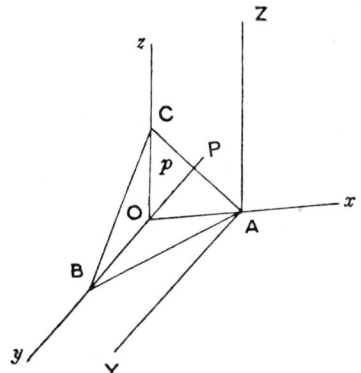

Can anyone tell me of an English book which contains a clear and intelligible account of the 'line at infinity'? Such accounts as are contained in the ordinary books on Conics, or in Miss Scott's *Modern Analytical Geometry*, appear to me confusing in the highest degree.

Most undergraduates seem to believe that there really are points at infinity, and that they really do lie on a line, and that if you could get there you would find that $1=0$. The fault lies in the books, which persist in treating conventions as if they were sober statements of fact.

I have found the following construction useful (see figure). Project p in the plane $x+y+z=1$ into P in the plane $x=1$. Taking axes AY, AZ, as shown, in the latter plane we find

$$Y = \frac{y}{x},\ Z = \frac{z}{x},$$

x, y, z being the Cartesian coordinates of p—or (what is the same thing) the areal coordinates of P in the plane $x+y+z=1$, referred to the triangle ABC—and Y, Z the Cartesian coordinates of P in the plane AYZ.

Thus if the locus of P is $f(1, Y, Z)=0$, that of p is $f(x, y, z)=0$.

G. H. HARDY.

MATHEMATICAL NOTES.

319. [D. 6. b.] *To find an approximation to the large positive root of the equation*
$$e^{e^x} = 10^{10} x^{10} e^{10^{10} x^{10}}.$$

We have
$$e^x = 23 \cdot 0 \ldots + 10 \log x + 10^{10} x^{10}$$
$$= 10^{10} x^{10} (1 + \epsilon),$$

where ϵ is quite small (certainly less than $40/10^{10}$, for example, since clearly $\log x < x^{10}$). Hence

(1) $\quad x = 23 \cdot 0 \ldots + 10 \log x + \epsilon',$

where ϵ' is positive and less than ϵ. This equation enables us to find a succession of lower and upper limits for x. Thus $x > 23$, $\log x > \log 23 > 3$; hence again $x > 23 + 30 = 53$, $\log x > \log 53 > 4$; and again $x > 23 + 40 = 63$; and so on. Again

(2) $\quad \log x = \log 10 + \log \log x + \log\left(1 + \dfrac{23 + \epsilon'}{10 \log x}\right).$

The last term is certainly less than $\dfrac{24}{40} = \dfrac{3}{5}$. Again, since $\dfrac{\log \log x}{\log x}$ decreases as x increases (at any rate when $\log x > e$, as is certainly the case), we have
$$\frac{\log \log x}{\log x} < \frac{\log \log 53}{\log 53} < \frac{\log 4}{4} < \frac{1 \cdot 4}{4} < \frac{2}{5}.$$

Hence $\qquad \log x < \log 10 + \tfrac{2}{5} \log x + \tfrac{3}{5},$

and so $\qquad \log x < 1 + \tfrac{5}{3} \log 10 < 5,$

and so $x < 170$. Substituting in (1) we get
$$x < 24 + 10 \log 170$$
$$< 24 + 56 = 80,$$

and so $\qquad x < 24 + 10 \log 80$
$$< 24 + 46 = 70,$$

and so $\qquad x < 24 + 10 \log 70$
$$< 24 + 43 = 67,$$

and so on. The value of the root certainly lies between 63 and 67: a closer approximation could be found with a little trouble.

I have purposely chosen a rather complicated equation of its type. The points to observe are (i) that the factor $10^{10} x^{10}$ proves to be of no importance whatever, and (ii) that it is futile to try to be very accurate in the early stages of the work. In all my inequalities I have left a good deal to spare, in order to work in round numbers as far as possible; and I have lost nothing by so doing. This is why examples of this sort are instructive, and teach a sense of proportion. The great weakness of boys confronted with a numerical problem is that they cannot see where accuracy is essential and where it is entirely useless.

All the logarithms are of course Napierian. As simpler examples take
$$e^x = 10^{10} x^{10}, \quad e^{x^2} = 100^{100} x^{100}, \quad e^{\sqrt{x}} = 10^{10} x (\log x)^{10},$$
etc., etc. Graphical methods may sometimes be used with advantage.

G. H. HARDY.

COMMENT

This equation appears as example (iii) on p. 62 of Hardy's Cambridge Tract, *Orders of infinity: the 'Infinitärcalcul' of Paul Du Bois-Reymond* (1st edn, Cambridge, 1910).

439. [I. 5. a.] *The Definition of a Complex Number.*

It is generally recognised by now that a "complex number" $a+ib$ is merely a symbolic equivalent for a pair of real numbers (a, b).

Writers of elementary text-books have sometimes attempted to introduce complex numbers somewhat as follows.

Let $O(a)$ denote the result of an operation performed on a real number a, and such that
$$O\{O(a)\} = -a. \quad \ldots\ldots\ldots\ldots\ldots\ldots\ldots\ldots\ldots\ldots(1)$$

Then ia is identified with $O(a)$.

This procedure is open to a multitude of objections. For one thing, $O(a)$ is *ex hypothesi* an operation on a *real* number. If then $O(a)$ is not itself real, what is $O\{O(a)\}$? But there is an even more fatal objection, and the revival of the procedure in question in a recent book (a book too of some merit) leads me to think that it may be worth while to develop the objection in detail.

The objection is that the property (1) does not suffice to define any unique operation. Let us suppose, for example, that O, operating on any complex number $a+ib$ or (a, b), produces
$$O(a, b) = (\lambda a + \mu b, \lambda' a + \mu' b),$$
where $\lambda, \mu, \lambda', \mu'$ are real.

Then
$$O\{O(a, b)\} = (A, B),$$
where
$$A = \lambda(\lambda a + \mu b) + \mu(\lambda' a + \mu' b),$$
$$B = \lambda'(\lambda a + \mu b) + \mu'(\lambda' a + \mu' b).$$

It will easily be verified that we shall have $A = -a, B = -b$, if the two conditions
$$\lambda + \mu' = 0, \quad \lambda^2 + \lambda'\mu = \lambda'\mu + \mu'^2 = -1$$
are satisfied. We can satisfy these equations by taking
$$\lambda = \sinh\theta, \quad \mu = -\rho\cosh\theta, \quad \lambda' = \frac{\cosh\theta}{\rho}, \quad \mu' = -\sinh\theta,$$
θ and ρ being any real numbers; so that
$$O(a, b) = \left(a\sinh\theta - b\rho\cosh\theta, \frac{a\cosh\theta}{\rho} - b\sinh\theta\right).$$

In particular the operation O, when performed on a real number a, produces the complex number
$$a\sinh\theta + \frac{ia\cosh\theta}{\rho}.$$

The special case required is of course that in which $\theta = 0, \rho = 1$; when
$$O(a, b) = (-b, a), \quad O(a) = ia. \qquad \text{G. H. Hardy.}$$

MATHEMATICAL NOTES.

1104. *A problem in elementary probability.*

(1) A batsman plays a large number of innings for which his average number of scoring strokes is a. There is a certain constant probability that he gets out to any ball bowled to him, and another constant probability that he makes a scoring stroke (and so a third constant probability that neither of these events happens). Prove that the commonest number of scoring strokes in an innings is 0, and, more generally, that the number of innings of exactly n scoring strokes is proportional to
$$\left(\frac{a}{a+1}\right)^n.$$

It may be assumed that all innings are completed and that the batsman is never run out.

(2) Discuss the assertion that "a batsman's average is governed primarily by his very large scores".

(3) Show that if we neglect the differences in the scoring values of different strokes, and psychological or other factors peculiar to the game of cricket, then a batsman who has a large average a should score $2a$ or over in about e^{-2} of his innings.

(4) Discuss the application of the preceding results to the case of J. B. Hobbs.

[Hobbs has, to a first approximation, scored 50,000 runs in 1000 innings, failed to score 50 times, made 200 centuries and 15 double centuries: more precise details will be found in the appropriate works of reference.]   G. H. HARDY and J. E. LITTLEWOOD.

1844. *A mathematical theorem about golf.*

It is usually held that steadiness, as against brilliancy, will tell more by strokes than by holes. It should follow that, if A is the steadier of two nearly equal players, B the more brilliant, and they are equal over a series of medal rounds, then B should have the advantage in a series of matches by holes. This seems to be the commonly accepted doctrine, but the evidence is inconclusive, since matches by holes between any given pair of leading golfers are comparatively rare. It may therefore be worth while to point out that a little mathematical analysis points, for what it is worth, to the opposite conclusion.

I construct a mathematical model as follows. Suppose that the course consists of 18 holes, all par fours (an immaterial simplification). Suppose that A is a completely mechanical player who does every hole in four (so that the conditions are similar to those of a bogey competition); and that B has equal chances x per stroke of making a *supershot* which gains a stroke or a *subshot* which costs one. It is plain that x cannot exceed $\frac{1}{2}$, and will be fairly small in any case at all corresponding to reality; that B will average par; and that the players will be equal over a series of medal rounds. According to the apparently accepted doctrine, B should win a long series of matches by holes.

To see the working of the model, suppose first that we may neglect terms of order x^2, *i.e.* consequences of B's playing more than one abnormal shot at the same hole. Then, to do a three, B must produce a supershot at one of his first *three* strokes, while he will take five if he makes a subshot at one of his first *four*. He will thus have a net expectation $4x - 3x$ or x of loss on the hole, and should lose the match, contrary to common expectation.

This approximation is too rough, but there is no difficulty in calculating the chances exactly. I find that B's chance of winning a hole is $3x - 9x^2 + 10x^3$, and his chance of losing $4x - 18x^2 + 40x^3 - 35x^4$, so that there is a balance

$$f(x) = x - 9x^2 + 30x^3 - 35x^4$$

against him. Mr. A. M. Binnie has plotted $f(x)$ for me. It increases to a maximum, about ·037, for x about ·09; and then decreases, vanishing for x about ·37 and falling to $-\frac{3}{16}$ for $x = \frac{1}{2}$. There are two inflexions, for x about ·16 and ·27, but these merely flatten part of the curve and do not affect its appearance seriously. The model has, of course, lost all its plausibility by the time that x is near $\frac{1}{2}$; a player with $x = \frac{1}{2}$ would be a "completely erratic" player, *all* of whose strokes are abnormal, and the conclusion that such a player would win, paradoxical as it may seem, need not disturb us. But for smaller x, and in particular for x about $\frac{1}{10}$, the model seems fairly reasonable.

If experience points the other way—and I cannot deny it, since I am no golfer—what is the explanation? I asked Mr. Bernard Darwin, who should be as good a judge as one could find, and he put his finger at once on a likely flaw in the model. To play a "subshot" is to give yourself an *opportunity* of a "supershot" which a more mechanical player would miss: if you get into a bunker you have an opportunity of recovering without loss, and one which you are naturally keyed up to take. Thus the less mechanical player's chance of a supershot is to some extent automatically increased. How far this may resolve the paradox, if it is one, I cannot say, and changes in the model make it unpleasantly complex.

I add one word about possible testing of the model. There are few data about matches between, say, Hagen and R. T. Jones. But the model, as I said, is more like that of a bogey competition, and there must be abundant information about them. If a player averages bogey, does he win or lose when he plays against bogey by holes? There should be plenty of data for testing that.

G. H. HARDY.

(e) Addresses, Invited Lectures, etc.

Mr S. Ramanujan's Mathematical Work in England*.

I shall confine myself in this report to the work which Mr Ramanujan has done since his arrival in England in April 1914, and to work which has been published, or is completed and on the point of publication. I need hardly say that he has in his possession, as he had before he came to England, a mass of unpublished material containing hundreds of most curious and interesting formulae, some definitely proved and others only conjectured. Many of the most remarkable of these results will no doubt be published in due course, as the process of sifting them continues.

In one respect Mr Ramanujan has been most unfortunate. The war has naturally had disastrous results on the progress of mathematical research. It has distracted three-quarters of the interest that would otherwise have been taken in his work, and has made it almost impossible to bring his results to the notice of the continental mathematicians most certain to appreciate it. It has moreover deprived him of the teaching of Mr Littlewood, one of the great benefits which his visit to England was intended to secure. All this will pass; and, in spite of it, it is already safe to say that Mr Ramanujan has justified abundantly all the hopes that were based upon his work in India, and has shown that he possesses powers as remarkable in their way as those of any living mathematician. His work is only the more valuable because his abilities and methods are of so unusual a kind, and so unlike those of a European mathematician trained in the orthodox school.

Mr Ramanujan's papers are as follows†:

(1) 'Some definite integrals', *Messenger of Mathematics*, vol. 44, 1914, pp. 10—18.
(2) 'Some definite integrals connected with Gauss's sums', *ibidem*, pp. 75—85.
(3) 'Modular equations and approximations to π', *Quarterly Journal of Mathematics*, vol. 45, 1914, pp. 350—372.
(4) 'New expressions for Riemann's functions $\zeta(s)$ and $\Xi(t)$', *ibidem*, vol. 46, 1915, pp. 253—261.
(5) 'On certain infinite series', *Messenger of Mathematics*, vol. 45, 1915, pp. 11—15.
(6) 'Summation of a certain series', *ibidem*, pp. 157—160.
(7) 'Highly composite numbers', *Proc. London Math. Soc.*, ser. 2, vol. 4, 1915, pp. 347—409.
(8) 'Some formulae in the analytic theory of numbers', *Messenger of Mathematics*, vol. 45, 1916, pp. 81—84.

* This report on Mr Ramanujan's work has been prepared for publication at the request of the authorities of the University of Madras.

† I do not include the various short notes published by him in India, as these are presumably already familiar to Indian readers.

2   Mr S. RAMANUJAN'S MATHEMATICAL WORK IN ENGLAND.

(9) 'On certain arithmetical functions', *Trans. Cambridge Phil. Soc.*, vol. 22, 1916, no. 9, pp. 159—184.

(10) 'Some series for Euler's constant' (*Messenger of Mathematics*, as yet unpublished).

(11) 'On the expression of numbers in the form $ax^2 + by^2 + cz^2 + dt^2$' (*Proc. Cambridge Phil. Soc.*, as yet unpublished).

(12) 'A problem in the analytic theory of numbers' (by G. H. Hardy and S. Ramanujan, communicated to the London Mathematical Society, as yet unpublished).

All of these papers, except the first three, represent work done for the most part in England. The most important are (3), (4), (7), and (9). I shall say a few words about the remaining published papers first, then give a rather fuller account of these four, and end by saying something about those still unpublished.

(1) This paper, published very shortly after Mr Ramanujan's arrival in England, contains the values of some curious and interesting definite integrals, such as

$$\int_0^\infty |\Gamma(a+ix)|^2 \cos 2mx \, dx = \tfrac{1}{2} \sqrt{\pi} \, \Gamma(a) \, \Gamma(a + \tfrac{1}{2}) (\operatorname{sech} m)^{2a} \quad \ldots (1\cdot1),$$

$$\int_0^\infty \left|\frac{\Gamma(a+ix)}{\Gamma(b+ix)}\right|^2 dx = \tfrac{1}{2} \sqrt{\pi} \, \frac{\Gamma(a) \Gamma(a+\tfrac{1}{2}) \Gamma(b-a-\tfrac{1}{2})}{\Gamma(b-\tfrac{1}{2}) \Gamma(b) \Gamma(b-a)} \quad \ldots (1\cdot2),$$

$$\int_0^\infty |\Gamma(a+ix) \Gamma(b+ix)|^2 dx = \tfrac{1}{2} \sqrt{\pi} \, \frac{\Gamma(a) \Gamma(a+\tfrac{1}{2}) \Gamma(b) \Gamma(b+\tfrac{1}{2}) \Gamma(a+b)}{\Gamma(a+b+\tfrac{1}{2})} \quad \ldots (1\cdot3),$$

where a, b, and m are positive, and, in (1·2), $a < b$; and some transformation formulae of which I quote

$$F(\alpha) = F\left(\frac{\pi^2}{\alpha}\right) \quad \ldots (1\cdot4),$$

where
$$F(\alpha) = \alpha^{-\tfrac{1}{4}} \left(1 + 4\alpha \int_0^\infty \frac{xe^{-\alpha x^2}}{e^{2\pi x} - 1} dx\right) \quad \ldots (1\cdot41).$$

(2) Mr Ramanujan shows that the integral

$$\int_0^\infty \frac{\cos 2\pi tx}{\cosh \pi x} e^{-n\pi i x^2} dx \quad \ldots (2\cdot1)$$

can be calculated in finite terms whenever n is rational. Only very special cases of this result, found by Kronecker and by myself, were known before. Similar results are proved for other classes of definite integrals. The method of proof depends upon an ingenious application of a well-known theorem of Lerch.

As typical formulae I quote:

$$\int_0^\infty \frac{\cos \pi x^2}{\cosh \pi x} \cos 2\pi tx \, dx = \frac{1 + \sqrt{2} \sin \pi t^2}{2\sqrt{2} \cosh \pi t} \quad \ldots (2\cdot2),$$

$$\int_0^\infty \frac{\sin \pi x^2}{\sinh \pi x} \sin 2\pi tx \, dx = \frac{\sin \pi t^2}{2 \sinh \pi t} \quad \ldots (2\cdot3),$$

and
$$\phi(0) = \frac{1}{24}, \quad \phi(1) = \frac{2-\sqrt{2}}{16}, \quad \phi(\tfrac{1}{2}) = \frac{1}{8\pi}, \quad \phi(\tfrac{2}{5}) = \frac{8 - 3\sqrt{5}}{32} \quad \ldots (2\cdot4),$$

where
$$\phi(n) = \int_0^\infty \frac{x \cos n\pi x^2}{e^{2\pi x} - 1} dx \quad \ldots (2\cdot41).$$

(5) This paper is a supplement to (2). Certain series, formally analogous to the integrals of (2), are expressible, for special values of a parameter, in terms of elliptic functions.

Thus for example, if a is a positive integer, the series

$$\frac{\sin(\pi/a)}{\cosh \tfrac{1}{2}\pi} - \frac{3\sin(9\pi/a)}{\cosh \tfrac{3}{2}\pi} + \frac{5\sin(25\pi/a)}{\cosh \tfrac{5}{2}\pi} - \ldots \qquad \ldots\ldots\ldots(5\cdot1)$$

is equal to

$$\frac{a\sqrt{a}}{8\sqrt{2}}\left\{\frac{1}{\cosh \tfrac{1}{8}\pi a} - \frac{3}{\cosh \tfrac{3}{8}\pi a} + \frac{5}{\cosh \tfrac{5}{8}\pi a} - \ldots\right\} = \frac{a\sqrt{a}}{4\sqrt{2}}(e^{-\tfrac{1}{32}\pi a} - e^{-\tfrac{9}{32}\pi a} + e^{-\tfrac{25}{32}\pi a} - \ldots)^4$$
$$\ldots\ldots\ldots(5\cdot21),$$

or to

$$\frac{a\sqrt{a}}{8\sqrt{2}}\left\{\frac{1}{\sinh \tfrac{1}{8}\pi a} + \frac{3}{\sinh \tfrac{3}{8}\pi a} + \frac{5}{\sinh \tfrac{5}{8}\pi a} + \ldots\right\} = \frac{a\sqrt{a}}{4\sqrt{2}}(e^{-\tfrac{1}{32}\pi a} + e^{-\tfrac{9}{32}\pi a} + e^{-\tfrac{25}{32}\pi a} + \ldots)^4$$
$$\ldots\ldots\ldots(5\cdot22),$$

according as a is even or odd. Another typical formula is

$$\frac{1}{8\pi} + \frac{\cos(2\pi/a)}{e^{2\pi}-1} + \frac{2\cos(8\pi/a)}{e^{4\pi}-1} + \frac{3\cos(18\pi/a)}{e^{6\pi}-1} + \ldots$$
$$= \int_0^\infty \frac{x\cos(2\pi x^2/a)}{e^{2\pi x}-1}dx + \tfrac{1}{4}a\sqrt{a}\left(\frac{1}{e^{\pi a}+1} + \frac{3}{e^{3\pi a}+1} + \frac{5}{e^{5\pi a}+1} + \ldots\right) \ldots(5\cdot3),$$

where a is odd.

(6) The series

$$\phi(s) = \sum_0^\infty \{\sqrt{(n+1)} - \sqrt{n}\}^s \qquad (s=3, 5, 7, \ldots) \quad \ldots\ldots(6\cdot1)$$

may be expressed finitely in terms of the Riemann ζ-function. Thus

$$\phi(3) = \frac{3}{2\pi}\zeta(\tfrac{3}{2}), \qquad \psi(5) = \frac{15}{2\pi^2}\zeta(\tfrac{5}{2}) \qquad \ldots\ldots\ldots\ldots(6\cdot11).$$

The paper also contains some interesting formulae concerning sums of square roots of integers, which have appeared also in the *Indian Mathematical Journal*.

(8) This paper contains a number of formulae, stated without proof, discovered incidentally by Mr Ramanujan in the course of other investigations. If $\sigma_a(n)$ denotes the sum of the a-th powers of the divisors of n, then

$$1^{-s}\sigma_a(1)\sigma_b(1) + 2^{-s}\sigma_a(2)\sigma_b(2) + \ldots = \frac{\zeta(s)\zeta(s-a)\zeta(s-b)\zeta(s-a-b)}{\zeta(2s-a-b)} \ldots(8\cdot1).$$

In particular

$$1^{-s}d^2(1) + 2^{-s}d^2(2) + \ldots = \frac{\zeta^4(s)}{\zeta(2s)} \qquad \ldots\ldots\ldots\ldots(8\cdot11),$$

$d(n)$ being the number of divisors of n. A similar formula is given in which $d(n)$ is replaced by $r(n)$, the number of representations of n as the sum of two squares. Asymptotic formulae are deduced for the sums

$$d^2(1) + d^2(2) + \ldots + d^2(n), \qquad r^2(1) + r^2(2) + \ldots + r^2(n) \qquad \ldots\ldots\ldots(8\cdot2).$$

Thus the dominant terms in the latter sum are

$$\tfrac{1}{4}n\log n + \tfrac{1}{4}Cn \qquad \ldots\ldots\ldots\ldots(8\cdot21),$$

where
$$C = 4\gamma - 1 + \tfrac{1}{3}\log 2 - \log \pi + 4\log \Gamma(\tfrac{3}{4}) - \frac{12}{\pi^2}\zeta'(2) \quad\ldots\ldots\ldots\ldots(8\cdot 211),$$

γ being Euler's constant. Among the other interesting formulae contained in this paper, I will mention only the formula

$$\frac{1}{1^s d(1)} + \frac{1}{2^s d(2)} + \ldots = \sqrt{\{\zeta(s)\}}\, \phi(s) \quad\ldots\ldots\ldots\ldots(8\cdot 3),$$

where $\phi(s)$ is a Dirichlet's series absolutely convergent when the real part of s is greater than $\tfrac{1}{2}$.

I come now to **Mr Ramanujan's** most important papers.

(3) This paper embodies the results of a great deal of Mr Ramanujan's Indian work. This work was done when he had very little knowledge of the progress made in Europe during the latter half of the nineteenth century, and it was inevitable that many of his results should have been anticipated by European mathematicians, in particular by Hermite, Kronecker, Weber, and Greenhill. None the less the paper is of the greatest interest and contains a large number of new results. I will state the principal problems as they present themselves from the author's individual point of view.

We have, in the ordinary notation of the theory of elliptic functions,

$$(1+q)(1+q^3)(1+q^5)\ldots = 2^{\tfrac{1}{6}} q^{\tfrac{1}{24}} (kk')^{-\tfrac{1}{12}} \quad\ldots\ldots\ldots\ldots(3\cdot 11),$$

$$(1-q)(1-q^3)(1-q^5)\ldots = 2^{\tfrac{1}{6}} q^{\tfrac{1}{24}} k^{-\tfrac{1}{12}} k'^{\tfrac{1}{6}} \quad\ldots\ldots\ldots\ldots(3\cdot 12).$$

Suppose now that $n = r/s$, where r and s are positive integers. The relation between k and l, which makes

$$n\frac{K'}{K} = \frac{L'}{L} \quad\ldots\ldots\ldots\ldots(3\cdot 2),$$

is given by the modular equation of the rs-th degree. If

$$k = l', \quad k' = l, \quad K = L', \quad K' = L \quad\ldots\ldots\ldots\ldots(3\cdot 31),$$

then

$$q = e^{-\pi K'/K} = e^{-\pi/\sqrt{n}}, \quad q' = e^{-\pi L'/L} = e^{-\pi\sqrt{n}} \quad\ldots\ldots\ldots\ldots(3\cdot 32),$$

and the corresponding values of k and l may be found by the solution of an algebraic equation.

Thus, if

$$G_n = 2^{-\tfrac{1}{4}} e^{\tfrac{1}{24}\pi\sqrt{n}}(1 + e^{-\pi\sqrt{n}})(1 + e^{-3\pi\sqrt{n}})\ldots \quad\ldots\ldots\ldots\ldots(3\cdot 41),$$

$$g_n = 2^{-\tfrac{1}{4}} e^{\tfrac{1}{24}\pi\sqrt{n}}(1 - e^{-\pi\sqrt{n}})(1 - e^{-3\pi\sqrt{n}})\ldots \quad\ldots\ldots\ldots\ldots(3\cdot 42),$$

G_n and g_n are algebraical numbers whenever n is rational. And since the ratio of either of them to

$$2^{-\tfrac{1}{4}} e^{\tfrac{1}{24}\pi\sqrt{n}}$$

is exceedingly near to 1 if n is at all large, we obtain an approximation to π of the form (e.g.)

$$\pi = \frac{24}{\sqrt{n}} \log(2^{\tfrac{1}{4}} G_n) \quad\ldots\ldots\ldots\ldots(3\cdot 43).$$

Mr Ramanujan has found G_n or g_n for a very large number of values of n, his results in this direction going far beyond those of any previous writer. Thus (to quote only new results)

$$G_{81}^3 = \frac{(2\sqrt{3}+2)^{\frac{1}{3}}+1}{(2\sqrt{3}-2)^{\frac{1}{3}}-1},$$

$$g_{98} + \frac{1}{g_{98}} = \tfrac{1}{2}\{\sqrt{2}+\sqrt{(14+4\sqrt{14})}\},$$

$$G_{553}^2 = \left\{\sqrt{\left(\frac{96+11\sqrt{79}}{4}\right)}+\sqrt{\left(\frac{100+11\sqrt{79}}{4}\right)}\right\}\left\{\sqrt{\left(\frac{141+16\sqrt{79}}{4}\right)}+\sqrt{\left(\frac{143+16\sqrt{79}}{4}\right)}\right\}.$$

Among his approximate formulae for π I may quote

$$e^{\pi\sqrt{58}} = 24591257751,$$

$$\pi = \frac{12}{\sqrt{190}} \log\{(2\sqrt{2}+\sqrt{10})(3+\sqrt{10})\},$$

correct to 6 and 18 places of decimals respectively. The most accurate of his approximations of this kind holds to 31 places.

In the later sections he develops a quite original method for approximating to π by *algebraical numbers only*. Let

$$\phi(q) = q^{\frac{1}{12}}(1-q^2)(1-q^4)(1-q^6)\ldots \quad\ldots\ldots\ldots\ldots(3\cdot 51),$$

$$\psi(q) = 12q\frac{\phi'(q)}{\phi(q)} = 1 - 24\left(\frac{q^2}{1-q^2} + \frac{2q^4}{1-q^4} + \ldots\right) \quad\ldots\ldots\ldots\ldots(3\cdot 52),$$

$$\chi(q) = 1 - 24\left(\frac{q}{1+q} + \frac{3q^3}{1+q^3} + \ldots\right) = \left(\frac{2K}{\pi}\right)^2(1-2k^2) \quad\ldots\ldots\ldots\ldots(3\cdot 53).$$

Then

$$\frac{\phi(q)}{\phi(q^n)} = \left(\frac{kk'}{ll'}\right)^{\frac{1}{6}}\sqrt{\left(\frac{K}{L}\right)} \quad\ldots\ldots\ldots\ldots(3\cdot 61),$$

if k and l are connected by the relation (3·2). Take $k = l'$, etc., as in (3·31), so that $q = e^{-\pi/\sqrt{n}}$; substitute in (3·61); and differentiate logarithmically with respect to n. We obtain

$$n\psi(e^{-\pi\sqrt{n}}) + \psi(e^{-\pi/\sqrt{n}}) = \frac{6\sqrt{n}}{\pi} \quad\ldots\ldots\ldots\ldots(3\cdot 62).$$

Now suppose n an integer. Then l is an algebraical function of k, determined by an equation with rational coefficients. Let us denote a function of this type, generally, by $A(k)$. We have

$$\frac{dk}{dl} = A(k).$$

But

$$\frac{n\,dk}{kk'^2 K^2} = \frac{dl}{ll'^2 L^2},$$

and so

$$\frac{K}{L} = A(k).$$

Again, differentiating (3·61) logarithmically with respect to k, and observing that

$$\frac{dq}{dk} = \frac{\pi^2 q}{2kk'^2 K^2},$$

we find

$$n\psi(q^n) - \psi(q) = \left(\frac{K}{\pi}\right)^2 A(k) \quad\ldots\ldots\ldots\ldots(3\cdot 71).$$

And if in this relation we suppose $k = l'$, etc., we obtain

$$n\psi(e^{-\pi\sqrt{n}}) - \psi(e^{-\pi/\sqrt{n}}) = \alpha\left(\frac{K}{\pi}\right)^2 \quad\ldots\ldots\ldots\ldots\ldots\ldots(3\cdot72),$$

where α is an algebraical number obtained by substituting in $A(k)$ the value of k drawn from the modular equation when $k = l'$ and $k' = l$.

From (3·62) and (3·72) we deduce

$$2n\psi(e^{-\pi\sqrt{n}}) - \frac{6\sqrt{n}}{\pi} = \alpha\left(\frac{K}{\pi}\right)^2 \quad\ldots\ldots\ldots\ldots\ldots\ldots(3\cdot81).$$

And from (3·53) we obtain

$$\chi(e^{-\pi\sqrt{n}}) = \beta\left(\frac{K}{\pi}\right)^2 \quad\ldots\ldots\ldots\ldots\ldots\ldots(3\cdot82),$$

where β is a number similar to α. Thus the quotient of the left-hand sides of (3·81) and (3·82) is an algebraical number.

If n is at all large, ψ and χ are both very nearly equal to unity. We thus obtain an approximate algebraical value for

$$2n - \frac{6\sqrt{n}}{\pi}$$

and so for π. Mr Ramanujan finds, for example, that

$$\frac{9}{5} + \sqrt{\frac{9}{5}} = 3\cdot14164\ldots, \qquad \frac{19}{16}\sqrt{7} = 3\cdot14180\ldots,$$

$$\frac{63}{25}\frac{17 + 15\sqrt{5}}{7 + 15\sqrt{5}} = 3\cdot14159265380\ldots.$$

This last number gives π correctly to 9 places of decimals.

The paper also includes a number of remarkable series for $1/\pi$ drawn from the theory of elliptic functions. I may mention the series

$$\frac{1}{2\pi\sqrt{2}} = \frac{1103}{99^2} + \frac{27493}{99^6}\cdot\frac{1}{2}\cdot\frac{1\cdot3}{4^2} + \frac{53883}{99^{10}}\cdot\frac{1\cdot3}{2\cdot4}\cdot\frac{1\cdot3\cdot5\cdot7}{4^2\cdot8^2} + \ldots.$$

The first term alone gives the sum to 8 places.

Finally Mr Ramanujan gives certain expressions (exact or approximate) for the perimeter of an ellipse. Thus, when the major semi-axis is a and the eccentricity is $\tan\tfrac{1}{8}\pi$, the perimeter is

$$a\sqrt{\left(\frac{\pi}{4}\right)}\left\{\frac{\Gamma(\tfrac{1}{8})}{\Gamma(\tfrac{5}{8})} + \frac{\Gamma(\tfrac{5}{8})}{\Gamma(\tfrac{9}{8})}\right\}.$$

(4) Riemann's function $\Xi(t)$ is defined by the equations

$$\xi(s) = \frac{s(s-1)}{2}\pi^{-\tfrac{1}{2}s}\Gamma(\tfrac{1}{2}s)\zeta(s), \qquad \xi(\tfrac{1}{2} + it) = \Xi(t),$$

and is an even integral function of t, which is real when t is real, and all of whose zeros are probably real. Mr Ramanujan supposes that α and β are two numbers whose real part is positive, that $\alpha\beta = \pi^2$, that

$$\phi(t, \alpha, x) = \frac{3}{3^2 + t^2} - \frac{\alpha}{1!}\frac{7x^2}{7^2 + t^2} + \frac{\alpha^2}{2!}\frac{11x^4}{11^2 + t^2} - \ldots,$$

and that
$$f(t, \alpha) = \alpha^{-\frac{1}{4}} \left\{ \frac{1}{1+t^2} - 4\alpha \int_0^\infty \phi(t, \alpha, x) \frac{x\, dx}{e^{2\pi x} - 1} \right\};$$

and shows that
$$\frac{1}{4} \pi^{-\frac{3}{4}} \Gamma\left(\frac{-1+it}{4}\right) \Gamma\left(\frac{1+it}{4}\right) \Xi\left(\frac{1}{2}t\right) \cos\left(\frac{t}{8} \log \frac{\alpha}{\beta}\right) = f(t, \alpha) + f(t, \beta) \quad \ldots\ldots\ldots (4\cdot 1).$$

In particular, if $\alpha = \beta = \pi$ and t is real, we have
$$\frac{1}{8\sqrt{\pi}} \left| \Gamma\left(\frac{-1+it}{4}\right) \right|^2 \Xi\left(\frac{1}{2}t\right) = \frac{1}{1+t^2} - 4\pi \int_0^\infty \phi(t, \pi, x) \frac{x\, dx}{e^{2\pi x} - 1} \quad \ldots\ldots\ldots (4\cdot 2).$$

From these formulae, and others of a similar character, he deduces, by means of Fourier's double integral theorem, the values of some remarkable definite integrals containing the function $\Xi(t)$ under the sign of integration.

It is difficult at present to estimate the importance of these results. The unsolved problems concerning the zeros of $\zeta(s)$ or of $\Xi(t)$ are among the most obscure and difficult in the whole range of pure mathematics. Any new formulae involving $\zeta(s)$ or $\Xi(t)$ are of very great interest, because of the possibility that they may throw new light on some of these outstanding questions. It is, as I have shown in a short note attached to Mr Ramanujan's paper, certainly possible to apply his formulae in this direction; but the results which can be deduced from them do not at present go beyond those obtained already by Mr Littlewood and myself in other ways. But I should not be at all surprised if still more important applications were to be made of Mr Ramanujan's formulae in the future.

(7) Mr Ramanujan's elaborate memoir on 'Highly composite numbers' contains an account of the longest and perhaps the most important connected piece of work which he has done since his arrival in England. A *highly composite* number n is a number which has more divisors than any smaller number, which is, so to say, as unlike a prime as a number can be. Thus 2, 4, 6, 12, 24, 36, 48, 60, 120, and 180 are the first ten such numbers.

Mr Ramanujan shows how, by reasoning of an elementary but highly ingenious character, we can obtain surprisingly accurate information as to the structure of highly composite numbers. There are two ways of stating the problem. We may write
$$n = 2^{a_2} 3^{a_3} 5^{a_5} \ldots p^{a_p},$$
and study the indices $a_2, a_3, \ldots$, which, as it is easy to see, must satisfy the inequalities
$$a_2 \geqslant a_3 \geqslant a_5 \ldots \geqslant a_p;$$
or we may write
$$n = 2 \cdot 3 \cdot 5 \ldots\ldots\ldots p$$
$$\times\, 2 \cdot 3 \cdot 5 \ldots\ldots p_1$$
$$\times\, 2 \cdot 3 \cdot 5 \ldots p_2$$
$$\times\, \ldots,$$

where $p \geqslant p_1 \geqslant p_2 \ldots$, and study the distribution of the primes $p, p_1, p_2, \ldots$. The second method proves in some ways the better: but it is easier to give an idea of the results by stating them in terms of the indices $a_2, a_3, a_5, \ldots$.

Mr Ramanujan proves, in the first place, that
$$a_p = 1 \quad\quad\quad\quad\quad\quad (7\cdot 1)$$
except for $n = 4$ and $n = 36$. He then shows that *near the beginning* the indices form a strictly decreasing sequence, that is to say that
$$a_2 > a_3 > a_5 \ldots > a_\lambda \quad\quad\quad\quad (7\cdot 2),$$
where λ is a certain function of n (or of p) which tends to infinity with n; but that later on groups of equal indices occur; and that, when n is very large, there are certainly groups of indices equal to $1, 2, 3, \ldots, \mu$, where μ again is a function of n (or p) which tends to infinity with n; and that the size of these later groups also becomes large with n. All these phenomena have begun to exhibit themselves, for example, in the largest highly composite number given by Mr Ramanujan, viz.:
$$6746328388800 = 2^6 \cdot 3^4 \cdot 5^2 \cdot 7^2 \cdot 11 \cdot 13 \cdot 17 \cdot 19 \cdot 23.$$
He shows further that, if q is fairly small in comparison with p, then
$$a_2 \log 2 \sim a_3 \log 3 \sim a_5 \log 5 \ldots \sim a_q \log q \sim \frac{\log p}{\log 2} \quad\quad (7\cdot 3);$$
and that a large group of indices near the end can be assigned with an error of at most unity.

He proves also that two successive highly composite numbers are asymptotically equivalent: from which it can be deduced that the number $N(x)$ of highly composite numbers less than x is of an order at any rate greater than that of $\log x$. The more precise determination of the order of $N(x)$ appears to be a problem of extreme difficulty. Mr Ramanujan has since found more definite results in this direction; but the problem is still unsolved.

Later in the paper Mr Ramanujan considers a special class of highly composite numbers which he calls *superior* highly composite numbers. The form of these numbers he determines *exactly*, and he bases on their properties a study of the 'maximum order' of the function $d(n)$ far more precise than any made before. In particular, assuming the Riemann hypothesis as to the zeros of $\zeta(s)$, he shows that the true maximum order of $d(n)$ is
$$2^{Li \log n + \phi(n)} \quad\quad\quad\quad\quad (7\cdot 41),$$
where
$$\phi(n) = \frac{\log \tfrac{3}{2}}{\log 2} Li \{(\log n)^{\log \tfrac{3}{2} / \log 2}\} - \frac{(\log n)^{\log \tfrac{3}{2} / \log 2}}{\log \log n} + \psi(n) \quad\quad (7\cdot 42),$$
$\psi(n)$ being at most of order
$$\frac{\sqrt{(\log n)}}{(\log \log n)^2} \quad\quad\quad\quad\quad (7\cdot 43).$$

The paper concludes with a study of the order of $d(n)$ for a number of special forms of n.

Mr Ramanujan has, since this paper was published, obtained a number of further results of the same kind. The problem is a very peculiar one, standing somewhat apart from the main channels of mathematical research. But there can be no question as to the extraordinary insight and ingenuity which he has shown in treating it, nor any doubt that his memoir is one of the most remarkable published in England for many years.

(9) This paper falls naturally into two parts. In the first part Mr Ramanujan considers the function

$$\Sigma_{a,b}(n) = \sigma_a(0)\sigma_b(n) + \sigma_a(1)\sigma_b(n-1) + \ldots + \sigma_a(n)\sigma_b(0) \ldots \ldots (9\cdot11),$$

where a and b are odd positive integers, and $\sigma_a(n)$ denotes the sum of the a-th powers of the divisors of n, except when $n=0$, when $\sigma_a(0)$ is to be interpreted as $\frac{1}{2}\zeta(-a)$. He writes

$$\Sigma_{a,b}(n) = S_{a,b}(n) + E_{a,b}(n) \ldots \ldots (9\cdot12),$$

where

$$S_{a,b}(n) = \frac{\Gamma(a+1)\Gamma(b+1)}{\Gamma(a+b+2)} \frac{\zeta(a+1)\zeta(b+1)}{\zeta(a+b+2)} \sigma_{a+b+1}(n) + \frac{\zeta(1-a)+\zeta(1-b)}{a+b} n\sigma_{a+b-1}(n)$$
$$\ldots \ldots (9\cdot13);$$

and his main problem is that of the determination of the maximum order of the 'error term' $E_{a,b}(n)$ as a function of n. He shows, in the first place, that $E_{a,b}(n)$ is zero if $a+b$ has one or other of the values 2, 4, 6, 8, and 12. In this case $\Sigma_{a,b}(n)$ may be expressed finitely in terms of the function σ: thus, e.g.,

$$\Sigma_{3,3}(n) = \frac{\sigma_7(n)}{120}, \quad \Sigma_{5,7}(n) = \frac{\sigma_{13}(n)}{10080}, \quad \Sigma_{3,9}(n) = \frac{\sigma_{13}(n)}{2640} \ldots \ldots (9\cdot2).$$

In all other cases the expression of $\Sigma_{a,b}(n)$ involves other arithmetical functions of a more obscure nature. Thus, for example, when $a+b=10$, we have

$$E_{a,b}(n) = F_{a,b}(1)\tau(n) \ldots \ldots (9\cdot31),$$

where $\tau(n)$ is defined by the equation

$$\sum_{1}^{\infty}\tau(n)x^n = x\{(1-x)(1-x^2)(1-x^3)\ldots\}^{24} \ldots \ldots (9\cdot32).$$

Mr Ramanujan regards it as highly probable that the maximum order of $E_{a,b}(n)$ is not less than that of

$$n^{\frac{1}{2}(a+b+1)},$$

and less than that of

$$n^{\frac{1}{2}(a+b+1)+\epsilon}$$

for every positive value of ϵ. Thus $\tau(n)$ should be about of the order of $n^{\frac{11}{2}}$. He can however only *prove* that its order is not less than that of n^5 and not greater than that of n^7, with corresponding results in the general case. In all cases we have

$$\Sigma_{a,b}(n) \sim S_{a,b}(n) \ldots \ldots (9\cdot4),$$

a formula remarkable in that it shows the asymptotic equivalence of two arithmetical functions, neither of which increases in a regular manner.

The truth of the more exact results concerning the maximum order of $\tau(n)$, which Mr Ramanujan is at present unable to prove, would follow without difficulty from a remarkable formula which he has conjectured, but of which he has at present no rigorous proof, viz.

$$\sum_1^\infty \frac{\tau(n)}{n^s} = \prod_p \left\{ \frac{1}{1 - \tau(p) p^{-s} + p^{11-2s}} \right\} \quad \ldots \ldots \ldots \ldots \ldots \ldots (9 \cdot 51),$$

where the product on the right-hand side extends over all primes p. The assertion of this formula is equivalent to the assertion that, if

$$n = p_1^{a_1} p_2^{a_2} \ldots p_r^{a_r} \quad \ldots \ldots \ldots \ldots \ldots \ldots (9 \cdot 52),$$

and

$$\cos \theta_p = \tfrac{1}{2} p^{-\tfrac{11}{2}} \tau(p) \quad \ldots \ldots \ldots \ldots \ldots \ldots (9 \cdot 53),$$

then

$$n^{-\tfrac{11}{2}} \tau(n) = \frac{\sin(1+a_1)\theta_{p_1}}{\sin \theta_{p_1}} \frac{\sin(1+a_2)\theta_{p_2}}{\sin \theta_{p_2}} \ldots \frac{\sin(1+a_r)\theta_{p_r}}{\sin \theta_{p_r}} \quad \ldots \ldots \ldots (9 \cdot 54).$$

Mr Ramanujan gives a number of equally remarkable conjectural formulae. It appears, for example, that the coefficient of x^n in

$$x \{(1 - x^{12})(1 - x^{24})(1 - x^{36}) \ldots\}^2$$

is zero unless n is of the form

$$(5^{a_5} 7^{a_7} 11^{a_{11}} 17^{a_{17}} \ldots)^2 13^{a_{13}} 37^{a_{37}} 61^{a_{61}} 73^{a_{73}} \ldots,$$

where $5, 7, 11, 17, \ldots$ are the primes of the forms $12k+5$, $12k+7$, and $12k+11$, and $13, 37, 61, 73, \ldots$ those of the form $12k+1$; and that when n is of this form the coefficient is

$$(-1)^{a_5 + a_{13} + a_{17} + a_{29} + a_{41} + \cdots} (1 + a_{13})(1 + a_{37})(1 + a_{61})(1 + a_{73}) \ldots,$$

where the sequence $13, 37, 61, 73, \ldots$ is defined as above, and the sequence $5, 13, 17, 29, 41, \ldots$ includes all primes of the form $4k+1$ which cannot be expressed in the form $x^2 + (6y)^2$.

In the second part of the paper Mr Ramanujan develops a corresponding series of results for the function $r_{2k}(n)$ which is equal to the number of representations of n as the sum of $2k$ squares, so that

$$1 + 2 \sum_1^\infty r_{2k}(n) q^n = (1 + 2q + 2q^4 + 2q^9 + \ldots)^{2k} \quad \ldots \ldots \ldots \ldots \ldots (9 \cdot 61).$$

He shows that

$$r_{2k}(n) = \delta_{2k}(n) + e_{2k}(n) \quad \ldots \ldots \ldots \ldots \ldots \ldots (9 \cdot 62),$$

where $\delta_{2k}(n)$ is a function which can be defined in finite terms by means of the divisors of n; and proves results concerning the order of $e_{2k}(n)$ analogous to those concerning that of $E_{a,b}(n)$ proved in the first part of the paper. In all cases

$$r_{2k}(n) \sim \delta_{2k}(n) \quad \ldots \ldots \ldots \ldots \ldots \ldots (9 \cdot 63).$$

These functions have of course been considered from a different point of view by earlier writers, notably Jacobi, Eisenstein, H. J. S. Smith, and Glaisher; but Mr Ramanujan's most characteristic results are quite new. To this part of the paper also Mr Ramanujan adds a number of remarkable formulae as yet unproved, containing expressions of the series

$$\sum \frac{e_{2k}(n)}{n^s}$$

as products analogous to that in (9·51), and corresponding explicit expressions for the functions $e_{2k}(n)$ in terms of the prime factors of n.

I may quote, for example, the formula

$$\sum_1^\infty \frac{e_{10}(n)}{n^s} = \frac{e_{10}(1)}{1+2^{2-s}} \Pi_1 \Pi_2 \quad\quad\quad\quad (9\cdot71),$$

where
$$\Pi_1 = \frac{1}{(1-3^{4-2s})(1-7^{4-2s})(1-11^{4-2s})\ldots} \quad\quad\quad\quad (9\cdot711),$$

$$\Pi_2 = \frac{1}{(1-2c_5\cdot 5^{-s}+5^{4-2s})(1-2c_{13}\cdot 13^{-s}+13^{4-2s})\ldots} \quad\quad\quad\quad (9\cdot712).$$

In these formulae 3, 7, 11, ... are the primes of the form $4k+3$, and 5, 13, ... the primes of the form $4k+1$; and

$$c_p = u^2 - (4v)^2 \quad\quad\quad\quad (9\cdot713),$$

where u and v are the unique pair of positive integers such that $u^2 + (4v)^2 = p^2$. The assertion of the formula (9·71) is equivalent to the assertion that $e_{10}(n)$ is zero unless

$$n = (3^{a_3} 7^{a_7} 11^{a_{11}} \ldots)^2 2^{a_2} 5^{a_5} 13^{a_{13}} \ldots \quad\quad\quad\quad (9\cdot72);$$

and that if n is of this form then

$$\frac{e_{10}(n)}{n^2 e_{10}(1)} = (-1)^{a_2} \frac{\sin 4(1+a_5)\theta_5}{\sin 4\theta_5} \frac{\sin 4(1+a_{13})\theta_{13}}{\sin 4\theta_{13}} \quad\quad\quad\quad (9\cdot73),$$

where θ_p is defined by the formulae

$$\tan \theta_p = \frac{u}{v}, \quad 0 < \theta_p < \tfrac{1}{2}\pi \quad\quad\quad\quad (9\cdot731);$$

u and v being the unique pair of positive integers such that $u^2 + v^2 = p$.

I conclude my analysis of Mr Ramanujan's work by a few words concerning the papers (10), (11), and (12), as yet unpublished. Paper (10) contains extensive generalisations of certain formulae of Dr Glaisher, and in particular of a formula which Dr Glaisher only conjectured, viz.

$$\gamma = \lambda_r - (r+1)(r+2)\ldots 2r \left\{ \frac{S_3}{3(r+3)(r+4)\ldots(2r+2)} + \frac{S_5}{5(r+5)(r+6)\ldots(2r+4)} + \ldots \right\}$$
$$\quad\quad\quad\quad (10\cdot1),$$

where
$$S_p = 1^{-p} + 2^{-p} + 3^{-p} + \ldots \quad\quad\quad\quad (10\cdot11),$$

and λ_r is a certain rational number. Mr Ramanujan finds that

$$\lambda_r = 1 - \frac{1}{2} + \frac{1}{3} - \ldots + \frac{1}{2r-1} \quad\quad\quad\quad (10\cdot12).$$

In paper (11) Mr Ramanujan shows that the form

$$ax^2 + by^2 + cz^2 + dt^2 \quad\quad\quad\quad (11\cdot1)$$

will represent *all* integers in exactly 55 cases; viz. when a, b, c, d have one or other of the sets of values

1, 1, 1, 1	1, 2, 3, 5	1, 2, 5, 8
1, 1, 1, 2	1, 2, 4, 5	1, 1, 2, 9
1, 1, 2, 2	1, 2, 5, 5	1, 2, 3, 9
1, 2, 2, 2	1, 1, 1, 6	1, 2, 4, 9
1, 1, 1, 3	1, 1, 2, 6	1, 2, 5, 9
1, 1, 2, 3	1, 2, 2, 6	1, 1, 2, 10
1, 2, 2, 3	1, 1, 3, 6	1, 2, 3, 10
1, 1, 3, 3	1, 2, 3, 6	1, 2, 4, 10
1, 2, 3, 3	1, 2, 4, 6	1, 2, 5, 10
1, 1, 1, 4	1, 2, 5, 6	1, 1, 2, 11
1, 1, 2, 4	1, 1, 1, 7	1, 2, 4, 11
1, 2, 2, 4	1, 1, 2, 7	1, 1, 2, 12
1, 1, 3, 4	1, 2, 2, 7	1, 2, 4, 12
1, 2, 3, 4	1, 2, 3, 7	1, 1, 2, 13
1, 2, 4, 4	1, 2, 4, 7	1, 2, 4, 13
1, 1, 1, 5	1, 2, 5, 7	1, 1, 2, 14
1, 1, 2, 5	1, 1, 2, 8	1, 2, 4, 14.
1, 2, 2, 5	1, 2, 3, 8	
1, 1, 3, 5	1, 2, 4, 8	

Of these 55 forms twelve, viz.

1, 1, 1, 2	1, 1, 2, 4	1, 2, 4, 8
1, 1, 2, 2	1, 2, 2, 4	1, 1, 3, 3
1, 2, 2, 2	1, 2, 4, 4	1, 2, 3, 6
1, 1, 1, 4	1, 1, 2, 8	1, 2, 5, 10

have been considered before by Liouville and Pepin. These writers have indeed, for these particular forms, gone further, and have found explicit expressions, in terms of other arithmetical functions such as $d(n)$, for the number of representations of a given number n.

Mr Ramanujan has also considered the more difficult problem of finding when the form will represent all numbers *with a finite number of exceptions*. He has considered, up to the present, only the special forms

$$a(x^2 + y^2 + z^2) + dt^2 \quad\quad\quad\quad\quad\quad\quad\quad(11\cdot2)$$

and

$$a(x^2 + y^2) + c(z^2 + t^2) \quad\quad\quad\quad\quad\quad\quad\quad(11\cdot3).$$

The results concerning the form (11·2) are included in this paper. He shows that a must be 1 or 2. If $a = 2$, d must be odd: the form then fulfils the requirements. If $a = 1$, there are but a finite number of exceptions if d has one of the values

$$1,\quad 4,\quad 9,\quad 17,\quad 25,\quad 36,\quad 68,\quad 100,$$

or is of one of the forms

$$4k+2,\quad 4k+3,\quad 8k+5,\quad 16k+12,\quad 32k+20:$$

in all other cases the number of exceptions is infinite.

The last paper (12) is a joint work of Mr Ramanujan and myself, and arose from some further consideration of the problem, already alluded to, of finding the order of the function $N(x)$, the number of highly composite numbers less than x. As we cannot solve this problem, it is of considerable interest to determine, as precisely as possible, the density of the distribution of any wider class of numbers possessing some, if not all, of the properties of highly composite numbers. The class which we consider is that of numbers of the form

$$2^{a_2} 3^{a_3} 5^{a_5} \ldots p^{a_p} \quad \ldots\ldots\ldots\ldots\ldots\ldots\ldots\ldots\ldots\ldots\ldots (12\cdot 11),$$

where

$$a_2 \geqslant a_3 \geqslant a_5 \ldots \geqslant a_p \quad \ldots\ldots\ldots\ldots\ldots\ldots\ldots\ldots\ldots\ldots\ldots (12\cdot 12).$$

If q denotes a number of this form, we have

$$\Sigma \frac{1}{q^s} = \Pi \frac{1}{1 - l_n^{-s}} \quad \ldots\ldots\ldots\ldots\ldots\ldots\ldots\ldots\ldots\ldots\ldots (12\cdot 2),$$

where l_n is the product of the first n primes; and by means of this formula, and the theory of Dirichlet's series in a real variable, we show that the number of q's less than x is

$$\exp\left\{\frac{2\pi}{\sqrt{3}} \sqrt{\left(\frac{\log x}{\log \log x}\right)} (1 + \epsilon)\right\} \quad \ldots\ldots\ldots\ldots\ldots\ldots\ldots\ldots\ldots (12\cdot 3),$$

where ϵ is small when x is large. Our proofs are based on certain general theorems concerning Dirichlet's series with positive coefficients, of the same general character as those proved by Mr Littlewood and myself in a number of recent memoirs. These theorems enable us to solve a number of other problems of a similar character: to prove, for example, that the number of partitions of n is of the form

$$\exp\left\{\pi \sqrt{\left(\frac{2n}{3}\right)} (1 + \epsilon)\right\} \quad \ldots\ldots\ldots\ldots\ldots\ldots\ldots\ldots\ldots\ldots\ldots (12\cdot 4),$$

where ϵ is small when n is large.

My account of Mr Ramanujan's work has been necessarily fragmentary and incomplete. I have said enough, I hope, to give some idea of its astonishing individuality and power. India has produced many talented mathematicians in recent years, a number of whom have come to Cambridge and attained high academical distinction. They will be the first to recognise that Mr Ramanujan's work is of a different category. In him India now possesses a pure mathematician of the first order, whose achievements suggest the brightest hopes for its scientific future.

G. H. HARDY, M.A., F.R.S.,
*Fellow and Lecturer of Trinity College
and Cayley Lecturer in Mathematics
in the University of Cambridge.*

COMMENT

Hardy's report on Ramanujan's work was published in the *Journal of the Indian Mathematical Society*. An identical version was privately printed by the Cambridge University Press on larger pages. The version reproduced here is the Cambridge one.

Sir George Stokes and the concept of uniform convergence. By G. H. HARDY, M.A., Trinity College.

[*Received* 1 Jan. 1918. *Read* 4 Feb. 1918.]

1. The discovery of the notion of uniform convergence is generally and rightly attributed to Weierstrass, Stokes, and Seidel. The idea is present implicitly in Abel's proof of his celebrated theorem on the continuity of power series; but the three mathematicians mentioned were the first to recognise it explicitly and formulate it in general terms*. Their work was quite independent, and it would be generally agreed that the debt which mathematics owes to each of them is in no way diminished by any anticipation on the part of the others. Each, as it happens, has some special claim to recognition. Weierstrass's discovery was the earliest, and he alone fully realised its far-reaching importance as one of the fundamental ideas of analysis. Stokes has the actual priority of publication; and Seidel's work is but a year later and, while narrower in its scope than that of Stokes, is even sharper and clearer.

My object in writing this note is to call attention to and, so far as I can, explain two puzzling features in the justly famous memoir† in which Stokes announces his discovery. The memoir is remarkable in many respects, containing a general discussion of the possible modes of convergence, both of series and of integrals, far in advance of the current ideas of the time. It contains also two serious mistakes, mistakes which seem at first sight almost inexplicable on the part of a mathematician of so much originality and penetration.

The first mistake is one of omission. It does not seem to have occurred to Stokes that his discovery had any bearing whatever on the question of term by term integration of an infinite series. The same criticism, it is true, may be made of Seidel's paper. But Seidel is merely silent on the subject. Stokes, on the other hand, quotes the false theorem that a convergent series may always be integrated term by term, and refers, apparently with approval, to the erroneous proof offered by Cauchy and Moigno‡.

Of this there is, I think, a fairly simple and indeed a double

---

* The idea was rediscovered by Cauchy, five or six years after the publication of the work of Stokes and Seidel. See Pringsheim, 'Grundlagen der allgemeinen Funktionenlehre', *Encykl. der Math. Wiss.*, II A 1, §17, p. 35.

† 'On the critical values of the sums of periodic series', *Trans. Camb. Phil. Soc.*, vol. 8, 1847, pp. 533–583 (*Mathematical and physical papers*, vol. 1, pp. 236–313).

‡ See p. 242 of Stokes's memoir (as printed in the collected papers).

explanation. In the first place it must be remembered that Stokes was primarily a mathematical physicist. He was also a most acute pure mathematician; but he approached pure mathematics in the spirit in which a physicist approaches natural phenomena, not looking for difficulties, but trying to explain those which forced themselves upon his attention. The difficulties connected with continuity and discontinuity are of this character. The theorem that a convergent series of continuous functions has necessarily a continuous sum is one whose falsity is open and aggressive: examples to the contrary obtrude themselves on analyst and physicist alike. The falsity of this theorem Stokes therefore observed and corrected. The falsity of the corresponding theorem concerning integration lies somewhat deeper. It is easy enough, when one's attention has been called to it, to see that the proof of Cauchy and Moigno is invalid. But there are no particularly obvious examples to the contrary: simple and natural examples are indeed somewhat difficult to construct*. And Stokes, his suspicions never having been excited, seems to have accepted the false theorem without examination or reflection.

This is half the explanation. The second half, I think, lies in the distinctions between different modes of uniform convergence which I shall consider in a moment.

Stokes's second mistake is more obvious and striking. He proves, quite accurately, that uniform convergence implies continuity†. He then enunciates and offers a proof‡ of the converse theorem, which is false. The error is not one merely of haste or inattention. The argument is as explicit and as clearly stated in one case as in the other; and, up to the last sentence, it is perfectly correct. He proves that continuity involves *something*, and then states, without further argument, that this something is what he has just defined as uniform convergence. It is merely this last statement that is false.

Stokes's mistake seems at first sight so palpable that I was for some time quite at a loss to imagine how he could have made it. A closer examination of his memoir, and a comparison of his work with other work of a very much later date, has made the lapse a good deal more intelligible to me; and my attempts to understand it have led me to a number of remarks which, although they contain very little that is really novel, are, I think, of some historical and intrinsic interest.

2. There are no less than *seven* different senses, all important, in which a series may be said to be uniformly convergent.

---

* See Bromwich, *Infinite series*, pp. 116–118; Hardy, 'Notes on some points in the integral calculus', XL, *Messenger of Mathematics*, vol. 44, 1915, pp. 145–149.
† p. 282. I use 'uniform' instead of Stokes's 'not infinitely slow'.
‡ p. 283.

I shall write the series in the form
$$\sum_{1}^{\infty} u_n(x);$$
and I shall suppose, for simplicity, that every term of the series is continuous, and the series convergent, for every x of the interval $a \leq x \leq b$. I shall denote the sum of the series by $s(x)$; and I shall write
$$s_n(x) = u_1(x) + u_2(x) + \ldots + u_n(x), \quad s(x) = s_n(x) + r_n(x).$$
The fundamental inequality in all my definitions will be of the type
$$|r_n(x)| \leq \epsilon \ldots\ldots\ldots\ldots\ldots\ldots\ldots(A).$$
I shall refer to this inequality simply as (A).

When we define uniform convergence, in one sense or another, we have to choose various numbers in a definite logical order, those which are chosen later being, in general, functions of those which are chosen before. I shall write each number in a form in which all the arguments of which it is a function appear explicitly: thus $n_0(\xi, \epsilon)$ is a function of ξ and ϵ, $n_0(\epsilon)$ one of ϵ alone.

It will sometimes happen that one of the later numbers depends upon several earlier numbers *already connected by functional relations*, so that it is really a function of a selection of these numbers only. Thus δ may have been determined as a function of ϵ; and n_0 may have to be determined as a function of ξ, ϵ, and δ, so that it is in reality a function of ξ and ϵ only. I shall express this by writing
$$n_0 = n_0(\xi, \epsilon, \delta) = n_0(\xi, \epsilon);$$
and I shall use a similar notation in other cases of the same kind.

3. The first three senses of uniform convergence are as follows.

A 1: Uniform convergence throughout an interval. *The series is said to be uniformly convergent throughout the interval* (a, b) *if to every positive ϵ corresponds an $n_0(\epsilon)$ such that* (A) *is true for* $n \geq n_0(\epsilon)$ *and* $a \leq x \leq b$.

This is the ordinary or 'classical', and most important, sense, the sense in which uniform convergence is defined in every treatise on the theory of series.

A 2: Uniform convergence in the neighbourhood of a point. *The series is said to be uniformly convergent in the neighbourhood of the point ξ of the interval* (a, b) *if an interval* $(\xi - \delta(\xi), \xi + \delta(\xi))$* *can be found throughout which it is uniformly convergent; that is to say if a positive $\delta(\xi)$ exists such that* (A) *is true for every positive ϵ, for* $n \geq n_0(\xi, \delta, \epsilon) = n_0(\xi, \epsilon)$, *and for* $\xi - \delta(\xi) \leq x \leq \xi + \delta(\xi)$.

* A trivial change is of course required in the definition if $\xi = a$ or $\xi = b$. The same point naturally arises in the later definitions.

A 3: Uniform convergence at a point. *The series is said to be uniformly convergent at the point $x = \xi$ (or for $x = \xi$) if to every positive ϵ correspond a positive $\delta(\xi, \epsilon)$ and an $n_0(\xi, \epsilon, \delta) = n_0(\xi, \epsilon)$ such that (A) is true for $n \geqslant n_0(\xi, \epsilon)$ and for $\xi - \delta(\xi, \epsilon) \leqslant x \leqslant \xi + \delta(\xi, \epsilon)$.*

4. Before proceeding further it will be well to make a few remarks concerning these definitions and their relations to one another.

The idea of uniform convergence *in the neighbourhood of a particular point* (Definition **A 2**) is substantially that defined by Seidel in 1848*. It is clear, however, that definitions **A 1** and **A 2** were both familiar to Weierstrass as early as 1841 or 1842†. It is obvious that a series uniformly convergent throughout an interval is uniformly convergent in the neighbourhood of every point of the interval. The converse theorem is important and by no means obvious, and was first proved by Weierstrass‡ in a memoir published in 1880. This theorem would now be proved by a simple application of the 'Heine-Borel Theorem', and is a particular case of a theorem which will be referred to in a moment.

Definition **A 3** appears first, in the form in which I state it, in a paper of W. H. Young published in 1903§; but the idea is present in an earlier paper of Osgood‖. The essential difference between definitions **A 2** and **A 3** is that in the latter δ is chosen *after* ϵ and is a function of ξ and ϵ, while in the former it is chosen before ϵ and is a function of ξ alone. In each case n_0 is a function of two independent variables, ξ and ϵ. It is plain that uniform convergence in the neighbourhood of ξ involves uniform convergence at ξ, and at (and indeed in the neighbourhood of) all points sufficiently near to ξ. But uniform convergence at ξ does not involve uniform convergence in the neighbourhood of ξ.

It is important, however, to observe that *uniform convergence at every point of an interval involves uniform convergence throughout the interval*. This important theorem is proved very simply by

---

* 'Note über eine Eigenschaft der Reihen, welche discontinuirliche Functionen darstellen', *Münchener Abhandlungen*, vol. 7, 1848, pp. 381–394. This memoir has been reprinted in Ostwald's *Klassiker der exakten Wissenschaften*, no. 116. The reference there given to vol. 5, 1847, is incorrect.

† For detailed references bearing on this and similar historical points, see Pringsheim's article already quoted.

‡ See the memoir 'Zur Functionenlehre' (*Abhandlungen aus der Funktionenlehre*, pp. 69–104 (pp. 71–72)).

§ 'On non-uniform convergence and term-by-term integration of series', *Proc. London Math. Soc.*, ser. 2, vol. 1, pp. 89–102.

‖ 'Non-uniform convergence and the integration of series', *American Journal of Math.*, vol. 19, 1897, pp. 155–190. See Prof. Young's remarks on this point at the beginning of his later paper 'On uniform and non-uniform convergence of a series of continuous functions and the distinction of right and left', *Proc. London Math. Soc.*, ser. 2, vol. 6, 1907, pp. 29–51.

Young, in his paper already quoted, by means of the Heine-Borel Theorem*; and it plainly includes, as a particular case, Weierstrass's theorem referred to above.

5. It seems to me that the definition given by Stokes is not any one of **A 1**, **A 2**, **A 3**; and that, if we are to understand him rightly, we must consider another parallel group of definitions. These definitions differ from those given above in that (A) is supposed to be satisfied, not for *all* sufficiently large values of n, but only for *an infinity of* values.

B 1: Quasi-uniform convergence throughout an interval. *The series is said to be quasi-uniformly convergent throughout (a, b) if to every positive ϵ and every N corresponds an $n_0(\epsilon, N)$ greater than N and such that* (A) *is true for* $n = n_0(\epsilon, N)$ *and* $a \leqslant x \leqslant b$.

B 2: Quasi-uniform convergence in the neighbourhood of a point. *The series is said to be quasi-uniformly convergent in the neighbourhood of ξ if an interval $(\xi - \delta(\xi), \xi + \delta(\xi))$ can be found throughout which it is quasi-uniformly convergent; i.e., if a positive $\delta(\xi)$ exists such that* (A) *is true for every positive ϵ, every N, an $n_0(\xi, \delta, \epsilon, N) = n_0(\xi, \epsilon, N)$ greater than N, and $\xi - \delta(\xi) < x \leqslant \xi + \delta(\xi)$.*

B 3: Quasi-uniform convergence at a point. *The series is said to be quasi-uniformly convergent for $x = \xi$ if to every positive ϵ and every N correspond a positive $\delta(\xi, \epsilon, N)$ and an*

$$n_0(\xi, \epsilon, \delta, N) = n_0(\xi, \epsilon, N),$$

greater than N, such that (A) *is true for $n = n_0(\xi, \epsilon, N)$ and for $\xi - \delta(\xi, \epsilon, N) \leqslant x \leqslant \xi + \delta(\xi, \epsilon, N)$.*

Definition **B 1** is to be attributed to Dini or to Darboux†. Another form of it has been given by Hobson‡. As Arzelà and Hobson§ have pointed out, a series is quasi-uniformly convergent throughout an interval if, and only if, it can be made uniformly convergent by an appropriate bracketing of its terms.

Definition **B 2** is for us at the moment of peculiar interest, for (as I shall show in a moment) it is really *this* definition that is given by Stokes.

Definition **B 3** is also of great interest, both in itself and in

---

* Choose ϵ and determine $\delta(\xi, \epsilon)$ and $n_0(\xi, \epsilon)$, as in definition **A 3**, for every ξ of the interval. Every point of (a, b) is included in an interval $(\xi - \delta, \xi + \delta)$. By the Heine-Borel Theorem, every point of (a, b) is included in one or other of a finite sub-set of these intervals. If $N(\epsilon)$ is the largest of the n_0's corresponding to each of the intervals of this finite sub-set, then (A) is true for $n \geqslant N$ and $a \leqslant x \leqslant b$.

This is the essence of the proof, though, like all proofs of the same character, it requires a somewhat more careful statement if all appearance of dependence upon Zermelo's *Auswahlsprinzip* is to be avoided.

† See Pringsheim, *l. c.*

‡ 'On modes of convergence of an infinite series of functions of a real variable', *Proc. London Math. Soc.*, ser. 2, vol. 1, 1903, pp. 373–387. Hobson (following Dini) uses the expression 'simply uniformly'.

§ *L. c.*, p. 375.

relation to Stokes's memoir. For *the necessary and sufficient condition that $s(x)$ should be continuous for $x = \xi$ is that the series should be quasi-uniformly convergent for $x = \xi$*. This theorem is in substance due to Dini*. I give the proof, as it is essential for the criticism of Stokes's memoir.

(1) *The condition is sufficient.* For
$$|s(x) - s(\xi)| \leqslant |s_n(x) - s_n(\xi)| + |r_n(x)| + |r_n(\xi)|.$$

Choose ϵ, N, $\delta(\xi, \epsilon, N)$, and $n = n_0(\xi, \epsilon, N)$ as in definition **B 3**. Then $|r_n(x)| < \epsilon$ for $\xi - \delta \leqslant x \leqslant \xi + \delta$. Now that n is fixed we can choose δ_1 less than δ and such that $|s_n(x) - s_n(\xi)| < \epsilon$ for $\xi - \delta_1 \leqslant x \leqslant \xi + \delta_1$. And thus
$$|s(x) - s(\xi)| < 3\epsilon$$
for $\xi - \delta_1 \leqslant x \leqslant \xi + \delta_1$, so that $s(x)$ is continuous for $x = \xi$.

It is plain that this argument proves, *a fortiori*, that **A 2**, **A 3**, and **B 2** all furnish sufficient conditions for continuity at a point, and **A 1** and **B 1** sufficient conditions for continuity throughout an interval.

(2) *The condition is necessary.* For
$$|r_n(x)| \leqslant |s(x) - s(\xi)| + |r_n(\xi)| + |s_n(x) - s_n(\xi)|.$$

Suppose that ϵ and N are given. Then we can choose $\delta(\xi, \epsilon)$ so that $|s(x) - s(\xi)| < \epsilon$ for $\xi - \delta \leqslant x \leqslant \xi + \delta$, and $n_0(\xi, \epsilon, N)$ so that $n_0 > N$ and $|r_{n_0}(\xi)| < \epsilon$. And, when n_0 has thus been fixed, we can choose $\delta_1(\xi, \epsilon, n_0) = \delta_1(\xi, \epsilon, N)$ so that $\delta_1 < \delta$ and $|s_{n_0}(x) - s_{n_0}(\xi)| < \epsilon$ for $\xi - \delta_1 \leqslant x \leqslant \xi + \delta_1$. Thus $|r_n(x)| < 3\epsilon$ for $n = n_0 > N$ and $\xi - \delta_1 \leqslant x \leqslant \xi + \delta_1$, so that the series is quasi-uniformly convergent for $x = \xi$.

6. If a series is uniformly convergent at every point ξ of an interval, it is (as we saw in § 4) uniformly convergent throughout the interval: definition **A 3** (and *a fortiori* definition **A 2**) passes over, in virtue of the Heine-Borel Theorem, into definition **A 1**. It is important to observe that this relation does not hold between **B 3** (or **B 2**) and **B 1**: a series quasi-uniformly convergent at every point of an interval (or in the neighbourhood of every such point) is not necessarily quasi-uniformly convergent throughout the interval. We can apply the Heine-Borel Theorem in the manner indicated in the first sentences of the footnote * to p. 152; but the last stage of the argument, in which every one of a finite number of different integers is replaced by the largest of them, fails. What we obtain is *the necessary and sufficient condition that $s(x)$ should be continuous throughout the interval*; and this is not

---

* *Fondamenti...*, p. 107 (German translation, *Grundlagen...*, pp. 143–145).

the condition **B 1** but a condition first formulated by Arzelà*, viz.:

C: Quasi-uniform convergence by intervals (*convergenza uniforme a tratti*). *The series is said to be quasi-uniformly convergent by intervals if to every positive ϵ and every N correspond a division of (a, b) into a finite number $\nu(\epsilon, N)$ of intervals $\delta_r(\epsilon, N)$, and a corresponding number of numbers $n_r(\epsilon, N)$, all greater than N, and such that* (A) *is true for* $n = n_r (r = 1, 2, ..., \nu)$ *and all values of x which belong to δ_r.*

The deduction of Arzelà's criterion from **B 3**, in the manner sketched above, was first made by Hobson†.

There is one further point which seems worth noticing here, although it is not directly connected with Stokes's memoir. Dini‡ proved that *if $u_n(x) \geq 0$ for all values of n and x, and $s(x)$ is continuous throughout (a, b), then the series is uniformly convergent throughout (a, b)*. This theorem is now almost intuitive. For it is obvious that, for series of positive terms, quasi-uniform convergence in any one of the senses **B 1**, **B 2**, or **B 3** involves uniform convergence in the corresponding sense **A 1**, **A 2**, or **A 3**. If then $s(x)$ is continuous throughout (a, b) it is continuous for every ξ of (a, b); and therefore the series is quasi-uniformly convergent for every ξ; and therefore uniformly convergent for every ξ; and therefore uniformly convergent throughout (a, b).

7. Let us now consider Stokes's definitions and proofs in the light of the preceding discussion.

It is clear, in the first place, that Stokes has in his mind some phenomenon characteristic of *a small, but fixed, neighbourhood of a point*.

'Let $u_1 + u_2 + \ldots$ (66)', he says§, 'be a convergent infinite series having U for its sum. Let $v_1 + v_2 + \ldots$ (67) be another infinite series of which the general term v_n is a function of the positive variable h and becomes equal to u_n when h vanishes. Suppose that *for a sufficiently small value of h and all inferior values* the series (67) is convergent, and has V for its sum. It might at first sight be supposed that the limit of V for $h = 0$ was necessarily equal to U. This however is not true....

'THEOREM. The limit of V can never differ from U unless the convergency of the series (67) becomes infinitely slow when h vanishes.

---

* 'Sulle serie di funzioni', *Memorie di Bologna*, ser. 5, vol. 8, 1900, pp. 131–186, 701–744.

† *L. c.*, pp. 380–382.

‡ *L. c.* (German edition), pp. 148–149. See also Bromwich, *Infinite series*, p. 125 (Ex. 3).

§ p. 279.

'The convergency of the series is here said to become infinitely slow when, if n be the number of terms which must be taken in order to render the sum of the neglected series numerically less than a given quantity e, which may be as small as we please, n increases beyond all limit as h decreases beyond all limit.

'DEMONSTRATION. If the convergency do not become infinitely slow it will be possible to find *a number n, so great that for the value of h we begin with and for all inferior values greater than zero the sum of the neglected terms shall be numerically less than e....*'

Stokes's words, and in particular those which I have italicised, seem to me to make two things perfectly clear.

(1) Stokes is considering neither a property of an interval (a, b) *im Grossen* (such as is contemplated in **A 1** or **B 1**), nor a property of a single point which (as in **A 3** or **B 3**) need not be shared by any neighbouring point, but a property of an interval *im Kleinen*, that is to say *a small but fixed interval chosen to include a particular point*. His definition is therefore one of the type of **A 2** or **B 2**.

Stokes's failure to perceive the bearing of his discovery on problems of integration is made much more natural when we realise that he is considering throughout a neighbourhood of a point and not an interval *im Grossen*. And this remark applies to Seidel as well.

(2) Stokes is considering an inequality satisfied for a special value of n, or at most an infinite sequence of values of n, and *not* necessarily for all values of n from a certain point onwards. In this respect there is a quite sharp distinction between Stokes's work and Seidel's. What Stokes defines is (to use the language of this note) a mode of *quasi-uniform* convergence and not one of strictly uniform convergence.

It seems to me, then, that what Stokes defines is what I have called *quasi-uniform convergence in the neighbourhood of a point* (**B 2**).

8. If we adopt this view, Stokes's mistake becomes very much more intelligible. He proves, quite correctly, that uniform convergence in his sense implies continuity: his proof, stated quite formally and by means of inequalities, is substantially that given in §5, under (1). He then continues* as follows.

'Conversely, if (66) is convergent, and if $U = V_0$†, the convergency of the series (67) cannot become infinitely slow when h

---

* p. 282. The italics are mine.

† V_0 is what Stokes calls 'the value of V for $h=0$', by which he means, of course, its limit when h tends to 0.

vanishes. For if U_n', V_n' represent the sums of the terms after the nth in the series (66), (67) respectively, we have
$$V = V_n + V_n', \ U = U_n + U_n';$$
whence
$$V_n' = V - U - (V_n - U_n) + U_n'.$$

Now $V - U$, $V_n - U_n$ vanish with h, and U_n' vanishes when n becomes infinite. Hence *for a sufficiently small value of h and all inferior values, together with a value of n sufficiently large and independent of h*, the value of V_n' may be made numerically less than any given quantity e however small; *and therefore, by definition, the convergency of the series* (67) *does not become infinitely slow when h vanishes.*'

Now this argument is, until we reach the last sentence, perfectly accurate, and indeed, if we translate it into inequalities, substantially identical with that given in §5, under (2). Stokes proves, in fact, that continuity at ξ involves quasi-uniform convergence at ξ. Where he falls into error is simply in his final assertion that this property is that which he has previously defined, the mistake being due to a failure to observe that his intervals of values of h depend upon a prior choice of ϵ. In a word, he confuses, momentarily, **B 2** and **B 3**. The ordinary view that Stokes defined uniform convergence in the same sense as Weierstrass compels us to suppose that he confused **B 3** with **A 1**, or at any rate with **A 2**: and this is hardly credible.

I add one final remark. If we could identify Stokes's idea with **B 3**, instead of with **B 2**, we could acquit him of having made any mistake at all, since **B 3** really is a necessary and sufficient condition for continuity. We could then regard Stokes as having anticipated Dini's theorem. This view, however, does not seem to me to be tenable.

The Theory of Numbers.[1]

By Prof. G. H. HARDY, M.A., F.R.S.

I FIND myself to-day in the same embarrassing position in which a predecessor of mine at Oxford found himself at Bradford in 1875, the president of a Section, probably the largest and most heterogeneous in the Association, which is absorbed by a multitude of divergent professional interests, none of which agree with his or mine.

There are two courses possible in such circumstances. One is to take refuge, as Prof. Henry Smith did then, with visible reluctance, in a series of general propositions to which mathematicians, physicists, and astronomers may all be expected to return a polite assent. The importance of science and scientific method, the need for better organisation of scientific education and research, are all topics on which I could no doubt say something without undue strain either on my own honesty or on your credulity. That there is no finer education and discipline than natural science; that it is, as Dr. Campbell has said, "the noblest of the arts"; that the crowning achievements of science lie in those directions with which this Section is professionally concerned: all this I could say with complete sincerity, and, if I were the head of a deputation approaching a Government Department, I suppose that I would not shirk even so unprofitable a task.

It is unfortunate that these essential and edifying truths, important as it is that they should be repeated as loudly as possible from time to time, are, to the man whose interest in life lies in scientific work and not in propaganda, unexciting, and in fact quite intolerably dull. I could, if I chose, say all these things, but, even if I wanted to, I should scarcely increase your respect for mathematics and mathematicians by repeating to you what you have said yourselves, or read in the newspapers, a hundred times already. I shall say them all some day; the time will come when we shall none of us have anything more interesting to say. We need not anticipate our inevitable end.

I propose therefore to adopt the alternative course suggested by my predecessor, and try to say something to you about the one subject about which I have anything to say. It happens, by a fortunate accident, that the particular subject which I love the most, and which presents most of the problems which occupy my own researches, is by no means overwhelmingly recondite or obscure, and indeed is sharply distinguished from almost every other branch of pure mathematics, in that it makes a direct, popular, and almost irresistible appeal to the heart of the ordinary man.

There is, however, one preliminary remark which I cannot resist the temptation of making. The present is a particularly happy moment for a pure mathematician, since it has been marked by one of the greatest recorded triumphs of pure mathematics. This triumph is the work, as it happens, of a man who probably would not describe himself as a mathematician, but who has done more than any mathematician to vindicate the dignity of mathematics, and to put that obscure and perplexing construction,

commonly described as "physical reality," in its proper place.

There is probably less difference between the methods of a physicist and a mathematician than is generally supposed. The most striking among them seems to me to be this, that the mathematician is in much more direct contact with reality. This may perhaps seem to you a paradox, since it is the physicist who deals with the subject-matter to which the epithet "real" is commonly applied. But a very little reflection will show that the "reality" of the physicist, whatever it may be (and it is extraordinarily difficult to say), has few or none of the attributes which common-sense instinctively marks as real. A chair may be a collection of whirling atoms, or an idea in the mind of God. It is not my business to suggest that one account of it is obviously more plausible than the other. Whatever the merits of either of them may be, neither draws its inspiration from the suggestions of common-sense.

Neither the philosophers, nor the physicists themselves, have ever put forward any very convincing account of what physical reality is, or of how the physicist passes, from the confused mass of fact or sensation with which he starts, to the construction of the objects which he classifies as real. We cannot be said, therefore, to know what the subject-matter of physics is; but this need not prevent us from understanding the task which a physicist is trying to perform. That, clearly, is to correlate the incoherent body of facts confronting him with some definite and orderly scheme of abstract relations, the kind of scheme, in short, which he can borrow only from mathematics.

A mathematician, on the other hand, fortunately for him, is not concerned with this physical reality at all. It is impossible to prove, by mathematical reasoning, any proposition whatsoever concerning the physical world, and only a mathematical crank would be likely now to imagine it his function to do so. There is plainly one way only of ascertaining the facts of experience, and that is by observation. It is not the business of a mathematician to suggest one view of the universe or another, but merely to supply the physicists with a collection of abstract schemes, which it is for them to select from, and to adopt or discard at their pleasure.

The most obvious example is to be found in the science of geometry. Mathematicians have constructed a very large number of different systems of geometry, Euclidean or non-Euclidean, of one, two, three, or any number of dimensions. All these systems are of complete and equal validity. They embody the results of mathematicians' observations of *their* reality, a reality far more intense and far more rigid than the dubious and elusive reality of physics. The old-fashioned geometry of Euclid, the entertaining seven-point geometry of Veblen, the space-times of Minkowski and Einstein, are all absolutely and equally real. When a mathematician has constructed, or, to be more accurate, when he has observed them, his professional interest in the matter ends. It may be the seven-point

[1] Presidential address delivered to Section A (Mathematics and Physics) of the British Association at Hull on Sept. 8.

geometry that fits the facts the best, for anything that mathematicians have to say. There may be three dimensions in this room and five next door. As a professional mathematician, I have no idea; I can only ask some competent physicist to instruct me in the facts.

The function of a mathematician, then, is simply to observe the facts about his own intricate system of reality, that astonishingly beautiful complex of logical relations which forms the subject-matter of his science, as if he were an explorer looking at a distant range of mountains, and to record the results of his observations in a series of maps, each of which is a branch of pure mathematics. Many of these maps have been completed, while in others, and these, naturally, are the most interesting, there are vast uncharted regions. Some, it seems, have some relevance to the structure of the physical world, while others have no such tangible application. Among them there is perhaps none quite so fascinating, with quite the same astonishing contrasts of sharp outline and mysterious shade, as that which constitutes the theory of numbers.

The number system of arithmetic is, as we know too well, not without its applications to the sensible world. The currency systems of Europe, for example, conform to it approximately; west of the Vistula, two and two make something approaching four. The practical applications of arithmetic, however, are tedious beyond words. One must probe a little deeper into the subject if one wishes to interest the ordinary man, whose taste in such matters is astonishingly correct, and who turns with joy from the routine of common life to anything strange and odd, like the fourth dimension, or imaginary time, or the theory of the representation of integers by sums of squares or cubes.

It is impossible for me to give you, in the time at my command, any general account of the problems of the theory of numbers, or of the progress that has been made towards their solution even during the last twenty years. I must adopt a much simpler method. I will merely state to you, with a few words of comment, three or four isolated questions, selected in a haphazard way. They are seemingly simple questions, and it is not necessary to be anything of a mathematician to understand them; and I have chosen them for no better reason than that I happen to be interested in them myself. There is no one of them to which I know the answer, nor, so far as I know, does any mathematician in the world; and there is no one of them, with one exception which I have included deliberately, the answer to which any one of us would not make almost any sacrifice to know.

1. *When is a number the sum of two cubes, and what is the number of its representations?* This is my first question, and first of all I will elucidate it by some examples. The numbers $2 = 1^3 + 1^3$ and $9 = 2^3 + 1^3$ are sums of two cubes, while 3 and 4 are not: it is exceptional for a number to be of this particular form. The number of cubes up to 1,000,000 is 100, and the number of numbers, up to this limit and of the form required, cannot exceed 10,000, one-hundredth of the whole. The density of the distribution of such numbers tends to zero as the numbers tend to infinity. Is there, I am asking, any simple criterion by which such numbers can be distinguished?

Again, 2 and 9 are sums of two cubes, and can be expressed in this form in one way only. There are numbers so expressible in a variety of different ways. The least such number is 1729, which is $12^3 + 1^3$ and also $10^3 + 9^3$. It is more difficult to find a number with *three* representations; the least such number is

$$175,959,000 = 560^3 + 70^3 = 552^3 + 198^3 = 525^3 + 315^3.$$

One number at any rate is known with *four* representations, namely,

$$19 \times 363510^3$$

(a number of 18 digits), but I am not prepared to assert that it is the least. No number has been calculated, so far as I know, with more than four, but theory, running ahead of computation, shows that numbers exist with five representations, or six, or any number.

A distinguished physicist has argued that the possible number of isotopes of an element is probably limited because, among the ninety or so elements at present under observation, there is none which has more isotopes than six. I dare not criticise a physicist in his own field; but the figures I have quoted may suggest to you that an arithmetical generalisation, based on a corresponding volume of evidence, would be more than a little rash.

There are similar questions, of course, for squares, but the answers to these were found long ago by Euler and by Gauss, and belong to the classical mathematics. Suppose, for simplicity of statement, that the number in question is *prime*. Then, if it is of the form $4m+1$, it is a sum of squares, and in one way only, while if it is of the form $4m+3$ it is not so expressible; and this simple rule may readily be generalised so as to apply to numbers of any form. But there is no similar solution for our actual problem, nor, I need scarcely say, for the analogous problems for fourth, fifth, or higher powers. The smallest number known to be expressible in two ways by two biquadrates is

$$635318657 = 158^4 + 59^4 = 134^4 + 133^4;$$

and I do not believe that any number is known expressible in three. Nor, to my knowledge, has the bare existence of such a number yet been proved. When we come to fifth powers, nothing is known at all. The field for future research is unlimited and practically untrodden.

2. I pass to another question, again about cubes, but of a somewhat different kind. *Is every large number* (every number, that is to say, from a definite point onwards) *the sum of five cubes?* This is another exceptionally difficult problem. It is known that every number, without exception, is the sum of nine cubes; two numbers, 23 (which is $2.2^3 + 7.1^3$) and 239, actually require so many. It seems that there are just fifteen numbers, the largest being 454, which need eight, and 121 numbers, the largest being 8042, which need seven; and the evidence suggests forcibly that the six-cube numbers also ultimately disappear. In a lecture which I delivered on this subject at Oxford I stated, on the authority of Dr. Ruckle, that there were two numbers, in the immediate neighbourhood of 1,000,000, which could not be resolved into fewer cubes than six; but Dr. A. E. Western has refuted this assertion by resolving each of them into five, and is of opinion,

I believe, that the six-cube numbers have disappeared entirely considerably before this point. It is conceivable that the five-cube numbers also disappear, but this, if it be so, is probably in depths where computation is helpless. The four-cube numbers must certainly persist for ever, for it is impossible that a number $9n+4$ or $9n+5$ should be the sum of three.

I need scarcely add that there is a similar problem for every higher power. For fourth powers the critical number is 16. There is no case, except the simple case of squares, in which the solution is in any sense complete. About the squares there is no mystery; every number is the sum of four squares, and there are infinitely many numbers which cannot be expressed by fewer.

3. I will next raise the question *whether the number $2^{137}-1$ is prime*. I said that I would include one question which does not interest me particularly; and I should like to explain to you the kind of reasons which damp down my interest in this one. I do not know the answer, and I do not care greatly what it is.

The problem belongs to the theory of the so-called "perfect" numbers, which has exercised mathematicians since the times of the Greeks. A number is perfect if, like 6 or 28, it is the sum of all its divisors, unity included. Euclid proved that the number

$$2^m(2^{m+1}-1)$$

is perfect if the second factor is prime; and Euler, 2000 years later, that all *even* perfect numbers are of Euclid's form. It is still unknown whether a perfect number can be odd.

It would obviously be most interesting to know generally in what circumstances a number 2^n-1 is prime. It is plain that this can be so only if n itself is prime, as otherwise the number has obvious factors; and the 137 of my question happens to be the least value of n for which the answer is still in doubt. You may perhaps be surprised that a question apparently so fascinating should fail to arouse me more.

It was asserted by Mersenne in 1644 that the only values of n, up to 257, for which 2^n-1 is prime are

2, 3, 5, 7, 13, 17, 19, 31, 67, 127, 257;

and an enormous amount of labour has been expended on attempts to verify this assertion. There are no simple general tests by which the primality of a number chosen at random can be determined, and the amount of computation required in any particular case may be appalling. It has, however, been imagined that Mersenne perhaps knew something which later mathematicians have failed to rediscover. The idea is a little fantastic, but there is no doubt that, so long as the possibility remained, arithmeticians were justified in their determination to ascertain the facts at all costs. "The riddle as to how Mersenne's numbers were discovered remains unsolved," wrote Mr. Rouse Ball in 1891. Mersenne, he observes, was a good mathematician, but not an Euler or a Gauss, and he inclines to attribute the discovery to the exceptional genius of Fermat, the only mathematician of the age whom any one could suspect of being hundreds of years ahead of his time.

These speculations appear extremely fanciful now, for the bubble has at last been pricked. It seems now that Mersenne's assertion, so far from hiding unplumbed depths of mathematical profundity, was a conjecture based on inadequate empirical evidence, and a somewhat unhappy one at that. It is now known that there are at least four numbers about which Mersenne is definitely wrong; he should have included at any rate 61, 89, and 107, and he should have left out 67. The mistake as regards 61 and 67 was discovered so long ago as 1886, but could be explained with some plausibility, so long as it stood alone, as a merely clerical error. But when Mr. R. E. Powers, in 1911 and 1914, proved that Mersenne was also wrong about 89 and 107, this line of defence collapsed, and it ceased to be possible to take Mersenne's assertion seriously.

The facts may be summed up as follows. Mersenne makes fifty-five assertions, for the fifty-five primes from 2 to 257. Of these assertions forty are true, four false, and eleven still doubtful. Not a bad result, you may think; but there is more to be said. Of the forty correct assertions many, half at least, are trivial, either because the numbers in question are comparatively small, or because they possess quite small and easily detected divisors. The test cases are those in which the numbers are prime, or Mersenne asserts that they are so; there are only four of these cases which are difficult and in which the truth is known; and in these Mersenne is wrong in every case but one.

It seems to me, then, that we must regard Mersenne's assertion as exploded; and for my part it interests me no longer. If he is wrong about 89 and 107, I do not care greatly whether he is wrong about 137 as well, and I should regard the computations necessary to decide as very largely wasted. There are so many much more profitable calculations which a computer could undertake.

I hope that you will not infer that I regard the problem of perfect numbers as uninteresting in itself; that would be very far from the truth. There are at least two intensely interesting problems. The first is the old problem, which so many mathematicians have failed to solve, whether a perfect number can be odd. The second is whether the number of perfect numbers is infinite or not. If we assume that all perfect numbers are even, we can state this problem in a still more arresting form. *Are there infinitely many primes of the form 2^n-1?* I find it difficult to imagine a problem more fascinating or more intricate than that. It is plain, though, that this is a question which computation can never decide, and it is very unlikely that it can ever give us any data of serious value. And the problem itself really belongs to a different chapter of the theory, to which I should like next to direct your attention.

4. *Are there infinitely many primes of the form n^2+1?* Let me first remind you of some well-known facts in regard to the distribution of primes.

There are infinitely many primes; their density decreases as the numbers increase, and tends to zero when the numbers tend to infinity. More accurately, the number of primes less than x is, to a first approximation,

$$\frac{x}{\log x}.$$

The chance that a large number n, selected at random, should be prime is, we may say, about $\frac{1}{\log n}$.

Still more precisely, the "logarithm-integral"

$$\operatorname{Li} x = \int_2^x \frac{dt}{\log t}$$

gives a very good approximation to the number of primes. This number differs from Li x by a function of x which oscillates continually, as Mr. Littlewood, in defiance of all empirical evidence to the contrary, has shown, between positive and negative values, and is sometimes large, of the order of magnitude $\sqrt{x}$ or thereabouts, but always small in comparison with the logarithm-integral itself.

Except for one lacuna, which I must pass over in silence now, this problem of the general distribution of primes, the first and central problem of the theory, is in all essentials solved. But a variety of most interesting problems remain as to the distribution of primes among numbers of special forms. The first and simplest of these is that of the arithmetical progressions: *How are the primes distributed among all possible arithmetical progressions* $an+b$? We may leave out of account the case in which a and b have a common factor; this case is trivial, since $an+b$ is then obviously not prime.

The first step towards a solution was made by Dirichlet, who proved for the first time, in 1837, that any such arithmetical progression contains an infinity of primes. It has since been shown that the primes are, to a first approximation at any rate, distributed evenly among all the arithmetical progressions. When we pursue the analysis further, differences appear; there are on the average, for example, more primes $4n+3$ than primes $4n+1$, though it is not true, as the evidence of statistics has led some mathematicians to conclude too hastily, that there is always an excess to whatever point the enumeration is carried.

The problem of the arithmetical progressions, then, may also be regarded as solved; and the same is true of the problem of the primes of a given quadratic form, say $am^2 + 2bmn + cn^2$, homogeneous in the two variables m and n. To take, for example, the simplest and most striking case, there is the natural and obvious number of primes $m^2 + n^2$. A prime is of this form, as I have mentioned already, if, and only if, it is of the form $4k+1$. The quadratic problem reduces here to a particular case of the problem of the arithmetical progressions.

When we pass to cubic forms, or forms of higher degree, we come to the region of the unknown. This, however, is not the field of inquiry which I wish now to commend to your attention. The quadratic forms of which I have spoken are forms in two independent variables m and n; the form n^2+1 of my question is a non-homogeneous form in a single variable n, the simplest case of the general form $an^2 + 2bn + c$. It is clear that one may ask the same question for forms of any degree: are there, for example, infinitely many primes n^3+2 or n^4+1? I do not choose n^3+1, naturally, because of the obvious factor $n+1$.

This problem is one in which computation can still play an important part. You will remember that I stated the same problem for perfect numbers. There a computer is helpless. For the numbers $2^n - 1$, which dominate the theory, increase with unmanageable rapidity, and the data collected by the computers appear, so far as one can judge, to be almost devoid of value. Here the data are ample, and, though the question is still unanswered, there is really strong statistical evidence for supposing a particular answer to be true. It seems that the answer is affirmative, and that there is a definite approximate formula for the number of primes in question. This formula is

$$\tfrac{1}{2}\operatorname{Li}\sqrt{x} \times \left(1+\tfrac{1}{3}\right)\left(1-\tfrac{1}{5}\right)\left(1+\tfrac{1}{7}\right)\left(1+\tfrac{1}{11}\right)\cdots,$$

where the product extends over all primes p, and the positive sign is chosen when p is of the form $4n+3$. Dr. A. E. Western has submitted this formula to a most exhaustive numerical check. It so happens that Colonel Cunningham some years ago computed a table of primes n^2+1 up to the value 15,000 of n, a limit altogether beyond the range of the standard factor tables, and Cunningham's table has made practicable an unusually comprehensive test. The actual number of primes is 1199, while the number predicted is 1219. The error, less than 1 in 50, is much less than one could reasonably expect. The formula stands its test triumphantly, but I should be deluding you if I pretended to see any immediate prospect of an accurate proof.

5. The last problem I shall state to you is this: *Are there infinitely many prime-pairs p, $p+2$?* One may put the problem more generally: *Does any group of primes, with assigned and possible differences, recur indefinitely, and what is the law of its recurrence?*

I must first explain what I mean by a "possible" group of primes. It is possible that p and $p+2$ should both be prime, like 3, 5, or 101, 103. It is not possible (unless p is 3) that p, $p+2$ and $p+4$ should *all* be prime, for one of them must be a multiple of 3: but p, $p+2$, $p+6$ or p, $p+4$, $p+6$ are possible triplets of primes. Similarly

$$p,\ p+2,\ p+6,\ p+8,\ p+12$$

can all be prime, so far as any elementary test of divisibility shows, and in fact 5, 7, 11, 13 and 17 satisfy the conditions. It is easy to define precisely what we understand by a "possible" group. We mean a group the differences in which, like 0, 2, 6, have at least one missing residue to every possible modulus. The "impossible" group 0, 2, 4 does not satisfy the condition, for the remainders after division by 3 are 0, 2, 1, a complete set of residues to modulus 3. There is no difficulty in specifying possible groups of any length we please.

We define in this manner, then, a "possible" group of primes, and we put the questions: Do all possible groups of primes actually occur, do they recur indefinitely often, and how often on the average do they recur? Here again it would seem that the answers are affirmative, that all possible groups occur, and continue to occur for ever, and with a frequency the law of which can be assigned. The order of magnitude of the number of prime-pairs, p, $p+2$, or p, $p+4$, or p, $p+6$, both members of which are less than a large number x, is, it appears,

$$\frac{x}{(\log x)^2}.$$

The order of magnitude of the corresponding number of triplets, of any possible type, is

$$\frac{x}{(\log x)^3},$$

and so on generally. Further, we can assign the relative frequencies of pairs or triplets of different types; there are, for example, about twice as many pairs the difference of which is 6 as there are pairs with the difference 2. All these results have been tested by actual enumeration from the factor tables of the first million numbers; and a physicist would probably regard them as proved, though we of course know very well that they are not.

There is a great deal of mathematics the purport of which is quite impossible for any amateur to grasp, and which, however beautiful and important it may be, must always remain the possession of a narrow circle of experts. It is the peculiarity of the theory of numbers that much of it could be published broadcast, and would win new readers for the *Daily Mail*. The positive integers do not lie, like the logical foundations of mathematics, in the scarcely visible distance, nor in the uncomfortably tangled foreground, like the immediate data of the physical world, but at a decent middle distance, where the outlines are clear and yet some element of mystery remains. There is no one so blind that he does not see them, and no one so sharp-sighted that his vision does not fail; they stand there a continual and inevitable challenge to the curiosity of every healthy mind. I have merely directed your attention for a moment to a few of the less immediately conspicuous features of the landscape, in the hope that I may sharpen your curiosity a little, and that some may feel tempted to walk a little nearer and take a closer view.

COMMENTS

This address, published by the British Association, was summarized in *Nature* 110 (1922), 352, and reprinted in full on pp. 381–5 of the same volume. In a letter to that journal dated 4 October, 1922 (p. 542), Hardy points out that, at the time when the address was given, it was already known that $2^{137}-1$ was composite and states that the number 139 should be substituted for 137 wherever it occurs in the address. The pages reproduced here are those in *Nature*, loc. cit., 381–5.

WHAT IS GEOMETRY?

(Presidential Address to the Mathematical Association, 1925.)

By Prof. G. H. Hardy, F.R.S.

I HAVE put the title of my address in the form of a definite question, to which I propose to return an equally definite answer. I wish to make it quite plain from the beginning that there will be nothing in the least degree original, still less anything paradoxical or sensational, in my answer, which will be the orthodox answer of the orthodox professional mathematician.

I expect that you, as members of an Association which stands half-way between the ordinary mathematical teacher and the professional mathematician in the narrower sense, will probably agree with me that I am wiser to avoid topics of what is usually called a "pedagogical" character. I am sorry to be compelled to use the unpleasant word "pedagogical," and I am sure that you will believe me when I say that I do not use it in any contemptuous sense, and that I am enough of a pedagogue myself to realise the very genuine interest of many "pedagogical" questions. But I do not regard it as the business of a professional mathematician to concern himself primarily with such questions, and, even if I did, I should have very little to say about them. It has always seemed to me that in all subjects, and most of all in mathematics, questions concerning methods of teaching, whether this should come before that, and how the details of a particular chapter are best presented, however interesting they may be, are of secondary importance; and that in mathematics at all events there is one thing only of primary importance, that a teacher should make an honest attempt to understand the subject he teaches as well as he can, and should expound the truth to his pupils to the limits of their patience and capacity. In a word, I do not think it matters greatly what you teach, so long as you are really certain what it is; and I feel that you might reasonably be impatient with me, whether you agreed with me or not, if I occupied your attention for an hour and had nothing more to say to you than that. It is obviously better that I should take some definite chapter of mathematical doctrine, a chapter which is at any rate of the most obvious and direct educational interest, and expound it to you as clearly as I can.

It is, however, quite likely that some of the more sophisticated of you, and particularly any genuine geometer who may be present, will criticise my choice of a subject in a manner which I might find a good deal more difficult to meet. You might object that it would be reasonable enough for me to try to expound the differential calculus, or the theory of numbers, to you, because the view that I might find something of interest to say to you about it is not *prima facie* absurd; but that geometry is, after all, the business of geometers, and that I know, and you know, and I know that you know, that I am not one; and that it is useless for me to try to tell you what geometry is, because I simply do not know. And here I am afraid that we are confronted with a regrettable but quite definite cleavage of opinion. I do not claim to know any geometry, but I do claim to understand quite clearly what geometry is.

I think that this claim is in reality not quite so impertinent as it may seem. The question, "What is geometry?" is not, in the ordinary sense of the phrase, a geometrical question, and I certainly do not think it absurd to suppose that a logician, or even an analyst, may be better qualified to answer it than a geometer. There have been very bad geometers who could have answered it quite well, and very great geometers, such as Apollonius, Poncelet, or Darboux, who would probably have answered it extremely badly. It is a comfort, at any rate, to reflect that my answer can hardly be worse than theirs would in all probability have been.

I propose, then, to cast doubts of this sort aside, and to proceed to answer my question to the best of my ability. There are two things, I think, which become quite clear the moment we reflect about the question seriously. In the first place, there is not one geometry, but an infinite number of geometries, and the answer must to some extent be different for each of them. In the second place, the elementary geometry of schools and universities is not this or that geometry, but a most disorderly and heterogeneous collection of fragments from a dozen geometries or more. These are, or should be, platitudes, and I have no doubt that they are to some extent familiar to all of you; but it is a small minority of teachers of geometry that has envisaged such platitudes clearly and sharply, and it is probably desirable that I should expand them a little.

I begin with the second. It is obvious, first, that a great part of what is taught in schools and universities under the title of geometry is not geometry, or at any rate mathematical geometry, at all, but physics or perhaps philosophy. It is an attempt to set up some kind of ordered explanation of what has been humorously called the real world, the world of physics and sensation, of sight and hearing, heat and cold, earthquakes and eclipses; and earthquakes and eclipses are plainly not constituents of the world of mathematics.

It is dangerous to repeat truisms in public, and the particular truism which I have just stated to you is one which I have often expressed before, and which has sometimes been received in a manner very different from that which I had anticipated. But I am not speaking now to an audience of rude and simple physicists, or of philosophers dazed by centuries of Aristotelian tradition, but to one of mathematicians familiar with common mathematical ideas. I find it difficult to believe that any mathematician of the twentieth century is quite so unsophisticated as to suppose that geometry is primarily concerned with the phenomena of spatial perception, or the physical facts of the world of common sense. It is, however, perhaps unwise to take too much for granted, and I will therefore try to drive home my point by a simple illustration.

Imagine that I am giving an ordinary mathematical lecture at Oxford, let us suppose on elementary differential geometry, and that I write out the proof of a theorem on the blackboard. John Stuart Mill would have maintained that the theorem was at the best approximately true, and that the closeness of the approximation depended on the quality of the chalk; and, though Mill was a man for whom I feel in many ways a very genuine admiration, I can hardly believe that there is anybody quite so innocent as that to-day. I want, however, to push my illustration a stage further. Let us imagine now that a very violent dynamo, or an extremely heavy gravitating body, is suddenly introduced into the room. Einstein and Eddington tell us, and I have no doubt that they are right, that the whole geometrical fabric of the room is changed, and every detail of the pattern to which it conforms is distorted. Does common sense really tell us that my theorem is no longer true, or that the strength or weakness of the arguments by which I have established it has been in the very slightest degree affected? Yet that is the glaring and intolerable paradox to which anyone is committed who supports the old-fashioned view that geometry is "the science of space.".

The simple view, then—the view which I will call for shortness the view of common sense, though there is uncommonly little common sense about it—the view that geometry is the science which tells us the facts about the space of physics and sensation, is one which will not stand a moment's critical examination; and this, of course, was plain enough before Einstein, though it is Einstein who, by enabling us to exhibit its paradoxes in so crude a form, has finally completed the demonstration. The philosophers, of course, have tried to restate the view of common sense in a more sophisticated form. Geometry, they have explained to us, tells us, not exactly the facts of physical or perceptual space, but certain general laws to which all spatial perception must conform. Philosophers have been singularly unhappy in their excursions into mathematics, and this is no exception. It is, as usual, an attempt to restrict the liberty of mathematicians, by proving that it is impossible for them to think except in some particular way; and the history of mathematics shows conclusively that mathematicians will never accept the tyranny of any philosopher. The moment a philosopher has demonstrated the impossibility of any mode of thought, some rebellious mathematician will employ it with unconquerable energy and conspicuous success. No sooner was the apodeictic certainty of Euclid firmly established, than the non-Euclidean geometries were constructed; no sooner were the inherent contradictions of the infinite finally exposed, than Cantor erected a coherent theory. I do not think, then, that we need trouble ourselves with the views of the philosophers concerning geometry. They are, indeed, of much less interest than those of the man in the street, which do possess some interest, since there are valid reasons for supposing that others may share them.

It will be more profitable to leave the philosophers alone, and to consider what the mathematicians themselves have to say. We shall then have reasonable hope of making some substantial progress, since mathematicians, or those of them who are at all interested in the logic of mathematics, hold fairly definite views, and views which are in tolerable agreement, concerning this question of the relation of geometry to the external world. The views of the mathematicians are also much more modest than those which the philosophers have tried to impose upon them.

A geometry, like any other mathematical theory, is essentially a map or scheme. It is a *picture*, and a picture, naturally, of *something*; and as to what that something is opinions do and well may differ widely. Some will say that it is a picture of something in our minds, or evolved from them or constructed by them, while others, like myself, will be more disposed to say that it is a picture of some independent reality outside them; and personally I do not think it matters very much which type of view you may prefer to adopt. What is much more important and much clearer is this, that there is one thing at any rate of which a geometry is *not* a picture, and that that is the so-called real world. About this, I think that almost all modern mathematicians would agree.

This is only common mathematical orthodoxy, but it is an orthodoxy which outsiders very frequently misunderstand or misrepresent. I need hardly say that it does not mean that mathematicians regard the world of physical reality as uninteresting or unimportant. That would be on a par with the view that mathematicians are peculiarly absent-minded, always lose at bridge, and are habitually unfortunate in their investments. Still less does it mean that they regard as uninteresting or unimportant the contribution which mathematics can make to the study of the real world. The Ordnance Survey suggests to me that Waterloo Station, and Piccadilly Circus, and Hyde Park Corner lie roughly in a straight line. That is a geometrical statement about reality, and it enables me to catch my train at Paddington. Einstein is more daring, and issues his orders to the stars, and the stars halt in their courses to obey him. Einstein, and the Ordnance Survey, and even I, can all of us, armed with our mathematics, put forward suggestions concerning the structure of physical reality, and our suggestions will continually prove to be not merely interesting, but of the most direct and practical importance. We can point to this or that mathematical model, Euclidean or Lobatschewskian or Einsteinian geometry, and suggest that perhaps the structure of the universe resembles it, or can be correlated with it in one way or another; that that is a possibility at any rate which the physicists may find it worth their while to consider. We can offer these suggestions, but, when we have offered them, our function as mathematicians is discharged. We cannot, do not profess to, and do not wish to *prove* anything whatsoever. There is not, and cannot be, any question of a mathematician proving anything about the physical world; there is one way only in which we can possibly discern its structure, that is to say the laboratory method, the method of direct observation of the facts.

I will venture here on an illustration which I have used before. If one of you were to tell me that there are three dimensions in this room, but five in Southampton Row, I should not believe him. I would not even suggest that we should adjourn our discussion and go outside to see. The assertion would, of course, be one of an exceedingly complicated character, and a very painstaking analysis might prove necessary before we were quite certain what it meant. However, I could attach a definite meaning to it. I should understand it to imply that, owing to particularities in the geography of London which had up to the present escaped my attention, the common three-dimensional model, sufficient for our purposes in here, becomes inadequate when we pass out into the street. And, however sceptical I might feel about such a theory, I should certainly not be so foolish as to advance mathematical arguments against it, for the all-sufficient reason that I am quite certain that there are none. I should be sceptical, not as a geometer but as a citizen of London, not because I am a mathematician, but in spite of it; and, indeed, I am sure that, if you appealed from me to the nearest policeman, you would find him not less but far more obstinately sceptical than me.

I must pass on, however, to what is really the proper subject-matter of my address. Geometries, I will ask you to agree provisionally, are *models*, and models of something which, whatever it may be in the last analysis, we may allow for our present purposes to be described as mathematical reality. The question which we have now to consider is that of the nature of these models, and the characteristics which distinguish one from another; and there is one great class of geometries for which the answer is immediate and easy, namely, that of the *analytical* geometries.

An analytical geometry, whether of one, two, three, four, or n dimensions, whether real or complex, projective or metrical, Euclidean or non-Euclidean, and it may, of course, be any of these, is a branch of analysis concerned with the properties of certain sets or classes of sets of numbers. I will take the simplest example, the two-dimensional Cartesian geometry which resembles very closely, though it is by no means the same as, the elementary "analytical geometry" taught in schools. I will call it, as I usually call it in lectures, *Common Cartesian Geometry*.

In Common Cartesian Geometry, a *point* is, by definition, a pair of real numbers (x, y), which we call its *coordinates*. A *line* is, again by definition, a certain class of points, viz. those which satisfy a linear relation $ax + by + c = 0$, where a, b, c are real numbers and a and b are not both zero. The relation itself is called the *equation* of the line. If the coordinates of a point satisfy the equation of a line, the line is said to *pass through* the point, and the point to *lie on* the line. And that is the end of Common Cartesian Geometry, in so far as it is projective, that is to say in so far as it does not use the so-called metrical notions of distance and angle, and in so far as it is concerned only with equations of the first degree. What remains is just algebraical deduction from the definitions.

Common Cartesian Geometry, as I have defined it, is a very simple and not a very interesting subject. It gains a great deal in interest, as you will readily imagine, when "metrical" concepts are introduced. We define the *distance* of two points (x_1, y_1) and (x_2, y_2) by the usual formula
$$d = \sqrt{\{(x_1 - x_2)^2 + (y_1 - y_2)^2\}},$$
and the *angle* between two lines by another common formula, which I need not repeat. We have still, however, only to explore the algebraical consequences of our definitions, and no new point of principle arises, so that I can illustrate what I want to say quite adequately from the projective and linear system. This system, trivial as it is, has certain features to which I wish to call your attention as characteristic of analytical geometries in general.

The first feature is this, that a point in Common Cartesian Geometry is *a definite thing*. This is so in all analytical geometries. Thus in any system of two-dimensional and homogeneous analytical geometry a point is a class of triads (x, y, z), those triads being classified together whose coordinates are proportional, and in the geometry of Einstein a point is a set of four numbers (x, y, z, t). This is a very obvious observation, but it is of fundamental importance, since it marks the most essential difference between analytical geometries and " pure " geometries, in which, as we shall see, a point is not a definite entity at all.

The next point which I ask you to observe is the absence of *axioms*. There are no axioms in any analytical geometry. An analytical geometry consists entirely of *definitions* and *theorems*; and this is only natural, since the object of axioms is, as we shall see, merely to limit our subject-matter, and in an analytical geometry our subject-matter is known.

It is most important to realise clearly that, in different geometrical systems, propositions verbally identical may occupy entirely different positions. What is an axiom in one system may be a definition in another, a true theorem in a third, and a false theorem in a fourth. You are accustomed, for example, to *proving* that the equation of a straight line is of the first degree, and I am not suggesting that the " proof " to which you are accustomed is meaningless, trivial, or false. You profess to be proving a theorem, and you are, in fact, genuinely proving *something*, though it might take us some time to ascertain exactly what it is. There is one thing, however, that is quite plain, and that is that the something which you are proving is not a theorem of analytical geometry, for your supposed theorem is, as a proposition of analytical geometry, not a theorem at all but the definition of a straight line.

Let us take another simple illustration, the " parallel postulate " of Euclid. *If L is a line, and P is a point which does not lie on L, then there is one and only one line through P which has no point in common with L.* This, in school geometry, is sometimes called an " axiom " and sometimes, I suppose, an " experimental fact." It cannot be either of these in analytical geometry, where there are neither axioms nor experimental facts, and it is obviously not a definition. It is, in fact, a theorem, which in Common Cartesian Geometry is true, though in other systems it may be false; and it is a theorem which any schoolboy can prove. It is the algebraical theorem that, given an equation $ax + by + c = 0$, and a pair of numbers, x_0, y_0, which do not satisfy this equation, then it is possible to find numbers A, B, C, such that

(i) $$Ax_0 + By_0 + C = 0$$

and (ii) the equations

$$ax + by + c = 0, \quad Ax + By + C = 0$$

are inconsistent with one another; and that the ratios $A : B : C$ are determined uniquely by these conditions.

These are the characteristics of Common Cartesian Geometry which it is most essential for us to observe at the moment. There are others which I should like to say something about if I had time. There is no infinite and no imaginary in this geometry; there are imaginaries, naturally, only in complex systems, and infinites in homogeneous systems. Further, the principle of duality is untrue. All these topics call for comment; and I should have liked particularly to say something on the subject of the geometrical infinite, since the tragical misunderstandings which have beset many writers of text-books of analytical geometry, and which have generated such appalling confusion in the minds of university students, are misunderstandings for which writers like myself of text-books on analysis have been largely though innocently responsible. The geometrical infinite, however, is a subject which would demand at least a lecture to itself. Apart from this, there is nothing in analytical geometry which presents any logical difficulty whatever, and I may pass to the slightly more delicate topic of pure geometry.

The nature of a system of pure geometry, such as the ordinary projective system, is most easily elucidated, I think, by contrast with analytical systems. The contrasts, which I have made by implication already, are sharp and striking, and when once they have been clearly observed the road to the understanding of the subject is open. I observed, first, that the points and lines of analytical geometry were *definite objects*, such as the pair of numbers (2, 3). Secondly, I observed that there were no *axioms* in an analytical geometry, which consists of definitions and theorems only; and that it is the definitions which differentiate one system of analytical geometry from another. The business of an analytical geometer is, in short, to investigate the properties of *particular systems of things*. The standpoint of a pure geometer is entirely different. He is not, except for incidental and subsidiary purposes, concerned with particular things at all. His function is always to consider *all things which possess certain properties*, and otherwise to be strictly indifferent to what they are. His "points" and "lines" are neither spatial objects, nor sets of numbers, nor this nor that system of entities, but *any* system of entities which are subject to a certain set of logical relations. The particular system of relations which he studies is that which is expressed by the *axioms* of his geometry. It is the axioms only which really matter; it is they which discriminate systems, and the definitions play an altogether subsidiary part.

Suppose, for example, if I may take a frivolous illustration, that a pure geometer and an analytical geometer were to go together to the Zoo. The analytical geometer might be interested in tigers, in their colour, their stripes, and in the fact that they eat meat. A point, he would say, is by definition a tiger, and the central theorems of my geometry are that "points are yellow," that "points are striped," and above all that "points eat meat." The pure geometer would reply that he was quite indifferent to tigers, except in so far as they possessed the properties of being yellow and striped; that *anything* yellow and striped was a point to him; that "points are yellow" and "points are striped" were the *axioms* of his geometry, and that all he wanted to know was whether "points eat meat" is a logical deduction from them.

You will, in fact, find, if you consult any standard work on pure geometry, such as Hilbert's *Grundlagen* or Veblen and Young's *Projective Geometry*, that a pure geometer begins somewhat as follows. We consider a system S of objects $A, B, C, \ldots$. We call these objects *points*, and their aggregate *space*; the *plane*, I may say, if I confine myself for simplicity to geometries of two dimensions. From the complete system S which constitutes space we pick out certain partial aggregates $L, M, N, \ldots$, which we call *lines*. If a point A belongs to the particular partial aggregate L, we say that A *lies on* L and that L *passes through* A. These are the *definitions*, and you will observe the quite subsidiary part they play. They are, in fact, purely verbal, and common to all systems; and they do not indicate or imply any special property whatever of the objects which they are said to define, which are indeed often called the *indefinables* of the geometry. The function of the definitions, in fact, is merely to point to the indefinables.

The serious business of the geometry begins when the axioms are introduced. We suppose next that our points and lines are subject to certain logical relations. These suppositions are assumptions, and we call them axioms. To construct a geometry is to state a system of axioms and to deduce all possible consequences from them.

Let us take an actual example. I select the following system of axioms:

AXIOM 1. *There are just three different points.*

AXIOM 2. *No line contains more than two points.*

AXIOM 3. *There is a line through any two points.*

These axioms are consistent with one another, for it is easy to construct a system of objects which satisfy them. We might, for example, take the numbers 1, 2, 3 as our points and the pairs of numbers 23, 31, 12 as our lines, in which case all our axioms are obviously satisfied. Further, the axioms are independent of one another. If the numbers 1, 2, 3 were still our points, but the pairs 2 3, 3 1 alone, and not the pair 1 2, were taken as lines, then the first and second axioms would be satisfied but not the third, and it naturally follows that Axiom 3 is incapable of deduction from the other two. You will have no difficulty in proving in a similar manner, if you care to do so, that each of the three axioms is logically independent of the others. I do not profess to have stated the axioms in the best form possible, but at any rate they are consistent and independent.

It is easy to deduce from our axioms:

THEOREM 1. *There are just three lines*;

and THEOREM 2. *There are just two lines through any point.*

The state of affairs in this geometry is, in short, that suggested by a figure consisting of three points on a blackboard and three lines joining them in pairs. With this, our geometry appears to be exhausted.

The geometry which I have constructed is not an interesting system, since it has no particular application and virtually no content. For our present purpose, however, that is an advantage, as it makes it possible for me to exhibit the system to you in its entirety. However little interest it may possess, it is a perfectly fair specimen of a pure geometry. All systems of pure geometry, projective geometry, metrical geometry Euclidean or non-Euclidean, are constructed in just this way. They are usually very much more complicated, for you must naturally be prepared to sacrifice simplicity to some extent if you wish to be interesting; but their differences from my trivial geometry are differences not at all of principle or of method, but merely of richness of content and variety of application.

I have now given to you the substance of the orthodox answer to the question which I started by asking. I might expand it indefinitely in detail, but I should add nothing essentially new. Geometry is a collection of logical systems. The number of systems is infinite, and any of you can invent as many new systems as he pleases; I have myself, with the aid of a few pupils, constructed seven or eight in the course of an hour. There are two kinds of systems, analytical geometries and pure geometries. An analytical geometry attaches the usual geometrical vocabulary to more or less complicated systems of numbers, and investigates their properties by means of the ordinary machinery of algebra and analysis. A pure geometry, on the other hand, considers all possible fields of certain logical relations, and explores their connections without reference to the nature of the objects among which they hold.

I said when I started that I did not propose to offer any very definite suggestions about the teaching of mathematics; but I should like to conclude with a few words about some of the practical problems with which members of this Association are primarily concerned. It should be obvious to you by now, I think, that school geometry is, as I stated early in my address, not a well-defined subject, a rational exposition of a particular geometrical system, but a collection of miscellaneous scraps, a selection of airs from different pieces, strung together in the manner which experience shows to be the most enlivening. It would be very easy for me to illustrate my thesis by examining a few passages from current text-books of geometry. What is taught as projective geometry, for example, is not projective geometry, and makes very little pretence of being so, since it is based quite frankly on ratios of lengths and other obviously metrical concepts. Indeed, so far as I know, no English book on projective geometry proper exists, except Mathews' *Projective Geometry*, Dr. Whitehead's tract, and parts of Prof. Baker's treatise. On the other hand, a great deal of what is taught as analytical geometry is not analytical geometry, but an attempt to apply the methods of analytical geometry in other fields, partly to some rough kind of physical geometry supposed to be given intuitively, partly to some system of hybrid pure geometry of which some previous knowledge is assumed. But I must not enter into detail, since detail would mean criticism, and criticism of particular books and particular passages, which I have no time for, and am in any case anxious to avoid.

It is not my object now to offer criticisms of the present methods of geometrical teaching. There are a good many very obvious criticisms suggested by the doctrines which I have tried to explain to you, but I recognise that most of these criticisms would be to a very great extent unfair. It is obvious that the teaching of geometry must be based on what is at best a very illogical compromise, and I am prepared to believe that the compromise evolved by experience, and applied by people who know a good deal more about the practical necessity of compromise than I do, is in substance as reasonable a compromise as the difficulties of the problem permit. My object, so far, has been one not of criticism but of explanation.

I do propose, however, to conclude with one word of criticism, directed only to those of you whose pupils are comparatively able and comparatively mature. There is no doubt that the standard of teaching of analysis has improved out of all knowledge during the last twenty years. The elements of the calculus, even the elements of what foreign mathematicians call algebraical analysis, are taught in a manner with which I personally have comparatively little fault to find. The stupid old superstition that falsehood is always easy and attractive, the truth inevitably repulsive and dull, is almost dead, and it is no longer supposed that ignorance of analysis is in itself a proof either of superior intelligence, or high moral character, or profound geometrical or physical intuition. The teaching of higher geometry does not seem to me to have advanced in the same degree.

I think that it is time that teachers of geometry became a little more ambitious. Geometry in its highest developments may be, for all I know, a more difficult subject than analysis; it is not for me as an analyst to deny it. But what may be true enough of the theory of deformation of surfaces, or of algebraical curves in space, is not even plausible of the elements of higher geometry. Those stages of the subject are surely very much easier than the corresponding stages of analysis. There is something hard and prickly about the basic difficulties of analysis, definite stages on the road where definite types of mind seem to come to an inevitable halt. The difficulties of geometry seem to me a little softer and vaguer; knowledge and general intelligence will carry a student appreciably further on the way. And, if this is so, it seems to me regrettable that students are not given the opportunity, while still at school, of learning a good deal more about the real subject-matter out of which modern geometrical systems are built. It is probably easier, and certainly vastly more instructive, than a great deal of what they are actually taught. Anyone who can investigate properties of six or eight points on a conic is capable of understanding what projective geometry is. Anyone who has the faintest hope of a scholarship at Oxford or Cambridge could learn the nature of an axiom, and how a system of axioms may be shown to be consistent with, or independent of, one another. And anyone who can be taught to project two arbitrary points into the circular points at infinity could learn, what he certainly does not learn at present, to attach some sort of definite meaning to the process he performs. Small as my own knowledge of geometry is, and slight as are my qualifications for teaching it to anybody, I have not yet encountered the student who finds difficulty with such ideas when once they are put before him clearly. I am well aware of the very great services which the Association has rendered in the improvement of geometrical teaching. I think that it might well now concentrate its efforts on a general endeavour to widen the horizon of knowledge, recognising, as regards niceties of logic, sequence, and exposition, that the elementary geometry of schools is a fundamentally and inevitably illogical subject, about whose details agreement can never be reached.

G. H. HARDY.

THE CASE AGAINST THE MATHEMATICAL TRIPOS.
(Presidential Address to the Mathematical Association, 1926.)

My address to-day is the result of an informal discussion which arose at our meeting last year after the reading of Mr. Bryon Heywood's paper. You may remember that Mr. Heywood put forward a number of suggestions, with whose general trend I found myself entirely in sympathy, for the improvement of the courses in higher pure mathematics in English universities. He did not criticise one university more than another; but Cambridge is admittedly the centre of English mathematics, so that it is almost inevitable that such suggestions should be considered from the Cambridge standpoint; and that, if my recollections are correct, is what actually happened in the discussion.

My own contribution to the discussion consisted merely in an expression of my feeling that the best thing that could happen to English mathematics, and to Cambridge mathematics in particular, would be that the Mathematical Tripos should be abolished. I stated this on the spur of the moment, but it is my considered opinion, and I propose to defend it at length to-day. And I am particularly anxious that you should understand quite clearly that I mean exactly what I say; that by "abolished" I mean "abolished", and not "reformed"; that if I were prepared to co-operate, as in fact I have co-operated in the past, in "reforming" the Tripos, it would be because I could see no chance of any more revolutionary change; and that my "reforms" would be directed deliberately towards destroying the traditions of the examination and so preparing the way for its extinction.

There are, however, certain possible grounds of misunderstanding which I wish to remove before I attempt to justify my view in detail. The first of these is unimportant and personal, but probably I shall be wise if I refer to it and deal with it explicitly. Our proceedings here do not as a rule attract a great deal of attention, but they are occasionally noticed in the press; and the writer of a well-known column in an evening paper, who was inspired last year to comment on these particular remarks of my own, observed that it was unnecessary to take such iconoclastic proposals seriously, since Cambridge mathematicians were very unlikely to be disturbed by the criticisms of an Oxford man. Perhaps, then, I had better begin by stating that the Mathematical Tripos is an institution of which I have an extensive and intimate knowledge. It is true that I have not taken any part in it during the last six years; but I was a candidate in both parts of it, I took my degree on it, I have examined in it repeatedly under both the old regulations and the new, and, when the old order of merit was abolished in 1910, I was a secretary of the committee which forced this and other changes through a reluctant Senate. I am not then a mere jealous outsider, itching to destroy an institution which I cannot comprehend, but a critic perfectly competent to express an opinion on a subject which I happen to know unusually well.

The second possible misapprehension which I am anxious to remove is decidedly more important. It is possible that some of you may have come here expecting me to deliver a general denunciation of examinations; and if so I am afraid that I shall disappoint you. Denunciation of examinations, like denunciation of lectures, is very popular now among educational reformers, and I wish to say at once that most of what they say, on the one topic and on the other, appears to me to be little better than nonsense. I judge such denunciations, naturally, as a mathematician; and it has always seemed to me that mathematics among all subjects is, up to a point, the subject most obviously adapted to teaching by lecture and to test by examination. If I wish to teach twenty pupils, for example, the exponential theorem, the product theorem for the sine, or any of the standard theorems of analysis or geometry, it seems to me that by far the best, the simplest, and the most economical course is to assemble them in a lecture-room and explain to them collectively the essentials of the proofs. It seems to me also that, if I wish afterwards to be certain that they have understood me, the obviously sensible way of finding out is to ask them to reproduce the substance of what I said, or to apply the theorems which I proved to simple examples. In short, up to a point, I believe in formal lectures, and I believe also in formal examinations.

There are in fact certain traditional purposes of examinations, purposes for which they always have been used, and for which they seem to me to be the obvious and the appropriate instrument. There are certain qualities of mind which it is often necessary to test, and which can be tested by examination much more simply and more effectively than in any other way. If a teacher wishes to test his pupils' industry, for example, or their capacity to understand something he has told them, something perhaps of no high order of difficulty, but difficult enough to require some little real intelligence and patience for its appreciation, it seems to me that his most reasonable course is to subject them to some sort of examination. Examinations have been used in this manner, from time immemorial, in every civilised country; there are, in England, quite a number of large, elaborately organised, and, so far as I can judge, quite sensibly conducted examinations of this type; and with such examinations I have no sort of quarrel.

There are, however, in England, and, so far as I know, in no other country in the world, a number of examinations, of which the Mathematical Tripos at Cambridge, and Greats at Oxford, have been the outstanding examples, which are of quite another type, and which fulfil, or purport to fulfil, quite different and very much more ambitious ends. These examinations originate in Oxford and Cambridge, and are found in their full development there only, though they have been copied to a certain extent by our modern universities. They are described as "honours" examinations, and pride themselves particularly on their traditions and their "standards". To these examinations are subjected a heterogeneous mass of students of entirely disparate attainments, and the examination professes to sort out the candidates and to label them according to the grade of their abilities. Thus in the old mathematical Tripos there were three classes, each arranged in order of merit, while in the new there are three classes and two degrees of marks of special distinction. It is evident that such an examination is not content with fulfilling the ends which I have admitted that an examination can fulfil so well; it is not, and prides itself that it is not, merely a useful test of industry, intelligence, and comprehension. It purports to appraise, and it must be admitted that to some extent, though very imperfectly, it does appraise, higher gifts than these. A "$b*$" in the Tripos, or a first in Greats, is taken to be, and in a measure is, an indication of a man quite outside the common run. It is these examinations and these only, these examinations with reputations and standards and traditions, which seem to me mistaken in their principle and useless or damaging in their effect, and which I would destroy if I had the power. An examination can do little harm, so long as its standard is low.

I suppose that it would be generally agreed that Cambridge mathematics, during the last hundred years, has been dominated by the Mathematical Tripos in a way in which no first-rate subject in any other first-rate university has ever been dominated by an examination. It would be easy for me, were the fact disputed, to justify my assertion by a detailed account of the history of the Tripos, but this is unnecessary, since you can find an excellent account, written by a man who was very much more in sympathy with the Tripos than I am, in Mr. Rouse Ball's *History of Mathematics in Cambridge*. I must, however, call your attention to cert   rather melancholy reflections which the history of Cambridge mathematics suggests. You will understand that when I speak of mathematics I mean primarily pure mathematics, not that I think that anything which I say about pure mathematics is not to a great extent true of applied mathematics also, but merely because I do not want to criticise where my competence as a critic is doubtful.

Mathematics at Cambridge challenges criticism by the highest standards. England is a first-rate country, and there is no particular reason for supposing that the English have less natural talent for mathematics than any other race; and if there is any first-rate mathematics in England, it is in Cambridge that it may be expected to be found. We are therefore entitled to judge Cambridge mathematics by the standards that would be appropriate in Paris or Göttingen or Berlin. If we apply these standards, what are the results? I will state them, not perhaps exactly as they would have occurred to me spontaneously—though the verdict is one which, in its essentials, I find myself unable to dispute—but as they were stated to me by an outspoken foreign friend.

In the first place, about Newton there is no question; it is granted that he stands with Archimedes or with Gauss. Since Newton, England has produced no mathematician of the very highest rank. There have been English mathematicians, for example Cayley, who stood well in the front rank of the mathematicians of their time, but their number has been quite extraordinarily small; where France or Germany produces twenty or thirty, England produces two or three. There has been no country, of first-rate status and high intellectual tradition, whose standard has been so low; and no first-rate subject, except music, in which England has occupied so consistently humiliating a position. And what have been the peculiar characteristics of such English mathematics as there has been? Occasional flashes of insight, isolated achievements sufficient to show that the ability is really there, but, for the most part, amateurism, ignorance, incompetence, and triviality. It is indeed a rather cruel judgment, but it is one which any competent critic, surveying the evidence dispassionately, will find it uncommonly difficult to dispute.

I hope that you will understand that I do not necessarily endorse my friend's judgment in every particular. He was a mathematician whose competence nobody could question, and whom nobody could accuse of any prejudice against England, Englishmen, or English mathematicians; but he was also, of course, a man developing a thesis, and he may have exaggerated a little in the enthusiasm of the moment or from curiosity to see how I should reply. Let us assume that it is an exaggerated judgment, or one rhetorically expressed. It is, at any rate, not a *ridiculous* judgment, and it is serious enough that such a condemnation, from any competent critic, should not be ridiculous. It is inevitable that we should ask whether, if such a judgment can really embody any sort of approximation to the truth, some share of the responsibility must not be laid on the Mathematical Tripos and the grip which it has admittedly exerted on English mathematics.

I am anxious not to fall into exaggeration in my turn and use extravagant language about the damage which the Tripos may have done, and it would no doubt be an extravagance to suggest that the most ruthless of examinations could destroy a whole side of the intellectual life of a nation. On the other hand it is really rather difficult to exaggerate the hold which the Tripos has exercised on Cambridge mathematical life, and the most cursory survey of the history of Cambridge mathematics makes one thing quite clear; the reputation of the Tripos, and the reputation of Cambridge mathematics stand in correlation with one another, and the correlation is large and negative. As one has developed, so has the other declined. As, through the early and middle nineteenth century, the traditions of the Tripos strengthened, and its importance in the eyes of the public grew greater and greater, so did the external reputation of Cambridge as a centre of mathematical learning steadily decay. When, in the years perhaps between 1880 and 1890, the Tripos stood, in difficulty, complexity, and notoriety, at the zenith of its reputation, English mathematics was somewhere near its lowest ebb. If, during the last forty years, there has been an obvious revival, the fortunes of the Tripos have experienced an equally obvious decline.

Perhaps you will excuse me if I interpolate here a few words concerning my own experience of the Tripos, which may be useful as a definite illustration of part of what I have said. I took the first part of the Tripos in 1898, and the second in 1900: you must remember that it was then the first part which produced wranglers and caught the public eye.

I am inclined to think that the Tripos had already passed its zenith in 1898. There had already been one unsuccessful attempt to abolish the order of merit, a reform not carried finally till 1910. When the first signs of decline might have been detected I cannot say, but the changes in the Smith's Prize examination, and the examination for Trinity Fellowships, must have been partly responsible, and these had been determined by dissertation for a considerable time. At any rate it was beginning to be recognised, by the younger dons in the larger colleges, and to some extent by undergraduates themselves, that the difference of a few places in the order of merit was without importance for a man's career. This, however, is comparatively unimportant, since it is less the examination itself than its effect on teaching in the university that I wish to speak of now.

The teaching at Cambridge when I was an undergraduate was, of course, quite good of its kind. There were certain definite problems which we were taught to solve; we could learn, for example, to calculate the potential of a nearly spherical gravitating body by the method of spherical harmonics, or to find the geodesics on a surface of revolution. I do not wish to suggest that the two years which I spent over the orthodox course of instruction—my second two years were occupied in a different way—were altogether wasted. It remains true that, when I look back on those two years of intensive study, when I consider what I knew well, what I knew slightly, and of what I had never heard, and when I compare my mathematical attainments then with those of a continental student of similar abilities and age, or even with those of a Cambridge undergraduate of to-day, it seems to me almost incredible that anyone not destitute of ability or enthusiasm should have found it possible to take so much trouble and to learn no more. For I was indeed ignorant of the rudiments of my profession. I can remember two things only that I had learnt. Mr. Herman of Trinity had taught me the elements of differential geometry, treated from the kinematical point of view; this was my most substantial acquisition, and I am grateful for it still. I had also picked up a few facts about analysis, towards the end of those two years, from Prof. Love. I owe, however, to Prof. Love something much more valuable than anything he taught me directly, for it was he who introduced me to Jordan's *Cours d'analyse*, the bible of my early years; and I shall never forget the astonishment with which I read that remarkable work, to which so many mathematicians of my generation owe their mathematical education, and learnt for the first time as I read it what mathematics really meant.

It has often been said that Tripos mathematics was a collection of elaborate futilities, and the accusation is broadly true. My own opinion is that this is the inevitable result, in a mathematical examination, of high standards and traditions. The examiner is not allowed to content himself with testing the competence and the knowledge of the candidates; his instructions are to provide a test of more than that, of initiative, imagination, and even of some sort of originality. And as there is only one test of originality in mathematics, namely the accomplishment of original work, and as it is useless to ask a youth of twenty-two to perform original research under examination conditions, the examination necessarily degenerates into a kind of game, and instruction for it into initiation into a series of stunts and tricks. It was in any case certainly true, at the time of which I am speaking, that an undergraduate might study mathematics diligently throughout the whole of his career, and attain the very highest honours in the examination, without having acquired, and indeed without having encountered, any knowledge at all of any of the ideas which dominate modern mathematical thought. His ignorance of analysis would have been practically complete. About geometry I speak with less confidence, but I am sure that such knowledge as he possessed would have been exceedingly one-sided, and that there would have been whole fields of geometrical knowledge, and those perhaps the most fruitful and fascinating of all, of which he would have known absolutely nothing. A mathematical physicist, I may be told, would on the contrary have received an appropriate and an excellent education. It is possible; it would no doubt be very impertinent for me to deny it. Yet I do remember Mr. Bertrand Russell telling me that he studied electricity at Trinity for three years, and that at the end of them he had never heard of Maxwell's equations; and I have also been told by friends whom I believe to be competent that Maxwell's equations are really rather important in physics. And when I think of this I begin to wonder whether the teaching of applied mathematics was really quite so perfect as I have sometimes been led to suppose.

I remember asking another friend, who was Senior Wrangler some years later, and has since earned a very high reputation by research of the most up-to-date and highbrow kind, how the Tripos impressed him in his undergraduate days, and his reply was approximately as follows. He had learnt a little about modern mathematics while he was still at school, and he understood perfectly while he was an undergraduate, as I certainly did not, that the mathematics he was studying was not quite the real thing. But, he continued, he regarded himself as playing a game. It was not exactly the game he would have chosen, but it was the game which the regulations prescribed, and it seemed to him that, if you were going to play the game at all, you might as well accept the situation and play it with all your force. He believed—and remember, if you think him arrogant, that his judgment was entirely correct —that he could play that game at least as well as any of his rivals. He therefore decided deliberately to postpone his mathematical education, and to devote two years to the acquisition of a complete mastery of all the Tripos technique, resuming his studies later with the Senior Wranglership to his credit and, he hoped, without serious prejudice to his career. I can only add —lost as I am in hopeless admiration of a young man so firmly master of his fate—that every detail of these precocious calculations has been abundantly justified by the event.

I feel, however, that I am laying myself open at this point to a challenge which I shall certainly have to meet sooner or later, and which I may as well deal with now. It will be said—I know from sad experience that such things are always said—that I am applying entirely wrong criteria to what is after all an examination for undergraduates. I shall be told that I am assuming that the principal object of the Cambridge curriculum is to increase learning and to encourage original discovery, and that this is false; that learning and research are admirable things, but that a great university must not allow itself to be overshadowed by them; and that, in short, a German professor of mathematics, however universal his reputation and profound his erudition, is not necessarily the noblest work of God. Indeed, at this point I seem to hear the voice of my opponent grow a little louder, as he points out to me that I am entirely misconceiving the function of an English university, that the universities of England are not at all intended as machines for the generation of an infinite sequence of professors, but as schools for the development of intellect and character, as training grounds of teachers, civil servants, statesmen, captains of industry, and proconsuls, in short as nurseries where every young Englishman may learn to add his quotum to the fulfilment of the destinies of an imperial race. I wish very heartily, I confess, that I was not going to be told all this, but I know very well that it is coming, for have I not heard it all a hundred times already, and did we not hear it all in 1910, from all the Justices who had been wranglers in their day?

Perhaps, however, I shall not be wasting your time entirely if I occupy a few minutes in an attempt to examine this indictment as dispassionately as I can. I find it very difficult to believe that most of the quite considerable body of quite intelligent people who continue to use this kind of language at the present day, and to turn it to the defence of our present university education, can have considered at all coolly some of the implications of what they say. On the other hand I recognise that it is a good deal easier to laugh at these people than to refute them, and that, if I were to attempt a reasoned reply to their contention, considered as a general principle to guide us in the construction of an educational system, then I should have a long and tiresome argument before me.

Fortunately, this is unnecessary. We are not now discussing educational systems generally, but the merits of a particular examination. We have not to undertake a general defence of mathematics and the position which is at present allowed to it in education, or to repel the very formidable onslaught which might be directed against it by Philistinism pure and simple. You and I and the Justices are after all agreed in wanting to see some sort of education in higher mathematics, and differ only in the kind of mathematical education which we prefer. The question is merely whether it is possible to defend the Mathematical Tripos on these lines, and we can appeal here, I think, to the method of *reductio ad absurdum*.

I have already put forward one test of a mathematical education, namely that it should produce mathematicians, as "mathematician" is understood by the leading mathematicians of the world; and this test, whatever its defects may be, has one merit at any rate, namely that it is clear and sharp and easily applied. It is also a test to which I suppose that everybody would agree in attaching *some* degree of importance, since it must be extraordinarily difficult for any English mathematician to maintain that it is of no importance whatever whether English mathematics be good or bad. The question therefore is not of the validity of the test, but only of its relative importance.

Now there is one obvious difference between my test, which I will call for shortness the professional test, and the slightly more orotund test which I have tried to state in general terms. My test has certainly this advantage, that I am testing a mathematical education as a means to one of the ends which a mathematical education may reasonably be expected to secure, and which it is hardly possible to secure in any other way. When, on the other hand, we attempt to test a training in higher mathematics, the highest such training the country offers, by its effects generally on the intelligence and character of those who submit to it, we are at once confronted with a question which is obviously more fundamental, whether intelligence and character of the type at which we aim are really developed very effectively by a training in higher mathematics. And as we are all mathematicians here, we need not indulge in humbug about it. We know quite well that the answer is No.

It is hardly likely that anybody here will accuse me of any lack of devotion to the subject which has after all been the one great permanent happiness of my life. My devotion to mathematics is indeed of the most extravagant and fanatical kind; I believe in it, and love it, and should be utterly miserable without it, and I have never doubted that, for any one who takes real pleasure in it and has a genuine talent for it, it is the finest intellectual discipline in the world. I believe also that a fair knowledge of mathematics is, even for those who have no pronounced mathematical talent, extremely useful and extremely stimulating, and that it should be part of the ordinary intellectual capital of all intelligent men. I am prepared indeed to go further, since I believe that a very large proportion of students abandon mathematics merely because it is often very badly taught, and might push their mathematical studies a good deal further than they do at present with very great profit to themselves. But I do not believe for a moment, and I do not believe that the majority of competent mathematicians believe, that the intensive study of higher mathematics, whether it be understood as it would be in a foreign university, or whether it be understood as it has in the past been understood in the Mathematical Tripos, forms a good basis of a general education. I am not at all sure that, among all possible subjects which might be selected as special courses of study, for an intelligent young man of no particular talent, mathematics is not the worst. Indeed, I think that this is being gradually recognised both by teachers of mathematics and by students themselves, and that it is for this reason that the Mathematical Tripos is more and more becoming, and rightly becoming, the special preserve of professional mathematicians. And if this be so, then surely it is quite obviously futile to judge the Tripos by anything but a professional standard.

It seems to me, then, that the opponents of the professional standard are committed from the beginning to a very paradoxical position, and yet it seems —such is the attraction of a paradox—that they are actually dissatisfied with its already sufficiently serious difficulties and determined to surround it by still more fantastic entrenchments. For they generally go on to maintain that mathematics may indeed be made the finest of intellectual disciplines, but only if it is taught in a manner which ignores or rejects every development of recent years. It will teach you to think, so long as you are not allowed to think quite correctly; it will widen your interests and stimulate your imagination, so long as you are carefully confined to problems in which mathematicians have lost interest for fifty years. In a word, the mathematics of the amateur is all right, and that precisely because it is so much more than a little wrong, but if we once allow mathematics to be dominated by the professionals, that is to say by the men who live in the subject and are familiar with its vital developments, then its energy will be sapped and its educational efficacy destroyed. And of all insane paradoxes, surely, this is one of the most portentous.

I have told you already that I am not much of a believer in the general educational efficacy of a specialised mathematical training. I do not believe that it is possible to build a character or an empire on a foundation of mathematical theories; but surely it must be still more impossible to build either on a foundation of Tripos problems. If I were compelled to undertake so crazy an enterprise, I would select the true theorems rather than the false, the fundamental facts of mathematics rather than its trivial excrescences, the problems which are alive to-day rather than those which perished in the mid-Victorian era.

I would suggest to you, then, that, when you have next to listen to the mathematical reactionary who laments the good old days, if you doubt your competence to judge for yourself the merits of his complaints, you should apply to what he says Hume's test of the greater improbability. It does not seem very likely that the modern experts are all wrong, but it is quite possible. It is also possible that the times have really left a conservatively-minded mathematician a little bit behind; that his lectures and his text-books have run out of date; that there is a good deal in modern mathematics which he finds it too great an effort to master; and that it gives him a good deal less trouble to abuse the modern tendencies than to repair the gaps in his own mathematical equipment. This also is, of course, extremely improbable; but you must ask yourself which is the greater improbability of the two.

I do not propose to waste further time on the discussion of this question: in what more I have to say about the Tripos I shall adopt a frankly professional view. I shall judge the Tripos by its real or apparent influence on English mathematics. I have already told you that in my judgment this influence has in the past been bad, that the Tripos has done negligible good and by no means negligible harm, and that, so far from being the great glory of Cambridge mathematics, it has gone a very long way towards strangling its development. There are further questions to consider. We may ask in the first place, if it be granted that what I have said about the past is roughly true, how far have things improved? Is it not true already that the Tripos means a great deal less, and English mathematics appreciably more, than forty years ago, and is it not extremely likely that, even if there be no further radical changes, this process will continue? Then, if we are not content to answer this question by a simple affirmative and leave it there, we may ask what really are the fundamental faults of an examination on the Tripos model, and whether it is not possible to make less drastic suggestions for its improvement.

I began my address with what was to a certain extent a defence of examinations. I said that under certain conditions I believed in examinations, that is to say in examinations of a sufficiently lowly type, which do not profess to be more than a reasonable test of certain rather humdrum qualities. The phrases which I used were vague, and I ought no doubt to attempt to define my own standard a little more precisely. This is naturally not quite easy, but I will risk some sort of definition. I should say, roughly, that the qualities which I have in mind—reasonable industry, reasonable intelligence, reasonable grasp—would be about sufficient to carry a candidate, in any of the orthodox Oxford or Cambridge examinations, into a decent second class. Beyond that, I do not believe in recognising differences of ability by examination.

I said this here last year, and I was at once challenged. I was asked, whatever could you do, if you could not tell the quality of a man by looking at his examination record? I wonder whether my questioner realised that these elaborate honours examinations, so far from being one of the fundamental necessities of modern civilisation, are a phenomenon almost entirely individual to Oxford and Cambridge, copied in a half-hearted fashion by other English universities, and, beyond that, having hardly a parallel in the world? Does Germany suffer from intellectual stagnation, because there are no honours examinations in her universities? Germany does not think in terms of firsts and seconds; we think in terms of them, so far as we do so think, and perhaps the practice is to some extent abating, merely because we have heard so much about them that they have become to us like bitter ale or eggs and bacon, and we have forgotten that we could get on quite happily without them.

I remember, if you will excuse my referring once more to the forgotten controversies of 1910, a curious saying of, I think, Mr. Justice Romer. Mr. Justice Romer circulated a flysheet to the Senate, deploring, of course, the proposal to abolish the Senior Wrangler. "What", he asked judicially, "is the function of the Tripos?", and he replied "Surely to examine and to make distinctions between young men". It would indeed be difficult to compress a larger quantity of vicious educational doctrine into a smaller number of words. The exactly opposite doctrine, that no distinctions should be made by examination except such as practical necessities may make imperative, is surely somewhere a little nearer to the truth.

Let us then consider, with the view of meeting the objection which was raised to what I said last year, whether the kind of distinctions made by the Mathematical Tripos are, in fact, of any particular practical utility. The evidence of ability provided by the Tripos is as follows. A candidate may obtain a first, second, or a third. He may obtain a mark—the "b" mark—

of adequate knowledge of some special subject, or a higher mark—the "$b*$" mark—of special distinction in that subject. The test case for us, and the only one I have time to consider now, is the highest mark. When a candidate has attained this mark, what has he gained?

In the first place, he has gained the natural feeling of satisfaction which everyone experiences when he is adjudged to have performed a definite task at least as well as anybody else. He will feel with pleasure and pride that the world is awarding honest work, and these entirely creditable feelings may spur him on to further effort. Has he gained anything of more tangible or permanent value?

A man who can attain the highest honours in the Tripos is generally a good enough mathematician to hope for a permanent academic career. How far will his "$b*$" assist him along this career? Will anyone give him a position, a fellowship or a lectureship, on the strength of it? If he thinks that, he will be very quickly disillusioned.

It is possible that there are positions, in the junior grades of the teaching staffs of certain universities, which are sometimes filled on the strength of an examination record. I have never come across such a post myself, but it is probable enough that they exist. Academic positions are usually bestowed, not on examination record, but from personal knowledge or on the strength of private recommendations from competent people. I have taken part myself in many such appointments. When applications are invited, the testimonials submitted by candidates contain statements of their academic qualifications, and often of their performance in examinations, and it would be an exaggeration to say that such records are never referred to. There are usually a fair proportion of the candidates whose qualifications seem obviously below the standard expected, and a glance at their examination records often provides useful evidence in confirmation of this view. I do not remember any case of any other kind in which such a record has played any part in the decision, or has been referred to in the discussion by any member of the board of electors.

I suppose that this is generally understood, and that candidates for such positions are not usually under any delusion about the attention paid to the records which they submit. It may, however, be urged that an examination record of high distinction might often determine the decision if the post were of a less purely academic kind, for example if it were a mastership in one of the big public schools. It may be so, but I must confess that—at any rate in the particular case which we are considering—I am uncommonly sceptical about it. In the first place, people who obtain "$b*$"s have usually scientific ambitions, and the last thing they want is a mastership in the most historic of public schools. It is not possible now for the richest or the most aristocratic school to obtain a really distinguished mathematician, even if it wants one, which of course in general it does not. Finally, even if the demand existed and the supply were there, the headmasters of the great public schools do not, so far as my experience has shown me, select their assistants in this way, but proceed much more in the spirit of a board of electors, though naturally in a more capricious and autocratic way.

My conclusion, then, is that the highest certificate of merit offered by the Tripos might just as well be scrapped, for all the influence it exerts on the careers of those who obtain it. I suppose, in fact, that the universities, and most of the other bodies in whose hands educational patronage is vested, have come in practice to very much the same conclusion as my own, that examinations are an admirable test of competence and industry, but ineffective and erratic as a test of any higher gift. The Government stands alone, so far as I know, in attaching a definite money value to an examination class, and even the Government stops short of rewarding the only mark which could plausibly pretend to be a mark of real distinction.

If such distinctions are in effect futile, why should we waste our time and our energies in making them, even if we were certain that they do no harm? If Einstein had taken the Mathematical Tripos, what would it matter what place he took? The world can recognise its Einsteins quickly enough when it gets the opportunity. If Einstein sits for an examination, let him have his degree, assuming that he can satisfy the examiners. What is the object of taking all these pains to make to-day, uncertainly and half-heartedly, distinctions which, if they have any foundation in reality, the world will make in its own much sharper fashion to-morrow?

I have left to the last the defence of the Tripos which I find myself most difficult to meet. It is a defence difficult to overcome, because it proceeds on what a chess-player would call close lines. This defence, which I have often heard from mathematicians whose judgment I value, and which I wish to treat with all respect, is simply this: that the examination has already been considerably relaxed, and that the effects of its relaxation can already be traced in a corresponding strengthening of English mathematics; that it may be indefensible in principle, but that the spirit of emulation which it fosters may conceivably do some slight positive good; and that, now that so many of its teeth have been successfully drawn, it is not very obvious that it does any very serious harm. This is undeniably the case for the Tripos in its strongest and sanest form.

I should admit that, up to a point, the defence is sound. I would go so far as to admit that the system now does little harm to men of what I may define roughly as fellowship standard. The truth is that the principles for which I am contending have been so far recognised that a man of this degree of ability need not really disturb himself very seriously about the examination. Such a man may pursue a course of serious mathematical study with every confidence that, unless he is wilfully neglectful, he can obtain without any intensive effort all such honours as the Tripos can bestow. The test, in short, does him no harm, because for him it has lost its meaning. This I admit, and, of course, I recognise that it is a very large admission, since it destroys a good deal of the case which could be urged so irresistibly against the Tripos thirty years ago on strictly professional grounds. It is no longer true that Cambridge is notably behind the times, or that its courses compare particularly unfavourably, at any rate in the subjects about which I am best qualified to judge, with those at any but the very best of continental universities. This, I think, Mr. Heywood did not recognise sufficiently; it was the only point in his address from which I particularly dissented. It is no longer true that the development of a decent school of English mathematics is being steadily throttled by the vices of its principal examination.

We must recognise this and rejoice that things have moved so far, and if they have moved just because the glamour of the Tripos has faded, we shall only rejoice the more. We need not rush to the conclusion that the whole case against the Tripos has been destroyed. We have to think of its effects, not only on students of the highest class, but also upon teachers of mathematics in the university, and upon students a little less gifted than those of whom I have spoken hitherto. I am afraid that it is still true that mathematical teaching is hampered very seriously by the examinations, both in Cambridge and in Oxford, where the system is different in detail, but in essentials the same.

In the first place, it is still true that a large proportion of students, either wilfully, because they exaggerate the importance of the examination, or from ignorance, because they have never heard of anything better, or (and I am afraid that this is the most common explanation) because they are driven to it by tutors who have to justify themselves in the eyes of college authorities greedy for firsts, for one or other of these reasons allow their mathematical ucation to b e stunted by absorption in examination technique. They spend hour after hour, which ought to be devoted to lectures or reading, in working through examination papers, or the collections of problems in which English text-books are so rich, exhausting themselves and their tutors in the struggle to turn a comfortable second into a marginal first. It is possible that the effect of all this mistaken exertion is more directly damaging to the tutors than to the students themselves; but a pupil cannot draw much inspiration from a tutor who is always tired, and there is hardly a tutor in Oxford, and not very many in Cambridge, who has not about twice as much teaching as any active mathematician should be asked to undertake. A professor at Oxford or Cambridge is very much his own master, but even a professor may be handicapped very seriously in his teaching by the recollection of the syllabus of the schools. It is often very hard to ask your pupils to go on listening to you when you know that what you tell them will gain them no credit in an examination to which they attach enormous importance and over which you have practically no control. I should always like to ignore the examination completely, and often summon up the courage to do so for a while, only to be pulled up short a few weeks later by the thought that after all it is hardly fair.

Whatever, then, may be said about the improvement of Cambridge teaching, and however much the dominion of the Tripos may have abated of its rigours of thirty years ago, I adhere to the view which I expressed to you last year, that the system is vicious in principle, and that the vice is too radical for what is usually called reform. I do not want to reform the Tripos, but to destroy it. And if you ask me whether the Tripos is a peculiar case, or whether what I have said applies to all other high-grade honours examinations, I can only answer that, so far as I can see, it does. The Tripos is the worst case. It is the oldest examination and the most famous, and generally the most strongly entrenched; and mathematics is a subject in which it is particularly easy to examine ferociously, so that the evils of the system stand out here in the clearest light. But, of course, the greater part of what I have said about the Tripos could be applied with almost equal force to Greats.

I wish, then, to abolish the Tripos, and as I know perfectly well that neither I nor anyone else will succeed in doing so, since the practical difficulties would be so serious, and the force of tradition is so terribly strong, I may reasonably be asked what seem to me the best practical steps for a more moderate reformer to take. You will probably have inferred from my remarks that I am not prepared with any very illuminating suggestions. I could, of course, suggest many changes of detail, both in the schedules and in the conduct of the examination; but my suggestions would be comparatively unimportant; and I should not be prepared to expend my energies in pressing them, since they would all be inspired by the same ideal, and that an ideal whose realisation will no doubt remain hopeless for many years. Indeed I am afraid that my advice to reformers might sound like a series of stupid jokes. I should advise them to let down the standard at every opportunity; to give first classes to almost every candidate who applied; to crowd the syllabus with advanced subjects, until it was humanly impossible to show reasonable knowledge of them under the conditions of the examination. In this way, in the course of years, they might succeed in corrupting the value of the prizes which they have to offer, and in all probability time would do the rest.

---

348. When David Ramsay was asked to consent to his daughter's wedding, "He multiplied six figures progressively, and reported the product—then gave his consent."—*The Fortunes of Nigel*, c. xxxv.

349. Dodd was one of the rare examples of an eloquent mathematician.—Sir Chartres Biron. *Johnson Club Papers*, 1920. ["Dodd, William, 1729-1777, forger" (*D.N.B.*). 15th Wrangler, 1749.]

COMMENTS

The paper by Mr Bryon Heywood referred to is:

H. B. Heywood, 'The reform of university mathematics', *Math. Gaz.* 12 (1925), 322–30.

On the final page there is a brief report on the subsequent discussion at which Hardy was present and declared his view that very difficult honours examinations, like the Mathematical Tripos or Oxford Greats, were fundamentally vicious and should be abolished.

Hardy's address was reprinted in the *Mathematical Gazette* 32 (1948), 134–45, as part of the 300th number of the *Gazette*.

AN INTRODUCTION TO THE THEORY OF NUMBERS*

BY G. H. HARDY

PART I

1. *Farey Series*. The theory of numbers has always occupied a peculiar position among the purely mathematical sciences. It has the reputation of great difficulty and mystery among many who should be competent to judge; I suppose that there is no mathematical theory of which so many well-qualified mathematicians are so much afraid. At the same time it is unique among mathematical theories in its appeal to the uninstructed imagination and in its fascination for the amateur. It would hardly be possible in any other subject to write books like Landau's *Vorlesungen* or Dickson's *History*, six great volumes of overwhelming erudition, better than the football reports for light breakfast table reading.

The excursions of amateur mathematicians into mathematics do not usually produce interesting results. I wish to draw your attention for a moment to one very singular exception. Mr. John Farey, Sen., who lived in the Napoleonic era, has a notice of twenty lines in the *Dictionary of National Biography*, where he is described as a geologist. He received as a boy "a good mathematical training". He was at one time agent to the Duke of Bedford, but afterwards came to London, where he acquired an extensive practice as a consulting surveyor, which led him to travel much about the country and "collect minerals and rocks". His principal work was a geological survey of Derbyshire, undertaken for the Board of Agriculture, but he also wrote papers in the *Philosophical Magazine*, on geology and on many other subjects, such as

---

* The sixth Josiah Willard Gibbs Lecture, read at New York City, December 28, 1928, before a joint session of the American Mathematical Society and the American Association for the Advancement of Science.

music, sound, comets, carriage wheels and decimal coinage. As a geologist, Farey is apparently forgotten, and, if that were all there were to say about him, I doubt that he would find his way into the *Dictionary of National Biography* today.

It is really very astonishing that Farey's official biographer should be so completely unaware of his subject's one real title to fame. For, in spite of the *Dictionary of National Biography*, Farey is immortal; his name stands prominently in Dickson's *History* and in the German encyclopaedia of mathematics, and there is no number-theorist who has not heard of "Farey's series". Just once in his life Mr. Farey rose above mediocrity and made an original observation. He did not understand very well what he was doing, and he was too weak a mathematician to prove the quite simple theorem he had discovered. It is evident also that he did not consider his discovery, which is stated in a letter of about half a page, at all important; the editor of the Philosophical Magazine printed a very stupid criticism in the next volume, and Farey, usually a rather acrid controversialist, ignored it completely. He had obviously no idea that this casual letter was the one event of real importance in his life. We may be tempted to think that Farey was very lucky; but a man who has made an observation that has escaped Fermat and Euler deserves any luck that comes his way.*

Farey's observation was this. The *Farey series of order* n is the series, in order of magnitude, of the irreducible rational fractions between 0 and 1 whose denominators do not exceed n. Thus

$$\frac{0}{1}, \frac{1}{7}, \frac{1}{6}, \frac{1}{5}, \frac{1}{4}, \frac{2}{7}, \frac{1}{3}, \frac{2}{5}, \frac{3}{7}, \frac{1}{2},$$
$$\frac{4}{7}, \frac{3}{5}, \frac{2}{3}, \frac{5}{7}, \frac{3}{4}, \frac{4}{5}, \frac{5}{6}, \frac{6}{7}, \frac{1}{1}$$

---

* It should be added that Farey's discovery had been anticipated 14 years before by C. Haros: see Dickson's *History*, vol. 1, p. 156. Cauchy happened to see Farey's note and attributed the theorem to him, and everyone else has followed Cauchy's example.

is the Farey series of order 7. There are two simple theorems about Farey series; (1) if p/q and p'/q' are two consecutive terms, then
$$p'q - pq' = 1,$$
and (ii) if p/q, p'/q', p''/q'' are three consecutive terms, then
$$\frac{p'}{q'} = \frac{p + p''}{q + q''}.$$
The second theorem (which is that actually stated by Farey) is an immediate consequence of the first, as we see by solving the equations
$$p'q - pq' = 1, \quad p''q' - p'q'' = 1,$$
for p' and q'.

The theorems are not of absolutely first class importance, but they are not trivial, and all of the many proofs have some feature of real interest. One of the simplest uses the language of elementary geometry. We consider the *lattice* or *Gitter L* in a plane formed by drawing parallels to the axes at unit distance from each other; the intersections, the points (x, y) with integral coordinates, are called the *points of the lattice*. It is obvious that the properties of the lattice are independent of the particular lattice point O selected as origin and symmetrical about any origin. The lattice is transformed into itself by the linear substitution
$$x' = \alpha x + \beta y, \quad y' = \gamma x + \delta y,$$
where $\alpha, \beta, \gamma, \delta$ are integers and $\Delta = \alpha\delta - \beta\gamma = 1$, since then there is a pair x, y which give any assigned integral values for x', y'.

The area of the parallelogram P based on the origin and two lattice points (x_1, y_1), and (x_2, y_2), not collinear with O, is
$$\delta = \pm (x_1 y_2 - x_2 y_1).$$
We can construct a lattice L' (an oblique lattice) by producing and drawing parallels to the sides of P. A necessary and sufficient condition that L' should be equivalent to L, that

is, that they should contain the same lattice points, is that $\delta=1$, that is, that δ should have its smallest possible value. It is clear that this is also a necessary and sufficient condition that *there should be no lattice point inside P*, and it is easy to see that if there is such a point inside P, there is one inside, or on the boundary of, the triangular half of P nearer to O.

We may call the lattice point (q,p) which corresponds to a fraction p/q in its lowest terms a *visible* lattice point; there is no other lattice point which obscures the view of it from O. Let us consider all the visible lattice points which lie inside, or on the boundary of, the triangle bounded by the lines $y=0$, $x=n$, $y=x$. It is plain that these points correspond one by one to the fractions of the Farey series of order n. When the ray R from O to (q, p) rotates from the x-axis to the line $y=x$, it passes through each of these points in turn. If we take two consecutive positions of R, corresponding to the points (q, p), (q', p'), the parallelogram based on these two points contains no lattice point inside it, since otherwise there would be a lattice point inside its nearer triangle, and therefore a Farey fraction between p/q and p'/q'. It follows that

$$\delta = p'q - pq' = 1,$$

which proves Farey's theorem.

2. *Purpose of this Lecture.* So much then for Farey's discovery; it is a curious theorem, and its history is still more curious; but I have no doubt allowed myself to dwell upon it a little longer than its intrinsic importance deserves. My discussion of it will, however, help me to explain what I am trying to do in this lecture.

I shall imagine my audience to be made up entirely of men like Farey. I know that most of them are very much better mathematicians, but I shall not assume so; I shall assume only that they possess the common school knowledge of arithmetic and algebra. But I shall also assume that, like Farey, they are curious about the properties of integral numbers; one need after all be no Ramanujan for that.

Let us then imagine such a man playing about with numbers (as so many retired officers in England do) and puzzling himself about the curious properties which they seem to possess. What odd properties would strike him? What are the first questions he would ask? We must not try to be very systematic; if we do, we shall make no progress in an hour. We must aim merely at a rough preliminary survey of the ground. If in the course of our survey, we find the opportunity for any illuminating remark, we may delay to make it, as I have already delayed over Mr. Farey, even if it does not seem to fall in quite its proper logical place. Then, if time permits, we may return to examine a little more closely any important difficulties which our preliminary survey has revealed.

3. *Congruences to a Modulus.* There is no doubt that the first general idea which we should have to explain is that of a *congruence.* Two numbers a and b are *congruent to modulus* m if they leave the same remainder when divided by m, that is, if m is a divisor of $a-b$. We write

$$a \equiv b \pmod{m}, \quad m \mid a - b.$$

It is obvious that congruences are of immense practical importance. Ordinary life is governed by them; railway time tables and lists of lectures are tables of congruences. The absolute values of numbers are comparatively unimportant; we want to know what time it is, not how many minutes have passed since the creation.

A great many problems both of arithmetic and of common life depend upon the solution of congruences involving an unknown x, such as

$$a_0 x^n + a_1 x^{n-1} + \cdots + a_n \equiv 0 \pmod{m}.$$

Such congruences may be classified like algebraical equations, as linear, quadratic, $\cdots$, according to the value of n.

Our first instinct in dealing with congruences is to follow up the analogy with algebra. In algebra a linear equation has one root, a quadratic two, and so on. We find at once that there are obvious and striking contrasts; even the linear congruence suggests a whole series of problems, and a full discussion of quadratic congruences involves quite an imposing body of general ideas.

Let us take the simplest case, the linear congruence, and suppose first that we are concerned only with one particular modulus, such as 7 or 24. We have then an example of a genuinely finite mathematics. Congruent numbers have exactly the same properties and cannot be distinguished, and our mathematics contains only *a finite number of things*. In such a mathematics any problem can be solved by enumeration; we can solve $2x \equiv 5 \pmod{7}$ by trying all possible values of x, and we find there is a unique solution, $x \equiv 6$. If we try to solve $2x \equiv 5 \pmod{24}$, we find that there is no solution; if I lecture every other day, I shall sooner or later lecture on Thursday, but if I lecture every other hour, I may never lecture at 5 P.M.

The difference is of course accounted for by the fact that 7 is *prime* and 24 is not. Here we encounter the notion of a prime, a number without factors, and all kinds of speculations suggest themselves. Can we tell, by any method short of trial of all possible divisors, whether any given number is prime or not? Are there formulas for primes? Are the primes infinite in number, and if so, what is the law of their distribution?

Again, it appears that all numbers are composed of primes, that primes are the ultimate material out of which the world of numbers is built up. We are bound to ask *how;* and here we meet our first big theorem, the "fundamental theorem of arithmetic," the theorem that factorization is unique. But we shall probably be wise to allow our enquirer to take this theorem for granted until he has acquired a little of the sophistication which comes with wider knowledge.

We may observe, however, before passing on, that the contrast between arithmetic and algebra becomes much more marked as soon as we consider congruences of higher degree. An equation of the fourth degree has, with appropriate conventions, just four roots. But

$$x^4 \equiv 1 \qquad (\text{mod } 13)$$

has 4 roots, 1, 5, 8, and 12;

$$x^4 \equiv 1 \qquad (\text{mod } 16)$$

has 8 roots, 1, 3, 5, 7, 9, 11, 13, and 15; and

$$x^4 \equiv 2 \qquad (\text{mod } 16)$$

has none.

4. *Regarding Decimals.* I pass to another subject that has an irresistible fascination for amateurs, the subject of *decimals*. Some decimals are finite and some recurring, but it is easy to write down decimals, such as

(a)   0.10100100010 · · ·   (b) 0.11010001000 · · ·

which are neither. Here (a) the number of 0's increases by one at each stage, (b) the ranks of the 1's are 1, 2, 4, 8, · · · . More amusing examples are

(c)   0.01101010001010 · · ·

(in which the 1's have prime rank) and

(d)   0.23571113171923 · · ·

(formed by writing down the prime numbers in order). The proof for (c) demands the knowledge that there is an infinity of primes, and that for (d) rather more.*

The answer to some of the obvious questions is immediate. A finite decimal represents a rational fraction $p/(2^\alpha 5^\beta)$, a pure recurring decimal a fraction p/q, where q is not divisible by 2 or 5, and a mixed recurring decimal a fraction in which q is divisible by 2 or 5 and also by some other number. The converses of these theorems are also true, but the proof demands a little genuinely arithmetical reasoning. I shall state the proof in the simplest case, since it depends upon

---

* See Pólya and Szegö's *Aufgaben aus der Analysis*, vol. 2, pp. 160, 383.

the logical principle which is perhaps our most effective weapon in the elementary parts of the theory, where we are dealing with so simple a subject matter that our choice of arguments is naturally very restricted.

Suppose $p<q$ and q prime to 10. If we divide all powers 10^ν by q, there are only q possible remainders, and one at least must be repeated. It follows that there are a ν_1 and a $\nu_2 > \nu_1$ such that

$$10^{\nu_2} \equiv 10^{\nu_1}, \quad 10^{\nu_1}(10^{\nu_2-\nu_1} - 1) \equiv 0$$

to modulus q. It follows that, if we write $\nu_2 - \nu_1 = N$, we have $10^N \equiv 1$, so that $q \mid 10^N - 1$ and

$$\frac{p}{q} = \frac{P}{10^N - 1} = P \cdot 10^{-N} + P \cdot 10^{-2N} + \cdots.$$

Since $P < 10^N$, this is a pure recurring decimal with a period of at most N. The principles which we have used are (a) that *if there are more than q things of at most q kinds, there must be two of them of the same kind*; (b) that if $10^\nu Q$ is divisible by q, and q is prime to 10, then Q is divisible by q. In the second we are of course appealing to the "fundamental theorem". The first is the general logical principle to which I referred just now.

Let us take a slightly more complicated variant of this principle. *If there are two sets of objects*

$$a_1, a_2, \cdots, a_m, \quad b_1, b_2, \cdots, b_m,$$

no two of either set being the same; and if every b is equal to an a; then the b's are the a's arranged in a different order. We may apply this principle to obtain further information about the period of our recurring decimal. I suppose now that q is prime. If q and a are given, and a is not a multiple of q, it is impossible that

$$ra \equiv sa \qquad (\mathrm{mod}\ q)$$

unless $r \equiv s$. If (ra) is the remainder when ra is divided by q, the two sets

$$r, \quad (ra) \qquad (r = 1, 2, \cdots, q-1)$$

satisfy the conditions of our principle and are therefore the same except in order. It follows that

$$(q-1)!\, a^{q-1} \equiv \prod (ra) \equiv \prod r = (q-1)! \quad (\mathrm{mod}\ q),$$

and therefore that

$$a^{q-1} \equiv 1 \quad (\mathrm{mod}\ q)\ ;$$

Fermat's Theorem. In the particular case in which we are interested, a is 10, and Fermat's Theorem shows that we may take $N = q-1$, so that the period of p/q cannot exceed $q-1$ figures. Observe that we have appealed to the fundamental theorem twice in the proof.

It is familiar to everyone that $\frac{1}{7}$ has 6 figures, the maximum number. We are bound to ask what other primes q possess this property; the values of q less than 50 are in fact 7, 17, 19, 23, 29, and 47, but here we begin to get into deeper water. I cannot stop to discuss this question now, but before passing on I must mention another familiar text-book theorem which I shall have to quote later. This is Wilson's Theorem, that

$$(q-1)! + 1 \equiv 0 \quad (\mathrm{mod}\ q)$$

if and only if q is prime. Of the mass of proofs catalogued by Dickson, that of Dirichlet depends most directly on principles which we have used already. It is an immediate consequence of these principles that, if x is any one of the set $1, 2, \cdots, q-1$, there is just one other, y, such that $xy \equiv 1\ (\mathrm{mod}\ q)$; we call y the *associate* of x. It is plain that 1 and $q-1$ are associated with themselves; and no other number can be, since $x_1^2 \equiv x_2^2$ implies $x_1 \equiv x_2$ or $x_1 \equiv q-x_2$. It follows that the numbers $2, 3, \cdots, q-2$ are composed of $\frac{1}{2}(q-3)$ distinct pairs the product of each of which is congruent to 1. Hence

$$2 \cdot 3 \cdots (q-2) \equiv 1^{(q-3)/2} = 1,$$
$$(q-1)! \equiv q-1 \equiv -1,$$

which is one half of Wilson's Theorem. The converse half is practically obvious, since $(q-1)!$ would be divisible by any factor of q.

5. *Algebraic and Transcendental Numbers*. The study of decimals leads directly to problems concerning *rationality and irrationality*. Our decimals such as $0.1010010001\cdots$ must represent irrational numbers. What criteria are there for deciding whether a given number is rational or irrational? To ask this question is to go a little outside the theory of numbers proper, which is concerned first with integers, and then with rationals or irrationals of special forms, such as the form $a+b\sqrt{2}$, and not with irrationals as a whole or general criteria for irrationality. The problem is, however, one about which an amateur will certainly demand information.

The famous argument of Pythagoras shows that $\sqrt{2}$ is irrational; if a/b is in its lowest terms and $a^2=2b^2$, then a and b must both be even, a contradiction. It is obvious to us now that the Pythagorean argument extends at once to $\sqrt{3}$, $\sqrt{5}$, $\cdots$, $2^{1/3}$, $\cdots$, and generally to $N^{1/m}$, where N is any number which is not a perfect mth power. There is a curious and very instructive historical puzzle connected with this argument. There is a passage in Plato's *Theaetetus*, discussed at length by Heath in his *History of Greek Mathematics*, about the attempt of Theodorus to generalize Pythagoras's proof. Theodorus, working some 50 years after Pythagoras, proved the irrationality of $\sqrt{N}$ for all values of N (except square values) up to 17 inclusive. Why, ask the historians did he stop? Why in any case should it have taken mathematicians like the Greeks 50 years to make so obvious an extension? Zeuthen in particular expended a great deal of ingenuity upon this question, but I think that the ingenuity was misplaced, and that the answer is obvious.

Theodorus *did not know the fundamental theorem of arithmetic;* there is something of a puzzle about the history of that theorem, but it cannot have been known to the Greeks before Euclid's time. The triviality of the generalization to us is due entirely to our knowledge of this theorem. Suppose, for example, we wish to prove that

$$a^2 = 60b^2,$$

where a and b are integers without common factor, is impossible. We argue that a^2 cannot be divisible by 3 unless a is divisible by 3; hence $a = 3c$, $a^2 = 9c^2$, $3c^2 = 20b^2$, and a repetition of the argument shows that b also is divisible by 3. We can prove that $3 \mid a^2$ implies $3 \mid a$ *without the fundamental theorem*, by enumeration of possible cases, considering separately the cases in which $a \equiv 0, 1, 2 \pmod 3$. If it were 17 instead of 3, the process would be a little tedious; and in any case such a classification of numbers would have been very novel in Theodorus's time. I am so far from being puzzled by the limitations of his work that I regard what he did as a very remarkable achievement.

There are very few types of numbers which present themselves at all naturally in analysis and which can be proved to be irrational. It is obvious that a number like $\log_{10} 2$ is irrational, for a power of 2 cannot be a power of 10. The proof for e, from the exponential series, is quite easy, and that for e^2 not very much more difficult. That for π is decidedly more so, and when we come to numbers like e^3 and π^2, it ceases to be worth while to worry about elementary proofs; we may as well go the whole way and prove e and π are transcendental. The most famous constant in analysis, after e and π, is Euler's constant γ; and the proof of the irrationality of γ is one of the classical unsolved problems of mathematics. It has never been proved that $2^{\sqrt{2}}$, $3^{\sqrt{2}}$, and similar numbers are irrational; no plausible method for attacking such problems has even been suggested. I am inclined to think that the number which holds out the best hopes for new discovery is the number e^{π}, which presents itself so naturally in the formulas of elliptic functions.

I said just now that e and π were "transcendental". I must not stop to talk at length about this famous theorem of Lindemann,* which contains the final proof that the

---

* See for example Hobson's *Trigonometry*, third edition, p. 305, or the same author's *Squaring the Circle*.

quadrature of the circle, in the classical sense, is impossible; but the *statement* of the theorem introduces a notion that we shall require, that of an *algebraic number*. An algebraic number is the root of an equation

$$a_0 x^n + a_1 x^{n-1} + \cdots + a_n = 0,$$

where the a's are integers. An *algebraic integer* is an algebraic number whose characteristic equation has unity for its leading coefficient. Thus $\sqrt{2}$ and $1+\sqrt{(-5)}$ are algebraic integers. A *transcendental* number is a number which is not algebraic; and Lindemann's Theorem is that π *is transcendental*. It is easy to show that all lengths which can be constructed by euclidean methods are algebraic, and indeed algebraic numbers of a quite special kind. If follows that the quadrature of the circle by any euclidean construction is impossible.

There is another direction in which we may be tempted to digress at this point, the theory of the approximation of irrationals by rationals, what is now called "diophantine approximation". There is just one theorem in this field that I shall mention, because it is connected so directly with what I have just been saying, and because it depends upon another of the stock arguments of number theory, the principle that *an integer numerically less than* 1 *is* 0. This is Liouville's theorem, that *there are transcendental numbers*. It is naturally much easier to prove this than to prove that a given number such as π is transcendental.

Liouville proves first that *it is impossible to approximate rationally to an algebraic number with more than a certain accuracy*. It is quite easy to see why. Suppose that ξ is an algebraic number defined by

$$f(\xi) = a_0 \xi^n + a_1 \xi^{n-1} + \cdots + a_n = 0.$$

We may suppose that the equation is irreducible, that is to say that $f(\xi)$ cannot be resolved into simpler algebraic factors of similar form; in this case we say that ξ is *of degree n*. We can obviously find a number M, depending only on ξ,

such that
$$|f'(x)| < M$$
for x near ξ. Suppose now that p/q is a rational, near ξ. Then
$$f\left(\frac{p}{q}\right) = \frac{N}{q^n},$$
where N is an integer not zero. It follows from our general principle that $|N| \geq 1$ and
$$\left|f\left(\frac{p}{q}\right)\right| \geq \frac{1}{q^n}.$$
But
$$f\left(\frac{p}{q}\right) = f\left(\frac{p}{q}\right) - f(\xi) = \left(\frac{p}{q} - \xi\right)f'(\eta),$$
where η lies between p/q and ξ. Hence, for all q,
$$\left|\frac{p}{q} - \xi\right| = \left|\frac{f(p/q)}{f'(\eta)}\right| > \frac{1}{Mq^n}.$$

It is impossible to approximate rationally to an algebraic number of degree n with an order of accuracy higher than q^{-n}.

On the other hand it is easy to write down numbers which have rational approximations of much higher accuracy than this; we have only to take a decimal of 0's and 1's in which the 1's are spaced out sufficiently widely. Thus
$$\xi = \frac{1}{10^{1!}} + \frac{1}{10^{2!}} + \frac{1}{10^{3!}} + \cdots = .11000100000 \cdots$$
is approximated by its first k terms, that is, by a fraction
$$\frac{p}{q} = \frac{p}{10^{k!}}$$
with an error of order $10^{-(k+1)!} = q^{-k-1}$. Hence it is not an algebraic number of degree k and, since k is arbitrary, it must be transcendental. Obviously Liouville's argument enables us to *construct* transcendental numbers as freely as we please.

6. *Arithmetic. Forms.* The theory of irrationals starts from Pythagoras, and there is another great branch of the theory of numbers which also starts from him and about which I must now say something. This is the theory of *forms*.

Our interest in the theory of forms begins when we observe that there are Pythagorean triangles with integral sides; thus $3^2+4^2=5^2$. The first problem which suggests itself is that of determining all such triangles, and the solution given in substance by Diophantus, is easy. All the integral solutions of
$$x^2 + y^2 = z^2$$
are given by
$$x = \lambda(\xi^2 - \eta^2), \qquad y = 2\lambda\xi\eta, \qquad z = \lambda(\xi^2 + \eta^2),$$
where the letters are integers and ξ and η are coprime and of opposite parity. This problem is trivial, but it suggests an infinity of others.

It is natural to begin by a generalization of the problem. Let us discard the hypothesis that the hypotenuse z is integral; then
$$n = z^2 = x^2 + y^2$$
is the sum of two squares, and we are led to ask what numbers n possess this property. This is the first and simplest problem in the theory of *quadratic forms*, and the answer to it shows that no such problem can be quite easy. Even linear forms are not quite trivial; the solution of $ax+by=n$ in integers is a quite interesting elementary problem. When we consider quadratic forms, we come up against difficulties of a different order.

The first theorem in the subject is another theorem of Fermat, that $x^2+y^2=n$ *is soluble when n is a prime* $p=4m+1$ and, apart from trivial variations of the sign and order of x and y, uniquely. It is to be observed that the equation is plainly insoluble when n is $4m+3$, since any square is congruent to 0 or 1 to modulus 4. This theorem is one of the most famous in the theory of numbers, and very rightly so, since it was the first really difficult theorem in the subject proved by any mathematician. There is no really simple proof, and the most natural, that which depends on the Gaussian numbers $a+bi$, introduces a whole series of ideas of revolutionary importance.

The first stage of the proof consists in proving that *there is a number x such that*

$$x^2 \equiv -1 \qquad (\bmod\ p),$$

or $p\,|\,1+x^2$. Let us go back for a moment to the proof I sketched of Wilson's Theorem. Let us associate the numbers $x=1, 2, \cdots, p-1$ in pairs x, y not, as then, so that $xy\equiv 1$, but so that

$$xy \equiv -1 \qquad (\bmod\ p).$$

If any x is associated with itself, our proposition is established. If not, we have arranged the numbers from 1 to $p-1$ in $\frac{1}{2}(p-1)$ pairs of different numbers each satisfying the condition. Hence

$$(p-1)! \equiv \prod xy \equiv (-1)^{(p-1)/2} = 1\ ;$$

which is false, since, by Wilson's Theorem,

$$(p-1)! \equiv -1.$$

We thus obtain our proposition by reductio ad absurdum.

The second stage of the proof depends on much more novel ideas. We are concerned with the simplest case of an *algebraic field*. The field $K(i)$ is the aggregate of numbers

$$\xi = r + si = r + s\sqrt{(-1)},$$

where r and s are rational. This number satisfies the equation

$$\xi^2 - 2r\xi + r^2 + s^2 = 0,$$

and is an algebraic integer, in the sense I defined before, when $2r$ and r^2+s^2 are integers, that is, when r and s are integers. We may denote by $K^*(i)$ the aggregate of all the integers

$$\alpha = a + bi$$

of $K(i)$; a and b are ordinary integers. The numbers of $K^*(i)$ reproduce themselves by addition and multiplication, and we can define division in this field just as we define it in ordinary arithmetic. We can also define a *prime* of $K^*(i)$, and factorization of numbers into primes. There are four numbers, ± 1 and $\pm i$, which play a part in the new arithmetic similar to that of 1 and -1 in ordinary arithmetic. These are the "unities" or divisors of 1. If we define the *norm* of $\alpha = a+bi$ as

$$N(\alpha) = a^2 + b^2,$$

then the unities are characterized by the fact that their norm is 1. We do not count them as primes, just as, in the ordinary theory, we do not count 1 as a prime.

We now make an assumption, namely that *the analog of the fundamental theorem holds in the field $K^*(i)$*, that is to say that, apart from any trivial complications which may be introduced by the unities, *the factorization of a number of $K^*(i)$ into primes is unique*. This assumption is in fact correct. Returning now to the first stage of our proof, there is an x such that

$$p \mid 1 + x^2 = (1 + ix)(1 - ix).$$

It is obvious that p does not divide $1+ix$ or $1-ix$, so that *p divides the product of two numbers without dividing either of them*. Hence p cannot be a prime in $K^*(i)$. We may therefore write

$$p = \pi\lambda,$$

where $N(\pi) > 1$ and $N(\lambda) > 1$. But

$$N(\pi)N(\lambda) = N(p) = p^2,$$

so that $N(\pi)$ and $N(\lambda)$ must each be p. If we write

$$\pi = a + ib,$$

it follows that

$$p = N(\pi) = a^2 + b^2,$$

which is Fermat's theorem.

We may be tempted by our success to further efforts in the same direction. It is easy to satisfy ourselves, by considering particular cases, that *any prime $p = 20m+1$ is of the form a^2+5b^2*: thus $61 = 4^2+5 \cdot 3^2$. Let us try to prove this theorem by a similar method. We must evidently consider now the field $K^*[\sqrt{(-5)}]$ formed of the algebraic integers of the form

$$\alpha = a + b\sqrt{(-5)} \; ;$$

it is easy to show that such a number is an algebraic integer if and only if a and b are ordinary integers. There is no difficulty in defining divisibility and primality in this field also.

The first step in our proof must plainly be to prove the existence of an x for which $p \mid 1+5x^2$. This is not difficult, but it demands a little more knowledge of quadratic congruences than I can assume, and I must take it for granted.

We define the norm $N(\alpha)$ of a number of this field as a^2+5b^2. We then argue as before; we have

$$p \mid 1 + 5x^2 = (1 + x\sqrt{(-5)})(1 - x\sqrt{(-5)}),$$

so that p divides a product without dividing either factor and is therefore not a prime. Hence, as before $p = \pi\lambda$, where $N(\pi) > 1$ and $N(\lambda) > 1$, and $N(\pi)$ and $N(\lambda)$ must each be p. It follows that

$$p = N(\pi) = a^2 + 5b^2,$$

the theorem we set out to prove.

At this point, however, there is a shock in store for us; we find that we can prove *too much*. The number

$$q = (2 + \sqrt{(-5)})(2 - \sqrt{(-5)})$$

is divisible by 3, while neither factor is so. Hence 3 is not a prime. Hence

$$3 = \pi\lambda, \quad 9 = N(\pi)N(\lambda),$$

and $N(\pi)$ and $N(\lambda)$ are each 3. It follows that

$$3 = N(\pi) = a^2 + 5b^2.$$

Similarly we can prove that
$$7 = a_2 + 5b_2;$$
and both of these theorems are obviously false.

There must therefore be a mistake somewhere in our argument, and if you examine it, and are prepared to believe that I have not been misleading you wilfully, you will see that there is only one step which can be questioned. In all three cases I concluded the argument by an appeal to the same theorem; *a number which divides the product of two numbers without dividing either of them cannot be prime.* This is true in ordinary arithmetic, because of the fundamental theorem; if 7 were a divisor of $15 = 3 \cdot 5$, 15 would be factorable into primes in two distinct manners. It follows that *the analog of the fundamental theorem in the field $K^*[\sqrt{(-5)}]$ must be false*; and this is easily verified when once our suspicions have been excited; thus

$$2 \cdot 3 = (1 + \sqrt{(-5)})(1 - \sqrt{(-5)}),$$
$$3 \cdot 7 = (1 + 2\sqrt{(-5)})(1 - 2\sqrt{(-5)}),$$

and all of these numbers are prime in $K^*[\sqrt{(-5)}]$. The proof which I gave of the theorem concerning primes $20m+1$ was therefore fallacious, although the theorem is true. The proof of Fermat's theorem, on the other hand, was correct, since factorization *is* unique in $K^*(i)$.

7. *Further Problems.* It is clear that we must go back to the beginning and study the theory of primes a little more closely; but before I do this I should like to call your attention to a series of further problems suggested by Fermat's theorem. We know now when a *prime* is the sum of two squares, and we have to consider the same problem for general n. Here in fact there are three different problems.

The first and most obvious problem is that of determining the necessary and sufficient conditions that n should be representable. This problem may be solved quite easily with the aid of the Gaussian numbers; n must be $2^\alpha M^2 N$, where α is 0 or 1 and N contains prime factors of the form

$4m+1$ only. We are then led naturally to the corresponding problem for other forms, first for the general binary quadratic form

$$ax^2 + bxy + cy^2,$$

then for quadratic forms in a larger number of variables, such as

$$x^2 + y^2 + z^2, \quad x^2 + y^2 + z^2 + t^2,$$

and then for forms of higher degree, such as x^3+y^3 and x^4+y^4. There is a highly developed theory of the general quadratic form; the most famous theorem is perhaps Lagrange's theorem, that *every number is the sum of four squares*. But as soon as we begin to consider cubic or higher forms we find ourselves on the boundary of knowedge. There is for example no criterion analogous to Fermat's by which we can decide whether a given number is the sum of two cubes.

The second problem about the form x^2+y^2 suggested by Fermat's theorem is that of determining the *number of representations*. This problem may be interpreted in two different ways. We may want an exact formula, in terms of the factors of n, and in this case the Gaussian theory again gives what we want; $r(n)$, the number of representations, is given by the formula

$$r(n) = 4\{d_1(n) - d_3(n)\},$$

where $d_1(n)$ and $d_3(n)$ are the numbers of divisors of n of the forms $4m+1$ and $4m+3$ respectively. This is, however, not the most interesting interpretation of the problem. We may want, not a formula like this, but information concerning the *order of magnitude* of $r(n)$, whether $r(n)$ is generally large when n is large, whether numbers are usually representable freely or with difficulty. In this case our formula gives us very little help, and the solution of the problem requires quite different methods.

It is here that we come into contact for the first time with a new branch of the theory, the modern "analytic" theory.

This theory has two special characteristics. The first is one of method; it uses, besides the methods of the classical theory, the methods of the modern theory of functions of a complex variable. The second is that it is concerned primarily with problems of order of magnitude and asymptotic distribution. The distinction is not a perfectly sharp one; there are "exact", "finite" theorems which have only been solved by "analytic" methods. For example, *every number greater than* $10^{10^{10}}$ *is expressible as the sum of* 8 *cubes*; this theorem includes no reference to "order of magnitude", and is a "finite" theorem in just the same sense as Fermat's theorem about the squares, but the only known proof is analytic. On the whole, however, it is the problems of asymptotic distribution which dominate the theory.

The answer given by the analytic theory to the special question which I raised is roughly as follows. The average value of $r(n)$ is π. It must be observed that representations which differ only trivially, that is, in the sign or order of x and y, are reckoned as distinct. If we allow for this, the average number of representations is rather less than a half; this is explained by the fact that, as we shall see, *most* numbers are not representable. On the other hand $r(n)$ tends to infinity with n with tolerable rapidity for numbers of appropriate forms, more rapidly for example than any power of $\log n$. The corresponding problems for cubes or higher powers present difficulties which are at present quite insuperable, and all that I can do is to mention a few curiosities. The smallest number representable by two cubes in two really distinct ways is

$$1729 = 1^3 + 12^3 = 9^3 + 10^3,$$

and the smallest representable in three ways is probably

$$175959000 = 70^3 + 560^3 = 198^3 + 552^3 = 315^3 + 525^3.$$

It can be proved that there exist numbers with as many different representations as we please. A. E. Western has carried out very heavy computations concerning representations by cubes; he has for example found 6 numbers, of which the smallest is 1,259,712, representable as the sum of *three* cubes in *six* different ways. The smallest number doubly representable by two fourth powers is probably

$$635318657 = 59^4 + 158^4 = 133^4 + 134^4 ;$$

there is, so far as I know, no known example of a number with three such representations, nor any proof that such a number exists.

The nature of the problems of the analytic theory becomes clearer when we consider the third problem suggested by Fermat's theorem. This is the problem of determining the *distribution* of the representable numbers. We want to know *how many numbers are representable*, or, to put it more precisely, how many numbers less than a large assigned number x are representable. If $Q(x)$ is the number of such numbers, what is the order of magnitude of $Q(x)$? Are nearly all numbers representable, or just a majority, or only a few? The answer is in fact that $Q(x)$ is approximately

$$\frac{Ax}{(\log x)^{1/2}},$$

where A is a constant; to put it roughly, quite a lot of numbers are representable, but strictly an infinitesimal proportion of the whole. This explains why the average number of representations turned out to be less than one.

This problem about $Q(x)$ is a very interesting one, but there is another of the same kind which is obviously still more interesting and much more fundamental. This is the problem of the distribution of the primes themselves; *how many primes are there less than x?* I shall say something about this problem in a moment; it is in any case time for us to return to the theory of primes, since all our enquiries have ended in questions about them, and it is obviously impossible to make serious progress until we know more both of their elementary properties and of the laws which govern their distribution.

Part II

8. *The Fundamental Theorem.* The *fundamental theorem* of arithmetic is the beginning of the theory of numbers, and it is plain that our first task must be to make this theorem secure.

There is another historical puzzle about the fundamental theorem. Who first stated the theorem, explicitly and generally? The natural answer is *Euclid*, since the *Elements* contain all the materials for the proof. Everything rests on Euclid's famous algorithm for the greatest common divisor. Given two numbers a, b, of which a is the greater, we form the table

$$a = bc + b_1, \quad b = b_1c_1 + b_2, \quad b_1 = b_2c_2 + b_3, \cdots$$

where $b_1, b_2, \cdots$, are the remainders in the ordinary sense of elementary arithmetic. Since

$$b > b_1 > b_2 > \cdots,$$

b_n must sooner or later be zero. The last positive remainder δ has the properties implied by the words *greatest common divisor*, and it follows from the process by which δ is formed that any number which divides both a and b divides δ.

Let us note in passing that there is an analogous process in $K^*(i)$, but that the analogy fails in $K^*[\sqrt{(-5)}]$. In ordinary arithmetic, given a and b, we can find a number congruent to a mod b and less than b. There is a similar theorem for the Gaussian numbers. Here there is no strict order of magnitude between different numbers, and we have to use the order of magnitude of their norms. Given α and β, there is a number, congruent to α mod β, whose norm is less than that of β. There is no such theorem in $K^*[\sqrt{(-5)}]$, and the process analogous to Euclid's fails.

When the existence of δ is once established, the proof of the fundamental theorem is easy. We write

$$\delta = (a, b)$$

and we say that a is prime to b when $(a, b) = 1$. The crucial lemma is that *if $(a, b) = \lambda$ and $b | ac$, then $b | c$*; in particular, *a prime cannot divide a product without dividing one or other of* *the factors.* This once granted, anybody can construct the proof of the fundamental theorem for himself; and you will remember that it was just this proposition which led to our troubles in $K^*[\sqrt{(-5)}]$.

The lemma itself may be proved as follows. We construct the euclidean algorithm for a and b, with the final remainder 1. If we multiply it throughout by c, we have the algorithm for ac and bc, and the final remainder is c. It follows that

$$(ac, bc) = c.$$

Since b divides ac, by hypothesis, and also bc, it divides c.

This is Euclid's own argument, and with it he had proved what is essential in the fundamental theorem. It is a very singular thing that he should then omit to state the magnificent theorem that he has proved. He is over the line and free, but apparently disdains the formality of touching down. I do not know of any formal statement of the theorem earlier than Gauss. The substance of the theorem, however, is in the *Elements*; it was plainly unknown, as I explained before, to the Greeks from 50 to 100 years before Euclid's time; and I see no particular reason for questioning the obvious view that it is Euclid's own.

As soon as we have proved the fundamental theorem our elementary knowledge falls into line. The theory of linear congruences, the theorems of Fermat and Wilson and all their consequences, the elementary theory of decimals and of the divisors of numbers, may be developed straightforwardly and without the introduction of essentially new ideas. I can now say something about the more modern side of the theory of primes.

9. *Problems Concerning Primes.* What are the most natural questions to ask about primes? I say deliberately the most *natural*; we must remember that a natural question does not always seem, on fuller reflection, to have been a *reasonable* one. It is natural to an engineer to ask us for a finite formula for

$$\int e^{-x^2} dx,$$

or for a solution of some simple looking differential equation in finite terms. If we fail to satisfy him, it is not because of our stupidity, but because the world does not happen to have been made that way.

So, if any one asks us (1) *to give a general formula for the nth prime p_n*, a formula in the sense in which

$$p_n = n^2, \quad p_n = n^2 + 1, \quad p_n = [e^n],$$

where $[x]$ denotes the integral part of x, would be a formula, I can only reply that it is not a reasonable question. It is, I will not say demonstrably impossible, but wildly improbable, that any such formula exists. The distribution of the primes is not like what it would have to be on any such hypothesis. I should make the same reply to a good many other questions which an amateur might be likely to ask, for example if he asked me (2) *to give a rule for finding the prime which immediately follows a given prime*. It would of course be perfectly reasonable that he should press me for the reasons why I gave so purely a negative a reply. On the other hand the problem (3) *to find the number of primes below a given limit* is, if interpreted properly, an entirely reasonable and a soluble problem. The problems (4) *to prove that there are infinitely many pairs of primes differing by 2*, and (5) *to prove that there are infinitely many primes of the form n^2+1*, are also entirely reasonable, and if (as is the case) we cannot solve them, it is quite reasonable to condemn our lack of ingenuity.

10. *The Distribution of Primes.* If we wish to classify these problems and to decide which of them are reasonable and which are not, the first essential is to understand broadly the present state of knowledge about the distribution, the distribution *in the large* or *asymptotic* distribution, of the primes. It is this theory which gives the solution of problem (3).

We denote by $\pi(x)$ the number of primes not exceeding

x. The first step is to prove that (a) *the number of primes is infinite*; $\pi(x)$ *tends to infinity with* x. This is another of Euclid's great contributions to knowledge, and Euclid's proof is perhaps the classical example of proof by reductio ad absurdum. If the theorem is false, we may denote the primes by $2, 3, 5, \cdots, P$, and all numbers are divisible by one of these. On the other hand the number

$$(2 \cdot 3 \cdot 5 \cdots P) + 1$$

is obviously not divisible by any of $2, 3, \cdots, P$, and this is a contradiction.

Another very interesting proof is due to Pólya.* It is easy to see that any two of the numbers

$$2 + 1, 2^2 + 1, 2^4 + 1, \cdots, u_n = 2^{2^n} + 1$$

are prime to each other. For suppose that p is an odd prime and that $p | u_n, p | u_{n+k}$. Then also

$$p | 2^{2^{n+k}} - 1 = u_{n+k} - 2,$$

since

$$x^{2^k m} - 1$$

is algebraically divisible by $x^m + 1$, and therefore

$$p | u_{n+k} - (u_{n+k} - 2) = 2,$$

which is absurd. It follows that the number of primes less than u_n is at least n, and therefore that the number of primes is infinite. In fact the argument shows not merely that $\pi(x) \to \infty$ but that

$$\pi(x) > A \log \log x,$$

where A is a constant. Something in this direction, though a little less, can be proved by a refinement of Euclid's argument.

There is a third line of argument which is a little less elementary but may be made to prove a good deal more.†

---

* See Pólya and Szegö, loc. cit., pp. 133, 342.

† See Dickson's *History*, vol. 1, p. 414, where the proof is attributed to Auric.

If 2, 3, 5, $\cdots$, P were the only primes, then every number would be of the form

$$2^a 3^b 5^c \cdots P^k.$$

If this number is less than x, then a fortiori 2^a is less than x, so that a is less than a constant multiple of $\log x$, and the same argument applies to $b, c, \cdots, k$. The number of possible choices of $a, b, \cdots, k$ is therefore less than a multiple of $(\log x)^\pi$, where π is the total number of primes. In other words the number of numbers less than x is less than

$$A(\log x)^\pi,$$

where A is a constant, and this is impossible, since x tends to infinity more rapidly than any power of $\log x$. A refinement of the argument leads to the inequality

$$\pi(x) > A \frac{\log x}{\log \log x};$$

and the underlying principle may be stated roughly thus, that *if the number of primes were finite, there would not be enough numbers to go round.*

We are still a very long way from the ultimate truth. It is in fact possible to prove, and by comparatively elementary methods, that *the order of magnitude of $\pi(x)$ is $x(\log x)^{-1}$*. This theorem, conjectured by Legendre and Gauss, was first proved by Tchebycheff in 1848.

There are two much earlier theorems of Euler which point in this direction. The first is the theorem that (b) *the series*

$$\sum \frac{1}{p}$$

extended over all prime numbers p, is divergent. The proof of this theorem depends upon an identity, also due to Euler, upon which the whole of the modern theory of primes is founded. The identity is

$$1^{-s} + 2^{-s} + 3^{-s} + \cdots = \sum n^{-s}$$
$$= \frac{1}{(1-2^{-s})(1-3^{-s})(1-5^{-s})\cdots}$$
$$= \prod\left(\frac{1}{1-p^{-s}}\right)$$

and is valid for $s > 1$; it is at bottom merely the analytical expression of the fundamental theorem, and its importance arises from the fact that it asserts the equivalence of two expressions of which one contains the primes explicitly while the other does not. From Euler's identity we deduce (b) roughly as follows: if $\sum p^{-1}$ were convergent, then

$$\prod\left(\frac{1}{1-p^{-1}}\right)$$

would be convergent, and therfore $\sum n^{-1}$ would be convergent, which is false. Of course the proof really needs a rather more careful statement.

Euler's second theorem is (c) *the quotient of $\pi(x)$ by x tends to zero*; or in symbols

$$\frac{\pi(x)}{x} \to 0,$$

or, as we write it now

$$\pi(x) = o(x).$$

The proportion of primes is ultimately infinitesimal, "almost all" *numbers are composite*. The theorem is a quite simple corollary of (b); roughly, if we remove from the numbers less than x all multiples of the primes $2, 3, \cdots, p$, other than these primes themselves, we are left something like

$$x\left(1-\frac{1}{2}\right)\left(1-\frac{1}{3}\right)\left(1-\frac{1}{5}\right)\cdots\left(1-\frac{1}{p}\right)$$

numbers. The product multiplying x tends to zero when

$p\to\infty$, because of (b), and from this we can deduce Euler's second theorem.

It is rather curious that, although Euler's second theorem is a corollary of the first, the lessons which we learn from the two theorems concerning the distribution of the primes have exactly opposite tendencies. The second theorem tells us that the number of primes below a given limit is *not too great*, that the primes are in the end rather liberally spaced out; it is in fact exactly equivalent to the theorem that (d) *the nth prime p_n has an order of magnitude greater than n*, or

$$\frac{p_n}{n} \to \infty.$$

If on the other hand the order of magnitude of p_n were *much* greater than n, if it were for example n^2 or $n^{10/9}$ or $n(\log n)^2$, then the series $\sum p_n^{-1}$ would be *convergent*, which is just what Euler's first theorem denies. What we learn from the two theorems together is something like this. If, as we hope, the true order of magnitude of p_n can be measured by some simple function $\phi(n)$, then that function must be of order higher than n, but somewhere near the boundary of convergence of the series

$$\sum \frac{1}{\phi(n)}.$$

The most obvious function which satisfies these requirements is $n \log n$, and to say that p_n is of order $n \log n$ is the same thing as to say that $\pi(x)$ is of order $x(\log x)^{-1}$. This is just what is asserted by Tchebycheff's theorem.

11. *Tchebycheff's Theorem.* The formal statement of Tchebycheff's theorem is (e) *the order of magnitude of $\pi(x)$ is $x(\log x)^{-1}$; there are constants A and B such that*

$$\frac{Ax}{\log x} < \pi(x) < \frac{Bx}{\log x}.$$

This theorem is precisely equivalent to (f) *the order of magnitude of p_n is $n \log n$; there are constants A and B such that*

$$An \log n < p_n < Bn \log n.$$

The proofs of these theorems given by Tchebycheff have been simplified a good deal by Landau, and I can give you a sketch of one half of the proof which should enable you to understand without much difficulty the general character of the whole.

We begin by replacing $\pi(x)$ by another function. We can write $\pi(x)$ in the form

$$\pi(x) = \sum_{p \leq x} 1 \ ;$$

count one for every prime up to x. A more convenient and really a more natural function is

$$\theta(x) = \sum_{p \leq x} \log p,$$

the logarithm of the product of all primes up to x. This function seems at first sight a more complicated function, but it is easy enough to see why it is more convenient to work with. The most natural operation to perform on primes is *multiplication*, and this is the operation which we employ in forming $\theta(x)$. It is because it is natural to multiply primes and not to add or subtract them that problems like the problem of the prime pairs $(p, p+2)$, or Goldbach's problem of expressing numbers as sums of primes, turn out to be so terribly difficult.

Since $x/x^{1-\delta}$ tends to infinity, for any positive value of δ, we may expect that nearly all the primes which contribute to $\theta(x)$ will lie in the interval $(x^{1-\delta}, x)$, so that their logarithms lie between $(1-\delta) \log x$ and $\log x$. Hence we may expect $\theta(x)$ to be very much the same function as $\pi(x) \log x$, and in fact there is no difficulty in proving that

$$\theta(x) \sim \pi(x) \log x,$$

that is, that the ratio of the two functions tends to 1. It follows that the inequalities in (e) are equivalent to

$$Ax < \theta(x) < Bx.$$

I shall sketch the proof of the second inequality, which is rather the simpler.

Suppose that x is a power of 2, say 2^m. The primes between $x/2$ and x divide $x!$ but not $(x/2)!$, so that

$$\prod_{x/2 < p \leq x} p \;\Big|\; \frac{x!}{(x/2)!(x/2)!}.$$

The expression on the right is *one term* in the binomial expansion of $(1+1)^x = 2^x$, and therefore

$$\prod_{x/2 < p \leq x} p \leq 2^x.$$

Replacing x by

$$x/2, \; x/4, \; x/8, \cdots$$

and multiplying the results, we find that

$$\prod_{p \leq x} p \leq 2^{x+x/2+x/4+\cdots} \leq 2^{2x},$$

and

$$\theta(x) \leq 2 \log 2 \cdot x.$$

This proves the theorem when $x = 2^m$. If

$$2^m < x < 2^{m+1}$$

we have

$$\theta(x) \leq \theta(2^{m+1}) \leq 4 \log 2 \cdot 2^m < 4 \log 2 \cdot x.$$

Hence we may take $B = 4 \log 2$. The proof of the second inequality is, as I said, not quite so simple, but does not involve essentially more difficult ideas. We have thus determined the order of magnitude of $\pi(x)$ and of p_n, and it is perhaps a little astonishing that a problem which sounds so abstruse should have so comparatively simple a solution.

12. *The Prime Number Theorem.* Tchebycheff's solution of the problem is, however, one with which it is impossible to remain content for long, since the whole trend of our discussion has been to suggest that much more is true than we have proved. In fact Tchebycheff's work, fine as it is, is the record of a failure; it is what survives of an unsuccessful attempt to prove what is now called the *Prime Number Theorem*.

This is the theorem that (g) *$\pi(x)$ and $x(\log x)^{-1}$ are asymptotically equivalent; the ratio of the two functions tends to unity.* We express this by writing

$$\pi(x) \sim \frac{x}{\log x}.$$

The Prime Number Theorem is equivalent to

$$p_n \sim n \log n,$$

and we may express it very roughly by saying that *the odds are* $\log x$ *to* 1 *that a large number x is not prime.*

The Prime Number Theorem, the central theorem of the analytic theory of numbers, was proved independently by Hadamard and by de la Vallée-Poussin in 1896. The empirical evidence for its truth had for long been overwhelming, and I suppose that every number-theorist since Legendre had tried to prove it. The theorem differs from all those which I have discussed so far in that it is apparently impossible to prove it by properly elementary methods; there is no proof known which does not depend essentially on complex function theory. I do not mean to imply that there is any terrible difficulty in the proof; there are considerable difficulties of detail, but the fundamental ideas on which it depends are tolerably straightforward. They are, however, quite unlike any of those of which I have spoken, and I should require a whole lecture to explain them even to a strictly mathematical audience. Actually, a good deal more is known; it can be proved that $\pi(x)$ is approximated still more closely by the "logarithm-integral" of x,

$$\operatorname{Li} x = \int_2^x \frac{dt}{\log t},$$

that in fact

$$\pi(x) = \operatorname{Li} x + O\left\{\frac{x}{(\log x)^k}\right\}$$

for every k, the error being of lower order than the quotient of x by *any* power of $\log x$; and it is probable, though not yet proved, that the order of the error does not very materially exceed that of $\sqrt{x}$.

13. *Formulas for Primes.* I return now for a moment to a question which I discussed shortly before, the question whether it was reasonable to expect an "elementary formula" for the nth prime p_n. Let us imagine that my questioner was obstinate in his desire for such a formula; how could I refute his successive suggestions? If he suggested

$$p_n = n \log n,$$

I should have the obvious reply that $n \log n$ is not an integer. Suppose then that he modified his formula to

$$p_n = [n \log n].$$

I should reply that his formula did not agree with the known facts of the asymptotic theory. It agrees with $p_n \sim n \log n$, the first and most obvious deduction from the Prime Number Theorem itself; but the theory carries us much further, it enables us, for example, to show that

$$p_n = n \log n + n \log \log n + O(n),$$

which contradicts the formula. If, becoming more cautious, he asked me what ground I had for denying that p_n might be *some* elementary combination of

$$n, \log n, \log \log n, \cdots,$$

I should naturally find it harder to refute him, but I could advance three arguments which are enough in the aggregate to make up a tolerably convincing case. (i) Since $\text{Li}\,x$ is a very good approximation to $\pi(x)$, the inverse function $\text{Li}^{-1}n$ must be a very good approximation to p_n. Now it is demonstrable that neither the logarithm integral nor its inverse* is an elementary function. It is therefore very unlikely that there should be an elementary formula for p_n. (ii) If the "elementary formula" does not involve the symbol $[\cdots]$ of the "integral part", the function which it defines will generally not be integral for integral n. If it does, it loses all its simplicity and all its plausibility. (iii) An elementary function may be expected to behave with tolerable regularity at infinity, and so may all its *differences*. Now extremely little is known about the difference $p_{n+1}-p_n$ of two successive primes, but everything that is known, or seems probable from the evidence of the tables, suggests *extreme irregularity* in its behavior. The Prime Number Theorem shows that the *average* value of $p_{n+1}-p_n$ must be $\log n$, and tend to infinity with n. On the other hand there is overwhelming evidence that the smallest possible values of $p_{n+1}-p_n$, namely, 2, 4, 6, $\cdots$, recur indefinitely. It seems practically certain, not merely that there are infinitely many prime pairs $(p, p+2)$ but that there are infinitely many triplets $(p, p+2, p+6)$, and so with any combination of successive primes that is arithmetically possible; such a combination as $(p, p+2, p+4)$ is naturally not possible, since one of these numbers must be divisible by 3. All this seems hopelessly inconsistent with the existence of such a formula as was suggested, and it is clear that speculation in this direction is a waste of time.

There are, however, questions which have a somewhat similar tendency and which cannot be dismissed so summarily. There is one, for example, mentioned in Carmichael's little book. The problem, as he states it, is *"to find a prime greater than a given prime,"* which might be interpreted as

---

* This may be deduced from general theorems proved recently by J. F. Ritt.

meaning either *"to find an elementary function $\phi(n)$ such that $\phi(n) \to \infty$ and $\phi(n)$ is prime for every n, or for all n beyond a certain limit"* or as meaning *"to find an elementary function $\phi(p)$ such that $\phi(p) > p$ and $\phi(p)$ is prime whenever p is prime."* With either interpretation, it is a reasonable challenge, and the problem has not been solved.

Let us take the first form of the problem, which is perhaps the more natural, and let us begin by demanding *less*, namely that $\phi(n)$ shall be prime only for *an infinity of values of n*. In this case the problem becomes trivial, since n is a solution, by Euclid's theorem. It is, however, very interesting to observe that even then n, and certain simple linear functions such as $4n-1$ and $6n-1$, are the *only* trivial solutions. Dirichlet proved that *any* linear function $an+b$ has the property required, provided only that b is prime to a, or in other words that *every arithmetical progression* (subject to the last reservation) *contains an infinity of primes*. This theorem is quite difficult, except in a few special cases such as those which I mentioned, and it exhausts our knowledge in this particular direction. No one has ever proved that any of the functions

$$n^2 + 1, \quad 2^n - 1, \quad 2^n + 1$$

is prime for an infinity of values of n. With functions of *two* variables we can progress a good deal farther; we know for example that every quadratic form $am^2 + bmn + cn^2$ contains an infinity of primes, provided of course that a, b, c have no common factor and that $b^2 \neq 4ac$, and we can study the law of their distribution.

To find a $\phi(n)$ prime for *every* n is naturally still more difficult. Here linear functions are obviously useless, and no solution of any kind is known. Fermat conjectured that

$$2^{2^n} + 1,$$

is always prime, but Euler proved that this is false, since

$$2^{32} + 1 = 4294967297 = 641 \cdot 6700417.$$

So far as I know, no one else has ever advanced any other suggestion which is even plausible.

In view of the apparently insuperable difficulties of this problem, there is a certain interest in *negative* results. It is plain, first, that $an+b$ cannot be prime for all n, or all large n. More generally, no polynomial

$$f(n) = a_0 n^k + a_1 n^{k-1} + \cdots + a_k$$

can be prime for all or all large n; for if $f(m) = M$ then $f(rn+m)$ is divisible by M for all r. There are entertaining curiosities in this field; thus

$$n^2 - n + 41$$

is prime for the first 41, and

$$n^2 - 79n + 1601$$

for the first 80 values of n. It is obvious that forms like

$$a^n - 1, \quad a^n + 1$$

cannot be prime for all large n, since, for example, $a^{3m}-1$ is divisible by a^m-1, and it is natural to suppose that the same is true for

$$P(n, 2^n, 3^n, 4^n, \cdots, k^n),$$

where P is any polynomial with integral coefficients.†

14. *The Fundamental Theorem in an Algebraic Field.* I must not allow myself to succumb to the temptation of talking too long about the theory of the distribution of primes, which is after all only one chapter in arithmetic. There are other topics about which our imaginary enquirer will certainly demand more information, and of these I think one stands out; it is certain that he will want fuller explanations about the field $K^*[\sqrt{(-5)}]$ and the other algebraic fields in which the analog of the fundamental theorem fails. All ordinary arithmetic depends, it seems, upon the fundamental theorem; how then can there *be* an arithmetic in a field in

---

† Morgan Ward of Pasadena has found a very simple proof of this theorem.

which it is false? It would seem that the arithmetic of such a field can bear no real resemblance to ordinary arithmetic. I shall spend the rest of my time in an attempt to explain, in the very broadest outline, how order is restored.

I shall begin by quoting a remark of Hilbert which is trivial in itself but which shows us at once the direction in which we must look for a solution. Consider the numbers

$$1, 5, 9, 13, 17, 21, \cdots$$

of the form $4m+1$. These numbers form a group for multiplication (though naturally not for addition), and we can define divisibility and primality in the group. The "primes" are the numbers

$$5, 9, 13, 17, 21, 29, 33, 37, 41, 49, \cdots$$

which are greater than 1, of the form $4m+1$, and not decomposable into factors of this form. Thus 21, 57, 77, and 209 are "primes"; but

$$4389 = 21 \cdot 209 = 57 \cdot 77,$$

so that a number of the group may be resolved into "prime" factors in different ways.

In this case the solution of the mystery is obvious. The "fundamental theorem" fails *because of the absence from the group of the numbers $4m+3$ of ordinary arithmetic.* In fact

$$21 = 3 \cdot 7, \quad 57 = 3 \cdot 19, \quad 77 = 7 \cdot 11, \quad 209 = 11 \cdot 19$$

and

$$21 \cdot 209 = (3 \cdot 7)(11 \cdot 19) = (3 \cdot 19)(7 \cdot 11) = 57 \cdot 77.$$

We cannot give a proper account of the properties of the numbers $4m+1$ so long as we insist on excluding the numbers $4m+3$; *the numbers $4m+1$ do not form by themselves an adequate basis for arithmetic.* This observation has of course no intrinsic interest, since no reasonable person would expect that they would do so. It is trivial in itself, but it is not at all trivial in its suggestion, since it suggests that the troubles of the field $K^*[\sqrt{(-5)}]$ may be remedied *by considering the field as part of some larger field.*

This is in fact the solution found by Kummer. We consider the field $L[\sqrt{(-5)}]$ of numbers

$$\xi = \sqrt{(a + b\sqrt{(-5)})},$$

where a and b are ordinary integers. This is only an approximate statement; we do not actually consider all such numbers, but only those satisfying certain further conditions; the greatest common divisor of a and b must be a square or five times a square, and a^2+5b^2 must be a square. The field L includes K^*. The numbers of L form a group for multiplication, and we can define divisibility and primality in the field. Finally, the analog of the fundamental theorem is valid; *factorization is unique in L.* The proof of this is quite simple, but requires a little attention to detail, and I must refer you for the details to Mordell's tract on *Fermat's Last Theorem.*

We can now give a simple account of the equations in $K^*(\sqrt{(-5)})$ which puzzled us before. Consider for example the equation

$$3 \cdot 7 = (1 + 2\sqrt{(-5)})(1 - 2\sqrt{(-5)}).$$

It is easily verified that

$$3^2 = (2+\sqrt{(-5)})(2-\sqrt{(-5)}),$$
$$7^2 = (2+3\sqrt{(-5)})(2-3\sqrt{(-5)}),$$
$$(1+2\sqrt{(-5)})^2 = -19+4\sqrt{(-5)} = -(2-\sqrt{(-5)})(2+3\sqrt{(-5)}),$$
$$(1-2\sqrt{(-5)})^2 = -19-4\sqrt{(-5)} = -(2+\sqrt{(-5)})(2-3\sqrt{(-5)}).$$

Hence, if we write

$$\alpha = \sqrt{(2 + \sqrt{(-5)})}, \quad \alpha' = \sqrt{(2 - \sqrt{(-5)})},$$
$$\beta = \sqrt{(2 + 3\sqrt{(-5)})}, \quad \beta' = \sqrt{(2 - 3\sqrt{(-5)})},$$

we have

$$3 = \alpha\alpha', 7 = \beta\beta', 1 + 2\sqrt{(-5)} = -\alpha'\beta, 1 - 2\sqrt{(-5)} = -\alpha\beta',$$
$$3 \cdot 7 = \alpha\alpha' \cdot \beta\beta' = \alpha'\beta \cdot \alpha\beta' = (1 + 2\sqrt{(-5)})(1 - 2\sqrt{(-5)});$$

and all of these equations are entirely natural. *In order to obtain a satisfactory theorem of factorization in K^*, we must conceive K^* as immersed in the larger field L.* The logic of the solution is exactly the same as that of the solution of the corresponding, but trivial, problem for the numbers $4m+1$.

On the other hand there is an obvious contrast between the two solutions. It is *natural* to think of the field "$4m+1$" as part of the field "m"; "m" is the more obvious and simpler field. It is not natural to think of K^* as part of L; K^* is a much simpler and more natural field than L, and we should like to do without the reference to the latter if we could. It will be very tiresome if, whenever we consider an algebraic field, we are to be compelled to construct some more elaborate field of which it is a part. We should prefer to tidy up the house without going out of doors.

We may look for a hint once more in the numbers $4m+1$. Some of these numbers are divisible by 7, a number outside the field; and these numbers stand in certain specific relations to one another inside the field. Could we give a rational account of these relations without explicit reference to the number 7? It is a very unnatural thing to try to do, since what is *important* about the numbers is precisely that they *are* divisible by 7, but we could do it; we could define the class

$$21, 49, 77, 105, \cdots,$$

of numbers $4m+1$ divisible by 7 in terms of the field $4m+1$ itself. For example, we could take the first two numbers 21 and 49, and say "the class in question is the class which begins with these two numbers and whose members recur at regular intervals in the field." It is of course an artificial definition, and it is impossible to conceal from ourselves what we are really doing.

It is often a very profitable exercise for a mathematician to force himself to solve some simple problem without the weapon obviously appropriate to the occasion, to throw away the key of the front door and insist on forcing himself in somehow through the window. The forced and unnatural solution of one problem will often turn out to contain the germ of a quite natural solution of another. So it proves in this case; it is natural to try to define the numbers of K^* divisible by ξ without going outside K^*; it is natural, and possible, and it gives us the key to what is, in the general case, the established method of constructing a satisfactory arithmetic.

It is obvious that, if α and β belong to K^*, and $\xi|\alpha$ and $\xi|\beta$, then $\xi|\lambda\alpha+\mu\beta$, where λ and μ are any numbers of K^*. The converse proposition is not true; it is not true that if I is any set of numbers of the field K^* which has the property "if α and β belong to I, then $\lambda\alpha+\mu\beta$ belongs to I, for every λ and μ of the field", then there exists a number ξ, belonging to K^* which divides every number of I. What *is* true is that every number of I is divisible by a ξ which belongs to L but not in general to K^*. The set I is identical with the set of numbers of K^* divisible by ξ. Such a set I, or the more general set based on any finite number of numbers $\alpha, \beta, \gamma, \cdots$, of K^*, is called an *ideal*, the numbers ξ, underlying K^* but not belonging to it, having been described by Kummer as "ideal numbers". In ordinary arithmetic ideals are simply the sets of numbers divisible by some special number such as 3, and there is nothing in particular to be gained by their introduction. In an algebraic field they are not, in general, the sets of numbers divisible by a number *of the field*, and their introduction is essential before arithmetic can get properly started. We can define multiplication and division of ideals, prime ideals, and so on, and when we have done this we find that the arithmetic of ideals has all the properties of ordinary multiplicative arithmetic. In particular, *every ideal can be resolved uniquely into prime ideals*; the fundamental theorem is true when stated in terms of ideals.

The proof of the fundamental theorem is not particularly difficult; Landau presents it, with all the preliminary definitions, in about a dozen pages of quite simple reasoning. But I would not commit the impertinence, even if I had the time, of assuming the airs of an expert in the algebraic theory of numbers, a subject which I admire only at a distance and in which I have never worked. It is ordinary rational arithmetic which attracts the ordinary man, and I have digressed outside it only because there is a good deal in it which it is impossible to appreciate properly without a little knowledge of the larger theory. It is impossible, for example, to appreciate Euclid's arithmetical achievements until we realize that there are arithmetics in which the most obvious analogs of his theorems are false.

15. *Conclusion. Pedagogy.* There are few things in the world for which I have less taste than I have for mathematical pedagogics, but I cannot resist the temptation of concluding with one pedagogic lesson. There was, and I fear still is, a popular English text book of algebra which I used at school and which contained a chapter on the theory of numbers. It might be expected that such a chapter would be among the most instructive in the book; we might suppose, for example, that Euclid's algorithm, with its elegance, its simplicity, and its far reaching consequences, would be an ideal text for the instruction of a bright young mathematician. In fact the algorithm was never mentioned; one was to find the highest common factor of 12091 and 14803, I suppose, by "trial"; and all that the authors had to say of the fundamental theorem was that "it is so evident that it may be regarded as a necessary law of thought." It is possible of course that all this may have been expunged from later editions. It is certain, however, that chapters on number theory in textbooks of algebra are usually quite intolerably bad, and it is conceivable that Oxford University may have been right in erasing the subject altogether from its more elementary examination schedule.

The elementary theory of numbers should be one of the very best subjects for early mathematical instruction. It demands very little previous knowledge; its subject matter is tangible and familiar; the processes of reasoning which it employs are simple, general and few; and it is unique among the mathematical sciences in its appeal to natural human curiosity. A month's intelligent instruction in the theory of numbers ought to be twice as instructive, twice as useful, and at least ten times as entertaining as the same amount of "calculus for engineers". It is after all only a minority of us who are going to spend our lives in engineering workshops, and there is no particular reason why most of us should feel any overpowering interest in machines; nor is it in the least likely that, on those occasions when machines are of real importance to us, we shall require the power of dealing with them by methods more elaborate than the simplest rule of thumb. It is not engineering mathematics that is wanted for the understanding of modern physics, and still less is it wanted by most of us for the ordinary needs of life; we do not actually drive cars by solving differential equations. There may be a case for subordinating mathematics to the linguistic and literary studies which are so much more obviously useful to ordinary men, but there is none for sacrificing a splendid subject to meet a quite imaginary need.

PRINCETON UNIVERSITY

MATHEMATICAL PROOF.[1]

By G. H. Hardy.

1. I HAVE chosen a subject for this lecture, after much hesitation, not from technical mathematics but from the doubtful ground disputed by mathematics, logic and philosophy; and I have done this deliberately, knowing that I shall be setting myself a task for which I have no sufficient qualifications. I have been influenced by three different motives. In the first place, the exercise will be good for me, since it will force me to think seriously about questions which a professional mathematician like myself is apt to neglect. Secondly, it is difficult to find a branch of pure mathematics suitable for popular exposition in an hour. Finally if, in a desperate attempt to be interesting, I lose myself in discussions where I am admittedly an amateur, then, whoever I may offend, I should certainly not have offended the founder of this lectureship and the Rouse Ball chair.

I do not regret my choice, but I am bound in self-defence to begin with a double apology. The first is to any real mathematical logicians who may be present. I am myself a professional pure mathematician in the narrow sense, and, in my own subject, quite as intolerant of amateurs as a self-respecting professional should be. I have therefore no difficulty in understanding that mathematical logic also is a subject for professionals; that it demands a detailed knowledge which I do not possess and, so long as I am active in my proper sphere, have hardly leisure to acquire; and that I am certain to be guilty of all sorts of confusions

[1] Rouse Ball Lecture in Cambridge University, 1928.

which would be impossible to a properly qualified logician. Indeed there is only one thought which gives me courage to proceed, and that is that I may be concerned less with strictly logical questions than with questions of general philosophy. However treacherous a ground mathematical logic, strictly interpreted, may be for an amateur, philosophy proper is a subject, on the one hand so hopelessly obscure, on the other so astonishingly elementary, that there knowledge hardly counts. If only a question be sufficiently fundamental, the arguments for any answer must be correspondingly crude and simple, and all men may meet to discuss it on more or less equal terms.

My second apology must be addressed to those mathematicians who dislike all discussions savouring of philosophy. But if I apologise to them, it is perhaps with less sincerity. I feel that this distaste is usually based on no better foundation than an unreasoning shrinking from anything unfamiliar, the distaste of the pragmatist for truth, of the engineer for mathematics, of the pavilion critic at Lords for the in-swinger and the two-eyed stance. It is reasonable to ask an audience like this to put aside this dislike of the fundamental for its own sake.

You must also remember that ordinary mathematics has a good deal at stake in some of these recent controversies. These controversies have seemed to threaten methods which we have used with confidence for nearly one hundred years. There are familiar elementary theorems—that any aggregate of real numbers has an upper bound, that any infinite aggregate has a point of condensation—the truth of which is simply denied by the 'intuitionist' school of logicians. There are also theorems of an apparently much less abstract or suspicious type, theorems for example in the theory of numbers, the only known proofs of which depend, in appearance at any rate, on principles which they reject.

2. It may not be possible to distinguish precisely between mathematics, mathematical logic, and philosophy, as the words are currently used. We can, however, by considering a few typical problems, recognise roughly the disputed tracts across which the boundaries must be drawn.

(i) *Is Goldbach's Theorem true?* Is any even number the sum of two primes? This is a strictly mathematical question to which all questions of logic or philosophy seem irrelevant.

(ii) *Is the cardinal number of the continuum the same as that of Cantor's second number class?* This again appears to be a mathematical question; one would suppose that, if a proof were found, its kernel would lie in some sharp and characteristically mathematical idea. But the question lies much nearer to the borderline of logic, and a mathematician interested in the problem is likely to hold logical and even philosophical views of his own.

(iii) *What is the best system of primitives for the logic of propositions?* This is a question of mathematical logic in the strict professional sense. A logician qualified to discuss it will probably belong to some more or less definite philosophical school, but it is hardly likely that his philosophical views will have any very noticeable influence on his choice.

(iv) *What is a proposition, and what is meant by saying that it is true?* This, finally, is a problem of simple philosophy.

It is often said that mathematics can be fitted on to any philosophy, and up to a point it is obviously true. Relativity does not (whatever Eddington may say) compel us to be idealists. The theory of numbers does not commit us to any particular view of the nature of truth. However that may be, there is no doubt that mathematics does create very strong philosophical *prejudices*, and that the tests which a philosophy must satisfy before a mathematician will look at it are likely to be very different from those imposed by a biologist or a theologian. I am sure that my own philosophical prejudices are as strong as my philosophical knowledge is scanty.

One may divide philosophies into *sympathetic* and *unsympathetic*, those in which we should like to believe and those which we instinctively hate, and into *tenable* and *untenable*, those in which it is possible to believe and those in which it is not. To me, for example, and I imagine to most mathematicians, Behaviourism and Pragmatism are both unsympathetic and untenable. The philosophy of Mr. Bradley may be just tenable, but it is highly unsympathetic. The Cambridge New Realism, in its cruder forms, is very sympathetic, but I am afraid that, in the forms in which I like it best, it may be hardly tenable. 'Thin' philosophies, if I may adopt the expressive classification of William James, are generally sympathetic to me, and 'thick' ones unsympathetic. The problem is to find a philosophy which is both sympathetic and tenable; it is not reasonable to hope for any higher degree of assurance.

3. The crucial test of a philosophy, for a mathematician, is that it should give some sort of rational account of *propositions* and of *proof*. A mathematical theorem is a proposition; a mathematical proof is clearly in some sense a collection or pattern of propositions. It is plain then that if I ask what are, to a mathematician, the most obvious characteristics of

a mathematical theorem or a mathematical proof, I am inviting philosophical discussion of the most fundamental kind. I wish to begin, however, by being as unsophisticated as I can, and I will therefore try to sketch what seems to be the view of mathematical common sense, the sort of view natural to a man who does not profess to be a logician but has spent his life in the search for mathematical truth. It is after all the misapprehensions of such a man that a logician may find the least fundamentally unreasonable and the least hopeless to remove.

I will begin then by enumerating some rough criteria which I think that a philosophy must satisfy if it is to be at all sympathetic to a working mathematician. I know too well how probable it is that just the most sympathetic philosophies will prove untenable.

(1) It seems to me that no philosophy can possibly be sympathetic to a mathematician which does not admit, in one manner or another, the immutable and unconditional validity of mathematical truth. Mathematical theorems are true or false; their truth or falsity is absolute and independent of our knowledge of them. In *some* sense, mathematical truth is part of objective reality.

'Any number is the sum of 4 squares'; 'any number is the sum of 3 squares'; 'any even number is the sum of 2 primes'. These are not convenient working hypotheses, or half-truths about the Absolute, or collections of marks on paper, or classes of noises summarising reactions of laryngeal glands. They are, in one sense or another, however elusive and sophisticated that sense may be, theorems concerning reality, of which the first is true, the second is false, and the third is either true or false, though which we do not know. They are not creations of our minds; Lagrange discovered the first in 1774; when he discovered it he discovered *something*; and to that something Lagrange, and the year 1774, are equally indifferent.

(2) When we know a mathematical theorem, there is something, some object, which we know; when we believe one, there is something which we believe; and this is so equally whether what we believe is true or false.

It is obvious that by this time we have escaped only too successfully from the domain of platitude and triviality. We have done no more than to make explicit a few of the instinctive prejudices of the 'mathematician in the street'. Yet with our first demand we have antagonised at least two-thirds of the philosophers in the world; and with the second we have reduced our first indiscretion to entire insignificance, since we have committed ourselves, in one form or another, to the objective reality of propositions, a doctrine rejected, I believe, not only by all philosophers, but also by all three of the current schools of mathematical logic.

(3) In spite of this I am going farther, and in a direction relevant to the recent controversies concerning 'transfinite' mathematics to which I shall return later. Mathematicians have always resented attempts by philosophers or logicians to lay down dogmas imposing limitations on mathematical truth or thought. And I am sure that the vast majority of mathematicians will rebel against the doctrine—even if it is supported by some of themselves, including mathematicians so celebrated as Hilbert and Weyl—that it is only the so-called 'finite' theorems of mathematics which possess a real significance. That 'the finite cannot understand the infinite' should surely be a theological and not a mathematical war-cry.

No one disputes that there are infinite processes which appear to be prohibited to us by the facts of the physical world. It is true, as Hilbert says, that no mathematician has completed an infinity of syllogisms. It is equally true that there is no mathematician who has never drunk a glass of water, and, so far as I can see, one of these facts has neither more nor less logical importance than the other. There is no more *logical* reason why a mathematician should not prove an infinity of theorems in this world than why he should not (as he has been so often encouraged to hope) emit an infinite sequence of musical notes in the next.

The history of mathematics shows conclusively that mathematicians do not evacuate permanently ground which they have conquered once. There have been many temporary retirements and shortenings of the line, but never a general retreat on a broad front. We may be confident that, whatever the precise issue of current controversies, there will be no general surrender of the ground which Weierstrass and his followers have won. 'No one', as Hilbert says himself, 'shall chase us from the paradise that Cantor has created': the worst that can happen to us is that we shall have to be a little more particular about our clothes.

4. Such then are the presuppositions and prejudices with which a working mathematician is likely to approach philosophical or logical systems. How far are they satisfied by the existing schools of mathematical logic? There are three such schools, the logisticians (represented at present by Whitehead, Russell, Wittgenstein, and Ramsey), the finitists or intuitionists (Brouwer and Weyl), and the formalists

(Hilbert and his pupils). I am primarily interested at the moment in the formalist school, first because it is perhaps the natural instinct of a mathematician (when it does not conflict with stronger desires) to be as formalistic as he can, secondly because I am sure that much too little attention has been paid to formalism in England, and finally because of the title of my lecture and because Hilbert's logic is above everything an explicit theory of mathematical proof. I must begin by a rapid summary of the most striking differences between these schools and of the difficulties which have brought them into existence. It is not my object to discuss these difficulties in detail, but what I have to say later can hardly be intelligible unless I give some sort of general explanation of their character. I can fortunately base this explanation on the extremely clear account of the situation given recently by Ramsey.

5. (1) I shall refer to the logisticians generally under the short title of 'Russell'. It is necessary to say that by 'Russell' I mean the Russell of *Principia Mathematica*. *Principia Mathematica* is not a treatise on philosophy, but it has a philosophical background, with which I am in general sympathy. I think that I can understand, in broad outline, how the logical edifice can support itself on *that* foundation. The problem of erecting it on the foundation of Russell's latest philosophical writings is one which I prefer to leave to bolder minds.

To Russell, then, logic and mathematics are substantial sciences which in some way give us information concerning the form and structure of reality. Mathematical theorems have *meanings*, which we can understand directly, and this is just what is important about them. In this, I may observe, Russell and the so-called 'intuitionists' are in complete agreement; and (since it is something of this sort which seems to me the natural implication of the word) I should prefer to avoid the use of 'intuitionism' as distinguishing one school from the other.

Mathematics is to Russell, up to a certain point at any rate, a branch of logic. It is concerned with particular kinds of assertions about reality, with particular logical concepts, propositions, classes, relations, and so forth. The propositions of logic and mathematics share certain general characteristics, in particular complete generality, though this is not an adequate description of them. There is no particular reason that I can see why any of this should be distasteful to us as mathematicians. It does not seem to conflict with the criteria which I suggested a moment ago; it seems likely at first sight even to indulge our desire for real propositions, though here we are ultimately disappointed.

There are certain definite points at which Russell's attempted reduction of mathematics to logic fails. In this, of course, there is nothing likely to astonish an unsophisticated mathematician. That mathematics should follow naturally, up to a point, from purely logical premisses, premisses to whose simplicity and 'self-evidence' no one can reasonably take exception, when proper allowance is made for the element of sophistication inevitable in a highly complex structure; but that it should then prove necessary to import fresh raw material and add new assumptions—all this is only what a mathematician might expect. In particular, I think that this is true of two of the three 'non-logical' axioms necessary in Russell's scheme; the Axiom of Infinity, that the universe contains an infinity of individuals, and the Multiplicative Axiom or Axiom of Zermelo, which is very famous but required only in particular theorems which might conceivably be discarded, and which I need not stop to explain, since I shall not refer to it further.

6. The situation is quite different with the third axiom, the notorious Axiom of Reducibility. The point here is much more important and also much more difficult. It is essential that I should say something about it, impossible that I should explain it fully. I cannot hope to find popular language clearer than Ramsey's, and I shall follow him very closely.

The theory of aggregates, in the classical form of Cantor and Dedekind, leads to certain antinomies, of which the most famous is Russell's paradox of the class of classes which are not members of themselves, a concept which may be shown to lead at once to flat contradiction. Russell met the difficulty by his Theory of Types.

Suppose that we are given a set S of properties, defined as being all properties of a certain kind K. Given an object x, we can ask whether x possesses *any* property of the kind K. If x has *any* such property, this is another property of x, say Σ; and we can then ask whether Σ can be itself a property of the set S, that is to say of the kind K. It is natural to suppose that the answer must be negative, since the idea of Σ already presupposes the totality S; and this is in fact Russell's answer. The property Σ is, he says, a property 'of higher order' than any property belonging to S; and so generally we must classify properties according to their orders, and any property defined by reference to all properties of a certain order must be a property of higher order. It is impossible to make any statement which is significant for properties of all orders simultaneously. Further, since, in Russell's logic, statements about classes are merely disguised statements about their defining properties, classes also must be divided into orders, and any statement about 'all' classes must really be confined to all classes of a certain order. This doctrine seems inherently plausible, and leads to an easy solution of Russell's and similar antinomies.

The theory of types has, however, very unfortunate mathematical consequences, since it appears to destroy some of the most fundamental theorems of analysis. The typical theorem is the theorem that any aggregate of numbers has an upper bound, a theorem which is substantially the same as what, in my *Pure Mathematics*, is called 'Dedekind's Theorem'. A real number is defined as a class of rationals. Suppose now that we are given a set S of real numbers x, i.e. a set of classes of rationals. The upper bound U of S is defined as the class of rationals which is the logical sum of the classes defining the various members of X, and it is taken for granted that this class stands on the same footing as the classes of which it is the sum. But a moment's consideration shows that this is not so. The classes which are the members of X are defined by certain properties of rationals, and the class which is U is defined by the property of belonging to *some one or other* of these classes, that is to say of possessing some one or other of these properties. Thus the defining characteristic of U involves a reference to *all* the defining characteristics of members of X, and is therefore a characteristic of higher order. It follows that, if we were to attempt to develop analysis without further assumptions, we should have to distinguish real numbers of different orders. We should have to say that the upper bound of an aggregate of real numbers of order n was a real number of order $n + 1$, and so on; and this, whether practicable or not (a point about which I express no opinion) would certainly be extremely inconvenient and probably intolerable.

Russell meets this difficulty by the Axiom of Reducibility, which asserts roughly that there is a property of the lowest order equivalent to any property of any order, not of course equivalent in *meaning*, but equivalent in *extension*, so that any object which possesses the one possesses the other, and they define the same class. The upper bound U may then be defined, not only by the property used to define it above, but also by the equivalent property of lower order, and it is thus a real number in the same sense as each of the numbers of which it is the upper bound. It is not disputed by anybody, so far as I know, that the axiom does yield a solution of the problem. Analysis can be developed in the classical manner and without further difficulty when once the truth of the axiom is granted; and there seems to be no ground for supposing that the axiom will lead to contradiction.

There are, however, objections to the axiom, about the force of which opinions may perhaps differ, but which have proved sufficient to prevent all other logicians from accepting it. It is complicated and (what is more important) very *unconvincing*. It has none of the 'self-evidence' of the properly logical assumptions; and it is obvious that Russell himself dislikes it very heartily and regards its presence in his system as a most regrettable necessity. Finally, an argument suggested in the rough by Ramsey, and developed in a more precise form by Waismann, appears to show conclusively that the axiom is definitely *not* a 'truth of logic' in the same sense as the other primitive propositions of *Principia Mathematica*. It is therefore impossible to regard Russell's solution as satisfactory, and this is about the only point on which the logicians, Russell himself included, are unanimous.

7. (2) I pass to the finitists, Brouwer and Weyl, and I shall dismiss them very shortly. Much as I admire the contributions of Brouwer and Weyl to constructive mathematics, I find their contribution to logic singularly unsympathetic. Finitism rejects, first, all attempts to push the analysis of mathematics beyond a certain point, and for this I see no sort of justification. I have no particular desire to be committed to the extreme Russellian doctrine, that all mathematics is logic and that mathematics has no fundamentals of its own. If it should turn out that there are parts of mathematics irreducible to logic, I do not see why I should be particularly distressed. On the other hand I see no reason for denying that, up to a point, the reduction has actually been made, and the arguments for denying in principle the possibility of a further reduction seem to me entirely inconclusive. That there is some particular sanctity about the notion of an integer which should protect it against the humiliation of further analysis, that general existential propositions have no real significance, that there is some peculiar certainty in knowledge based, in some sense, in immediate perception of a finite number of sensible things— all these are dogmas to which the finitists seem to be committed; and all of them seem to be founded on philosophical doctrines with which I have no sympathy, which indeed I find it extremely difficult to understand, and which seem to me, so far as I can understand them, to rest on all

sorts of questionable assumptions, and in particular on an impossibly naïve attitude towards our knowledge of the physical world.

This, however, is a minor point for a mathematician. What is much more serious to a mathematician is that the mathematical consequences of finitism involve rejection not (like those of denying the Multiplicative Axiom) of particular isolated outworks of mathematics but of integral regions of ordinary analysis. It is no use trying to deny that the finitists have the better of the argument up to a point; the parts of analysis which they admit are unquestionably, at present, in a more secure position than the rest; and so long as finitism merely insists on this its position is unassailable. I cannot believe that mathematicians generally will be so ready to accept a check as final, so anxious to find metaphysical reasons for supposing that the prettiest path is that which passes on the side of the hedge away from the bull.

8. (3) I go on then to consider the logic of Hilbert and his school; and here I find it very necessary to distinguish between Hilbert the philosopher and Hilbert the mathematician. I dislike Hilbert's philosophy quite as much as I dislike that of Brouwer and Weyl, but I see no reason for supposing that the importance of his logic depends in any way on his philosophy.

I am sure that the Hilbert logic has been unreasonably neglected by English logicians. 'The formal school', says Ramsey, 'have concentrated on the propositions of mathematics, which they have pronounced to be meaningless formulæ to be manipulated according to certain rules, and mathematical knowledge they hold to consist in knowing what formulæ can be derived from what others consistently with the rules. Such being the propositions of mathematics, the account of its concepts, for example the number 2, immediately follows: "2" is a meaningless mark occurring in these meaningless formulæ. But, whatever may be thought of this as an account of mathematical propositions, it is obviously hopeless as a theory of mathematical concepts; for these occur not only in mathematical propositions, but also in those of everyday life. Thus "2" occurs not merely in "$2 + 2 = 4$", but also in "it is 2 miles to the station", which is not a meaningless formula but a significant proposition, in which "2" cannot conceivably be a meaningless mark. Nor can there be any doubt that "2" is used in the same sense in the two cases, for we can use "$2 + 2 = 4$" to infer from "it is 2 miles to the station and 2 miles on to the

Gogs" to "it is 4 miles to the Gogs *via* the station", so that these ordinary meanings of "2" and "4" are clearly involved in "2 + 2 = 4".'

Let me say at once that this argument seems to me to be unanswerable and that, if I thought that this really was the beginning and the end of formalism, I should agree with Ramsey's rather contemptuous rejection of it. But is it really credible that this is a fair account of Hilbert's view, the view of the man who has probably added to the structure of significant mathematics a richer and more beautiful aggregate of theorems than any other mathematician of his time? I can believe that Hilbert's philosophy is as inadequate as you please, but not that an ambitious mathematical theory which he has elaborated is trivial or ridiculous. It is impossible to suppose that Hilbert denies the significance and reality of mathematical concepts, and we have the best of reasons for refusing to believe it: 'the axioms and demonstrable theorems,' he says himself, ' which arise in our formalistic game, are the images of the ideas which form the subject-matter of the ordinary mathematics '.

I must, however, begin with a few remarks about the philosophical background which seems to lie behind Hilbert's views; and here of course I need not be alarmed if I find myself disagreeing with him as hopelessly as with the finitists. Hilbert's philosophy appears indeed to be in broad outline much the same as Weyl's, as Weyl himself has very fairly pointed out. There is the same rejection of the possibility of any purely logical analysis of mathematics: ' mathematics is occupied with a content given independently of all logic, and cannot in any way be founded on logic alone.' There is the same insistence on some sort of concrete, perceptible basis, for which Hilbert (with what justice I have no idea) claims the support of ' the philosophers and especially Kant'. ' in order that we should be able to apply logical forms of reasoning, it is necessary that there should first be something given in presentation, some concrete, extra-logical object, immediately present to intuition and perceived independently of all thought. . . . In particular, in mathematics, the objects of our study are the concrete signs themselves.' There is, I think, no doubt at all that Hilbert does assert, quite unambiguously, that the subject matter of mathematics proper is the actual physical mark, not general formal relations between the marks, properties which one system of marks may share with another, but the black dots on paper which we see.

I had better state at once what is to me a fatal objection to this view. If Hilbert has made the Hilbert mathematics with a particular series of marks on a particular sheet of paper, and I copy them on another sheet, have I made a *new* mathematics? Surely it is *the same* mathematics, and that even if he writes in pencil and I in ink, and his marks are black while mine are red. Surely the Hilbert mathematics must be in some sense something which is common to all such sets of marks. I make this point here, because there are two questions which suggest themselves at once about Hilbert's marks. The first is whether we are studying the physical signs themselves or general formal relations in which they stand, and the second is whether these signs or relations have 'meaning' in the sense in which the symbols of mathematics are usually supposed to have meaning. It seems to me that the two questions are quite distinct.

9. It is no doubt this philosophical outlook, and this consequent insistence on the importance of the physical mark or sign, that inspire Hilbert's finitism, which appears at first sight as extreme as that of Brouwer and Weyl themselves. I naturally find this attitude very disappointing; it seems to me that formalism is bound to die for want of air within the narrow confines of a finitistic system. But on the face of it Hilbert is entirely uncompromising: 'there is no infinite anywhere in reality', he says, and again ' is it not clear that, when we think we can recognise the reality of the infinite in any sense, we are merely allowing ourselves to be deceived by the enormity of the largeness or smallness which confronts us everywhere . . . ?'.

Hilbert says that 'infinite theorems', theorems such as 'there are infinitely many primes', are not genuine propositions but 'ideal' propositions. I am not at all sure what he means by an 'ideal proposition', but I suppose that one thing at any rate that he would say (if he used Russell's language) is that the infinite is essentially *incomplete*. We know that mathematics is full of 'incomplete symbols', symbols which have no meaning in themselves, though larger collections of symbols of which they are parts have perfectly definite meanings. There are, for example, the ordinary 'operational' symbols; '$\frac{d}{dx}$', '∇^2', '$\int_a^b \ldots dx$'. The most striking example is the '∞' of elementary analysis; we define '$\sum_0^\infty$' and '$f(x) \to \infty$', but (at any rate in the ordinary presentations of the subject) we never define '∞' standing by itself. There is, in the classical analysis, no number ∞ standing on all fours with e or π; there is a sharp contrast here between the infinite of analysis and the infinite of geometry, in which 'the line at infinity', say $z = 0$, is on just the same footing as any other line.

It is one of Russell's admitted achievements to have recognised in a precise and explicit manner the immense importance of 'incomplete symbolism' in logic and philosophy also, and so to have shown how widely the correct analysis of a proposition may diverge from the analysis of unreflecting common sense. The standard example is that of propositions containing denoting phrases or *descriptions*, 'the so-and-so', 'the murderer', 'the author of Waverley'. The 'Waverley' argument applies to all propositions of the form 'a is the b', and shows that the proposition cannot be analysed, as the words expressing it suggest, into an assertion of identity between 'a' and 'the b'. I wish to know whether a is the b, whether Dr. Sheppard was the murderer of Roger Ackroyd; and in fact he was. If 'a' and 'the b' are the same object, I can substitute one for the other in any proposition without destroying its sense or its truth; and therefore it appears that what I really wanted to know was whether Dr. Sheppard was Dr. Sheppard, which is obviously false. It follows that the analysis was wrong, and that there is no such object in reality as 'the b'; 'a is the b' must be analysed in an entirely different manner.

I am not suggesting that Hilbert would accept the statement that the infinite is incomplete as an adequate account of his attitude towards it. No doubt he would want to go very much further. I have inserted this explanation merely (1) because I shall need it later and (2) because rival views about the infinite are apt to differ more violently in expression than reality, and the notion of an incomplete symbol might in some cases be a basis for a reconciliation between them. I have the less hope that it would do so in this case because Hilbert uses, as instances in support of his thesis that all 'infinite theorems' are in some sense 'ideal theorems', such divergent illustrations as (a) the infinite of analysis, (b) the infinite of geometry, and (c) the ideal numbers of higher arithmetic, and it seems to me quite impossible to regard all these as inspired by the same logical motive, the first representing a *purification* of mathematics by an agreement to regard certain notions as 'incomplete', the others an *enlargement* of it by the introduction of new elements as 'complete' as those which they generalise.

10. It is time, however, to proceed to some description of Hilbert's system, and I do this in language based upon that

of v. Neumann, a pupil of Hilbert's whose statement I find sharper and more sympathetic than Hilbert's own.

(1) Hilbert's logic is a theory of proof. Its object is to provide a system of formal axioms for logic and mathematics, and a formal theory of logical and mathematical proof, which (a) is sufficiently comprehensive to generate the whole of recognised mathematics, and (b) can be proved to be consistent. The system of *Principia Mathematica* fulfils the first but not the second criterion.

(2) If we can do this, we shall be troubled by antinomies no more. But for this end the whole existing apparatus of axioms, proofs and theorems must first be formalised strictly, so that to every mathematical theorem a formula will correspond. The structure of the formal system will of course be *suggested* by the current logic and mathematics. Every formula will *seem* to have a meaning, a meaning which we must afterwards forget.

(3) For example, we have the 'logical' formula

$$a \to (b \to a).$$

This is suggested by an obvious 'logical truth', the truth that (in Russell's symbolism) $a \mathbin{.} \supset \mathbin{.} b \supset a$, that a true proposition is a consequence of any hypothesis. This formula is an 'axiom', which means *simply* that it is one of the formulæ with which we start.

Similarly we have the formula (again an axiom)

$$Za \to Z(a+1),$$

which is suggested by the 'mathematical truth' that $a + 1$ is an integer if a is one. We thus start with a finite system of axioms or 'given formulæ'. They are, so to say, the chessmen, the bat, ball and stumps, *the material with which we play*.

(4) We also need *rules for the game*, of which there are two. Rule (1) is that we may substitute one formula inside another, in the first instance inside an axiom, while Rule (2) is embodied in the 'scheme of demonstration'

$$\begin{array}{c} a \\ a \to b \\ \hline b \end{array} \quad \ldots \ldots \text{(A)}$$

(which corresponds to the 'non-formal principle of inference' in *Principia Mathematica*). Such a scheme is called a *demonstration*, a the *hypothesis*, b the *conclusion*. A formula is said to be *demonstrable* (1) if it is an axiom, or (2) it is b, a and $a \to b$ being axioms, or (3) it is b, a and

$a \to b$ being demonstrable, or (4) it is derivable from an axiom or a demonstrable formula by substitution. We have thus a quite precise concept of 'demonstration'. To use Weyl's illustration, we are playing chess. The *axioms* correspond to the given position of the pieces; the *process of proof* to the rules for moving them; and the *demonstrable formulæ* to all possible positions which can occur in the game.

(5) Let us observe in passing that there are far more axioms in Hilbert's scheme than in such a scheme as that of *Principia Mathematica*, and *no definitions* in the sense of *Principia Mathematica*. This is inevitable, since it is cardinal in Hilbert's logic that, however the formulæ of the system may have been suggested, the 'meanings' which suggested them lie entirely outside the system, so that the 'meaning' of a formula is to be forgotten immediately it is written down. The definitions of *Principia Mathematica* are the most important elements of the system, and embody 'philosophical' analyses of the meanings of the symbols used. The definition of a cardinal number, for example, presents to us at any rate one possible meaning of number, and tells us that that is the meaning with which Russell proposes to use the word. Hilbert is not concerned with that, or any, 'meaning' of 'number', and the only conceivable sense of a definition in his system is that of a symbolic convention which instructs us to replace a prolix formula by a more concise one.

11. (6) Mathematics proper, then, is reduced to a game like chess. We can, however, regard a game like chess from two quite different standpoints. In the first place we can *inspect*, or *construct*, chess, by reading the games whose aggregate constitutes chess, or playing new ones. Secondly, we can think and theorise *about* chess; we make judgements about it, and these judgements contain theorems which are in no sense part of the game. To take a definite illustration, which is in one form or another essential to the understanding of the Hilbert logic, we can judge, and in a sense *prove*, that *certain positions cannot occur*. There cannot be more than eighteen queens on the board; two knights cannot mate; these are true and provable theorems, not theorems *of* chess —the theorems *of* chess are the actual positions—but theorems *about* chess.

Similarly there is the Hilbert mathematics on the one hand, and what Hilbert calls 'metamathematics' on the other, the metamathematics being the aggregate of theorems *about* the mathematics; and of course it is the metamathematics which is the exciting subject and affords the real justification for our interest in this particular sort of mathematics. Suppose, for example, that we could find a finite system of rules which enabled us to say whether any given formula was demonstrable or not. This system would embody a theorem of metamathematics. There is of course no such theorem, and this is very fortunate, since if there were we should have a mechanical set of rules for the solution of all mathematical problems, and our activities as mathematicians would come to an end.

Such a theorem is not to be expected or desired, but there are metamathematical theorems of a different kind which it is entirely reasonable to expect and which it is in fact Hilbert's dominating aim to prove. These are the negative theorems of the kind which I illustrated a moment ago; they assert, for example, in chess, that two knights cannot mate, or that some other combination of the pieces is impossible, in mathematics that certain theorems cannot be demonstrated, that certain combinations of symbols cannot occur. In particular we may hope (and it is this hope that has inspired the whole construction of the logic) to show the impossibility of the combination

$$a . - a,$$

where - is the symbol corresponding to the 'negation' of *Principia Mathematica*.

Let us suppose that our analysis of the game has established this, and then recur to the 'meanings' which suggested the game but were afterwards discarded. We may think about meanings *now*, because we are engaged in metamathematics, *outside* the game. It will plainly follow that the concepts and propositions which we symbolised cannot lead to contradiction. If this has been done, and for a formal system rich enough to be correlated with the whole of mathematics, the purpose of the Hilbert logic will have been achieved.

12. It is now time for me to interpolate a remark which gives the justification for the title of my lecture. It is obvious that to Hilbert *proof* means two quite different things. I have tried to anticipate the point in my choice of words: we fortunately have two words, *proof* and *demonstration*.

'Proof' has always meant at least two different things, even in ordinary mathematics. *We* distinguish vaguely and halfheartedly; in the Hilbert logic the distinction becomes absolutely sharp and clear. First, there is the *formal, mathematical, official* proof, the proof inside the system, the pattern (A), what I called the *demonstration*. These inside official proofs are, in the mathematics, the actual formulæ or patterns, in the metamathematics, the subject matter for discussion.

Secondly there are the proofs of the theorems of the metamathematics, the proof that two knights cannot mate. These are *informal, unofficial, significant* proofs, in which we reflect on the meaning of every step. The structure of these proofs is not dictated by our formal rules; in making them we are guided, as in ordinary life, by 'intuition' and common sense. 'Prof. Hardy will lecture at 12.0 to-day, because it says so in the *Reporter*, and because statements in the *Reporter* are always true.'

You must not imagine that the unofficial, metamathematical, non-formal, intuitionist proof is in any sense slacker or less 'rigorous' than the formal mathematical proof. The subject matter is abstract and complicated, and every step has to be scrutinised with the utmost care. We may even find it necessary to guide our thoughts by the introduction of new formalism, and it is quite likely that, if we do, we shall use over again the same symbols that we have used already. And here, of course, lies a danger; for we may be tempted to forget that we are using the same symbols in different contexts and with different aims; even Russell has been accused of making this mistake by logicians of the more formal schools. In the Hilbert logic at any rate the distinction is quite precise; the unofficial proof lies entirely outside the official system, and its object is simply *to produce conviction*, unofficial conviction of the absence of official contradiction —which is what we want.

13. At this point I should like to leave the Hilbert logic for a moment, and make a few general remarks about mathematical proof as we working mathematicians are familiar with it. It is generally held that mathematicians differ from other people in *proving* things, and that their proofs are in some sense *grounds* for their beliefs. Dedekind said that 'what is provable, ought not to be believed without proof'; and it is undeniable that a decent touch of scepticism has generally (and no doubt rightly) been regarded as some indication of a superior mind.

But if we ask ourselves why we believe particular mathematical theorems, it becomes obvious at once that there are very great differences. I believe the Prime Number Theorem because of de la Vallée-Poussin's proof of it, but I do not believe that $2 + 2 = 4$ because of the proof in *Principia Mathematica*. It is a truism to any mathematician that the 'obviousness' of a conclusion need not necessarily affect the interest of a proof.

I have myself always thought of a mathematician as in the first instance an *observer*, a man who gazes at a distant range of mountains and notes down his observations. His object is simply to distinguish clearly and notify to others as many different peaks as he can. There are some peaks which he can distinguish easily, while others are less clear. He sees A sharply, while of B he can obtain only transitory glimpses. At last he makes out a ridge which leads from A, and following it to its end he discovers that it culminates in B. B is now fixed in his vision, and from this point he can proceed to further discoveries. In other cases perhaps he can distinguish a ridge which vanishes in the distance, and conjectures that it leads to a peak in the clouds or below the horizon. But when he sees a peak he believes that it is there simply because he sees it. If he wishes someone else to see it, he *points to it*, either directly or through the chain of summits which led him to recognise it himself. When his pupil also sees it, the research, the argument, the *proof* is finished.

The analogy is a rough one, but I am sure that it is not altogether misleading. If we were to push it to its extreme we should be led to a rather paradoxical conclusion; that there is, strictly, no such thing as mathematical proof; that we can, in the last analysis, do nothing but *point*; that proofs are what Littlewood and I call *gas*, rhetorical flourishes designed to affect psychology, pictures on the board in the lecture, devices to stimulate the imagination of pupils. This is plainly not the whole truth, but there is a good deal in it. The image gives us a genuine approximation to the processes of mathematical pedagogy on the one hand and of mathematical discovery on the other; it is only the very unsophisticated outsider who imagines that mathematicians make discoveries by turning the handle of some miraculous machine. Finally the image gives us at any rate a crude picture of Hilbert's metamathematical proof, the sort of proof which is a *ground* for its conclusion and whose object is to *convince*.

On the other hand it is not disputed that mathematics is full of proofs, of undeniable interest and importance, whose purpose is not in the least to secure conviction. Our interest in these proofs depends on their formal and æsthetic properties. This is almost always so with *logical* proofs; Theorem 3·24 of *Principia Mathematica* is the law of contradiction, and it is certainly not because we require to be convinced of its truth that we are prepared to study its elaborate deduction from equally 'self-evident' premisses. Here we are interested in the pattern of proof *only*. In our practice as mathematicians, of course, we cannot distinguish so sharply, and our proofs are neither the one thing nor the other, but a more or less rational compromise between the two. Our object is *both* to exhibit the pattern and to obtain assent. We cannot exhibit the pattern completely, since it is far too elaborate; and we cannot be content with mere assent from a hearer blind to its beauty.

14. Let us return to the Hilbert logic. The very structure of the logic, its mere existence, are enough, I think, to prove two propositions of great importance. The first is that it is possible to establish the consistency of a system of axioms *internally*, that is to say by direct examination of its structure; and the second is that it is possible to prove a system consistent even when the axioms embody logical principles such as the law of contradiction itself. Each of these propositions has been disputed.

Consider for a moment the ordinary procedure of axiomatic geometry. In abstract geometry we consider unspecified systems of things, a class S of objects A, B, C, . . . which we call *points*, and sub-classes of these objects which we call *lines*. We make certain assumptions about these points and lines, which we call *axioms*, such as that there is a line which contains any given pair of points, that there is only one such line, and so on. To lay down a system of axioms in geometry is simply to limit the subject matter, to say that we propose to consider only objects of certain kinds. Thus, in a geometry which contains the two axioms I have mentioned, our 'points' might be the players in a tournament, and our 'lines' the opponents in a game, but the points and lines could not be undergraduates and colleges, because then the axioms would be untrue.

In a geometry we are not concerned with any *particular* meaning of 'point' or 'line'. We may say, if we like, that we are concerned with *all possible* meanings, or that we are not concerned with meanings at all; we might accept Hilbert's language, and say that we are concerned simply with *marks*, or we might say (what would, I think, be at any rate one stage nearer to the truth) that we are concerned with what Wittgenstein calls *forms*. It is possible that the question is mainly one of words. We assume merely that our unspecified subject matter possesses the properties stated in our axioms, and we set out to investigate its other properties, the theorems of our geometry, by the usual processes of logical inference.

Every geometry demands a *consistency theorem*, which is naturally not a theorem of the geometry. We have to prove that the axioms do not contradict one another. We produce an example, an 'interpretation', of the geometry, a set of objects which actually have the properties attributed by the axioms to our points and lines. In general in these discussions we take arithmetic or analysis for granted, and our example is one in which points and lines are sets of numbers. Thus our points might be the numbers 1, 2, and 3, and our lines the classes 23, 31, 12: these objects do in fact satisfy the particular axioms which I mentioned. It was by this process, for example, that the old difficulties about the possibility of non-Euclidean Geometry were ultimately settled. It has always been held, and no doubt correctly, that in Geometry, where only the 'subject-matter' is symbolised, and there is no attempt to symbolise the process of inference itself, there is no other possible method.

If we try to apply a similar process to arithmetic, we are met by a difficulty. It is natural that a mathematician should wish to treat arithmetic axiomatically, to say not (with *Principia Mathematica*) that a number is such or such a particular object, but that numbers are any set of objects which have certain properties: there are so many plausible definitions of a number, and the reasons for selecting one rather than another seem so purely technical. There is, however, an obvious difficulty about the inevitable proof of consistency. When we wanted such a proof for a geometry, we could appeal to arithmetic; but there is nothing in ordinary mathematics which comes before arithmetic, and it is not easy at first to see where any 'example' is to be found. There seems only one possibility, if we are to pursue the established method, and that is to find an example in which the rôle of number is played by some logical construct, such as the Frege-Russell class of similar classes, which can be shown to have the properties required. If we approach the subject from a standpoint different from that of *Principia Mathematica*, we may say that this is what the authors of that work have actually done.

Finally, if we have established consistency in geometry and arithmetic, can we do so in logic, or in a subject which includes logic? It has been held, and I think by Russell, that we cannot, because our formulæ symbolise, among other things, the logical processes which we use in examining it, because the rules of the game are required in forming the judgement that what purports to be an instance of the game really is one. Other logicians, with whom here I agree, have held that this is a misunderstanding, due to a failure to distinguish between the use of our symbolism inside and outside the formal system.

My own view is that even here the classical method, the method of instances, is available in principle, and that, in restricted subjects such as the logic of classes or of propositions, it can be and has been successfully carried through. If, however, we are as ambitious as Hilbert, so that our system is to cover the whole field of abstract thought, I imagine that the attempt to do what we want on these lines is hopeless. I cannot imagine where we could find an adequate image of so comprehensive a symbolism, except in the whole field of thought which it was actually constructed to symbolise. There remains only the 'internal' method followed by Hilbert, based on study of the formal properties of the rules themselves. Whatever we may think about the philosophical basis on which Hilbert has erected his system, and with whatever success he or his followers may pursue it, it seems to me unquestionable that this method is valid in principle, in mathematics in exactly the same sense as in chess. And in this case Hilbert is entirely justified in his claim that he has found a necessary condition for all systems of mathematical logic, and that 'even the assertions of intuitionism, however modest they may be, require first a certificate of authorisation from this tribunal'.

15. My remarks up to this point have been mainly explanations of things which I think I understand. The rest of what I have to say amounts to little more than a confession of a series of perplexities.

The first question which you will naturally ask is this: granted that Hilbert's method is valid in principle, what has it *done*? How far has the proof of consistency progressed? *Does* it establish freedom from contradiction in a domain co-extensive with mathematics? So far as I know the answer is, up to the present, *No*. There has been very substantial progress, and consistency has been proved up to a point beyond the point up to which success might be expected to be easy. The region accounted for includes the mathematics of the finitists, and that part of *Principia Mathematica* which is independent of the Axiom of Reducibility; but this region does not cover analysis.

It would be very reasonable to ask me, as an analyst, to explain my own attitude towards this hiatus in the foundations of analysis, and I do not profess to be able to give any satisfactory answer. I could only say this: in the first place, I am no finitist; I believe that the analysis of the text-books is true. Secondly, Ramsey has advanced a solution, which he does not profess to regard as entirely satisfactory, but in which I can find a good deal of encouragement. Ramsey makes a distinction, which seems to me obviously valid, between the properly mathematical antinomies, those which (like Russell's) would appear, unless precautions against them were taken, in the structure of mathematics itself, and those which appear to arise from some epistemological or psychological confusion concerning 'meaning' or 'definition'. He observes that Russell's theory of types can be divided into two parts, of which only the first, which is harmless, is required in order to dispose of the first category of antinomies, the second, from which all the trouble arises, being needed only for the antinomies of the second kind. He then puts forward a new theory, which might be described roughly as a revival, with appropriate safeguards, of the old-fashioned theory of classes in extension. In this theory there is no need for any axiom of reducibility; and this is at any rate the *sort* of solution that I should like to see. I cannot really doubt that there is a class which is the logical sum of any given set of classes, and this, or something like it, is all that is required by the Dedekind theory.

16. I will return for a moment in conclusion to the properly 'philosophical' question to which I referred at the beginning, about the reality or 'completeness' of propositions. I am entirely unable to exorcise my craving for real propositions, a weakness which is after all only natural in a mathematician, to whom mathematical theorems ought to be the first basic reality of life. But I can find no sort of encouragement wherever I turn.

Our first instinct is to suppose that a judgement, whether true or false, must be analysable into a mind and an object in relation. In a sense this is admitted to be true by everybody; it is undisputed that there is something objective, what Russell and Wittgenstein call the 'proposition as fact', which enters into any judgement. When we judge, we form a picture of the reality about which we are judging, a form of words, a set of marks or noises, which we suppose, rightly or wrongly, to afford an image of the facts. This is the 'proposition as fact'; the question is, what, if anything, is there more?

It can hardly be questioned that there is *something* more, something which is common to a whole class of factual propositions. If I say that 'George is the father of Edward', I create a factual proposition. If I and all other men say it, in all languages printed, written, or spoken, and formalise it in every conceivable symbolism, we create a class of facts, and there will plainly be something common to all these facts. This also is admitted; all such factual propositions have something in common, something which may be called their *form*. This, however, is by no means enough to satisfy me, since ' Edward is the father of George ' has *the same* form as ' George is the father of Edward ', while the propositions, if such there be, are plainly different.

In Russell's 'multiple relation' theory, the theory of truth accepted provisionally in the first edition of *Principia Mathematica*, no such entity as the proposition is recognised. A judgement is a complex of objects, of which a mind is one, my mind and ' George ' and ' Edward ' and ' fatherhood ', if we treat all these for simplicity as simple objects. If George *is* the father of Edward (so that the judgement is true) then there is a smaller complex, the 'fact' that George is the father of Edward, which is a part of the larger complex which is the judgement. If the judgement is false, there is no such subordinate complex. In neither case is there anything which can be called the ' proposition '. First descriptions, then classes, then propositions have been washed away into the ocean of the incomplete.

I have myself always detested this theory of truth. Apart from my bewilderment about how a structure such as that of *Principia Mathematica* could possibly be built up on so bottomless a foundation, Russell's theory has always seemed to me to banish entirely the element of correspondence which I have felt to be essential in any theory. My own difficulty has always been this, that I find it impossible not to believe in false or uncertain propositions and almost equally difficult to believe in true ones. When we judge truly, there is something which is admitted, namely the fact; and it seems unreasonable to insist on the independent existence of the proposition as something distinct from either the judgement or the fact. When we judge falsely, there is no fact, and, unless we admit the proposition, there seems to be no foundation for our judgement. It seems, therefore, that there must be some subsidiary complex present in *any* judgement, and this is just what Russell's theory denies.

It was therefore with great relief that I found that Wittgenstein rejects Russell's theory, for a variety of reasons of which the most convincing seems to be that Russell's theory leaves it entirely unexplained why it should be impossible to judge a *nonsense*. It would seem, on Russell's theory, that if you can judge that Edward is the father of George, you should be equally capable of judging that Edward is the father of blue.

Wittgenstein's own theory, if I understand it correctly, is something like this. We begin with reality, the facts. Of these facts we construct pictures, the factual propositions. A factual proposition consists of objects, words, noises, chairs or tables, arranged in a certain *form*. This form is *the same* form as that of reality; it is only because the picture and the facts have the same form that they can be compared with one another. If the fact is that George is the father of Edward, then the picture 'Edward is the father of George' has the same form, and it is just because of this that we can say that the picture is a bad one, the proposition is false. 'The picture can represent every reality whose form it has. . . . The picture, however, cannot represent its form of representation; it shows it forth. . . . The picture has the logical form of representation in common with what it pictures. . . . It agrees with reality or not; it is right or wrong, true or false. . . .'

There is, however, something beside the picture or factual proposition, namely the proposition in the sense which is relevant to logic. What is relevant to logic is not the factual proposition but what is common to all the factual propositions that can be pictures of a given state of affairs. A proposition is thus, in some sense, a *form*. The propositions of Hilbert's logic are also forms, but Wittgenstein's forms are more substantial than Hilbert's, since they contain what Russell and Wittgenstein call the 'logical constants', 'and', 'or', 'not' and so forth, whereas Hilbert's can hardly be said to 'contain' anything at all. These logical constants do not represent and are not represented, but are present in the proposition (that is to say the factual proposition) as in the fact. The proposition (that is to say here the logical proposition) is thus a form of logical constants, whereas Hilbert's propositions are so to say *pure* form.

I ask then, finally, whether there is anything in the proposition, as relevant to logic and as Wittgenstein seems to conceive it, which affords any justification for my belief in 'real' propositions, my invincible feeling that, if Littlewood and I both believe Goldbach's theorem, then there is something, and that the same something, in which we both believe, and that that same something will remain the same something when each of us is dead and when succeeding generations of more skilful mathematicians have proved our belief to be right or wrong. I hoped to find support for such a view, when I read that 'the essential in a proposition is that which is common to all propositions which can express the same sense' and that 'the proposition is the propositional sign in its projective relation to the world'. When I read further, both in the book itself and in what

Russell says about it, I concluded that I had been deceived. I can find nothing, in Wittgenstein's theory, that is common to all the ways in which I can say that something is true and is not common also to many of the ways in which I can say that it is false. So here I can find no support for my belief; and if not here, where am I likely to find it? Yet my last remark must be that I am still convinced that it is true.

POSTSCRIPT.

I have left this lecture as it was delivered, but I should like to add two remarks.

(1) My quotation from Mr. Ramsey at the beginning of § 8 may lead to a misinterpretation of his general view of formalism. I understand from what Mr. Ramsey has written later, and from conversation with him, that his attitude towards Hilbert's logic is, up to a point at any rate, somewhat like my own, that is to say that he accepts the logic without accepting its philosophical foundation. In saying this, of course, I must not be interpreted as claiming Mr. Ramsey's approval for anything in particular that I say in the lecture.

(2) Prof. J. W. Alexander of Princeton has made the following remark to me concerning Hilbert's 'ideal theorems'. The fact that a great part of a formalism has been suggested by 'significant' concepts and propositions does not show that *all* its theorems must be capable of interpretation; there will generally be formulæ to which 'no meaning' can be attributed, and the study of these 'meaningless' formulæ may well advance our understanding of the relations of those which can be interpreted. Indeed (as v. Neumann has pointed out) the formalism *must* contain formulæ of this kind, since (*e.g.*) we can substitute a 'numerical' symbol inside a 'logical' formula, 2 for a and b in '$a \to (b \to a)$': no one has suggested any 'meaning' for '$2 \to (2 \to 2)$'.

It is natural to interpret Hilbert as meaning that his 'ideal theorems' are all of this kind; and that his logic does contain theorems 'ideal' in this sense is obvious after what I have just said. It is one thing to admit this, and another to admit that a particular proposition such as 'there are infinitely many primes' is 'ideal'. If I cannot admit that 'there are infinitely many primes' has no 'meaning', it is simply because it seems evident to me what the 'meaning' is.

CORRECTION

p. 8, *line* 15. For x read X.

COMMENTS

This paper gives a lucid and readable account of some of the basic problems of mathematical logic, though it could be argued that Hardy does less than justice to the case for intuitionism. Some of the later sections would probably have been modified if the paper had been written after Gödel's incompleteness results became known.

MATHEMATICS

MATHEMATICS is one of the traditional subjects of study in Oxford as elsewhere, but it probably attracts less notice in Oxford than in any other considerable university. The first thing which a Cambridge man notices when he migrates to Oxford is that mathematics and physics, the 'ranking' subjects in his old University, are overshadowed in Oxford not merely by the literary schools but even by other sciences. The friend who conveyed to me the invitation to write this article remarked to me (I believe quite innocently) that 'the Editor wondered whether it would be worth while to include in this series an article on the School of Mathematics.' I take the invitation as a compliment in the circumstances, but I cannot help reflecting that a similar question would seem a very odd one in Paris or Göttingen or Harvard or Princeton. I have encountered a number of curious instances of the ordinary Oxford estimate of mathematics. One is in Jowett's delightfully naive memoir of Henry Smith, printed in Smith's *Collected Mathematical Papers*. Smith, says Jowett, 'was not the author of any considerable work'; he 'lived and died almost unknown to the world at large.' I wonder how many Germans knew the work of Smith, for one who had ever heard of Benjamin Jowett.

Another example was in an amusing little undergraduate book called *Isis*, which I picked up last Term in Blackwell's shop. The thesis of the book was, briefly, 'to Hell with Science.' Stamp out these insufferable scientists, and these uncouth and illiterate scholars from the secondary schools, and Oxford may be once again a home for a gentleman and a Hellenist. Remembering that mathematics ranks as a science in some universities, but as an art in others, and hoping against hope that we mathematicians might escape condemnation, I turned hastily over the pages. I am sorry to say that what I found was more humiliating than the worst that I had dreaded. Mathematics was not mentioned; the author did not know that such a subject existed.

What are the reasons for this ignorance and indifference? Mathematics is an ancient, and one would have supposed quite a gentlemanly, study. If Oxford desires above all things to be Hellenistic, it could hardly find a more Hellenistic activity; Pythagoras and Archimedes, after all, were Greeks, and Archimedes could have given Aristotle a good fifteen. Why then should Oxford tolerate its mathematicians but be so totally uninterested in what they do?

It is usually supposed that mathematics has languished in Oxford because the School here has been overshadowed by the Cambridge School. Cambridge has had its Tripos, the most difficult, the most notorious, and the best advertised examination in the world, and all the best mathematicians in England have flocked to Cambridge to distinguish themselves in this ordeal, so that Oxford has been starved of ability. This is the common explanation, and no doubt it contains a certain amount of truth. Oxford has produced a good many more distinguished mathematicians than is generally realised, but the majority of English mathematicians are Cambridge men, and they almost monopolise the more important chairs; and this is naturally discouraging to an Oxford student who wishes to make a name and a career for himself as a mathematician.

I do not think that it ought to be very difficult for the Oxford School of Mathematics to eradicate this rather humiliating inferiority complex. We may begin by reflecting (and it is easy for me as a Cambridge man to say so) that English opinion has overrated very grossly the status of Cambridge mathematics; if we have lagged behind them, they have lagged still further behind the great continental centres, and we have only to beat an opponent who has been beaten very badly a great many times already.

We can then remember that there is no subject in which it would be easier to develop a fine school than in mathematics, if once we had decided that we really wanted to do so. Harvard has created a school of the first rank in thirty years, Princeton in twenty, Hamburg in ten. 'The circumstances of the University,' says Jowett, 'hardly admitted' of Smith's 'raising up a school of mathematical pupils.' It is possible, but, if so, it was the fault of Jowett and people like him. There is no good reason at all why we should not do what these other universities have done. Mathematics, indeed, has here two advantages which no other subject combines; it is indispensable and it is comparatively inexpensive. It is indispensable because every other science (however little it may like it) finds itself compelled, sooner or later, to use mathematical methods; and it is inexpensive because it does not require either an enormous library or a costly laboratory equipment. No mathematician in the world needs very much more than brains and leisure.

In any case, we need not distress ourselves about Cambridge and its Tripos. If we cannot hope to rival those faded glories, so much the better; it is not possible to found a school of learning on an examination. I must not succumb to the temptation of airing views on 'honours schools' which most of my colleagues might reject. Still, the mathematical world at large has never cared two pins about the Tripos, Cambridge mathematicians themselves have long since ceased to attach any exaggerated importance to it, and, if we have here no examination of equal status and fame, our freedom from such shackles should be an inestimable advantage to us in the development of our School.

If then Oxford wishes to have a first-rate School of Mathematics, it can have one; the difficulty for us is to persuade the University (or rather the Colleges) that such a School is an asset worth possessing. It is a question, primarily, of buying the men; mathematicians are reasonably cheap, but they cannot be had for nothing, and at present there is a very serious lack of openings for rising men. It is here that we find the really serious contrast with Cambridge. The University itself is liberal enough; the number of professorships here and at Cambridge is the same; but we find a distressing disparity as soon as we look at college appointments. In this respect mathematics falls between two stools. It is not a 'modern' subject, with a centre of its own independent of college life, and it is not one of the favoured subjects on which scholarships and fellowships are showered. At Cambridge, Trinity has four mathematical lecturers, St. John's three, and every college one at least. Here Exeter, Lincoln, Magdalen, Oriel, Pembroke, Trinity, University, Wadham, and Worcester have not a mathematician between them; in none of those great centres of philosophy is there a man who can hope to understand Hilbert or Russell, Einstein or Schrödinger.

This is the great handicap with which Oxford mathematics has to contend, and I imagine that it can be overcome only gradually and indirectly. We can raise the prestige of Oxford mathematics in Oxford only by raising it first outside, and it is plain that we can do this only by making our school primarily a school of advanced study and research. There is no word which excites such bitter feeling in a 'teaching university' as the word 'research,' and I do not wish to suggest that the advancement of knowledge is the only worthy outlet for a don's activities. The fact remains that it is quite impossible to erect a school of any science, and in particular a School of Mathematics, on any other foundation. The world, when asked to admire the Oxford School, will inquire 'what it has done,' and will be quite uninterested unless it can show that it has done something substantial. It is already practically impossible for a mathematician who has not 'added to knowledge' even to start on an academical career.

It is fortunate that we can claim that our School already stands this test quite well. The current periodicals certainly show no lack of substantial contributions by Oxford men. The new *Quarterly Journal*, captured from Cambridge by the enterprise of the Clarendon Press, should put our activities still more clearly in evidence and add materially to our prestige abroad. We have now a journal, and we demand next a home. Why should Oxford, almost alone among great Universities, have no centre of any kind for its mathematical work? Why should a professor have to grovel before a College Bursar when he wants a room and a decent blackboard?

We are, therefore, asking for a Mathematical Institute as a centre for Oxford mathematical life. It is an obviously reasonable, and a modest, demand, remarkable only because it has been delayed so long, and sooner or later it must be met. Whether it is met sooner or later, the proposal is in itself a most encouraging sign, as evidence of a new spirit of enterprise and assertiveness on the part of Oxford mathematicians, and proof that they are beginning to rebel against the subjection, unworthy of a splendid subject, in which they have acquiesced too long.

G. H. HARDY.

COMMENT

Hardy's plea for a Mathematical Institute was ultimately granted when in 1953 the Oxford Mathematical Institute was established at 10 Parks Road. The Institute, which moved to its present location at 22–29 St Giles in 1966, is under the general supervision of the Board of the Faculty of Mathematics.

THE J-TYPE AND THE S-TYPE AMONG MATHEMATICIANS

BY

PROF. G. H. HARDY, F.R.S.
New College, Oxford

MATHEMATICIANS in England and America have been recently intrigued by reports of a lecture delivered by Prof. L. Bieberbach, of the University of Berlin, to the Verein zur Förderung des mathematisch-naturwissenschaftlichen Unterrichts. They have, however, found difficulty in judging the lecture fairly from secondhand reports. It is now possible to form a more reasoned estimate, Prof. Bieberbach having published a considerable extract, under the title "Persönlichkeitsstruktur und mathematisches Schaffen", in the issue of *Forschungen und Fortschritte* of June 20.

Prof. Bieberbach begins by explaining that his exposition will make clear by examples the influence of nationality, blood and race upon the creative style. For a National Socialist, the importance of this influence requires no proof. Rather is it intuitive that all our actions and thoughts are rooted in blood and race and receive their character from them. Every mathematician can recognise such influences in different mathematical styles. Blood and race determine our choice of problems, and so influence even the assured content of science (den Bestand der Wissenschaften an gesicherten Ergebnissen); but naturally do not go so far as to affect the value of π or the validity of Pythagoras' theorem in Euclidean geometry. . . .

Our nature becomes conscious of itself in the malaise (in dem Unbehagen) produced by alien ways. There is an example in the manly rejection (mannhafte Ablehnung) of a great mathematician, Edmund Landau, by the students of Göttingen. The unGerman style of this man in teaching and research proved intolerable to German sensibilities. A people which has understood how alien lust for dominance has gnawed into its vitals . . . must reject teachers of an alien type. . . .

Prof. Bieberbach proceeds to distinguish between the 'J-type' and the 'S-type' among mathematicians. Broadly, the J-type are Germans, the S-type Frenchmen and Jews. The differences of type appear quite clearly in the varying treatments by different mathematicians of the theory of imaginary numbers. For example, in Gauss (an outstanding instance of the J-type) one finds above all insistence on the 'anschauliche Bedeutung von $\sqrt{-1}$'. . . . On the other hand, there are expositions of the theory by mathematicians of the S-type (for example, Cauchy) which produce a malaise (die Unbehagen verursachen) in one belonging to the J-type. . . . Technical virtuosity and juggling with conceptions are signs betraying the S-type, hostile to life and inorganic (dem Lebensfeindlichen unorganischen S-typus). . . .

Typical of the J-type are the 'nordisch-falische' Gauss, the 'nordisch-dinarische' Klein, and the 'ostbaltisch-nordische' Hilbert. . . . One of the crowning achievements of the J-type is Hilbert's work on axiomatics, and it is particularly regrettable that abstract Jewish thinkers of the S-type should have succeeded in distorting it into an intellectual variety performance (intellektuelles Variété). . . .

But perhaps I have quoted enough; and I feel disposed to add one comment only. It is not reasonable to criticise too closely the utterances, even of men of science, in times of intense political or national excitement. There are many of us, many Englishmen and many Germans, who said things during the War which we scarcely meant and are sorry to remember now. Anxiety for one's own position, dread of falling behind the rising torrent of folly, determination at all costs not to be outdone, may be natural if not particularly heroic excuses. Prof. Bieberbach's reputation excludes such explanations of his utterances; and I find myself driven to the more uncharitable conclusion that he really believes them true.

COMMENTS

The circumstances that forced Landau to resign his chair in Göttingen in 1933 are referred to in the obituary notice O 14 (p. 762 of this volume) and in Constance Reid's *Hilbert* (Allen & Unwin, Springer, 1970). Bieberbach was not the first to link mathematical creative style with race. See, for example, Hardy's review (R18) of a book by Hadamard, where reference is made to Felix Klein's views on the subject (p. 837 of this volume).

Hardy's letter is dated 20 July, 1934. After he returned to Cambridge in 1931 he used to go over to Oxford for several weeks in the Long Vacation to captain the New College Senior Common Room at cricket against the choir school and against the college servants.

THE INDIAN MATHEMATICIAN RAMANUJAN*

By G. H. HARDY, Cambridge University

I have set myself a task in these lectures which is genuinely difficult and which, if I were determined to begin by making every excuse for failure, I might represent as almost impossible. I have to form myself, as I have never really formed before, and to try to help you to form, some sort of reasoned estimate of the most romantic figure in the recent history of mathematics; a man whose career seems full of paradoxes and contradictions, who defies almost all the canons by which we are accustomed to judge one another, and about whom all of us will probably agree in one judgment only, that he was in some sense a very great mathematician.

The difficulties in judging Ramanujan are obvious and formidable enough. Ramanujan was an Indian, and I suppose that it is always a little difficult for an Englishman and an Indian to understand one another properly. He was, at the best, a half-educated Indian; he never had the advantages, such as they are, of an orthodox Indian training; he never was able to pass the "First Arts Examination" of an Indian university, and never could rise even to be a "Failed B.A." He worked, for most of his life, in practically complete ignorance of modern European mathematics, and died when he was a little over 30 and when his mathematical education had in some ways hardly begun. He published abundantly—his published papers make a volume of nearly 400 pages—but he also left a mass of unpublished work which had never been analysed properly until the last few years. This work includes a great deal that is new, but much more that is rediscovery, and often imperfect rediscovery; and it is sometimes still impossible to distinguish between what he must have rediscovered and what he may somehow have learnt. I cannot imagine anybody saying with any confidence, even now, just how great a mathematician he was and still less how great a mathematician he might have been.

These are genuine difficulties, but I think that we shall find some of them less formidable than they look, and the difficulty which is the greatest for me has nothing to do with the obvious paradoxes of Ramanujan's career. The real difficulty for me is that Ramanujan was, in a way, my discovery. I did not invent him—like other great men, he invented himself—but I was the first really competent person who had the chance to see some of his work, and I can still remember with satisfaction that I could recognise at once what a treasure I had found. And I suppose that I still know more of Ramanujan than any one else, and am still the first authority on this particular subject. There are other people in England, Professor Watson in particular, and Professor Mordell, who know parts of his work very much better than I do, but neither Watson nor Mordell knew Ramanujan himself as I did. I saw him and talked with him almost every

---

* A lecture delivered at the Harvard Tercentenary Conference of Arts and Sciences, August 31, 1936.

day for several years, and above all I actually collaborated with him. I owe more to him than to any one else in the world with one exception, and my association with him is the one romantic incident in my life. The difficulty for me then is not that I do not know enough about him, but that I know and feel too much and that I simply cannot be impartial.

I rely, for the facts of Ramanujan's life, on Seshu Aiyar and Ramachaundra Rao, whose memoir of Ramanujan is printed, along with my own, in his *Collected Papers*. He was born in 1887 in a Brahmin family at Erode near Kumbakonam, a fair-sized town in the Tanjore district of the Presidency of Madras. His father was a clerk in a cloth-merchant's office in Kumbakonam, and all his relatives, though of high caste, were very poor.

He was sent at 7 to the High School of Kumbakonam, and remained there nine years. His exceptional abilities had begun to show themselves before he was 10, and by the time that he was 12 or 13 he was recognised as a quite abnormal boy. His biographers tell some curious stories of his early years. They say for example that, soon after he had begun the study of trigonometry, he discovered for himself "Euler's theorems for the sine and cosine" (by which I understand the relations between the circular and exponential functions), and was very disappointed when he found later, apparently from the second volume of Loney's *Trigonometry*, that they were known already. Until he was 16 he had never seen a mathematical book of any higher class. *Whittaker's Modern Analysis* had not yet spread so far, and Bromwich's *Infinite Series* did not exist. There can be no doubt that either of these books would have made a tremendous difference to him if they could have come his way. It was a book of a very different kind, Carr's *Synopsis*, which first aroused Ramanujan's full powers.

Carr's book (*A synopsis of elementary results in pure and applied mathematics*, by George Shoobridge Carr, formerly Scholar of Gonville and Caius College, Cambridge, published in two volumes in 1880 and 1886) is almost unprocurable now. There is a copy in the Cambridge University Library, and there happened to be one in the library of the Government College of Kumbakonam, which was borrowed for Ramanujan by a friend. The book is not in any sense a great one, but Ramanujan has made it famous, and there is no doubt that it influenced him profoundly and that his acquaintance with it marked the real starting point of his career. Such a book must have had its qualities, and Carr's, if not a book of any high distinction, is no mere third-rate textbook, but a book written with some real scholarship and enthusiasm and with a style and individuality of its own. Carr himself was a private coach in London, who came to Cambridge as an undergraduate when he was nearly 40, and was 12th Senior Optime in the Mathematical Tripos of 1880 (the same year in which he published the first volume of his book). He is now completely forgotten, even in his own college, except in so far as Ramanujan has kept his name alive; but he must have been in some ways rather a remarkable man.

I suppose that the book is substantially a summary of Carr's coaching notes. If you were a pupil of Carr, you worked through the appropriate sections of the

Synopsis. It covers roughly the subjects of Schedule A of the present Tripos (as these subjects were understood in Cambridge in 1880), and is effectively the "synopsis" it professes to be. It contains the enunciations of 6165 theorems, systematically and quite scientifically arranged, with proofs which are often little more than cross-references and are decidedly the least interesting part of the book. All this is exaggerated in Ramanujan's famous note-books (which contain practically no proofs at all), and any student of the note-books can see that Ramanujan's ideal of presentation had been copied from Carr's.

Carr has sections on the obvious subjects, algebra, trigonometry, calculus and analytical geometry, but some sections are developed disproportionally, and particularly the formal side of the integral calculus. This seems to have been Carr's pet subject, and the treatment of it is very full and in its way definitely good. There is no theory of functions; and I very much doubt whether Ramanujan, to the end of his life, ever understood at all clearly what an analytic function is. What is more surprising, in view of Carr's own tastes and Ramanujan's later work, is that there is no elliptic functions. However Ramanujan may have acquired his very peculiar knowledge of this theory, it was not from Carr.

On the whole, considered as an inspiration for a boy of such abnormal gifts, Carr was not too bad, and Ramanujan responded amazingly.

"Through the new world thus opened to him," say his Indian biographers,* "Ramanujan went ranging with delight. It was this book which awakened his genius. He set himself to establish the formulae given therein. As he was without the aid of other books, each solution was a piece of research so far as he was concerned ... Ramanujan used to say that the goddess of Namakkal inspired him with the formulae in dreams. It is a remarkable fact that frequently, on rising from bed, he would note down results and rapidly verify them, though he was not always able to supply a rigorous proof. ..."

I have quoted the last sentences deliberately, not because I attach any importance to them—I am no more interested in the goddess of Namakkal than you are—but because we are now approaching the difficult and tragic part of Ramanujan's career, and we must try to understand what we can of his psychology and of the atmosphere surrounding him in his early years.

I am sure that Ramanujan was no mystic and that religion, except in a strictly material sense, played no important part in his life. He was an orthodox high-caste Hindu, and always adhered (indeed with a severity most unusual in Indian residents in England) to all the observances of his caste. He had promised his parents to do so, and he kept his promises to the letter. He was a vegetarian in the strictest sense—this proved a terrible difficulty later when he fell ill—and all the time he was in Cambridge he cooked all his food himself, and never cooked it without first changing into pyjamas.

Now the two memoirs of Ramanujan printed in the *Papers* (and both written

---

* Quotations (except those from my own memoir of Ramanujan) are from Seshu Aiyar and Ramachaundra Rao.

by men who, in their different ways, knew him very well) contradict one another flatly about his religion. Seshu Aiyar and Ramachaundra Rao say

"Ramanujan had definite religious views. He had a special veneration for the Namakkal goddess.... He believed in the existence of a Supreme Being and in the attainment of Godhead by men.... He had settled convictions about the problem of life and after . . . ";

while I say

" . . . his religion was a matter of observance and not of intellectual conviction, and I remember well his telling me (much to my surprise) that all religions seemed to him more or less equally true . . . ".

Which of us is right? For my part I have no doubt at all; I am quite certain that I am.

Classical scholars have, I believe, a general principle, *difficilior lectio potior*—the more difficult reading is to be preferred—in textual criticism. If the Archbishop of Canterbury tells one man that he* believes in God, and another that he does not, then it is probably the second assertion which is true, since otherwise it is very difficult to understand why he should have made it, while there are many excellent reasons for his making the first whether it be true or false. Similarly, if a strict Brahmin like Ramanujan told me, as he certainly did, that he had no definite beliefs, then it is 100 to 1 that he meant what he said.

This was no sufficient reason why Ramanujan should outrage the feelings of his parents or his Indian friends. He was not a reasoned infidel, but an "agnostic" in its strict sense, who saw no particular good, and no particular harm, in Hinduism or in any other religion. Hinduism is, far more for example than Christianity, a religion of observance, in which belief counts for extremely little in any case, and, if Ramanujan's friends assumed that he accepted the conventional doctrines of such a religion, and he did not disillusion them, he was practising a quite harmless, and probably necessary, economy of truth.

This question of Ramanujan's religion is not itself important, but it is not altogether irrelevant, because there is one thing which I am really anxious to insist upon as strongly as I can. There is quite enough about Ramanujan that is difficult to understand, and we have no need to go out of our way to manufacture mystery. For myself, I liked and admired him enough to wish to be a rationalist about him; and I want to make it quite clear to you that Ramanujan, when he was living in Cambridge in good health and comfortable surroundings, was, in spite of his oddities, as reasonable, as sane, and in his way as shrewd a person as anyone here. The last thing which I want you to do is to throw up your hands and exclaim "here is something unintelligible, some mysterious manifestation of the immemorial wisdom of the East!" I do not believe in the immemorial wisdom of the East, and the picture I want to present to you is that of a man who had his peculiarities like other distinguished men, but a man in whose

---

* The Archbishop.

society one could take pleasure, with whom one could take tea and discuss politics or mathematics; the picture in short, not of a wonder from the East, or an inspired idiot, or a psychological freak, but of a rational human being who happened to be a great mathematician.

Until he was about 17, all went well with Ramanujan.

"In December 1903 he passed the Matriculation Examination of the University of Madras, and in the January of the succeeding year he joined the Junior First in Arts class of the Government College, Kumbakonam, and won the Subrahmanyam scholarship, which is generally awarded for proficiency in English and Mathematics . . . ",

but after this there came a series of tragic checks.

"By this time, he was so absorbed in the study of Mathematics that in all lecture hours—whether devoted to English, History, or Physiology—he used to engage himself in some mathematical investigation, unmindful of what was happening in the class. This excessive devotion to mathematics and his consequent neglect of the other subjects resulted in his failure to secure promotion to the senior class and in the consequent discontinuance of the scholarship. Partly owing to disappointment and partly owing to the influence of a friend, he ran away northward into the Telugu country, but returned to Kumbakonam after some wandering and rejoined the college. As owing to his absence he failed to make sufficient attendances to obtain his term certificate in 1905, he entered Pachaiyappa's College, Madras, in 1906, but falling ill returned to Kumbakonam. He appeared as a private student for the F. A. examination of December 1907 and failed . . .".

Ramanujan does not seem to have had any definite occupation, except mathematics, until 1912. In 1909 he married, and it became necessary for him to have some regular employment, but he had great difficulty in finding any because of his unfortunate college career. About 1910 he began to find more influential Indian friends, Ramaswami Aiyar and his two biographers, but all their efforts to find a tolerable position for him failed, and in 1912 he became a clerk in the office of the Port Trust of Madras, at a salary of about £30 a year. He was then nearly 25. The years between 18 and 25 are the critical years in a mathematician's career, and the damage had been done. Ramanujan's genius never had again its chance of full development.

There is not much to say about the rest of Ramanujan's life. His first substantial paper had been published in 1911, and in 1912 his exceptional powers began to be understood. It is significant that, though Indians could befriend him, it was only the English who could get anything effective done. Sir Francis Spring and Sir Gilbert Walker obtained a special scholarship for him, £60 a year, sufficient for a married Indian to live in tolerable comfort. At the beginning of 1913 he wrote to me, and Professor Neville and I, after many difficulties, got him to England in 1914. Here he had three years of uninterrupted activity, the results of which you can read in his *Papers*. He fell ill in the summer of 1917, and never really recovered, though he continued to work, rather spasmodically, but with no real sign of degeneration, until his death in 1920. He became a Fellow of the Royal Society early in 1918, and a Fellow of Trinity College, Cambridge, later in the same year (and was the first Indian elected to either society). His last mathematical letter on "Mock-Theta functions", the subject of Professor Watson's presidential address to the London Mathematical Society last year, was written about two months before he died.

The real tragedy about Ramanujan was not his early death. It is of course a disaster that any great man should die young, but a mathematician is often comparatively old at 30, and his death may be less of a catastrophe than it seems. Abel died at 26 and, although he would no doubt have added a great deal more to mathematics, he could hardly have become a greater man. The tragedy of Ramanujan was not that he died young, but that, during his five unfortunate years, his genius was misdirected, side-tracked, and to a certain extent distorted.

I have been looking again through what I wrote about Ramanujan 16 years ago, and, although I know his work a good deal better now than I did then, and can think about him more dispassionately, I do not find a great deal which I should particularly want to alter. But there is just one sentence which now seems to me indefensible. I wrote

"Opinions may differ about the importance of Ramanujan's work, the kind of standard by which it should be judged, and the influence which it is likely to have on the mathematics of the future. It has not the simplicity and the inevitableness of the very greatest work; it would be greater if it were less strange. One gift it shows which no one can deny, profound and invincible originality. He would probably have been a greater mathematician if he could have been caught and tamed a little in his youth; he would have discovered more that was new, and that, no doubt, of greater importance. On the other hand he would have been less of a Ramanujan, and more of a European professor, and the loss might have been greater than the gain . . ."

and I stand by that except for the last sentence, which is quite ridiculous sentimentalism. There was no gain at all when the College at Kumbakonam rejected the one great man they had ever possessed, and the loss was irreparable; it is the worst instance that I know of the damage that can be done by an inefficient and inelastic educational system. So little was wanted, £60 a year for five years, occasional contact with almost anyone who had real knowledge and a little imagination, for the world to have gained another of its greatest mathematicians.

Ramanujan's letters to me, which are reprinted in full in the *Papers*, contain the bare statements of about 120 theorems, mostly formal identities extracted from his note-books. I quote fifteen which are fairly representative. They include two theorems, (14) and (15), which are as interesting as any but of which one is false and the other, as stated, misleading. The rest have all been verified since by somebody; in particular Rogers and Watson found the proofs of the extremely difficult theorems (10)–(12).

(1) $$1 - \frac{3!}{(1!2!)^3}x^2 + \frac{6!}{(2!4!)^3}x^4 - \cdots$$
$$= \left(1 + \frac{x}{(1!)^3} + \frac{x^2}{(2!)^3} + \cdots\right)\left(1 - \frac{x}{(1!)^3} + \frac{x^2}{(2!)^3} - \cdots\right).$$

(2) $$1 - 5\left(\frac{1}{2}\right)^3 + 9\left(\frac{1.3}{2.4}\right)^3 - 13\left(\frac{1.3.5}{2.4.6}\right)^3 + \cdots = \frac{2}{\pi}.$$

(3) $$1 + 9\left(\frac{1}{4}\right)^4 + 17\left(\frac{1.5}{4.8}\right)^4 + 25\left(\frac{1.5.9}{4.8.12}\right)^4 + \cdots = \frac{2^{3/2}}{\pi^{1/2}\{\Gamma(\frac{3}{4})\}^2}.$$

(4) $$1 - 5\left(\frac{1}{2}\right)^5 + 9\left(\frac{1.3}{2.4}\right)^5 - 13\left(\frac{1.3.5}{2.4.6}\right)^5 + \cdots = \frac{2}{\{\Gamma(\frac{3}{4})\}^4}.$$

(5) $$\int_0^\infty \frac{1 + \left(\frac{x}{b+1}\right)^2}{1 + \left(\frac{x}{a}\right)^2} \cdot \frac{1 + \left(\frac{x}{b+2}\right)^2}{1 + \left(\frac{x}{a+1}\right)^2} \cdots dx = \frac{1}{2}\pi^{1/2}\frac{\Gamma(a+\frac{1}{2})\Gamma(b+1)\Gamma(b-a+\frac{1}{2})}{\Gamma(a)\Gamma(b+\frac{1}{2})\Gamma(b-a+1)}.$$

(6) $$\int_0^\infty \frac{dx}{(1+x^2)(1+r^2x^2)(1+r^4x^2)\cdots} = \frac{\pi}{2(1 + r + r^3 + r^6 + r^{10} + \cdots)}.$$

(7) If $\alpha\beta = \pi^2$, then
$$\alpha^{-1/4}\left(1 + 4\alpha\int_0^\infty \frac{xe^{-\alpha x^2}}{e^{2\pi x} - 1}dx\right) = \beta^{-1/4}\left(1 + 4\beta\int_0^\infty \frac{xe^{-\beta x^2}}{e^{2\pi x} - 1}dx\right).$$

(8) $$\int_0^a e^{-x^2}dx = \frac{1}{2}\pi^{1/2} - \frac{e^{-a^2}}{2a+}\frac{1}{a+}\frac{2}{2a+}\frac{3}{a+}\frac{4}{2a+}\cdots.$$

(9) $$4\int_0^\infty \frac{xe^{-x\sqrt{5}}}{\cosh x}dx = \frac{1}{1+}\frac{1^2}{1+}\frac{1^2}{1+}\frac{2^2}{1+}\frac{2^2}{1+}\frac{3^2}{1+}\frac{3^2}{1+}\cdots.$$

(10) If $$u = \frac{x}{1+}\frac{x^5}{1+}\frac{x^{10}}{1+}\frac{x^{15}}{1+}\cdots, \quad v = \frac{x^{1/5}}{1+}\frac{x}{1+}\frac{x^2}{1+}\frac{x^3}{1+}\cdots,$$

then $$v^5 = u\frac{1 - 2u + 4u^2 - 3u^3 + u^4}{1 + 3u + 4u^2 + 2u^3 + u^4}.$$

(11) $$\frac{1}{1+}\frac{e^{-2\pi}}{1+}\frac{e^{-4\pi}}{1+}\cdots = \left\{\sqrt{\left(\frac{5+\sqrt{5}}{2}\right)} - \frac{\sqrt{5}+1}{2}\right\}e^{2\pi/5}.$$

(12) $$\frac{1}{1+}\frac{e^{-2\pi\sqrt{5}}}{1+}\frac{e^{-4\pi\sqrt{5}}}{1+}\cdots = \left[\frac{\sqrt{5}}{1+\sqrt[5]{\left\{5^{3/4}\left(\frac{\sqrt{5}-1}{2}\right)^{5/2} - 1\right\}}} - \frac{\sqrt{5}+1}{2}\right]e^{2\pi/\sqrt{5}}.$$

(13) If $F(k) = 1 + \left(\frac{1}{2}\right)^2 k + \left(\frac{1.3}{2.4}\right)^2 k^2 + \cdots$ and $F(1-k) = \sqrt{(210)} F(k)$, then

$k = (\sqrt{2}-1)^4 (2-\sqrt{3})^2 (\sqrt{7}-\sqrt{6})^4 (8-3\sqrt{7})^2 (\sqrt{10}-3)^4 (4-\sqrt{15})^4 (\sqrt{15}-\sqrt{14})^2 (6-\sqrt{35})^2.$

(14) The coefficient of x^n in $(1 - 2x + 2x^4 - 2x^9 + \cdots)^{-1}$ is the integer nearest to

$$\frac{1}{4n}\left(\cosh(\pi\sqrt{n}) - \frac{\sinh(\pi\sqrt{n})}{\pi\sqrt{n}}\right).$$

(15) The number of numbers between A and x which are either squares or sums of two squares is

$$K \int_A^x \frac{dt}{\sqrt{(\log t)}} + \theta(x),$$

where $K = 0.764\cdots$ and $\theta(x)$ is very small compared with the previous integral.

I should like you to begin by trying to reconstruct the immediate reactions of an ordinary professional mathematician who receives a letter like this from an unknown Hindu clerk.

The first question was whether I could recognize anything. I had proved things rather like (7) myself, and seemed vaguely familiar with (8). Actually (8) is classical; it is a formula of Laplace first proved properly by Jacobi; and (9) occurs in a paper published by Rogers in 1907. I thought that, as an expert in definite integrals, I could probably prove (5) and (6), and did so, though with a good deal more trouble than I had expected. On the whole the integral formulas seemed the least impressive.

The series formulas (1)–(4) I found much more intriguing, and it soon became obvious that Ramanujan must possess much more general theorems and was keeping a great deal up his sleeve. The second is a formula of Bauer well known in the theory of Legendre series, but the others are much harder than they look. The theorems required in proving them can all be found now in Bailey's Cambridge Tract on hypergeometric functions.

The formulas (10)–(13) are on a different level and obviously both difficult and deep. An expert in elliptic functions can see at once that (13) is derived somehow from the theory of "complex multiplication", but (10)–(12) defeated me completely; I had never seen anything in the least like them before. A single look at them is enough to show that they could only be written down by a mathematician of the highest class. They must be true because, if they were not true, no one would have had the imagination to invent them. Finally (you must remember that I knew nothing whatever about Ramanujan, and had to think of every possibility), the writer must be completely honest, because great mathematicians are commoner than thieves or humbugs of such incredible skill.

The last two formulas stand apart because they are not right and show Ramanujan's limitations, but that does not prevent them from being additional evidence of his extraordinary powers. The function in (14) is a genuine approximation to the coefficient, though not at all so close as Ramanujan imagined, and Ramanujan's false statement was one of the most fruitful he ever made, since it ended by leading us to all our joint work on partitions. Finally (15), though literally "true", is definitely misleading (and Ramanujan was under a real misapprehension). The integral has no advantage, as an approximation, over the simpler function

$$\text{(16)} \qquad \frac{Kx}{\sqrt{(\log x)}},$$

found in 1908 by Landau. Ramanujan was deceived by a false analogy with the problem of the distribution of primes. I must postpone till later what I have to say about Ramanujan's work on this side of the theory of numbers.

It was inevitable that a very large part of Ramanujan's work should prove on examination to have been anticipated. He had been carrying an impossible handicap, a poor and solitary Hindu pitting his brains against the accumulated wisdom of Europe. He had had no real teaching at all; there was no one in India from whom he had anything to learn. He can have seen at the outside three or four books of good quality, all of them English. There had been periods in his life when he had access to the library in Madras, but it was not a very good one; it contained very few French or German books; and in any case Ramanujan did not know a word of either language. I should estimate that about two-thirds of Ramanujan's best Indian work was rediscovery, and comparatively little of it was published in his life-time, though Watson, who has worked systematically through his notebooks, has since disinterred a good deal more.

The great bulk of Ramanujan's published work was done in England. His mind had hardened to some extent, and he never became at all an "orthodox" mathematician, but he could still learn to do new things, and do them extremely well. It was impossible to teach him systematically, but he gradually absorbed new points of view. In particular he learnt what was meant by proof, and his later papers, while in some ways as odd and individual as ever, read like the works of a well-informed mathematician. His methods and his weapons, however, remained essentially the same. One would have thought that such a formalist as Ramanujan would have revelled in Cauchy's Theorem, but he practically never used it,* and the most astonishing testimony to his formal genius is that he never seemed to feel the want of it in the least.

It is easy to compile an imposing list of theorems which Ramanujan rediscovered. Such a list naturally cannot be quite sharp, since sometimes he found a part only of a theorem, and sometimes, though he found the whole theorem, he was without the proof which is essential if the theorem is to be properly understood. For example, in the analytic theory of numbers he had,

---

* Perhaps never. There is a reference to "the theory of residues" on p. 129 of the *Papers*, but I believe that I supplied this myself.

in a sense, discovered a great deal, but he was a very long way from understanding the real difficulties of the subject. And there is some of his work, mostly in the theory of elliptic functions, about which some mystery still remains; it is not possible, after all the work of Watson and Mordell, to draw the line between what he may have picked up somehow and what he must have found for himself. I will take only cases in which the evidence seems to me tolerably clear.

Here I must admit that I am to blame, since there is a good deal which we should like to know now and which I could have discovered quite easily. I saw Ramanujan almost every day, and could have cleared up most of the obscurity by a little cross-examination. Ramanujan was quite able and willing to give a straight answer to a question, and not in the least disposed to make a mystery of his achievements. I hardly asked him a single question of this kind; I never even asked him whether (as I think he must have done) he had seen Cayley's or Greenhill's *Elliptic Functions*.

I am sorry about this now, but it does not really matter very much, and it was entirely natural. In the first place, I did not know that Ramanujan was going to die. He was not particularly interested in his own history or psychology; he was a mathematician anxious to get on with the job. And after all I too was a mathematician, and a mathematician meeting Ramanujan had more interesting things to think about than historical research. It seemed ridiculous to worry him about how he had found this or that known theorem, when he was showing me half a dozen new ones almost every day.

I do not think that Ramanujan discovered much in the classical theory of numbers, or indeed that he ever knew a great deal. He had no knowledge at all, at any time, of the general theory of arithmetical forms. I doubt whether he knew the law of quadratic reciprocity before he came here. Diophantine equations should have suited him, but he did comparatively little with them, and what he did do was not his best. Thus he gave solutions of Euler's equation

(17) $$x^3 + y^3 + z^3 = w^3,$$

such as

(18) $\quad x = 3a^2 + 5ab - 5b^2, \ y = 4a^2 - 4ab + 6b^2, \ z = 5a^2 - 5ab - 3b^2, \ w = 6a^2 - 4ab + 4b^2;$

and

(19) $\quad x = m^7 - 3m^4(1+p) + m(2+6p+3p^2), \ y = 2m^6 - 3m^3(1+2p) + 1 + 3p + 3p^2,$

$\qquad z = m^6 - 1 - 3p - 3p^2, \ w = m^7 - 3m^4 p + m(3p^2 - 1);$

but neither of these is the general solution.

He rediscovered the famous theorem of von Staudt about the Bernoullian numbers:

(20) $$(-1)^n B_n = G_n + \frac{1}{2} + \frac{1}{p} + \frac{1}{q} + \cdots \frac{1}{r},$$

where $p, q, \cdots$ are those odd primes such that $p-1, q-1, \cdots$ are divisors of

$2n$, and G_n is an integer. In what sense he had proved it it is difficult to say, since he found it at a time of his life when he had hardly formed any definite concept of proof. As Littlewood says "the clear-cut idea of what is *meant* by a proof, nowadays so familiar as to be taken for granted, he perhaps did not possess at all; if a significant piece of reasoning occurred somewhere, and the total mixture of evidence and intuition gave him certainty, he looked no further". I shall have something to say later about this question of proof, but I postpone it to another context in which it is much more important. In this case there is nothing in the proof that was not obviously within Ramanujan's powers.

There is a considerable chapter of the theory of numbers, in particular the theory of the representation of integers by sums of squares, which is closely bound up with the theory of elliptic functions. Thus the number of representations of n by two squares is

$$(21) \qquad r(n) = 4\{d_1(n) - d_3(n)\},$$

where $d_1(n)$ is the number of divisors of n of the form $4k+1$ and $d_3(n)$ the number of divisors of the form $4k+3$. Jacobi gave similar formulas for 4, 6 and 8 squares. Ramanujan found all these, and much more of the same kind.

He also found Gauss's theorem that n is the sum of 3 squares except when it is of the form

$$(22) \qquad 4^a(8k+7),$$

but I do not attach much importance to this. The theorem is quite easy to guess and difficult to prove. All known proofs depend upon the general theory of ternary forms, of which Ramanujan knew nothing, and I agree with Professor Dickson in thinking it very unlikely that he possessed one. In any case he knew nothing about the number of representations.

Ramanujan, then, before he came to England, had added comparatively little to the theory of numbers; but no one can understand him who does not understand his passion for numbers in themselves. I wrote before

"He could remember the idiosyncrasies of numbers in an almost uncanny way. It was Littlewood who said that every positive integer was one of Ramanujan's personal friends. I remember going to see him once when he was lying ill in Putney. I had ridden in taxi-cab No. 1729, and remarked that the number seemed to me rather a dull one, and that I hoped that it was not an unfavorable omen. 'No,' he replied, 'it is a very interesting number; it is the smallest number expressible as a sum of two cubes in two different ways.'[*] I asked him, naturally, whether he could tell me the solution of the corresponding problem for fourth powers; and he replied, after a moment's thought, that he knew no obvious example, and supposed that the first such number must be very large."[†]

---

[*] $1729 = 12^3 + 1^3 = 10^3 + 9^3$.

[†] The smallest known is Euler's example
$$635318657 = 158^4 + 59^4 = 134^4 + 133^4.$$

In algebra, Ramanujan's main work was concerned with hypergeometric series and continued fractions (I use the word algebra, of course, in its old-fashioned sense). These subjects suited him exactly, and here he was unquestionably one of the great masters. There are three now famous identities, the "Dougall-Ramanujan identity"

$$(23) \quad \sum_{n=0}^{\infty} (-1)^n (s+2n) \frac{s^{(n)}}{1^{(n)}} \frac{(x+y+z+u+2s+1)^{(n)}}{(x+y+z+u+s)_{(n)}} \prod_{x,y,z,u} \frac{x_{(n)}}{(x+s+1)^{(n)}}$$

$$= \frac{s}{\Gamma(s+1)\Gamma(x+y+z+u+s+1)} \prod_{x,y,z,u} \frac{\Gamma(x+s+1)\Gamma(y+z+u+s+1)}{\Gamma(z+u+s+1)},$$

where

$$a^{(n)} = a(a+1) \cdots (a+n-1), \qquad a_{(n)} = a(a-1) \cdots (a-n+1),$$

and the "Rogers-Ramanujan identities"

$$(24) \quad 1 + \frac{q}{1-q} + \frac{q^4}{(1-q)(1-q^2)} + \frac{q^9}{(1-q)(1-q^2)(1-q^3)} + \cdots$$

$$= \frac{1}{(1-q)(1-q^6) \cdots (1-q^4)(1-q^9) \cdots},$$

$$1 + \frac{q^2}{1-q} + \frac{q^6}{(1-q)(1-q^2)} + \frac{q^{12}}{(1-q)(1-q^2)(1-q^3)} + \cdots$$

$$= \frac{1}{(1-q^2)(1-q^7) \cdots (1-q^3)(1-q^8) \cdots},$$

in which he had been anticipated by British mathematicians, and about which I shall speak in other lectures. As regards hypergeometric series one may say, roughly, that he rediscovered the formal theory, set out in Bailey's tract, as it was known up to 1920. There is something about it in Carr, and more in Chrystal's *Algebra*, and no doubt he got his start from that. The four formulas (1)–(4) are highly specialized examples of this work.

His masterpiece in continued fractions was his work on

$$(25) \quad \frac{1}{1+} \frac{x}{1+} \frac{x^2}{1+} \cdots,$$

which includes the theorems (10)–(12). The theory of this fraction depends upon the Rogers-Ramanujan identities, in which he had been anticipated by Rogers, but he had gone beyond Rogers in other ways and the theorems which I have quoted are his own. He had many other very general and very beautiful formulas, of which formulas like Laguerre's

$$(26) \quad \frac{(x+1)^n - (x-1)^n}{(x+1)^n + (x-1)^n} = \frac{n}{x+} \frac{n^2-1}{3x+} \frac{n^2-2^2}{5x+} \cdots$$

are extremely special cases. Watson* has recently published a proof of the most imposing of them.

It is perhaps in his work in these fields that Ramanujan shows at his very best. I wrote before

"It was his insight into algebraical formulae, transformation of infinite series, and so forth, that was most amazing. On this side most certainly I have never met his equal, and I can compare him only with Euler or Jacobi. He worked, far more than the majority of modern mathematicians, by induction from numerical examples; all his congruence properties of partitions, for example, were discovered in this way. But with his memory, his patience, and his power of calculation he combined a power of generalization, a feeling for form, and a capacity for rapid modification of his hypotheses, that were often really startling, and made him, in his own peculiar field, without a rival in his day."

I do not think now that this extremely strong language is extravagant. It is possible that the great days of formulas are finished, and that Ramanujan ought to have been born 100 years ago; but he was by far the greatest formalist of his time. There have been a good many more important, and I suppose one must say greater, mathematicians than Ramanujan during the last 50 years, but not one who could stand up to him on his own ground. Playing the game of which he knew the rules, he could give any mathematician in the world fifteen.

In analysis proper Ramanujan's work is inevitably less impressive, since he knew no theory of functions, and you cannot do real analysis without it, and since the formal side of the integral calculus, which was all that he could learn from Carr or any other book, has been worked over so repeatedly and so intensively. Still, Ramanujan rediscovered an astonishing number of the most beautiful analytic identities. Thus the functional equation for the Riemann Zeta-function

$$\zeta(s) = \sum_{n=1}^{\infty} \frac{1}{n^s},$$

namely

(27) $\qquad \zeta(1-s) = 2(2\pi)^{-s} \cos \tfrac{1}{2}s\pi \; \Gamma(s)\zeta(s),$

stands (in an almost unrecognizable notation) in the notebooks. So does Poisson's summation formula

(28) $\quad \alpha^{1/2}\{\tfrac{1}{2}\phi(0) + \phi(\alpha) + \phi(2\alpha) + \cdots\} = \beta^{1/2}\{\tfrac{1}{2}\psi(0) + \psi(\beta) + \psi(2\beta) + \cdots\}$

where

$$\psi(x) = \sqrt{\left(\frac{2}{\pi}\right)} \int_0^{\infty} \phi(t) \cos xt\, dt$$

and $\alpha\beta = 2\pi$; and so also does Abel's† functional equation

---

* G. N. Watson, *Proceedings of the Cambridge Philosophical Society*, vol. 31, 1935, p. 7.

† The equation was rediscovered by Rogers and is attributed to him in the *Papers* (p. 337); but it is to be found in a posthumous fragment of Abel (*Œuvres*, t.2., p. 193).

$$
(29) \quad L(x) + L(y) + L(xy) + L\left(\frac{x(1-y)}{1-xy}\right) + L\left(\frac{y(1-x)}{1-xy}\right) = 3L(1)
$$

for

$$
L(x) = \frac{x}{1^2} + \frac{x^2}{2^2} + \frac{x^3}{3^2} + \cdots.
$$

He had most of the formal ideas which underlie the recent work of Watson and of Titchmarsh and myself on "Fourier kernels" and "reciprocal functions"; and he could of course evaluate any evaluable definite integral. There is one particularly interesting formula, viz.

$$
(30) \quad \int_0^\infty x^{s-1} \{\phi(0) - x\phi(1) + x^2\phi(2) - \cdots\} dx = \frac{\pi \phi(-s)}{\sin s\pi},
$$

of which he was especially fond and made continual use. This is really an "interpolation formula", which enables us to say, for example, that, under certain conditions, a function which vanishes for all positive integral values of its argument must vanish identically. I have never seen this formula stated explicitly by any one else, though it is closely connected with the work of Mellin and others.

I have left till last the two most intriguing sides of Ramanujan's early work, his work on elliptic functions and in the analytic theory of numbers. The first is probably too specialized and intricate for anyone but an expert to understand, and I shall say nothing about it now. The second subject is still more difficult (as anyone who has read Landau's book on primes or Ingham's tract will know), but anyone can understand roughly what the problems of the subject are, and any decent mathematician can understand roughly why they defeated Ramanujan. For this was Ramanujan's one real failure; he showed, as always, astonishing imaginative power, but he proved next to nothing, and a great deal even of what he imagined was false.

Here I am obliged to interpolate some remarks on a very difficult subject, *proof* and its importance in mathematics. All physicists, and a good many quite respectable mathematicians, are contemptuous about proof. I have heard Professor Eddington, for example, maintain that proof, as pure mathematicians understand it, is really quite uninteresting and unimportant, and that no one who is really certain that he has found something good should waste his time looking for a proof. It is true that Eddington is inconsistent, and has sometimes even descended to proof himself. It is not enough for him to have direct knowledge that there are exactly

$$
136 \cdot 2^{256}
$$

protons in the universe; he cannot resist the temptation of proving it; and I cannot help thinking that the proof, whatever it may be worth, gives him a certain amount of intellectual satisfaction. His apology would no doubt be that "proof" means something quite different for him from what it means for a pure mathematician, and in any case we need not take him too literally. But the opinion which I have attributed to him, and with which I am sure that almost all physicists agree at the bottom of their hearts, is one to which a mathematician ought to have some reply.

I am not going to get entangled in the analysis of a particularly prickly concept, but I think that there are a few points about proof where nearly all mathematicians are agreed. In the first place, even if we do not understand exactly what proof is, we can, in ordinary analysis at any rate, recognise a proof when we see one. Secondly, there are two different motives in any presentation of a proof. The first motive is simply to secure conviction. The second is to exhibit the conclusion as the climax of a conventional pattern of propositions, a sequence of propositions whose truth is admitted and which are arranged in accordance with rules. These are the two ideals, and experience shows that, except in the simplest mathematics, we can hardly ever satisfy the first ideal without also satisfying the second. We may be able to recognise directly that 5, or even 17, is prime, but nobody can convince himself that

$$2^{127} - 1$$

is prime except by studying a proof. No one has ever had an imagination so vivid and comprehensive as that.

A mathematician usually discovers a theorem by an effort of intuition; the conclusion strikes him as plausible, and he sets to work to manufacture a proof. Sometimes this is a matter of routine, and any well-trained professional could supply what is wanted, but more often imagination is a very unreliable guide. In particular this is so in the analytic theory of numbers, where even Ramanujan's imagination led him very seriously astray.

There is a striking example, which I have very often quoted, of a false conjecture which seems to have been endorsed even by Gauss and which took about 100 years to refute. The central problem of the analytic theory of numbers is that of the distribution of the primes. The number $\pi(x)$ of primes less than a large number x is approximately

$$(31) \qquad \frac{x}{\log x};$$

this is the "Prime Number Theorem", which had been conjectured for a very long time, but was never established properly until Hadamard and de la Vallée-Poussin proved it in 1896. The approximation errs by defect, and a much better one is

$$(32) \qquad \mathrm{Li}\, x = \int_2^x \frac{dt}{\log t}.$$

In some ways a still better one is

(33) $\quad \operatorname{Li} x - \tfrac{1}{2}\operatorname{Li} x^{1/2} - \tfrac{1}{3}\operatorname{Li} x^{1/3} - \tfrac{1}{5}\operatorname{Li} x^{1/5} + \tfrac{1}{6}\operatorname{Li} x^{1/6} - \tfrac{1}{7}\operatorname{Li} x^{1/7} + \cdots.$

(we need not trouble now about the law of formation of the series). It is extremely natural to infer that

(34) $\quad\quad\quad\quad\quad\quad\quad \pi(x) < \operatorname{Li} x,$

at any rate for large x, and Gauss and other mathematicians commented on the high probability of this conjecture. The conjecture is not only plausible but is supported by *all* the evidence of the facts. The primes are known up to 10,000,000, and their number at intervals up to 1,000,000,000, and (34) is true for every value of x for which data exist.

In 1912 Littlewood proved that the conjecture is false, and that there are an infinity of values of x for which the sign of inequality in (34) must be reversed. In particular, there is a number X such that (34) is false for some x less than X. Littlewood proved the existence of X, but his method did not give any particular value, and it is only very recently that an admissible value, viz.

$$X = 10^{10^{10^{34}}},$$

was found by Skewes.* I think that this is the largest number which has ever served any definite purpose in mathematics.

The number of protons in the universe is about

$$10^{80}.$$

The number of possible games of chess is much bigger, perhaps

$$10^{10^{50}}$$

(in any case a second order exponential). If the universe were the chessboard, the protons the chessmen, and any interchange in the position of two protons a move, then the number of possible games would be something like the Skewes number. However much the number may be reduced by refinements on Skewes' argument, it does not seem at all likely that we shall ever know a single instance of the truth of Littlewood's theorem.

This is an example in which the truth has defeated not only all the evidence of the facts and of common sense but even a mathematical imagination so powerful and profound as that of Gauss; but of course it is taken from the most difficult parts of the theory. No part of the theory of primes is really easy, but up to a point simple arguments, although they will prove very little, do not actually mislead us. For example, there are simple arguments which might lead any good mathematician to the conclusion†

(35) $\quad\quad\quad\quad\quad\quad\quad \pi(x) \sim \dfrac{x}{\log x}$

---

* S. Skewes, *Journal of the London Mathematical Society*, vol. 8, 1933, p. 277.
† $f(x) \sim g(x)$ means that the ratio f/g tends to unity.

of the Prime Number Theorem, or, what is the same thing, to the conclusion that

(36) $$p_n \sim n \log n,$$

where p_n is the n-th prime number.

In the first place, we may start from Euler's identity

(37) $$\prod_p \frac{1}{1-p^{-s}} = \frac{1}{(1-2^{-s})(1-3^{-s})(1-5^{-s})\cdots} = \frac{1}{1^s} + \frac{1}{2^s} + \frac{1}{3^s} + \cdots = \sum_n \frac{1}{n^s}.$$

This is true for $s>1$, but both series and product become infinite for $s=1$. It is natural to argue that, when $s=1$, the series and the product should diverge in the same sort of way. Also

(38) $$\log \prod \frac{1}{1-p^{-s}} = \sum \log \frac{1}{1-p^{-s}} = \sum \frac{1}{p^s} + \sum \left(\frac{1}{2p^{2s}} + \frac{1}{3p^{3s}} + \cdots \right),$$

and the last series remains finite for $s=1$. It is natural to infer that

$$\sum \frac{1}{p}$$

diverges like

$$\log \left(\sum \frac{1}{n} \right),$$

or, more precisely, that

(39) $$\sum_{p \leq x} \frac{1}{p} \sim \log \left(\sum_{n \leq x} \frac{1}{n} \right) \sim \log \log x$$

for large x. Since also

$$\sum_{n \leq x} \frac{1}{n \log n} \sim \log \log x,$$

formula (39) indicates that p_n is about $n \log n$.

There is a slightly more sophisticated argument which is really simpler. It is easy to see that the highest power of a prime p which divides $x!$ is

$$\left[\frac{x}{p}\right] + \left[\frac{x}{p^2}\right] + \left[\frac{x}{p^3}\right] + \cdots ,$$

where $[y]$ denotes the integral part of y. Hence

$$x! = \prod_{p \leq x} p^{[x/p]+[x/p^2]+\cdots},$$

(40) $$\log x! = \sum_{p \leq x}\left(\left[\frac{x}{p}\right]+\left[\frac{x}{p^2}\right]+\cdots\right)\log p.$$

The left-hand side of (40) is practically $x \log x$, by Stirling's Theorem. As regards the right-hand, one may argue; squares, cubes, ... of primes are comparatively rare, and the terms involving them should be unimportant, and it should also make comparatively little difference if we replace $[x/p]$ by x/p. We thus infer that

$$x \sum_{p \leq x} \frac{\log p}{p} \sim x \log x, \qquad \sum_{p \leq x} \frac{\log p}{p} \sim \log x,$$

and this again just fits the view that p_n is approximately $n \log n$.

This is broadly the argument used, naturally in a less naïve form, by Tchebychef, who was the first to make substantial progress in the theory of primes, and I imagine that Ramanujan began by arguing in the same sort of way, though there is nothing in the note-books to show. All that is plain is that Ramanujan found the form of the Prime Number Theorem for himself. This was a considerable achievement; for the men who had found the form of the theorem before him, like Legendre, Gauss, and Dirichlet, had all been very great mathematicians; and Ramanujan found other formulas which lie still further below the surface. Perhaps the best instance is (15). The integral is better replaced by the simpler function (16), but what Ramanujan says is correct as it stands and was proved by Landau in 1909; and there is nothing obvious to suggest its truth.

The fact remains that hardly any of Ramanujan's work in this field had any permanent value. The analytic theory of numbers is one of those exceptional branches of mathematics in which proof really is everything and nothing short of absolute rigour counts. The achievement of the mathematicians who found the Prime Number Theorem was quite a small thing compared with that of those who found the proof. It is not merely that in this theory (as Littlewood's theorem shows) you can never be sure of the facts without the proof, though this is important enough. The whole history of the Prime Number Theorem, and the other big theorems of the subject, shows that you cannot reach any real understanding of the structure and meaning of the theory, or have any sound instincts to guide you in further research, until you have mastered the proofs. It is comparatively easy to make clever guesses; indeed there are theorems, like "Goldbach's Theorem",* which have never been proved and which any fool could have guessed.

The theory of primes depends upon the properties of Riemann's function $\zeta(s)$, considered as an analytic function of the complex variable s, and in particu-

---

* "Any even number greater than 2 is the sum of two primes."

lar on the distribution of its zeros; and Ramanujan knew nothing at all about the theory of analytic functions. I wrote before

"Ramanujan's theory of primes was vitiated by his ignorance of the theory of functions of a complex variable. It was (so to say) what the theory might be if the Zeta-function had no complex zeros. His method depended upon a wholesale use of divergent series. . . . That his proofs should have been invalid was only to be expected. But the mistakes went deeper than that, and many of the actual results were false. He had obtained the dominant terms of the classical formulae, although by invalid methods; but none of them are such close approximations as he supposed.

"This may be said to have been Ramanujan's one great failure . . .",

and if I had stopped there I should have had nothing to add, but I allowed myself again to be led away by sentimentalism. I went on to argue that "his failure was more wonderful than any of his triumphs", and that is an absurd exaggeration. It is no use trying to pretend that failure is something else. This much perhaps we may say, that his failure is one which, on the balance, should increase and not diminish our admiration for his gifts, since it gives us additional, and surprising, evidence of his imagination and versatility.

But the reputation of a mathematician cannot be made by failures or by rediscoveries; it must rest primarily, and rightly, on actual and original achievement. I have to justify Ramanujan on this ground, and that I hope to do in my later lectures.

COMMENT

This is the first of two lectures delivered in the autumn of 1936 at the Harvard Tercentenary Conference of Arts and Sciences and forms Lecture I of Hardy's book (B9) *Ramanujan; twelve lectures on subjects suggested by his life and work* (Cambridge, 1940). The second lecture was gradually expanded to fill the remaining eleven lectures.

The Harvard lectures no doubt grew out of two public lectures on 'the life and scientific work of Srinivasa Ramanujan' given by Hardy on 5 May and 20 May of the same year in Cambridge, England. In the following year (Lent Term, 1937) he gave a course of 24 lectures on 'Mathematical problems connected with the work of Ramanujan'. The frontispiece to the present volume is reproduced from his lecture notes for this course.

MATHEMATICS IN WAR-TIME

THE editor asked me at the beginning of term to write an article for EUREKA, and I felt that I ought to accept the invitation; but all the subjects which he suggested seemed to me at the time quite impossible. "My views about the Tripos"—I have never really been much interested in the Tripos since I was an undergraduate, and I am less interested in it now than ever before. "My reminiscences of Cambridge"—surely I have not yet come to that. Or, as he put it, "something more topical, something about mathematics and the war"—and that seemed to me the most impossible subject of all. I seemed to have nothing at all to say about the functions of mathematics in war, except that they filled me with intellectual contempt and moral disgust.

I have changed my mind on second thoughts, and I select the subject which seemed to me originally the worst. Mathematics, even my sort of mathematics, has its "uses" in war-time, and I suppose that I ought to have something to say about them; and if my opinions are incoherent or controversial, then perhaps so much the better, since other mathematicians may be led to reply.

I had better say at once that by "mathematics" I mean *real* mathematics, the mathematics of Fermat and Euler and Gauss and Abel, and not the stuff which passes for mathematics in an engineering laboratory. I am not thinking only of "pure" mathematics (though that is naturally my first concern); I count Maxwell and Einstein and Eddington and Dirac among "real" mathematicians. I am including the whole body of mathematical knowledge which has permanent aesthetic value, as for example the best Greek mathematics has, the mathematics which is eternal because the best of it may, like the best literature, continue to cause intense emotional satisfaction to thousands of people after thousands of years. But I am not concerned with ballistics or aerodynamics, or any of the other mathematics which has been specially devised for war. That (whatever one may think of its purposes) is repulsively ugly and intolerably dull; even Littlewood could not make ballistics respectable, and if he could not, who can?

Let us try then for a moment to dismiss these sinister by-products of mathematics and to fix our attention on the real thing. We have to consider whether real mathematics serves any purposes of importance in war, and whether any purposes which it serves are good or bad. Ought we to be glad or sorry, proud or ashamed, in war-time, that we are mathematicians?

It is plain at any rate that the real mathematics (apart from the elements) has no *direct* utility in war. No one has yet found any war-like purpose to be served by the theory of numbers or relativity or quantum mechanics, and it seems very unlikely that anybody will do so for many years. And of that I am glad, but in saying so I may possibly encourage a misconception.

It is sometimes suggested that pure mathematicians glory in the "uselessness" of their subject, and make it a boast that it has no "practical" applications.[1] The imputation is usually based on an incautious saying attributed to Gauss,[2] which has always seemed to me to have been rather crudely misinterpreted. If the theory of numbers could be employed for any practical and honourable purpose, if it could be turned directly to the furtherance of human happiness or the relief of human suffering (as for example physiology and even chemistry can), then surely neither Gauss nor any other mathematician would have been so foolish as to decry or regret such applications. But if on the other hand the applications of science have made, on the whole, at least as much for evil as for good—and this is a view which must always be taken seriously, and most of all in time of war—then both Gauss and lesser mathematicians are justified in rejoicing that there is one science at any rate whose very remoteness from ordinary human activities should keep it gentle and clean.

It would be pleasant to think that this was the end of the matter, but we cannot get away from the mathematics of the workshops so easily. Indirectly, we are responsible for its existence. The gunnery experts and aeroplane designers could

[1] I have been accused of taking this view myself. I once stated in a lecture, which was afterwards printed, that "a science is said to be useful if its development tends to accentuate the existing inequalities in the distribution of wealth, or more directly promotes the destruction of human life"; and this sentence, written in 1915, was quoted in the *Observer* only a few months ago. It was, of course, a conscious rhetorical flourish (though one perhaps excusable at the time when it was written).

[2] To the effect that, if mathematics is the queen of the sciences, then the theory of numbers is, because of its supreme "uselessness", the queen of mathematics. I cannot find an accurate quotation.

not do their job without quite a lot of mathematical training, and the best mathematical training is training in real mathematics. In this indirect way even the best mathematics becomes important in war-time, and mathematicians are wanted for all sorts of purposes. Most of these purposes are ignoble and dreary—what could be more soul-destroying than the numerical solution of differential equations?—but the men chosen for them must be mathematicians and not laboratory hacks, if only because they are better trained and have the better brains. So mathematics is going to be really important now, whether we like it or regret it; and it is not so obvious as it might seem at first even that we ought to regret it, since that depends upon our general view of the effect of science on war.

There are two sharply contrasted views about modern "scientific" war. The first and the most obvious is that the effect of science on war is merely to magnify its horror, both by increasing the sufferings of the minority who have to fight and by extending them to other classes. This is the orthodox view, and it is plain that, if this view is just, then the only possible defence lies in the necessity for retaliation. But there is a very different view which is also quite tenable. It can be maintained that modern warfare is *less* horrible than the warfare of pre-scientific times, so far at any rate as combatants are concerned; that bombs are probably more merciful than bayonets; that lachrymatory gas and mustard-gas are perhaps the most humane weapons yet devised by military science; and that the "orthodox" view rests solely on loose-thinking sentimentalism. This is the case presented with so much force by Haldane in *Callinicus*.[1] It may also be urged that the equalisation of risks which science was expected to bring would be in the long run salutary; that a civilian's life is not worth more than a soldier's, or a woman's than a man's; that anything is better than the concentration of savagery on one particular class; and that, in short, the sooner war comes "all out" the better. And if this be the right view, then scientists in general and mathematicians in particular may have a little less cause to be ashamed of their profession.

It is very difficult to strike a balance between these extreme opinions, and I will not try to do so. I will end by putting to myself, as I think every mathematician ought to, what is perhaps an easier question. Are there *any* senses in which we can say,

[1] J. B. S. Haldane, *Callinicus ; a defence of chemical warfare* (Kegan Paul, 1924).

with any real confidence, that mathematics "does good" in war? I think I can see two (though I cannot pretend that I extract a great deal of comfort from them).

In the first place it is very probable that mathematics will save the lives of a certain number of young mathematicians, since their technical skill will be applied to "useful" purposes and will keep them from the front. "Conservation of ability" is one of the official slogans; "ability" means, in practice, mathematical, physical, or chemical ability; and if a few mathematicians are "conserved" then that is at any rate something gained. It may be a bit hard on the classics and historians and philosophers, whose chances of death are that little much increased; but nobody is going to worry about the "humanities" now. It is better that some should be saved, even if they are not necessarily the most worthy.

Secondly, an older man may (if he is not *too* old) find in mathematics an incomparable anodyne. For mathematics is, of all the arts and sciences, the most austere and the most remote, and a mathematician should be of all men the one who can most easily take refuge where, as Bertrand Russell says, "one at least of our nobler impulses can best escape from the dreary exile of the actual world." But he must not be too old—it is a pity that it should be necessary to make this very serious reservation. Mathematics is not a contemplative but a creative subject; no one can draw much consolation from it when he has lost the power or the desire to create; and that is apt to happen to a mathematician rather soon. It is a pity, but in that case he does not matter a great deal anyhow, and it would be silly to bother about him.

COMMENTS

This article appeared in January 1940 in *Eureka*, the journal of the Archimedeans, by which short title the Cambridge University (undergraduate) Mathematical Society is known. It formed the basis of §28 of Hardy's book *A mathematician's apology*, which was published later in the same year.

3. QUESTIONS FROM *THE EDUCATIONAL TIMES*

INTRODUCTION TO THE QUESTIONS FROM *THE EDUCATIONAL TIMES*

The *Educational Times and Journal of the College of Preceptors* was a monthly periodical devoted to educational matters. Each issue contained a list of mathematical questions posed by contributors, and the solutions sent in by the proposer and others were published in due course, in curtailed form if too lengthy. A separate volume of complete solutions was published annually under the title of *Mathematical Questions and Solutions from the 'Educational Times'*. Where the brief title *Educational Times* is used in the list of Hardy's works and in the following papers, it is this latter publication that is referred to and not the monthly periodical in which the Questions were first proposed.

Hardy became a contributor, to both journals, in 1898 and continued as such until they became defunct at the end of the 1914–18 war. As one would expect from Hardy's more substantial publications during this period, the bulk of the questions are concerned with the evaluation of definite integrals. They display his extraordinary ingenuity in this field, which was one in which he retained an interest throughout his life. The other questions which he proposed or solved cover a wide field, including divergent series, elliptic functions, two-dimensional geometry, and a not entirely successful excursus into Galois theory.

The list of Hardy's papers given in earlier volumes of his Collected Papers contained 26 contributions from the *Educational Times*. Since then several previously undetected Questions, that he proposed or solved, have come to light bringing the total to 44.

R.A.R.

EDUCATIONAL TIMES

13848. (G. H. Hardy.)—Give, in a symmetrical form, the general equation of a circle through two fixed points.

Solution by the Proposer.

Let the two points be $(\alpha\beta)$, $(\alpha'\beta')$. Then the equation of the circle described on the line joining them as diameter is
$$(x-\alpha)(x-\alpha') + (y-\beta)(y-\beta') = 0 ;$$
the line joining them is $\quad \dfrac{x-\alpha}{\alpha'-\alpha} = \dfrac{y-\beta}{\beta'-\beta} ;$

and therefore any circle through them may be represented by
$$(x-\alpha)\left(x-\alpha' - \frac{\lambda}{\alpha'-\alpha}\right) + (y-\beta)\left(y-\beta' + \frac{\lambda}{\beta'-\beta}\right) = 0 ;$$
or $(x-\alpha)(x-\alpha') + (y-\beta)(y-\beta')$ is equal to each of the quantities
$$\lambda\left(\frac{x-\alpha}{\alpha'-\alpha} - \frac{y-\beta}{\beta'-\beta}\right), \quad \lambda\left(\frac{x-\alpha'}{\alpha'-\alpha} - \frac{y-\beta'}{\beta'-\beta}\right),$$
and, therefore, to half their sum. Hence the general equation required is
$$(x-\alpha)(x-\alpha') + (y-\beta)(y-\beta') = \pm\lambda\frac{2x-\alpha-\alpha'}{\alpha-\alpha'} \mp \lambda\frac{2y-\beta-\beta'}{\beta-\beta'}$$
(writing λ for $\tfrac{1}{2}\lambda$, λ being arbitrary); and this form is symmetrical in xy, $\alpha\beta$, $\alpha'\beta'$.

[Mr. Curjel solves the Question as follows:—Let equation be
$$x^2 + y^2 + 2ax + 2by + c = 0.$$
Since it passes through the two given points (x_1, y_1), (x_2, y_2),
$$x_1^2 + y_1^2 + 2ax_1 + 2by_1 + c = 0 ;$$
$$x_2^2 + y_2^2 + 2ax_2 + 2by_2 + c = 0 ;$$
therefore equation becomes, on eliminating a and b,
$$\begin{vmatrix} x^2 + y^2 - c, & x, & y \\ x_1^2 + y_1^2 - c, & x_1, & y_1 \\ x_2^2 + y_2^2 - c, & x_2, & y_2 \end{vmatrix} = 0,$$
where c is arbitrary.]

COMMENTS

This appears to be Hardy's first mathematical publication. The question was posed in the issue of the *Educational Times and Journal of the College of Preceptors* that appeared on 1 May, 1898. Hardy's solution appeared in the issue of 1 August, 1898 (Vol. 51, pp. 225, 327) and, together with Curjel's solution, the following year in Volume 70 of the *Mathematical Questions and Solutions*.

1899, 1 *Educational Times*, 70, 43–4.

The simpler and more elegant solution by H. W. Curjel has been included for comparison. Harold W. Curjel obtained First Class Honours both in Mathematical Moderations in 1887, and in the Honours School of Mathematics and Physics in 1890, as an undergraduate at Exeter College, Oxford. He resided in Southport and was a frequent contributor to the *Questions*.

13917. (G. H. Hardy.)—A parabola is described in a triangle. If its focus is $\alpha' : \beta' : \gamma'$, its vertex is

$$\alpha'^3 \left\{ \frac{\gamma' \cos B - \beta' \cos C}{a} \right\}^2 : \beta'^3 \left\{ \frac{\alpha' \cos C - \gamma' \cos A}{b} \right\}^2 : \ldots .$$

Solution by the Proposer.

Let the equation of the parabola be
$$\sqrt{(l\alpha)} + \sqrt{(m\beta)} + \sqrt{(n\gamma)} = 0;$$
let its foci be $\alpha', \beta', \gamma'; \ 1/\alpha', 1/\beta', 1/\gamma';$
then the latter is at infinity and satisfies
$$a/\alpha' + \ldots = 0, \quad \sqrt{(l/\alpha')} + \ldots = 0 \quad\quad\quad\quad (1),$$
i.e., $\quad \alpha' : \beta' : \gamma :: a^2/l : b^2/m : c^2/n,$
since these values satisfy equations (1) in virtue of $l/a + m/b + n/c = 0$,
i.e., $l : m : n :: a^2/\alpha' : \ldots,$ and the parabola is
$$a\sqrt{(\alpha/\alpha')} + b\sqrt{(\beta/\beta')} + c\sqrt{(\gamma/\gamma')} = 0;$$
also, its axis is
$$a\alpha' (\beta'^2 - \gamma'^2) + \ldots + \ldots = 0,$$
and the vertex is obtained by solving these equations for $\alpha : \beta : \gamma$ with the condition
$$a\beta'\gamma' + b\gamma'\alpha' + c\alpha'\beta' = 0 \quad\quad\quad\quad (2).$$

Eliminating α, we get an equation of the form
$$p\beta + q\gamma - r\sqrt{(\beta\gamma)} = 0,$$
one root of which is given by
$$\beta : \gamma :: 1/\beta' : 1/\gamma' \quad \text{or} \quad \sqrt{(\beta\beta')} - \sqrt{(\gamma\gamma')} = 0,$$
and, therefore, the other by
$$p\sqrt{(\beta/\beta')} - q\sqrt{(\gamma/\gamma')} = 0 \quad\quad\quad\quad (3).$$

Now $\quad p = b^2/\beta' + a^2/\alpha' \cdot \dfrac{\beta' (\gamma'^2 - \alpha'^2)}{\alpha' (\beta'^2 - \gamma'^2)},$

and, by use of equation (2), and of the identity
$$ac \cos A \cdot \beta'^2 - bc \cos B \cdot \alpha'^2 + (b^2 - a^2) \alpha'\beta' = (\beta' \cos A - \alpha' \cos B)(bc\alpha' + ac\beta')$$
$$= -(c^2\alpha'\beta'/\gamma')(\beta' \cos A - \alpha' \cos B),$$
we find $\quad p = \dfrac{2ab\gamma' (\alpha' \cos B - \beta' \cos A)}{\alpha' (\beta'^2 - \gamma'^2)}$

and a similar expression for q; and, by (3),
$$\alpha : \beta : \gamma :: \alpha'^3 \left(\frac{\beta' \cos C - \gamma' \cos B}{a} \right)^2 : \ldots : \ldots .$$

CORRECTION

Line 4 after equation (1). Insert closing bracket after β'.

COMMENT

The coordinates used are trilinear.

13889. (G. H. Hardy.)—Sum $1 - \frac{2}{5} + \frac{3}{9} - \frac{4}{13} + \ldots$ ad inf.

Solution by Annie Chartres; Rev. T. Roach, M.A.; H. W. Curjel, M.A.; Professor Sanjána, M.A.; *and others.*

$$1 - \tfrac{2}{5} + \tfrac{3}{9} - \tfrac{4}{13} + \ldots = \int_0^1 \frac{dx}{(1+x^4)^2} = \tfrac{1}{8} + \tfrac{3}{4}\int_0^1 \frac{1}{1+x^4}dx$$
$$= \tfrac{1}{8} + \tfrac{3}{16}\sqrt{2}\left\{\log(\sqrt{2}+1) + \tfrac{1}{2}\pi\right\}.$$

See Todhunter's *Integral Calculus*, p. 36.

[It is uncertain which sign precedes the term at infinity. The respective results, however, differ by $\tfrac{1}{4}$. The value given in the above solution is the mean.]

COMMENT

This question is the earliest evidence of Hardy's interest in divergent series. The value given would now be described less picturesquely as the Abel or $(C, 1)$ sum of the series.

1899, 2a *Educational Times*, 71, 26.

13979. (R. Chartres.)—Show that
$$\int_0^\infty \frac{x^{k-1}\,dx}{1+x} = \frac{\pi}{\sin k\pi}$$
without Demoivre's theorem or imaginaries ($k < 1$); evaluate
$$\int_0^\infty \frac{\log x \, dx}{1+x^3};$$
and sum the series $\quad \dfrac{1}{1^2 \cdot 2^2} - \dfrac{3}{4^2 \cdot 5^2} + \dfrac{5}{7^2 \cdot 8^2} - \&\text{c., ad infinitum}.$

Solution by Professor Nanson, G. H. Hardy, *and the* Proposer.

(1) $\qquad \displaystyle\int_0^\infty \frac{x^{k-}}{1+x}\,dx = \int_0^1 \frac{x^{k-1} + x^{-k}}{1+x}\,dx = \pi \operatorname{cosec} k\pi,$

on expanding $(1-x)^{-1}$ in a G.P., integrating each term separately, and summing by the known formula

$$\pi \operatorname{cosec} k\pi = \frac{1}{k} + \sum_{r=1}^{r=\infty}\left(\frac{1}{k+r} + \frac{1}{k-r}\right)(-1)^r,$$

which can be proved without using Demoivre's theorem, or imaginary quantities.

(2) $\qquad \displaystyle\int_0^\infty \frac{\log x}{1+x^3}\,dx = \int_0^1 \frac{1-x}{1+x^3} \log x \, dx$

$\qquad\qquad\qquad = \displaystyle\int_0^1 \{1 - x - x^3 + x^4 + x^6 - x^7 - \ldots\} \log x \, dx$

$\qquad\qquad\qquad = -\dfrac{1}{1^2} + \dfrac{1}{2^2} + \dfrac{1}{4^2} - \dfrac{1}{5^2} - \dfrac{1}{7^2} + \dfrac{1}{8^2} + \ldots$

$\qquad\qquad\qquad = -\dfrac{2\pi^2}{27}$, as in next part.

(3) Let $\qquad S = \dfrac{1}{1^2 \cdot 2^2} - \dfrac{3}{4^2 \cdot 5^2} + \dfrac{5}{7^2 \cdot 8^2} - \ldots;$

therefore $\qquad 3S = \dfrac{1}{1^2} - \dfrac{1}{2^2} - \dfrac{1}{4^2} + \dfrac{1}{5^2} + \dfrac{1}{7^2} - \dfrac{1}{8^2} - \ldots .$

Now $\qquad \dfrac{\pi^2}{12} = \dfrac{1}{1^2} - \dfrac{1}{2^2} + \dfrac{1}{3^2} - \dfrac{1}{4^2} + \dfrac{1}{5^2} - \dfrac{1}{6^2} - \dfrac{1}{7^2} + \dfrac{1}{8^2} - \ldots ;$

therefore $\qquad 3S - \dfrac{\pi^2}{12} = -\dfrac{1}{9}\dfrac{\pi^2}{12}$; therefore $S = \dfrac{2\pi^2}{81}.$

[The Proposer remarks that (2) follows at once from (1) by differentiating with respect to k and putting $k = \tfrac{1}{3}$.]

CORRECTIONS

The numerator of the integrand on the left of (1) should be x^{k-1}. In the next line, replace $(1-x)^{-1}$ by $(1+x)^{-1}$; the integration term by term is easily justified.

1899, 2b *Educational Times*, 71, 61–2.

13964. (G. H. Hardy.)—Prove

$$\int_0^{\frac{1}{4}\pi} \tanh^{-1}(p\tan^2 x)\,dx = \tan^{-1}\sqrt{p}\,\tanh^{-1}\sqrt{p} - \tfrac{1}{4}\pi\tanh^{-1}p,$$

$$\int_0^{\lambda} \log(1 + \tan\lambda \tan x)\,dx = \lambda \log\sec\lambda,$$

$$\int_0^{\frac{1}{2}\pi}(x\cot x)^2\,dx = \tfrac{2}{3}\int_0^1 \left(\frac{\sin^{-1}x}{x}\right)^3 dx = \pi\log 2 - \frac{\pi^3}{24}.$$

Solution by Professor Umes Chandra Ghosh, M.A.

(i.) Let $\quad u = \int_0^{\frac{1}{4}\pi} \tanh^{-1}(p\tan^2 x)\,dx,$

therefore $\quad 2\dfrac{du}{dp} = \int_0^{\frac{1}{4}\pi} \dfrac{\tan^2 x}{1+p\tan^2 x}dx + \int_0^{\frac{1}{4}\pi} \dfrac{\tan^2 x}{1-p\tan^2 x}dx$

$= \dfrac{1}{\sqrt{p}}\dfrac{1}{1-p}\tan^{-1}\sqrt{p} + \dfrac{1}{\sqrt{p}}\dfrac{1}{1+p}\tfrac{1}{2}\log\left(\dfrac{1+\sqrt{p}}{1-\sqrt{p}}\right) - \dfrac{2}{1-p^2}\tfrac{1}{4}\pi.$

Let $\quad \tan^{-1}\sqrt{p} = \theta;$

then $\quad 2u = \int \dfrac{2\theta}{\cos 2\theta}d\theta + \int \log\left(\dfrac{1+\tan\theta}{1-\tan\theta}\right)d\theta - \pi\int \dfrac{d(\tan\theta)}{1-\tan^2\theta}$

$= \theta \log\left(\dfrac{1+\tan\theta}{1-\tan\theta}\right) - \int \log\left(\dfrac{1+\tan\theta}{1-\tan\theta}\right)d\theta + \int \log\left(\dfrac{1+\tan\theta}{1-\tan\theta}\right)d\theta$

$\quad - \tfrac{1}{4}\pi \log\left(\dfrac{1+\sqrt{p}}{1-\sqrt{p}}\right)$

$= 2\tanh^{-1}\sqrt{p}\,\tan^{-1}\sqrt{p} - \tfrac{1}{2}\pi\tanh^{-1}\sqrt{p};$

therefore $\quad u = \tanh^{-1}\sqrt{p}\,\tan^{-1}\sqrt{p} - \tfrac{1}{4}\pi\tanh^{-1}p,$
the constant being zero.

(ii.) $\mathrm{I} = \displaystyle\int_0^{\lambda}\log(1+\tan\lambda\tan x)\,dx = \int_0^{\lambda}\log\{1+\tan\lambda\tan(\lambda-x)\}\,dx$

$= \displaystyle\int_0^{\lambda}\log\left(\dfrac{\sec^2\lambda}{1+\tan\lambda\tan x}\right)dx = \int_0^{\lambda}\log(\sec^2\lambda)\,dx - \mathrm{I};$

therefore $\quad 2\mathrm{I} = 2\lambda\log(\sec\lambda);\quad$ therefore $\mathrm{I} = \lambda\log(\sec\lambda).$

$\displaystyle\int_0^1\left(\dfrac{\sin^{-1}x}{x}\right)^3 dx = \tfrac{1}{2}\int_0^1 \dfrac{3(\sin^{-1}x)^2}{x^2\sqrt{(1-x^2)}}dx\quad$ [by integration by parts]

$= \tfrac{3}{2}\displaystyle\int_0^{\frac{1}{2}\pi}\theta^2\operatorname{cosec}^2\theta\,d\theta - \tfrac{1}{2}\cdot\tfrac{1}{8}\pi^3\quad$ [assuming $\sin\theta = x$];

therefore $\quad \tfrac{2}{3}\displaystyle\int_0^1\left(\dfrac{\sin^{-1}x}{x}\right)^3 dx = \int_0^{\frac{1}{2}\pi}\theta^2(\operatorname{cosec}^2\theta - 1)\,d\theta = \int_0^{\frac{1}{2}\pi}x^2\cot^2 x\,dx,$

and $\quad \displaystyle\int_0^{\frac{1}{2}\pi}x^2\cot^2 x\,dx = \int_0^{\frac{1}{2}\pi}x^2(\operatorname{cosec}^2 x - 1)\,dx = \int_0^{\frac{1}{2}\pi}x^2\operatorname{cosec}^2 x\,dx - \dfrac{\pi^3}{24}.$

But $\int_0^{\frac{1}{2}\pi} x^2 \operatorname{cosec}^2 x\, dx = \int_0^{\frac{1}{2}\pi} \cot x \cdot 2x\, dx$ [by integration by parts]

$$= -2\int_0^{\frac{1}{2}\pi} \log(\sin x)\, dx = \pi \log 2\,;$$

therefore $\int_0^{\frac{1}{2}\pi} x^2 \cot^2 x\, dx = \pi \log 2 - \dfrac{\pi^3}{24}.$

CORRECTIONS
On the previous page (iii.) is missing on the third line after (ii.).
On the same line the term $-\pi^3/16$ should be added on the right-hand side.

COMMENT
The solver was Professor of Mathematics at Allahabad.

14124. (G. H. Hardy.)—Prove that
$$\int_0^{\frac{1}{12}\pi} \log \tan \phi \, d\phi = \tfrac{2}{3} \int_0^{\frac{1}{4}\pi} \log \tan \phi \, d\phi = -4 \int_0^{\frac{1}{6}\pi} \tanh^{-1} \tan \phi \, d\phi.$$

Solution by the Proposer.

We have
$$\int_0^u \frac{du}{\cos^2 u - p} = \frac{1}{\sqrt{\{p(1-p)\}}} \sinh^{-1} \frac{\sqrt{p} \sin u}{\sqrt{(\cos^2 u - p)}};$$

i.e., putting $p = \cos^2 x$,

$$\int_0^u \frac{2 \cos x \sin x}{\cos^2 u - \cos^2 x} du = 2 \tanh^{-1}(\tan u \cot x),$$

whence $\int_{x_0}^{x'_0} \int_0^{u_0} \frac{2 \cos x \sin x}{\cos^2 u - \cos^2 x} du \, dx = 2 \int_{x_0}^{x'_0} \tanh^{-1}(\tan u_0 \cot x) \, dx.$

Let $u_0 = \tfrac{1}{4}\pi$; then

$$\int_0^{\frac{1}{4}\pi} \log \frac{\cos^2 u - \cos^2 x'_0}{\cos^2 u - \cos^2 x_0} du = \int_{x_0}^{x'_0} \log \frac{1+\cot x}{1-\cot x} dx.$$

Put $\quad x'_0 = \tfrac{1}{3}\pi, \quad x_0 = \tfrac{1}{2}\pi, \quad$ and $\quad \cos^2 u - \tfrac{1}{4} = \dfrac{\sin 3u}{4 \sin u},$

and $\quad \int_0^{\frac{1}{4}\pi} \log \tan x \, dx = -\sum_0^\infty \dfrac{1}{(2n+1)^2} = -\Delta,$ say,

when we get, after some reduction,

$$\int_0^{\frac{1}{12}\pi} \log \tan x \, dx = -\tfrac{2}{3}\Delta,$$

and, writing $\tfrac{1}{4}\pi - y = x$,

$$-\int_{\frac{1}{4}\pi}^{\frac{1}{6}\pi} \log \frac{1-\tan y}{1+\tan y} dy = -\tfrac{2}{3}\Delta,$$

whence $\int_0^{\frac{1}{6}\pi} \log \frac{1+\tan y}{1-\tan y} dy = 2 \int_0^{\frac{1}{6}\pi} \tanh^{-1} \tan y \, dy = \tfrac{1}{3}\Delta,$

which proves the result.

14005. (G. H. Hardy.)—Evaluate the definite integrals

$$\int_0^{\frac{1}{2}\pi} \tan^{-1}(p\tan^2 t)\, dt, \quad \int_0^{\frac{1}{2}\pi} \tan^{-1}(p\sin^2 t)\, dt,$$

$$\int_0^a \frac{(a_0 a_1 a_2 \ldots a_n \emptyset \log x, 1)^n}{\sqrt{(\log a - \log x)}}\, dx, \quad \text{and} \quad \int_0^a \binom{\sin}{\sinh} \kappa \sqrt{(\log a - \log x)}\, dx.$$

Solution by the Proposer.

(1) $\quad \int_0^{\frac{1}{2}\pi} \log(1 + n\sin^2\theta)\, d\theta = \pi \log \tfrac{1}{2}\{1 + \sqrt{(1+n)}\}.$

Putting $n = ip$ and equating imaginary parts,

$$\int_0^{\frac{1}{2}\pi} \tan^{-1}(p\sin^2\theta)\, d\theta = \pi \tan^{-1} \frac{\sqrt{\{\sqrt{(1+p^2)} - 1\}}}{\sqrt{2} + \sqrt{\{\sqrt{(1+p^2)} + 1\}}}.$$

More generally, $\int_0^{\frac{1}{2}\pi} \tan^{-1}(p\sin^2\theta / 1 + q\sin^2\theta)\, d\theta$

$$= \pi \tan^{-1} \frac{\sqrt{[\sqrt{\{(1+q)^2 + p^2\}} - (1+q)]}}{\sqrt{2} + \sqrt{[\sqrt{\{(1+q)^2 + p^2\}} + (1+q)]}};$$

and, similarly, $\int_0^{\frac{1}{2}\pi} \tan^{-1}(p\tan^2\theta)\, d\theta = \pi \cot^{-1}\{1 + \sqrt{(2/p)}\}.$

(2) It is easy to show that

$$\int_0^a \frac{dx}{\sqrt{(\log a - \log x)}} = a\sqrt{\pi} = \frac{2^{n+1}}{1.3\ldots 2n+1}\int_0^a (\log a - \log x)^{n+\frac{1}{2}}\, dx,$$

and so, expanding $(a_0 \ldots a_n \emptyset \log x, 1)^n$ in powers of $\log a - \log x$, we find

$$\int_0^a \frac{(a_0 \ldots a_n \emptyset \log x, 1)^n}{\sqrt{(\log a - \log x)}}\, dx = a\sqrt{\pi}\left\{A_n - \tfrac{1}{2}nA_{n-1} + \frac{1.3}{2^2}\frac{n.n-1}{1.2}A_{n-2} - \ldots\right\},$$

where $A_0 = a_0$, $A_1 = a_0 \log a + a_1$, $A_2 = a_0(\log a)^2 + 2a_1 \log a + a_2$, &c.

Again, $\int_0^a \sin \kappa \sqrt{(\log a - \log x)}\, dx$

$$= \tfrac{1}{2}\kappa a \sqrt{\pi}\{1 - (\tfrac{1}{2}\kappa)^2 (1/1!) + (\tfrac{1}{2}\kappa)^4 (1/2!) - \ldots\} = \tfrac{1}{2}\kappa a e^{-\frac{1}{4}\kappa^2}\sqrt{\pi},$$

and $\int_0^a \sinh \kappa \sqrt{(\log a - \log x)}\, dx = \tfrac{1}{2}\kappa a e^{\frac{1}{4}\kappa^2}\sqrt{\pi}.$

CORRECTION

First line after (2). Read $\alpha\sqrt{\pi}$ for $a\sqrt{\pi}$.

14155. (G. H. Hardy.)—Prove, by contour integration, or otherwise, that
$$\int_0^\infty e^{n \cos \alpha x} \sin(\beta x + n \sin \alpha x) \, (dx)/x = \tfrac{1}{2}\pi e^n.$$

What conditions, if any, must α, β, n satisfy?

COMMENTS

The result, as stated, is valid for $\alpha \geq 0$, $\beta > 0$ and all real n, as can be shown by integrating $z^{-1} \exp(n e^{-\alpha z} - \beta z)$ over a large indented semicircle centred at and to the right of the origin. If α and β are replaced by their negatives, the value of the integral becomes $-\tfrac{1}{2}\pi e^n$. The solution, by G. D. Muggeridge, is not reproduced, nor is it free from minor errors.

1899, 5 *Educational Times*, 71, 117.

14089. (D. Edwardes, B.A.)—Prove that
$$\int_0^{\frac{1}{2}\pi} \sin x \log \sin x \log \cos x \, dx = 2 - \log 2 - \tfrac{1}{8}\pi^2.$$

Solution by G. H. Hardy, B.A.; J. O. Watts; *and many others.*

$\int_0^{\frac{1}{2}\pi} \sin x \log \sin x \log \cos x \, dx$

$= \tfrac{1}{2} \int_0^1 \log t \log(1-t^2) \, dt$

$= \dfrac{1}{2 \cdot 3^2} + \dfrac{1}{4 \cdot 5^2} + \ldots = \left(\dfrac{1}{2 \cdot 3} - \dfrac{1}{3^2}\right) + \left(\dfrac{1}{4 \cdot 5} - \dfrac{1}{5^2}\right) + \ldots$

$= \left(\dfrac{1}{2} - \dfrac{1}{3} + \dfrac{1}{4} - \dfrac{1}{5} \ldots\right) - \left(\dfrac{1}{3^2} + \dfrac{1}{5^2} + \ldots\right)$

$= 1 - \log 2 - (\tfrac{1}{8}\pi^2 - 1) = 2 - \log 2 - \tfrac{1}{8}\pi^2.$

Similarly we may find
$$\int_0^{\frac{1}{2}\pi} \sin x \log \cos x \log \tfrac{\sin}{\cos} \tfrac{1}{2}x \, dx.$$

COMMENT

The steps in the proof are easily justified.

14090. (G. H. Hardy, B.A.)—Evaluate

$$\int_0^\infty \tan^{-1}(p \operatorname{sech} u)\, du \quad \text{and} \quad \int_0^\infty \tan^{-1}\left\{ \tfrac{1}{2}\left(a - \frac{1}{a}\right) \operatorname{sech}^2 u \right\} du.$$

[The Proposer solves the Question as follows:—Let

$$P = \int_0^\infty \log(1 + p \operatorname{sech} u)\, du \quad (p < 1), \qquad \frac{dP}{dp} = \int_0^\infty \frac{du}{\cosh u + p},$$

and putting $(\cosh u + p)(\cos x - p) = 1 - p^2$,

this becomes $\dfrac{dP}{dp} = \dfrac{\cos^{-1} p}{\sqrt{(1-p^2)}}$,

whence $P = \tfrac{1}{2}\{(\tfrac{1}{2}\pi)^2 - (\cos^{-1} p)^2\}$ .... (1).

Now it is easy to justify our putting $p = iq$, and then, equating real and imaginary parts, we find

$$\int_0^\infty \tan^{-1}(q \operatorname{sech} u)\, du = \tfrac{1}{2}\pi \sinh^{-1} q \quad \ldots\ldots\ldots (2),$$

$$\int_0^\infty \log(1 + q^2 \operatorname{sech}^2 u)\, du = \{\sinh^{-1} q\}^2 \ldots\ldots\ldots (3).$$

In (3) put $q = \sqrt{(ik)}$, as is permissible, and equate imaginary parts; we get

$$\int_0^\infty \tan^{-1}(k \operatorname{sech}^2 u)\, du = \operatorname{R}\left[-i\{\sinh^{-1}\sqrt{(ik)}\}^2\right] \ldots\ldots (4),$$

or $= \operatorname{R}\left[-i\{\sin^{-1}\sqrt{(ik)}\}^2\right]$,

agreeing with Mr. Webb's solution.

If, moreover, we put $k = \tfrac{1}{2}(a - 1/a)$, this may be expressed very simply; for

$$\sinh^{-1}\sqrt{(ik)} = \log\left[\sqrt{(ik)} + \sqrt{(1+ik)}\right] = \log(u + iv),$$

where $u = (1/\sqrt{2})\left[\sqrt{k} + \sqrt{\{\sqrt{(1+k^2)} + 1\}}\right]$,
$v = (1/\sqrt{2})\left[\sqrt{k} + \sqrt{\{\sqrt{(1+k^2)} - 1\}}\right]$;

and, if $a = e^\theta$, $k = \sinh\theta$, $(a > 1)$,

$u = \sqrt{(\cosh\tfrac{1}{2}\theta)}\left[\sqrt{(\sinh\tfrac{1}{2}\theta)} + \sqrt{(\cosh\tfrac{1}{2}\theta)}\right]$,
$v = \sqrt{(\sinh\tfrac{1}{2}\theta)}\left[\sqrt{(\sinh\tfrac{1}{2}\theta)} + \sqrt{(\cosh\tfrac{1}{2}\theta)}\right]$,

$\log(u^2 + v^2) = \log\{e^\theta + \sqrt{(e^{2\theta} - 1)}\} = \cosh^{-1} a$

$\tan^{-1} v/u = \tfrac{1}{2}\cos^{-1}(e^{-\theta}) = \tfrac{1}{2}\cos^{-1}(1/a)$,

and $\int_0^\infty \tan^{-1}\{\tfrac{1}{2}(a - 1/a)\operatorname{sech}^2 u\}\, du = \log(u^2 + v^2)\tan^{-1}(v/u)$
$= \tfrac{1}{2}\cosh^{-1} a \cos^{-1}(1/a)$,

at any rate when $a > 1$, $a - 1/a < 1$.]

CORRECTIONS

In equation (1) the right-hand side is $\tfrac{1}{2}\{(\tfrac{1}{2}\pi)^2 - (\cos^{-1} p)^2\}$.
On *line* 8 from the end the equation $a = e^\theta$ is unclear.

COMMENTS

The first solution printed, by H. A. Webb, is not reproduced, although simpler; the integrals are evaluated by making the substitution $\operatorname{sech} u = \cos x$ and expanding the inverse tangent. It is assumed that $0 \leq p < 1$.

1900, 1b *Educational Times*, 72, 57–9.

14243. (G. H. Hardy, B.A.)—If p, m are integers, $p > m$, then, if m is odd,

$$\int_0^\pi x \cos px \sin^m x \, dx = (-)^{\frac{1}{2}[m+(-1)^p]} \left\{ \frac{m! \, \pi}{(p^2-1^2)(p^2-3^2) \ldots (p^2-m^2)} \right\};$$

and, if m is even,

$$\int_0^\pi x \sin px \sin^m x \, dx = (-)^{\frac{1}{2}[m+1+(-1)^p]} \left\{ \frac{m! \, \pi}{p(p^2-2^2)(p^2-4^2) \ldots (p^2-m^2)} \right\}.$$

Solution by J. H. Dibb, B.Sc.

We know that $\int_0^\pi \sin px \sin^m x \, dx$ vanishes when m is odd, and $\int_0^\pi \cos px \sin^m x \cos x \, dx$, when m is even. Let $f(m)$ denote

$$\int_0^\pi x \cos px \sin^m x \, dx.$$

Then, integrating by parts, we get

$$f(m) = -\frac{1}{p} \int_0^\pi \sin px \, (\sin^m x + mx \sin^{m-1} x \cos x) \, dx$$

$$= -\frac{m}{p} \int_0^\pi x \sin^{m-1} x \cos x \sin px \, dx.$$

Again integrating by parts,

$$f(m) = -\frac{m}{p^2} \int_0^\pi \cos px \left\{ \sin^{m-1} x \cos x + (m-1) x \sin^{m-2} x - mx \sin^m x \right\} dx$$

$$= -\frac{m(m-1)}{p^2} f(m-2) + \frac{m^2}{p^2} f(m);$$

therefore $f(m) = -\frac{m(m-1)}{p^2 - m^2} f(m-2)$

$$= (-)^{\frac{1}{2}(m-1)} \frac{m(m-1) \ldots 2}{(p^2-m^2)\{p^2-(m-2)^2\} \ldots (p^2-3^2)} f(1).$$

But $f(1) = \int_0^\pi x \cos px \sin x \, dx$

$$= \tfrac{1}{2} \int_0^\pi x \sin(p+1) x \, dx - \tfrac{1}{2} \int_0^\pi x \sin(p-1) x \, dx = \frac{(-1)^{p+1} \pi}{p^2 - 1^2}.$$

Considering separately the cases of p odd and even, we see that

$$(-1)^{\frac{1}{2}(m-1)} (-1)^{p+1} = (-1)^{\frac{1}{2}[m+(-1)^p]};$$

therefore $f(m) = (-1)^{\frac{1}{2}[m+(-1)^p]} \left\{ \dfrac{m! \, \pi}{(p^2-1^2)(p^2-3^2) \ldots (p^2-m^2)} \right\}.$

The second formula is established in a precisely similar manner.

N.B.—*When p is even* the first formula is easily obtained by the well known transformation

$$f(m) = \int_0^\pi (\pi-x)\cos p(\pi-x)\sin^m(\pi-x)\,dx$$
$$= \pi\int_0^\pi \cos px \sin^m x\,dx - f(m).$$

A similar method can be applied to the second formula *when p is odd.*

COMMENT

Hardy graduated in 1899; this is the first of his published works to record his B.A. The author of the solution, John Hustwick Dibb, graduated from London University in 1888.

14164. (Professor ELLIOTT, F.R.S.)—If a, b are positive moduli, and $\alpha - \beta$ lies between 0 and 2π, and if $f(z)$ is a function of the complex variable $z = re^{i\theta}$ which approaches definite limits $f(0)$ and $f(\infty)$ as z comes infinitely small and infinitely great within the sector from $\theta = \beta$ to $\theta = \alpha$, and is holomorphic within and on the bounding lines of the sector except for simple poles at $c_1, c_2, \ldots, c_n$ within the sector, prove the FRULLANIAN formula

$$\int_0^\infty \frac{f(ae^{i\alpha}x) - f(be^{i\beta}x)}{x} dx = \{f(\infty) - f(0)\}\{\log a/b + i(\alpha - \beta)\} - 2\pi i \sum_{r=1}^{r=n} \left[(z - c_r)\frac{f(z)}{z}\right]_{z = c_r},$$

the variable x being real.

Solutions (1) by G. H. HARDY, B.A.; (2) by H. W. CURJEL, M.A.

(1) Integrate $\int \frac{f(z)}{z} dz$ round the contour bounded by $\theta = \alpha$, $\theta = \beta$, $r = a\rho_0$, $r = aR$; where ρ_0 decreases, and R increases beyond limit.

$$\{f(\infty) - f(0)\}\int_\beta^\alpha i\,d\theta + \int_{b\rho_0 e^{i\beta}}^{bRe^{i\beta}} \frac{f(z)}{z} dz - \int_{a\rho_0 e^{i\alpha}}^{aRe^{i\alpha}} \frac{f(z)}{z} dz$$

$$-f(0)\int_{b\rho_0 e^{i\beta}}^{a\rho_0 e^{i\alpha}} \frac{dz}{z} + f(\infty)\int_{bRe^{i\beta}}^{aRe^{i\alpha}} \frac{dz}{z}$$

$$= 2\pi i \sum_{r=1}^{r=n} \left[(z - c_r)\frac{f(z)}{z}\right]_{z = c_r},$$

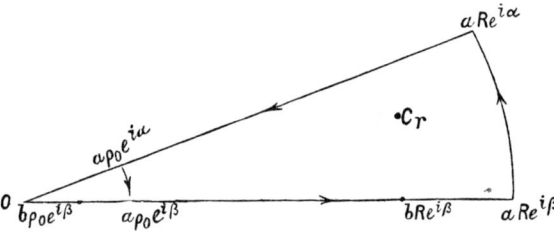

which reduces to

$$\int_0^\infty \frac{f(ae^{i\alpha}x) - f(be^{i\beta}x)}{x} dx = \{f(\infty) - f(0)\}\{i(\alpha - \beta) + \log a/b\}$$
$$- 2\pi i \sum_{r=1}^{r=n} \left[(z - c_r)\frac{f(z)}{z}\right]_{z = c_r},$$

x being real.

CORRECTION
In the third line of Hardy's solution there is a missing parenthesis.

COMMENT
The solution by H. W. Curjel is not reproduced.

The Proposer, Edwin Bailey Elliott (1851–1937), was Waynflete Professor of Pure Mathematics at Oxford from 1892 to 1921. He was President of the London Mathematical Society during the period 1896–8.

1900, 2a *Educational Times*, 72, 89–90.

14332. (Professor M. W. Crofton, F.R.S.)—(1) If $4n-1$ is a prime, $1^2 2^2 3^2 \ldots (2n-1)^2 - 1$ is a multiple of $4n-1$. (2) If $4n+1$ is a prime, $1^2 2^2 3^2 \ldots (2n)^2 + 1$ is a multiple of $4n+1$. It hence follows that $4n+1$ is the sum of two squares.

Solution by G. H. Hardy, B.A.; *and* Lt.-Col. Allan Cunningham, R.E.

If $2p-1$ be prime, then, by Wilson's theorem,
$$2p-2\,!+1 \equiv 0 \pmod{2p-1}.$$
Now $\quad 2p-2-s \equiv -(s+1), \quad (s = 0, 1, \ldots p-2),$
therefore $p \cdot p+1 \ldots 2p-2 \equiv (-)^{p-1} p-1\,!, \quad 2p-2\,! \equiv (-)^{p-1}(p-1\,!)^2.$

So, if $\quad p = 2n, \quad (2n-1\,!)^2 - 1 \equiv 0 \pmod{4n-1}$;
if $\quad\quad p = 2n+1, \quad (2n\,!)^2 + 1 \equiv 0 \pmod{4n+1}.$

Thus a prime of the form $4n+1$ is a divisor of a number which is the sum of two squares, and is therefore, by a known theorem, itself the sum of two squares.

CORRECTION

In *line* 3 of the Question there is an unclear plus sign. The number is $1^2 2^2 3^2 \ldots (2n)^2 + 1$.

COMMENTS

The 'known theorem', referred to in the last sentence of the solution, is Euler's lemma, which states that every divisor of the sum of two relatively prime squares is itself the sum of two squares; see L. E. Dickson, *History of the Theory of Numbers* (New York, 1934), Vol. II, p. 230. It is important that the two squares should be, as in this application, relatively prime.

The proposer of the Question, Professor Morgan William Crofton (1826–1915), was Professor of Mathematics and Mechanics at the Royal Military Academy.

1900, 2b *Educational Times*, 72, 105–6.

14271. (G. H. Hardy, B.A.)—Prove

(1) $a^{\frac{1}{4}}\int_0^\infty \dfrac{e^{-ax^2}}{\cosh \pi x}\,dx = b^{\frac{1}{4}}\int_0^\infty \dfrac{e^{-bx^2}}{\cosh \pi x}\,dx$ if $ab = \pi^2$;

(2) $a^{\frac{1}{4}}\int_0^\infty \dfrac{e^{-ax^2}}{1+2\cosh \pi x}\,dx = b^{\frac{1}{4}}\int_0^\infty \dfrac{e^{-bx^2}}{1+2\cosh \pi x}\,dx$ if $ab = (\tfrac{3}{4}\pi)^2$;

(3) $a^{\frac{1}{4}}\int_0^\infty \dfrac{\cosh \tfrac{1}{2}\pi x}{\cosh \pi x}e^{-ax^2}\,dx = b^{\frac{1}{4}}\int_0^\infty \dfrac{\cosh \tfrac{1}{2}\pi x}{\cosh \pi x}e^{-bx^2}\,dx$ if $ab = (\tfrac{1}{2}\pi)^2$.

Solution by the Proposer.

It is known that
$$\int_0^\infty \frac{\sinh \mu\xi}{\sinh \tfrac{1}{2}\xi}\cos\beta\xi\,d\xi = \frac{\pi \sin 2\mu\pi}{\cos 2\mu\pi + \cosh 2\beta\pi}.$$

Multiply both sides by $e^{-a\beta^2}$ and integrate from $\beta = 0$ to ∞. Since
$$\int_0^\infty e^{-a\beta^2}\cos\beta\xi\,d\beta = \tfrac{1}{2}\sqrt{(\pi/a)}\,e^{-\xi^2/4a},$$

(a) $\displaystyle\int_0^\infty \frac{\sinh \mu\xi}{\sinh \tfrac{1}{2}\xi}e^{-\xi^2/4a}\,d\xi = 2\sqrt{(a\pi)}\sin 2\mu\pi \int_0^\infty \frac{e^{-a\beta^2}\,d\beta}{\cos 2\mu\pi + \cosh 2\beta\pi}.$

Let $\mu = \tfrac{1}{4}$, $\xi = 4\pi x$, $\beta = \tfrac{1}{2}y$, $4\pi^2/a = a$, and $\tfrac{1}{4}a = b$, so that $ab = \pi^2$; then (a) reduces to

(1) $\qquad a^{\frac{1}{4}}\displaystyle\int_0^\infty \frac{e^{-ax^2}}{\cosh \pi x}\,dx = b^{\frac{1}{4}}\int_0^\infty \frac{e^{-by^2}}{\cosh \pi y}\,dy.$

Let $\mu = \tfrac{1}{6}$; then $\dfrac{\sinh \tfrac{1}{6}\xi}{\sinh \tfrac{1}{2}\xi} = \dfrac{1}{1 + 2\cosh \tfrac{1}{3}\xi}$,

and if $\xi = 3\pi x$, $\beta = \tfrac{1}{2}y$, $a = 9\pi^2/4a$, $b = \tfrac{1}{4}a$, so that $ab = (\tfrac{3}{4}\pi)^2$, we find

(2) $\qquad a^{\frac{1}{4}}\displaystyle\int_0^\infty \frac{e^{-ax^2}}{1+2\cosh \pi x}\,dx = b^{\frac{1}{4}}\int_0^\infty \frac{e^{-by^2}}{1+2\cosh \pi y}\,dy.$

Again, $\displaystyle\int_0^\infty \frac{\cosh \mu\xi}{\cosh \tfrac{1}{2}\xi}\cos\beta\xi\,d\xi = 2\pi\frac{\cos \mu\pi \cosh \beta\pi}{\cos 2\mu\pi + \cosh 2\beta\pi}.$

Using the same integrating factor as before,

(b) $\displaystyle\int_0^\infty \frac{\cosh \mu\xi}{\cosh \tfrac{1}{2}\xi}e^{-\xi^2/4a}\,d\xi = 4\sqrt{(a\pi)}\cos \mu\pi \int_0^\infty \frac{e^{-a\beta^2}\cosh \beta\pi\,d\beta}{\cos 2\mu\pi + \cosh 2\beta\pi}.$

Put $\mu = \tfrac{1}{4}$, $\xi = 2\pi x$, $\beta = \tfrac{1}{2}y$, $a = \pi^2/a$, $b = \tfrac{1}{4}a$, so that $ab = (\tfrac{1}{2}\pi)^2$;

then (3) $\quad a^{\frac{1}{4}}\displaystyle\int_0^\infty \frac{\cosh \tfrac{1}{2}\pi x}{\cosh \pi x}e^{-ax^2}\,dx = b^{\frac{1}{4}}\int_0^\infty \frac{\cosh \tfrac{1}{2}\pi y}{\cosh \pi y}e^{-by^2}\,dy.$

If we put $\mu = 0$ in (b), we find (1) again.

[The integrals assumed in the above proof will be found in Harkness and Morley's *Introduction to the Theory of Analytic Functions*, pp. 226-9.]

14179. (G. H. Hardy, B.A.)—Find a root of $x^3 + 3x(\beta - 1/\beta)^{\frac{2}{3}} = 4$. If A, B, C denote the quantities $8/(1+3x^2)$, $\frac{1}{8}\{x(3+x^2)\}$, $\frac{1}{4}\{(4-x^3)/3x\}^3$, then, if A = B, each is $= \pm \sqrt{\{\pm\sqrt{(1+C)}\pm\sqrt{C}\}}$. Fix the signs when x is real and positive.

[The Proposer solves the Question as follows:—A root of
$$x^3 + 3x(\beta - 1/\beta)^{\frac{2}{3}} = 4$$
is $\qquad x = \{\sqrt{\beta} + \sqrt{(1/\beta)}\}^{\frac{2}{3}} - \{\sqrt{\beta} - \sqrt{(1/\beta)}\}^{\frac{2}{3}}$.

Now, if $\quad A \equiv 8/(1+3x^2) = B \equiv \frac{1}{8}\{x(3+x^2)\} = 1/y$, say,
$$y + 1/y = \{\tfrac{1}{2}(1+x)\}^3, \quad y - 1/y = \{\tfrac{1}{2}(1-x)\}^3,$$
and therefore $\qquad (y+1/y)^{\frac{2}{3}} - (y-1/y)^{\frac{2}{3}} = x$;

so x is a root of $\qquad x^3 + 3x(y^2 - 1/y^2)^{\frac{2}{3}} = 4$,

i.e., $\qquad y^2 - 1/y^2 = \{(4-x^3)/3x\} = \pm 2\sqrt{C}$,

and $\qquad y = \pm\sqrt{\{\pm\sqrt{(1+C)}\pm\sqrt{C}\}}$.

To fix the signs, the equation $64 = (1+3x^2)(3x+x^3)$

has evidently one, and only one, real positive root, which lies between 1 and $4^{\frac{1}{3}}$; thus C is positive; also y is real and positive; so
$$y = +\sqrt{\{+\sqrt{(1+C)}\pm\sqrt{C}\}}, \quad 1/y = +\sqrt{\{+\sqrt{(1+C)}\mp\sqrt{C}\}},$$
and, as $x > 1$, $\{\tfrac{1}{2}(1-x)\}^3$ and $y - 1/y$ are negative, and the lower sign must be taken.]

CORRECTIONS

The equation for $y^2 - 1/y^2$ should read:
$$y^2 - 1/y^2 = \{(4-x^2)/3x\}^{3/2} = \pm 2\sqrt{C}.$$

Two lines later a closing parenthesis is required.

COMMENT

The first solution by K. G. Panchapagasa Aiyar is omitted; he has misread the cubic equation to be solved.

1900, 4 *Educational Times*, 73, 53–4.

14317. (G. H. Hardy, B.A.)—Prove Frullani's theorem

$$\int_0^\infty \frac{\phi(ax)-\phi(bx)}{x}\,dx = \{\phi(\infty)-\phi(0)\}\log a/b,$$

extending it to apply to the cases in which (1) $\phi(x)$, while possessing a determinate integral near any value of x, has a discrete set of infinities in the range $(0, \infty)$; (2) only the *principal value* of the integral is determinate. Consider also the case in which $\phi(x)$ oscillates at infinity between finite or infinite limits. As examples, evaluate

$$\int_0^\infty \log\left(\frac{1-pe^{-ax}}{1-pe^{-bx}}\right)^2 \frac{dx}{x}, \quad \int_0^\infty \left(\frac{e^{-ax}-e^{-bx}}{x}\right)^2 dx,$$

$$\int_0^\infty \left(\frac{1}{\cos ax} - \frac{1}{\cos bx}\right)\frac{dx}{x}, \quad \int_0^\infty \left(\frac{1}{\cos ax} - \frac{1}{\cosh bx}\right)\frac{dx}{x}.$$

[The last two integrals are defined by their principal values.]

Solution by the Proposer.

The usual proof of the formula is, if $b > a$,

$$\lim_{h=0,\,H=\infty}\int_h^H \frac{\phi(ax)-\phi(bx)}{x}\,dx = \lim_{h=0,\,H=\infty}\left(\int_{ah}^{bh} + \int_{aH}^{bH}\right)\phi(x)\frac{dx}{x}$$

$$= \{\phi(\infty)-\phi(0)\}\log a/b.$$

Now, provided $\phi(\infty)$, $\phi(0)$ be determinate, this remains valid under the conditions (1); for we can choose h, H initially so that all the infinities of $\phi(ax)$, $\phi(bx)$ fall in (h, H). And the same is true under conditions (2).

Thus, $\quad\displaystyle\int_0^\infty \log\left(\frac{1-pe^{-ax}}{1-pe^{-bx}}\right)^2 \frac{dx}{x} = \log(1-p)^2 \log\frac{b}{a},$

whether p be $\gtrless 1$. And

$$P\int_0^\infty \left(\frac{1}{1-ax}-\frac{1}{1-bx}\right)\frac{dx}{x} = \log\left(\frac{b}{a}\right).$$

If $\phi(x)$ oscillates at infinity, and $\displaystyle\int_h^\infty \frac{\phi(x)}{x}dx$ $(h>0)$ be determinate,

$$\lim \int_{aH}^{bH} \phi(x)\frac{dx}{x} = 0;$$

and then the correct formula is $\phi(0)\log(b/a)$.

The same holds if only $P\displaystyle\int_h^\infty \frac{\phi(x)}{x}dx$ be determinate. But in this case, if there be infinities of $\phi(x)$ extending up to ∞ [as, *e.g.*, if $\phi(x)=1/\cos x$], we must fix a quantity δ, and make H tend to infinity through a series of values such that aH, bH differ by not less than δ from any of the infinities in question. Thus it is not difficult to prove that

$$P\int_h^\infty \frac{1}{\cos x}\frac{dx}{x} \quad (0 < h < \tfrac{1}{2}\pi)$$

is determinate, and hence

$$P\int_0^\infty \left(\frac{1}{\cos ax}-\frac{1}{\cos bx}\right)\frac{dx}{x} = \log\frac{b}{a}.$$

Again, it is easy to prove by contour integration that
$$P\int_0^\infty \left(\frac{1}{\cos bx} - \frac{1}{\cosh bx}\right)\frac{dx}{x} = 0,$$
and so
$$P\int_0^\infty \left(\frac{1}{\cos ax} - \frac{1}{\cosh bx}\right)\frac{dx}{x} = \log\frac{b}{a}.$$

This affords a verification of the formula
$$\int_0^\infty dy \left\{ P\int_0^\infty \left(\frac{1}{\cos x} - \frac{1}{\cos yx}\right)\frac{dx}{x(1+y^2)} \right\}$$
$$= \int_0^\infty dx \left\{ P\int_0^\infty \left(\frac{1}{\cos x} - \frac{1}{\cos yx}\right)\frac{dy}{x(1+y^2)} \right\}.$$

For the left-hand is
$$\int_0^\infty \frac{\log y \, dy}{1+y^2} = 0,$$

and the right-hand is
$$\frac{\pi}{2}\int_0^\infty \left(\frac{1}{\cos x} - \frac{1}{\cosh x}\right)\frac{dx}{x} = 0.$$

Finally, if
$$u = \int_0^\infty \left(\frac{e^{-ax} - e^{-bx}}{x}\right)^2 dx,$$
$$\frac{du}{da} = 2\int_0^\infty \frac{e^{-2ax} - e^{-(a+b)x}}{x} dx = 2\log\frac{a+b}{2a};$$

whence
$$u = 2a\log\frac{a+b}{2a} + 2b\log\frac{a+b}{2b}.$$

CORRECTION
In the second line of the solution + should be −.

COMMENTS
It is assumed that $\phi(\infty)$ and $\phi(0)$ are 'determinate', i.e. that
$$\lim_{x\to\infty} \phi(x) \quad \text{and} \quad \lim_{x\to 0+} \phi(x)$$
exist as finite numbers. This justifies the statement that h and H may be chosen so that all the infinities of $\phi(ax)$, $\phi(bx)$ fall in (h, H). Note that, as in 1901, 1, Hardy uses the word 'discrete' to describe a set of zero Jordan content.

The initial statement of the cases (1) and (2) to be considered is not sufficiently precise. It is not the integral of $\phi(x)$, but the integral of $\phi(x)/x$ that must be 'determinate', as is shown by the example
$$\phi(x) = (x-1)^{-3} \quad (x \neq 1),$$
where $\int \phi(x)\,dx$ is determinate as a principal value 1, but $\int \{\phi(x)/x\}\,dx$ is not.

For a more detailed discussion of Frullani integrals see 1902, 3 (Vol. V, pp. 195–228); see also the earlier solution to Question 14164, in 1900, 2a.

14263. (D. Biddle.)—T is a multiple of $(N-T)^{\frac{1}{2}}$: prove that N is factorizable, unless $N-T = 1$; and, conversely, that N is either a prime or the square of a prime, when no lower value of T than $N-1$ will fulfil the conditions.

Solution by H. A. Webb; Lt.-Col. Allan Cunningham, R.E.; *and* G. H. Hardy, B.A.

Let $(N-T)^{\frac{1}{2}} = P$, and $T = PQ$, P and Q being positive integers. Then $N = P^2 + T = P^2 + PQ = P(P+Q)$. Hence N is factorizable, unless
$$N - T = P^2 = 1. \quad \quad \quad (1)$$

In the second case, the condition $T = Q \times (N-T)^{\frac{1}{2}}$ is not satisfied by a lower value of T than $(N-1)$. Hence $T = N-1$, or $T = N$, or $Q = 0$, and only under one of these three conditions does the equation $N = P(P+Q)$ hold good. The first condition gives $Q = N-1$, and, since $P(P+N-1) = N$, we must have $P = 1$. The second condition gives $N = 0$. The third condition gives $N = P^2$. Hence the only possible factors of N are N and 1, or P and P. N is therefore a prime, or the square of a prime.

[Denoting the half-sum of the factors of N by H, and the half-difference by h, we have $Q = 2h$ and $P = H-h$; therefore $T = 2h(H-h)$. In cases where it is known that the factors of N are of form $2\Delta m + 1$, it is clear that $Q \equiv 0 \pmod{2\Delta}$, and that T, besides being the difference between N and some square below it (namely, P^2), is of form $2\Delta m'(2\Delta m + 1)$, where $2\Delta m + 1 = P$. If m' be zero, $N = P^2$; if m be zero, $P = 1$, and N is prime. The former requires but little trouble to discover; the latter is determined by our failing to factorize, after trials conducted as follows :—Beginning with the first integer below $\sqrt{N}$ that is of form $2\Delta m + 1$, for P_1, and finding therefrom $N - P_1^2 = T_1$, we can readily set down in pairs P_n and T_n, because $P_n = P_{n-1} - 2\Delta$ and $T_n = T_{n-1} + 2\Delta(P_{n-1} + P_n)$. Or we can set down in pairs P_n and K_n, where $K_n = T_n/(2\Delta)$, and where $K_n = K_{n-1} + (P_{n-1} + P_n)$, which is easier. Excluding those pairs in which K_n is not a multiple of P_n, we at length find K_x and P_x, so that $N = P_x(P_x + 2\Delta K_x/P_x)$, or N is determined a prime. The excluding process can be facilitated by well known guides. Moreover, $K_1 + 2\Delta(x-1)^2$ is divisible by $P_1 - 2\Delta(x-1)$.]

1900, 6 *Educational Times*, 73, 76.

14496. (G. H. Hardy, B.A.)—Prove that

$$\Sigma\, \sigma^2(u)\, \sigma_1^2(u)\, \sigma_2^2(v)\, \sigma_3^2(v) = \tfrac{3}{2}\{\wp(u+v) + \wp(u-v)\}\, \sigma^2(u+v)\, \sigma^2(u-v)\,;$$

the notation being that of Weierstrass's theory of elliptic functions, and the summation applying to the six possible divisions into pairs of the functions $\sigma, \sigma_1, \sigma_2, \sigma_3$.

Solution by H. W. Curjel, M.A.

Consider the function

$$f(u) = \frac{\Sigma\, \sigma^2(u)\, \sigma_1^2(u)\, \sigma_2^2(v)\, \sigma_3^2(v)}{\sigma^2(u+v)\, \sigma^2(u-v)} = \frac{N(u)}{D(u)}\ \text{(say)}.$$

Since $\sigma(u)$, $\sigma_1(u)$, &c., are integral transcendental functions, all the poles of $f(u)$ are zeros of $D(u)$, i.e., $\pm v + 2m\omega_1 + 2n\omega_2$ (each a double one). Also, if we change u into $u + 2\omega_a$ ($a = 1, 2, 3$), we get

$$f(u + 2\omega_a) = \frac{N(u+2\omega_a)}{D(u+2\omega_a)} = \frac{e^{8\eta_a(u+\omega_a)}\,N(u)}{e^{8\eta_a(u+\omega_a)}\,D(u)} = f(u)\,;$$

therefore $f(u)$ is an elliptic function of periods $2\omega_1, 2\omega_2$. Again,

$$\lim_{u = \mp v}\, (u \pm v)^2 f(u) = \frac{6\sigma^2(v)\, \sigma_1^2(v)\, \sigma_2^2(v)\, \sigma_3^2(v)}{\sigma^2(2v)} = \tfrac{3}{2}\,;$$

therefore the infinite parts of the developments of $f(u)$ near the poles $\pm v$ are

$$\frac{3}{2(u-v)^2} + \frac{B}{u-v},\quad \frac{3}{2(u+v)^2} + \frac{A}{u+v}\,;$$

therefore $f(u) = A\zeta(u+v) + B\zeta(u-v) - \tfrac{3}{2}\{\zeta'(u+v) + \zeta'(u-v)\}$

$$= A\zeta(u+v) + B\zeta(u-v) + \tfrac{3}{2}\{\wp(u+v) + \wp(u-v)\},$$

where $A + B = 0$. When $u = 0$, we get

$$3\wp(v) = A\zeta(v) + B\zeta(-v) + 3\wp(v)\,;\quad \text{therefore}\ (A-B)\,\zeta(v) = 0\,;$$

therefore, if $\zeta(v) \neq 0$, $A - B = 0$; therefore $A = B = 0$; therefore

$$f(u) = \tfrac{3}{2}\{\wp(u+v) + \wp(u-v)\}\,;$$

if $\zeta(v) = 0$, substitute $v + 2\omega_a$ for v; this does not change $f(u)$, and, as $\zeta(v + 2\omega_a)$, $\zeta(v)$ cannot both be zero, the above proof can then be applied.

14511. (John C. Malet, M.A., F.R.S.)—If, in the sextic algebraic equation $x^6 - p_1 x^5 + p_2 x^4 - p_3 x^3 + p_4 x^2 - p_5 x + p_6 = 0$, the sum of three roots is equal to the sum of the other three, (1) prove
$$4p_6 Q_4 - Q_4 Q_3^2 - p_6 Q_2^2 + p_5 Q_2 Q_3 - p_5^2 = 0,$$
where $Q_2 \equiv p_2 - \tfrac{1}{4} p_1^2$, $Q_3 \equiv p_3 - \tfrac{1}{2} p_1 p_2 + \tfrac{1}{8} p_1^3$, $Q_4 \equiv p_4 - \tfrac{1}{2} p_1 p_3 + \tfrac{1}{4} p_1^2 p_2 - \tfrac{1}{16} p_1^4$; (2) solve the equation.

I. *Solution by the* Proposer; H. W. Curjel, M.A.; *and* Prof. Sanjána.

(1) Let $x_1, x_2, x_3, x_4, x_5, x_6$ be the roots of the equation; then
$$x_1 + x_2 + x_3 = x_4 + x_5 + x_6 = \tfrac{1}{2} p_1.$$
Let now $\quad x_1 x_2 + x_2 x_3 + x_3 x_1 = u_1, \quad x_4 x_5 + x_5 x_6 + x_6 x_4 = u_2,$
$$x_1 x_2 x_3 = v_1, \quad x_4 x_5 x_6 = v_2;$$
and we find $\quad u_1 + u_2 = Q_2 \dotfill$ (i.).
$v_1 + v_2 + \tfrac{1}{2} p_1 (u_1 + u_2) = p_3$; therefore $v_1 + v_2 = Q_3$ ...... (ii.).
$$u_1 u_2 + \tfrac{1}{2} p_1 (v_1 + v_2) = p_4;$$
therefore $\quad u_1 u_2 = Q_4, \quad v_1 u_2 + v_2 u_1 = p_5, \quad v_1 v_2 = p_6 \ldots$ (iii., iv., v.).
From (i.) and (iii.), we have
$$2u_1 = Q_2 + \sqrt{(Q_2^2 - 4Q_4)}, \quad 2u_2 = Q_2 - \sqrt{(Q_2^2 - 4Q_4)};$$
and, from (ii.) and (v.),
$$2v_1 = Q_3 + \sqrt{(Q_3^2 - 4p_6)}, \quad 2v_2 = Q_3 - \sqrt{(Q_3^2 - 4p_6)};$$
Substituting for u_1, u_2, v_1, v_2 in (v.) and rationalizing, we get the required condition.

(2) The roots of the sextic are the roots of the cubics
$$2x^3 - p_1 x^2 + \{Q_2 \pm \sqrt{(Q_2^2 - 4Q_4)}\} x - \{Q_3 \pm \sqrt{(Q_3^2 - 4p_6)}\} = 0.$$

II. *Solution by* G. H. Hardy, B.A.

It is interesting to consider this equation from the point of view of the Galois theory. The function $\phi_1 \equiv x_1 + x_2 + x_3 \equiv [123]$ has, in general, 20 values, viz.:
[123], [124], [125], [126], [134], [135], [136], [145], [146], [156], [234], [235], [236], [245], [246], [256], [345], [346], [356], [456];
and satisfies an equation $\quad g_{20}(\phi) = 0 \dotfill$ (i.).

The group of ϕ_1 is of order $(3!)^2 = 36$; and the solution of (i.) involves the complete solution of the sextic, for, if $x_1 + x_2 + x_3, \ldots$ are known, so are $x_1, x_2, x_3, \ldots$.

If we know *one* root of (i.), e.g., ϕ_1, and *adjoin* it, we can determine x_1, x_2, x_3 by the solution of a cubic, for $x_2 x_3 + x_3 x_1 + x_1 x_2$ and $x_1 x_2 x_3$ are rational functions of ϕ_1. As we also know $x_4 + x_5 + x_6$ rationally, we can determine x_4, x_5, x_6 by the solution of a second cubic.

Thus, when the coefficients of the sextic are so conditioned that ϕ_1 is a rational function of them, the equation can be solved by means of square and cube roots. This remains true in the present case, although two values of ϕ, viz., [123], [456], are numerically equal.

Let $\quad x^6 - p_1 x^5 + p_2 x^4 - p_3 x^3 + p_4 x^2 - p_5 x + p_6$
$$\equiv (x^3 - \tfrac{1}{2} p_1 x^2 + Ax - B)(x^3 - \tfrac{1}{2} p_1 x^2 + A'x - B');$$

1901, 5a *Educational Times*, 74, 42–3.

then
$$p_2 = \tfrac{1}{4}p_1^2 + A + A', \quad p_3 = \tfrac{1}{2}p_1(A+A') + B + B'$$
$$p_4 = \tfrac{1}{2}p_1(B+B') + AA', \quad p_5 = AB' + A'B, \quad p_6 = BB' \quad \Big\} \ldots (a).$$
$$A + A' = p_2 - \tfrac{1}{4}p_1^2 = Q_2,$$
$$B + B' = p_3 - \tfrac{1}{2}p_1 p_2 + \tfrac{1}{8}p_1^3 = Q_3, \quad AA' = p_4 - \tfrac{1}{2}p_1 p_3 + \tfrac{1}{4}p_1^2 p_2 - \tfrac{1}{16}p_1^4 = Q_4,$$
$$(A-A')(B-B') = Q_2 Q_3 - 2p_5, \quad (Q_2 Q_3 - 2p_5)^2 = (Q_2^2 - 4Q_4)(Q_3^2 - 4p_6);$$
i.e.,
$$4p_6 Q_4 - Q_4 Q_3^2 - p_6 Q_2^2 + p_5 Q_2 Q_3 - p_5^2 = 0 \ldots\ldots\ldots (\text{ii.}).$$

To solve the equation, we have to calculate
$$A, A' = \tfrac{1}{2}\left\{Q_2 \pm \sqrt{(Q_2^2 - 4Q_4)}\right\},$$
which must, in virtue of (ii.), be rational functions of the coefficients; and then determine B, B' by equations (*a*). The roots are then found by solving the two cubics.

COMMENTS

John C. Malet was Assistant Commissioner of Intermediate Education in Ireland, having formerly been Professor of Mathematics at Queen's College, Cork; he died in 1901.

Hardy's comments are neither helpful nor correct. Consider the sextic
$$x^6 - 2(x+1)^2 = (x^3 - \sqrt{2}x - \sqrt{2})(x^3 + \sqrt{2}x + \sqrt{2}),$$
which has rational coefficients. If x_1, x_2, x_3 are the zeros of the first cubic factor, then $x_2 x_3 + x_3 x_1 + x_1 x_2$ and $x_1 x_2 x_3$ are not rational functions of $\phi_1 (=0)$ and of the coefficients of the sextic. Moreover, Hardy's parameters A, A' take the values $\pm \sqrt{2}$ and are not, as claimed, rational functions of these coefficients.

14447. (H. W. Curjel, M.A.)—If $f(x)$ is finite and continuous for all positive finite values of x except a finite number of values, then

$$\int_0^\infty \sin\{f(x)\}\,dx \quad \text{and} \quad \int_0^\infty \cos\{f(x)\}\,dx$$

are convergent or divergent according as $\lim\limits_{x=\infty} \dfrac{df(x)}{dx}$ is infinite or finite; except in the case where $\lim\limits_{x=\infty} f(x) = 0$ or $n\pi$, when $\int_0^\infty \sin\{f(x)\}\,dx$ may be convergent, and the case where $\lim\limits_{x=\infty} f(x) = \tfrac{1}{2}\pi$ or $(2n+1)\tfrac{1}{2}\pi$, when $\int_0^\infty \cos\{f(x)\}\,dx$ may be convergent.

Solution by the Proposer.

First consider the case where $\lim\limits_{x=\infty} \dfrac{df(x)}{dx}$ is infinite. Here it is obvious that to any assigned small quantity ϵ there corresponds a finite quantity m such that $f(x+\epsilon) - f(x) > \pi$ if $x > m$, and that the smaller ϵ is the greater the least value of m is. Hence, if $x > m$, as x increases $\genfrac{}{}{0pt}{}{\sin}{\cos}\{f(x)\}$ changes signs at intervals which continually decrease and the greatest of which is less than ϵ. Hence $\left| \int_m^{m+n} \genfrac{}{}{0pt}{}{\sin}{\cos} \{f(x)\}\,dx \right| <$ a series of terms whose signs are alternately positive and negative, of which the first and second are less than ϵ, and the third, fourth, fifth, &c., are each less than the preceding one; therefore

$$\left| \int_m^{m+n} \genfrac{}{}{0pt}{}{\sin}{\cos} \{f(x)\}\,dx \right| < \epsilon;$$

therefore the integral is convergent.

Again, if $\lim\limits_{x=\infty} \dfrac{df(x)}{dx} = a$, a finite quantity not zero, then the interval between the changes of sign of $\genfrac{}{}{0pt}{}{\sin}{\cos}\{f(x)\}$ is always finite, and the integral can be reduced to a series of terms which are alternatively positive and negative, but never become indefinitely small; hence the integral is clearly divergent. This reasoning can be applied *a fortiori* to the case where $\lim\limits_{x=\infty} \dfrac{df(x)}{dx} = 0$, except in the cases where $\lim\limits_{x=\infty} \genfrac{}{}{0pt}{}{\sin}{\cos} \{f(x)\} = 0$, *i.e.*, when $\lim\limits_{x=\infty} f(x) = \begin{cases} -0 \text{ or } n\pi \\ \tfrac{1}{2}\pi \text{ or } n\pi + \tfrac{1}{2}\pi \end{cases}$, when the convergency of the integral clearly depends on the form of the function $f(x)$.

Note on Quest. 14447. *By* G. H. Hardy, B.A.

Mr. Curjel's solution does not seem to me to be entirely satisfactory. "The smaller ϵ is, the greater the least value of m is"; but it does not follow, so far as I can see, that, when m is fixed, the intervals between the points a_i at which $\sin f(x)$ changes sign in (m, ∞) will diminish continually as i increases, though they will certainly all be less than ϵ; and, even if they did, it would not follow that $\left| \int_{a_i}^{a_{i+1}} \sin\{f(x)\}\, dx \right|$ diminished continually as i increased. And, in fact, the theorem is untrue.

For, if $\dfrac{df(x)}{dx}$ be continuous and >0 for all finite values of x,
$$\int^{H} \sin\{f(x)\}\, dx = \int^{f(H)} \sin y\, \frac{dy}{f'(x)} \quad \{y = f(x)\}.$$

Now suppose $\quad f'(x) = y^\mu - a \sin y \quad (0 < \mu < \tfrac{1}{2})$;

i.e., $\quad x = \int_0^y \dfrac{dy}{y^\mu - a \sin y} \quad \{y = f(x)\}.$

Then $\displaystyle\int_0^\infty \sin\{f(x)\}\, dx$ converges or diverges with
$$\int_0^\infty \frac{\sin y\, dy}{y^\mu - a \sin y} \quad (0 < \mu < \tfrac{1}{2}),$$
as x, y become infinite together. But it is not difficult to show that this integral *diverges*. It would take too long to enter into the proof at present; but it is easy to see the reason. The oscillations of the denominator *coincide in phase* with those of the numerator; so that, to speak roughly, all the positive elements are increased by the presence of the oscillating term in the denominator, and all the negative elements decreased.

The second part of Mr. CURJEL's theorem is true. But the integral may be determinate if $f'(x)$ does not tend to a limit at all, for $x = \infty$. But it *is* a sufficient condition that $f'(x)$ tend to ∞ *steadily*; i.e., without oscillations.

[Mr. CURJEL having seen the above, and having been requested to send any further remarks he might deem requisite, says: "I have nothing to add, as Mr. HARDY's criticism is quite correct; my proof only applies when $f'(x)$ increases *steadily* to ∞, and consequently the theorem to a comparatively limited class of functions; i.e., the condition given in the Question is not by itself sufficient for convergence, but that for divergence is sufficient."]

COMMENTS

In Hardy's example, where
$$y = f(x), \quad f'(x) = y^\mu - a \sin y \quad (0 < \mu < \tfrac{1}{2}),$$
he tacitly assumes that $f(0) = 0$; presumably, also, a must be taken sufficiently small so that $f'(x) \neq 0$ for $y > 0$.

14467. (G. H. Hardy, B.A.)—Prove that

$$\int_{-\infty}^{\infty} \{\phi(x-a)-\phi(x-b)\}\, dx = (b-a)\{\phi(\infty)-\phi(-\infty)\},$$

provided each side of the equation represents a determinate quantity. Deduce the values of

$$\int_{-\infty}^{\infty} \frac{dx}{\cosh(x-a)\cosh(x-b)}, \quad P\int_{-\infty}^{\infty} \frac{dx}{\sinh(x-a)\sinh(x-b)}.$$

Solution by the Proposer.

$$\int_{-\infty}^{\infty} \{\phi(x-a)-\phi(x-b)\}\, dx = \lim_{H=\infty} \int_{-H}^{H}$$

$$= \lim_{H=\infty}\left[\int_{-H-a}^{H-a} \phi(u)\,du - \int_{-H-b}^{H-b} \phi(u)\,du\right]$$

$$= \lim_{H=\infty}\left[\int_{H-b}^{H-a} \phi(u)\,du - \int_{-H-b}^{-H-a} \phi(u)\,du\right]$$

$$= (b-a)\{\phi(\infty)-\phi(-\infty)\},$$

if both sides of the equation be determinate.

If $\phi(u) = \tanh u$,

$$\phi(x-a)-\phi(x-b) = \frac{\sinh(b-a)}{\cosh(x-a)\cosh(x-b)},$$

and

$$\int_{-\infty}^{\infty} \frac{dx}{\cosh(x-a)\cosh(x-b)} = \frac{b-a}{\sinh(b-a)}.$$

If $\phi(u) = \coth u$, we find

$$P\int_{-\infty}^{\infty} \frac{dx}{\sinh(x-a)\sinh(x-b)} = -\frac{b-a}{\sinh(b-a)}.$$

It is easy to see that the proof remains valid, although in the latter case only the principal value of the integral is determinate.

1901, 7 *Educational Times*, 74, 111–12.

14028. (G. H. Hardy.)—Reduce the evaluation of $\int_0^\pi \frac{\cos(p/q)\phi\, d\phi}{1 + 2t\cos\phi + t^2}$, where p, q are integers, $p < q$ and $t < 1$, to the integration of a rational fraction. Prove, in particular, that

$$\int_0^\pi \frac{\cos\tfrac{1}{2}\phi\, d\phi}{1 + 2t\cos\phi + t^2} = \frac{2}{1+t}\frac{\tanh^{-1}\sqrt{t}}{\sqrt{t}};$$

and deduce (and also prove independently) that

$$\int_0^\pi \tan^{-1}\left(\frac{2t\sin\phi}{1-t^2}\right)\frac{d\phi}{\sin\tfrac{1}{2}\phi} = 8\tan^{-1}\sqrt{t}\,\tanh^{-1}\sqrt{t}.$$

Solution by the Proposer.

Since
$$\frac{1-t^2}{1+2t\cos\phi+t^2} = 1 + 2\sum_1^\infty (-t)^n \cos n\phi,\quad t<1,$$

$$\int_0^\pi \frac{\cos a\phi\, d\phi}{1+2t\cos\phi+t^2} = \frac{1}{1-t^2}\int_0^\pi \left\{\cos a\phi + 2\sum_1^\infty(-t)^n \cos a\phi \cos n\phi\right\}d\phi$$

$$= \frac{\sin a\pi}{1-t^2}\left\{\frac{1}{a} + 2\sum_1^\infty\left(\frac{1}{n+a} - \frac{1}{n-a}\right)t^n\right\},$$

if $a < 1$. That is to say,

$$= \frac{\sin a\pi}{1-t^2}\left\{t^{-a}\int_0^t \frac{t^{a-1}\,dt}{1-t} - t^a\int_0^t\frac{t^{-a}\,dt}{1-t}\right\}.$$

If $a = p/q$, this is

$$\frac{q\sin(p/q)\pi}{1-t^2}\left\{t^{-(p/q)}\int_0^{t^{1/q}}\frac{\tau^{p-1}\,d\tau}{1-\tau^q} - t^{p/q}\int_0^{t^{1/q}}\frac{\tau^{q-p-1}\,d\tau}{1-\tau^q}\right\}.$$

If $p=1$, $q=2$, we get

$$\frac{2}{1-t^2}\left\{t^{-\frac{1}{2}}\int_0^{t^{\frac{1}{2}}}\frac{d\tau}{1-\tau^2} - t^{\frac{1}{2}}\int_0^{t^{\frac{1}{2}}}\frac{d\tau}{1-\tau^2}\right\} = \frac{1}{1-t^2}(t^{-\frac{1}{2}} - t^{\frac{1}{2}})\log\frac{1+t^{\frac{1}{2}}}{1-t^{\frac{1}{2}}}$$

$$= \frac{2}{1+t}\frac{\tanh^{-1}\sqrt{t}}{\sqrt{t}}.$$

(This integral and the corresponding integral with a sine in the numerator may also be easily obtained by contour integration applied to $\int\frac{z^{a-1}\,dz}{1+z}$.) That is to say,

$$\int_0^\pi \frac{\cos\tfrac{1}{2}\phi\, d\phi}{1+2t\cos\phi+t^2} = 2\tanh^{-1}\sqrt{t}\,\frac{d}{dt}(2\tan^{-1}\sqrt{t}).$$

Similarly, $\int_0^\pi \frac{\cos\tfrac{1}{2}\phi\, d\phi}{1-2t\cos\phi+t^2} = 2\tan^{-1}\sqrt{t}\,\frac{d}{dt}(2\tanh^{-1}\sqrt{t}).$

Adding and integrating from 0 to t, we find

$$\int_0^\pi \tan^{-1}\left(\frac{2t\sin\phi}{1-t^2}\right)\frac{d\phi}{\sin\tfrac{1}{2}\phi} = 8\tan^{-1}\sqrt{t}\,\tanh^{-1}\sqrt{t}.$$

This may be verified independently, as follows:—

$$\int_0^\pi \tan^{-1}\left(\frac{2t\sin\phi}{1-t^2}\right)\frac{d\phi}{\sin\frac{1}{2}\phi} = 2\sum_1^\infty \frac{t^{2n+1}}{2n+1}\int_0^\pi \frac{\sin(2n+1)\phi}{\sin\frac{1}{2}\phi}d\phi$$

$$= 8\sum_1^\infty \left(1-\frac{1}{3}\ldots+\frac{1}{4n+1}\right)\frac{t^{2n+1}}{2n+1}$$

$$= 8\left[t+(1-\frac{1}{3}+\frac{1}{5})\frac{1}{3}t^3+(1-\frac{1}{3}+\frac{1}{5}-\frac{1}{7}+\frac{1}{9})\frac{1}{5}t^5\ldots\right]$$

$$= 8\left[(t^{\frac{1}{2}}-\frac{1}{3}t^{\frac{3}{2}}+\frac{1}{5}t^{\frac{5}{2}}\ldots)(t^{\frac{1}{2}}+\frac{1}{3}t^{\frac{3}{2}}+\frac{1}{5}t^{\frac{5}{2}}\ldots)\right]$$

$$= 8\tan^{-1}\sqrt{t}\tanh^{-1}\sqrt{t}.$$

CORRECTIONS

The word 'fraction' in the third line should be replaced by 'function'. In the third line of the solution the 2 before the summation sign should be omitted.

COMMENT

The parameters p, q, α, and t are assumed to satisfy the conditions

$$0<p<q,\quad 0<\alpha<1,\quad 0<t<1.$$

14369. (G. H. Hardy, B.A.—If $\sum_{0}^{\infty} u_n(x)$ be a *summable divergent* series, having the sum $\int_0^\infty e^{-a} u(x, a)\, da,$

where $u(x, a) = \sum_{0}^{\infty} \frac{a^n}{n!} u(x),$

prove that it may be integrated term by term in (x_0, X), provided (1) $u(x, a)$ may be integrated term by term, and (2) the order of integration in $\int_0^\infty e^{-a} da \int_{x_0}^{X} u(x, a)\, dx$

may be changed. Deduce and verify independently that, if
$$\phi(x, a) = e^{-a(1+\cos x)} \cos(x - a \sin x),$$
$$\psi(x, a) = e^{-a(1-\cos 2mx)} \cos(mx + a \sin 2mx)/(1+x^2),$$

then $\int_0^{2\pi} dx \int_0^\infty \phi\, da = \pi, \quad \int_0^\infty da \int_0^{2\pi} \phi\, dx = 0;$

$\int_0^\infty dx \int_0^\infty \psi\, da = 0, \quad \int_0^\infty da \int_0^\infty \psi\, dx = \tfrac{1}{2} \dfrac{\pi}{e^m - e^{-m}}.$

Solution by the Proposer.

According to Borel's definition of the sum of a divergent series,
$$\sum_{0}^{\infty} u_n(x) = \int_0^\infty e^{-a} u(x, a)\, dx.$$

Hence $\int_{x_0}^{X} dx \sum_{0}^{\infty} u_n(x) = \int_{x_0}^{X} dx \int_0^\infty e^{-a} u(x, a)\, dx = \int_0^\infty e^{-a} da \int_{x_0}^{X} u(x, a)\, dx$

$= \int_0^\infty e^{-a} da \int_{x_0}^{X} \sum_{0}^{\infty} \frac{a^n}{n!} u_n(x)\, dx$

$= \int_0^\infty e^{-a} da \sum_{0}^{\infty} \frac{a^n}{n!} \int_{x_0}^{X} u_n(x)\, dx = \sum_{0}^{\infty} \int_{x_0}^{X} u_n(x)\, dx;$

provided conditions (1), (2) be satisfied. Here X may be finite or infinite. It follows that, if integration of the divergent series term by term leads to a false result, one or other of (1), (2) cannot be satisfied. Generally it is (2).

Thus consider the divergent series $\cos x - \cos 2x \ldots$.

In this case $u(x, a) = \sum_{0}^{\infty} (-)^n \frac{a^n}{n!} \cos(n+1)x = e^{-a\cos x} \cos(x - a \sin x),$

and $\int_0^\infty e^{-a} u(x, a)\, da = \dfrac{\cos x(1 + \cos x) + \sin^2 x}{(1 + \cos x)^2 + \sin^2 x} = \tfrac{1}{2};$

so that $\cos x - \cos 2x + \ldots = \tfrac{1}{2}.$

If we integrate from 0 to 2π, we find $0 = \pi$.

Now it is easy to see that in this case (1) is satisfied. It follows that, if
$$\phi(x, a) = e^{-a(1+\cos x)} \cos(x - a \sin x),$$
$\int_0^{2\pi} dx \int_0^\infty \phi\, da = \pi, \quad \int_0^\infty da \int_0^{2\pi} \phi\, dx = 0.$

The second equation is easily verified, for
$$\int_0^{2\pi} \phi\, dx = -\frac{e^{-a}}{a} \left(e^{-ae^{ix}} \right)_0^{2\pi} = 0.$$

Again, it is easy to show that $\cos mx + \cos 3mx + \ldots = 0.$

1901, 9 *Educational Times*, 75, 135–6.

Multiply by $\frac{1}{1+x^2}$, and integrate from 0 to ∞. We find

$$\tfrac{1}{2} \frac{\pi}{e^m - e^{-m}} = 0;$$

and it follows that, if

$$\psi(x, \alpha) = e^{-\alpha(1-\cos 2mx)} \frac{\cos(mx + \alpha \sin 2mx)}{1+x^2},$$

$$\int_0^\infty dx \int_0^\infty \psi \, d\alpha = 0, \qquad \int_0^\infty d\alpha \int_0^\infty \psi \, dx = \tfrac{1}{2} \frac{\pi}{e^m - e^{-m}}.$$

The second equation is also easily verified, for

$$\int_0^\infty \psi \, dx = e^{-\alpha} \int_0^\infty \frac{dx}{1+x^2} \sum_0^\infty \frac{\alpha^n}{n!} \cos(2n+1)mx$$

$$= \tfrac{1}{2}\pi e^{-\alpha} \sum_0^\infty \frac{\alpha^n}{n!} e^{-(2n+1)m} = \tfrac{1}{2}\pi e^{-m-\alpha(1-e^{-2m})};$$

and

$$\tfrac{1}{2}\pi e^{-m} \int_0^\infty e^{-\alpha(1-e^{-2m})} d\alpha = \tfrac{1}{2} \frac{\pi}{e^m - e^{-m}}.$$

CORRECTIONS

Line 1. Add closing parenthesis after B.A.
Line 3. This should read:

$$u(x, \alpha) = \sum_0^\infty \frac{\alpha^n}{n!} u_n(x).$$

Line 14. In integrand replace dx by $d\alpha$.
Line 15. The third dx should be $d\alpha$.

COMMENT

Note that, in *line* 17, the last series is to be interpreted in Borel's sense.

1423, 2316, 3941, 4794. (The late Professor CLIFFORD, F.R.S.)—
Prove that $\int_0^{\frac{1}{2}\pi} \cos(a \tan x) e^{\beta \tan x} dx = \frac{1}{2}\pi e^{-a}(\cos \beta + \sin \beta)$.

Note by G. H. HARDY, B.A.

This question has been reproposed so often that it is curious that no one should have pointed out that the formula is obviously untrue. If $\beta > 0$, the integral is divergent. If $\beta < 0$, $\int_0^\infty \frac{\cos ax \, e^{\beta x}}{1+x^2} dx$ certainly cannot be evaluated in terms of elementary functions. If, for instance, $a = 0$, $\beta = -b$,
$$\int_0^\infty \frac{e^{-bx}}{1+x^2} dx = \int_0^\infty \frac{\sin bx}{1+x} dx,$$

(CAUCHY, *Œuvres*, I., p. 377), *i.e.*,

$$= \int_1^\infty \frac{\sin b(u-1)}{u} du = \cos b \int_b^\infty \frac{\sin t}{t} dt - \sin b \int_b^\infty \frac{\cos t}{t} dt$$

$$= \cos b \left(\tfrac{1}{2}\pi - \operatorname{Si} b\right) + \sin b \operatorname{Ci} b \qquad \text{(DE HAAN, *Tables* 91, 7)}.$$

The solution given in Vol. LIII. of the *Reprint* (p. 107) is altogether wrong. The integrals $\int_0^\infty \frac{x^r \cos ax}{1+x^2} dx$, $\int_0^\infty \frac{x^r \sin ax}{1+x^2} dx$ are divergent if $r \geq 2$. And, in any case, there would be no more reason for supposing them equal than for supposing that

$$\int_0^\infty e^{-x} \cos ax \, dx = \int_0^\infty e^{-x} \sin ax \, dx,$$

which is obviously untrue.

COMMENT

If the stated result were true, the integral would vanish for $\beta = -\tfrac{1}{4}\pi$ and $a = 0$, which is clearly false.

William Kingdon Clifford (1845–1879), the proposer, was Second Wrangler in 1867, and held the chair of Applied Mathematics at University College, London from 1871 until his death.

14851. (G. H. HARDY, B.A.)—Prove that, if m, n are positive integers whose greatest common divisor is r, and $m = r\mu$, $n = r\nu$; and $0 < \alpha < \pi$, $0 < \beta < \pi$; and p, q are numerically less than unity; then

$$\int_0^\pi \frac{dx}{(1-2p\cos mx + p^2)(1-2q\cos nx + q^2)} = \frac{\pi}{(1-p^2)(1-q^2)} \frac{1+p^\nu q^\mu}{1-p^\nu q^\mu} \quad \text{(i.)},$$

$$P\int_0^\pi \frac{dx}{(1-2p\cos mx + p^2)(\cos\beta - \cos nx)} = -\frac{2\pi}{(1-p^2)\sin\beta} \frac{p^\nu \sin\mu\beta}{1-2p^\nu \cos\mu\beta + p^{2\nu}} \quad \text{(ii.)},$$

$$P\int_0^\pi \frac{dx}{(\cos\alpha - \cos mx)(\cos\beta - \cos nx)} = 0 \quad \text{(iii.)},$$

unless $n\alpha = m\beta$, in which case (iii.) is not determinate.

Solution by the PROPOSER.

(i.) $\int_0^\pi \dfrac{dx}{(1-2p\cos mx + p^2)(1-2q\cos nx + q^2)}$

$$= \frac{1}{1-p^2}\left[\int_0^\pi \frac{dx}{1-2q\cos nx + q^2} + 2\sum_1^\infty p^\lambda \int_0^\pi \frac{\cos\lambda mx}{1-2q\cos nx + q^2}dx\right].$$

Now $\int_0^\pi \dfrac{\cos\lambda mx}{1-2q\cos nx + q^2}dx = \dfrac{\pi q^\kappa}{1-q^2}$

if $\lambda m = \kappa n$; $= 0$ otherwise. And, if $\lambda m = \kappa n$, $\lambda\mu = \kappa\nu$; and λ is a multiple of ν, κ of μ: say $\lambda = \xi\nu$, $\kappa = \xi\mu$. Then the original integral is

$$\frac{\pi}{(1-p^2)(1-q^2)}\left[1 + 2\sum_1^\infty (q^\mu p^\nu)^\xi\right] = \frac{\pi}{(1-p^2)(1-q^2)} \frac{1+p^\nu q^\mu}{1-p^\nu q^\mu}.$$

(ii.) $P\int_0^\pi \dfrac{dx}{(1-2p\cos mx + p^2)(\cos\beta - \cos nx)}$

$$= \frac{1}{1-p^2}\left[P\int_0^\pi \frac{dx}{\cos\beta - \cos nx} + 2\sum_1^\infty p^\lambda P\int_0^\pi \frac{\cos\lambda mx}{\cos\beta - \cos nx}dx\right].$$

Now $P\int_0^\pi \dfrac{\cos\lambda mx}{\cos\beta - \cos nx}dx = -\pi\dfrac{\sin\kappa\beta}{\sin\beta}$

if $\lambda m = \kappa n$, $= 0$ otherwise. Hence the original principal value is

$$-\frac{2\pi}{(1-p^2)\sin\beta}\sum_1^\infty p^{\xi\mu}\sin\xi\nu\beta = -\frac{2\pi}{(1-p^2)\sin\beta}\frac{p^\nu \sin\mu\beta}{1-2p^\nu\cos\mu\beta + p^{2\nu}}.$$

(iii.) To evaluate this, we expand $\dfrac{1}{\cos\alpha - \cos mx}$ as a *divergent* series

$$-\frac{2}{\sin\alpha}\sum_1^\infty \sin\lambda\alpha\cos\lambda mx,$$

and take the principal value term by term. We obtain

$$\frac{\pi}{\sin\alpha\sin\beta}\sum_1^\infty\{\cos\xi(\mu\beta - \nu\alpha) - \cos\xi(\mu\beta + \nu\alpha)\};$$

and the "sum" of this *divergent* series is 0 unless $\mu\beta = \nu\alpha$, in which case (iii.) is plainly indeterminate.

The methods used in evaluating (ii.) and (iii.) are strictly justifiable; but to prove this would require a long discussion into which we cannot enter at present. See solution to Quest. 14639.

CORRECTION

At the end of Solution (i.) the denominator should be $1-p^\nu q^\mu$ and not $-p^\nu q^\mu$.

The Question referred to at the end of Solution (iii.) is presumably **14369** (i.e. 1901, 9) and not **14639**.

14055. (G. H. HARDY.)—Express

$$\int_0^{\frac{1}{2}\pi} \log \sin x \sqrt{(\sin x)}\, dx, \quad \int_0^{\frac{1}{2}\pi} \frac{\log \sin x}{\sqrt{(\sin x)}}\, dx, \quad \int_0^{\frac{1}{2}\pi} \frac{\log \sin x}{\sqrt{(1+\sin^2 x)}}\, dx,$$

$$\int_0^{\pi} \frac{dx}{(\sec \frac{1}{2}x + \tan \frac{1}{2}x)^m \sqrt{(\sin x)}} \quad (m \text{ a positive integer}),$$

$$\int_0^{\pi} \frac{\log(\sec \frac{1}{2}x + \tan \frac{1}{2}x)}{\sqrt{(\sin x)}}\, dx, \quad \text{and} \quad \int_0^{\infty} \frac{x}{\sqrt{(\sinh x)}}\, dx$$

as multiples of $[\Gamma(\tfrac{1}{4})]^{\pm 2}$; and prove that

$$\int_0^{\frac{1}{2}\pi} \sin x (\log \sin x)^2\, dx = (\log 2 - 1)^2 + 1 - \tfrac{1}{12}\pi^2.$$

[The last integral is given in Quest. 8423 (D. EDWARDES); another proof is asked for.]

Solution by the PROPOSER.

(*a*) Let $\quad f(n, p) = \int_0^{\frac{1}{2}\pi} \sin^n x (\log \sin x)^p\, dx \quad (-1 < n,\ -1 < p).$

Then it is easy to prove that (WOLSTENHOLME's *Problems*, 1918, 51), if
$-1 < n < 1, \qquad f(n, 0) f(-n, 0) = (2\pi/n) \tan \tfrac{1}{2} n\pi \ \dots\dots\dots\dots\dots\dots (1),$
and, if $0 < n, \qquad f(n, 0) f(n-1, 0) = 2\pi/n \ \dots\dots\dots\dots\dots\dots(2).$

Differentiating these equations with respect to n,

$$f(n, 1) f(-n, 0) - f(n, 0) f(-n, 1) = -(2\pi/n^2) \tan \tfrac{1}{2} n\pi + (\pi^2/n) \sec^2 \tfrac{1}{2} n\pi,$$
$$f(n, 1) f(n-1, 0) + f(n, 0) f(n-1, 1) = -(2\pi/n^2).$$

If $n = \tfrac{1}{2}, \quad f(\tfrac{1}{2}, 1) f(-\tfrac{1}{2}, 0) - f(\tfrac{1}{2}, 0) f(-\tfrac{1}{2}, 1) = -8\pi + 4\pi^2,$
$$f(\tfrac{1}{2}, 1) f(-\tfrac{1}{2}, 0) + f(\tfrac{1}{2}, 0) f(-\tfrac{1}{2}, 1) = -8\pi.$$

Therefore $\quad f(\tfrac{1}{2}, 1) = \dfrac{2\pi(\pi-4)}{f(-\tfrac{1}{2}, 0)}, \quad f(-\tfrac{1}{2}, 1) = -\dfrac{2\pi^2}{f(\tfrac{1}{2}, 0)}.$

Now $\quad f(-\tfrac{1}{2}, 0) = \sqrt{\pi}\, \dfrac{\Gamma(\tfrac{1}{4})}{\Gamma(\tfrac{3}{4})} = \dfrac{[\Gamma(\tfrac{1}{4})]^2}{\sqrt{(2\pi)}},$

$$f(\tfrac{1}{2}, 0) = \sqrt{\pi}\, \dfrac{\Gamma(\tfrac{3}{4})}{\Gamma(\tfrac{5}{4})} = \dfrac{4\pi\sqrt{(2\pi)}}{[\Gamma(\tfrac{1}{4})]^2}.$$

Hence $\quad \int_0^{\frac{1}{2}\pi} \sqrt{(\sin x)} \log \sin x\, dx = \dfrac{(\pi-4)(2\pi)^{\frac{3}{2}}}{[\Gamma(\tfrac{1}{4})]^2},$

$$\int_0^{\frac{1}{2}\pi} \frac{\log \sin x\, dx}{\sqrt{(\sin x)}} = -2^{-\frac{3}{2}} \sqrt{\pi}\, [\Gamma(\tfrac{1}{4})]^2.$$

(*b*) Suppose $\quad \operatorname{cosec} x = \sec \tfrac{1}{2}\theta + \tan \tfrac{1}{2}\theta.$

Then $\quad \dfrac{dx}{\sqrt{(\sin x)}} = -\dfrac{(\sin x)}{\cos x} \dfrac{1 + \sin \tfrac{1}{2}\theta}{2 \cos^2 \tfrac{1}{2}\theta}\, d\theta = -\tfrac{1}{2} \dfrac{\sqrt{(\sin x)}}{\cos x}\, \dfrac{d\theta}{\cos \tfrac{1}{2}\theta}$

$$= -\tfrac{1}{2} \sec \tfrac{1}{2}\theta \sqrt{\left(\dfrac{\sec \tfrac{1}{2}\theta + \tan \tfrac{1}{2}\theta}{(\sec \tfrac{1}{2}\theta + \tan \tfrac{1}{2}\theta)^2 - 1} \right)}\, d\theta = -\tfrac{1}{2} \dfrac{d\theta}{\sqrt{(\sin \theta)}}.$$

1902, 14 *Educational Times*, (2), 2, 41-2.

As x goes from 0 to $\tfrac{1}{2}\pi$, θ goes from π to 0. Hence
$$\int_0^\pi \phi(\sec \tfrac{1}{2}\theta + \tan \tfrac{1}{2}\theta) \frac{d\theta}{\sqrt{(\sin \theta)}} = 2 \int_0^\pi \phi(\operatorname{cosec} x) \frac{dx}{\sqrt{(\sin x)}}.$$

This formula gives
$$\int_0^\pi \frac{\log(\sec \tfrac{1}{2}x + \tan \tfrac{1}{2}x)}{\sqrt{(\sin x)}} dx = -2 \int_0^{\frac{1}{2}\pi} \frac{\log \sin x}{\sqrt{(\sin x)}} dx = 2^{\frac{5}{2}} \sqrt{\pi} [\Gamma(\tfrac{1}{4})]^2.$$

And
$$\int_0^\pi \frac{(\sec \tfrac{1}{2}x + \tan \tfrac{1}{2}x)^{-m}}{\sqrt{(\sin x)}} dx = -2 \int_0^{\frac{1}{2}\pi} (\sin x)^{m-\frac{1}{2}} dx;$$

which is easily expressed in terms of $[\Gamma(\tfrac{1}{4})]^2$ when m is an integer.

(c) If we put $\sin x = e^{-u}$, $dx = -\dfrac{e^{-u} du}{\sqrt{(1-e^{-2u})}}$,
$$\int_0^{\frac{1}{2}\pi} \frac{\log \sin x}{\sqrt{(\sin x)}} dx = -\int_0^\infty \frac{u e^{-\frac{1}{2}u} du}{\sqrt{(1-e^{-2u})}} = -\frac{1}{\sqrt{2}} \int_0^\infty \frac{u\, du}{\sqrt{(\sinh u)}}.$$

If we put $\sin x = \sin^2 y$, $dx = \dfrac{2 \sin y\, dy}{\sqrt{(1+\sin^2 y)}}$,
$$\int_0^{\frac{1}{2}\pi} \frac{\log \sin x\, dx}{\sqrt{(\sin x)}} = 4 \int_0^{\frac{1}{2}\pi} \frac{\log \sin y\, dy}{\sqrt{(1+\sin^2 y)}}.$$

(d) We may observe that, if
$$\int_0^\infty e^{-x^4} dx = k, \quad k e^{-a^4} = \int_0^\infty e^{-a^4(1+x^4)} a\, dx,$$
$$k^2 = \int_0^\infty da \int_0^\infty e^{-a^4(1+x^4)} a\, dx = \tfrac{1}{2} \int_0^\infty dx \int_0^\infty e^{-a^2(1+x^4)} da$$
$$= \tfrac{1}{4} \sqrt{\pi} \int_0^\infty \frac{dx}{\sqrt{(1+x^4)}};$$

so that
$$[\Gamma(\tfrac{1}{4})]^2 = 4\sqrt{\pi} \int_0^\infty \frac{dx}{\sqrt{(1+x^4)}}.$$

Thus, e.g.,
$$\int_0^{\frac{1}{2}\pi} \frac{\log \sin x\, dx}{\sqrt{(\sin x)}} = -\pi \sqrt{2} \int_0^\infty \frac{dx}{\sqrt{(1+x^4)}}.$$

14988. (The late Professor CAYLEY, F.R.S.)—If
$$l^2 + m^2 + n^2 + \ldots, \quad l'^2 + m'^2 + n'^2 + \ldots$$
are each $= 0$, and s, s' denote positive integers, show that the integral
$$\int (lx + my + nz + \ldots)^s (l'x + m'y + n'z + \ldots)^{s'} dx\,dy\,dz \ldots,$$
over the interior of the hyper-sphere $x^2 + y^2 + z^2 + \ldots = 1$ is $= 0$ or $=$ a numerical multiple of $(ll' + mm' + nn' + \ldots)^s$, according as s' is not, or is, $= s$.

Solution by G. H. HARDY, B.A.

It is convenient to write $x_1, x_2, \ldots, l_1, l_2, \ldots, l'_1, l'_2, \ldots$ for $x, y, \ldots, l, m, \ldots$, $l', m', \ldots$; Λ for $l_1 x_1 + l_2 x_2 + \ldots$; and Λ' for $l'_1 x_1 + l'_2 x_2 + \ldots$. And we suppose that n is the number of variables. Now

$$\int \Lambda^s \Lambda'^{s'} dv = \frac{1}{s + s' + n} \int \Lambda^s \Lambda'^{s'} dS,$$

the latter integral being extended over the surface of the unit sphere. Also
$$\Lambda^s \Lambda'^{s'} = (l_1 x_1 + l_2 x_2 + \ldots) \Lambda^{s-1} \Lambda'^{s'},$$
and, as $x_1, x_2, \ldots$ are the direction cosines of the outward normal at $(x_1, x_2, \ldots)$,
$$\int \Lambda^s \Lambda'^{s'} dS = \int \left(\frac{\partial}{\partial x_1} (l_1 \Lambda^{s-1} \Lambda'^{s'}) + \frac{\partial}{\partial x_2} (l_2 \Lambda^{s-1} \Lambda'^{s'}) \ldots \right) dv.$$

But $\quad \Sigma \partial/\partial x_i (l_i \Lambda^{s-1} \Lambda'^{s'}) = (s-1) \Lambda^{s-2} \Lambda'^{s'} \Sigma l_i^2 + s' \Lambda^{s-1} \Lambda'^{s'-1} \Sigma l_i l'_i$;

so that $\quad \int \Lambda^s \Lambda'^{s'} dv = \frac{s'}{s+s'+n} \Sigma l_i l'_i \int \Lambda^{s-1} \Lambda'^{s'-1} dv.$

The same reasoning shows that it is also equal to
$$\frac{s}{s+s'+n} \Sigma l_i l'_i \int \Lambda^{s-1} \Lambda'^{s'-1} dv ;$$

it is therefore equal to 0 unless $s = s'$. And, if $s = s'$,

$$\int \Lambda^s \Lambda'^s dv = \frac{s}{2s+n} \Sigma l_i l'_i \int \Lambda^{s-1} \Lambda'^{s-1} dv = \frac{s}{2s+n} \frac{s-1}{2s+n-2} \cdots \frac{1}{n+2} (\Sigma l_i l'_i)^s \int dv$$

$$= \frac{s(s-1)\ldots 1}{(2s+n)(2s+n-2)\ldots(n+2)} \frac{[\Gamma(\tfrac{1}{2})]^n}{\Gamma(\tfrac{1}{2}n+1)} (\Sigma l_i l'_i)^s$$

$$= \frac{\pi^{\tfrac{1}{2}n}}{2^s} \frac{\Gamma(s+1)}{\Gamma(s+1+\tfrac{1}{2}n)} (\Sigma l_i l'_i)^s.$$

This result may easily be verified for small values of n. If $n = 2$, we obtain
$$\int (lx+my)^s (l'x+m'y)^{s'} dv = 0$$
or
$$= \frac{\pi}{2^s (s+1)} (ll'+mm')^s.$$

The volume integral is
$$\frac{1}{s+s'+2} \int (lx+my)^s (l'x+m'y)^{s'} dS = \frac{l^s l'^{s'}}{s+s'+2} \int_0^{2\pi} e^{\pm si\theta \pm s'i\theta} d\theta$$

(since $m = \pm il$, $m' = \pm il'$). This is 0 unless $s = s'$ and the signs are different. In this case it is
$$\frac{2\pi (ll')^s}{2(s+1)} = \frac{\pi (ll' + mm')^s}{2^s (s+1)}.$$
If the signs are the same, $ll' + mm' = 0$; so that the formula still holds.

CORRECTIONS

Line 12. Read Λ^s for Λ_s in integrand on right.
Line 19. Read $l_i l_i'$ for $l_i l'$.

9969. (J. Brill, M.A.)—A family of equipotential curves is drawn on a plane, and the locus of the points of contact of a system of parallel tangents is drawn. Prove that the loci corresponding to different directions of these parallel tangents form an equipotential family.

Solution by G. H. Hardy, B.A.

Suppose that the equipotential curves are $u = $ const. and their orthogonal trajectories $v = $ const., and that $f(z) \equiv f(x+iy) = u+iv$. The locus of the points of contact with the equipotentials of tangents making an angle θ with the axis of x is

$$\cos\theta \, \partial u/\partial x + \sin\theta \, \partial u/\partial y = 0.$$

But $\quad e^{i\theta} f'(z) = e^{i\theta} \partial f/\partial x = (\cos\theta + i\sin\theta)(\partial u/\partial x + i\, \partial v/\partial x),$

and the real part of this

$$\cos\theta \, \partial u/\partial x - \sin\theta \, \partial v/\partial x = \cos\theta \, \partial u/\partial x + \sin\theta \, \partial u/\partial y.$$

The loci are therefore equipotentials.

COMMENT

When $\dfrac{\partial u}{\partial x}$ and $\dfrac{\partial u}{\partial y}$ are not both zero the family of loci can be written in the form

$$\operatorname{Im}\{-\log f'(z)\} = \phi \quad \left(\phi = \theta \pm \frac{\pi}{2}\right),$$

which, perhaps, makes it more evident that it is an equipotential family.

The proposer, John Brill (born 1858), was fourth Wrangler in 1882 and became a Board of Education Inspector.

14989. (G. H. Hardy, B.A.)—Prove that, if $\phi(x)$ is continuous and tends to a definite limit for $x = \infty$,

$$\lim_{t=\infty} \int_0^\infty e^{-x} \phi(tx)\, dx = \phi(\infty).$$

If $\phi(\infty)$ is not determinate, but the limit on the left is, we *define* $\phi(\infty)$ or the *generalized limit* $\operatorname{G}\lim_{x=\infty} \phi(x)$ as equal to it. Prove that

$$\operatorname{G}\lim_{x=\infty} e^{aix} f(\sin^2 x) = 0,$$

unless a is an even integer, in which case

$$\operatorname{G}\lim_{x=\infty} e^{aix} f(\sin^2 x) = \frac{1}{\pi} \int_0^\pi e^{aix} f(\sin^2 x)\, dx.$$

Solution by the Proposer.

(1) We can choose a constant K such that $|\phi(x)| < K$ $(0 \leq x \leq \infty)$, and, if σ be an arbitrarily small positive quantity, we can determine H so that $|\phi(x) - \phi(\infty)| < \sigma$, $(x \geq H)$. Now

$$\int_0^\infty e^{-x} \phi(tx)\, dx = \int_0^{1/\sqrt{t}} e^{-x} \phi(tx)\, dx + \int_{1/\sqrt{t}}^\infty e^{-x}\{\phi(tx) - \phi(\infty)\}\, dx + \phi(\infty) e^{-1/\sqrt{t}}.$$

The first term is in absolute value $< K/\sqrt{t}$; and, if $\sqrt{t} > H$, the second is in absolute value $< \sigma e^{-1/\sqrt{t}}$. The theorem follows.

(2) By definition,

$$\operatorname{G}\lim_{x=\infty} e^{aix} f(\sin^2 x) = \lim_{t=\infty} \int_0^\infty e^{(tai-1)x} f(\sin^2 tx)\, dx$$

$$= \lim \frac{1}{t} \int_0^\infty e^{(ai-1/t)u} f(\sin^2 u)\, du$$

$$= \lim \tau \sum_{n=0}^\infty \int_{n\pi}^{(n+1)\pi} e^{(ai-\tau)u} f(\sin^2 u)\, du$$

$$= \lim \int_0^\pi \frac{\tau e^{(ai-\tau)\theta}}{1 - e^{(ai-\tau)\pi}} f(\sin^2 \theta)\, d\theta = 0,$$

unless $e^{ai\pi} = 1$, in which case it is plainly $\dfrac{1}{\pi} \int_0^\pi e^{ai\theta} f(\sin^2 \theta)\, d\theta.$

CORRECTIONS

Line 3 of (1). Read $\phi(\infty)$ for $\phi(\ )$.
Line 4 of (1). Read $e^{-1/\sqrt{t}}$ at end of line.

15019. (G. H. Hardy, B.A.)—If
$$F(m, n) = m - \frac{n}{2.3!} m(m^2 - 1^2) + \frac{n^2}{3.5!} m(m^2 - 1^2)(m^2 - 2^2) - \ldots,$$

prove that (1) $F(m, 1) = 0, 1/m, 3/m, 4/m$,
according as $m \equiv 0, \pm 1, \pm 2, 3 \pmod 6$;

(2) $F(m, 2) = 0, 1/m, 2/m$,
according as $m \equiv 0, \pm 1, 2 \pmod 4$;

(3) $F(m, 3) = 0, 1/m$,
according as $m \equiv 0, \pm 1 \pmod 3$;

and (4) $F(m, 4) = 0, 1/m$,
according as $m \equiv 0, 1 \pmod 2$.

Solution by the Proposer.

The coefficient of x^m in $-\log\{1 - [nx/(1+x)^2]\}$ or
$$[nx/(1+x)^2] + [n^2 x^2/2 (1+x)^4] + \ldots$$
is easily seen to be $(-)^{m-1} nF(m, n)$. Hence (1) $(-)^{m-1} F(m, 1)$ is the coefficient of x^m in
$$\log[(1+x^2)/1+x+x^2)] = 2\log(1+x) + \log(1-x) - \log(1-x^3)$$
$$= 2(x - \tfrac{1}{2}x^2 + \tfrac{1}{3}x^3 - \tfrac{1}{4}x^4 + \tfrac{1}{5}x^5 - \tfrac{1}{6}x^6 \ldots)$$
$$- x - \tfrac{1}{2}x^2 - \tfrac{1}{3}x^3 - \tfrac{1}{4}x^4 - \tfrac{1}{5}x^5 - \tfrac{1}{6}x^6 \ldots$$
$$+ x^3 + \tfrac{1}{2}x^6 \ldots .$$

Thus $F(m, 1) = 0, 1/m, 3/m, 4/m$,
according as $m \equiv 0, \pm 1, \pm 2, 3 \pmod 6$.

Again, (2) $(-)^{m-1} 2F(m, 2)$ is the coefficient of x^m in
$$\log[(1+x)^2/(1+x^2)] = 2(x - \tfrac{1}{2}x^2 + \tfrac{1}{3}x^3 - \tfrac{1}{4}x^4 \ldots)$$
$$- x^2 + \tfrac{1}{2}x^4 \ldots .$$

Thus $F(m, 2) = 0, 1/m, 2/m$,
according as $m \equiv 0, \pm 1, 2 \pmod 4$.

Again, (3) $(-)^{m-1} 3F(m, 3)$ is the coefficient of x^m in
$$\log[(1+x)^3/(1-x+x^2)] = 3\log(1+x) - \log(1+x^3) = 3(x - \tfrac{1}{2}x^2 + \tfrac{1}{3}x^3 \ldots)$$
$$- x^3 \ldots ;$$
so that $F(m, 3) = 0, 1/m$, according as $m \equiv 0, \pm 1 \pmod 3$.

Finally, (4) $(-)^{m-1} 4F(m, 4)$ is the coefficient of x^m in
$$\log[(1+x)/(1-x)]^2 = 2(x - \tfrac{1}{2}x^2 + \ldots + x + \tfrac{1}{2}x^2 + \ldots);$$
so that $F(m, 4) = 0, 1/m$, according as $m \equiv 0, 1 \pmod 2$.

1903, 12 *Educational Times*, (2), 4, 75.

15265. (G. H. Hardy, B.A.)—(See Question 14988.)—If
$$l^2 + m^2 + n^2 + \ldots, \quad l'^2 + m'^2 + n'^2 + \ldots, \quad l''^2 + m''^2 + n''^2 + \ldots$$
are each $= 0$, and s, s', s'' are positive integers which satisfy the conditions $s' + s'' - s \geq 0$, $s'' + s - s' \geq 0$, $s + s' - s'' \geq 0$, $s + s' + s'' \equiv 0 \pmod 2$, the integral
$$\int (lx + my + nz + \ldots)^s \, (l'x + m'y + n'z + \ldots)^{s'} \, (l''x + m''y + n''z + \ldots)^{s''} \, dx\,dy\,dz \ldots,$$
taken over the interior of the hyper-sphere $x^2 + y^2 + z^2 + \ldots = 1$, is equal to
$$\frac{\pi^{\frac{1}{2}\nu}}{2^\sigma} \frac{s!\, s'!\, s''!}{(\sigma-s)!\,(\sigma-s')!\,(\sigma-s'')!} \frac{(\Sigma l'l'')^{\sigma-s}(\Sigma l''l)^{\sigma-s'}(\Sigma ll')^{\sigma-s''}}{\Gamma(\sigma+1+\tfrac{1}{2}\nu)}.$$
where $\sigma = \tfrac{1}{2}(s + s' + s'')$ and ν is the number of dimensions of the sphere. If s, s', s'' do not satisfy the conditions stated, the value of the integral is 0. [If $s'' = 0$, we obtain the result of Question 14988.]

Solution by the Proposer.

Schlömilch has shown that, if ϕ is continuous,
$$\int \phi(\alpha x + \beta y + \gamma z + \ldots)\, dx\,dy\,dz \ldots \quad (x^2 + y^2 + z^2 + \ldots < 1)$$
$$= \frac{2\pi^{\frac{1}{2}(\nu-1)}}{\Gamma[\tfrac{1}{2}(\nu-1)]} \int_{-1}^{1} \phi[x\sqrt{(\alpha^2+\beta^2+\gamma^2+\ldots)}]\, dx \int_0^{\sqrt{(1-x^2)}} y^{\nu-2}\, dy.$$

Suppose $\alpha = lt + l't' + l''t''$, $\beta = mt + m't' + m''t''$, $\ldots$,

so that $\alpha^2 + \beta^2 + \gamma^2 = (2\Sigma l'l'')\, t't'' + (2\Sigma l''l)\, t''t + (2\Sigma ll')\, tt'$,

and $\phi(u) = e^{-u}$. Expanding and equating the coefficients of $t^s t'^{s'} t''^{s''}$, we find for the value of the integral in the Question
$$\frac{s!\, s'!\, s''!}{s + s' + s''!} \frac{2\pi^{\frac{1}{2}(\nu-1)}}{\Gamma[\tfrac{1}{2}(\nu-1)]} \int_{-1}^{1} \frac{(1-x^2)^{\frac{1}{2}(\nu-1)}}{\nu-1} x^{s+s'+s''}\, dx,$$
multiplied by the coefficient of $t^s t'^{s'} t''^{s''}$ in
$$[(2\Sigma l'l'')\, t't'' + (2\Sigma l''l)\, t''t + (2\Sigma ll')\, tt']^\sigma.$$

This is
$$\frac{\sigma!\, 2^\sigma (\Sigma l'l'')^{\sigma-s} (\Sigma l''l)^{\sigma-s'} (\Sigma ll')^{\sigma-s''}}{(\sigma-s)!\,(\sigma-s')!\,(\sigma-s'')!}.$$

Also
$$\int_{-1}^{1} (1-x^2)^{\frac{1}{2}(\nu-1)} x^{2\sigma}\, dx = \int_0^1 (1-u)^{\frac{1}{2}(\nu-1)} u^{\sigma-\frac{1}{2}}\, du$$
$$= \frac{\Gamma[\tfrac{1}{2}(\nu+1)]\, \Gamma(\sigma+\tfrac{1}{2})}{\Gamma(\sigma+1+\tfrac{1}{2}\nu)}.$$

But $\sigma!\, \Gamma(\sigma+\tfrac{1}{2}) = \Gamma(\sigma+1)\, \Gamma(\sigma+\tfrac{1}{2}) = 2^{-2\sigma} 2\sigma!\, \sqrt{\pi}$;

so that we obtain, finally,
$$\frac{\pi^{\frac{1}{2}\nu}}{2^\sigma} \frac{s!\, s'!\, s''!}{(\sigma-s)!\,(\sigma-s')!\,(\sigma-s'')!} \frac{(\Sigma l'l'')^{\sigma-s}(\Sigma l''l)^{\sigma-s'}(\Sigma ll')^{\sigma-s''}}{\Gamma(\sigma+1+\tfrac{1}{2}\nu)}.$$

CORRECTION

In *line* 5 of the solution $\alpha^2+\beta^2+\gamma^2$ should be replaced by $\alpha^2+\beta^2+\gamma^2+\cdots$.

COMMENT

For Question 14988 see 1903, 10.

9515. (R. Holmes, B.A.)—If P_n is Legendre's coefficient of order n, and J_0 is the Bessel's function of order zero, show that

$$P_0(\mu) - rP_1(\mu) + \frac{r^2}{2!}P_2(\mu) - \frac{r^3}{3!}P_3(\mu) + \ldots + (-1)^n \frac{r^n}{n!}P_n(\mu) + \ldots$$
$$= e^{-\mu r} J_0\{r(1-\mu^2)^{\frac{1}{2}}\}.$$

Solution by G. H. Hardy, M.A., *and* R. F. Whitehead.

If $\nu = (1-\mu^2)^{\frac{1}{2}}$, $\quad P_n(\mu) = \int_0^\pi (\mu + i\nu \cos\phi)^n \, d\phi$,

$$\sum_0^\infty \frac{(-)^n r^n}{n!} P_n(\mu) = \int_0^\pi d\phi \left[\sum_0^\infty \frac{(-1)^n r^n (\mu + i\nu \cos\phi)^n}{n!}\right] = \int_0^\pi e^{-r(\mu + i\nu \cos\phi)} d\phi$$
$$= e^{-r\mu} \int_0^\pi \cos(r\nu \cos\phi) \, d\phi = e^{-r\mu} J_0(r\nu).$$

COMMENTS

A generalization of this formula can be found in L. Gegenbauer (1884), 'Zur Theorie der Functionen $C_n^\nu(x)$', *Wiener Akad. Denkschriften* 48, 293–316; see formula (2) on p. 294.

Hardy took his M.A. in 1903.

1904, 12a *Educational Times*, (2), 5, 46.

15300. (G. H. Hardy, M.A.)—If $0 < \alpha < \tfrac{1}{2}\pi$, $0 < \beta < \tfrac{1}{2}\pi$, and $0 < s < 1$,
$$\int_0^{\frac{1}{2}\pi} \cos^{-s}(\theta-\alpha)\cos^{s-2}(\theta-\beta)\,d\theta = \operatorname{cosec}(\beta-\alpha)\left\{\left(\frac{\cos\alpha}{\cos\beta}\right)^{1-s} - \left(\frac{\sin\alpha}{\sin\beta}\right)^{1-s}\right\}.$$

Solution by the Proposer.

Suppose $a_1, a_2, c_1, c_2 > 0$, and $0 < s < 1$. Then
$$\frac{\Gamma(s)}{(a_1 x + a_1 y)^s} = \int_0^\infty e^{-(ax_1 + a_2 y)t}\, t^{s-1}\, dt,$$
$$\Gamma(s)\int_0^\infty\int_0^\infty \frac{e^{-(c_1 x + c_2 y)}}{(a_1 x + a_2 y)^s}\, dx\, dy = \int_0^\infty t^{s-1}\, dt \left(\int_0^\infty\int_0^\infty e^{-(a_1 t + c_1)x - (a_2 t + c_2)y}\, dx\, dy\right)$$
$$= \int_0^\infty \frac{t^{s-1}\, dt}{(a_1 t + c_1)(a_2 t + c_2)}$$
$$= \frac{1}{a_1 c_2 - a_2 c_1}\left\{a_1 \int_0^\infty \frac{t^{s-1}}{a_1 t + c_1}\, dt - a_2 \int_0^\infty \frac{t^{s-1}}{a_2 t + c_2}\, dt\right\}$$
$$= \frac{\pi}{(a_1 c_2 - a_2 c_1)\sin s\pi}\left\{\left(\frac{a_1}{c_1}\right)^s - \left(\frac{a_2}{c_2}\right)^s\right\}.$$

But the left-hand side is
$$\Gamma(s)\int_0^{\frac{1}{2}\pi} \frac{d\theta}{(a_1\cos\theta + a_2\sin\theta)^s}\int_0^\infty e^{-(c_1\cos\theta + c_2\sin\theta)r}\, r^{1-s}\, dr$$
$$= \frac{\pi}{\sin s\pi}\int_0^{\frac{1}{2}\pi} \frac{d\theta}{(a_1\cos\theta + a_2\sin\theta)^s (c_1\cos\theta + c_2\sin\theta)^{2-s}};$$
so that
$$\int_0^{\frac{1}{2}\pi}(a_1\cos\theta + a_2\sin\theta)^{-s}(c_1\cos\theta + c_2\sin\theta)^{2-s}\, d\theta$$
$$= \frac{1}{a_1 c_2 - a_2 c_1}\left\{\left(\frac{a_1}{c_1}\right)^{1-s} - \left(\frac{a_2}{c_2}\right)^{1-s}\right\}.$$

The formula in the Question follows on putting
$$a_1 = \cos\alpha, \quad a_2 = \sin\alpha, \quad c_1 = \cos\beta, \quad c_2 = \sin\beta.$$

15282. (Hon. G. R. Dick, M.A.)—Prove that, if a is any real quantity, positive or negative,

$$\int_0^\infty \frac{\sin ax}{1+x^2} dx = \tfrac{1}{2} e^{-a} \operatorname{li} e^a - \tfrac{1}{2} e^a \operatorname{li} e^{-a},$$

where $\operatorname{li} e^a$ is the logarithmic integral and may be defined by

$$\operatorname{li} e^a = \operatorname{Ei}(a) = \int_0^1 \frac{e^{az}}{z} dz - \int_0^\infty \frac{e^{-z\sqrt{a^2}}}{z} dz.$$

for all real values of a, positive and negative.

[N.B.—The error in the above definition of the li function is noticed in each of the solutions given below.—Editor.]

Solutions (I.) *by* θ. β.; (II.) *by* G. H. Hardy, M.A.; (III.) *by the* Proposer.

(II.) The expression given in the Question for $\operatorname{li} e^a$ is not correct, as

$$\int_0^1 \frac{e^{az}}{z} dz, \quad \int_0^\infty \frac{e^{-z\sqrt{a^2}}}{z} dz$$

are both divergent. It should be

$$\int_0^1 \frac{e^{az} - e^{-z\sqrt{a^2}}}{z} dz - \int_1^\infty \frac{e^{-z\sqrt{a^2}}}{z} dz.$$

For this is
$$\int_0^a \frac{e^u - e^{-u \operatorname{sgn} a}}{u} du + \int_{-\infty}^{-\sqrt{a^2}} \frac{e^u}{u} du = \int_{-\infty}^a \frac{e^u}{u} du \quad (a < 0),$$

or
$$= \lim_{\epsilon = 0} \int_\epsilon^a \frac{e^u - e^{-u}}{u} du + \int_{-\infty}^{-a} \frac{e^u}{u} du$$

$$= \lim_{\epsilon = 0} \left(\int_{-a}^{-\epsilon} + \int_\epsilon^a \right) \frac{e^u}{u} du + \int_{-\infty}^{-a} \frac{e^u}{u} du = \operatorname{P} \int_{-\infty}^a \frac{e^u}{u} du \quad (a > 0).$$

These are the definitions usually adopted. The symbol P implies that the principal value of the integral is to be taken.

To solve the Question, integrate $\int \frac{e^{axi}}{1+x^2} dx$ $(a > 0)$ round a contour formed by the positive parts of the real and imaginary axes, and a very large quadrant whose centre is O. The point $x = i$ must be avoided by a small semicircle round it. We easily find that

$$\int_0^\infty \frac{\sin ax}{1+x^2} dx = \operatorname{P} \int_0^\infty \frac{e^{-ay}}{1-y^2} dy = \tfrac{1}{2} \left(\operatorname{P} \int_0^\infty \frac{e^{-ay}}{1-y} dy + \operatorname{P} \int_0^\infty \frac{e^{-ay}}{1+y} dy \right)$$

$$= \tfrac{1}{2} \left(e^{-a} \operatorname{P} \int_{-\infty}^a \frac{e^u}{u} du - e^a \int_\infty^{-a} \frac{e^u}{u} du \right)$$

$$= \tfrac{1}{2} (e^{-a} \operatorname{li} e^a - e^a \operatorname{li} e^{-a}).$$

COMMENTS

Only Hardy's solution is reproduced. The first solution employs a similar method; the solver θ.β. points out that the result is to be found in L. Kronecker's *Vorlesungen über Mathematik*,

1904, 13a *Educational Times*, (2), 5, 89–92.

Band I (Leipzig, 1894), p. 212, formula (17) and that Kronecker refers to a 'more elementary solution' in O. X. Schlömilch's *Compendium der höheren Analysis*, Band II (Braunschweig, 1868–9), pp. 201–3. This is, in fact, a mistranslation, Kronecker's words being 'auf nicht ganz einfache Weise abgeleitet'. This is corroborated by Hardy's statement in 1904, 14 that Schlömilch's solution is far from elementary.

See 1904, 14 for further remarks on the two other solutions.

George Roger Dick (1845–1913), a Mauritian, was ninth Wrangler in 1866, became Professor of Mathematics at the Royal College Mauritius in 1880 and later Registrar General there. The identity of $\theta.\beta.$ is not known. It is possible that he may have been T. J. I'A. Bromwich.

Additional Note on Question 15282.

[*V. Educational Times* for November, 1903, and *Reprint*, Vol. v. (New Series), p. 89.]

By G. H. Hardy, M.A.

I agree with $\theta.\ \beta.$ that the value of the integral was in all probability known to Cauchy. The result of the question and the corresponding formula for $\int_0^\infty \frac{x \cos ax}{1+x^2}\, dx$ correspond exactly to the two formulæ

$$\int_0^\infty \frac{\cos ax}{1+x}\, dx = \int_0^\infty \frac{xe^{-ax}}{1+x^2}\, dx, \quad \int_0^\infty \frac{\sin ax}{1+x}\, dx = \int_0^\infty \frac{e^{-ax}}{1+x^2}\, dx, \quad (a>0)$$

which express the integrals on the right in terms of the functions $\int \frac{\sin x}{x}\, dx$, $\int \frac{\cos x}{x}\, dx$, and these formulæ are given in Cauchy's "Mémoire sur les Intégrales définies" (*Œuvres*, t. I., p. 377).

The question is such an interesting one that I may perhaps be allowed to add a few remarks on some other ways in which it can be solved. In the first place, Schlömilch's solution is far from "elementary." He proves first that

$$\int_0^\infty \left\{ e^{-t} \operatorname{li}(e^t) - e^t \operatorname{li}(e^{-t}) \right\} \sin ut\, dt = \pi/(1+u^2),$$

and deduces the result by means of Fourier's double-integral theorem (*Compendium*, Vol. II., pp. 201-3). I gave another solution recently in a paper "The Theory of Cauchy's Principal Values" (*Proc. Lond. Math. Soc.*, Vol. xxxv., p. 96). Since $\sin ay = -\frac{2a}{\pi} \operatorname{P} \int_0^\infty \frac{\cos xy}{x^2 - a^2}\, dx$,

$$\int_0^\infty \frac{\sin ay}{1+y^2}\, dy = -\frac{2a}{\pi} \int_0^\infty \frac{dy}{1+y^2} \operatorname{P} \int_0^\infty \frac{\cos xy}{x^2 - a^2}\, dx = -\frac{2a}{\pi} \operatorname{P} \int_0^\infty \frac{dx}{x^2 - a^2} \int_0^\infty \frac{\cos xy}{1+y^2}\, dy$$

$$= -a \operatorname{P} \int_0^\infty \frac{e^{-x}\, dx}{x^2 - a^2} = \tfrac{1}{2} \left\{ e^{-a} \operatorname{li}(e^a) - e^a \operatorname{li}(e^{-a}) \right\}.$$

This argument of course needs exceedingly careful theoretical justification.

Other solutions may be deduced from the theory of divergent series and integrals. Thus, if the integral of the question be denoted by u,

$$\frac{du}{da} = \int_0^\infty \frac{y \cos ay}{1+y^2}\, dy, \quad \frac{d^2u}{da^2} = -\operatorname{G} \int_0^\infty \frac{y^2 \sin ay}{1+y^2}\, dy,$$

$$\frac{d^2u}{da^2} - u = -\operatorname{G} \int_0^\infty \sin ay\, dy = -\frac{1}{a},$$

which is the differential equation found otherwise by Mr. Dick. The meaning of the symbol G is explained in a paper in the last issue of the *Quarterly Journal*.

Or again expand $1/(1+y^2)$ in the series $1-y^2+y^4-\ldots$, divergent if $y>1$. Take the divergent integral term by term. Since

$$\mathrm{G}\int_0^\infty \sin ay\, dy = \frac{1}{a}, \qquad \mathrm{G}\int_0^\infty y^{2n} \sin ay\, dy = \frac{2n!}{a^{2n+1}},$$

we obtain $u = 1/a - 2!/a^3 + 4!/a^5 - \ldots$, a divergent series which gives the result of the question on summation by M. BOREL's method.

The last two methods are of course really symbolic, the argument being really a shorthand representation of another which is much more involved.

Finally, I would add that Mr. DICK's definition of the function $\mathrm{li}\, e^a$, though slightly inaccurate in form, as at first stated, is really very interesting, as (if modified as suggested in my solution) it gives a formula applicable to all positive or negative values of a and involving only *unconditionally* convergent integrals.

COMMENTS

The paper on Cauchy's principal values referred to is 1903, 1 (Vol. V, pp. 259–86). The symbol G is defined in 1904, 4, p. 49 (Vol. VI, p. 64). This generalized limit is also used in Question 14989 (1903, 11).

15361. (G. H. Hardy, M.A.)—The area of the diagonal triangle of the quadrangle whose vertices are (x_1, y_1, z_1), (x_2, y_2, z_2), (x_3, y_3, z_3), (x_4, y_4, z_4) (in areal co-ordinates) is numerically $\dfrac{54\Delta\, D_{234}\, D_{314}\, D_{124}\, D_{123}}{D_{23,14}\, D_{31,24}\, D_{12,34}}$,

where $D_{\lambda\mu\nu} = \begin{vmatrix} x_\lambda & y_\lambda & z_\lambda \\ x_\mu & y_\mu & z_\mu \\ x_\nu & y_\nu & z_\nu \end{vmatrix}$, $D_{\alpha\beta,\gamma\delta} = \begin{vmatrix} 1 & x_\alpha - x_\beta & x_\gamma - x_\delta \\ 1 & y_\alpha - y_\beta & y_\gamma - y_\delta \\ 1 & z_\alpha - z_\beta & z_\gamma - z_\delta \end{vmatrix}$,

and Δ is the area of the triangle of reference.

Solution by the Proposer.

If we denote (x_1, y_1, z_1), (x_2, y_2, z_2), ..., by (1), (2), ..., the co-ordinates of the lines (12), (34) are

$y_1 z_2 - y_2 z_1 : z_1 x_2 - z_2 x_1 : x_1 y_2 - x_2 y_1$ and $y_3 z_4 - y_4 z_3 : z_3 x_4 - z_4 x_3 : x_3 y_4 - x_4 y_3$;

and their point of intersection is

$(z_1 x_2 - z_2 x_1)(x_3 y_4 - y_4 x_3) - (x_1 y_2 - x_2 y_1)(z_3 x_4 - z_4 x_3) : \dots : \dots.$

Hence the area of the diagonal triangle is $\dfrac{K f_{12}}{D'_{23,14}\, D'_{31,24}\, D'_{12,34}}$, where K is a constant, f_{12} a homogeneous function of degree 12, and

$D'_{\alpha\beta,\gamma\delta} = \begin{vmatrix} 1 & y_\alpha z_\beta - y_\beta z_\alpha & y_\gamma z_\delta - y_\delta z_\gamma \\ 1 & z_\alpha x_\beta - z_\beta x_\alpha & z_\gamma x_\delta - z_\delta x_\gamma \\ 1 & x_\alpha y_\beta - x_\beta y_\alpha & x_\gamma y_\delta - x_\delta y_\gamma \end{vmatrix}.$

It is easy to prove (using the equations $x_1 + y_1 + z_1 = 1, \dots$) that $D'_{\alpha\beta,\gamma\delta} = \tfrac{1}{3} D_{\alpha\beta,\gamma\delta}$. Moreover, it is evident that the diagonal triangle degenerates into a straight line when three of the vertices of the quadrangle are collinear: and therefore f_{12} is a constant multiple of $D_{234}\, D_{314}\, D_{124}\, D_{123}$. We may suppose the constant multiplier absorbed in K. To determine K we may suppose either (i.) that the quadrangle is formed by the vertices of the triangle of reference and the orthocentre, in which case the diagonal triangle is the pedal triangle; or (ii.) that the quadrangle is formed by the four in-centres, in which case the diagonal triangle is the triangle of reference. In either case we find $K = -2\Delta$, and the result follows.

COMMENT

This question appeared in the *Educational Times and Journal of the College of Preceptors* on 1 June, 1903. The following variant appeared as Question 15488 on 1 January, 1904: Find the area of the diagonal triangle of a quadrilateral, given the equations of its sides in areal coordinates. No separate solution to this question was published.

1904, 15 *Educational Times*, (2), 5, 118.

15125. (G. H. Hardy, M.A.)— Prove directly that
$$\lim_{k=1} \int_1^{1/k} \frac{du}{\sqrt{[(u^2-1)(1-k^2u^2)]}} = \tfrac{1}{2}\pi.$$

Solution by the Proposer.

Put $u^2 = \cos^2\phi + (1/k^2)\sin^2\phi$, $u\,du = (1/k^2 - 1)\cos\phi\sin\phi\,d\phi$, when the integral beomes $\dfrac{1}{k}\int_0^{\frac{1}{2}\pi} \dfrac{d\phi}{\sqrt{[\cos^2\phi + (1/k^2)\sin^2\phi]}}$, the limit of which for $k = 1$ is $\tfrac{1}{2}\pi$. More generally, if $f(u)$ is continuous,
$$\lim_{k=1} \int_1^{1/k} \frac{f(u)\,du}{\sqrt{\{(u^2-1)(1-k^2u^2)\}}} = \tfrac{1}{2}\pi f(1).$$

[Indirectly the result is obvious, for the integral is known to be equal to K'.]

COMMENT

It is assumed that $k < 1$.

1904, 16 *Educational Times*, (2), 6, 31.

15357. (A. M. NESBITT, M.A.) — Evaluate $\int_0^\infty \frac{x\, dx}{e^{cx} \cosh x}$, c being between $+1$ and -1.

Solution by G. H. HARDY, M.A.

Since $e^{-cx} \operatorname{sech} x = 2 \sum_0^\infty (-)^n e^{-(c+2n+1)x}$ and $\int_0^\infty e^{-px} x\, dx = \frac{1}{p^2}\ (p > 0)$,

$$\int_0^\infty \frac{x\, dx}{e^{cx} \cosh x} = 2 \sum_0^\infty \frac{(-)^n}{(c+2n+1)^2} = 2 \frac{d^2}{dc_2} \log\left[\Gamma\tfrac{1}{4}(c+1)/\Gamma\tfrac{1}{4}(c+3)\right].$$

By integration with respect to c we can deduce the values of

$$\int_0^\infty \frac{e^{-cx}}{\cosh x}\, dx,\quad \int_0^\infty \frac{e^{-ax} - e^{-bx}}{\cosh x} \frac{dx}{x}.$$

It is not necessary that $c < 1$; the condition $c > -1$ is sufficient.

CORRECTION

In second line of solution read $\frac{d^2}{dc^2}$.

COMMENT

Alfred Mortimer Nesbitt obtained First Class Honours in Mathematical Moderations in 1874 and in the Honours School of Mathematics in 1876, as an undergraduate at Corpus Christi College, Oxford.

1904, 17 *Educational Times*, (2), 6, 45.

15686. (G. H. HARDY, M.A.)—The area Δ and semi-perimeter s of a triangle are fixed. Show that the maximum and minimum values of one of the sides are roots of the equation $sx^2(x-s) + 4\Delta^2 = 0$. Discuss the existence of real maximum and minimum values.

Solution by the PROPOSER.

If a is stationary, and $a+b+c = 2s = $ const. .... (i.),

$$s(s-a)(s-b)(s-c) = \Delta^2 = \text{const.} \quad \text{................. (ii.),}$$

then $\quad da = 0, \quad db + dc = 0 \quad \text{and} \quad (s-c)db + (s-b)dc = 0,$

or $\quad b = c, \quad s - b = \tfrac{1}{2}(a+c-b) = \tfrac{1}{2}a,$

and so (ii.) takes the form $sa^2(a-s) + 4\Delta^2 = 0$. It is worth while to note the fact that a and R attain stationary values simultaneously; hence we cannot satisfy $da/dR = 0$. For R is stationary if abc is so, and this gives

$$bc\, da + ca\, db + ab\, dc = 0, \quad da + db + dc = 0,$$

$$(s-b)(s-c)da + (s-c)(s-a)db + (s-a)(s-b)dc = 0;$$

so that
$$\begin{vmatrix} (s-b)(s-c) & bc & 1 \\ (s-c)(s-a) & ca & 1 \\ (s-a)(s-b) & ab & 1 \end{vmatrix} = 0,$$

or $(b-c)(c-a)(a-b)s = 0$. If $b = c$, we have the case already discussed. If $a = b$ (say), then $s - c = 2a - s$, and so (ii.) gives

$$s(s-2a)(s-a)^2 + \Delta^2 = 0.$$

Let us consider the equation $sx^3 - s^2x^2 + 4\Delta^2 = 0$. If $x = 1/y$,

$$y^3 - s^2y/(4\Delta^2) + s/(4\Delta^2) = 0.$$

The condition that all the roots should be real is that

$$s^2/(16\Delta^4) - s^6/(432\Delta^6) < 0, \quad s^4 > 27\Delta^2.$$

If this condition is not satisfied, the values of s and Δ are impossible for a real triangle. If it is satisfied, two roots of $sx^2(x-s) + 4\Delta^2 = 0$ are positive, and there is one real maximum and one real minimum, as is geometrically obvious.

COMMENT

The equation $4R\Delta = abc$ is used.

15498. (G. H. Hardy, M.A.)—Prove that the reciprocal of the product of two determinants is the product of their reciprocals.

Solution by C. M. Ross.

Take the determinant

$$\Delta = \begin{vmatrix} x_{11} & x_{12} & \ldots & x_{1n} \\ x_{21} & x_{22} & \ldots & x_{2n} \\ \ldots & \ldots & & \ldots \\ x_{n1} & x_{n2} & \ldots & x_{nn} \end{vmatrix} \quad \ldots\ldots(1).$$

Its reciprocal is

$$\Delta' = \begin{vmatrix} X_{11} & X_{12} & \ldots & X_{1n} \\ X_{21} & X_{22} & \ldots & X_{2n} \\ \ldots & \ldots & & \ldots \\ X_{n1} & X_{n2} & \ldots & X_{nn} \end{vmatrix} \quad \ldots\ldots(2),$$

X_{rs} in (2) being the co-factor of x_{rs} in (1).

Multiplying together (1) and (2), the result is

$$\Delta\Delta' = \begin{vmatrix} x_{11}X_{11}+\ldots+x_{1n}X_{1n} & x_{11}X_{21}+\ldots+x_{1n}X_{2n} & \ldots & x_{11}X_{n1}+\ldots+x_{1n}X_{nn} \\ x_{21}X_{11}+\ldots+x_{2n}X_{1n} & x_{21}X_{21}+\ldots+x_{2n}X_{2n} & \ldots & x_{21}X_{n1}+\ldots+x_{2n}X_{nn} \\ \ldots & \ldots & & \ldots \\ x_{n1}X_{11}+\ldots+x_{nn}X_{1n} & x_{n1}X_{21}+\ldots+x_{nn}X_{2n} & \ldots & x_{n1}X_{n1}+\ldots+x_{nn}X_{nn} \end{vmatrix}$$

$$= \begin{vmatrix} \Delta & 0 & \ldots & 0 \\ 0 & \Delta & \ldots & 0 \\ \ldots & \ldots & & \ldots \\ 0 & 0 & \ldots & \Delta \end{vmatrix} = \Delta^n;$$

therefore $\Delta' = \Delta^{n-1}$; *i.e.*, the reciprocal of any determinant of the n-th order is equal to the $(n-1)$-th power of the latter. Hence, if Δ denote the product of two n-th order determinants Δ_1 and Δ_2, and D the reciprocal of their product, then

$$D = \Delta^{n-1} = \{\Delta_1\Delta_2\}^{n-1} = \Delta_1^{n-1}\Delta_2^{n-1}$$

= product of the reciprocals of Δ_1 and Δ_2.

16041. (Professor Sanjána, M.A., and M. V. Arunachalan, M.A.)—If $R_n = 1 + \frac{1}{2} + \frac{1}{3} + \ldots + 1/n$, prove that $R_1 - R_2 + R_3 - R_4 + \ldots$ ad inf. $= \frac{1}{2} \log 2$.

Note by G. H. Hardy, M.A.

The series $R_1 - R_2 + R_3 - \ldots$ where
$$R_n = 1 + \tfrac{1}{2} + \tfrac{1}{3} + \ldots + 1/n$$
is obviously not convergent. If, however, S_n denotes the sum of the first n terms, $s_{2n} = -\tfrac{1}{2} - \tfrac{1}{4} - \ldots - 1/2n$, $s_{2n-1} = 1 + \tfrac{1}{3} + \ldots + 1/(2n+1)$,
and
$$(s_1 + s_2 + \ldots + s_{2n})/2n = (\sigma_2 + \sigma_4 + \ldots + \sigma_{2n})/2n,$$
where
$$\sigma_{2n} = 1 - \tfrac{1}{2} + \tfrac{1}{3} - \ldots + 1/(2n-1) - 1/2n.$$
Since $\lim \sigma_{2n} = \log 2$, it follows, by a well known theorem, that
$$\lim (s_1 + s_2 + \ldots + s_{2n})/2n = \tfrac{1}{2} \log 2;$$
and, since $\lim s_n/n = 0$, we can say simply that
$$\lim (s_1 + s_2 + \ldots + s_n)/n = \tfrac{1}{2} \log 2.$$
Thus the series, though not convergent, is summable, in the sense defined by Frobenius and Césaro, and its "sum" is $\tfrac{1}{2} \log 2$.

CORRECTION

In third line of solution read s_n for S_n.

COMMENT

Professor Sanjána was Professor of Mathematics at the University of Bombay.

16257. (G. H. HARDY, M.A.)—Show that the series
$$\log 1 - \log 2 + \log 3 - \ldots$$
is summable, *i.e.*, that, if s_n denotes the sum of the first n terms, then $(s_1 + s_2 + \ldots + s_n)/n$ tends to a limit as n tends to infinity: and find its sum. Show also that, if $\phi(n)$ tends steadily to 0 as n tends to infinity, the series $R_1 - R_2 + R_3 - \ldots$ where $R_n = \phi(1) + \phi(2) + \ldots + \phi(n)$ is summable, and its sum is equal to that of the convergent series
$$\phi(1) - \phi(2) + \phi(3) - \ldots.$$

Solution, with remarks, by V. RAMASWAMI AIYAR, M.A.

There is a slight inaccuracy in the statement which will be rectified. Apart from this, Mr. Hardy's conditions as to ϕ are narrower than they need be. Let $\phi_1 - \phi_2 + \phi_3 - \ldots$ be *any* convergent series, its sum being s. Let s_n denote the sum of its first n terms; so that
$$\lim_{n = \infty} s_n = s.$$
Let t_{2n-1} mean $s_1 + s_3 + \ldots + s_{2n-1}$; and t_{2n} mean $s_2 + s_4 + \ldots + s_{2n}$. Observing that t_{2n-1} and t_{2n} consist each of n terms tending to the limit s when $n = \infty$, we infer that
$$\lim_{n = \infty} t_{2n-1}/n = s \quad \text{and} \quad \lim_{n = \infty} t_{2n}/n = s.$$
Hence, whether n be odd or even, we have
$$\lim_{n = \infty} t_n/n = \tfrac{1}{2} s \quad \ldots \ldots \ldots \ldots (1).$$
Now let $R_n = \phi_1 + \phi_2 + \ldots + \phi_n$, $S_n = R_1 - R_2 + R_3 - \ldots \pm R_n$,
$$T_n = S_1 + S_2 + \ldots + S_n.$$
We shall show that the series $R_1 - R_2 + R_3 - \ldots$ is "summable" in the sense of Frobenius and Césaro given by Mr. Hardy, and that the sum $= \tfrac{1}{2} S$. This will be proved if we show that
$$\lim_{n = \infty} T_n/n = \tfrac{1}{2} s.$$
Now, the series $\phi_1 - \phi_2 + \phi_3 - \ldots$ being convergent, the power series $u \equiv \phi_1 - \phi_2 x + \phi_3 x^2 - \ldots$ is absolutely convergent ($|x| < 1$). Hence, in succession,
$$\left. \begin{array}{l} u(1+x)^{-1} = R_1 - R_2 x + R_3 x^2 - \ldots \\ u(1+x)^{-1}(1-x)^{-1} = S_1 + S_2 x + S_3 x^2 + \ldots \\ u(1+x)^{-1}(1-x)^{-2} = T_1 + T_2 x + T_3 x^2 + \ldots \end{array} \right\} \ldots \ldots (2).$$
Now the left-hand side of the above can be written $u(1-x)^{-1}(1-x^2)^{-1}$; and we have, in succession,
$$\left. \begin{array}{l} u(1-x)^{-1} = s_1 + s_2 x + s_3 x^2 + \ldots \\ u(1-x)^{-1}(1-x^2)^{-1} = t_1 + t_2 x + t_3 x^2 + \ldots \end{array} \right\} \ldots \ldots \ldots (3).$$
Now, comparing (2) with (3), we get $T_n = t_n$. Hence
$$\lim_{n = \infty} T_n/n = \lim_{n = \infty} t_n/n = \tfrac{1}{2} s, \text{ by (1)}.$$

In the above, let $\phi_1 = \log 1$, $\phi_2 = \log \frac{2}{1}$, $\phi_3 = \log \frac{3}{2}$, ... Then $\phi_2, \phi_3, \phi_4, \ldots$ form a decreasing series of positive terms tending to vanish when continued indefinitely. Hence $\phi_1 - \phi_2 + \phi_3 - \phi_4 + \ldots$ is convergent. The series $R_1 - R_2 + R_3 - \ldots$ becomes

$$\log 1 - \log 2 + \log 3 \ldots .$$

Hence this series is summable, and the sum

$$= \tfrac{1}{2}[\phi_1 - \phi_2 + \phi_3 - \phi_4 + \ldots] = \tfrac{1}{2}[\log 1 - \log \tfrac{2}{1} + \log \tfrac{3}{2} - \log \tfrac{4}{3} + \ldots]$$
$$= \tfrac{1}{2}\log[\tfrac{1}{2} \cdot \tfrac{3}{2} \cdot \tfrac{3}{4} \cdot \tfrac{5}{4} \ldots] = \tfrac{1}{2}\log(2/\pi),$$

by Wallis's theorem.

CORRECTIONS

In the last line of the question the series should be $\tfrac{1}{2}\phi(1) - \tfrac{1}{2}\phi(2) + \tfrac{1}{2}\phi(3) - \cdots$.
In the fifth line after (1) read $\tfrac{1}{2}s$ for $\tfrac{1}{2}S$.

COMMENT

Mr V. Ramaswami Aiyar was the founder of the Indian Mathematical Society. He appears to have been the first person to have recognized the genius of Srinivasa Ramanujan when, in 1910, the latter came to Tirukoilur near Madras, where Ramaswami Aiyar was Deputy Collector, seeking a clerical post; see *Collected Papers of S. Ramanujan* (Cambridge, 1927), p. xiii.

16402. (Professor NANSON.)—If $u_1, u_2, \ldots, u_n, v$ are $n+1$ quadratic functions of x, show that the value of

$$\int \frac{u_1 u_2 \ldots u_n}{v^{n+1}} dx$$

is algebraic, provided v is harmonically related to one of the u's.

Note by G. H. HARDY, M.A.

The result stated in this Question is not quite accurate, as is easily seen by an example. Let

$$u_1 = a_1 x^2 + c_1, \quad u_2 = a_2 x^2 + 2b_2 x + c_2, \quad v = x.$$

Then the coefficient of $1/x$ in $u_1 u_2/v^3$ is $a_1 c_2 + a_2 c_1$, and $\int u_1 u_2 v^{-3} dx$ will be rational only if $a_1 c_2 + a_2 c_1 = 0$, *i.e.*, if u_2 as well as v is harmonically related to u_1.

The Proposer's argument, "Now operate on $\int u_1 v^{-2} dx$ with $a_r \partial/\partial a + b_r \partial/\partial b + c_r \partial/\partial c, \ldots,$" is unsound because (if a_1, b_1, c_1 are supposed given) a, b, c are not independent, being, in fact, connected by the relation $ac_1 + ca_1 = 2bb_1$. The correct form of the result, viz.,

"$\int \dfrac{u_1 u_2 \ldots u_n}{v^{n+1}} dx$

is rational if *all* the quadratics are harmonically related to any one in the numerator," is stated in my pamphlet, "The Integration of Functions of a Single Variable" (*Camb. Math. Tracts*, No. 2, 1905, p. 16).

COMMENTS

The proposer uses the word *algebraic* to mean *rational*.

The statement that two quadratic functions

$$u_1 = a_1 x^2 + 2b_1 x + c_1, \quad u_2 = a_2 x^2 + 2b_2 x + c_2$$

are harmonically related means that their roots are harmonically conjugate. A necessary and sufficient condition for this to be the case is that $a_1 c_2 + a_2 c_1 = 2b_1 b_2$.

The proposer's argument (ibid., p. 52) is as follows: 'If the quadratic functions u_2, v of x are harmonically related, then, by a linear transformation $\int u_1 v^{-2} dx$ can be reduced to the form $\int (p + qX^2) X^{-2} dX$ and is therefore algebraic. Now operate on $\int u_1 v^{-2} dx$ with $a_r \partial/\partial a + b_r \partial/\partial b + c_r \partial/\partial c$, where a, b, c are the coefficients of v and r has the values $2, 3, \ldots, n$ and the result stated follows.'

Edward John Nanson (1850–1936; see also 1899, 2b) was Professor of Mathematics at the University of Melbourne from 1875 until 1922; he was Second Wrangler in 1873.

1909, 8 *Educational Times*, (2), 16, 66.

14587. (G. H. Hardy, B.A.)—Prove that the line element of a sphere of unit radius may be expressed in the form
$$ds^2 = (\mathrm{dn}^2 \alpha - \mathrm{dn}^2 \beta)(d\alpha^2 - d\beta^2);$$
that the equation of a circle on the sphere is
$$p \operatorname{sn}^2 \tfrac{1}{2}(\alpha - \beta) \operatorname{sn}^2 \tfrac{1}{2}(\alpha + \beta) + q \operatorname{sn}^2 \tfrac{1}{2}(\alpha + \beta) + r \operatorname{sn}^2 \tfrac{1}{2}(\alpha - \beta) + s = 0;$$
and that the condition that two circles should cut at right angles is
$$ps' + p's = qr' + q'r.$$

Solution by Maurice A. Giblett, B.Sc.

The point $x = k \operatorname{sn}\alpha \operatorname{sn}\beta$, $y = ik/k' \operatorname{cn}\alpha \operatorname{cn}\beta$, $z = 1/k' \operatorname{dn}\alpha \operatorname{dn}\beta$, all the functions being formed for modulus k, lies on the unit sphere, for
$$x^2 + y^2 + z^2 = k^2 \operatorname{sn}^2\alpha \operatorname{sn}^2\beta - k^2/k'^2 \operatorname{cn}^2\alpha \operatorname{cn}^2\beta + 1/k'^2 \operatorname{dn}^2\alpha \operatorname{dn}^2\beta$$
$$= 1/k'^2 [k^2 k'^2 \operatorname{sn}^2\alpha \operatorname{sn}^2\beta - k^2(1 - \operatorname{sn}^2\alpha)(1 - \operatorname{sn}^2\beta)$$
$$\qquad + (1 - k^2 \operatorname{sn}^2\alpha)(1 - k^2 \operatorname{sn}^2\beta)]$$
$$= 1 \quad (\text{since } k^2 + k'^2 = 1).$$

Now
$$dx = k c_1 d_1 s_2 d\alpha + k s_1 c_2 d_2 d\beta,$$
$$dy = -i \cdot k/k' \; s_1 d_1 c_2 d\alpha - ik/k' \; c_1 s_2 d_2 d\beta,$$
$$dz = -k^2/k' \; s_1 c_1 d_2 d\alpha - k^2/k' \; d_1 s_2 c_2 d\beta,$$
writing $s_1 = \operatorname{sn}\alpha$, $s_2 = \operatorname{sn}\beta$, ... Therefore
$$ds^2 = dx^2 + dy^2 + dz^2$$
$$= (k^2 c_1^2 d_1^2 s_2^2 - k^2/k'^2 s_1^2 d_1^2 c_2^2 + k^4/k'^2 s_1^2 c_1^2 d_2^2) \, d\alpha^2$$
$$\qquad + (k^2 s_1^2 c_2^2 d_2^2 - k^2/k'^2 c_1^2 s_2^2 d_2^2 + k^4/k'^2 d_1^2 s_2^2 c_2^2) \, d\beta^2.$$

Coefficient of
$$d\alpha^2 = 1/k'^2 [k'^2 k^2 (1 - s_1^2)(1 - k^2 s_1^2) s_2^2 - k^2 s_1^2 (1 - k^2 s_1^2)(1 - s_2^2)$$
$$\qquad + k^4 s_1^2 (1 - s_1^2)(1 - k^2 s_2^2)]$$
$$= 1/k'^2 [k^2 k'^2 (s_2^2 - s_1^2)] \quad (\text{since } k^2 + k'^2 = 1)$$
$$= k^2 (s_2^2 - s_1^2) = d_1^2 - d_2^2.$$

Coefficient of $d\beta^2$ is obtained by interchanging α and β in coefficient of $d\alpha^2$; and therefore equals $d_2^2 - d_1^2$. Thus
$$ds^2 = (\mathrm{dn}^2 \alpha - \mathrm{dn}^2 \beta)(d\alpha^2 - d\beta^2).$$

Write $\quad u = \tfrac{1}{2}(\alpha + \beta), \quad v = \tfrac{1}{2}(\alpha - \beta).$

Then $\quad x = k \operatorname{sn}(u+v) \operatorname{sn}(u-v) = k [\operatorname{sn}^2 u - \operatorname{sn}^2 v] \;(\div),$
$$y = ik/k' \operatorname{cn}(u+v) \operatorname{cn}(u-v) = ik/k' [1 - \operatorname{sn}^2 u - \operatorname{sn}^2 v + k^2 \operatorname{sn}^2 u \operatorname{sn}^2 v] \;(\div),$$
$$z = 1/k' \operatorname{dn}(u+v) \operatorname{dn}(u-v)$$
$$= 1/k' [1 - k^2 \operatorname{sn}^2 u - k^2 \operatorname{sn}^2 v + k^2 \operatorname{sn}^2 u \operatorname{sn}^2 v] \;(\div).$$

Denominator in each case $1 - k^2 \operatorname{sn}^2 u \operatorname{sn}^2 v$. A circle on the sphere is the curve in which it is intersected by the plane
$$lx + my + nz = r;$$

and therefore, replacing the Cartesian by curvilinear co-ordinates, we have as the equation of a circle on the sphere,

$$p \operatorname{sn}^2 u \operatorname{sn}^2 v + q \operatorname{sn}^2 u + r \operatorname{sn}^2 v + s = 0,$$

or in the original co-ordinates

$$p \operatorname{sn}^2 \tfrac{1}{2}(\alpha+\beta) \operatorname{sn}^2 \tfrac{1}{2}(\alpha-\beta) + q \operatorname{sn}^2 \tfrac{1}{2}(\alpha+\beta) + r \operatorname{sn}^2 \tfrac{1}{2}(\alpha-\beta) + s = 0.$$

Now $\quad ds^2 = (\operatorname{dn}^2 \alpha - \operatorname{dn}^2 \beta)(d\alpha^2 - d\beta^2) = 2\mathrm{F}\, du\, dv,$

where F is a function of u and v, since $\alpha = u+v$, $\beta = u-v$, and therefore $d\alpha = du + dv$, $d\beta = du - dv$.

Suppose du, dv; $\delta u, \delta v$ refer to two small arcs drawn from a point on the sphere, and let their lengths be ds and δs. Then, if

$$ds^2 = \mathrm{E}\, du^2 + 2\mathrm{F}\, du\, dv + \mathrm{G}\, dv^2;$$

the condition that the arcs should be at right angles is

$$\mathrm{E}\, du\, \delta u + \mathrm{F}\, (du\, \delta v + dv\, \delta u) + \mathrm{G}\, dv\, \delta v = 0.$$

In the case considered $\mathrm{E} = \mathrm{G} = 0$, so that the condition is

$$du\, \delta v + dv\, \delta u = 0.$$

Now let the arcs be taken along the circles

$$p \operatorname{sn}^2 u \operatorname{sn}^2 v + q \operatorname{sn}^2 u + r \operatorname{sn}^2 v + s = 0 \quad \ldots\ldots\ldots\ldots (\mathrm{i}),$$
$$p' \operatorname{sn}^2 u \operatorname{sn}^2 v + q' \operatorname{sn}^2 u + r' \operatorname{sn}^2 v + s' = 0 \quad \ldots\ldots\ldots (\mathrm{ii}),$$

at a point of intersection. From (i),

$$\operatorname{sn} u \operatorname{cn} u \operatorname{dn} u\, (p \operatorname{sn}^2 v + q)\, du + \operatorname{sn} v \operatorname{cn} v \operatorname{dn} v\, (p \operatorname{sn}^2 u + r)\, dv = 0.$$

From (ii),

$$\operatorname{sn} u \operatorname{cn} u \operatorname{dn} u\, (p' \operatorname{sn}^2 v + q')\, \delta u + \operatorname{sn} v \operatorname{cn} v \operatorname{dn} v\, (p' \operatorname{sn}^2 u + r')\, \delta v = 0.$$

Therefore, if $\quad du\, \delta v + dv\, \delta u = 0,$

$$(p' \operatorname{sn}^2 u + r')(p \operatorname{sn}^2 v + q) + (p' \operatorname{sn}^2 v + q')(p \operatorname{sn}^2 u + r) = 0,$$

i.e., $\quad 2pp' \operatorname{sn}^2 u \operatorname{sn}^2 v + (p'q + pq') \operatorname{sn}^2 u + (p'r + pr') \operatorname{sn}^2 v$
$$+ (q'r + qr') = 0 \quad \ldots\ldots\ldots (\mathrm{iii}),$$

where the co-ordinates u, v refer to the point considered, common to (i) and (ii). Eliminate $\operatorname{sn} u, \operatorname{sn} v$ between (i), (ii), (iii).

Multiply (i) by p', and (ii) by p, and add

$$2pp' \operatorname{sn}^2 u \operatorname{sn}^2 v + (p'q + pq') \operatorname{sn}^2 u + (p'r + pr') \operatorname{sn}^2 v$$
$$+ (p's + ps') = 0 \quad \ldots\ldots\ldots (\mathrm{iv}).$$

Comparing (iii) and (iv), we obtain the condition that the circles should cut orthogonally, viz., $p's + ps' = q'r + qr'$.

COMMENT

Maurice A. Giblett (1894–1930), a graduate of the University of London, became Superintendent of the Airship Services Division of the Meteorological Office and lost his life in the *R101* airship disaster at Beauvais on 5 October 1930.

4. OBITUARY NOTICES BY G. H. HARDY

INTRODUCTION TO THE OBITUARY NOTICES BY G. H. HARDY

Hardy's felicitous literary style is well illustrated by these obituary notices. For Ramanujan, Mittag-Leffler, Bromwich, Hobson, and Young he wrote obituary notices both for the London Mathematical Society and for the Royal Society; those published by the former body are reproduced here, since in several cases they contain slightly more information. Thus, for Ramanujan, O1 is the fuller version and is the one that appeared in *Collected Papers of Srinivasa Ramanujan* (Cambridge, 1927), xxi–xxxvi; the briefer version O2 differs mainly from O1 in the selection of formulae illustrating Ramanujan's work. Hardy also wrote shorter notices on Ramanujan, Hobson, and W. H. Young for *Nature* (O3, O13 and O17); these are not reproduced here.

The Hilbert obituary notice O18 is a provisional one written during the 1939–45 war in the unfulfilled expectation of a fuller notice to be published later.

A full list of the obituary notices by Hardy is given at the end of this Volume.

R.A.R.

SRINIVASA RAMANUJAN.

I.

SRINIVASA RAMANUJAN, who died at Kumbakonam on April 26th, 1920, had been a member of the Society since 1917. He was not a man who talked much about himself, and until recently I knew very little of his early life. Two notices, by P. V. Seshu Aiyar and R. Ramachandra Rao, two of the most devoted of Ramanujan's Indian friends, have been published recently in the *Journal of the Indian Mathematical Society*; and Sir Francis Spring has very kindly placed at my disposal an article which appeared in the *Madras Times* of April 5th, 1919. From these sources of information I can now supply a good many details with which I was previously unacquainted. Ramanujan (Srinivasa Iyengar Ramanuja Iyengar, to give him for once his proper name) was born on December 22nd, 1887, at Erode in southern India. His father was an accountant (*gumasta*) to a cloth merchant at Kumbakonam, while his maternal grandfather had served as *amin* in the Munsiff's (or local judge's) Court at Erode. He first went to school at five, and was transferred before he was seven to the Town High School at Kumbakonam, where he held a "free scholarship", and where his extraordinary powers appear to have been recognised immediately. "He used", so writes an old schoolfellow to Mr. Seshu Aiyar, "to borrow Carr's *Synopsis of Pure Mathematics* from the College library, and delight in verifying some of the formulæ given there. . . . He used to entertain his friends with his theorems and formulæ, even in those early days. . . . He had an extraordinary memory and could easily repeat the complete lists of Sanscrit roots (*atmanepada* and *parasmepada*); he could give the values of $\sqrt{2}$, π, e, ... to any number of decimal places. . . . In manners, he was simplicity itself. . . ."

He passed his matriculation examination to the Government College at Kumbakonam in 1904; and secured the "Junior Subraniam Scholarship". Owing to weakness in English, he failed in his next examination and lost his scholarship; and left Kumbakonam, first for Vizagapatam and then for Madras. Here he presented himself for the "First Examination in Arts" in December 1906, but failed and never tried again. For the next few years he continued his independent work in mathematics, "jotting down his results in two good-sized notebooks": I have one of these notes books in my possession still. In 1909 he married, and it became necessary for him to find some permanent employment. I quote Mr. Seshu Aiyar:

> To this end, he went to Tirukoilur, a small sub-division town in South Arcot District, to see Mr. V. Ramaswami Aiyar, the founder of the Indian Mathematical Society, but Mr. Aiyar, seeing his wonderful gifts, persuaded him to go to Madras. It was then after some four years' interval that Mr. Ramanujan met me at Madras, with his two well-sized note-books referred to above. I sent Ramanujan with a note of recommendation to that true lover of Mathematics, Dewan Bahadur R. Ramachandra Rao, who was then District Collector at Nellore, a small town some eighty miles north of Madras. Mr. Rao sent him back to me saying it was cruel to make an intellectual giant like Ramanujan rot at a mofussil station like Nellore, and recommended his stay at Madras, generously undertaking to pay Mr. Ramanujan's expenses for a time. This was in December 1910. After a while, other attempts to obtain for him a scholarship having failed, and Ramanujan himself being unwilling to be a burden on anybody for any length of time, he decided to take up a small appointment under the Madras Port Trust in 1911.
> But he never slackened his work at Mathematics. His earliest contribution to the *Journal of the Indian Mathematical Society* was in the form of questions communicated by me in Vol. III (1911). His first long article on 'Some Properties of Bernoulli's Numbers' was published in the December number of the same volume. Mr. Ramanujan's methods were so terse and novel and his presentation was so lacking in clearness and precision, that the ordinary reader, unaccustomed to such intellectual gymnastics, could hardly follow him. This particular article was returned more than once by the Editor before it took a form suitable for publication. It was during this period that he came to me one day with some theorems on Prime Numbers, and when I referred him to Hardy's Tract on *Orders of Infinity*, he observed that Hardy had said on p. 36 of his Tract 'the exact order of $\rho(x)$ [defined by the equation
> $$\rho(x) = \pi(x) - \int_2^x \frac{dt}{\log t},$$
> where $\pi(x)$ denotes the number of primes less than x], has not yet been determined', and that he himself had discovered a result which gave the order of $\rho(x)$. On this I suggested that he might communicate his result to Mr. Hardy, together with some more of his results.

This passage brings me to the beginning of my own acquaintance with Ramanujan. But before I say anything about the letters which I received from him, and which resulted ultimately in his journey to England, I must add a little more about his Indian career. Dr. G. T. Walker, F.R.S., Head of the Meteorological Department, and formerly Fellow and Mathematical Lecturer of Trinity College, Cambridge, visited Madras for some official purpose some time in 1912; and Sir Francis Spring, K.C.I.E., the Chairman of the Madras Port Authority, called his attention to Ramanujan's work. Dr. Walker was far too good a mathematician not to recognise its quality, little as it had in common with his own. He brought Ramanujan's case to the notice of the Government and the University of Madras. A research studentship, "Rs. 75 *per mensem* for a period of two years", was awarded him; and he became, and remained for the rest of his life, a professional mathematician.

II.

Ramanujan wrote to me first on January 16th, 1913, and at fairly regular intervals until he sailed for England in 1914. I do not believe that his letters were entirely his own. His knowledge of English, at that stage of his life, could scarcely have been sufficient, and there is an occasional phrase which is hardly characteristic. Indeed I seem to remember his telling me that his friends had given him some assistance. However, it was the mathematics that mattered, and that was very emphatically his.

Madras, 16th January 1913

"Dear Sir

I beg to introduce myself to you as a clerk in the Accounts Department of the Port Trust Office at Madras on a salary of only £20 per annum. I am now about 23 years of age. I have had no university education but I have undergone the ordinary school course. After leaving school I have been employing the spare time at my disposal to work at Mathematics. I have not trodden through the conventional regular course which is followed in a university course, but I am striking out a new path for myself. I have made a special investigation of divergent series in general and the results I get are termed by the local mathematicians as 'startling'.

Just as in elementary mathematics you give a meaning to a^n when n is negative and fractional to conform to the law which holds when n is a positive integer, similarly the whole of my investigations proceed on giving a meaning to Eulerian Second Integral for all values of n. My friends who have gone through the regular course of university education tell me that $\int_0^\infty x^{n-1} e^{-x} dx = \Gamma(n)$ is true only when n is positive. They say that this integral relation is not true when n is negative. Supposing this is true only for positive values of n and also supposing the definition $n\Gamma(n) = \Gamma(n+1)$ to be universally true, I have given meanings to these integrals and under the conditions I state the integral is true for all values of n negative and fractional. My whole investigations are based upon this and I have been developing this to a remarkable extent so much so that the local mathematicians are not able to understand me in my higher flights.

Very recently I came across a tract published by you styled *Orders of Infinity* in page 36 of which I find a statement that no definite expression has been as yet found for the no of prime nos less than any given number. I have found an expression which very nearly approximates to the real result, the error being negligible. I would request you to go through the enclosed papers. Being poor, if you are convinced that there is anything of value I would like to have my theorems published. I have not given the actual investigations nor the expressions that I get but I have indicated to the lines on which I proceed. Being inexperienced I would very highly value any advice you give me. Requesting to be excused for the trouble I give you.

I remain Dear sir Yours truly

S. Ramanujan

P.S. My address is S. Ramanujan, Clerk Accounts Department, Port Trust, Madras, India."

I quote now from the "papers enclosed," and from later letters :—

"In page 36 it is stated that 'the no of prime nos less than $x = \int_2^x \frac{dt}{\log t} + \rho(x)$ where the precise order of $\rho(x)$ has not been determined. . . .'

I have observed that $\rho(e^{2\pi x})$ is of such a nature that its value is very small when x lies between 0 and 3 (its value is less than a few hundreds when $x = 3$) and rapidly increases when x is greater than 3. . . .

The difference between the no of prime nos of the form $4n-1$ and which are less than x and those of the form $4n+1$ less than x is infinite when x becomes infinite. . . .

The following are a few examples from my theorems :—

(1) The nos of the form $2^p 3^q$ less than $n = \frac{1}{2} \frac{\log(2n) \log(3n)}{\log 2 \log 3}$ where p and q may have any positive integral value including 0.

(2) Let us take all nos containing an odd no of dissimilar prime divisors viz.

2, 3, 5, 7, 11, 13, 17, 19, 23, 29, 30, 31, 37, 41, 42, 43, 47 &c

(a) The no of such nos less than $n = \frac{3n}{\pi^2}$.

(b) $\frac{1}{2^2} + \frac{1}{3^2} + \frac{1}{5^2} + \frac{1}{7^2} + \cdots + \frac{1}{30^2} + \frac{1}{31^2} + \cdots = \frac{9}{2\pi^2}$.

(c) $\frac{1}{2^4} + \frac{1}{3^4} + \frac{1}{5^4} + \frac{1}{7^4} + \&c. = \frac{15}{2\pi^4}$.

(3) Let us take the no of divisors of natural nos viz.

1, 2, 2, 3, 2, 4, 2, 4, 3, 4, 2 &c (1 having 1 divisor, 2 having 2, 3 having 2, 4 having 3, 5 having 2, &c).

The sum of such nos to n terms

$$= n(2\gamma - 1 + \log n) + \tfrac{1}{2} \text{ of the no of divisors of } n$$

where $\gamma = \cdot 5772156649\ldots$, the Eulerian Constant.

(4) 1, 2, 4, 5, 8, 9, 10, 13, 16, 17, 18 &c are nos which are either themselves sqq. or which can be expressed as the sum of two sqq.

The no of such nos greater than A and less than B

$$= K \int_A^B \frac{dx}{\sqrt{\log x}} + \theta(x)^* \quad \text{where} \quad K = \cdot 764 \ldots$$

and $\theta(x)$ is very small when compared with the previous integral. K and $\theta(x)$ have been exactly found though complicated. ...."

Ramanujan's theory of primes was vitiated by his ignorance of the theory of functions of a complex variable. It was (so to say) what the theory might be if the Zeta-function had no complex zeros. His methods of proof depended upon a wholesale use of divergent series. He disregarded entirely all the difficulties which are involved in the interchange of double limit operations; he did not distinguish, for example, between the sum of a series Σa_n and the value of the Abelian limit

$$\lim_{x \to 1} \Sigma a_n x^n,$$

or that of any other limit which might be used for similar purposes by a modern analyst. There are regions of mathematics in which the precepts of modern rigour may be disregarded with comparative safety, but the Analytic Theory of Numbers is not one of them, and Ramanujan's Indian work on primes, and on all the allied problems of the theory, was definitely wrong. That his proofs should have been invalid was only to be expected. But the mistakes went deeper than that, and many of the actual results were false. He had obtained the dominant terms of the classical formulæ, although by invalid methods; but none of them are such close approximations as he supposed.

This may be said to have been Ramanujan's one great failure. And yet I am not sure that, in some ways, his failure was not more wonderful than any of his triumphs. Consider, for example, problem (4). The dominant term, which Ramanujan gives correctly, was first obtained by

---

* This should presumably be $\theta(B)$.

Landau in 1908. The correct order of the error term is still unknown. Ramanujan had none of Landau's weapons at his command; he had never seen a French or German book; his knowledge even of English was insufficient to enable him to qualify for a degree. It is sufficiently marvellous that he should have even dreamt of problems such as these, problems which it has taken the finest mathematicians in Europe a hundred years to solve, and of which the solution is incomplete to the present day.

"... IV. Theorems on integrals. The following are a few examples

(1) $\int_0^\infty \dfrac{1+\left(\frac{x}{b+1}\right)^2}{1+\left(\frac{x}{a}\right)^2} \cdot \dfrac{1+\left(\frac{x}{b+2}\right)^2}{1+\left(\frac{x}{a+1}\right)^2} \cdots \&c\ dx$

$= \dfrac{\sqrt{\pi}}{2} \cdot \dfrac{\Gamma(a+\frac{1}{2})}{\Gamma(a)} \cdot \dfrac{\Gamma(b+1)}{\Gamma(b+\frac{1}{2})} \cdot \dfrac{\Gamma(b-a+\frac{1}{2})}{\Gamma(b-a+1)}.$

...

(3) If $\int_0^\infty \dfrac{\cos nx}{e^{2\pi\sqrt{x}}-1}\, dx = \phi(n),$

then $\int_0^\infty \dfrac{\sin nx}{e^{2\pi\sqrt{x}}-1}\, dx = \phi(n) - \dfrac{1}{2n} + \phi\left(\dfrac{\pi^2}{n}\right)\sqrt{\dfrac{2\pi^3}{n^3}}.$

$\phi(n)$ is a complicated function. The following are certain special values

$\phi(0) = \dfrac{1}{12};\quad \phi\left(\dfrac{\pi}{2}\right) = \dfrac{1}{4\pi};\quad \phi(\pi) = \dfrac{2-\sqrt{2}}{8};\quad \phi(2\pi) = \dfrac{1}{16};$

$\phi\left(\dfrac{2\pi}{5}\right) = \dfrac{8-8\sqrt{5}}{16};\quad \phi\left(\dfrac{\pi}{5}\right) = \dfrac{6+\sqrt{5}}{4} - \dfrac{5\sqrt{10}}{8};\quad \phi(\infty) = 0;$

$\phi\left(\dfrac{2\pi}{3}\right) = \dfrac{1}{3} - \sqrt{3}\left(\dfrac{3}{16} - \dfrac{1}{8\pi}\right).$

(4) $\int_0^\infty \dfrac{dx}{(1+x^2)(1+r^2x^2)(1+r^4x^2)\cdots\&c} = \dfrac{\pi}{2(1+r+r^3+r^6+r^{10}+\&c)}$

where 1, 3, 6, 10 &c are sums of natural nos.

(5) $\int_0^\infty \dfrac{\sin 2nx}{x(\cosh \pi x + \cos \pi x)}\, dx = \dfrac{\pi}{4} - 2\left(\dfrac{e^{-n}\cos n}{\cosh\frac{\pi}{2}} - \dfrac{e^{-3n}\cos 3n}{3\cosh\frac{3\pi}{2}} \cdots \&c\right).$

...

V Theorems on summation of series;* *e.g.*

(1) $\dfrac{1}{1^3}\cdot\dfrac{1}{2}+\dfrac{1}{2^3}\cdot\dfrac{1}{2^2}+\dfrac{1}{3^3}\cdot\dfrac{1}{2^3}+\dfrac{1}{4^3}\cdot\dfrac{1}{2^4}+\&\text{c}$

$$=\tfrac{1}{6}(\log 2)^3-\dfrac{\pi^2}{12}\log 2+\left(\dfrac{1}{1^3}+\dfrac{1}{3^3}+\dfrac{1}{5^3}+\&\text{c}\right).$$

(2) $1+9\cdot(\tfrac{1}{4})^4+17\cdot\left(\dfrac{1.5}{4.8}\right)^4+25\cdot\left(\dfrac{1.5.9}{4.8.12}\right)^4+\&\text{c}=\dfrac{2\sqrt{2}}{\sqrt{\pi}\cdot\{\Gamma(\tfrac{3}{4})\}^2}.$

(3) $1-5\cdot(\tfrac{1}{2})^3+9\cdot\left(\dfrac{1.3}{2.4}\right)^3-\&\text{c}=\dfrac{2}{\pi}.$

(4) $\dfrac{1^{13}}{e^{2\pi}-1}+\dfrac{2^{13}}{e^{4\pi}-1}+\dfrac{3^{13}}{e^{6\pi}-1}+\&\text{c}=\dfrac{1}{24}.$

(5) $\dfrac{\coth\pi}{1^7}+\dfrac{\coth 2\pi}{2^7}+\dfrac{\coth 3\pi}{3^7}+\&\text{c}=\dfrac{19\pi^7}{56700}.$

(6) $\dfrac{1}{1^5\cosh\tfrac{\pi}{2}}-\dfrac{1}{3^5\cosh\tfrac{3\pi}{2}}+\dfrac{1}{5^5\cosh\tfrac{5\pi}{2}}-\&\text{c}=\dfrac{\pi^5}{768}.$

. . .

VI. Theorems on transformation of series and Integrals, *e.g.*

(1) $\pi\left(\dfrac{1}{2}-\dfrac{1}{\sqrt{1}+\sqrt{3}}+\dfrac{1}{\sqrt{3}+\sqrt{5}}-\dfrac{1}{\sqrt{5}+\sqrt{7}}+\&\text{c}\right)$

$$=\dfrac{1}{1\sqrt{1}}-\dfrac{1}{3\sqrt{3}}+\dfrac{1}{5\sqrt{5}}-\&\text{c}.$$

. . .

(3) $1-\dfrac{x^2\lfloor 3}{(\lfloor 1\ \lfloor 2)^3}+\dfrac{x^4\lfloor 6}{(\lfloor 2\ \lfloor 4)^3}-\dfrac{x^6\lfloor 9}{(\lfloor 3\ \lfloor 6)^3}+\&\text{c}$

$$=\left\{1+\dfrac{x}{(\lfloor 1)^3}+\dfrac{x^2}{(\lfloor 2)^3}+\&\text{c}\right\}\left\{1-\dfrac{x}{(\lfloor 1)^3}+\dfrac{x^2}{(\lfloor 2)^3}-\&\text{c}\right\}.$$

. . .

(6) If $\alpha\beta=\pi^2$, then $\dfrac{1}{\sqrt[4]{\alpha}}\left\{1+4\alpha\displaystyle\int_0^\infty\dfrac{xe^{-\alpha x^2}}{e^{2\pi x}-1}dx\right\}$

$$=\dfrac{1}{\sqrt[4]{\beta}}\left\{1+4\beta\int_0^\infty\dfrac{xe^{-\beta x^2}}{e^{2\pi x}-1}dx\right\}.$$

---

* There is always more in one of Ramanujan's formulæ than meets the eye, as anyone who sets to work to verify those which look the easiest will soon discover. In some the interest lies very deep, in others comparatively near the surface; but there is not one which is not curious and entertaining.

7) $n \left(e^{-n^2} - \dfrac{e^{-\frac{n^2}{3}}}{3\sqrt{3}} + \dfrac{e^{-\frac{n^2}{5}}}{5\sqrt{5}} - \&\text{c} \right)$

$\qquad = \sqrt{\pi} \left(e^{-n\sqrt{\pi}} \sin n\sqrt{\pi} - e^{-n\sqrt{3\pi}} \sin n \sqrt{3\pi} + \&\text{c} \right).$

(8) If n is any positive integer excluding 0

$$\dfrac{1^{4n}}{(e^{\pi}-e^{-\pi})^2} + \dfrac{2^{4n}}{(e^{2\pi}-e^{-2\pi})^2} \ldots \&\text{c} = \dfrac{n}{\pi} \left\{ \dfrac{B_{4n}}{8n} + \dfrac{1^{4n-1}}{e^{2\pi}-1} + \dfrac{2^{4n-1}}{e^{4\pi}-1} \ldots \&\text{c} \right\}$$

where $B_2 = \tfrac{1}{6}$, $B_4 = \tfrac{1}{30}$, &c.

VII. Theorems on approximate integration and summation of series.

. . .

(2) $1 + \dfrac{x}{\lfloor 1} + \dfrac{x^2}{\lfloor 2} + \dfrac{x^3}{\lfloor 3} + \ldots + \dfrac{x^x}{\lfloor x} \theta = \dfrac{e^x}{2}$

where $\theta = \dfrac{1}{3} + \dfrac{4}{135(x+k)}$ where k lies between $\dfrac{8}{45}$ and $\dfrac{2}{21}$.

(3) $1 + \left(\dfrac{x}{\lfloor 1} \right)^5 + \left(\dfrac{x^2}{\lfloor 2} \right)^5 + \left(\dfrac{x^3}{\lfloor 3} \right)^5 + \&\text{c} = \dfrac{\sqrt{5}}{4\pi^2} \cdot \dfrac{e^{5x}}{5x^2 - x + \theta}$

where θ vanishes when $x = \infty$.

(4) $\dfrac{1^2}{e^x - 1} + \dfrac{2^2}{e^{2x} - 1} + \dfrac{3^2}{e^{3x} - 1} + \dfrac{4^2}{e^{4x} - 1} + \&\text{c}$

$\qquad = \dfrac{2}{x^3} \left(\dfrac{1}{1^3} + \dfrac{1}{2^3} + \dfrac{1}{3^3} + \&\text{c} \right) - \dfrac{1}{12x} - \dfrac{x}{1440} + \dfrac{x^3}{181440}$

$\qquad + \dfrac{x^5}{7257600} + \dfrac{x^7}{159667200} + \&\text{c}$ when x is small.

(*Note.*—x may be given values from 0 to 2).

(5) $\dfrac{1}{1001} + \dfrac{1}{1002^2} + \dfrac{3}{1003^3} + \dfrac{4^2}{1004^4} + \dfrac{5^3}{1005^5} + \&\text{c}$

$\qquad = \dfrac{1}{1000} - 10^{-440} \times 1{\cdot}0125$ nearly.

(6) $\displaystyle\int_0^a e^{-x^2} dx = \dfrac{\sqrt{\pi}}{2} - \dfrac{e^{-a^2}}{2a} + \dfrac{1}{a} + \dfrac{2}{2a} + \dfrac{3}{a} + \dfrac{4}{2a} + \&\text{c}.$

(7) The coefficient of x^n in $\dfrac{1}{1 - 2x + 2x^4 - 2x^9 + 2x^{16} - \&\text{c}}$

$\qquad =$ the nearest integer to $\dfrac{1}{4n} \left\{ \cosh(\pi\sqrt{n}) - \dfrac{\sinh(\pi\sqrt{n})}{\pi\sqrt{n}} \right\}.$*

---

* This is quite untrue. But the formula is extremely interesting for a variety of reasons.

IX. Theorems on continued fractions, a few examples are :—

(1) $\dfrac{4}{x} + \dfrac{1^2}{2x} + \dfrac{3^2}{2x} + \dfrac{5^2}{2x} + \dfrac{7^2}{2x} + \&c = \left\{ \dfrac{\Gamma\left(\dfrac{x+1}{4}\right)}{\Gamma\left(\dfrac{x+3}{4}\right)} \right\}^2 .$

. . .

(4) If $\quad u = \dfrac{x}{1} + \dfrac{x^5}{1} + \dfrac{x^{10}}{1} + \dfrac{x^{15}}{1} + \dfrac{x^{20}}{1} + \&c$

and $\quad v = \dfrac{\sqrt[5]{x}}{1} + \dfrac{x}{1} + \dfrac{x^2}{1} + \dfrac{x^3}{1} + \&c$

then $\quad v^5 = u \cdot \dfrac{1 - 2u + 4u^2 - 3u^3 + u^4}{1 + 3u + 4u^2 + 2u^3 + u^4} .$

(5) $\dfrac{1}{1} + \dfrac{e^{-2\pi}}{1} + \dfrac{e^{-4\pi}}{1} + \dfrac{e^{-6\pi}}{1} + \&c = \left(\sqrt{\dfrac{5+\sqrt{5}}{2}} - \dfrac{\sqrt{5}+1}{2} \right) \sqrt[5]{e^{2\pi}} .$

(6) $\dfrac{1}{1} - \dfrac{e^{-\pi}}{1} + \dfrac{e^{-2\pi}}{1} - \dfrac{e^{-3\pi}}{1} + \&c = \left(\sqrt{\dfrac{5-\sqrt{5}}{2}} - \dfrac{\sqrt{5}-1}{2} \right) \sqrt[5]{e^{\pi}} .$

(7) $\dfrac{1}{1} + \dfrac{e^{-\pi\sqrt{n}}}{1} + \dfrac{e^{-2\pi\sqrt{n}}}{1} + \dfrac{e^{-3\pi\sqrt{n}}}{1} + \&c$ can be exactly found if n be any positive rational quantity. . . ."

27 February 1913

". . . I have found a friend in you who views my labours sympathetically. This is already some encouragement to me to proceed. . . . I find in many a place in your letter rigorous proofs are required and you ask me to communicate the methods of proof. . . . I told him* that the sum of an infinite no of terms of the series $1+2+3+4+\ldots = -\tfrac{1}{12}$ under my theory. If I tell you this you will at once point out to me the lunatic asylum as my goal. . . . What I tell you is this. Verify the results I give and if they agree with your results . . . you should at least grant that there may be some truths in my fundamental basis. . . .

To preserve my brains I want food and this is now my first consideration. Any sympathetic letter from you will be helpful to me here to get a scholarship either from the University or from Government. . . .

1. The no of prime nos. less than $e^a = \displaystyle\int_0^\infty \dfrac{a^x \, dx}{x S_{x+1} \Gamma(x+1)}$

where $\quad S_{x+1} = \dfrac{1}{1^{x+1}} + \dfrac{1}{2^{x+1}} + \ldots .$

---

* Referring to a previous correspondence.

2. The no of prime nos. less than $n =$
$$\frac{2}{\pi}\left\{\frac{2}{B_2}\left(\frac{\log n}{2\pi}\right)+\frac{4}{3B_4}\left(\frac{\log n}{2\pi}\right)^3+\frac{6}{5B_6}\left(\frac{\log n}{2\pi}\right)^5+\text{etc}\right\}$$

where $B_2 = \frac{1}{6}$; $B_4 = \frac{1}{30}$ &c, the Bernoullian nos. ....

For practical calculations
$$\int_\mu^n \frac{dx}{\log x} = n\left(\frac{1}{\log n}+\frac{\lfloor 1}{(\log n)^2}+\cdots+\frac{\lfloor k-1}{(\log n)^k}\theta\right)$$

where $\quad \theta = \frac{2}{3}-\delta+\frac{1}{\log n}\left\{\frac{4}{135}-\frac{\delta^2(1-\delta)}{3}\right\}$
$$+\frac{1}{(\log n)^2}\left\{\frac{8}{2835}+\frac{2\delta(1-\delta)}{135}-\frac{\delta(1-\delta^2)(2-3\delta^2)}{45}\right\}+\text{etc}$$

where $\delta = k - \log n$. ...

The order of $\theta(x)$ which you asked in your letter is $\sqrt{\left(\frac{x}{\log x}\right)}$.

...

(1) If $F(x) = \dfrac{1}{1+}\dfrac{x}{1+}\dfrac{x^2}{1+}\dfrac{x^3}{1+}\dfrac{x^4}{1+}\dfrac{x^5}{1+}\&c$

then $\left\{\dfrac{\sqrt{5}+1}{2}+e^{-\frac{2\alpha}{5}}F(e^{-2\alpha})\right\}\left\{\dfrac{\sqrt{5}+1}{2}+e^{-\frac{2\beta}{5}}F(e^{-2\beta})\right\} = \dfrac{5+\sqrt{5}}{2},$

with the conditions $\alpha\beta = \pi^2$. ...

e.g. $\dfrac{1}{1+}\dfrac{e^{-2\pi\sqrt{5}}}{1+}\dfrac{e^{-4\pi\sqrt{5}}}{1+}+\text{etc}\cdots = e^{\frac{2\pi}{\sqrt{5}}}\left\{\dfrac{\sqrt{5}}{1+\sqrt{5^{\frac{3}{4}}\left(\frac{\sqrt{5}-1}{2}\right)^{\frac{5}{2}}-1}}-\dfrac{\sqrt{5}+1}{2}\right\}$

The above theorem is a particular case of a theorem on the c.f.
$$\frac{1}{1+}\frac{ax}{1+}\frac{ax^2}{1+}\frac{ax^3}{1+}\frac{ax^4}{1+}\frac{ax^5}{1+}\&\text{c}.$$

which is a particular case of the c.f.
$$\frac{1}{1+}\frac{ax}{1+bx+}\frac{ax^2}{1+bx^2+}\frac{ax^3}{1+bx^3+}\&\text{c}$$

which is a particular case of a general theorem on c.f.

(2) i. $4\int_0^\infty \dfrac{xe^{-x\sqrt{5}}}{\cosh x}dx = \dfrac{1}{1+}\dfrac{1^2}{1+}\dfrac{1^2}{1+}\dfrac{2^2}{1+}\dfrac{2^2}{1+}\dfrac{3^2}{1+}\dfrac{3^2}{1+}\&\text{c}$

ii. $4\int_0^\infty \dfrac{x^2 e^{-x\sqrt{3}}}{\cosh x}dx = \dfrac{1}{1+}\dfrac{1^3}{1+}\dfrac{1^3}{3+}\dfrac{2^3}{1+}\dfrac{2^3}{5+}\dfrac{3^3}{1+}\dfrac{3^3}{7+}\&\text{c}$

(3) $1 - 5\cdot\left(\dfrac{1}{2}\right)^5 + 9\cdot\left(\dfrac{1\cdot 3}{2\cdot 4}\right)^5 - 13\cdot\left(\dfrac{1\cdot 3\cdot 5}{2\cdot 4\cdot 6}\right)^5 + \&\text{c} = \dfrac{2}{\{\Gamma(\frac{3}{4})\}^4}$

...

(6) If $v = \dfrac{x}{1+}\dfrac{x^3+x^6}{1+}\dfrac{x^6+x^{12}}{1+}\dfrac{x^9+x^{18}}{1}+\&c.$

then
i. $x\left(1+\dfrac{1}{v}\right) = \dfrac{1+x+x^3+x^6+x^{10}+\&c}{1+x^9+x^{27}+x^{54}+x^{90}+\&c}$

ii. $x^3\left(1+\dfrac{1}{v^3}\right) = \left(\dfrac{1+x+x^3+x^6+x^{10}+\&c}{1+x^3+x^9+x^{18}+x^{30}+\&c}\right)^4$

(7) If n is any odd integer,

$$\dfrac{1}{\cosh\dfrac{\pi}{2n}+\cos\dfrac{\pi}{2n}} - \dfrac{1}{3\left(\cosh\dfrac{3\pi}{2n}+\cos\dfrac{3\pi}{2n}\right)} + \dfrac{1}{5\left(\cosh\dfrac{5\pi}{2n}+\cos\dfrac{5\pi}{2n}\right)} \ldots \&c = \dfrac{\pi}{8}.$$

...

(10) If $F(\alpha, \beta, \gamma, \delta, \epsilon) = 1 + \dfrac{\alpha}{\underline{1}} \cdot \dfrac{\beta}{\delta} \cdot \dfrac{\gamma}{\epsilon} + \dfrac{\alpha(\alpha+1)}{\underline{2}} \cdot \dfrac{\beta(\beta+1)}{\delta(\delta+1)}$

$$\times \dfrac{\gamma(\gamma+1)}{\epsilon(\epsilon+1)} + \&c.$$

then $F(\alpha, \beta, \gamma, \delta, \epsilon) = \dfrac{\Gamma(\delta)\,\Gamma(\delta-\alpha-\beta)}{\Gamma(\delta-\alpha)\,\Gamma(\delta-\beta)} \cdot F(\alpha, \beta, \epsilon-\gamma, \alpha+\beta-\delta+1, \epsilon)$

$$+ \dfrac{\Gamma(\delta)\,\Gamma(\epsilon)\,\Gamma(\alpha+\beta-\delta)\,\Gamma(\delta+\epsilon-\alpha-\beta-\gamma)}{\Gamma(\alpha)\,\Gamma(\beta)\,\Gamma(\epsilon-\gamma)\,\Gamma(\delta+\epsilon-\alpha-\beta)}$$

$$\times F(\delta-\alpha, \delta-\beta, \delta+\epsilon-\alpha-\beta-\gamma, \delta-\alpha-\beta+1, \delta+\epsilon-\alpha-\beta).$$

...

(13) $\dfrac{a}{1+n}+\dfrac{a^2}{3+n}+\dfrac{(2a)^2}{5+n}+\dfrac{(3a)^2}{7+n}+\ldots$

$$= 2a\int_0^1 z^{\tfrac{n}{\sqrt{(1+a^2)}}} \dfrac{dz}{\{\sqrt{(1+a^2)}+1\}+z^2\{\sqrt{(1+a^2)}-1\}}.$$

(14) If $F(\alpha, \beta) = \alpha + \dfrac{(1+\beta)^2+k}{2\alpha} + \dfrac{(3+\beta)^2+k}{2\alpha} + \dfrac{(5+\beta)^2+k}{2\alpha}+\ldots,$

then $F(\alpha, \beta) = F(\beta, \alpha).$

(15) If $F(\alpha, \beta) = \dfrac{\alpha}{n+}\dfrac{\beta^2}{n+}\dfrac{(2\alpha)^2}{n+}\dfrac{(3\beta)^2}{n}+\ldots$

then $F(\alpha, \beta)+F(\beta, \alpha) = 2F\{\tfrac{1}{2}(\alpha+\beta), \sqrt{(\alpha\beta)}\}$

...

(17) If $F(k) = 1 + \left(\dfrac{1}{2}\right)^2 k + \left(\dfrac{1.3}{2.4}\right)^2 k^2 + \ldots$ and $F(1-k) = \sqrt{(210)}\,F(k),$

then $k = (\sqrt{2}-1)^4(2-\sqrt{3})^2(\sqrt{7}-\sqrt{6})^4(8-3\sqrt{7})^2(\sqrt{10}-3)^4(4-\sqrt{15})^4$
$$\times (\sqrt{15}-\sqrt{14})^2(6-\sqrt{35})^2.$$

. . .

(20) If $\quad F(a) = \int_0^{\frac{1}{2}\pi} \dfrac{d\phi}{\sqrt{\{1-(1-a)\sin^2\phi\}}} \Big/ \int_0^{\frac{1}{2}\pi} \dfrac{d\phi}{\sqrt{\{1-a\sin^2\phi\}}}$

and $\quad F(a) = 3F(\beta) = 5F(\gamma) = 15F(\delta),$

then i. $[(a\delta)^{\frac{1}{2}} + \{(1-a)(1-\delta)\}^{\frac{1}{2}}][(\beta\gamma)^{\frac{1}{2}} + \{(1-\beta)(1-\gamma)\}^{\frac{1}{2}}] = 1$

. . .

v. $(a\beta\gamma\delta)^{\frac{1}{2}} + \{(1-a)(1-\beta)(1-\gamma)(1-\delta)\}^{\frac{1}{2}}$
$$+ \{16a\beta\gamma\delta(1-a)(1-\beta)(1-\gamma)(1-\delta)\}^{\frac{1}{4}} = 1$$

. . .

(21) If $\quad F(a) = 3F(\beta) = 13F(\gamma) = 39F(\delta)$

or $\quad F(a) = 5F(\beta) = 11F(\gamma) = 55F(\delta)$

or $\quad F(a) = 7F(\beta) = 9F(\gamma) = 63F(\delta)$

then $\quad \dfrac{\{(1-a)(1-\delta)\}^{\frac{1}{2}} - (a\delta)^{\frac{1}{2}}}{\{(1-\beta)(1-\gamma)\}^{\frac{1}{2}} - (\beta\gamma)^{\frac{1}{2}}} = \dfrac{1+\{(1-a)(1-\delta)\}^{\frac{1}{4}}+(a\delta)^{\frac{1}{4}}}{1+\{(1-\beta)(1-\gamma)\}^{\frac{1}{4}}+(\beta\gamma)^{\frac{1}{4}}}$

. . .

(23) $(1+e^{-\pi\sqrt{1353}})(1+e^{-3\pi\sqrt{1353}})(1+e^{-5\pi\sqrt{1353}})\ldots$
$$= \sqrt[4]{2}\, e^{-\frac{3}{4}\pi\sqrt{1353}} \times \sqrt{\left\{\sqrt{\left(\dfrac{569+99\sqrt{33}}{8}\right)} + \sqrt{\left(\dfrac{561+99\sqrt{33}}{8}\right)}\right\}}$$
$$\times \sqrt{\left\{\sqrt{\left(\dfrac{25+3\sqrt{33}}{8}\right)} + \sqrt{\left(\dfrac{17+3\sqrt{33}}{8}\right)}\right\}} \times \sqrt[4]{\left(\dfrac{\sqrt{123}+11}{\sqrt{2}}\right)}$$
$$\times \sqrt[8]{(10+3\sqrt{11})} \times \sqrt[8]{(26+15\sqrt{3})} \times \sqrt[12]{\left(\dfrac{6817+321\sqrt{451}}{\sqrt{2}}\right)}$$

. . ."

17 April 1913

". . . I am a little pained to see what you have written. . . .* I am not in the least apprehensive of my method being utilized by others. On the contrary my method has been in my possession for the last eight years and I have not found anyone to appreciate the method. As I wrote in my last letter I have found a sympathetic friend in you and I am willing to place unreservedly in your hands what little I have. It was on

---

* Ramanujan might very reasonably have been reluctant to give away his secrets to an English mathematician, and I had tried to reassure him on this point as well as I could.

account of the novelty of the method I have used that I am a little diffident even now to communicate my own way of arriving at the expressions I have already given. . . .

. . . I am glad to inform you that the local University has been pleased to grant me a scholarship of £60 per annum for two years and this was at the instance of Dr. Walker, F.R.S., Head of the Meteorological Department in India, to whom my thanks are due. . . . I request you to convey my thanks also to Mr. Littlewood, Dr Barnes, Mr. Berry and others who take an interest in me. . . ."

III.

It is unnecessary to repeat the story of how Ramanujan was brought to England. There were serious difficulties; and the credit for overcoming them is due primarily to Prof. E. H. Neville, in whose company Ramanujan arrived in April 1914. He had a scholarship from Madras of £250, of which £50 was allotted to the support of his family in India, and an exhibition of £60 from Trinity. For a man of his almost ludicrously simple tastes, this was an ample income; and he was able to save a good deal of money which was badly wanted later. He had no duties and could do as he pleased; he wished indeed to qualify for a Cambridge degree as a research student, but this was a formality. He was now, for the first time in his life, in a really comfortable position, and could devote himself to his researches without anxiety.

There was one great puzzle. What was to be done in the way of teaching him modern mathematics? The limitations of his knowledge were as startling as its profundity. Here was a man who could work out modular equations, and theorems of complex multiplication, to orders unheard of, whose mastery of continued fractions was, on the formal side at any rate, beyond that of any mathematician in the world, who had found for himself the functional equation of the Zeta-function, and the dominant terms of many of the most famous problems in the analytic theory of numbers; and he had never heard of a doubly periodic function or of Cauchy's theorem, and had indeed but the vaguest idea of what a function of a complex variable was. His ideas as to what constituted a mathematical proof were of the most shadowy description. All his results, new or old, right or wrong, had been arrived at by a process of mingled argument, intuition, and induction, of which he was entirely unable to give any coherent account.

It was impossible to ask such a man to submit to systematic instruction, to try to learn mathematics from the beginning once more. I was afraid too that, if I insisted unduly on matters which Ramanujan found irksome, I might destroy his confidence or break the spell of his inspiration. On the other hand there were things of which it was impossible that he should remain in ignorance. Some of his results were wrong, and in particular those which concerned the distribution of primes, to which he attached the greatest importance. It was impossible to allow him to go through life supposing that all the zeros of the Zeta-function were real. So I had to try to teach him, and in a measure I succeeded, though obviously I learnt from him much more than he learnt from me. In a few years' time he had a very tolerable knowledge of the theory of functions and the analytic theory of numbers. He was never a mathematician of the modern school, and it was hardly desirable that he should become one; but he knew when he had proved a theorem and when he had not. And his flow of original ideas showed no symptom of abatement.

I should add a word here about Ramanujan's interests outside mathematics. Like his mathematics, they showed the strangest contrasts. He had very little interest, I should say, in literature as such, or in art, though he could tell good literature from bad. On the other hand, he was a keen philosopher, of what appeared, to followers of the modern Cambridge school, a rather nebulous kind, and an ardent politician, of a pacifist and ultra-radical type. He adhered, with a severity most unusual in Indians resident in England, to the religious observances of his caste; but his religion was a matter of observance and not of intellectual conviction, and I remember well his telling me (much to my surprise) that all religions seemed to him more or less equally true. Alike in literature, philosophy, and mathematics, he had a passion for what was unexpected, strange, and odd; he had quite a small library of books by circle-squarers and other cranks.

It was in the spring of 1917 that Ramanujan first appeared to be unwell. He went into the Nursing Home at Cambridge in the early summer, and was never out of bed for any length of time again. He was in sanatoria at Wells, at Matlock, and in London, and it was not until the autumn of 1918 that he showed any decided symptom of improvement. He had then resumed active work, stimulated perhaps by his election to the Royal Society, and some of his most beautiful theorems were discovered about this time. His election to a Trinity Fellowship was a further encouragement; and each of those famous societies may well congratulate themselves that they recognised his claims before it was too late. Early in 1919 he had recovered, it seemed, sufficiently for the voyage home to India, and the best medical opinion held out hopes of a permanent restoration. I was rather alarmed by not hearing from him for a considerable time; but a letter reached me in February 1920, from which it appeared that he was still active in research.

> University of Madras
> 12th January 1920
>
> "I am extremely sorry for not writing you a single letter up to now. ... I discovered very interesting functions recently which I call 'Mock' ϑ-functions. Unlike the 'False' ϑ-functions (studied partially by Prof. Rogers in his interesting paper) they enter into mathematics as beautifully as the ordinary ϑ-functions. I am sending you with this letter some examples. ...
>
> *Mock ϑ-functions*
>
> $$\phi(q) = 1 + \frac{q}{1+q^2} + \frac{q^4}{(1+q^2)(1+q^4)} + \ldots$$
>
> $$\psi(q) = \frac{q}{1-q} + \frac{q^4}{(1-q)(1-q^3)} + \frac{q^9}{(1-q)(1-q^3)(1-q^5)} + \ldots$$
>
> ...
>
> *Mock ϑ-functions (of 5th order)*
>
> $$f(q) = 1 + \frac{q}{1+q} + \frac{q^4}{(1+q)(1+q^2)} + \frac{q^9}{(1+q)(1+q^2)(1+q^3)} + \ldots$$
>
> ...
>
> *Mock ϑ-functions (of 7th order)*
>
> $$\text{(i)} \quad 1 + \frac{q}{1-q^2} + \frac{q^4}{(1-q^3)(1-q^4)} + \frac{q^9}{(1-q^4)(1-q^5)(1-q^6)} + \ldots$$
>
> ..."

He said little about his health, and what he said was not particularly discouraging; and I was quite unprepared for the news of his death.

IV.

Ramanujan published the following papers in Europe:—

(1) "Some definite integrals", *Messenger of Mathematics*, Vol. 44 (1914), pp. 10–18.
(2) "Some definite integrals connected with Gauss's sums", *ibid.*, pp. 75–85.
(3) "Modular equations and approximations to π", *Quarterly Journal of Mathematics*, Vol. 45 (1914), pp. 350–372.
(4) "New expressions for Riemann's functions $\zeta(s)$ and $\Xi(t)$", *ibid.*, Vol. 46 (1915) pp. 253–261.
(5) "On certain infinite series". *Messenger of Mathematics*, Vol. 45 (1915), pp. 11–15.
(6) "Summation of a certain series", *ibid.*, pp. 157–160.
(7) "Highly composite numbers", *Proc. London Math. Soc.*, Ser. 2, Vol. 14 (1915) pp. 347–409.

(8) "Some formulæ in the analytic theory of numbers", *Messenger of Mathematics*, Vol. 45 (1916), pp. 81–84.

(9) "On certain arithmetical functions", *Trans. Cambridge Phil. Soc.*, Vol. 22 (1916), No. 9, pp. 159–184.

(10) "Some series for Euler's constant", *Messenger of Mathematics*, Vol. 46 (1916), pp. 73–80.

(11) "On the expression of numbers in the form $ax^2 + by^2 + cz^2 + dt^2$", *Proc. Cambridge Phil. Soc.*, Vol. 19 (1917), pp. 11–21.

*(12) "Une formule asymptotique pour le nombre des partitions de n", *Comptes Rendus*, 2 Jan. 1917.

*(13) "Asymptotic formulæ concerning the distribution of integers of various types", *Proc. London Math. Soc.*, Ser. 2, Vol. 16 (1917), pp. 112–132.

*(14) "The normal number of prime factors of a number n", *Quarterly Journal of Mathematics*, Vol. 48 (1917), pp. 76–92.

*(15) "Asymptotic formulæ in Combinatory Analysis", *Proc. London Math. Soc.*, Ser. 2, Vol. 17 (1918), pp. 75–115.

*(16) "On the coefficients in the expansions of certain modular functions", *Proc. Roy. Soc.*, (A), Vol. 95 (1918), pp. 144–155.

(17) "On certain trigonometrical sums and their applications in the theory of numbers", *Trans. Camb. Phil. Soc.*, Vol. 22 (1918), pp. 259–276.

(18) "Some properties of $p(n)$, the number of partitions of n", *Proc. Camb. Phil. Soc.*, Vol. 19 (1919), pp. 207–210.

(19) "Proof of certain identities in Combinatory Analysis", *ibid.*, pp. 214–216.

(20) "A class of definite integrals", *Quarterly Journal of Mathematics*, Vol. 48 (1920), pp. 294–309.

(21) "Congruence properties of partitions", *Math. Zeitschrift*, Vol. 9 (1921), pp. 147–153.

Of these those marked with an asterisk were written in collaboration with me, and (21) is a posthumous extract from a much larger unpublished manuscript in my possession.† He also published a number of short notes in the *Records of Proceedings* at our meetings, and in the *Journal of the Indian Mathematical Society*. The complete list of these is as follows:

Records of Proceedings at Meetings.

*(22) "Proof that almost all numbers n are composed of about $\log \log n$ prime factors", 14 Dec. 1916.

*(23) "Asymptotic formulæ in Combinatory Analysis", 1 March, 1917.

(24) "Some definite integrals", 17 Jan., 1918.

(25) "Congruence properties of partitions", 13 March, 1919.

(26) "Algebraic relations between certain infinite products", 13 March, 1919.

Journal of the Indian Mathematical Society.

(A) Articles and Notes.

(27) "Some properties of Bernoulli's numbers", Vol. 3 (1911), pp. 219–235.

(28) "On Q. 330 of Prof. Sanjaná", Vol. 4 (1912), pp. 59–61.

(29) "A set of equations", Vol. 4 (1912), pp. 94–96.

---

† All of Ramanujan's manuscripts passed through my hands, and I edited them very carefully for publication. The earlier ones I rewrote completely. I had no share of any kind in the results, except of course when I was actually a collaborator, or when explicit acknowledgment is made. Ramanujan was almost absurdly scrupulous in his desire to acknowledge the slightest help.

(30) "Irregular numbers", Vol. 5 (1913), pp. 105–107.
(31) "Squaring the circle", Vol. 5 (1913), pp. 132–133.
(32) "On the integral $\int_0^x \arc\tan t . \frac{dt}{t}$", Vol. 7 (1915), pp. 93–96.
(33) "On the divisors of a number", Vol. 7 (1915), pp. 131–134.
(34) "The sum of the square roots of the first n natural numbers", Vol. 7 (1915), pp. 173–175.
(35) "On the product $\pi \left[1 + \frac{x^2}{(a+nd)^2} \right]$", Vol. 7 (1915), pp. 209–212.
(36) "Some definite integrals", Vol. 11 (1919), pp. 81–88.
(37) "A proof of Bertrand's postulate", Vol. 11 (1919), pp. 181–183.
(38) (Communicated by S. Narayana Aiyar), Vol. 3 (1911), p. 60

(B) Questions proposed and solved.

Nos. 260, 261, 283, 289, 294, 295, 298, 308, 353, 358, 386, 427, 441, 464, 489, 507, 541, 546, 571, 605, 606, 629, 642, 666, 682, 700, 723, 724, 739, 740, 753, 768, 769, 783, 785.

(C) Questions proposed but not solved as yet.

Nos. 284, 327, 359, 387, 441, 463, 469, 524, 525, 526, 584, 661, 662, 681, 699, 722, 738, 754, 770, 784, 1049, 1070, and 1076.

Finally, I may mention the following writings by other authors, concerned with Ramanujan's work.

"Proof of a formula of Mr. Ramanujan", by G. H. Hardy (*Messenger of Mathematics*, Vol. 44, 1915, pp. 18–21).

"Mr. S. Ramanujan's mathematical work in England", by G. H. Hardy (Report to the University of Madras, 1916, privately printed).

"On Mr. Ramanujan's empirical expansions of modular functions", by L. J. Mordell (*Proc. Camb. Phil. Soc.*, Vol. 19, 1917, pp. 117–124).

"Life sketch of Ramanujan" (editorial in the *Journal of the Indian Math. Soc.*, Vol. 11, 1919, p. 122).

"Note on the parity of the number which enumerates the partitions of a number", by P. A. MacMahon (*Proc. Camb. Phil. Soc.*, Vol. 20, 1921, pp. 281–283).

"Proof of certain identities and congruences enunciated by S. Ramanujan", by H. B. C. Darling (*Proc. London Math. Soc.*, Ser. 2, Vol. 19, 1921, pp. 350–372).

"On a type of modular relation", by L. J. Rogers (*ibid.*, pp. 387–397).

It is plainly impossible for me, within the limits of a notice such as this, to attempt a reasoned estimate of Ramanujan's work. Some of it is very intimately connected with my own, and my verdict could not be impartial; there is much too that I am hardly competent to judge; and there is a mass of unpublished material, in part new and in part anticipated, in part proved and in part only conjectured, that still awaits analysis. But it may be useful if I state, shortly and dogmatically, what seems to me Ramanujan's finest, most independent, and most characteristic work.

His most remarkable papers appear to me to be (3), (7), (9), (17), (18), (19), and (21). The first of these is mainly Indian work, done before he came to England; and much of it had been anticipated. But there is much that is new, and in particular a very remarkable series of algebraic approximations to π. I may mention only the formulæ

$$\pi = \frac{63}{25}\frac{17+15\sqrt{5}}{7+15\sqrt{5}}, \quad \frac{1}{2\pi\sqrt{2}} = \frac{1103}{99^2},$$

correct to 9 and 8 places of decimals respectively.

The long memoir (7) represents work, perhaps, in a backwater of mathematics, and is somewhat overloaded with detail; but the elementary analysis of "highly composite" numbers—numbers which have more divisors than any preceding number—is exceedingly remarkable, and shows very clearly Ramanujan's extraordinary mastery over the algebra of inequalities. Papers (9) and (17) should be read together, and in connection with Mr. Mordell's paper mentioned above; for Mr. Mordell afterwards proved a great deal that Ramanujan conjectured. They contain, in particular, exceedingly remarkable contributions to the theory of the representation of numbers by sums of squares. But I am inclined to think that it was in the theory of partitions, and the allied parts of the theories of elliptic functions and continued fractions, that Ramanujan shows at his very best. It is in papers (18), (19), and (21), and in the papers of Prof. Rogers and Mr. Darling that I have quoted, that this side of his work (so far as it has been published) is to be found. It would be difficult to find more beautiful formulæ than the "Rogers-Ramanujan" identities, proved in (19); but here Ramanujan must take second place to Prof. Rogers; and, if I had to select one formula from all Ramanujan's work, I would agree with Major MacMahon in selecting a formula from (18), viz.

$$p(4)+p(9)x+p(14)x^2+\ldots = 5\frac{\{(1-x^5)(1-x^{10})(1-x^{15})\ldots\}^5}{\{(1-x)(1-x^2)(1-x^3)\ldots\}^6},$$

where $p(n)$ is the number of partitions of n.

I have often been asked whether Ramanujan had any special secret; whether his methods differed in kind from those of other mathematicians; whether there was anything really abnormal in his mode of thought. I cannot answer these questions with any confidence or conviction; but I do not believe it. My belief is that all mathematicians think, at bottom, in the same kind of way, and that Ramanujan was no exception. He had, of course, an extraordinary memory. He could remember the idiosyncrasies of numbers in an almost uncanny way. It was Mr. Littlewood (I believe) who remarked that "every positive integer was one of his personal friends." I remember once going to see him when he was lying ill at Putney. I had ridden in taxi-cab No. 1729, and remarked that the number (7.13.19) seemed to me rather a dull one, and that I hoped it was not an unfavourable omen. "No," he replied, "it is a very interesting number; it is the smallest number expressible as a sum of two cubes in two different ways." I asked him, naturally, whether he knew the answer to the corresponding problem for fourth powers; and he replied, after a moment's thought, that he could see no obvious example, and thought that the first such number must be very large.* His memory, and his powers of calculation, were very unusual, but they could not reasonably be called "abnormal". If he had to multiply two large numbers, he multiplied them in the ordinary way; he would do it with unusual rapidity and accuracy, but not more rapidly or more accurately than any mathematician who is naturally quick and has the habit of computation. There is a table of partitions at the end of our paper (15). This was, for the most part, calculated independently by Ramanujan and Major MacMahon; and Major MacMahon was, in general, slightly the quicker and more accurate of the two.

It was his insight into algebraical formulæ, transformations of infinite series, and so forth, that was most amazing. On this side most certainly I have never met his equal, and I can compare him only with Euler or Jacobi. He worked, far more than the majority of modern mathematicians, by induction from numerical examples; all of his congruence properties of partitions, for example, were discovered in this way. But with his memory, his patience, and his power of calculation, he combined a power of generalisation, a feeling for form, and a capacity for rapid modification of his hypotheses, that was often really startling, and made him, in his own peculiar field, without a rival in his day.

It is often said that it is much more difficult now for a mathematician to be original than it was in the great days when the foundations of modern analysis were laid; and no doubt in a measure it is true. Opinions may differ as to the importance of Ramanujan's work, the kind of standard by which it should be judged, and the influence which it is likely to have on the mathematics of the future. It has not the simplicity and the inevitableness of the very greatest work; it would be greater if it were less strange. One gift it has which no one can deny, profound and invincible originality. He would probably have been a greater mathematician if he had been caught and tamed a little in his youth; he would have discovered more that was new, and that, no doubt, of greater importance. On the other hand he would have been less of a Ramanujan, and more of a European professor, and the loss might have been greater than the gain.

G. H. H.

---

* Euler gave $542^4 + 103^4 = 359^4 + 514^4$ as an example. See Sir T. L. Heath's *Diophantus of Alexandria*, p. 380.

CAMILLE JORDAN, 1838–1922.

Camille Jordan, who died on January 20, 1922, at the age of 84, was one of the great French mathematicians of the nineteenth century. His life was not notably more eventful than that of most mathematicians. He was professor at the École Polytechnique from 1876 to 1912, and also at the Collège de France; a member of the Institute from 1881, and editor of the 'Journal de Mathématiques' ('Liouville's Journal') from 1885. He was elected a Foreign Member of the Society in 1919.

Jordan was a mathematician of great profundity and originality, and left behind him contributions of the first importance to every field in which he worked. He is, no doubt, best known now for his 'Cours d'Analyse,' the celebrated treatise in which, according to the admirable custom of the best French mathematicians, he embodied the substance of his professorial lectures. This work, 'Cours' though it is, is no mere text-book, but contains the substance of the main researches of Jordan's later life, and it is a rare tribute to the fundamental character of these researches that, in so general a treatise, they should seem so perfectly in place.

Jordan's 'Cours' was the first systematic treatise on analysis in which the fundamental problems of the theory of functions were envisaged from a really modern point of view, and it has accordingly played a great part in the education of most of the leading analysts of the day. It was in its second edition, of 1893–1896, that it assumed substantially its final form. The first edition had been comparatively undistinguished. "Dans la précedente édition, où nous tenions à conserver toute la simplicité possible, nous avions glissé un peu rapidement sur les premiers principes . . .", Jordan remarks himself: but now "nous les exposons avec toute la précision et la généralité que nous avons pu, dût-il en résulter quelque complication . . ." The result was a new book and the rise of a new school; for it is fair to attribute to the inspiration of Jordan the beginnings of the movement which, carried on by Hadamard, Borel and Lebesgue, has revolutionised the foundations of modern analysis.

The book, for all its masterfulness, is by no means a very easy one to read. Jordan, I imagine, was no believer in easy roads to the understanding of mathematical truth; he could not shirk a difficulty himself, and he had no intention of allowing his readers to do so. A writer of a text-book, if he be ingenious and competent enough, can often turn a dangerous mathematical obstacle, and lead his army of readers in perfect order to an easy victory, while formidable fortresses lie still unreduced behind him. Jordan invariably scorns such temporary triumphs, knowing too well the disastrous consequences of any later check. Every difficulty, as it arises, must be fairly faced and definitely conquered, and all the ground consolidated in the rear.

Jordan's ideal is, unquestionably, the right one, and it is because he followed it so consistently that his book has been, to the best of mathematical students, so signal an inspiration. To have read it and mastered it is a mathematical education in itself; and it is hardly possible to overstate the influence which it has had on those who, coming to it as I did from the elaborate futilities of 'Tripos' mathematics, have found themselves at last in presence of the real thing. But the ideal is a difficult one, and it is just possible to be too uncompromising in its pursuit. There are times when a little pedagogic ingenuity is innocuous and even useful, and Jordan is apt to push his scorn of it a little too far. He tends to neglect simplicity and symmetry of presentation, even when it might be attained quite easily and without any real surrender of the end in view. It was said of him, by his pupils at the École Normale, that "lorsque M. Jordan rencontre dans un raisonnement quatres quantités jouant enactement le même rôle, il les designe par u, A'', λ, e_3."* It is of course a humorous exaggeration, but it has a foundation of justice: "nous avions du moins senti nettement combien M. Jordan se soucie peu de certaines précautions pédagogiques vulgaires . . ."

Among the many individual investigations embodied in the book, there are two which stand out now for originality and for a fruitfulness still quite unexhausted. There is a good deal in the early chapters, the theory of "content," and so on, that has now been superseded; but we shall certainly never forget the function of bounded variation or the continuous Jordan curve. A function $f(x)$, defined in an interval (a, b), is of bounded variation if, when we divide the interval into parts in any manner by points $x_0 = a, x_1, x_2, \ldots, x_n = b$, the sum of the absolute values of the increments $f(x_{i+1}) - f(x_i)$ of the function is less than a number independent of the number or nature of the divisions. It is characteristic of such a function that it can be expressed as the difference of two steadily increasing functions, so that the concept is a generalisation of one of the simplest in mathematics. The generalisation may seem an obvious one, but it is one of the great generalisations which carry simplicity and symmetry in their train. The "monotonic" (increasing or decreasing) function has, for analytical purposes, a fatal defect, namely, that its characteristic property is not invariant for the elementary operations; the sum, for example, of two monotonic functions is not in general monotonic. Functions of bounded variation, while retaining the most essential characteristics of monotonic functions, form a group for these operations, and their introduction gives to the solutions of a large number of important problems a symmetry and unity which would otherwise be quite unattainable. The classical example is that of Dirichlet's conditions for the convergence of a Fourier's series. There is something anomalous in Dirichlet's demonstration, admirable as it is. It is not true that, if Dirichlet's conditions are satisfied by f and g individually,

---

* I quote from a review of the third edition of the 'Cours' (1915), by M. Lebesque.

they are also satisfied by $f + g$; while the *result*, if true for two functions, is obviously true also for their sum; the conditions break down and yet the result remains. It was this anomaly that Jordan set himself to remove, and from which his discovery originated.

There is another familiar example of the use of the concept of bounded variation, which shows in a striking manner the connection between different parts of Jordan's work. What is the condition that a continuous curve should have a length? A continuous curve is a set of points (x, y) defined by equations

$$x = x(t), \quad y = y(t),$$

where $x(t)$ and $y(t)$ are continuous functions of a parameter t. The precise definition is Jordan's, and a continuous curve, as so defined, still carries the name of a *Jordan curve*. The *length* of a curve is the limit of the perimeter of an inscribed polygon, and the problem is that of finding the conditions that this limit should exist. There is a complete solution: the curve is rectifiable (possesses a length) if and only if $x(t)$ and $y(t)$ are functions of bounded variation. So perfect is the solution, and so inevitably does the concept of bounded variation intervene, that it is difficult to imagine how the concept came to be invented for an entirely different purpose.

Any mention of the *Jordan curve* suggests inevitably the most famous of all the theorems with which Jordan's name is connected. What is meant by the *inside* or the *outside* of a closed curve? How can these notions be defined precisely, and the rough intuitions of geometrical "common sense" translated into accurate analytical terms? Jordan's solution of the problem is not absolutely complete, but it was he who put the question, stated the solution, and provided the essentials for the first accurate proof. The problem was no doubt suggested to him by his early "topological" researches. A simple closed continuous curve C is a set of points (x, y) in two dimensional space, defined by $x = x(t)$, $y = y(t)$, where t is a parameter which varies, say, from 0 to 2π, and $x(t)$, $y(t)$ are continuous functions such that

$$x(t_1) = x(t_2), \quad y(t_1) = y(t_2)$$

if and only if t_2 is congruent to t_1 to modulus 2π. Such a curve divides the plane into two regions, D_1 and D_2, of which C is the common boundary. If P and Q are points of the same region, then there is a continuous curve PQ which has no point in common with C; but if they lie one in D_1 and one in D_2, then any such curve must meet C once at least. This is Jordan's theorem, the most famous theorem in 'Analysis Situs,' and the source of a whole branch of modern mathematics. Jordan's proof is incomplete, since he assumes the truth of the theorem for a simple polygon, and the success of his attempt to free himself from the pre-suppositions of geometrical intuition is therefore incomplete. The gap has been filled by his successors by the construction, both of alternative demonstrations in which no such preliminary assumption is required, and of elementary inductive arguments specially applicable to the polygonal case; but the theorem remains justly and permanently associated with his name.

I have written of Jordan as an analyst, and it is no doubt as an analyst that he won the most permanent fame. But even his analysis bears continual witness, in his selection of problems and in the spirit in which he pursues them, to the interests which occupied him in his earlier life. The notion of a group, whether abstract or a group of algebraical or geometrical transformations, is never long absent from his mind, as anyone who has studied his admirable account of the elliptic functions will remember; and his early researches are almost entirely dominated by the theory of groups.

In his 'Traité des Substitutions et des Équations Algébriques,' published by Gauthier-Villars in 1870, Jordan gave the first comprehensive account of the theory of Galois ("dont tout ceci n'est qu'un commentaire") and its applications to the theory of algebraic equations, with chapters, on a less extensive scale, on geometry and elliptic functions. The treatise is occupied for the most part with groups of linear substitutions. In the last book he solves a problem put by Abel, that of finding all equations of a given degree which are soluble by radicals, and of deciding whether a given equation belongs to this class or not. But there are also very important original contributions to the general theory, in particular concerning primitive groups and factors of composition. A series of memoirs published later complete and extend his results. Only an expert in the theory could venture to pass judgment on Jordan's contributions to it, and I must be content to repeat the verdict of one of the most distinguished of his countrymen, that in his treatise he showed himself "un grand algébriste," with "une rare profondeur d'esprit et une extraordinaire puissance d'abstraction," who "se jouait au milieu des discussions les plus subtiles, se plaisant à aborder les questions dans toute leur généralité, comme s'il craignait que quelque particularité l'empêchat de voir les vraies raisons des choses." It is at any rate a verdict which no one who knows Jordan as an analyst will be inclined to dispute.

G. H. H.

GÖSTA MITTAG-LEFFLER*

G. H. HARDY.

MAGNUS GÖSTA MITTAG-LEFFLER, who died last year at the age of eighty-one, the oldest of our honorary members, was for many years the recognized leader of the Scandinavian school of mathematics.

Mittag-Leffler was a remarkable man in many ways. He was a mathematician of the front rank, whose contributions to analysis had become classical, and had played a great part in the inspiration of later research; he was a man of strong personality, fired by an intense devotion to his chosen study; and he had the persistence, the position, and the means to make his enthusiasms count. He thus won for himself a position unlike that of any other Scandinavian mathematician, and one which it was a little difficult for anyone who had not met him in his own country to realize. I can remember well the occasion when he lectured for the last time to a Scandinavian Congress, at Copenhagen in 1925, and the whole audience rose and stood as he entered the room. It was a reception rather astonishing at first to a visitor from a less ceremonious country; but it was an entirely spontaneous expression of the universal feeling that to him, more than to any other single man, the great advance in the status of Scandinavian mathematics during the last fifty years was due.

Mittag-Leffler had been overwhelmed with honours; he was a doctor, for example, of four English or Scottish universities; he had been a president or vice-president of endless congresses; and he was an honorary member of almost every scientific society in the world. He travelled continually, and there was no mathematician who was so familiar a figure abroad. But to the outside world he was, above everything, the editor of the *Acta Mathematica*†, the famous journal which he founded, and,

---

* Honorary member of the society 1892-1927.

† A fuller account of Mittag-Leffler's activities in connexion with the *Acta* will be found in Nörlund's notice of him in vol. 50. Prof. Nörlund very kindly allowed me to see the proof-sheets of this notice, and I have drawn from it freely.

with the co-operation of a committee of the four Scandinavian countries*, edited for forty-five years.

Periodicals are the most important material facts in the mathematical world. The *Acta* took its place in the front rank of mathematical periodicals from the beginning, and its rank and standard have never declined. It has always been the most completely international of all mathematical journals. All such journals are international to some extent; *Crelle's Journal* made its reputation on the works of Abel; the leading German periodicals of the present day, the *Mathematische Annalen* and the *Mathematische Zeitschrift*, are very largely international; and we know, even from our own short experience, how essential to a new journal is some sort of international connexion. The *Acta* was always international in a more thoroughgoing sense than this. It would hardly have been practicable, in 1881, to support a mathematical journal of the first rank on Scandinavian mathematics alone; and Mittag-Leffler understood this, and set himself from the beginning a standard which, without the co-operation of mathematicians of all countries, it would have been quite impossible to maintain.

Two great mathematicians in particular, Poincaré and Cantor, contributed freely to the early volumes. It is difficult to realize to-day how Cantor had to struggle for recognition. The mathematicians of his time, however, were singularly slow to understand the overwhelming importance of his work; and Mittag-Leffler, a younger man, was one of the first to recognize it, and to apply Cantor's ideas successfully in the region of "ordinary" analysis. In Mittag-Leffler's last published paper he recalls this with very justifiable pride†. But Mittag-Leffler was always a good judge of the quality of the work submitted to him for publication. Even in his later years, when most of the editorial work was delegated to others, he retained that curious sense which enables the great editor to feel the value of work at which he has hardly glanced; and it was this gift which enabled him to carry the heavy responsibility which he had assumed in a way which would hardly have been possible to any other mathematician. It was a heavy responsibility in more ways than one, for the finances of the *Acta*, like those of other journals, were imperilled by the war, and there was a period when Mittag-Leffler had to supplement the Government subventions heavily himself. Like other journals,

---

* Including Finland.

† A short note on a very interesting historical article by Schönflies ["Die Krisis in Cantor's mathematischem Schaffen", *Acta Math.*, 50 (1927), 1–23]. This article reproduces a number of Cantor's letters to Mittag-Leffler.

the *Acta* has overcome its troubles, and Mittag-Leffler could no doubt feel, when the last three volumes were dedicated to him as a *Festschrift* on his eightieth birthday, that the future of his great foundation was secure.

Mittag-Leffler's own work is occupied almost entirely with the general theory of functions analytic in Weierstrass's sense. He was an excellent writer. His account of the foundations of the theory of elliptic functions, modelled, no doubt, on Weierstrass's lectures which he had attended in Berlin two years before, published first in Helsingfors in 1876, and translated into English in 1923 in the *Annals of Mathematics*, is a really admirable piece of exposition, and makes one regret that he never wrote a treatise on a larger scale. He occupied himself on various occasions with fundamentals in the theory of functions. For example, he gave a proof of Cauchy's theorem, in 1875, which marked at the time a very definite advance. Like Goursat after him, he divides the contour into elementary contours by a network, and argues directly from the definition of the derivative in every mesh; but he assumes what amounts substantially to uniform differentiability, and so misses what is really characteristic in Goursat's proof. He also wrote a number of memoirs on linear differential equations.

His best known and most characteristic work, however, is concerned with the problem of the analytic representation of a one-valued function (or one-valued branch of a function) with assigned singularities, and falls into two halves associated with the periods of his life following on the years 1877 and 1900. It was in 1877 that Mittag-Leffler first published the classical "Mittag-Leffler's theorem" of the text-books, a theorem suggested, of course, by Weierstrass's factor-theorem, but in no sense a corollary of it. If $a_1, a_2, \ldots$ is any sequence of numbers such that $|a_n|$ tends to infinity with n, and

$$g_n\left(\frac{1}{z-a_n}\right) = \frac{A_{n,1}}{z-a_n} + \ldots + \frac{A_{n,\nu_n}}{(z-a_n)^{\nu_n}}$$

is an arbitrary polynomial associated with a_n, then there is a one-valued analytic function $f(z)$ which is regular in any part of the plane free from the points a_n and has at a_n a pole of the type g_n. Such a function may be defined by a series

$$f(z) = \Sigma \left\{ g_n\left(\frac{1}{z-a_n}\right) - h_n(z) \right\}$$

where h is an appropriate polynomial in z; and the most general $f(z)$ differs from this one by an arbitrary integral function. In a later memoir

—his first contribution to the *Acta*, in 1884—Mittag-Leffler gave various generalizations of his theorem, of which treatises such as those of Forsyth and Osgood give a full account.

In the second series of memoirs, the first of which appeared in 1900, we find the problem of analytic continuation envisaged in an entirely modern form. Mittag-Leffler considers the general problem of continuation for an element of any analytic function regular, say, at $z = a$. He introduces for the first time the idea, now well established as of fundamental importance in all such investigation, of the "star" (*Hauptstern*) associated with a; the region obtained by removing from the plane the further segments of all radii issuing from a and passing through singular points. He solves in various manners and in explicit form the problem of representing the function by an analytical expression convergent throughout the whole interior of the star. Some of these solutions are series of polynomials in z, such as

$$\lim_{n\to\infty} \sum_{m_1=0}^{n^2} \sum_{m_2=0}^{n^4} \cdots \sum_{m_n=0}^{n^{2n}} \frac{f^{(m_1+m_2+\ldots+m_n)}(a)}{m_1!\, m_2!\, \ldots m_n!} \left(\frac{z-a}{n}\right)^{m_1+m_2+\ldots+m_n};$$

while others are limits of associated functions, such as

$$\lim_{a\to 0} \sum_{n=0}^{\infty} \frac{c_n z^n}{\Gamma(1+an)},$$

where we suppose that a is zero and that $\Sigma c_n z^n$ is the Taylor's series of $f(z)$. Representations similar to the last, but involving slightly different convergence factors, such as n^{-an}, have also been used with much effect by other writers, such as Lindelöf and Le Roy. In particular, the analytic continuation of $(1-z)^{-1}$ over the plane cut along $(1, \infty)$ is given by

$$\lim_{a\to 0} E_a(z) = \lim_{a\to 0} \Sigma \frac{z^n}{\Gamma(1+an)},$$

a very interesting special function of whose properties Mittag-Leffler himself and Wiman made a detailed study, which may have suggested to the latter his well known theorem concerning integral functions of order less than $\frac{1}{2}$. A glance at Bieberbach's article in the *Enzyklopädie** is enough to show the range of later investigations which have their root in Mittag-Leffler's researches.

Mittag-Leffler had two homes between which he divided his time in Sweden. His principal residence was in Djursholm, in the suburbs of Stockholm. Here was his library, the finest mathematical library in the

---

* "Neuere Untersuchungen über Funktionen von komplexen Variabeln", *Enzykl. der Math. Wiss.*, II C 4.

world, perhaps, at any rate for a working mathematician, and now in the hands of the Mittag-Leffler Institut and the Swedish Academy of Sciences. All books and periodicals were there (or all, at any rate, that any sane man could want), and a colossal collection of "separata"; and if one got tired one could read the correspondence of all the mathematicians in the world, or enjoy the view of Stockholm from the roof. He had also a country house at Tällberg, some two hundred miles to the north, in the beautiful and rather primitive Dalarne country. It was a delightful experience to stay there, and there perhaps Mittag-Leffler appeared at his best, a most entertaining mixture of the great international mathematician and the rather naive country squire. He was a strong nationalist, in spite of his internationalism, as anyone who lived in so beautiful a country well might be; and he loved his house and his garden and his position as the landowner of the countryside. Even his tennis court, "the most northerly tennis court in the world"—that shapeless mass of disintegrating rubble, with holes in which the ball remained stationary where it pitched—he regarded with an entirely disarming pride. I am sure there can be no one who has ever stayed at Tällberg who is not saddened by the thought that he will never stay there again.

Mittag-Leffler played a very great part in the history of the mathematics of his time. He had exceptional opportunities, and exactly the blend of qualities required to take advantage of them to the full. There have been greater mathematicians during the last fifty years, but no one who has done in his way more for mathematics.

Dr. GLAISHER AND THE "MESSENGER OF MATHEMATICS".

By G. H. *Hardy*.

JAMES WHITBREAD LEE GLAISHER, who died on 7 Dec. 1928, at the age of 80, had been editor of the *Messenger of Mathematics* since the foundation of the journal in its present form in 1871. He was helped by other editors (W. Allen Whitworth, C. Taylor, W. J. Lewis, and R. Pendlebury) until 1877 (? 8), after which he had sole control. He contributed to the second number of Vol. 1 in May 1871, and his last contribution appeared in Vol. 57 in July 1927. Few mathematicians can have expected that the *Messenger* would survive Glaisher's death, and in fact it dies with him. The present volume, which he had begun, is the last; and in future the *Messenger* will be absorbed in the new Oxford series of the *Quarterly Journal*, to begin in April of this year.

Various notices of Glaisher have appeared already, the most important being those by Prof. Forsyth in the *Journal of the London Math. Soc.* (Vol. 3, pp. 101—112) and by the Master of Trinity (Sir J. J. Thomson) in the *Cambridge Review*. I do not propose to repeat what has been said about him there, but only to add a few words about him as a mathematician, and in particular as an editor.

Glaisher was, I suppose, the last of the old school of mathematical editors, the men who, like Liouville, contrived to run mathematical journals practically unaided. At any rate it is safe to say that no one will again attempt to run two simultaneously, as Glaisher did. Glaisher got a little assistance from friends, but for the most part he accepted or rejected contributions on his own responsibility. The best editors have of course a wonderful power of judging the quality of a manuscript without attempting to read it; but the task becomes steadily more difficult as mathematics grows and specialises, and a man cannot always be worrying his friends for opinions. The position of an editor, without a properly constituted body of experts behind him, grows more and more thankless, and it is hardly likely that the experiment will be repeated.

At the same time a private journal, controlled autocratically by a kindly and discriminating editor, can serve many useful purposes, and particularly that of giving early encouragement to beginners, on whom societies, with councils and referees, are sometimes rather severe. As Glaisher said to me himself when I submitted my own first paper to him, he 'never discouraged a young man from rushing into print'. It is then only natural that a very large proportion of well known

English mathematicians should have begun their career as authors in the *Messenger*. Baker, Barnes, Burnside, Elliott, and Forsyth, for example, all began there. It is a little more surprising to find that J. J. Thomson and Jeans both published their first mathematical papers in the *Messenger*, and that both of them started as experts in the theory of numbers.

The *Messenger* was of course always a 'minor' journal; it did not aim at the standards of the *Acta* or the *Annalen*, or even at that of the *Proceedings* of the London Mathematical Society. It occupied a comparatively humble position in the mathematical world, but a useful, individual, and honourable position, and we must all regret its extinction even if we accept it as inevitable.

I should like to end with a few words about Glaisher's position as a mathematician, because I think that he has generally been underestimated. He wrote a great deal, of very uneven quality, and he was 'old fashioned' in a sense which is most unusual now; but the best of his work is really good. This work is almost all arithmetical, but it belongs to a peculiar region of the theory of numbers; neither to the 'classical' theory on the one hand nor to the full-blooded 'analytic' theory on the other, but to the 'semi-analytic' theory of Kronecker, Liouville, Ramanujan, Mordell, or Bell, in which we apply to arithmetic not general principles of function theory but special properties of particular functions such as the elliptic modular functions.

The standard problem of this theory is that of the representation of numbers by sums of squares. Glaisher studied this problem in a series of elaborate papers in Vols. 36–39 of the *Quarterly Journal*, in which he considers not merely representations by any even number of squares from 2 to 18 but also 'classified' representations in which stated numbers of the squares are odd or even. Two of his most important results, concerning the representation of numbers $4k+3$ by 10 squares, and even numbers by 12, had been anticipated, as he points out himself, by Eisenstein and by Liouville respectively; but he proves many other very interesting theorems, and there are still unsolved problems about the new arithmetical functions which he introduced. This was probably Glaisher's most important work, but he did much more, and anyone who will take the trouble to work through the index of Dickson's *History* will probably be surprised at the number of striking theorems associated with Glaisher's name.

THOMAS JOHN I'ANSON BROMWICH

G. H. Hardy†.

Thomas John I'Anson Bromwich, who died, by suicide, on 24 August, 1929, was one of the most accomplished and most versatile among English mathematicians of the last fifty years. He was born in Wolverhampton on 8 February, 1875, but spent his youth in Natal, and was educated in Durban. He came to Cambridge, as a Pensioner of St John's College, in October, 1892. A brilliant career as an undergraduate ended when he was Senior Wrangler in 1895, in an exceptionally strong year which included also E. T. Whittaker and J. H. Grace. He obtained a Fellowship in 1897, but left Cambridge in 1902 to be Professor of Mathematics in Galway, returning in 1907 when appointed a permanent lecturer at St. John's. He was also a University Lecturer from 1909 to 1926. He became a Fellow of the Royal Society in 1906 and a Doctor of Science in 1909. He was for many years a most enthusiastic and energetic member of the Society, and one of the men who have done most to consolidate its present position. He became a member in 1899, served on the Council in 1907 and from 1909 to 1922, was Secretary (in succession to A. E. H. Love) from 1911 to 1919, and Vice-President in 1919 and 1920. During all these years he took a leading part in the management of the Society, and would undoubtedly, but for the breakdown of his health, have succeeded to the Presidency in due course. He married in 1901, and leaves a widow and one son.

As an undergraduate and a Fellow of St. John's, Bromwich was very popular. He had a number of interests outside his work, being an active lawn tennis player, fond of music, and an accomplished dancer. He was sensible and kindly, and always willing to take trouble to oblige his friends or pupils; and whatever he did was done in the most business-like way imaginable. No one could have seemed more sane; he was the last man whom anyone would have suspected of any mental instability, and when this developed later it was a great surprise as well as a great shock to his many friends.

---

† Bromwich's work covers so wide a field that it is hardly possible for any one person to deal with it competently. His later work in mathematical physics is discussed in Dr. Jeffreys' notice which follows this. Prof. H. W. Turnbull and Prof. A. E. H. Love have very kindly provided me with notes concerning Bromwich's early work, in algebra and in applied mathematics respectively, and what I say about these subjects is very largely based on them.

Bromwich was the author of one large treatise, *An introduction to the theory of infinite series* (Macmillan, 1908), a second edition of which appeared in 1926†; of one of the Cambridge Tracts, *Quadratic forms and their classification by means of invariant-factors* (1906), long out of print; of two very useful pamphlets, *Elementary integrals and Examples in geometrical optics* (Bowes and Bowes, 1911 and 1921); and of from seventy to eighty papers in mathematical and physical journals. It is noteworthy that over fifty of these papers had appeared by the end of 1906, and that after 1908 he published comparatively little of importance, except for the work on "normal coordinates" and the operational calculus which was his primary interest in his later years of activity. After the war he was never in normal health, but his period of greatest fertility had ended a good many years before. There is no doubt that he had for long been overworked. He was engaged in original work in several different fields; he put a great deal of energy into his college and university lectures, where his passion for working out every point in detail must have added enormously to his labours; and to all this he added a considerable amount of examining and private coaching. The best pure mathematician among the applied mathematicians of Cambridge, and the best applied mathematician among the pure mathematicians, he must have been an ideal coach for a Tripos candidate of sufficient strength, and if there could have been a don's Tripos, we would all have laid odds on him to be first; but the cost of all this was heavy both for mathematics and for Bromwich himself, and he never quite fulfilled his early promise. He would have had a happier life, and been a greater mathematician, if his mind had worked with less precision. As it was, even the best of his work is a little wanting in imagination. For mastery of technique in a wide variety of subjects, it would be difficult to find his superior, but he lacked the power of "thinking vaguely".

Bromwich plunged enthusiastically into research immediately after taking his degree. He was, in the first instance, an applied mathematician; his special subjects in Part II of the Tripos had been Hydrodynamics, Elasticity, Optics, and Electricity and Magnetism. He was, however, an applied mathematician of an extremely "analytical" type. He had no physical training, and does not (in his early work at any rate) show any very great interest in physics for its own sake. He does not seem really to care much about the physical world or the light which his

---

† Bromwich had been working at a new edition in collaboration with A. E. Jolliffe; but he was unable to finish the work, and the volume was ultimately edited by T. M. MacRobert.

analysis may throw on it, but rather to regard the world as created to illustrate his analysis. What he always wanted was the correct answer to the physical question when reduced to a problem of analysis, a convincing proof, and a logical presentation. It is only fair to add that in his later years, when applied mathematics became again the principal object of his thoughts, he showed a good deal more of the spirit of a genuine physicist than ever before.

The result of this was that Bromwich soon found himself engaged primarily in pure mathematical research. He began, no doubt, by finding a good deal of the current physics deficient in accuracy and in technique, and set himself to remedy some of the deficiencies. This made him, first an algebraist, and then an "analyst" in the technical sense. As soon as he began to study analysis seriously he found, as well he might, that the teaching of analysis in England, in Cambridge and elsewhere, was slovenly and incompetent, and he set to work with characteristic energy to improve it. This was the period of his *Infinite series*, when he was in the first place an analyst, more from force of circumstances than from any spiritual necessity. It was however with algebra that he occupied himself first, and I must begin by saying something about his work in this field, which he abandoned altogether later.

Bromwich's most important contributions to algebra are to be found in his Tract and in the papers **5-8, 10, 32, 37,** and **38**†. The most striking of these papers to the ordinary reader is certainly **37**, and the result is easily stated and stands apart from the rest of his work, though the proof depends on theorems first proved explicitly by Bromwich himself in **8, 10,** and **38.** The problem is the old problem of finding bounds for the real and imaginary parts of the roots of the characteristic equation of a linear substitution. If the substitution is

(A) $$\xi_s = \sum_r a_{sr} x_r \quad (r, s = 1, 2, \ldots, n),$$

then the equation is

$$|A - \lambda E| \equiv \begin{vmatrix} a_{11}-\lambda & a_{12} & \ldots \\ a_{21} & a_{22}-\lambda & \ldots \\ \ldots & \ldots & \ldots \end{vmatrix} = 0,$$

where $|A|$ is the matrix of A and $|E|$ is the unit matrix. The classical results are as follows. The roots are all real when A is real and symmetrical (Cauchy, 1829), or, more generally, when A is Hermitian,

---

† See the references at the end of the notice.

i.e. when a_{rs} and a_{sr} are conjugate (Christoffel, 1864). They have unit modulus when A is real and orthogonal (Brioschi, 1854). Finally, they are pure imaginaries when A is real and alternate, *i.e.* when $a_{rs} = -a_{sr}$ (Weierstrass, 1879). The general case was considered first by Bendixson (1900) and Hirsch (1902), by the former for real and by the latter for complex A. Bromwich, using a different method, completes their results and arrives at a final theorem. Suppose that $\lambda = \mu + i\nu$ is a root, and write

$$b_{rs} = \tfrac{1}{2}(a_{rs} + \bar{a}_{sr}), \quad b_{sr} = \tfrac{1}{2}(a_{sr} + \bar{a}_{rs}),$$

$$ic_{rs} = \tfrac{1}{2}(a_{rs} - \bar{a}_{sr}), \quad ic_{sr} = \tfrac{1}{2}(a_{sr} - \bar{a}_{rs}),$$

where the bar denotes the conjugate, so that $|B|$ and $|C|$ are Hermitian. Then, after Christoffel, the roots of $|B - \lambda E| = 0$ and $|C - \lambda E| = 0$ are real, say $\mu_1^*, \mu_2^*, \ldots$, and $\nu_1^*, \nu_2^*, \ldots$, respectively; and Bromwich's theorem is that *μ lies between the least and the greatest of the μ^* and ν between the least and the greatest of the ν^*.*

The content of the other algebraical papers is more difficult to characterise. They are not at all easy to read, for Bromwich shows defects as an expositor which he never wholly overcame. He is clear enough in detail, but, as Berry said very justly later, in a review of his *Infinite series*, he does not help the reader by laying emphasis at the right moment on the really fundamental idea. His first object in all this work is to expound and extend the ideas and methods of Kronecker in the theory of quadratic and bilinear forms, which he was the first to introduce to English readers. Yet one will search in vain for any quite clear and explicit statement of the fundamental difference between Kronecker's methods and those of writers before him.

The difference shows itself particularly clearly in the very simplest case, that of the reduction of a single quadratic form

$$A(x_1, x_2, \ldots, x_n) = \Sigma a_{rs} x_r x_s$$

to a sum of squares. Lagrange and Kronecker each reduce step by step, making

$$A(x_1, x_2, \ldots, x_n) = c\xi_1^2 + B(\xi_2, \xi_3, \ldots, \xi_n),$$

to take the simplest case. In Lagrange's reduction $\xi_2, \xi_3, \ldots$ are $x_2, x_3, \ldots$; the coefficients in B have changed but not the variables. In Kronecker's, $B = \Sigma a_{rs} \xi_r \xi_s$, the coefficients are the same but the variables have changed†. This divergence of procedure naturally persists and

---

† See Bôcher's *Introduction to higher algebra*, 131.

widens throughout the whole theory which Bromwich is occupied in developing.

In detail Bromwich's work is impressive, and his contributions to the theory are substantial. In **5** he considers the reduction of the linear substitution A to a "canonical" substitution†, by a method similar to that of Netto. In **6** he begins by considering the corresponding problem for the more general system

$$\sum_r b_{sr} \xi_r = \sum_r a_{sr} x_r \quad (r, s = 1, 2, \ldots, n),$$

with special reference to the "singular" case in which $|A - \lambda B| = 0$ for all λ. Then, after a dynamical application, he passes to the central problem of the theory, that of the simultaneous reduction of the two bilinear forms

$$A(x, y) = \sum a_{rs} x_r y_s, \quad B(x, y) = \sum b_{rs} x_r y_s.$$

It is familiar that a single bilinear form A of "rank" p is reducible to the form

$$A^* = X_1 Y_1 + X_2 Y_2 + \ldots + X_p Y_p,$$

by an infinity of substitutions $x = PX$, $y = QY$. The most interesting case is that in which the substitution is "congruent", *i.e.* when the form of the substitutions for X and Y is the same. The necessary and sufficient condition that A should be reducible to A^* by a congruent substitution is that A should be symmetric. An alternate form may be reduced by such a substitution to‡

$$X_1 Y_2 - X_2 Y_1 + X_3 Y_4 - X_4 Y_3 + \ldots.$$

Kronecker gave a general process for the congruent reduction of any A, which Bromwich, in **38**, works out for symmetric and alternate forms.

The next problem is that of two forms A, B, or a "Schaar" $A - \lambda B$. If $|B| \neq 0$, then $A - \lambda B$ can be reduced to

$$C(X, Y) - \lambda(X_1 Y_1 + X_2 Y_2 + \ldots + X_n Y_n),$$

where C is a sum of forms like §

$$a X_1 Y_1 + (a X_2 + X_1) Y_2 + \ldots + (a X_r + X_{r-1}) Y_r.$$

Thus, in the simple case when the roots of $|A - \lambda B| = 0$ are distinct,

---

† See Hilton's *Linear substitutions*, 29.
‡ For all these theorems, see Hilton, 69, 73, 75.
§ Hilton, 173.

there are n such forms each consisting of the first term only. The general problem has been attacked by many writers. In **6**, Bromwich gives a method of his own, again paying special attention to the singular case when $|A-\lambda B| \equiv 0$, and illustrating it on Darboux's example

$$A = a_1(x_1y_1+\ldots+x_{n-1}y_{n-1})+x_2y_1+\ldots+x_ny_{n-1},$$

$$B = x_1y_1+\ldots+x_{n-1}y_{n-1}.$$

In all this there is no question of congruent reduction, a special problem which Bromwich considers in **8** and **38**. Kronecker gave a process for the congruent reduction of two quadratic forms, and Bromwich, in **38**, applies this method to the four cases in which each form is symmetric or alternate, and also works out the reduction of a pair of Hermitian forms by conjugate substitutions.

In his Tract (and in **32**, which is practically a sketch for it) Bromwich confines himself to *quadratic* forms, again using Kronecker's methods. He does not follow Kronecker at all slavishly, and, indeed, the very friendly critic (W. F. Meyer) in the *Jahrbuch* takes him to task for his neglect of Kronecker's rational invariants. In many ways the Tract is exceptionally clear and illuminating, especially for readers who like general processes illustrated by particular applications at every stage, and it has been widely read and quoted while his more general work has been neglected. The geometrical illustrations are very full and clear; there is a complete statement on pages 46–47 of the results for two quadratics in four variables, in terms of the geometry of quadrics; and the applications in the last chapter include, besides the obvious applications to the metrical classification of quadrics (in point or line coordinates), applications to bicircular quartics, to cyclides, to quadratic line complexes, and to dynamics. This richness in detail, within the narrow limits of a Tract, may well excuse some neglect of certain sides of the theory.

It was perhaps as an algebraist that Bromwich showed the highest natural talent, but it is no doubt as an analyst, as the author of *Infinite series*, that he is most widely known. This book, and the various investigations into which he was led when writing it, must have occupied the principal energies of a good many years of his life.

It is to misjudge Bromwich to imagine him an "analyst" in the fullest sense of the word. He was a very able analyst, but an analyst made and not born. He was no *Funktionentheoretiker*. He knew the theory of series extremely well, and the theory of functions very well up to a point, but there both his knowledge and his inclination stopped short rather abruptly, and none of his work shows any interest in the most modern developments of the theory of functions either on the real or the complex side. No one will blame for that a man whose fault was to be interested in too many things rather than too few.

The book is unquestionably a very fine one. It is not merely a good and an interesting book; it has a character and a distinction which show at once that it is written by an exceptional mathematician. It is easy to criticise it, and many people have done so. It is excellent in detail; there are singularly few actual mistakes, when the richness of the content is considered, and it is easy to find a clear and convincing proof of any particular theorem; but the general plan of the book is open to obvious objections, and, considered as a reasoned and comprehensive text-book, it is not so satisfactory as Knopp's. There is a clear and temperate statement of these objections in the review by Berry which I have already quoted. The continual forward references to the elaborate appendices are, as Berry points out, very trying, confusing to a student and irritating to an expert; the arrangement suggests that Bromwich "originally intended to confine himself almost entirely to series and to use only the elements of the infinitesimal calculus, but that in the course of writing he became interested in the allied theories of definite integrals, and added his Appendix III rather as an afterthought". One might add that it is really impossible to justify the position and arrangement of Appendix I; part of it should come at the beginning, and the rest should be split up and incorporated in different chapters. It is also very difficult to defend the logical attitude adopted by Bromwich towards the elementary transcendental functions. To use the properties of logarithms from the beginning, justifying their use later by methods based on the integral calculus, may perhaps be defensible in an elementary course of analysis, but not in a systematic treatise on the theory of infinite series. "Algebraical analysis" as such has a right, in a book like this, to a more serious treatment.

It would be unfair to lay too much stress on these undeniable defects, in view of the great and striking merits of the book. It has two supreme merits, rare indeed in English text-books of its time; it is thoroughly interesting, and the detailed analysis is almost always sound and clear. Whatever one may think of the general scheme, individual sections are hardly ever obscure. Finally, among all treatises on infinite series, this book stands by itself as a work of reference and a storehouse of information; it really is "a book that no mathematician can do without".

One particular merit is that, when Bromwich has to prove a rather difficult theorem, he is so careful to separate any general difficulty of principle from the small complications due to the special functions considered. A very good example of this is his treatment of the product for $\sin x$. The difficulties of principle are disposed of by a general theorem ("Tannery's theorem" of § 49), and all that remains is to verify that the conditions are satisfied in the special case. In such proofs as those in Hobson's *Trigonometry* or Chrystal's *Algebra* the two sources of trouble are encountered simultaneously, with the result of a much heavier strain on the comprehension of a moderate student.

The best chapters have always seemed to me to be Ch. 5, on double series, and Ch. 11, on divergent and asymptotic series. In each case Bromwich was practically introducing the subject to English readers. Ch. 5 is quite short, but it contains all the essentials and is admirably clear. The examples are most instructive, and it cannot be said here, as it can in some parts of the book, that they are so numerous and so difficult that they overshadow the general theory.

Ch. 11 must be judged by a different standard. The subject matter of Ch. 5 was already almost "classical", and there would be no reason even now for any very substantial change; while Ch. 11 is essentially a "pioneer" account of a theory which has changed almost out of recognition since. It would be difficult now, for example, to justify the proportion of space allotted to Borel's method, while the account of the "Cesàro" methods is obviously inadequate. For these reasons much of this chapter has been struck out of the second edition, a decision which is perhaps correct, but which takes away a good deal from the attractiveness of the book.

The book contains the results of a great deal of original research, especially of the papers which Bromwich published in the *Proceedings* from 1903 to 1908. Large parts of these papers are incorporated in the text, the appendices, or the examples, and give the book a refreshing air of liveliness and actuality. If we contrast it, for example, with Chrystal's *Algebra*, there can be no doubt that it is incomparably a better book. There are many directions in which it has suggested problems for further research. Thus in Ch. 11 Bromwich raises for the first time the problem of the relations between Borel's and Euler's methods of summation; it was not till 1922 that Knopp and Rademacher gave a definite solution†. What Bromwich actually proved was very

---

† See Knopp, *Math. Zeitschrift*, 15 (1922), 238–242. Actually the region of Borel summability includes that of Euler summability (as is suggested by Bromwich's examples), and is the limit of the region covered by iteration of Euler's method.

little, but it is only one of many cases in which his book initiated a very interesting discussion.

I may add a few remarks about one or two of Bromwich's individual contributions to analysis. They begin with **13**, in which he makes a remark of considerable importance in the theory of double and repeated limits. The question is that of the equality of the limits

(1) $$\lim_{x \to 0} \lim_{y \to 0} f(x, y), \quad \lim_{y \to 0} \lim_{x \to 0} f(x, y);$$

let us denote the inner (simple) limits, which we assume to exist, by $g(x)$ and $h(y)$ respectively. "It is the merit of Bromwich", says W. H. Young†, "to have recognized the importance" of the equation

(2) $$\lim_{y \to 0} \lim_{x \to 0} \{f(x, y) - g(x)\} = 0$$

in this problem. It is evident that (2) is true whenever the limits (1) exist and are equal. What is less obvious is that (2) implies the existence of the limits (1) and not merely their equality when they exist. Bromwich makes a number of interesting applications of this principle.

Another paper on double limits is **14**, about the extension of Abel's continuity theorem to power series in several variables. This is a joint paper, but I am concerned here only with a remark which is Bromwich's, and which illustrates his quickness in seeing curious connections between pure and applied mathematics. The force between two equal electrified spheres of radius a, in contact and at potential V, is $V^2(\frac{1}{6} \log 2 - \frac{1}{24})$. This result was found by Kelvin; Bromwich gives a direct proof in **34**. Kelvin, however, uses the method of images, which leads to the series

$$V^2 \Sigma \Sigma (-1)^{i+j} \frac{ij}{(i+j)^2},$$

a series which is obviously not convergent as a double series, and does not even converge when summed by squares, though the *repeated* series are convergent and have the value which Kelvin finds. The method of images, in short, fails; it is not true that the force between the spheres is "the force between the two sets of images", which could only be regarded as the limit of the sum over a square.

I mention, in passing, the very interesting papers **15** and **16**. The first concerns the theory of the logarithmic potential, and reveals some

---

† *Trans. Camb. Phil. Soc.*, 21 (1910), 368.

curious contrasts with Petrini's results for the Newtonian case. In the second Bromwich studies the analogues for series of zonal harmonics of Abel's continuity theorem (a field in which it seems that there is still something to be done). Another important paper on a similar topic is **39**, in which Bromwich considers the limit

$$\lim_{x \to 0} \Sigma a_n v_n(x),$$

where $v_n(x) \to 1$ for every n, and Σa_n is a divergent series summable by Cesàro's or Hölder's means. The results which he obtains have been mostly superseded by more comprehensive theorems, but the paper played a considerable part in its time in the development of the theory of divergent series; it is here, for example, that we first find an explicit statement of the critical condition

$$\Sigma n^k \, | \, \Delta^{k+1} v_n \, | < K.$$

It seems odd now that so skilful an algebraist as Bromwich should not have been able to prove the general equivalence of the two kinds of means, but there is a great deal in the theory of series which seems easy now and seemed very difficult in 1908.

It remains to say something about Bromwich's earlier work in applied mathematics. His first published paper was **1**, on a subject suggested by Larmor. This paper has become a classic. Later Bromwich found his own subjects, though he often makes it clear that they have been suggested by the work of others. Thus **3** was suggested by a paper of Forsyth, and Bromwich's interest in Forsyth's work arose from a passage in Darboux's chapters on minimal surfaces, referred to in **2**. The last paper contains the elegant theorem that a surface defined by a tangential equation $p = p(l, m, n)$, where p is homogeneous and of degree one, is minimal when $\nabla^2 p = 0$†. This was one of Bromwich's occasional contributions to geometry. There are others in **9**, which contains a new proof and generalizations of Liouville's theorem concerning conformal space-transformations, and in **33**, a joint paper with Hudson, in which there is a very clear account of some of the more delicate points in the theory of envelopes, and a number of results concerning osculating envelopes difficult to find elsewhere. But Bromwich would never have

---

† The theorem was discovered independently by Richmond. See H. W. Richmond, *Trans. Camb. Phil. Soc.*, 18 (1899), 324–332, where the idea is applied more systematically to the theory of minimal surfaces.

pretended to be a "geometer" in any very real sense of that difficult word.

Other early papers, such as **11** and **12**, are concerned with physical applications of his algebraical work. Bromwich, when a student, worked out for himself simple methods of finding the wave surface in a medium from the general equations of electricity or electrodynamics. A note on this subject by Macdonald appeared in Vol. 32 of the *Proceedings*, and Bromwich, having acquired a wide knowledge of algebra in the interval, was in a position to construct, in **12**, a much more comprehensive theory.

Another development of his early work on the electromagnetic field is shown in **36** and **31**. Bromwich had found a new solution of the general equations as early as 1899, when there was a good deal of discussion of the theory of scattering by a spherical obstacle. This was first published partially as a Tripos question in 1910, and then much later in **36** ; its value as a means of investigation is exhibited more fully in **31**.

21 is another very interesting and valuable paper in which Bromwich returned in later years to his work as a student. He had already published one paper (**35**) on the determination of the potential of a symmetrical distribution from its values on the axis of symmetry. He returns to this problem in **21**, and finds a new solution

$$X_n = r^n P_n(\mu) \log \frac{r+z}{2a} - 2r^n \left\{ \frac{2n-1}{1 \cdot 2n} P_{n-1} - \frac{2n-3}{2(2n-1)} P_{n-2} \cdots + \frac{(-1)^{n-1}}{n(n+1)} \right\}$$

of Laplace's equation with an axis of symmetry, and applies it to determine potentials of circular discs of variable density. A more obvious form of the solution, viz.

$$X_n = a^n \frac{\partial}{\partial n} \left\{ \left(\frac{r}{a}\right)^n P_n(\mu) \right\} - 2r^n P_n(\mu) \left(1 - \frac{1}{2} + \frac{1}{3} - \cdots - \frac{1}{2n}\right),$$

was pointed out afterwards by Watson.

It is, no doubt, on Bromwich's later work, which is dealt with by Dr. Jeffreys, that our judgment of him as a mathematical physicist must primarily depend. His earlier work reveals him not as a physicist but as a very powerful and accomplished mathematician, always on the watch for any application of his analysis to physics, and ready to turn it in almost any direction with singular quickness and versatility. However, we may judge him as a physicist, or as a pure mathematician of any particular stamp, there can be no doubt about the substance and the distinction of his contributions to mathematics as a whole. We may feel that a mathematician of Bromwich's powers should have done more than he did, but what he did is a great deal more than the performance of any ordinary man, and it would be very hard to find anyone among his contemporaries who did so many things so well.

References.

The following list contains (1) all papers published by the Society and (2) all other papers referred to in the notice.

Proc. London Math. Soc. (1): **1**, 30 (1898), 98–120; **2**, 30 (1899), 276–281; **3**, 30 (1899), 281–290; **4**, 31 (1900), 4–15; **5**, 31 (1900), 289–297; **6**, 32 (1901), 79–118; **7**, 32 (1901), 158–163; **8**, 32 (1901), 321–352; **9**, 33 (1901), 185–192; **10**, 33 (1901), 197–216; **11**, 33 (1901), 325–342; **12**, 34 (1902), 307–321.

Proc. London Math. Soc. (2): **13**, 1 (1904), 176–201; **14**, 2 (1905), 161–189 (with G. H. Hardy); **15**, 3 (1905), 345–370; **16**, 4 (1907), 204–222; **17**, 6 (1908), 58–76; **18**, 6 (1908), 119–121; **19**, 6 (1908), 327–338; **20**, 7 (1909), 101–106; **21**, 12 (1913), 100–125; **22**, 13 (1914), 222–235; **23**, 14 (1915), 450–463; **24**, 15 (1916), 401–448; **25**, 25 (1926), 103–114; **26**, 28 (1928), 438–475; **27**, 30 (1930), 165–173; **28**, 31 (1930), 209–216.

Proc. London Math. Soc. (2), *Records of Proceedings at Meetings*: **29**, 18 (1920), ix–xii.

Journal London Math. Soc.: **30**, 5 (1930), 10–13.

Phil. Trans. Royal Soc. (A): **31**, 220 (1920), 175–206.

Quarterly Journal: **32**, 33 (1902), 85–112; **33**, 34 (1903), 98–116 (with R. W. H. T. Hudson).

Messenger of Math.: **34**, 35 (1906), 1–12.

Phil. Mag. (6): **35**, 2 (1901), 237–240; **36**, 38 (1919), 143–164.

Acta Mathematica: **37**, 30 (1906), 297–304.

American Journal of Math.: **38**, 23 (1901), 235–258.

Math. Annalen: **39**, 65 (1908), 350–369.

RAYMOND EDWARD ALAN CHRISTOPHER PALEY

G. H. Hardy.

Raymond Edward Alan Christopher Paley, who was killed in an avalanche near Banff, Alberta, on April 7, 1933, was generally recognized as one of the two or three best English mathematicians who have made their reputation since the War. He was born on January 8, 1907, at Weymouth, and was the son of the late Capt. R. E. Paley, R.A. He was elected to a scholarship at Eton in 1920, and to one at Trinity College, Cambridge, in 1925. He was a Wrangler in 1928, and in 1930 he was awarded a Smith's Prize and elected a Fellow of Trinity. In the autumn of 1932 he was appointed to a Rockefeller Fellowship in the Massachusetts Institute of Technology, and he was working there with Prof. Norbert Wiener until a few days before the ski-ing expedition which resulted in his death.

Paley's mathematical career began in 1929, when he was working for a Fellowship under Littlewood's direction, and it was in the autumn of that year that, as a Fellowship referee, I first became acquainted with his work. It was hardly the kind of reading one would select as a holiday task, being rough and untidy and sometimes really difficult to understand, but it had all the signs of originality and strength; and it was obvious that, even if there were no room for him that year, he must certainly be elected before long. Next year, when I read the dissertation again, it had doubled, but (though Paley was never easy reading) a good deal of the roughness had gone, and his election was as certain as any such election can be. There is something very intimidating to an older man in such youthful quickness and power, and, of all the people who frightened me when I came back to Cambridge, Paley was the man who frightened me most.

In one respect Paley was exceptionally lucky. Quite a series of well-known foreign mathematicians came to Cambridge during the next few years, and several whose main interests lay very near his own. Paley was exactly the man to profit by the opportunity, since he was astonishingly "quick in the uptake", and more gregarious in his mathematical habits than most English mathematicians. He was never too tired for a troublesome discussion; people who tired other people, he could tire himself. He was thus always ready to turn to account and improve upon the suggestions of any good mathematician whom he met. He had also the confident and cheerful temperament best suited for work in common; and the result was an extreme productivity. He wrote some twenty-five papers, many of them long and full of heavy analysis, in about four years, and collaborated with quite a number of first-rate mathematicians; and his career, short as it was, can be divided into periods, corresponding naturally to the men with whom he worked.

Paley was, of course, primarily a pupil of Littlewood, and Littlewood's influence dominates nearly all his earliest work. His papers **1**, **2**, **3**, **8**, **9**, **12**, and **14**, for example, are all concerned with problems arising from work of Littlewood's (or joint work of Littlewood's and mine to which Littlewood had introduced him). Later he was very much influenced by the Polish mathematicians, and in particular by Zygmund; while during his last period, which had only begun, the dominating influence was Wiener's. His interests naturally widened, and changed to some extent as they widened, but the earliest of them were never lost, and the only two papers in which he actually collaborated with Littlewood were published in 1931 and 1932.

When a man dies very young like Paley (and no doubt only then), it is his latest work which is the most interesting of all. There is some of this, and in particular the "Littlewood-Paley" papers, about which I am qualified to speak myself, but for most of it I shall rely on Wiener*. The collaboration of Paley and Wiener developed, in the first instance, out of that of Paley and Zygmund. Paley and Zygmund on the one hand,

---

* Prof. Wiener allows me to quote from his notice of Paley, in vol. 39 of the *Bull. Amer. Math. Soc.*, which he has supplemented by letters.

Wiener on the other, were interested in the region where "Fourier series and probability come together". Paley and Zygmund, for example, had studied the properties of "almost all" functions

$$\Sigma \pm c_n e^{ni\theta},$$

with a given (c_n) and a "random distribution" of the ambiguous signs, to which they had applied with great success an "averaging principle" due in the first instance to Littlewood; and Wiener had studied "random functions" of a more general kind. The attempt to unify the two points of view led to a triangular collaboration and a paper (**18**) which has just appeared in the *Math. Zeitschrift*. But Paley and Wiener soon found themselves engaged in collaboration with a wider range. Paley had, in **4**, done valuable work on certain special systems of orthogonal functions (the "Rademacher" and "Walsh" systems), and this led, with Wiener's collaboration, to a general theory of orthogonal functions associated with Abelian groups. It was primarily the desire to pursue these common investigations, the first results of which have appeared in **17** and **23**, that led Paley to apply for a Rockefeller Fellowship and to visit America.

In America, at the Massachusetts Institute, Paley was intensely active, both in collaboration with Wiener and independently. The results of his latest work with Wiener are in course of publication in the *Trans. Amer. Math. Soc.*, and, fragmentary as they inevitably are, show an extraordinarily varied range. The seven notes published already, under the general title "Notes on the theory and application of Fourier transforms", deal with the following topics: (1) quasi-analytic functions (a new proof of the fundamental Denjoy-Carleman theorem), (2) conjugate functions, (3) the zeros of an analytic function regular in a circle or half-plane, (4) closure of systems of exponential functions, (5) integral functions of order $\tfrac{1}{2}$, (6) closure of generalized trigonometrical systems, and (7) Volterra integral equations. Each of these notes contains at least one clear-cut and striking theorem, and there are more of them to come. One need only read these notes, and the work which Paley published independently while he was in America, to be convinced (if conviction were wanted) that Paley had not at all reached the limit of his powers; that his actual contributions to mathematics, substantial as they are, should have been only a preface to bigger achievements; and that he had in him all the stuff required for the development of a great mathematician.

A mathematician is to be judged by the theorems which he proves, and probably the best tribute to be paid to Paley is to quote half a dozen in which he had a share. And here (good as the theorems are) I am bound to do a little injustice to Paley and his collaborators, since I am forced to quote the most immediately striking rather than the deepest theorems, since some of his best work (*e.g.* in **1**) was shown in new proofs of known results, and since some of the best new theorems are too sophisticated to bear statement without lengthy preliminary explanations.

(1) If the Fourier series of a bounded function has positive coefficients, then the series is bounded. If the function is continuous (and the end values fit), then it is uniformly convergent (Paley, **11**).

(2) If $\chi(m)$ is a primitive character (mod k), then

$$\sum_{m=1}^{n} \chi(m) = \Omega(k^{\frac{1}{2}} \log \log k):$$

that is to say, there is a positive P such that

$$\left| \sum_{m=1}^{n} \chi(m) \right| > P k^{\frac{1}{2}} \log \log k,$$

for an infinity of k and suitable χ, n (Paley, **8**).

(3) If $\Sigma c_n z^n$ has unit radius of convergence, and $\phi_n(t)$ is Rademacher's orthogonal function, then

$$f(z) = \Sigma c_n \phi_n(t) z^n$$

is non-continuable for almost all t (Paley-Zygmund, **15**; a precise form of the principle "almost all power series are non-continuable").

(4) An odd function, regular and schlicht in the unit circle, has bounded coefficients (Littlewood-Paley, **10**).

(5) If f is of the class L^r, where $r > 1$, then the Fourier polynomial of f, of rank 2^n, converges almost everywhere (Littlewood-Paley, **7**).

(6) If Jacobi's continued fraction algorithm for $1 : a : \beta$ is periodic, then the characteristic cubic of the algorithm is irreducible (Paley-Ursell, **13**).

(7) The conjugate of an odd, monotonic, integrable function is integrable (Paley-Wiener, **20**).

(8) If $M(r)$ and $m(r)$ are the maximum and minimum moduli of an integral function $f(z)$, and $n(r)$ the number of zeros in $|z| < r$; and if

$$\log M(r) = O(r^{\frac{1}{2}}), \quad \int_0^\infty \frac{\log^+ m(r)}{r^{\frac{3}{2}}} dr < \infty;$$

then $n(r) \sim A r^{\frac{1}{2}}$ (Paley-Wiener, **20**).

Some of these theorems are very difficult, others comparatively easy to prove, granted the right idea, but none of them is "cheap"; no mathematician would reply, when any of the theorems was explained to him, "why, naturally, that must be so". In Paley's work there is almost always some element of surprise, and it is for that reason that no one would have ventured to put a definite limit to what he might have done.

A list of Paley's papers.

Proc. London Math. Soc. (2).
1. "A proof of a theorem on averages", 31 (1930), 289–300.
2. "On some theorems connected with Weierstrass's non-differentiable function", 31 (1930), 301–378.
3. "On the k-analogues of some theorems in the theory of the Riemann ζ-function", 32 (1931), 273–311.
4. "A remarkable series of orthogonal functions", 34 (1932); (I) 241–264, (II) 265–279.

Journal London Math. Soc.
5. "A uniqueness theorem on trigonometrical series", 5 (1930), 182–189.
6. "A proof of a theorem on bilinear forms", 6 (1931), 226–230.
7. [with J. E. Littlewood] "Theorems on Fourier series and power series", 6 (1931), 230–233.
8. "A theorem on characters", 7 (1932), 28–32.
9. "A note on power series", 7 (1932), 122–130.
10. [with J. E. Littlewood] "A proof that an odd schlicht function has bounded coefficients", 7 (1932), 167–169.
11. "On Fourier series with positive coefficients", 7 (1932), 205–208.

Proc. Camb. Phil. Soc.
12. "On the Cesàro summability of Fourier series and allied series", 26 (1930), 173–203.
13. [with H. D. Ursell] "Continued fractions in several dimensions", 26 (1930), 127–145.
14. "On the strong summability of Fourier series", 26 (1930), 429–437.
15. [with A. Zygmund] "On some series of functions"; (I) 26 (1930), 337–357, (II) 26 (1930), 458–474, (III) 28 (1932), 190–205.
16. "A note on analytic functions in the unit circle", 28 (1932), 266–272.

Quarterly Journal of Math. (Oxford).
17. "Theorems on polynomials in a Galois field", 4 (1933), 52–63.

Math. Zeitschrift.
18. [with N. Wiener and A. Zygmund] "Notes on random functions", 37 (1933), 647–668.
19. "Note on a theorem of Kolmogoroff and Menchoff", 37 (1933), 669–673.

Trans. Amer. Math. Soc.
20. [with N. Wiener] "Notes on the theory and application of Fourier transforms", 35 (1933), (I–II), 348–355, (III–VII), 761–791.
21. "A special integral function", 35 (1933), 709–715.

Bull. Amer. Math. Soc.
22. "A note on bilinear forms", 39 (1933), 259–260.

Trans. Nat. Acad. of Sciences (Washington).
23. [with N. Wiener] "Characters of Abelian groups", 19 (1933), 253–257.
24. "On lacunary power series", 19 (1933), 271–272.

Studia Math.
25. [with A. Zygmund] "On the partial sums of Fourier series", 2 (1930), 221–227.
26. "Some theorems on orthogonal functions", 3 (1931), 226–238.

ERNEST WILLIAM HOBSON

G. H. Hardy.

Ernest William Hobson, who was born at Derby on 27 October, 1856, and died rather suddenly, after a short illness, on 19 April, 1933, had been for many years one of the first of English mathematicians. Although he lived to be 76, he was active almost up to his death; his last book (and perhaps in some ways his best) was published when he was 74. He was a singular exception to the general rule that good mathematicians do their best work when they are young.

Hobson was the son of William Hobson, who was editor and part proprietor of the *Derbyshire Advertiser* and was prominent in municipal affairs. He was the eldest of a family of six, J. A. Hobson, the well-known economist, being one of his brothers. His early education was at Derby School. Derby had a mathematical master* of more than usual ability, and Hobson's mathematical talents were very soon noticed and encouraged. At 13 he had his first opportunity of distinguishing himself

---

* The Rev. H. J. Hose, formerly Scholar of Trinity College, Cambridge, and Ninth Wrangler in 1849.

in competition, and was first in all England in the old "Junior Local". It is interesting to observe that he also attained "distinction" in French, music, and natural science; such lists usually show nothing but general ability, but in Hobson's case the subjects represented interests which survived. He was a good linguist (though German was his language in later life rather than French); was definitely musical; and, as he showed in his Gifford lectures, had an exceptional all-round knowledge of science.

When he was 15 Hobson obtained a "Whitworth Scholarship" at what is now the Royal College of Science, and studied physics in London for a short time under Prof. F. G. Guthrie. Two years later he was elected a mathematical scholar of Christ's. What is perhaps more interesting, in view of his later career, is that he had been previously rejected by Trinity, and that he was "not sure that, on the evidence before them, the examiners were wrong": these were his own words, but he may have been idealising the situation a little. However, he went into residence at Christ's in October, 1874, "coached" with Routh, and was Senior Wrangler in January, 1878. But he was unsuccessful in the Smith's Prize examination which followed, and does not seem to have stood out as one would have expected among the mathematicians of his year.

A Senior Wrangler of those days succeeded almost as of right to a Fellowship, and Hobson became a Fellow of Christ's, and a lecturer in mathematics, in the autumn of the same year. His duties as a lecturer were not difficult and, like a good many of the best Cambridge mathematicians of the time, he did a good deal of private coaching, which he did not abandon until about 1903. He was very successful as a coach, though not so notorious as Routh or Webb, and he enjoyed one theatrical triumph, Miss Fawcett, the woman Senior Wrangler of 1890, being one of his pupils*.

In 1883 Hobson was made one of the first University lecturers in mathematics. But "research" meant much less for a college and even a university lecturer then than it does now, and Hobson wrote very little, and that of little importance, in his early years. His first paper (**3**) in our *Proceedings* did not appear until 1888, and before that he had written nothing but one or two notes in the *Messenger of Mathematics*. It is interesting, to those who know his later work, to notice that one of these notes (**2**) contains a highly heuristic and "unrigorous" proof of Fourier's theorem†. His Royal Society paper (**12**) on spherical

---

* J. M. Keynes was another pupil.

† Another (**1**) contains a very simple proof of "Rodrigues' theorem" and (as Prof. Watson points out to me) may mark the beginning of Hobson's interest in spherical harmonics.

harmonics, the first of the papers on which his reputation rests, was not published until 1896. This classical memoir must have occupied him for years, and several of the papers which he had published in the interval were really preliminary sketches.

In 1891 Hobson wrote his only "text-book", the well-known *Trigonometry*. This book has run through many editions, and was for long the only English book, apart from Chrystal's *Algebra*, which gave any serious account of the elements of "algebraical analysis". It is instructive to observe the development of the book in successive editions, for Hobson, when first he wrote it, was not yet a professional analyst, and his standard of accuracy was by no means that of his later years.

Hobson's development as an original mathematician seems now to have been strangely slow. By 1903, however, he had moved a very long way; he had (largely, I believe, as the result of intercourse with W. H. Young) acquired his interest in the modern theory of functions; and he had abandoned coaching in order to win leisure for research. From this time onward he changed rapidly into the Hobson whom we knew. In 1903 he became "Stokes Lecturer", a position which is now associated definitely with applied mathematics, and has been occupied, since Hobson held it, by Jeans, Fowler, Dirac, and Born; but Hobson was by then very plainly a pure mathematician. The first edition of his *Theory of functions of a real variable* appeared in 1907. In 1910, at the age of 53, he succeeded Forsyth as Sadleirian Professor, and he held this office until his retirement in 1931. He was still surprisingly vigorous, but, as well he might be, tired, and he admitted that he found retirement a great relief.

Hobson took a great part in the development of this society. He joined in 1887, served on the Council for twenty five years in all, was President during 1900–2, and very frequently Vice-President. The de Morgan medal was awarded to him in 1920, and would no doubt have been awarded to him years before, had it not been for his habitual membership of the Council. He was most regular in his attendance at meetings before the war. I can still see him, at 2.10 p.m. on the second Thursday of every month, walking rapidly along the Station Road to catch the old 2.15, his hands clasped firmly behind his back and the tail of his overcoat flying in the wind. He was always a distinguished and conspicuous figure, but wore, to an observer who hardly knew him, a rather angular and "hostile" air. It was only on closer acquaintance that one could learn his extreme generosity and kindliness of heart.

He received many honours from other quarters. He was elected to the Royal Society in 1893, served twice on the Council, and was a Royal Medallist in 1907. He went with the British Association to Winnipeg in

1909: this was the only time he visited America. In 1910, at Sheffield, he was President of Section A. He represented Cambridge at the Abel Centenary in Oslo in 1902. He was an honorary doctor of six universities, and a member of various foreign academies. But he said, and I have no doubt he meant it, that no honour paid to him pleased him more than the dinner organized in his honour by the mathematical faculty of Cambridge less than a year before his death.

The significance of Hobson, as of other mathematicians, lay in his mathematics, and it was as a mathematician that he would have wished to have been judged; but his extra-mathematical interests were, for a mathematician, unusually keen and strong, and he had always a decided view about any question of the day, whether it were abstract or practical, and whether it concerned the University or the outside world. As was natural in these circumstances, he was active in university affairs. He was Proctor twice, served on the Council of the Senate, the Press Syndicate, and the Financial Board, and was a frequent speaker in discussions. His views were roughly those to be expected of a man of his capacities and upbringing, a man of keen intellect and strong independence, brought up in a narrowly religious atmosphere against which he had rebelled, mellowed in some ways and hardened in others by advancing age. He was, for example, a strong "pro-Boer"; orthodox, though sane, throughout the war of 1914; a little difficult to move when, after the war, the time for reconciliation had come. Academically he was radical, as radicalism was understood in university politics twenty years ago, and one usually expected to find him on the *Placet* side. "Women's degrees" was an exception, but Hobson had a curious touch of anti-feminism, and in any case that question cut across all ordinary party lines. Whatever side he took, his opinion was a definite and a fixed one. If you agreed with him, so much the better; you could be absolutely certain that he would never let you down. If you differed, then, in his later years at any rate, you could only acknowledge the disagreement.

As befitted a man of his origin and training, and an intimate and long-standing friend of James Ward, Hobson was a keen philosopher. Philosophy, indeed, was his strongest abstract interest outside mathematics, as one could judge from passages of his great book. He was not a "mathematical logician", but he was attracted by fundamentals, and was the first English mathematician to see the point of the discussions of the "antinomies" and to recognise the importance of "Zermelo's axiom". It was therefore quite appropriate that he should have been one of the two or three mathematicians who have been invited to deliver Gifford lectures. These lectures were published, as *The domain of natural science*, in 1926.

I must, however, confess that I find the book, scholarly and well written as it is, rather disappointing. Hobson naturally did not profess to be an original philosopher, and his modesty in this respect protects him very effectively from any danger of crudeness or overstatement; but his modesty has its drawbacks; he is overconscious of the limitations of what he has to say, and suppresses his personality too successfully. The position which he defends, a rather extreme and rather abstract form of the "descriptive" view of natural science, is definite enough, and he is doubtless right in thinking that this view demands precise and explicit statement. But, in Gifford lectures, one looks for something more; one wants, in short, either a theistic or an anti-theistic moral; and Hobson's moral, that religion has nothing to fear from science if only it will leave science alone, may be a very reasonable one but is not at all exciting. The reader itches to know what Hobson himself really thinks about it, and that is just what he will not say. Here, and in some others of his "unofficial" discourses, he would have been much more interesting if he could have brought himself to be more egotistical. Hobson belonged to the minority who are genuinely more interested in principles and movements than in their own relations to them, and was a little too abstract and too impersonal to be egotistical, even when egotism was wanted. This touch of impersonality was illustrated very curiously at the Faculty dinner which I have already mentioned. Then, if ever, was the time when Hobson should have talked about himself, and every encouragement was given to him to do so. But Hobson, though he spoke at length and obviously with careful preparation, would speak only about mathematics and about the university, and largely about mathematics in the university in times when he was not yet born.

The position of a mathematical professor in Cambridge is very different now from what it was fifty years ago, when Cayley lectured to classes of two or three. Now, as then, a professor has little or nothing to do with undergraduate work, and the majority of mathematical undergraduates are unaware of his existence; but the development of graduate work, the organisation of a rational faculty system, the frequent visits of foreign mathematicians, and the general preoccupation with research, have changed his functions profoundly. He can lecture, if he chooses, to a good-sized class, and he can have as many pupils as he wants to; there is no lack of recognised machinery through which he can influence younger mathematicians. Hobson represented the period of transition. He was a good lecturer, of a rather formal and "old-fashioned" type, but there were few who could be described properly as his "pupils", and his influence on Cambridge mathematics was exerted mainly through his books. His lectures were indeed, in style and substance, hardly distinguishable from books*. His books also contain a systematic account of most of his own discoveries, and give an unusually complete picture of him as a mathematician.

Hobson wrote five books in all. The Gifford lectures and the *Trigonometry* I have mentioned already. *Squaring the circle*, a reprint of six lectures delivered in 1913, is a "popular" book which may be compared with Klein's *Vorträge über ausgewahlte Fragen der Elementargeometrie*. It is more solid than Klein, but is full of interesting information and most agreeably written, and makes one regret that Hobson did so little in the way of popular exposition. The two remaining books, the great treatise *The theory of functions of a real variable*, which occupied him from before 1907 to 1926, and the *Spherical and ellipsoidal harmonics*, published only in 1931, though a great deal of it was written more than thirty years before, contain the record of most of the chief work of his life.

It is the later book which, one chapter apart, gives us the picture of the earlier Hobson. The *Harmonics* certainly seems to me the most attractive book that he ever wrote. It may be less important, but it is fresher and more lively than the bigger book; it is the work of a younger man. Most of it dates (and all of it, except Chapter 7, might date) from 1896, when Hobson wrote his Royal Society memoir, and when his interests were very different from what they became later. There is comparatively little about "convergence"; there might indeed well be more. In most of the book we see, not Hobson the "Funktionentheoretiker" of 50 or 60, but Hobson the Senior Wrangler, the accomplished all-round mathematician and problem solver, with a lively interest in formulae and in physical applications. Or rather, perhaps, in what were called physical applications in Cambridge forty years ago; for physics, to Hobson, meant generalizations of Laplace's equation.

The book contains, for example, many elegant theorems in formal differentiation and integration, and it may be worth while to quote two or three. Thus

$$(1) \quad f_n\left(\frac{\partial}{\partial x_1}, \ldots, \frac{\partial}{\partial x_p}\right) F(x_1^2 + \ldots + x_p^2)$$
$$= \left\{ 2^n \frac{d^n F}{(dr^2)^n} + \frac{2^{n-2}}{1!} \frac{d^{n-1} F}{(dr^2)^{n-1}} \nabla^2 + \frac{2^{n-4}}{2!} \frac{d^{n-2} F}{(dr^2)^{n-2}} \nabla^4 + \ldots \right\} f_n(x_1, \ldots, x_p).$$

---

* Hobson did not lecture on modern pure mathematics until after my undergraduate days, and I never attended one of his courses. I can only judge from hearsay, and from the notes which he lent me of his lectures on the calculus of variations. There is a story (which no doubt contains an element of truth) that he once began a lecture with the words "This being so".

if f_n is a homogeneous polynomial, F is an arbitrary function, $r^2 = \Sigma x^2$, and ∇^2 is Laplace's operator (p. 126);

(2) $\iint f(x, y, z) Y_n(x, y, z) dS = 4\pi R^{2n+2} \dfrac{2^n n!}{(2n+1)!} \left\{ 1 + \dfrac{R^2 \nabla^2}{2(2n+3)} \right.$
$\left. + \dfrac{R^4 \nabla^4}{2 \cdot 4(2n+3)(2n+5)} + \cdots \right\} Y_n\left(\dfrac{\partial}{\partial x}, \dfrac{\partial}{\partial y}, \dfrac{\partial}{\partial z}\right) f(x, y, z),$

if Y_n is a spherical harmonic, f is an arbitrary function, and the integration is over the surface of the sphere of radius R (p. 161);

(3) the most general harmonic of degree $-n-1$ is of the form

$$\dfrac{\partial^n}{\partial z^n} \left\{ \dfrac{1}{r} f\left(\dfrac{x \pm iy}{r+z}\right) \right\}$$

(p. 165). All of these theorems were first proved by Hobson in papers (**4, 5, 6, 7**) in our *Proceedings*.

Chapters 1–4 contain the formal theory of ordinary spherical harmonics. Chapter 5 is occupied by a systematic account of the functions $P_n{}^m(\mu)$ and $Q_n{}^m(\mu)$ for unrestricted values of m, n, and μ. The general theory of these functions is Hobson's own, and the chapter is largely a restatement of the Royal Society memoir. The theory is no trivial "exercise in generalization"; the functions really are needed in all sorts of problems, and it is not in the least obvious how they should be defined. In the introduction to his memoir Hobson enumerates some of the potential problems for which the functions are required. In the usual spherical problems m and n are positive integers, and μ is real and numerically less than 1. In spheroidal problems, μ may be greater than 1. In those connected with the anchor-ring, n is half an odd integer; in those connected with two spherical bowls, n is of the form $-\tfrac{1}{2} + ip$; and there are problems in which general values of m also are required. There is thus every need for a more general theory; and it is necessary to generalize so far that it would be silly not to generalize comprehensively.

Several writers before Hobson, such as Thomson and Tait, Oldbricht, Schläfli, and particularly Heine, had attempted to supply some of the necessary generalizations, but their results had been fragmentary and in a good many cases erroneous. Thus Heine, in his standard *Kugelfunktionen*, had defined $P_n{}^m(\mu)$, for $m = 0$ and unrestricted n and μ, in a manner which Hobson shows to be radically defective and to lead to false results. Hobson, using the double-circuit integrals introduced by Pochhammer and Jordan, gave a new definition of the functions, and was thus able to work out a complete and satisfactory theory, which is reproduced in the book with the expansions necessitated by later work. A number of the intermediate generalizations arising in physical problems are discussed in special detail in later chapters.

There is one obvious criticism of this chapter, a criticism of which Hobson himself was quite aware. Barnes, in a well-known paper published in the *Quarterly Journal* in 1908, developed a rival theory, in which the functions are defined, not by double-circuit integrals with algebraical integrand, but by ordinary contour integrals with gamma-functions under the sign of integration. Thus Hobson defines $P_n{}^m(\mu)$ by

$$P_n{}^m(\mu) = \frac{e^{-n\pi i}}{4\pi \sin n\pi} \frac{\Gamma(n+m+1)}{\Gamma(n+1)} 2^{-n}(\mu^2-1)^{\frac{1}{2}n} \int (t^2-1)^n (t-\mu)^{-n-m-1} dt$$

round a certain Pochhammer circuit, whereas Barnes defines it by

$$P_n{}^m(\mu) = \frac{\sin n\pi}{\pi} \frac{(\mu^2-1)^{\frac{1}{2}n}}{2\pi i} \int \frac{\Gamma(s-n)\,\Gamma(n+1+s)\,\Gamma(-s)}{2^s\,\Gamma(1-m-s)} (x-1)^{s-m} ds$$

along a simple contour running across the plane of s from south to north. Each definition has its origin in one of the ordinary representations of $P_n(\mu)$ as a hypergeometric series; and Barnes proves that they are equivalent, and that all of Hobson's results may be developed directly from his own definition. Hobson's Pochhammer integrals are, no doubt, more elementary, but Barnes's methods are often more intuitive, and certainly lead to a great economy in formal calculation. It may be regretted that Hobson could not include some account of Barnes's theory in the book, but it was perhaps too much to ask. Only part of Barnes's paper has been published, and this, in spite of the "economy", occupies 108 pages of the *Quarterly Journal*. Hobson said himself that to complete the paper, to coordinate all Barnes's theorems with his own, and to condense the results within printable compass, was a task which, as a man of 70, he simply could not face.

After Chapter 5, Chapter 7 is the most important in the book, and is the only chapter devoted to "convergence theory". This chapter, which is all too short, is in Hobson's later manner, and contains the standard theorems on the convergence and summability of Legendre's and Laplace's series. The results concerning "classical convergence" are very largely Hobson's own. Broadly, when $(1-x^2)^{-\frac{1}{4}} f(x)$ is integrable in $(-1, 1)$, and x lies inside the interval, the convergence properties of the Legendre series of $f(x)$ are the same as those of the Fourier series of $f(\cos\theta)$. At the ends of the interval (the singular points of Legendre's equation) the parallelism fails and supplementary conditions are required. The general lines of all this were laid down in Hobson's papers **8** and **9**, and completed later by work of Haar and Young; Young indeed went further by proving similar theorems for "general" series of Legendre functions, series which are not "Legendre series" because their coefficients are not expressible as integrals of the standard type. Here, and in the account which follows of the summability of Legendre series by Cesàro means, based on the work of Fejèr, Chapman, and Gronwall, one wishes that the chapter had been more complete. Hobson confines himself, for example, to summability (C, k) with $k > \frac{1}{2}$, and a discussion of the interval $0 < k \leqslant \frac{1}{2}$ is required to link up the results in a satisfying manner with his own concerning ordinary convergence.

The *Functions of a real variable* was published in 1907, at first as a single volume. Young's *Theory of sets of points* had appeared one year before. The modern theories of measure and integration were then almost new, and Hobson and Young were the first to introduce them to English readers. The classical theory of functions of a complex variable had been introduced into Cambridge by Forsyth, but real function theory was practically unknown. To-day it is the part of pure mathematics that has been studied most intensively, and it is to Hobson and Young that the revolution is due.

The book, in its various editions, occupied Hobson for twenty years, and it was no doubt the central fact in Hobson's life, both for himself and for English mathematics. The whole theory has expanded out of recognition, and very little of the first edition survives unchanged. In particular, nearly all of Hobson's own contributions to the subject were made after 1907 and appear only in the later editions. Of these there are two, both in the theory of orthogonal series, which obviously surpass the rest.

It was in 1908 that Hobson published the first (8) of his series of papers on the representation of an arbitrary function by a series of normal orthogonal functions. In this paper he aimed at obtaining conditions for the validity of such a representation "comparable in generality with the known sufficient conditions for Fourier series". The series in question include Sturm-Liouville series, Legendre series, and Bessel-Fourier series (and also Hermite and Laguerre series, which Hobson does not consider). The theory of integral equations, as developed by Hilbert and Schmidt, had led to a certain unification in the theory of these series, but only for functions of a severely restricted type; Kneser alone had obtained, for Sturm-Liouville series, "reasonably general" conditions. Hobson makes a big advance all along the line.

Hobson based much of his work on a "general convergence theorem which is very important in itself. The theorem is a generalization of the "Riemann-Lebesgue theorem" in the theory of Fourier series. In its simplest form it asserts that, if (i) $f(x)$ is integrable in (a, b), (ii) $\phi(x, y, \lambda)$ is

bounded, and (iii)
$$\int_a^\beta \phi(x, y, \lambda)\,dx \to 0$$
when $\lambda \to \infty$, for every (α, β) included in (a, b) and every y of a set Y, then
$$\int_a^b f(x)\,\phi(x, y, \lambda)\,dx \to 0,$$
and boundedly in y. The conditions are satisfied, for example, if $\phi(x, y, \lambda)$ is $\cos\lambda(x-y)$, when the theorem reduces to the "Riemann-Lebesgue" theorem. This is an example of the most important case of the general theorem, in which ϕ is of the form $\phi(x-y, \lambda)$. In this form the theorem was found a little later, but independently, by Lebesgue, in his famous memoir "Sur les intégrales singulières". Lebesgue's work is less general in this respect, but in others he goes further. In particular, he shows that the conditions on ϕ are in a sense necessary as well as sufficient; if they are not satisfied, we can find an $f(x)$ for which the conclusion is untrue. Hobson does not consider this problem in his original paper, but he incorporates and extends Lebesgue's result in the book, and also works out, in part in generalization of Lebesgue and in part independently, a large number of variations of the theorem in which $f(x)$ belongs to some narrower class of functions and the conditions on ϕ are modified appropriately.

This paper is the first of a series, all published in the *Proceedings*, which contain what is, perhaps, Hobson's most important original work. In **8** he applied his general theorem to Sturm-Liouville and Legendre series; in the latter case the results are not quite definitive and are superseded by those of his later paper **9**. In **10** he works out the more difficult application to Bessel-Fourier series. These results are not included in his books, which in this respect do less than justice to Hobson's achievement. The "general convergence theorem" is a fine theorem, with a character rather unlike that of the classical theorems concerning passage to the limit under the integral sign; theorems depending as it does on the theory of approximation are much more familiar now than they were in 1908, and a modern reader is likely to under-estimate the originality which it shows. Still, interesting and valuable as the theorem is, one can get along without it (as Hobson himself shows in **9**). The applications of the theorem, and the very substantial advances to which it led in the theory of orthogonal series, were more of an achievement than the theorem itself.

Hobson's second outstanding contribution to the theory of orthogonal series is one of a quite different kind. The familiar "Riesz-Fischer theorem" states that, if (ϕ_n) is a normal orthogonal system, and Σc_n^2 is convergent, then the system of equations

$$\int f(x)\phi_n(x)\,dx = c_n \quad (n = 1, 2, \ldots)$$

has a solution $f(x)$ of the class L^2, so that c_n is the "Fourier constant" of such a function. It would be natural to expect the series $\Sigma c_n \phi_n$, the "Fourier series" of $f(x)$, to converge "almost everywhere", and so to provide an "effective representation" of the function. This is untrue, even with the ordinary generalizations of the notion of convergence, and the question arises: how much more must we postulate about c_n in order that the conclusion should follow? More precisely, for what k_n is it true that, if $\Sigma(c_n k_n)^2$ is convergent, the series $\Sigma c_n \phi_n$ is convergent for almost all values of x?

This problem is one of "convergence in the large"; the problems considered in **8** and the papers which followed it were problems of "point convergence". When (ϕ_n) is a *special* system, such as the trigonometrical system, we can pass from the one theory to the other; we can show that, when the condition on c_n is satisfied, the series converges at all points of a certain type, and that the set complementary to the set of these points is nul. This kind of argument is impossible with general systems, since there is then no "point convergence" theory. We have to approach the problem quite differently, and Hobson has the credit of having introduced, in **11**, the "grouping" method which is now the standard line of attack.

The "truth" is that we can take $k_n = \log n$; the convergence of $\Sigma(c_n \log n)^2$ is a sufficient condition, and this condition is the best possible of its kind. These theorems, which settle the question definitely, were proved by Rademacher and Menchoff in 1922 and 1923. Hobson did not succeed in getting so far, but he made what was probably the greatest step towards the final solution. Weyl had shown in 1908, by a difficult method due in principle to Jerosch, that we can take $k_n = n^{\frac{1}{2}}$; Hobson showed, and very much more simply, that we can take $k_n = n^\delta$ for any positive δ, and this was the first condition bordering on the ultimate truth. Plancherel, a little later, replaced n^δ by $(\log n)^{\frac{3}{2}}$. This demanded only a refinement of Hobson's method*, but it proved decidedly more difficult to take the last step to $\log n$. The final result is set out in the book in the last chapter on orthogonal functions, which occupies pp. 752–772 of the

---

* Hobson takes $k_n = n^{1/\lambda}$, where λ is a large positive integer, and considers subsequences of type s_{n^λ}, while Plancherel and later writers use subsequences of type s_{2^n}. The simplest proof of the complete theorem is that given by Kaczmarz in Vol. 1 of *Studia Mathematica*.

second volume. It is the best tribute to the vitality of the subject and the author that, after 1,500 pages, the reader should still be capable of feeling that the chapter is much too short.

Hobson's book is, if any book ever was, a "standard treatise". There are parts of the theory which are developed much further in other books. For example, Hobson does not carry the abstract theory of sets of points beyond the point where he needs it for applications, and a reader interested in this theory primarily for its own sake may find his account inadequate. And there are certainly very many books superior to Hobson's in elegance and æsthetic appeal. It has the æsthetic faults inevitable in a book written in instalments and continually revised; and Hobson, though he always wrote like a scholar, had not, as a mathematician, a very flexible or graceful style. But it was an astonishing feat to create such a monster, and if the monster sometimes treads heavily, the marvel is that it can move at all. For a researching mathematician, the book is really "indispensable"; there is no other book in the field of "real variable" which is at all so comprehensive, or contains a reasoned account of such a proportion of modern knowledge. In particular it is, up to the present—the situation will no doubt be changed by Zygmund's forthcoming book—the only book which even professes to give an adequate account of the theory of Fourier series. I think that there is no doubt that it is the most important book written by a modern English mathematician.

The modern theory of functions of a real variable was in its infancy when Hobson began his work. In England it was practically unknown, and rather derided. There may, perhaps, have been a little excuse for the people who, like Greenhill, regarded it as a monstrosity; for there was still a faint air of mystery hanging about the elements, and much of the superstructure was inelegant and more than a little tiresome. Hobson and Young were the first English mathematicians to see the significance of the new ideas, and fought what must often have been a rather disheartening fight for their recognition. Hobson lived to see real function theory the most highly developed mathematical discipline in Cambridge, a subject recognized even as "a good Tripos subject", the most popular and paying subject in "Schedule B". The most commonplace Cambridge mathematician now has forgotten the superstition that it is impossible to be "rigorous" without being dull, and that there is some mysterious terror in exact thought: now we go to the opposite extreme, and say that "rigour is of secondary importance in analysis because it can be supplied, granted the right idea, by any competent professional". All this we owe very largely to Hobson, but Hobson never quite understood how completely he had won his fight. He always retained something of the air of the protagonist of an unpopular cause; he was a little too old to understand fully that everything that he had been fighting for had been achieved.

References.

[Complete lists of Hobson's papers in the *Proceedings* will be found in the index parts published in 1900 and 1930. This list contains only papers referred to in the notice.]

Messenger of Mathematics.

1. " Proof of Rodrigues' theorem ", 9 (1880), 53–54.
2. " On Fourier's theorems ", 11 (1882), 11–14.

Proc. London Math. Soc. (1).

3. " Synthetical solutions in the conduction of heat ", 19 (1888), 279–294.
4. " Systems of spherical harmonics ", 22 (1891), 431–439.
5. " On a theorem in differentiation and its application to spherical harmonics ", 24 (1893), 55–67.
6. " On the evaluation of a certain surface integral, and its application to the expansion in series of the potential of ellipsoids ", 24 (1893), 80–96.
7. " On the most general solution of given degree of Laplace's equation ", 26 (1895), 492–494.

Proc. London Math. Soc. (2).

8. " On a general convergence theorem, and the theory of the representation of a function by series of normal functions ", 6 (1908), 349–398.
9. " On the representation of a function by a series of Legendre functions ", 7 (1909), 14–23.
10. " On the representation of a function by series of Bessel's functions ", 7 (1909), 359–388.
11. " On the convergence of series of orthogonal functions ", 12 (1913), 297–308.

Phil. Trans. Roy. Soc. (*A*).

12. " On a type of spherical harmonics of unrestricted degree, order, and argument ", 178 (1896), 443–531.

EDMUND LANDAU

G. H. Hardy *and* H. Heilbronn*.

Edmund Landau, an honorary member of this Society since 1924, was born in Berlin on 14 February 1877, and died there on 19 February 1938. He was the son of Professor Leopold Landau, a well-known gynaecologist. After passing through the "French Gymnasium" in Berlin, he entered the University of Berlin as a student of mathematics, and remained there, apart from two short intervals in Munich and Paris, until 1909. His favourite teacher was Frobenius, who lectured on algebra and the theory of numbers. Landau worked through these lectures very thoroughly, and used his notes of them throughout his life. He took his doctor's degree in 1899†, and obtained the "venia legendi", or right to give lectures, in 1901.

In 1909 Landau succeeded Minkowski as ordinary professor in Göttingen. The University of Göttingen was then in its mathematical prime; Klein and Hilbert were Landau's colleagues, and young mathematicians came to Göttingen for inspiration from every country. After the war the University recovered its position quickly, so that Landau had always ample opportunities of training able pupils, and often had a decisive influence on their careers.

In 1933 the political situation forced him to resign his chair. He retired to Berlin, but still lectured occasionally outside Germany. In 1935 he came to Cambridge as Rouse Ball Lecturer, and gave the lectures which he developed later into a Cambridge Tract‡. He continued to take an active interest in mathematics, and his last lectures were in Brussels in November 1937, only a few months before his sudden death.

Landau's first and most abiding interest was the analytic theory of numbers, and in particular the theory of the distribution of primes and prime ideals. In his "doctor-dissertation" (**1**) he gave a new proof of the identity

$$(1) \qquad \sum_{1}^{\infty} \frac{\mu(n)}{n} = 0$$

(conjectured by Euler and first proved by von Mangoldt), and this was

---

* We have to thank Dr. Alfred Brauer for information concerning the facts of Landau's career.

† His thesis was **1** in the list of papers on p. 310.

‡ G in the list of books on p. 307.

the first of a long series of papers on the zeta-function and the theory of primes.

The "prime number theorem"

$$\pi(x) \sim \frac{x}{\log x} \tag{2}$$

was first proved by Hadamard and de la Vallée-Poussin in 1896. De la Vallée-Poussin went further, and proved that

$$\pi(x) = \int_2^x \frac{dt}{\log t} + O\{xe^{-A\sqrt{(\log x)}}\} = \mathrm{li}\, x + O\{xe^{-A\sqrt{(\log x)}}\}, \tag{3}$$

for a certain positive A. The proofs of both Hadamard and de la Vallée-Poussin depended upon Hadamard's theory of integral functions, and in particular on the fact that $\zeta(s)$, apart from a simple pole at $s = 1$, is regular all over the complex plane.

In 1903 Landau found (in **2**) a new proof of the prime number theorem which does not depend upon the general theory of Hadamard. For this proof we need know only that $\zeta(s)$ can be continued "a little way over" the line $\sigma = 1$; we do not need the functional equation, the Weierstrass product, and the other machinery used in the earlier proofs. On the other hand we do not obtain quite so precise a formula as (3).

This discovery of Landau's was very important, since it permitted a decisive step in the theory of the prime ideals of an algebraic field κ. This theory depends upon the properties of the Dedekind zeta-function

$$\zeta_\kappa(s) = \Sigma \frac{1}{(N\mathfrak{a})^s}, \tag{4}$$

where $\mathfrak{a}$ runs through the integer ideals of κ (except 0), and $N\mathfrak{a}$ is the norm of $\mathfrak{a}$. It was not proved until much later (by Hecke in 1917) that $\zeta_\kappa(s)$ can be continued all over the plane, but Landau had no difficulty in showing that it has properties like those of $\zeta(s)$ used in his proof of the prime number theorem. He thus obtained the "prime ideal theorem": if $\pi_\kappa(x)$ is the number of prime ideals of κ whose norm is less than x, then

$$\pi_\kappa(x) \sim \frac{x}{\log x}. \tag{5}$$

Later*, using Hecke's discoveries, he proved the formula for $\pi_\kappa(x)$ corresponding to (3).

---

* C, § 20.

The logic of prime number theory has developed a good deal since 1903 and even since 1917, and Landau kept fully in touch with all these developments. Thus his paper **12** contains the shortest and most direct proof of the prime number theorem known today (a proof based on the ideas of Wiener). He was also intensely interested in the logical relations between different propositions in the theory. Thus he first proved (in **6** and **10**) that (1) and (2) are "equivalent", that each can be deduced from the other by "elementary" reasoning, although there is no "elementary" proof of either*. It was Landau who first enabled experts to classify the theorems of prime number theory according to their "depths".

Landau's second big discovery was in an entirely different direction. Picard's theorem states that an integral function, not a constant, assumes all values with at most one exception. Picard deduced his theorem in 1879 from the properties of the "modular function", and it was not until 1896 that Borel found the first elementary proof.

In 1904 Landau, studying Borel's proof, made a most important, and then very unexpected, extension (**3**). If a_0 and a_1 are given, $a_1 \neq 0$, and

$$f(x) = a_0 + a_1 x + \ldots$$

is regular at the origin, then there is a number $\Omega = \Omega(a_0, a_1) > 0$, depending on a_0 and a_1 only, such that $f(x)$, if regular in the circle $|x| < \Omega$, must assume one of the values 0 and 1 somewhere in the circle. It is obvious that this theorem includes Picard's theorem.

A few weeks later Schottky developed Landau's theorem further. Suppose that $a_0 \neq 0$, $a_0 \neq 1$, and $0 < \vartheta < 1$. Then there is a number $\Phi = \Phi(a_0, \vartheta)$ with the following property: if $f(x)$ is regular, and never 0 or 1, for $|x| < 1$, and $f(0) = a_0$, then

$$|f(x)| < \Phi(a_0, \vartheta)$$

for $|x| < \vartheta$. Landau's theorem is a simple corollary†.

Schottky's theorem was imperfect in one important respect, since his function Φ was unbounded near $a_0 = 0$ and $a_0 = 1$. Landau, in **13**,

---

* The proof that (2) implies (1) is given in A, § 156, but that of the converse implication is later.

† See B (ed. ii), p. 103.

removed this imperfection. Suppose that $a > 0$, $0 < \vartheta < 1$. Then there is a number $\Psi = \Psi(a, \vartheta)$ with the property: if $f(x)$ is regular, and never 0 or 1, for $|x| < 1$, and $|f(0)| \leqslant a$, then

$$|f(x)| < \Psi(a, \vartheta)$$

for $|x| < \vartheta$. In **13** he makes an important application to the theory of $\zeta(s)$ and $\zeta_\kappa(s)$.

These theorems have inspired a great amount of later work. Carathéodory, for example, found the "best" Ω in terms of the modular function. The elementary proofs have also been transformed by the discovery of "Bloch's theorem", and are developed in this way in the second edition of Landau's *Ergebnisse* (B).

This theorem and the prime ideal theorem were probably the most striking of Landau's original discoveries. We state a few more with the minimum of comment.

(1) Every large positive integer is a sum of at most 8 positive integral cubes (**9**).

This 8 is the only number which has resisted the later analytic attacks on Waring's problem.

(2) Every positive definite polynomial with rational coefficients can be represented as a sum of 8 squares of polynomials with rational coefficients. In particular, every positive definite quadratic with rational coefficients is a sum of 5 squares of rational linear functions; and 5 is the best possible number (**5**).

Mordell has since proved the corresponding theorem for quadratics with *integral* coefficients.

(3) If $f(x) \sim x$ when $x \to \infty$, and $xf'(x)$ increases with x, then $f'(x) \to 1$ (**8**, 218).

This theorem (which contains the kernel of a differencing process used by de la Vallée-Poussin and others in the analytic theory of numbers) is, perhaps, the first genuine example of a "O-Tauberian" theorem.

(4) If $f(x) = a_0 + a_1 x + a_2 x^2 + \ldots$ is regular, and $|f(x)| < 1$, for $|x| < 1$, then

$$|a_0 + a_1 + \ldots + a_n| \leqslant 1 + \left(\frac{1}{2}\right)^2 + \left(\frac{1.3}{2.4}\right)^2 + \ldots + \left(\frac{1.3\ldots 2n-1}{2.4\ldots 2n}\right)^2.$$

There is equality, for every n, with an appropriate $f(x)$ depending on n (**11**).

(5) A Dirichlet's series $\Sigma a_n e^{-\lambda_n s}$, with non-negative coefficients, has a singularity at the real point of its line of convergence (**4**).

This had been proved before for power-series by Vivanti and Pringsheim, but their method of proof cannot be extended to the general case.

(6) If $N(\sigma_0, T)$ is the number of zeros of $\zeta(s)$ in the domain $\sigma \geqslant \sigma_0 > \frac{1}{2}, |t| \leqslant T$, then

$$N(\sigma_0, T) = o(T).$$

This was proved, first with O and then as stated, by Bohr and Landau in their joint papers **14** and **15**. It is known that $N(\frac{1}{2}, T)$ is of order $T \log T$, so that most of the zeros of $\zeta(s)$ lie very near $\sigma = \frac{1}{2}$. This was the first successful attempt to show that the Riemann hypothesis is at any rate "approximately" true. Carlson proved later that $o(T)$ may be replaced by $O(T^a)$, where $0 < a < 1$, and Titchmarsh and Ingham have since improved Carlson's value of a.

Landau also published a very large number of new, shorter, and simpler proofs of known theorems. We mention only his well-known proof of Weierstrass's approximation Theorem (**7**). This depends upon the singular integral

$$\int_{-\frac{1}{2}}^{\frac{1}{2}} \{1-(u-x)^2\}^n f(u) \, du,$$

and was perhaps the first proof in which the approximating functions are "visibly" polynomials. It is reproduced, in a more general form, in Hobson's book (vol. ii, ed. 2, 459–461).

Landau wrote over 250 papers, but it is possible that he will be remembered first for his books, of which he wrote seven.

A. *Handbuch der Lehre von der Verteilung der Primzahlen* (Leipzig and Berlin, Teubner, 1909: 2 vols., 961 pp.).

B. *Darstellung und Begründung einiger neuerer Ergebnisse der Funktionentheorie* (Berlin, Springer, 1916; second edition, 1929: 122 pp.).

C. *Einführung in die elementare und analytische Theorie der algebraischen Zahlen und Ideale* (Leipzig and Berlin, Teubner, 1918; second edition, 1927: 147 pp.).

D. *Vorlesungen über Zahlentheorie* (Leipzig, Hirzel, 1927: 3 vols., 1009 pp.).

E. *Grundlagen der Analysis* (Leipzig, Akademische Verlagsgesellschaft, 1930: 134 pp.).

F. *Einführung in die Differentialrechnung und Integralrechnung* (Groningen, Noordhoff, 1934: 368 pp.).

G. *Über einige neuere Fortschritte der additiven Zahlentheorie* (Cambridge Tracts in Mathematics, No. 35, 1937: 94 pp.).

Of these books, E and F are elementary, and we say nothing about them, interesting and individual as they are. All the rest are works of first-rate importance and high distinction.

Landau was the complete master of a most individual style, which it is easy to caricature (as some of his pupils sometimes did in an amusing way*), but whose merits are rare indeed. It has two variations, the "old Landau style", best illustrated by the *Handbuch*, which sweeps on majestically without regard to space, and the "new Landau style" of his post-war days, in which there is an incessant striving for compression. Each of these styles is a model of its kind. There are no mistakes—for Landau took endless trouble, and was one of the most accurate thinkers of his day—no ambiguities, and no omissions; the reader has no skeletons to fill, but is given every detail of every proof. He may, indeed, sometimes wish that a little more had been left to his imagination, since half the truth is often easier to picture vividly than the whole of it, and the very completeness of Landau's presentation sometimes makes it difficult to grasp the "main idea". But Landau would not, or could not, think or write vaguely, and a reader has to read as precisely and conscientiously as Landau wrote. If he will do so, and if he will then compare Landau's discussion of a theorem with those of other writers, he will be astonished to find how often Landau has given him the shortest, the simplest, and in the long run the most illuminating proof.

The *Handbuch* was probably the most *important* book he wrote. In it the analytic theory of numbers is presented for the first time, not as a collection of a few beautiful scattered theorems, but as a systematic science. The book transformed the subject, hitherto the hunting ground of a few

---

* For example in a mock *Festschrift* written on the occasion of his declining an invitation to leave Göttingen for another university.

adventurous heroes, into one of the most fruitful fields of research of the last thirty years. Almost everything in it has been superseded, and that is the greatest tribute to the book.

Landau would not publish a second edition of the *Handbuch* (which must necessarily have been a new book), but preferred to incorporate the results of later researches in his *Vorlesungen*, which is no doubt his *greatest* book. This remarkable work is complete in itself; he does not assume (as he had done in the *Handbuch*) even a little knowledge of number-theory or algebra. It stretches from the very beginning to the limits of knowledge, in 1927, of the "additive", "analytic", and "geometric" theories. Thus part 6 (vol. i, pp. 235–269) carries the solution of Waring's problem to where it stood before Vinogradov's recent work. Part 12 (vol. iii, pp. 201–328) contains practically everything then known about "Fermat's last theorem", and the rest of the book is conceived on the same scale. In spite of this enormous programme, Landau never deviates an inch from his ideal of absolute completeness. For example, he never refers to his *Algebraische Zahlen*, but proves from the beginning everything he needs.

The richness of content of the book, and the power of condensation it shows, are astonishing. Thus the classical theorems about decompositions into two, three, and four squares are proved in twenty-eight pages (vol. i, pp. 97–125). And Landau can find room (vol.i, pp. 153–171) for four different evaluations of Gauss's sums.

The *Vorlesungen* is not only Landau's finest book but also, in spite of the great difficulty and complexity of some of the subject matter, the most agreeably written. The style here is the rather informal style of his lectures, which he was persuaded by his friends to leave unchanged.

The *Algebraische Zahlen* gives a short and self-contained account (pp. 1–54) of the theory of algebraic numbers and ideals, intended as an introduction to the proofs of the prime ideal theorem and its refinements which occupy the remainder of the book. He does not go so deeply into the algebraic theory as, for example, Hecke, being content with what is required for his applications.

The *Ergebnisse* is probably Landau's most *beautiful* book. It contains a collection of elegant, significant, and entertaining theorems of modern function theory: Hadamard's and Fabry's "gap theorems", Fatou's theorem, the most striking "Tauberian" theorems, Bloch's theorem, the Picard-Landau group of theorems, and the fundamental theorems concerning "schlicht" functions. It is one of the most attractive little volumes in recent mathematical literature, and the most effective answer to any one who suggests that Landau's mathematics was dull.

Finally, his last work, the Cambridge Tract originating from his Rouse Ball lecture, gives an account of Vinogradov's "Waring" and Schnirelmann's "Goldbach" theorems, and of a group of half solved "elementary" problems of additive number theory which open a new field of research for young and unprejudiced mathematicians. There is a review of this tract by Mr. Ingham in the current volume of the *Mathematical Gazette*, to which we should have little to add.

Landau was certainly one of the hardest workers of our times. His working day often began at 7 a.m. and continued, with short intervals, until midnight. He loved lecturing, more perhaps even than he realized himself; and a lecture from Landau was a very serious thing, since he expected his students to work in the spirit in which he worked himself, and would never tolerate the tiniest rough end or the slightest compromise with the truth. His enforced retirement must have been a terrible blow to him; it was quite pathetic to see his delight when he found himself again in front of a blackboard in Cambridge, and his sorrow when his opportunity came to an end.

No one was ever more passionately devoted to mathematics than Landau, and there was something rather surprisingly impersonal, in a man of such strong personality, in his devotion. Everybody prefers to do things himself, and Landau was no exception; but most of us are at bottom a little jealous of progress by others, while Landau seemed singularly free from such unworthy emotions. He would insist on his own rights, even a little pedantically, but he would insist in the same spirit and with the same rigour on the rights of others.

This was all part of his passion for order in the world of mathematics. He could not stand untidiness in his chosen territory, blunders, obscurity, or vagueness, unproved assertions or half substantiated claims. If X had proved something, it was up to X to print his proof, and until that happened the something was nothing to Landau. And the man who did his job incompetently, who spoilt Landau's world, received no mercy; that was the unpardonable sin in Landau's eyes, to make a mathematical mess where there had been order before.

Landau received many honours in his lifetime. He was a member of the Academies of Berlin, Göttingen, Halle, Leningrad, and Rome; but no honour seemed to please him quite so much as his election to honorary membership of this society, and he came specially from Germany to attend our sixtieth anniversary dinner. This was natural, since there was no country where his reputation stood quite so high as in England, and none where his work has borne more fruit.

References.

[This list contains only papers referred to in the notice.]

1. " Neuer Beweis der Gleichung $\Sigma\,\mu(n)/n = 0$ ", *Inaugural-Dissertation* (Berlin, 1899).
2. " Neuer Beweis des Primzahlsatzes und Beweis des Primidealsatzes " *Math. Annalen*, 56 (1903), 645–670.
3. " Über eine Verallgemeinerung des Picardschen Satzes ", *Berliner Sitzungsberichte* (1904), 1118–1133.
4. " Über einen Satz von Tschebyschef ", *Math. Annalen*, 61 (1905), 527–550.
5. " Über die Darstellung definiter Funktionen durch Quadrate ", *Math. Annalen*, 62 (1906), 272–285.
6. " Über den Zusammenhang einiger neuerer Sätze der analytischen Zahlentheorie ", *Wiener Sitzungsberichte*, 115, 2a (1906), 589–632.
7. " Über die Approximation einer stetigen Funktion durch eine ganze rationale Funktion " *Rend. di Palermo*, 25 (1908), 337–345.
8. " Beiträge zur analytischen Zahlentheorie ", *Rend. di Palermo*, 26 (1908), 169–302.
9. " Über eine Anwendung der Primzahltheorie auf das Waringsche Problem in der elementaren Zahlentheorie ", *Math. Annalen*, 66 (1909), 102–105.
10. " Über die Äquivalenz zweier Hauptsätze der analytischen Zahlentheorie ", *Wiener Sitzungsberichte*, 120, 2a (1911), 973–988.
11. " Abschätzung der Koeffizientensumme einer Potenzreihe ", *Archiv d. Math. u. Phys*, (3), 24 (1916), 250–260.
12. " Über den Wienerschen neuen Weg zum Primzahlsatz. ", *Berliner Sitzungsberichte* (1932), 514–521.

With Harald Bohr.

13. " Über das Verhalten von $\zeta(s)$ und $\zeta_\kappa(s)$ in der Nähe der Geraden $\sigma = 1$ ", *Göttinger Nachrichten* (1910), 303–330.
14. " Ein Satz über Dirichletsche Reihen mit Anwendungen auf die ζ-Funktion und die L-Funktionen ", *Rend. di Palermo*, 37 (1914), 269–272.
15. " Sur les zéros de la fonction $\zeta(s)$ de Riemann ", *Comptes rendus*, 158 (1914), 106–110.

WILLIAM HENRY YOUNG

G. H. HARDY†.

William Henry Young, who died at Lausanne on 7 July, 1942, at the age of 78, was one of the most profound and original of the English mathematicians of the last fifty years.

He was born in London on 20 October, 1863. His ancestors were Ipswich people, but had been bankers in the City for some generations. His early education was at the City of London School; the headmaster, Edwin A. Abbott, had been a schoolfellow of Young's father. Abbott was the author of the entertaining mathematical fantasy *Flatland*, and, though he left the actual teaching in mathematics to others, was enough of a mathematician to recognize Young's exceptional talents. Young seems indeed to have been understood much better at school than at home, and he always spoke of Abbott with gratitude and admiration.

He came up to Cambridge, as a Scholar of Peterhouse, in 1881. He came with a reputation to sustain and, if we are to judge him as an undergraduate and by the standards of the time, he hardly lived up to it. He was expected to be Senior Wrangler, but was fourth, Sheppard, Workman and Bragg being above him; and he did not send in an essay for a Smith's Prize (though the new regulations, which should have suited him exactly, had just come into force).

It is easy now to see reasons for Young's comparative failure. The whole system of mathematical education in Cambridge was deplorable. The college teaching was negligible, the professors were inaccessible, and an undergraduate's only chance of learning some mathematics was from a private coach. Young, like nearly all the best mathematicians of his time, coached with Routh, from whom he could learn a lot. But he had many other interests, and no doubt he wasted much of his time. He was a good, though unsystematic, chess player, and an enthusiastic swimmer and rower; and his greatest disappointment as an undergraduate seems to have been his failure to get a place in the college boat. He had always immense physical as well as mental energy, and remained an ardent oarsman all his life.

Whatever disappointments he may have had, Young seems to have been happy as an undergraduate: he said afterwards that "he never

---

† I have received much help in writing this notice from Mrs. Grace Chisholm Young and Dr. J. C. Burkill.

began to live until he went to Cambridge", and he formed one friendship which was important throughout his career. This was with George and Foss Westcott, the sons of Professor Westcott, afterwards Bishop of Durham. Both of the Westcotts became bishops themselves later, and their influence led Young to turn his attention to theology. His family were Baptists, but he was baptized into the Church of England, and became for a time superintendent of a Sunday school. These preoccupations help to explain his neglect of the Smith's Prize competition; he was occupied at the time in winning a College theological prize.

Young was elected a Fellow of Peterhouse in 1886. There was no dissertation or examination; it was a matter of course because of his place in the Tripos. He remained a Fellow until 1892, but was never given any permanent position either by the College or the University. It was not until he was an old man that his College, in 1939, elected him an Honorary Fellow.

The next thirteen years of Young's life were spent, almost exclusively, in teaching and examining. It was common enough then for Cambridge mathematicians to earn quite large sums by private coaching, and Young set himself resolutely to do so. Here the Westcott connection was a help, bringing him a good many pupils. He also went twice to Charterhouse as a temporary assistant master. It is difficult for anyone who knew Young only later to imagine him as a schoolmaster, but he seems to have enjoyed the experience. He did much examining, at Eton (where he awarded the "Tomline Prize" to P. H. Cowell) and at other big schools; but primarily, through all these years, he was a coach. His position became a little more official in 1888, when Girton made him a lecturer in mathematics. "Lecturing" at Girton meant, in effect, more coaching, and after this coaching absorbed practically all his time and energy. He was working from early morning till late at night, sometimes taking two classes simultaneously in adjacent rooms, and often going without lunch.

There is general agreement that Young was an excellent coach, and here again I find something surprising in the testimony to his merits. It is easy enough to imagine him a great inspiration to any first-rate pupil, but he did not have the chance of teaching more than a very few. He had the monopoly of Girton, but Girton wranglers were rare; the second whom he taught was his future wife. I should hardly have expected him to have had the patience necessary for success with less gifted pupils, but apparently it was just there that he excelled.

His preferences in mathematics also seem very surprising to anyone familiar with his later work. He had read widely, and could teach anything in reason, but astronomy was his pet subject. "Astronomy" was the mathematical astronomy required for the Tripos of those days, and a man who could make that stimulating must have been a teacher indeed. This interest lasted, and his first suggestion for "independent work", in the early days of his marriage, was one of a text-book of astronomy to be written in collaboration with his wife.

And all this time we hear not one word of research. Young was the most original of the younger Cambridge mathematicians; twenty years later he was the most prolific. Yet no one suggested to him that he might have it in him to be a great mathematician; that the years between twenty-five and forty should be the best of a mathematician's life; that he should set to work and see what he could do. The Cambridge of those days would seem a strange place to a research student transplanted into it from to-day.

I still find it difficult to visualize Young's own attitude during these early years of unproductivity. The productivity, when it did come, was so astonishing; it seems at first as if it must have been the sequel to years of preparation, by a man who had succeeded at last in finding his subject and himself. One would have supposed that anyone so original, however he might be occupied, must surely have found something significant to say, but actually the idea of research seems hardly to have occurred to Young. Mr. Cowell says that Young once told him that he "deliberately accepted ten years of drudgery", that he "fancied his knowledge of the Stock Exchange", and that he thought that he could "win his leisure" by thirty-five; but "leisure" meant freedom, comfort, reading, and travel, not a life of mathematical research. The truth seems to be that Young had really no time to think of much but his teaching; that the atmosphere of Cambridge was mathematically stifling; that no one was particularly anxious to look out for or encourage originality; and that he was too much absorbed in his routine, in his pupils and their performance, to dream of higher ambitions.

However that may be, the dreams were to come and the "drudgery" to end, and the end came quickly after Young's marriage. In 1896 he married Grace Chisholm, the second of his wrangler pupils. Mrs. Young's father was H. W. Chisholm, for many years Warden of the Standards†; and her brother, Hugh Chisholm, was editor of the *Encyclopaedia Britannica*. The family carries on this tradition of distinction, and two of Young's six children are well known to us as mathematicians. The eldest son, Frank, was killed as an airman in France in 1917.

---

† The post afterwards occupied, though with less responsibilities, by MacMahon.

The great break in Young's life came, quite suddenly, in 1897; and here perhaps I had better quote Mrs. Young's own words. "At the end of our first year together he proposed, and I eagerly agreed, to throw up lucre, go abroad, and devote ourselves to research": it seems to imply a revolution in Young's whole attitude to mathematics. But Mrs. Young had studied in Göttingen before her marriage, and knew what the air of a centre of research was like, so that possibly the revolution was a little less abrupt than it appears. At any rate, the Youngs left Cambridge for Göttingen in September. "Of course all our relations were horrified, but we succeeded in living without help, and indeed got the reputation of being well off": Young's "banker's instincts" had served him well.

Young's permanent home was abroad for the rest of his life, in Göttingen until 1908 and then in Switzerland, first in Geneva and afterwards in Lausanne; but the continuity of home life was much broken by his many activities. In 1901 he came back to Cambridge, and had rooms in Peterhouse during term time for some years, returning home for vacations. During 1902–5 he was Chief Examiner to the Central Welsh Board, and seems to have thrown himself into the work with all his usual enthusiasm. His reputation was now rising, and he became a Fellow of the Royal Society in 1907, but it was not until 1913 that he obtained any definitely academic position. He was still not properly appreciated, and I can remember that, when he was a candidate for the Sadleirian chair in 1910, no one in Cambridge seemed to take his candidature very seriously.

The next few years were his years of greatest activity. He wrote a great deal—there are forty papers of his in Vols. 6–18 of our *Proceedings* only; and in 1913 he at last became a professor. His first posts were of an "occasional", though honourable, kind. He was the first Hardinge Professor of Mathematics in Calcutta; this involved residence in India for the three winter months of the next three years. He also became Professor of the Philosophy and History of Mathematics in Liverpool, and lectured there during the summer. This was a special post created for him, and he held it until 1919, when he was appointed Professor of Pure Mathematics at Aberystwyth. Here he was as energetic as ever, but his residence abroad was sometimes a source of trouble, and disagreements about this, and about appointments to the staff, led to his resignation in 1923.

By this time Young had almost ceased his activity as an original mathematician. It had slackened for a time about 1915, but the death of his son in 1917 caused him great distress and drove him back to mathematics "as a drug". It was in this year that the Society awarded him the De Morgan Medal. He did good work after, but none quite equal to his best, and after 1923 he wrote little. He was President during 1922–4, and his last papers (including his Presidential address) were printed in the *Proceedings* in 1925. It was a little later, in 1928, that the Royal Society gave him the Sylvester Medal in recognition of a life of invincible mathematical activity.

Young was well over sixty now and regarded his career as a constructive mathematician as finished, but there were other openings for his activities. He had always been keenly interested in the international organization of mathematics; and this brings me to the one controversy where I found myself in active opposition to him. We had no quarrel and, so far as I know, lost no respect for one another: our differences concerned means rather than ends but, within their limits, they were irreconcilable. They had first appeared in 1922–24, when he was President and I was Secretary, and the Society refused to send delegates to the mathematical congress at Toronto.

In 1929 Young became President of the "International Union of Mathematicians", one of the unions formed, under the aegis of the "International Research Council", in 1919–20. It seemed an honourable position, and there is no doubt that Young thought that he could use it to do real service to the cause of international co-operation, and worked whole-heartedly to that end. I am afraid (though my judgment is no doubt biassed) that it brought him little but worry and disappointment.

The truth seems to me to be that Young, though the objects of his activity were irreproachable, was carrying them on under an impossible handicap. The Union had been so much prejudiced by its previous history that all his efforts were foredoomed to failure. It had been founded too soon, before the passions of the war had had any time to cool; it was shackled by the statutes of the I.R.C.; and these statutes had been largely inspired by men anxious to direct them towards a boycott of ex-enemy nations. These feelings (with which Young had never sympathized) gradually weakened, and the opposition was always strong; so that in 1926 the Council declared itself ready to open its ranks. But then, as might have been anticipated, the "enemy nations" showed no desire to accept favours from such a quarter.

It was in these discouraging circumstances that Young succeeded to the Presidency of the Union. He did his best, but the case was hopeless; the majority of mathematicians had made up their minds to scrap all this machinery and to start again. The Zürich Congress of 1932 broke away from the Union, and that was effectively its end.

I think that Young himself ended by feeling that he was well quit of a thankless job. He turned back to his other old interests: law (he had been a member of Lincoln's Inn from early days), finance, and above all languages. These included Jugoslav and Polish, two of the most intricate of Central Europe.

The end of Young's life was rather tragic, since he was cut off from his family completely by the war. Mrs. Young had left him, as they meant for a few days only, and the collapse of France prevented her return; and his children were settled in London, South Africa, and Paris. He had always been the centre of a family, and found himself imprisoned in Switzerland and practically alone. Little had been heard from him, but it is known that he died quite suddenly, that he "just went out", and that the University of Geneva, of which he was an honorary doctor, did him every honour.

It is not particularly difficult to estimate Young's rank as a mathematician. There is no mystery about him; his work, three books and over 200 papers, is entirely characteristic; one may get a little lost in it at times, but both appreciation and criticism are straightforward tasks. Two features of it stand out on almost every page, intense energy and a profusion of original ideas. Indeed it is obvious to any reader that Young has a superabundance of ideas, far too many for any one man to work out exhaustively. One feels that he should have been a professor at Göttingen or Princeton, surrounded by research pupils eager to explore every bypath to the end. It may be that he had hardly the temperament or the patience to lead a school in this way, but he never had a chance to try.

His style is better in his books than in his papers, which are sometimes rather rambling and diffuse—faults natural in the writing of a man with many ideas, anxious to press on in a field which is developing rapidly and where there are many rivals. He makes astonishingly few mistakes, and the critical passage will almost always be found to be accurate and clear; but his repetitions are sometimes rather trying to a reader anxious to dig out the kernel of what he has to say. A theorem will be proved, in varying degrees of generality, in half a dozen different papers, with continual cross-references, and promises of further developments not always fulfilled. It is not surprising that a good many of Young's theorems should have been missed and rediscovered. At his best, however, he can be as sharp and concise as any reader could desire; and he (or he and his wife together) could write an excellent historical and critical résumé, with just the right spice of originality†.

---

† For example, **19**.

There is one particular compliment which I find it easy to pay to Young. His work stands up stoutly to critical examination. There are men who seem to me admirable mathematicians so long as they write about geometry or physics—it is easy to be impressed by what one does not understand very well. Young's best work seems to me to be his work on the subjects which I myself know best, on the theory of Fourier and other orthogonal series, on the differential calculus, and on certain parts of the theory of integration.

I will say something first about the last of these subjects, not because his work here (except that on the "Stieltjes integral" in **20**) seems to me his very best, but because it is the most widely known, and because it was the occasion of a disappointment which, coming as it did right at the beginning of his active career, might easily have broken the spirit of a weaker man.

The theory of functions of a real variable has been written afresh during the last forty or fifty years. In particular, the foundations of the integral calculus have been entirely remodelled; and it is acknowledged by everyone that, among those who have reconstructed them, Lebesgue stands first. The "Lebesgue integral" opens the blocked passages and smooths the jagged edges which disfigured the older theories, and gives the integral calculus the aesthetic outlines of the best "classical" mathematics. In particular it brings integration and differentiation into harmony with one another. It is Lebesgue's theorems about integrals and derivatives, the core of any modern treatment of the subject, which are his greatest achievement.

Young, working independently, arrived at a definition of the integral different in form from, but essentially equivalent to, Lebesgue's. He had not made Lebesgue's applications: the great theorems about integrals and derivatives are Lebesgue's and his alone. But naturally Young's integral, being equivalent to Lebesgue's, "has them in it". If Lebesgue had never lived, but the mathematical world had been presented with Young's definition, it would have found Lebesgue's theorems before long. In the definition itself Young was anticipated by about two years, and it must have been a heavy blow to a man who was just beginning to find himself as a mathematician; but he recognized the anticipation magnanimously, and set himself whole-heartedly to work at the further development of the theory. The phrase "the Lebesgue integral" is Young's.

It may seem a paradox, but it is possible that Young's work on integration, fine as it was, actually impeded his recognition. These subjects were not popular, even in France, with conservatively minded mathematicians. In England they were regarded almost as a morbid growth in mathematics, and it was convenient for men out of sympathy with Young's interests, and perhaps a little jealous of his growing reputation, to dismiss him as "the man who was anticipated by Lebesgue". It is easy enough now to recognize the absurdity of such a view: if Young had never given his definition of an integral, his reputation would not be very materially affected.

Most of this work is set out in **4, 5,** and **6**; it is in **6** that the "Young integral" is actually defined. All these papers were written in ignorance of Lebesgue's work, and recast when Young discovered it. This spoils their continuity a little, and it is perhaps a pity that he did not leave them as they stood and add the acknowledgments necessary in appendices; the genesis and progress of his own ideas would then have stood out more clearly. It is plain that Young, in his first essays at a theory of integration, was rather hypnotized by the familiar Darboux sums. He wished, at first, to *preserve* their properties, and came only gradually to see that the essential thing is to get away from them.

I confine myself to the ordinary integral over an interval, leaving aside integrals over sets and extensions to many dimensions. Then the essential results of **5** are (*a*) that the Darboux upper integral is unchanged by replacing $f(x)$ by its "upper function"; and (*b*) that if f is "upper semi-continuous", when the set in which $f \geqslant k$ is closed and has (in the old sense) a content $I(k)$, then $I(k)$ is (again in the old sense) integrable, and

$$(1) \qquad \overline{\int} f\,dx = SK + \int_{K}^{K'} I\,dk,$$

where S is the length of the interval and K and K' any lower and upper bounds for f. There are similar results, of course, for the lower integral. Here already we have the germs of the famous sums of Lebesgue.

In **6** he treats the whole subject more systematically. Starting from the point of view of Riemann and Darboux, he considers the effect of the most obvious changes which might be made in their definitions. In § 6 he suggests a "tentative" definition (which is substantially that which he ultimately adopted) and shows that its results are inconsistent with those of Darboux. He therefore dismisses it temporarily, and considers other definitions which secure agreement but seem otherwise unsatisfactory. Here he repeats a good deal of the analysis of **5**.

It is only in § 24 that he takes the decisive step. "We now return to the tentative definitions of § 6, which we saw did not agree with the usual definitions. On the other hand, the definitions we have since constructed seem more artificial than these. It suggests itself, therefore, that the most logical plan is to throw overboard the Riemann and Darboux definitions altogether and to define an integral as follows...". Divide (a, b) in any way into measurable sets e_ν, and let F_ν and f_ν be the upper and lower bounds of f in e_ν. Then (using the same symbol for a set and its measure) the upper and lower integrals of f are respectively the lower bound of $\Sigma F_\nu e_\nu$ and the upper bound of $\Sigma f_\nu e_\nu$. If f is measurable, in Lebesgue's sense, then the upper and lower integrals are each equal to the Lebesgue integral, and can be expressed in the form (1). The definitions presuppose Young's definition of measure, which had been worked out in 4 and which is also equivalent to Lebesgue's. In one way they are more general than Lebesgue's, since they do not presuppose the measurability of f, and Young's procedure seems at first a good deal more straightforward. He does not change a great deal in the older definitions; there is no such inversion of the roles of x and y as Lebesgue makes so dramatically.

In later papers, 9, 18 and 20, Young developed the whole theory differently, by the "method of monotone sequences". In this we begin by defining the integral for some particularly simple class of functions, such as continuous functions or step-functions, and agree that monotone sequences are integrable term-by-term (this convention supplying the definition of the integrals of their limits). This procedure† is particularly well adapted to *proofs* of theorems in the integral calculus. But I mention it here less for this reason than because it led Young to one of his admitted triumphs.

There is another important generalization of the Riemann integral, the integral first defined by Stieltjes in 1894. The Stieltjes integral covers sums as well as ordinary integrals, and has come rapidly into vogue since about 1909, primarily because of its outstanding importance in the theory of "linear functionals". In Stieltjes integration we integrate one function f with respect to another function g: the classical case is that in which f is continuous and g monotone. It was inevitable, after Lebesgue's work, that mathematicians should try to combine the two generalizations, and define the integral of any Lebesgue-integrable f with respect to any monotone g. In particular, Lebesgue had tried, but his results were not altogether satisfactory. Young solved the problem with complete success: he showed that his method of monotone sequence

---

† Of which there is a very clear account in L. C. Young's Cambridge Tract *The theory of integration*.

could be applied to this more general problem with little more than verbal changes.

The best tribute to Young's work that I can quote is that of Lebesgue himself†. Referring to his own attempt, he says "En réalité, je n'avais que très imparfaitement compris ce rôle [that of monotone sequences], sans quoi je n'aurais pas écrit . . . qu'il serait très difficile d'étendre la notion d'intégrale de Stieltjes par un procédé différent de celui que j'employais. Peu de temps après que j'eus commis cette imprudence, M. W. H. Young montrait que mon procédé était loin d'être indispensable, et que l'intégrale de Stieltjes se définit exactement comme l'intégrale ordinaire par le procédé des suites monotones. . . . Ce travail de M. Young est le premier de ceux qui ont finalement bien fait *comprendre* ce que c'est qu'une intégrale de Stieltjes . . . ".

Young solves the problem by two, slightly different, methods. In the first (which is rather more fundamental) he starts with "simple u- or l-functions": a simple u-function, for example, being a step function whose value at a jump is not less than its neighbouring values. If f is such a function, and g is increasing, then he defines $\int f dg$ by

$$\int_a^b f dg = \sum_{i=0}^{n} f(x_i)\{g(x_i+0)-g(x_i-0)\} + \sum_{i=0}^{n-1} f(x_i+0)\{g(x_{i+1}-0)-g(x_i+0)\},$$

where the x_i are the points of discontinuity‡. The definition is then generalized progressively, the final results being

$$\overline{\int} f dg = \underline{\text{bound}} \int f^* dg, \quad \underline{\int} f dg = \overline{\text{bound}} \int f_* dg,$$

where $f^*(f_*)$ is a lower (upper) semi-continuous function greater (less) than f. The Stieltjes integral $\int f dg$ is the common value of $\overline{\int} f dg$ and $\underline{\int} f dg$, when these two agree.

It is now a commonplace that any definition of the integral with respect to x may be generalized to cover integration with respect to g: we have only to replace the measure of a set e by the variation of g over e. If we do this either to Lebesgue's definition, or to that adopted originally by Young, we obtain a definition of $\int f dg$ which agrees with that just quoted.

Young's work on integration, which reaches its peak in the paper on the Stieltjes integral, was preceded and accompanied by a whole flood of

---

† *Leçons sur l'intégration*, ed. 2, p. 263.
‡ And $g(a-0) = g(a)$, $g(b+0) = g(b)$, by definition.

papers on the theory of sets of points and its application to the general theory of functions. A considerable part of the contents of these papers is incorporated in the Youngs' book 2, published in 1906 and, unfortunately, never revised and reprinted. But I must pass on to the other two fields in which Young seems to me to show his powers at their highest. These are the theory of Fourier series (and other special orthogonal series), and the elementary differential calculus of functions of several variables.

(1) There is a long series of papers on Fourier series in Vols. 9–18 of our *Proceedings* and Vols. 85–95 of the *Proceedings of the Royal Society*. Of his many contributions to this theory, I suppose that most of us would put first the theorems now known, in their final form, as the "Young-Hausdorff" theorems. I state them as they appear, for example, in Zygmund's book.

Suppose that $1 < p \leqslant 2$ and that p' is the "conjugate index $p/(p-1)$ (so that $p' \geqslant 2$); and that J_r and S_r are defined, for any $r > 1$, by

$$J_r(f) = \left(\frac{1}{2\pi}\int_{-\pi}^{\pi}|f|^r\,d\theta\right)^{1/r}, \quad S_r(c) = \left(\sum_{-\infty}^{\infty}|c_n|^r\right)^{1/r}.$$

If J_p is finite, and the c_n are the (complex) Fourier constants of f, then $S_{p'} \leqslant J_p$. If S_p is finite, then there is an f of which the c_n are the Fourier constants, and $J_{p'} \leqslant S_p$. Both theorems become false if $p > 2$. If $p = 2$ (and so $p' = 2$), then they reduce to the famous "Parseval" and "Riesz-Fischer" theorems.

Young (14, 15, 17) put the problem and found the solution, but it was not until eleven years later that Hausdorff completed the proof. Young proved the theorem only for the infinite sequence of special values of p, viz. 2, $\frac{4}{3}$, $\frac{6}{5}$, $\frac{8}{7}$, ..., for which p' is an even integer. It was a fine theorem even then, and the first of the 'p, p'' theorems of which the subject is now so full.

Young's proof depends upon an inequality which has other important applications. If $p > 1$ and $1 < q < p'$ (so that $pq < p+q$), x_n and y_n are positive, and

$$z_n = \sum_{s+t=n} x_s y_t$$

is the "Faltung" or "resultant" of x_n and y_n, then

$$\sum z_n^{pq/(p+q-pq)} \leqslant (\sum x_s^p)^{q/(p+q-pq)} (\sum y_t^q)^{p/(p+q-pq)}.$$

More elegantly, if $\lambda > 0$, $\mu > 0$ and $\lambda + \mu < 1$, then

(2) $$S_{1/(1-\lambda-\mu)}(z) \leqslant S_{1/(1-\lambda)}(x)\,S_{1/(1-\mu)}(y).$$

There are naturally many extensions and "integral analogues".

The simplest cases of the theorems are those in which $p = \frac{4}{3}$, $p' = 4$. Suppose, for example, that $\Sigma |c_n|^{\frac{4}{3}} < \infty$, and that C_n is the Faltung of c_n with itself. Then C_n is the Fourier constant of f^2. But, taking $\lambda = \frac{1}{4}$ in (2), we obtain $\Sigma |C_n|^2 \leqslant (\Sigma |c_n|^{\frac{4}{3}})^3$. Hence, by the Riesz-Fischer theorem, f^2 is L^2, i.e. f is L^4, and

$$\frac{1}{2\pi}\int_{-\pi}^{\pi} |f|^4 d\theta = \Sigma |C_n|^2 \leqslant (\Sigma |c_n|^{\frac{4}{3}})^3,$$

which is $J_4(f) \leqslant S_{\frac{4}{3}}(c)$. The inequalities when $p = \frac{6}{5}, \frac{8}{7}, \ldots$ may be deduced similarly from the appropriate generalizations of (2). But the hypothesis that p' is an even integer is essential for the success of this type of argument, and Hausdorff had to add a quite different technique.

There is another, and simpler, inequality of Young's which has also proved very fruitful in the theory of Fourier series. Young observed in **10** (as no doubt many others had done before him) that

$$xy \leqslant \frac{x^p}{p} + \frac{y^{p'}}{p'}:$$

a very useful inequality which leads, for example, to a simple proof of Hölder's. Later **(14)** he noticed that (4) is a special case of a simple but important general theorem: if $\phi(0) = 0$, $\phi(x)$ increases with x, ψ is the inverse of ϕ, and a and b are positive, then

$$ab \leqslant \int_0^a \phi(x)\,dx + \int_0^b \psi(x)\,dx,$$

with equality if, and only if, $b = \phi(a)$. Simple as it is, it is very important, since it enables us to set up a classification of integrable functions into pairs of "conjugate classes", of which the pair of "Lebesgue classes" L^p and $L^{p'}$ is the most fundamental.

(2) It will be seen from these examples that Young's work can be as simple, as terse, and as "snappy" as any mathematician could desire: it is quite absurd to picture him as a man habitually absorbed in a cloud of abstractions. I will give one more example: this is a theorem which seems to me perhaps the most *beautiful* that he ever proved.

Trigonometrical series of the special types $\Sigma \lambda_n \cos n\theta$ and $\Sigma \lambda_n \sin n\theta$, where λ_n is *positive and decreasing*, have many striking properties which illuminate the general theory. It was observed, for example, by Fatou

that the series
$$\frac{\sin 2\theta}{\log 2} + \frac{\sin 3\theta}{\log 3} + \cdots,$$
although it converges for every θ, *is not a Fourier series*: its sum is in fact not integrable. Young (**16**) showed that the corresponding cosine series (which looks at first the less promising, since it diverges for $\theta = 0$) *is a Fourier series*; more generally, $\Sigma \lambda_n \cos n\theta$ is a Fourier series whenever λ_n is positive, decreases to 0, and is *convex*. The kernel of the proof lies in the simple observation that

$$f(\theta) = \tfrac{1}{2}\lambda_0 + \overset{\infty}{\underset{1}{\Sigma}} \lambda_n \cos n\theta = \tfrac{1}{2} \overset{\infty}{\underset{0}{\Sigma}} (\lambda_n - 2\lambda_{n+1} + \lambda_{n+2}) \frac{\sin^2 \tfrac{1}{2}(n+1)\theta}{\sin^2 \tfrac{1}{2}\theta},$$

by two partial summations, and so that $f(\theta)$ is positive for all θ. The proof may be completed in a number of ways, of which that chosen by Young is not quite the most concise.

(3) I have only space to mention a few more of Young's many contributions to the theory of Fourier series, most of which can be found somewhere in Zygmund.

(i) First, there is "Young's convergence test" (**12, 21, 22**): of this there are many generalizations, and I state it only in its simplest and most significant form. If, in the usual notation, $\phi(t) = \tfrac{1}{2}\{f(\theta+t)+f(\theta-t)\}$, $\phi(t) \to \phi(+0)$ when $t \to 0$, and the variation of $u\phi(u)$ in $(0, t)$ is $O(t)$, then the Fourier series of $f(t)$, for $t = \theta$, converges to $\phi(+0)$. This is an important extension of the famous test of Jordan, and includes one particularly interesting special case, in which ϕ is continuous and $\phi'(t) = O(t^{-1})$ for small t.

(ii) Young was the first to attack seriously the convergence problems presented by the series $\Sigma (b_n \cos n\theta - a_n \sin n\theta)$ "conjugate" to the Fourier series of f. The "natural" sum for this series is the "conjugate function"

(3) $$\tilde{f}(\theta) = \frac{1}{2\pi} \int_0^\pi \{f(\theta+t) - f(\theta-t)\} \cot \tfrac{1}{2}t \, dt;$$

and all that was known before Young was that it converges to $\tilde{f}(\theta)$ when the integral (3) is absolutely convergent: this test corresponds to "Dini's test" for the Fourier series. Young (**11, 12, 22**) found tests for convergence corresponding to Jordan's, de la Vallée-Poussin's, and his own, and tests for summability corresponding to Féjer's and Lebesgue's. Suppose, for example, to take the simplest case, that f is of bounded variation; then a necessary and sufficient condition for the convergence of the conjugate series is the existence of $\tilde{f}(\theta)$ as a "Cauchy integral", an elementary generalized integral down to 0. This is the *natural* condition.

(4) I will say something next about Young's work on other orthogonal series. The Legendre series of a function $f(x)$, integrable in $(-1, 1)$, is the series $\Sigma a_n P_n(x)$, where $P_n(x)$ is Legendre's polynomial and

$$(4) \qquad a_n = (n+\tfrac{1}{2}) \int_{-1}^{1} f(x) P_n(x) \, dx.$$

Since $\qquad P_n(\cos\theta) \sim \left(\dfrac{2}{n\pi \sin\vartheta}\right)^{\frac{1}{2}} \cos\{(n+\tfrac{1}{2})\theta - \tfrac{1}{4}\pi\},$

for a fixed θ in $(0, \pi)$, it is to be expected that the convergence theory of Legendre series will, in broad outline, resemble that of Fourier series; but there are important differences, due at bottom to the fact that the points ± 1 are singular points of Legendre's equation. In particular it is not true that the general term of the series necessarily tends to 0, or (what is the same thing) that

$$(5) \qquad a_n = o(n^{\frac{1}{2}}).$$

This is naturally a *necessary* condition for the convergence of the series.

The first "reasonably satisfactory" tests for the convergence of a Legendre series were found by Hobson in 1909. A sufficient condition for (5) is that

$$(6) \qquad \int_{-1}^{1} (1-x^2)^{-\frac{1}{4}} |f(x)| \, dx < \infty;$$

and Hobson showed that, if this condition is satisfied, that the series is convergent for any x, interior to $(-1, 1)$, near which f is of bounded variation. At the ends of the interval, where the analogy with Fourier series fails, much more stringent conditions are required.

Hobson's theorem was an important advance, but it is not a theorem of the ideal type. Like any other test for the convergence of a Legendre series, it contains two clauses; a "broad" condition, here (6), involving all the values of f; and a "local" condition involving only those assumed near the particular x in question. For Fourier series, the "broad" condition disappears, since it is simply that f is integrable, and this is implied in the definition of a Fourier series. For Legendre series, some "broad" condition is essential, but the "local" condition should consist simply of a reduction of the problem: the Legendre series converges if, and only if, the corresponding Fourier series converges. The first theorem of this kind was proved in 1918 by Haar, his "broad" condition being that f^2

is integrable; this is less general than Hobson's condition (6), though Haar's theorem is better in other ways.

Young attacked the problem in **23**, and may be said to have solved it definitively. His fundamental idea is that of changing the "broad" condition into a condition *on the coefficients* a_n. The natural condition is plainly (5), and Young shows that, if this condition is satisfied, then *a necessary and sufficient condition for the convergence of the Legendre series, at any internal point x of* $(-1, 1)$, *is the convergence of the Fourier series of* f (or of any other function which agrees with f near x). The theorem includes all known previously and, since (5) is in any case *necessary* for convergence, it is difficult to suggest any improvement.

In **24** Young treats the same problem for "Bessel-Fourier" series: the technique is more difficult, but the essential ideas are the same, and the final theorem is equally satisfactory.

(5) It was probably in the theory of Fourier series (and these connected series) that Young did his finest work, but it is arguable that his most *important* work was in a much more elementary field.

I cannot introduce the subject better than in Young's own words in **7**. "As soon as we leave the domain of a single variable in the applications of the fundamental definitions of the calculus, we begin to feel that we are on less secure ground. There cannot, in the nature of things, be a theory applicable to functions of two or more variables so elegant and simple as that of the differential coefficient. A knowledge of the partial differential coefficients is in no way equivalent. . . . To understand and to characterize the behaviour of a function in the neighbourhood of a point, we must have recourse to *differentials*. But we have only to turn to the pages of Stolz, which still . . . constitute the last word on the subject, to see in what an imperfect form the theory of differentials has remained . . .". Young set out to create a more coherent theory, and succeeded completely: his theory is simple, elegant, and symmetrical, and runs strictly parallel to the classical theory with a single variable. The best tribute to it is perhaps that de la Vallée-Poussin should, in the light of it, have rewritten his whole account of partial derivatives and differentials in the third edition of his *Cours d'analyse*.

The first step had been taken by Stolz. In the old theory we begin by defining f_x and f_y, the partial derivatives of $f(x, y)$, and define the differential of f by $df = f_x dx + f_y dy$, where dx and dy are increments of the independent variables. But this definition is far too general, since the existence of f_x and f_y at (x, y) tells us nothing about the behaviour of f except on two particular lines through (x, y); the existence of df does not even involve the continuity of f. On the other hand, if we insist throughout that f_x and f_y shall be continuous functions of (x, y), we are restricting our analysis far too severely.

Stolz had therefore given a new definition. Just as we say that $f(x)$ is differentiable, and that its derivative is A, if $\delta f = (A+\epsilon)\delta x$, where $\epsilon \to 0$ when $\delta x \to 0$, so now we say that $f(x, y)$ is differentiable if

$$\delta f = (A+\epsilon)\delta x + (B+\eta)\delta y,$$

where ϵ and η tend to 0 with δx and δy, and write $df = A dx + B dy$ in these circumstances only. It is plain that f_x and f_y then exist and are equal to A and B. On the other hand, it is easy to prove that f is differentiable when f_x and f_y are continuous. Thus our hypothesis lies between the two "classical" hypotheses, and it proves to be exactly the right compromise. As de la Vallée-Poussin says, its superiority is incontestable.

Stolz had not seen the full implications of his idea. In Young's hands it develops simply and elegantly, and the parallelism with the case of one variable is preserved strictly. He begins with a theorem about the reversibility of the order of partial differentiations: *if f_x and f_y are differentiable at (x, y), then*

$$f_{xy} = \frac{\partial}{\partial x}\left(\frac{\partial f}{\partial y}\right) = \frac{\partial}{\partial y}\left(\frac{\partial f}{\partial x}\right) = f_{yx}.$$

The road is then clear for the definition of the second differential d^2f. We suppose that $df = f_x dx + f_y dy$ exists *near* (x, y) and (regarded as a function of x and y, with dx and dy treated as constants) has a differential $d(df)$ at (x, y). We define d^2f as $d(df)$, and it is easy to show that

$$d^2f = f_{xx} dx^2 + 2f_{xy} dx\, dy + f_{yy} dy^2.$$

The higher differentials are then defined in the same way, $d^n f$ being $d(d^{n-1}f)$.

With these definitions, the whole theory runs with admirable smoothness; we have only to take the known theorems for one variable and make the natural verbal changes. Thus, if f has an n-th differential at (x, y),

$$f(x+h, y+k) = f + df + \frac{1}{2!}d^2f + \ldots + \frac{1}{n!}(d^n f + R),$$

where the differentials on the right are formed with $dx = h$, $dy = k$,

$$R = \sum_{m=0}^{n} \binom{n}{m} h^{n-m} k^m \epsilon_m,$$

and the ϵ_m tend to 0 with h and k. This is an exact analogue of the "o" form of Taylor's Theorem for one variable.

The theory is particularly well adapted to the treatment of implicit functions, which beginners usually find so troublesome. All this is set out in detail in **7** and **8** and, in a more condensed form, in Young's Cambridge Tract **3**. The exposition is excellent: Young seems to have written here with unusual care.

The tract covers more ground, though the main theorems of the papers form its central and most striking feature. It is one of the best of the series, and it is astonishing that it should never have run into a second edition, since the subjects treated have been so singularly neglected by modern English writers. It is possible that Young's preface, and the rather intimidating air of his two short appendices, may have discouraged readers in 1910. "Rigidity of proof and novelty of treatment have been aimed at rather than simplicity of treatment . . . the subject cannot be completely mastered without some knowledge of the theory of sets of points . . .": an intending reader might be frightened by Young's language and reputation, and expect something much more "highbrow" than what he gets. For actually very little knowledge is wanted, and the "rigour" would not trouble any clever undergraduate now.

It is curious that Young should never have written a really successful book. He wrote three, alone or in collaboration with his wife; the *Sets of points*, this tract (both "classics" which somehow hung fire), and a third book which must have seemed to have every chance of popularity, but which was, to the publisher at any rate, a complete disappointment. This was the *First book of geometry* (**1**), written by the Youngs jointly, and published by Dent in 1905.

The book is a genuine "book for children" of a very interesting and original kind. The central idea is that children should be encouraged to think of geometrical objects in three dimensions, to think of a plane, for example, as a boundary of a solid, and of a line as an intersection of two planes, or as a fold in one. I am no authority on such a matter, but I should have thought that the idea was sound, and that a book based on it should be more concrete and more stimulating, for most learners, than those of the more conventional and abstract pattern. The authors, however, were asking too much of English teachers. It appeared that they could not, or would not, fold paper, and the book fell absolutely flat in England. It was much more successful abroad, has been translated into German, Italian, Magyar, and Swedish, and used with success in German schools.

(6) I must end by saying something about the series of papers (**25–29**) which dominated Young's last period of activity, though here I am not

a very well qualified critic†. The central problem is that of the area of surfaces, a problem of notorious difficulty still not completely solved.

The most interesting and characteristic paper is the short paper **26**. "No one", Young says, "has hitherto succeeded in giving a definition of the area of a surface"

(7) $$x = x(u, v), \quad y = y(u, v), \quad z = z(u, v),$$

"which allows us to state a sufficient condition of a general character that the surface should possess an area . . .", given by the familiar formula

(8) $$S = \iint \sqrt{(J_1^2 + J_2^2 + J_3^2)}\, du\, dv,$$

where the J are the three Jacobians of x, y and z. Lebesgue had attacked the problem in his thesis, defining the area as the lower bound of that of polyhedra which tend to the surface, without necessarily being inscribed in it; but the results were not altogether satisfactory, and Young suggests a new procedure.

He takes a network in the plane (u, v); this defines a network of curved quadrilaterals on the surface. He then defines *the area of a skew curve* (and this is his most characteristic contribution to the problem). Inscribe polygons in the curve, and imagine that (to use the language of statics) their sides represent forces. The forces are equivalent to a couple, and the area of the curve is the limit, if it exists, of the magnitude of the couple. For the surface, sum the areas of the network of curved quadrilaterals, and again proceed to the limit; if this limit exists, it is the area of the surface. The definition differs fundamentally from Lebesgue's in two respects; it uses only auxiliary figures *inscribed in* the surface, and it is based on limits and not on lower bounds.

The test of the definition lies in its results, and these are in some ways very satisfactory. Thus Young's main theorem (stated only in its simplest and most striking form) is that *if all of x_u, x_v, y_u, ... are of integrable square, then the area exists and is given by* (8). The theorem demands a detailed preliminary study of the formula

(9) $$A = \iint J\, du\, dv$$

for the plane area defined by $x = x(u, v)$, $y = y(u, v)$ when (u, v) varies over a rectangle. The first results concerning the auxiliary formula (9)

---

† What I say about them is based on notes given me by Dr. Burkill.

are set out in **25**: thus (9) is true whenever $x_u, \ldots$ are of integrable square. Young returned to them and generalized them in the later papers.

Young's definition has drawbacks. It is insufficiently intrinsic, depends too much on the particular parametric representation, and does not lend itself to proofs of invariance. His work was developed a little later by Burkill. Burkill's *definition* is much the same, but his theory of (non-additive) functions of intervals enables him to define the J quite differently and so to reach more satisfactory results.

Since 1926 Lebesgue's definition has come again into favour, as a result of Tonelli's work. Tonelli solves the problem definitely for surfaces $z = z(x, y)$†, but not in the general case. This has attracted many writers, but no final solution has been found.

This was Young's last work, and very remarkable work for a man of sixty, though it may not have quite the quality of the best work of his best years. He knew that it was his last, and said so rather dramatically in his Presidential Address (**30**) to us a little later. "What I have been able to do, I have done—

> this rough magic
> I here abjure".

There is a touch of grandiloquence in it, but it was excusable, for few of our Presidents could claim to have done more.

References.

[This list of books and papers contains only those referred to in the notice. The letters *PLMS* and *PRS* refer to the *Proc. London Math. Soc.* (2) and the *Proc. Royal Soc.* (A).]

1. *The first book of geometry* (with Grace Chisholm Young; London, Dent, 1905).
2. *The theory of sets of points* (with Grace Chisholm Young; Cambridge, University Press, 1906).
3. *The fundamental theorems of the differential calculus* (Cambridge Tracts in Mathematics, No. 11, 1910).
4. "Open sets and the theory of content", *PLMS*, 2 (1905), 16–51.
5. "On upper and lower integration", *PLMS*, 2 (1905), 52–66.
6. "On the general theory of integration", *Phil. Trans. Royal Soc.* (A), 204 (1905), 221–252.
7. "On differentials", *PLMS*, 7 (1909), 157–180.
8. "On implicit functions and their differentials", *PLMS*, 7 (1909), 397–421.
9. "On a new method in the theory of integration", *PLMS*, 9 (1911), 15–50.
10. "On a class of parametric integrals and their application in the theory of Fourier series", *PRS*, 85 (1911), 401–414.
11. "Konvergenzbedingungen für die verwandte Reihe einer Fouriersche Reihe", *Münchener Sitzungsberichte*, 41 (1911), 361–371.
12. "On the convergence of a Fourier series and of its allied series", *PLMS*, 10 (1912), 254–272.

---

† See Saks, *Théorie de l'intégrale*, ed. 2, ch. 5.

13. " On classes of summable functions and their Fourier series ", *PRS*, 87 (1912), 225–229.
14. " On the multiplication of successions of Fourier constants ", *PRS*, 87 (1912), 331–339.
15. " Sur la géneralisation du théorème de Parseval ", *Comptes rendus*, 155 (1912), 30–33.
16. " On the Fourier series of bounded functions ", *PLMS*, 12 (1913), 41–70.
17. " On the determination of the summability of a function by means of its Fourier constants ", *PLMS*, 12 (1913), 71–88.
18. " On the new theory of integration ", *PRS*, 88 (1913), 170–178.
19. " On the theorem of Riesz-Fischer " (with Grace Chisholm Young), *Quarterly Journal*, 44 (1913), 49–88.
20. " On integration with respect to a function of bounded variation ", *PLMS*, 13 (1914), 109–150.
21. " Sur la convergence des séries de Fourier ", *Comptes rendus*, 163 (1916), 187–190.
22. " On the convergence of the derived series of Fourier series ", *PLMS*, 17 (1918), 195–236.
23. " On the connexion between Legendre series and Fourier series ", *PLMS*, 18 (1920), 141–162.
24. " On series of Bessel functions ", *PLMS*, 18 (1920), 163–200.
25. " On a formula for an area ", *PLMS*, 18 (1920), 339–374.
26. " On the area of surfaces ", *PRS*, 96 (1920), 71–81.
27. " On the triangulation method of defining the area of a surface ", *PLMS*, 19 (1921), 117–252.
28. " On a new set of conditions for a formula for an area ", *PLMS*, 21 (1923), 75–94.
29. " Integration over the area of a curve and transformation of the variables in a multiple integral ", *PLMS*, 21 (1923), 161–190.
30. " The progress of mathematical analysis in the twentieth century ", *PLMS*, 24 (1926), 421–434.

DAVID HILBERT

David Hilbert, who had been an Honorary Member of the Society since 1901, died during the summer, and the Council feel that the death of so great a mathematician should not pass, even momentarily, unnoticed; but the difficulties of obtaining any adequate account of his work and influence are at present insuperable. They have therefore decided to print this short notice and defer the publication of a more appropriate obituary until later.

Hilbert was born in 1862. He was Privatdocent in Königsberg from 1886 to 1892, when he succeeded Hurwitz as Professor Extraordinarius. He was called to Göttingen as Ordinarius in 1895, and remained there for the rest of his life. His profoundly original researches, his inspiring teaching, and his very individual personality had far-reaching effects on the development of mathematics both in his own and in other countries; it was he who did most to make Göttingen rank so long with Paris as one of the first two mathematical centres of the world.

Almost all of Hilbert's greatest contributions to constructive mathematics were made before 1911, and are recounted in Poincaré's *Rapport sur le prix Bolyai*, published in 1911 in the *Acta Mathematica*. After 1918 he was occupied almost exclusively with logic and the foundations of mathematics. It had always been his habit to concentrate on one subject at a time, to make or renew it by his discoveries, and then to leave it and forget it. He began with algebra and invariant theory (1888–92); and passed successively to algebraic theory of numbers (1892–9), foundations of geometry (1898–1903), calculus of variations and Dirichlet's principle (1899–1905), and integral equations (1901–12). There is little overlapping: thus he published nothing on number-theory after 1899, except his solution of "Waring's Problem", an isolated effort in a quite different part of the subject. There was a short period (1912–7) during which he occupied himself with the mathematical foundations of physics, but after this he became a "logician", and remained one for the rest of his active life.

There is none of these fields in which Hilbert did not make discoveries of outstanding importance, and several which he transformed entirely. In three cases he gave a systematic account of his work, in the *Bericht über die Theorie der algebraischen Zahlkörper* (1897), the *Grundlagen der Geometrie* (of which many editions and translations have been published), and the *Grundzügen eine allgemeinen Theorie der linearen Integralgleichungen* (1912). The first of these is reprinted in Vol. 1 of his *Gesammelte Abhandlungen* (1932–6)*. Vol. 3 of the *Abhandlungen* contains an account of

Hilbert's life, and a reasoned estimate of his work, by O. Blumenthal, and this and Poincaré's *Rapport* between them give a fairly comprehensive picture of one of the greatest mathematicians of our time.

---

* Reviewed by Prof. Mordell in Vols. 16, 18, and 20 of the *Math. Gazette*. There is a short notice of Hilbert by Mrs. J. A. Todd (Olga Taussky) in *Nature*, 14 August, 1943.

Prof. H. L. Lebesgue, For.Mem.R.S.

HENRI LÉON LEBESGUE, professor in the Collège de France, a foreign member of the Royal Society and an honorary member of the London Mathematical Society, was born at Beauvais, Oise, on June 28, 1975; news of his death during 1941 recently reached Great Britain.

Lebesgue was one of the greatest mathematicians of recent times, but more of a specialist than most great mathematicians. His work has not the variety and versatility of Poincaré's or Hilbert's; he did not, as they did, enrich almost every branch of mathematics with his contributions. He was rather a man with one outstanding claim to fame. He had other things to his credit; for example, he made one important contribution (the *Pflastersatz* or 'pavement theorem') to topology; but that, and all his secondary work, of which there is not much, is overshadowed by his work on integration. There, he was first: the 'Lebesgue integral' is one of the supreme achievements of modern analysis.

It is no exaggeration to say that Lebesgue remade the integral calculus. The theory of functions of a real variable, of which the differential and integral calculus are parts, has been rewritten since 1900. The revolution had been initiated by Borel, whose famous monograph "Leçons sur la théorie des fonctions" was avowedly the starting point of Lebesgue's researches; but it was Lebesgue himself who took the decisive steps. The older theories of integration, in spite of all that Riemann, Darboux and others had done, were radically defective at several vital points, and in particular in their relations to differentiation. It was not true that differentiation and integration were 'usually' inverse operations, that integration and differentiation, or differentiation and integration, usually restored the original function. Indeed the whole theory was æsthetically unsatisfying; it tended to be cumbrous, longwinded and full of untidy exceptions; and the conservative mathematicians who detested it had a certain amount of excuse for their distaste.

All this is changed; the theory is now one of the most beautiful in mathematics. It unfolds itself in a series of terse and comprehensive theorems, with all the smoothness and elegance of the best 'classical' analysis. In particular, differentiation and integration dovetail harmoniously together. It is a readily intelligible, almost a 'popular' subject, and for this, by the unanimous testimony of every worker in these fields, it is to Lebesgue first that honour is due.

Lebesgue was a fine writer, with an admirably vigorous and lucid style. His two great memoirs, "Intégrale, longueur, aire" (1902) and "Sur les intégrales singulières" (1910), and his two books, "Leçons sur l'intégration" (1904 and 1928) and "Leçons sur les séries trigonométriques" (1906), rank among the classics of mathematical literature. It is sad that the second book should never have been re-edited, since the subject gives perhaps the finest illustration of the inevitability and vitality of Lebesgue's ideas.
G. H. HARDY.

JOHN RAYMOND WILTON

H. S. CARSLAW *and* G. H. HARDY.

John Raymond Wilton, Professor of Mathematics in the University of Adelaide, South Australia, died on 12 April, 1944.

He was born on 4 May, 1884. His father was on the literary staff of an Adelaide newspaper. He entered the University early in 1901 and obtained his B.Sc. at the end of 1903, with first class honours in both mathematics and physics. Prof. Bragg, later Sir William, then held the combined chairs of Mathematics and Physics, and he once described him as having had the greatest natural genius for mathematics among any of his students during his more than twenty years in Adelaide. On his advice Wilton proceeded to Cambridge, entering Trinity in October 1904. In March 1905 he was awarded a sizarship, later converted into a major scholarship. In the Tripos of 1907 he was fifth wrangler, the first five places being filled by Trinity men: G. N. Watson was senior, H. W. Turnbull second, and A. V. Hill third. Even in so good a year Wilton might have gained a higher place.

In the following year, on the advice of his private tutor Herman, he took Part II of the Natural Science Tripos, gaining a first class in physics. The next twelve months he worked in the Cavendish, did some lecturing in mathematics to students of physics, and some demonstrating. There can be no doubt that in forsaking mathematics for natural science he made a serious mistake. In spite of his first class he was not interested in physics, and he was not an applied mathematician. He would have attained distinction as an analyst much sooner had he taken Part II of the Mathematical Tripos in 1908 and devoted the next year to advanced study in mathematics.

In October 1909 he was appointed Assistant Lecturer in Mathematics at Sheffield. From 1912 to 1916 he published many papers in the *Messenger of Mathematics*, *Philosophical Magazine* and *Quarterly Journal of Mathematics*. They were mainly concerned with partial differential equations, hydrodynamics and elliptic functions. His first contributions to the *Proceedings* of this Society were his papers **22** and **23**. In 1916 (**17**) he had already turned his attention to the ζ-function of Riemann.

From 1916 to 1922 he published nothing. The war affected him very deeply, and his connection with Sheffield ended in 1916. He had been doing some X-ray work in a hospital there before conscription was introduced, and he hoped to be allowed to continue it; but the tribunal before which he appeared directed him to St. George's Hospital, London, where for two years, till the end of the war, he was kept copying "cases" from one book into another. His financial resources were small and both he and his wife suffered many privations. He was teaching in the Friends School at Saffron Walden for a short time after the war and about this time he became a member of the Society of Friends.

In 1919 he was appointed to a Lectureship in Mathematics in Manchester under Horace Lamb, and at the end of that year he was offered the chair at Adelaide of which Lamb had been the first holder, during 1876–1885.

When Bragg left Adelaide in 1908, a Professor of Physics was appointed, but mathematics, regarded as the handmaiden of engineering and physics, was placed under the supervision of the Professor of Engineering. This arrangement, which we may assume to have been intended to be temporary, lasted till a year after the end of the war. In 1919 it was decided that a Professor of Mathematics should be appointed, and the position was offered to Wilton, who arrived in Adelaide in January 1920 to take up his duties. He found the mathematical courses at the University much the same as those which he had attended some twenty years earlier, and for thirty years there had been little change in the mathematics taught in the schools. In South Australia secondary education had not made as much progress as in New South Wales and Victoria. His first task was a complete renovation of the work in mathematics at the University and a revision of the programmes in mathematics for the examinations controlled by the University, which then governed to a great extent the curricula of the schools in the state. Though he was thus able materially to improve the standard of the pass work of the first year, it remained for many years lower than he would have liked.

The pass classes in his department were large in the first year, 80–110, and much smaller in the second and third years, 30–50 and 10–20. His programme for both pass and honours was an ambitious one, stronger naturally in pure mathematics than in applied. In 1923 H. W. Sanders was appointed Lecturer and took over all the applied work. To him Wilton also handed over the large pass class of the first year; and he was thus able to devote himself to the pure mathematics of the second and third years, and to the courses offered to such students as were willing to undertake at least part of the programme for honours. This covered four years, but from 1920 to 1940 only eight persons graduated with honours in mathematics. In Sydney it was not unusual in these years to have more than that number of honours graduates in mathematics annually. Wilton envied N.S.W. its High Schools scattered through the state, the large endowments of its university, and its many graduate scholarships: Adelaide had only the 1851 Exhibition and the Rhodes Scholarships. Of one of its Rhodes Scholars, who promises to be the most distinguished mathematician Australia has produced, he had good reason to be proud.

He visited England only once after his appointment to Adelaide (December 1924–August 1925), but he corresponded with a number of fellow-workers, British and foreign. Among the latter were Landau, Walfisz and Hasse, and by the last named he was asked to contribute to *Crelle's Journal*. He had by now found his proper field, and from 1922 to 1934 he was really productive; but he was attacked by ill-health in 1934 and wrote practically nothing after. He explains this in a letter of 6 June 1941. He had made no attempt, he said, to do anything mathematical for years, the effect of overstrain, from which it had taken him a long time to recover; but in spite of a severe illness three years earlier he felt himself in better health than for years and he had been trying to pick up the threads again. In 1934 he had sent a long paper "On the ζ-function of Riemann", in two parts, to the *Quarterly Journal* and the *Proceedings of the Cambridge Philosophical Society*. Both parts had been accepted but returned to him for revision, owing to his ignorance of Siegel's analysis, in 1931, of Riemann's surviving manuscripts. He had put his paper aside and had not looked at it again till the beginning of 1941; but he had since been working through it and thought he would be able to make something of it when he got into his stride again. Two days later he told his wife that he had "seen something that he had been looking for for years", and that he felt he could now go straight ahead with the paper, which he was re-writing. That evening he had a paralytic stroke, which left him with speech and memory seriously impaired. He made a partial recovery, and after two years began again to do some work at his department, but he was never the same man again. On 9 April 1944 he had a second stroke and he died three days later.

He had taken his D.Sc. in Adelaide in 1912 and he was given a Cambridge Sc.D. in 1930. In 1934 he received the Lyle Medal of the Australian National Research Council, awarded for the best work in mathematics or physics produced in Australia in the five-year period, 1928–1933.

He had become one of the most influential leaders in Australia of the Society of Friends. He was a great lover of Dante and before his illness he was in the habit of reading through the *Divina Commedia* in the original several times a year. He used to say that this work had exercised the greatest single influence on his life. Late in life he discovered a love and deep understanding of classical music.

He was married twice. First, in England in 1910, to Annie Martha Gladstone of Forest Hill, London: she died in 1932 after a long illness. There were no children of this marriage. Second, in 1936, to Winifred Welbourn of Adelaide: she and a daughter born in 1937 survive him.

The most important of Wilton's papers are no doubt those (**30–1, 33–4, 36–7, 46, 50**) concerning "Gauss's circle problem", "Dirichlet's divisor problem", and their extensions; but the analysis in them is intricate and it is not easy to summarize his chief results. It has been familiar since Gauss that, if $r(n)$ is the number of representations of n by two squares, so that

$$R(x) = \sum_{n \leqslant x} r(n) = \sum_{u^2+v^2 \leqslant x} 1$$

is the number of integral lattice points in the circle $u^2+v^2 \leqslant x$, then $R(x) = \pi x + O(\sqrt{x})$, the area of the circle with an error of the order of the circumference. If we write

$$R(x) = \pi x + 1 + P(x),$$

then a variety of problems arise concerning $P(x)$. These are of two kinds, problems of order, and problems of identity, concerning the exact expression of $P(x)$ as a convergent series of Bessel functions. These form the content of the "circle problem", and the "divisor problem" covers the similar problems for the divisor function $d(n)$ and the rectangular hyperbola $uv \leqslant x$. There are a multitude of generalizations associated with more general figures, such as m-dimensional spheres and ellipsoids, with weighted lattice points, and with functions defined by sums of powers of divisors.

Wilton's work concerns the identities connected with these problems. It is known, for example, that

$$P(x) = \sqrt{x} \sum_1^\infty \frac{r(n)}{\sqrt{n}} J_1\{2\pi \sqrt{(nx)}\},$$

the series being convergent; and each generalization of the problem introduces a similar identity, generally with a summable divergent series. The widest generalizations are those of Oppenheim, Walfisz, and Wilton himself. It is hardly possible to state them without elaborate explanations, but there is a good historical account of the subject, up to 1929, in **31**, and this makes Wilton's own contributions clear.

Another arithmetical function which had an invincible attraction for Wilton was Ramanujan's $\tau(n)$. In **38**, for example, he investigates the "identities" connected with this function, and proves that

$$\frac{1}{\Gamma(a+1)} \sum_{n \leq x} (x-n)^a \tau(n) = \frac{1}{(2\pi)^a} \sum_1^\infty \left(\frac{x}{n}\right)^{6+\frac{1}{2}a} \tau(n) J_{12+a}\{4\pi \sqrt{(nx)}\}$$

if $a > 0$; the result was afterwards extended by Hardy [*PCPS*, 34 (1938)] to cover the case $a = 0$. This paper also contains other results of considerable interest; the functional equation

$$(2\pi)^{-s} \Gamma(s) F(s) = (2\pi)^{s-12} \Gamma(12-s) F(12-s)$$

for $F(s) = \Sigma n^{-s} \tau(n)$, and a proof that $F(s)$ has an infinity of zeros on $\Re s = 6$ (an analogue of a known theorem concerning Riemann's ζ-function). He was also much interested in the curious congruence properties of $\tau(n)$, and **40** contains what are perhaps the most general results known about them. He did further work in this direction which has never been published, and there is a long manuscript about the coefficients in

$$\{x^{\frac{1}{24}}(1-x)(1-x^2)\ldots\}^q,$$

for general q, still in the possession of the Society: the coefficient is $\tau(n)$ when $q = 24$.

A man with these interests was inevitably attracted by Poisson's formula, and Voronoï's more recondite formula for sums $\Sigma d(n) f(n)$. In **47** he extends the work of Dixon and Ferrar on Voronoï's formula, and in **48** he applies the results to series such as

$$\Sigma n^{-a} d(n) \genfrac{}{}{0pt}{}{\cos}{\sin} 2n\pi\theta,$$

already considered by Chowla and Walfisz. His various short notes on Fourier analysis, and Poisson's formula in particular, not of great importance in themselves, may be regarded as studies preliminary to this work.

Finally, another interesting pair of papers is 51–2, which contain almost his last work. Here he simplifies some intricate analysis of van der Corput and applies it to trigonometrical series. In particular he proves that if ϕ and ψ are sufficiently regular functions such that $\psi' \to \infty$ and $\phi^2 > A\psi''$, where A is constant, then

$$\Sigma \phi(n) e^{2\pi i \{n\theta - \psi(n)\}}$$

diverges for all θ. If $\psi = n \log n$, $\phi = n^{-\frac{1}{2}}$, we obtain a theorem of Hardy and Littlewood; if $\psi = n \log \log n$, $\phi = (\log n)^{-1}$, one of Steinhaus.

It will be plain from this summary that Wilton was a fine mathematician, with admirable taste and a natural inclination towards deep and difficult problems. He may have left nothing, strictly, of major importance, but his record is genuinely impressive. He might perhaps have made a bigger name if his taste had been less fine, and he had been content to work in fields which offer cheaper rewards.

We have to thank Prof. H. W. Sanders, who has succeeded him at Adelaide, for providing a list of Wilton's published papers and information about the mathematical courses in the University.

List of papers by J. R. Wilton.

[*JLMS*, Journal London Math. Soc.; *PLMS*, Proc. London Math. Soc. (2); *M*, Messenger of Math.; *QJ*, Quarterly Journal of Math.; *PCPS*, Proc. Camb. Phil. Soc.; *PRS*, Proc. Royal Soc. (A); *PM*, Phil. Mag.]

1. Note on the solution of a certain partial differential equation. *M*, 42 (1912).
2. Note on the solution of the equation $r = f(t)$. *M*, 43 (1913).
3. On the solution of an equation of the form $F(r, s, t) = 0$. *M*, 43 (1913).
4. Some simple transformations of Stokes' current function. *M*, 43 (1913).
5. Note on the equation $s = f(z)$. *M*, 43 (1913).
6. On plane waves of sound. *PM* (6), 26 (1913).
7. On the highest wave in deep water. *PM* (6), 26 (1913).
8. On the potential and force function of an electrified spherical bowl. *M*, 44 (1914).
9. Note on a certain partial differential equation of the second order possessing an intermediate integral of the first order. *M*, 44 (1914).
10. A simple transformation of certain partial differential equations. *M*, 44 (1914).
11. The seven-fourteen section of the Zeta function. *M*, 44 (1914).
12. The nine-eighteen section of the Zeta function. *M*, 44 (1914).
13. The eleven-twentytwo section of the Zeta function. *M*, 44 (1914).
14. On deep water waves. *PM* (6), 27 (1914).
15. Figures of equilibrium of rotating fluid under the restriction that the figure is to be a surface of revolution. *PM* (6), 28 (1914).
16. A transformation of the partial differential equation of the second order. *M*, 45 (1915).
17. Note on the zeros of Riemann's ζ-function. *M*, 45 (1915).
18. On ripples. *PM* (6), 29 (1915).
19. On the solution of certain problems of two-dimensional physics. *PM* (6), 30 (1915).

20. On the conditions that certain partial differential equations of the second order have an intermediate integral of the same order. *QJ*, 46 (1915).
21. A continued fraction solution of the linear differential equation of the second order. *QJ*, 46 (1915).
22. On Darboux's method of solution of partial differential equations of the second order. *PLMS*, 14 (1915).
23. A pseudo-sphere whose equation is expressible in terms of elliptic functions. *PLMS*, 14 (1915).
24. A formula in zonal spherical harmonics. *M*, 46 (1916).
25. Certain criteria for the success of Darboux's method when applied to the equation $s = f(x, y, z, p, q)$. *QJ*, 47 (1916).
26. A proof of Burnside's formula for $\log \Gamma(x+1)$ and certain allied properties of Riemann's Zeta function. *M*, 52 (1923).
27. The Gibbs phenomenon in series of the Schlömilch type. *M*, 56 (1927).
28. A note on the coefficients in the expansion of $\zeta(s, z)$ in powers of $(s-1)$. *QJ*, 50 (1927).
29. The approximate functional formula for the Theta function. *JLMS*, 2 (1927).
30. The lattice points of an n-dimensional ellipsoid. *JLMS*, 2 (1927).
31. The lattice points of a circle; an historical account of the problem. *M*, 58 (1928).
32. Some applications of a transformation of series. *PLMS*, 27 (1928).
33. The average value of an exponential function over the lattice points of a circle. *PCPS*, 24 (1928).
34. A series of Bessel functions connected with the lattice points of an n-dimensional ellipsoid. *PRS*, 120 (1928).
35. The Gibbs phenomenon in Fourier-Bessel series. *Journal f. Math.*, 159 (1928).
36. A series of Bessel functions connected with the theory of lattice points. *PLMS*, 29 (1929).
37. The lattice points of a circle. *Proc. Royal Soc. Edinburgh*, 48 (1929).
38. A note on Ramanujan's arithmetical function $\tau(n)$. *PCPS*, 25 (1929).
39. On Ramanujan's arithmetical function $\Sigma_{r,s}(n)$. *PCPS*, 25 (1929).
40. Congruence properties of Ramanujan's function $\tau(n)$. *PLMS*, 31 (1930).
41. An approximate functional equation for the product of two ζ-functions. *PLMS*, 31 (1930).
42. The mean value of the Zeta function on the critical line. *JLMS*, 5 (1930).
43. A proof of Poisson's summation formula. *JLMS*, 5 (1930).
44. Congruence properties of Ramanujan's function $\tau(n)$ to the modulus 11. *Boletin. Mat., Buenos Aires* (June 1930).
45. A proof of Fourier's Theorem. *JLMS*, 6 (1931).
46. On Dirichlet's divisor problem. *PRS*, 134 (1931).
47. Voronoï's summation formula. *QJ (Oxford)*, 3 (1932).
48. An approximate functional equation with applications to a problem of Diophantine approximation. *Journal f. Math.*, 169 (1933).
49. A note on Stirling's theorem. *Mathematical Notes* (Edinburgh Math. Soc., No. 28, 1933).
50. An extended form of Dirichlet's divisor problem. *PLMS*, 36 (1934).
51. An approximate functional equation of a simple type (I). *JLMS*, 9 (1934).
52. An approximate functional equation of simple type (II): applications to certain trigonometrical series. *JLMS*, 9 (1934).

Also in South Australia.

Certain Diophantine problems. *Proc. Royal Soc. of South Australia*, 44 (1920).
Notes on the mathematical syllabus of the Public Examinations (1921).

5. BOOK REVIEWS BY G. H. HARDY

INTRODUCTION TO THE BOOK REVIEWS BY G. H. HARDY

Hardy's reviews for the *Mathematical Gazette*, *Nature*, the *Times Literary Supplement* and the *Cambridge Review* are reproduced. In many of the earlier reviews he is concerned with the exposition of the concepts of the infinitesimal calculus; the improvement in the quality of elementary textbooks on analysis owes much to his reforming zeal and to his own book *Pure Mathematics*, in particular. See also, in this connexion, 1906, 9 [Vol. V, p. 284] and 1907, 8 [p. 481 of this volume]. Unlike E. W. Hobson (see O11, p. 749 of this volume) Hardy was young enough to realize by 1920 'that everything that he had been fighting for had been achieved', and he rarely returned to the subject after that date.

Of his reviews of works on other subjects R18, R22, R23 and R24, are possibly the most interesting.

R.A.R.

(a) Reviews from *The Mathematical Gazette*

REVIEWS.

Theoretische Arithmetik. (II. Abtheilung). By O. STOLZ und J. A. GMEINER. Leipzig, Teubner. 1902. Pp. 402.

This book is a new enlarged edition of certain chapters of Dr. Stolz' well-known *Allgemeine Arithmetik*. The merits of the latter are universally recognised, and no praise could be higher than to say that in lucidity and thoroughness the present volume is an improvement on it.

The first two chapters are not essential to the general plan of the book, the systematic development of the arithmetic theory of number, real and complex; but their inclusion is fully justified by their intrinsic interest. The first (chapter V. of the complete work) deals with the mathematical theory of *magnitudes*, and has no direct connection with number. Mr. Russell has shown in his recent work *The Principles of Mathematics* that the simplicity or complexity of a mathematical theory of magnitudes depends very much on the philosophical view which underlies it. The authors do not explicitly raise any philosophical questions, but what Mr. Russell calls the relative theory of magnitude is presupposed in their treatment of the subject. The result is an exceedingly complicated system of axioms. The working out of the theory on this basis, however, leaves nothing to be desired. The second chapter contains a very clear account of Euclid's theory of ratio. It is not a little remarkable that the part of Euclid's work which is perhaps least open to criticism is the part which most of us were brought up to despise.

The theory of real numbers is then developed from Cantor's point of view. A noteworthy feature of the authors' treatment of this part of the subject is that the fundamental theorems concerning limits are proved first for the particular cases in which the limits are rational numbers. This is much to be commended, although it adds considerably to the length of the chapter. Another feature of the book is the prominence given here and elsewhere to the theory of approximation by decimals. It is a pity that the authors have not seen their way to include at any rate an outline of the elements of the theory of aggregates and transfinite cardinals and ordinals. This would have added considerably to the value of the book.

The next two chapters deal with the elementary theory of powers, roots, logarithms, and series. These chapters do not present many very novel features, but are admirably thorough. I may perhaps mention particularly the sections on approximation to roots, on the solution of the functional equation
$$f(x+y) = f(x) f(y)$$
and on the development of the theory of logarithms from the equation
$$\log a = \operatorname*{L}_{n=\infty} n(\sqrt[n]{a} - 1).$$

It is characteristic of the spirit of the book that a proof should be given that if $a_n = b_n (n = 0, 1, 2, \ldots)$ the series Σa_n, Σb_n are *gleichartig*, i.e. converge, diverge, or oscillate together. Perhaps some mention of the theory of divergent series might be given: it is a little old-fashioned to say that the equation
$$1 - 1 + 1 - \ldots = \tfrac{1}{2}$$
has 'no meaning,' although, as the authors mean it, the remark is of course quite true. It is inevitable that in a book of this character most of the most interesting theorems about non-absolutely convergent series should be omitted, or relegated to the examples. And I may mention here that the examples (it is refreshing to see examples in a German book) are extremely well selected; they contain a large number of important theorems for which there is no space in the text, and nothing which is not of some interest, all 'tricks' being rigidly excluded. For instance, the development of the theory of e^x *from* its expression as a power series is given as an example. It is perhaps doubtful whether it is best to make the theory of e^x and $\log x$ entirely prior to the theory of series.

R1 *Mathematical Gazette*, 2 (1903), 312-13.

Of the remaining chapters only X. calls for special notice. Here the different possible varieties of complex numbers are considered, and it is shown that the ordinary complex numbers alone obey all the ordinary rules of arithmetic, and (after Frobenius) that only quaternions obey all but the commutative rule of multiplication.

Chapter XI. deals with the geometric theory of complex numbers; some very interesting geometric problems are solved as examples. The last chapters contain the theory of complex powers, roots, logarithms, and series.

The great merit of the book lies in the complete absence of sketchiness from any part of it; in this it affords a welcome contrast to many books written now. It is also unusually free from printers' errors. G. H. HARDY.

Einleitung in die Funktionentheorie, (1 Abteilung), von O. Stolz und I. A. Gmeiner. 1904.

This volume, the fourteenth of Teubner's new mathematical series, is a continuation of the same authors' *Theoretische Arithmetik* :* the two books will when completed form a new and much enlarged edition of the well-known *Allgemeine Arithmetik* of Prof. Stolz, and supersede that excellent work.

The range of the present volume is more like that of Harkness and Morley's *Introduction to the Theory of Analytic Functions* than that of any other English book. But it begins farther on and ends sooner, and the intervening ground is naturally covered with much greater thoroughness. So far as English readers are concerned it is only too likely that this book will fall between two stools. It is altogether too solid and systematic for one who is only beginning the theory of functions. On the other hand it contains nothing which an expert in the subject ought not to know perfectly well already. Still for a student who has acquired a superficial knowledge of a number of different branches of the theory, one, let us say, who has just taken Part II. of the Mathematical Tripos, and is about to begin to study the subject seriously, it is difficult to imagine a type of book more profitable to read. The authors develop the theory in a thoroughly consistent and logical way; they do not jump from one theory of functions to another, from Cauchy to Riemann, and from Riemann to Weierstrass, but proceed on strictly Weierstrassian lines throughout. So much is this the case that, in spite of the wealth of matter which the book contains, differential coefficients are only alluded to once or twice, and that quite incidentally, while the word 'integral' does not occur once.

As in their former book, the authors include numerous examples. This custom, which seems, happily, to be spreading rapidly abroad, is no doubt a result of the influence of English books. But there is nothing 'English' about the examples themselves. The authors adopt the excellent method of including among the examples all obvious corollaries and extensions for which there is no room in the text; and, whenever any example presents the slightest difficulty, hints are given which remove it. There is not an example in the book which would puzzle a Cambridge mathematician, who knows anything about the subject, for five minutes. And, although no doubt we sin in the opposite direction, most people will probably think that in this respect the authors go too far; it would be better to exercise the ingenuity of the reader a little more. However, of their kind the examples are excellent, and include some of types which ought to be but are not in every English book upon the subject.

There is not very much in the way of detail to which it is worth while to call attention. There is a serious logical error in the very first line, which recurs on p. 114. A variable, whatever it may be, is not a *sign* : it is what the sign denotes. How to define a 'variable' in strictly logical terms is no doubt a by no means easy question. But it is no more a sign than Prof. Stolz is his signature. On p. 26 the authors have missed an excellent opportunity of giving a short account

---

*See *Math. Gazette*, vol. II., p. 312.

of the general notion of 'orders of infinity.' Heine's theorem concerning uniform continuity (pp. 52-55) may be deduced from a general theorem first explicitly stated by Dr. Baker. 'If every point P of a certain region (in space of any number of dimensions) possesses the property that it can be enclosed in a small region such that, if Q is any other point within it, a relation $(P-Q)$ holds, then the whole region can be divided into a *finite* number of regions, each containing *at least one* point P satisfying the relation $(P-Q)$ with respect to all points Q of the sub-region within which it lies.' This point occurs, for instance, in Goursat's proof of Cauchy's theorem; and to state and prove it, and some similar general theorems, early in every book on the theory of functions would save a good deal of unnecessary repetition later on. The treatment of 'fonctions à variation bornée' (pp. 57-60) is particularly simple and lucid. On p. 62 the authors are not very clear as to what they mean by the 'possibility of geometrical representation' of a function. The theorems proved on pp. 89-95 concerning multiple and repeated limits are very interesting, although they do not go very far. The sections on complex functions of a *real* variable and their geometrical representation (pp. 108-114) are interesting and might well be drawn upon by elementary English books, as might the examples (pp. 133-134). The brief account of 'Arithmetic Series' and Interpolation is an excellent feature (pp. 160-171). The last chapter, giving an account of the Weierstrassian theory of Power Series, is really admirable; about the theory of Multiple Power series (pp. 219-236) just enough is said to whet the appetite for more.

It is to be hoped that in the concluding volume the authors will have something to say concerning the fundamental relations between Cauchy's and Weierstrass's theories of functions. It is certain that the last word on this most interesting subject has not yet been said. Cauchy's theory shews that the assumptions of the Weierstrassian theory are redundant. For in Cauchy's theory we need only assume that a function has a derivative at every point; we need not even assume that this derivative is continuous, much less that higher derivatives exist. All these things can be *proved*. But in the Weierstrassian theory we assume far more than this, in another language it is true. Why should this be? What is there in the notion of the integral in virtue of which its introduction simplifies the assumptions of the theory so immensely?
G. H. HARDY.

N. J. Lobatschefskij's Imaginäre Geometrie und Anwendung der Imaginären Geometrie auf einige Integrale. Herausgegeben von H. LIEBMANN. (Teubner, 1904; pp. xi, 188.)

These two memoirs are not of the same fundamental importance as some of Lobatschewsky's work, but they are very interesting none the less, and Herr Liebmann has performed a real service to mathematics by translating them from the Russian and editing them in an intelligible form. One would gather from the editor's elaborate notes and from some remarks in the Introduction, which have an almost pathetic ring, that, if ever a mathematician needed careful editing, then, if not Sylvester, it was Lobatschewsky. His style, at any rate in these memoirs, is obscure and condensed to the verge of unintelligibility, and the details of his analysis often inaccurate; the editor's task must indeed have been arduous and somewhat thankless.

In order to understand these memoirs it is essential to realise the peculiar point of view from which they were written. The author's earlier work upon the subject, written in the synthetic and geometrical spirit more usually associated with his name, had been imperfectly understood, and to some extent called in question, in part on the ground of supposed absurd analytical consequences. Nettled by these criticisms Lobatschewsky returns to the subject from a different side. Laying down *a priori* certain relations supposed to hold between the sides and angles of a right-angled triangle, he deduces a complete and consistent system of trigonometrical formulae which he shows to lead by an imaginary transformation to the ordinary formulae of spherical trigonometry. He then devotes himself to applications of the differential geometry of non-Euclidean space to analysis, and in particular to the evaluation of definite integrals, just as one might apply the elementary geometry of the Euclidean circle to the proof of the formula

$$\int_0^a \sqrt{a^2 - x^2}\,dx = \tfrac{1}{4}\pi a^2.$$

In each formula obtained by such quasi-geometrical methods, and verifiable by pure analysis, he sees a new confirmation of the logical possibility of his geometrical theory. This theory would not be called in question by anybody now, so that a good deal of his work is only of historical interest.

At the same time, many of Lobatschewsky's results are interesting in themselves. This is most true, perhaps, of his investigations of the volumes of certain figures in non-Euclidean space, cones, cylinders, and tetrahedra.

But some of the integrals which he evaluates are interesting too; such, for instance, as

$$\int_0^\pi \frac{\sin^n x\,dx}{(a+b\cos x)^{n+1}} = \frac{\sqrt{\pi}}{(a^2-b^2)^{\frac{n+1}{2}}} \frac{\sqrt{\left(\frac{n+1}{2}\right)}}{\sqrt{\left(\frac{n}{2}+1\right)}},$$

$$\int_0^{\frac{1}{2}\pi} \log\left(\frac{1+\cos\beta\cos x}{1-\cos\beta\cos x}\right) \frac{dx}{1-\cos^2\alpha\cos^2 x} = \frac{\pi}{\sin\alpha} \log\left(\frac{1+\sin\alpha}{\sin\alpha+\sin\beta}\right),$$

$$\int_0^\infty \frac{x\,dx}{\cosh 2x - \cos 2a} = \frac{a\log 2 - L(a)}{2\sin a \cos a},$$

where
$$L(x) = -\int_0^x \log \cos x\,dx,$$

and so on.

G. H. Hardy.

Leçons sur les fonctions de variables réelles, par E. BOREL;
Leçons sur les fonctions discontinues, par RENÉ BAIRE; Le calcul
des residus et ses applications à la théorie des fonctions, par E.
LINDELÖF. (Paris, Gauthier-Villars, 1905, 3 f. 50 c. each.)

M. Borel's book is the sixth of his series of monographs on the theory
of functions, of which the first appeared as recently as 1898. M. Borel
is only human, and by now he has decided to leave to his pupils the
work of preparing his lectures for publication. It cannot be said that
this method has proved in every case an unqualified success. The first
few volumes, prepared by M. Borel himself, and particularly the
admirable *Leçons sur les fonctions entières*, were remarkable alike for
their originality, for the judgment shewn in the selection of material,
and for the lucidity and proportion of the exposition. This high
standard has not been maintained in all of the later volumes, some of
which have been rather scrappy, and have given the impression of
hasty, and at times perfunctory composition. In these respects, however, the present volume is an improvement upon its immediate
predecessors, M. Maurice Fréchet having performed his task unusually
well. But I cannot help thinking that M. Borel would be fortunate if
he could find the time to write his books himself.

The principal problem with which M. Borel deals in this volume is
that of the representation of functions by means of series of polynomials, a form of representation the importance of which was first
shown by Weierstrass's well-known theorem that every continuous
function of a real variable can be expanded in such a series. M. Borel
confines himself to functions of real variables, reserving the complex
theory for the next volume of his series; and for the most part he is
concerned with continuous functions only, the short chapter on the
representation of discontinuous functions containing little more than a
discussion of the comparatively simple case in which the aggregate of
points of discontinuity is enumerable, and a reference to the results
obtained by M. Baire. The consequence is that the book is rather
disconnected; for the first two chapters, which deal with the theory of
aggregates and continuity and discontinuity in general, contain a good
deal which, though very interesting in itself, is really not required for
M. Borel's purpose. The central chapters (3 and 4) are excellent,[1] and
the long note added by M. Painlevé is perhaps the most interesting
feature of the book, though its natural place would not be in this
volume, but in the next one.

There is one criticism which will occur to every regular reader of
this series, in which M. Borel has now enlisted the collaboration of a
number of other eminent mathematicians. There is an amount of
repetition which a judicious general editor should be able to diminish;
and that the plan of the series is open to this criticism M. Borel, to
judge from his remarks in the preface, appears to recognise: 'Il a
paru préférable d'admettre parfois quelques brèves redites plutôt que de
renoncer à l'indépendance des Volumes de la Collection, chacun d'eux
devant pouvoir être lu isolément par un lecteur ayant des connaissances

---

[1] The argument of p. 66 has become inverted in some curious way. The function
is '*more* continuous' when $\phi(\epsilon)$ decreases *less* quickly.

générales d'Analyse. Sans ce principe d'indépendance, on aurait eu tous les inconvénients d'un grand Traité, sans en avoir les avantages.'

In principle, no doubt M. Borel is right. Each author should be asked to deal with some definite question, and he should have full liberty to preface his discussion of it with a general account of those modern developments of analysis which are necessary for his purpose, and with which a reader who has not read the other volumes of the series cannot be expected to be familiar. But he should be very careful to make this general account as short as is consistent with clearness, and to limit it strictly to results which will afterwards be required. This is the course adopted by M. Baire, who, if he is at times a little diffuse, is careful not to encumber his book with unnecessary matter. M. Borel has not set his colleagues so good an example. Why, for example, should he think it necessary to introduce a few pages concerning M. Lebesgue's generalisation of the notion of the definite integral? Most interesting and most important this generalisation certainly is; but it has already been expounded by M. Lebesgue himself in an earlier volume of the series, and no allusion whatever is made to it throughout the remainder of the book.

It will not be necessary to say much about the volume contributed by M. Baire. It is in substance a popular edition of his remarkable memoir, *Sur les fonctions de variables réelles*, published as a thesis in 1899, and afterwards in the *Annali di Matematica*. A good deal of introductory matter has been added, and the argument has been simplified and condensed. The problem of finding the *necessary and sufficient* conditions that a function whose points of discontinuity are given should be capable of representation as the sum of a series of continuous functions is one which most mathematicians would have regarded as hopeless if M. Baire had not completely solved it; and that M. Baire's researches should be made more accessible to the ordinary reader was much to be desired. But it is difficult to resist the impression that these two volumes might well have been condensed into one. M. Baire's results are particularly interesting when applied to differential coefficients of continuous functions. If $f(x)$ is continuous, so is

$$\frac{f(x+h)-f(x)}{h}$$

considered as a function of x. Now, if the differential coefficient exists for every value of x under consideration, and we denote by $h_1, h_2, h_3 \ldots$ a series of positive quantities whose limit is zero,

$$f'(x) = \lim_{n=\infty} \frac{f(x+h_n)-f(x)}{h_n},$$

and is therefore representable as the limit of a sequence of continuous functions, or (what is the same thing) the sum of a series of such functions. M. Baire's results, therefore, give us much important information as regards the possible discontinuities of the differential coefficients of continuous functions.

For M. Lindelöf's *Calcul des Residus* I have nothing but praise. The applications of Cauchy's "calculus" to the theory of functions, and in particular to the summation of series and the theory of analytic continuation, are of the most far-reaching character, and, so far as I know, no one before M. Lindelöf has attempted to give a systematic account of them. Laurent, it is true, published in 1865 a *Théorie des Residus*, which hardly deserved to have been so soon forgotten, but the applications of the theory have multiplied ten times since then. Some new account was urgently necessary, and M. Lindelöf has given us exactly what was wanted. One admirable feature of his book is the thoroughness and exactitude of his historical references, especially to the writings of Cauchy. M. Lindelöf is one of the few mathematicians who have found life long enough to make 'une étude détaillée' of Cauchy's works.

M. Lindelöf very wisely does not trouble himself with all the difficulties as to the *minimum* of assumption required to establish Cauchy's theory, which centre round Goursat's proof of Cauchy's theorem. These difficulties, of course, have absolutely no bearing on the applications with which M. Lindelöf is chiefly concerned. Assuming the continuity of the differential coefficient, he proceeds to show that any analytic function $f(x)$ is itself the differential coefficient of an analytic function $F(x)$ determinate save for an additive constant, a conclusion from which, combined with the definition of the definite integral along a curvilinear path, Cauchy's theorem immediately follows. There is certainly a great deal to be said for presenting the proof of the theorem in this way.

A short but clear account follows of some familiar applications of the formula

$$f(x) = \frac{1}{2\pi i} \int \frac{f(t)}{t-x} dt$$

and of some others which are not so familiar, such as occur in the proof of Jensen's theorem, the theory of the Bernoullian and Eulerian functions, the factorisation of such functions as $\sin x - ax \cos x$, and the transformation of slowly convergent series. In Chapter III. he comes to the ground which he has made particularly his own. He proves a whole series of general formulae, of which the formula of Plana and Abel,

$$\Sigma f(x) = -\tfrac{1}{2} f(x) + \int^x f(x) dx + \frac{1}{i} \int_0^\infty \frac{f(x+it) - f(x-it)}{e^{2\pi t} - 1} dt,$$

was historically the first. The applications of these formulae to different regions of analysis—Gauss's sums, the Zeta and Gamma functions, the continuation of power series, and the asymptotic behaviour of integral functions—are so numerous that it is impossible to enumerate them here. I will only cite the beautiful formula which defines the behaviour of the function

$$F(x, s) = \frac{x}{1^s} + \frac{x^2}{2^s} + \frac{x^3}{3^s} + \ldots$$

near $x = 1$, viz.,

$$F(x, s) = \Gamma(1-s) \left(\log \frac{1}{x}\right)^{s-1} + \sum_0^\infty \zeta(s-\nu) \frac{(\log x)^\nu}{\nu!},$$

which solves completely a problem which has exercised the ingenuity of many mathematicians.

M. Borel's series of monographs is almost indispensable for anyone who is engaged in research in the theory of functions, and with the possible exception of M. Borel's own first two volumes, M. Lindelöf's contribution certainly seems to me the best.   G. H. HARDY.

Introduction to the Infinitesimal Calculus. By H. S. Carslaw. Pp. 103. (Longmans, Green & Co., 1905.)

Kurze Einleitung in die Differential- und Integral Rechnung. Von Irving Fisher. Pp. 72. (Teubner, 1904.)

These two little books are very much alike in character. Each is designed to serve as a short introduction to the Calculus sufficient for the purposes of a special class of students. Mr. Carslaw is thinking of engineering students, Dr. Fisher of students of economics, and the former naturally require a good deal more mathematics than the latter; and so, although the range of the subjects treated is much the same in the two books, Mr. Carslaw generally gives us a good deal more in the way of detailed applications. Each author cherishes a hope that his book may be useful as an introductory course even for mathematical students. In Mr Carslaw's case this hope is certainly justified. I have myself subjected his book to the best of all possible tests, actual use in teaching backward students, and find it admirably adapted for the purpose. But I am afraid there is hardly enough detail in Dr. Fisher's book to make it very useful in such cases.

A great merit of Mr. Carslaw's book is his treatment of the elements of "Conics" in the only sensible way, viz.: as an easy illustration of the processes of the Calculus. He only devotes one chapter to the Conic Sections, but a great many of their elementary properties for which there is no room in the text are introduced in admirably chosen examples. It is a very great help to have an easy book from which one can teach Analytical Geometry and the Calculus at the same time. I wish that Mr. Carslaw had seen his way to include the elements of the theory of e^x and $\log x$ as well. These functions ought only to be introduced after the notions of the Calculus have been mastered, and Mr. Carslaw might have done it all in a very few pages. He says "we assume a knowledge of the properties of the following series:—$e^x = \ldots$, $a^x = \ldots$, $\log(1+x) = \ldots$." But the kind of person who will use this book has really no such knowledge at all: at least that is my experience—and there really is no book in which exactly what is wanted can be found. It is the treatment of the exponential and the logarithm which seems to me the least satisfactory feature in both books. Both authors assume too much, and neither makes it very clear precisely what he is assuming. On the other hand both authors handle the troublesome (though of course not really difficult) subject of "differentials" in a lucid and unobjectionable way. G. H. Hardy.

Einleitung in die Funktionentheorie. Von O. Stolz und J. A. Gmeiner (II. Abtheilung, pp. 243-598, Leipzig, Teubner, 1905).

The two parts of Stolz and Gmeiner's *Theoretische Arithmetik*, and the first part of the *Einleitung in die Funktionentheorie*, have been already noticed in these columns, and it is sufficient praise of the present volume to say that it is in every way as excellent as its predecessors. The 'second edition' of Prof. Stolz's well known *Vorlesungen ueber allgemeine Arithmetik* is now complete. No part of the original work has been more thoroughly revised or more judiciously enlarged than that which furnishes the subject matter of this volume.

A new chapter has been added on the Weierstrassian notion of the analytic function. After reading it one is disposed at first to regret that the authors do not go further into the developments of the theory, but it is probable that they have gone as far as it is wise to go without invoking the aid of the differential coefficient and the integral.

It is not easy to find anything to criticise adversely. The authors cannot lay claim to the conciseness of Jordan or the sprightliness of Picard: but in lucidity, thoroughness, and consistency of purpose they yield to no one, and although the book is not exactly easy reading it is never unreasonably difficult or heavy. It is enlivened with a large number of excellently chosen examples, many of them interesting and important theorems in themselves.

If there is one chapter which seems to me better than the rest it is chapter viii., " Die Kreisfunktionen für komplexe Werte des Arguments." The sections dealing with the equation
$$e^{\xi i} = \cos \xi + i \sin \xi,$$
the binomial theorem for a complex variable and exponent, and the functional character of the logarithm, are particularly good. The excursion into the theory of the Bernoullian numbers, and their use in asymptotic approximations, is very welcome. The chapter on Infinite Products contains a number of very useful theorems concerning products whose factors are functions of a variable, which have so far only appeared in memoirs, if at all. The chapters on continued fractions, it must be confessed, are a little less attractive; but continued fractions, indispensable and interesting as they are, are particularly difficult to treat in an attractive manner. G. H. Hardy.

REVIEWS.

A First Course in the Differential and Integral Calculus. By W. F. Osgood. Pp. xv, 423. New York: The Macmillan Company. 1907.

A Treatise on the Integral Calculus, founded on the Method of Rates. By W. Woolsey Johnson. Pp. v, 440. New York: John Wiley & Sons. 1907.

No one who wishes to realise the enormous advance that has been made during the last ten or twenty years in the teaching of the Calculus could do better than to read and contrast these two books, published in the same country and in the same year, but differing completely in their objects and methods and representing one the old school and the other the new.

It is no disparagement of Prof. Johnson's book to say that it might have been written thirty years ago. It is a book of a frankly old-fashioned type, another *Todhunter* or *Williamson*. It may be better or worse than *Todhunter*; on the whole we prefer it to *Todhunter*: but it is with such books as *Todhunter* that we must compare it.

We must not be understood as implying that Prof. Johnson's book may not be of considerable utility. There is plenty of room for an improved *Todhunter*. It is, as Prof. Osgood himself remarks, of the utmost importance that the formal side of the Calculus should be taught thoroughly in a first course; no one can be a good mathematician who cannot differentiate

$$\arccos\left(\frac{a\cos x + b}{a + b\cos x}\right),$$

or show a little ingenuity in devices of integration. And it is at anyrate open to argument that the best way to impart this most necessary training is by means of a course in which the theoretical difficulties clustering around the foundations of the Calculus are deliberately or unconsciously ignored. That some of these difficulties should, at first, be passed over in silence, or almost in silence, is indeed obviously necessary. It would be ridiculous to insist that nobody should learn how to determine maxima and minima until he can prove that a continuous function attains its upper and lower limits, or to calculate areas until he can prove their existence. But there are different ways of passing over difficulties. We may simply and absolutely ignore them: that is a course for which there is often much to be said. We may point them out and avowedly pass them by; or we may expand a little about them and endeavour to make our conclusions plausible without professing to make our reasoning exact. There is only one course for which no good defence can ever be found. This course is to give what profess to be proofs and are not proofs, reasoning which is ostensibly exact, but which really misses all the essential difficulties of the problem. This was Todhunter's method, and it is one which Prof. Johnson too often adopts.

Thus, when he professes to prove that "an integral can be differentiated with respect to a quantity independent of the current variable and the limits by differentiating the expression under the integral sign," he is professing to prove what he himself later on acknowledges to be untrue by arguments entirely destitute of validity. And what is gained? Why not simply say, "it is natural to suppose, and can be proved to be in fact often true, that an integral ..."? The reader would find the applications of the principle just as interesting, and the author, when confronted with the necessity of explaining why what he says he has proved is not true of the integral

$$\int_0^\infty \frac{\sin ax}{x}\,dx,$$

would not have been obliged to invent the untrue and absurd explanation that "this results from the fact that, although a occurs in it, the expression is not really a function of a."

R7 *Mathematical Gazette*, 4 (1907), 307–9.

But let us attempt to judge the book by more appropriate standards. The first two chapters, on methods of integration, seem to us fairly good and interesting, though far from ideal. There is a lack of system about the treatment. Surely, in 125 pages, more might be said as to what are the general classes of functions whose integrals can always be found. It is not even pointed out that
$$\int R(\cos x, \sin x)\,dx$$
can always be evaluated by the substitution $\tan \tfrac{1}{2}x = t$, although this substitution is used in particular cases; and the treatment of algebraical functions is equally unsystematic.

The most interesting sections are those on approximate integration (in Ch. III.) and on applications to probability (in Ch. IV.). The chapter on definite integrals (Ch. V.) is deplorable theoretically, but contains many interesting results and examples. But we could wish that Ch. VI. had been omitted. The end of a text-book on the Integral Calculus is not the right place for an introduction to the theory of functions of a complex variable, and the introduction itself is neither sound nor particularly clear.

Prof. Osgood's book is, of course, a work of an entirely different character. His task is a much more difficult one, for he is showing the way over a road the roughness of which he knows. On the whole, his book seems to us the best elementary treatise on the Calculus in the English language; only *Lamb* and *Gibson* seriously compete with it, and of the three books the latest seems to us the most lucid, the best arranged, and (what is very important) the most *level* in its standard of interest and difficulty. Unfortunately there is hardly enough technical and analytical complication to suit a Cambridge course. If there is one chapter which we would single out for praise it is Ch. XIV., on "partial differentiation," "the total differential," "small errors," and so on; or perhaps Ch. X. on "Mechanics," the 30 odd pages of which might well supersede the whole of some well-known text-books of "Dynamics."

The book is one, we fear, likely to make considerable demands on the knowledge and discrimination of the teacher; for Prof. Osgood is a firm believer in a *gradual* introduction to fundamental difficulties, and anyone using the book as a text-book must be ready to parry the occasional awkward question which comes a little before its time. For this reason we sometimes wish for more detail in the text. It is a big jump from the rather innocent attitude towards limits adopted at the beginning to the formal definition for functions of two variables given in Ch. XIV. In the case of one variable no such definition is ever given. And we should have liked to have seen included at anyrate a short and popular sketch of Dedekind's theory of number. It is not very difficult to explain in general terms, and enables a good many gaps to be filled in, notably in connection with the 'Fundamental Principle' (p. 246) that an increasing function approaches a limit or tends to infinity.

One minor point suggests itself. We do not like the notation "$\lim_{x \doteq a}$." Messrs. Leathem and Bromwich's "$x \to a$" is much clearer and much less ugly. And why "$\lim_{n=\infty}$"? It seems to us even more essential to use a different symbol here. For Prof. Osgood does not need to be told that nothing is ever "equal to ∞."

G. H. HARDY.

An Introduction to the Study of Integral Equations. M. BÔCHER. (Cambridge Tracts in Mathematics and Mathematical Physics, No. 10.)

Prof. Bôcher's tract is rather different in character from any of its predecessors in the Cambridge Series. Most of these have dealt with subjects selected from "classical" theories, with the general outlines of which every mathematician is familiar. The general theory of Integral Equations, on the other hand, is a product of the last ten years; it is still possible to be a mathematician of the highest reputation and to know next to nothing about it.

It has hitherto been very difficult for anyone who has not kept pace with the development of the theory from the beginning to make good his deficiencies afterwards. The subject is intrinsically difficult, and has been approached from different points of view, which have at first sight but little in common. The literature is scattered, and new contributions appear almost every month. It has been a severe handicap to anyone who wished to "get up" the subject that there has been no connected account of those parts of the theory which may be regarded as tolerably complete. At the same time, so much still remains confessedly incomplete that a treatise on the subject would probably be superseded almost as soon as it was written, and it is not surprising that no one has been willing to write one. The publication of a "Tract" is an ideal compromise, and Prof. Bôcher's should be one of the most valuable of the series.

The most striking characteristic of Prof. Bôcher's exposition is the closeness with which he follows the historical order of the development of the theory. Here, we think, he has probably been wise, though the method is one that has its disadvantages as well as its advantages. The advantages are obvious; the historical order is always the most interesting and the easiest to follow, and the reader is better able to preserve his sense of proportion when different sides of the theory are presented to him in turn. The chief disadvantage is that, when the author's space is as strictly limited as it is here, and when so much of it is devoted to the work of the pioneers, the more systematic modern theories are apt to receive less than their due. And in this tract we certainly wish that less space had been given to Volterra and more to Hilbert and his followers, and in particular that room had been found for some general account of the connection between Integral Equations and Differential Equations, and of the application of this theory to expansions such as Fourier's—to us at any rate the most remarkable and interesting of all its applications.

Still, it would be unreasonable to expect everything in 70 pages, and the wonder is that Prof. Bôcher has been able to give us so much as he has without compressing his argument beyond the limits of intelligibility. He is, indeed, considering the amount of information he contrives to give us in so short a space, extraordinarily lucid and readable throughout, and almost succeeds in making a difficult subject seem easy; and it is in many ways an excellent thing that he should have given us so full an account of the work of Abel and the other precursors of the theory—work which those who are familiar only with the writings of the German school might be in danger of forgetting.

May we make one remark of a technical character with reference to the formula (p. 6, f.n.)

$$\frac{d}{dx}\int_a^x \psi(x-\xi)\phi(\xi)\,d\xi = \int_a^x \psi(x-\xi)\phi'(\xi)\,d\xi,$$

where $\psi(x-\xi)$ is a function such as $(x-\xi)^{-s}$ $(0 < s < 1)$ and $\phi(a) = 0$? The simplest method of procedure seems to be to put $x - \xi = u$ before differentiating, when the result follows at once. This method also leads to the more general formula

$$\frac{d}{dx}\int_a^x \psi(x-\xi)\phi(\xi, x)\,d\xi = \psi(x-a)\phi(a, x) + \int_a^x \psi(x-\xi)\left\{\frac{\partial\phi}{\partial\xi} + \frac{\partial\phi}{\partial x}\right\}d\xi.$$

G. H. HARDY.

Leçons sur la théorie de la croissance. E. BOREL. 1910. (Gauthier-Villars.)

It is now twelve years since the appearance of M. Borel's *Leçons sur la théorie des fonctions*, the first of the admirable series of monographs on the Theory of Functions published under his direction. The object of the series was to provide connected and reasonably compact accounts of modern developments in analysis that have not yet found their way into the standard books. They were to occupy a position half-way between the *Traité d'Analyse* and the original memoir. Beyond question the series has been a most brilliant success, and many of its volumes are quite indispensable to any serious student of analysis: one has only to think of M. Lebesgue's *Leçons sur l'intégration*, of M. Baire's *Fonctions discontinues*, of M. Lindelöf's *Calcul des résidus*, or, above all, of M. Borel's own *Fonctions entières* and *Séries divergentes*. We are therefore naturally reluctant to use any language save that of praise in connection with an addition to the series. But we must confess that we are disappointed with M. Borel's latest volume.

Its cardinal defect is that it falls asunder into two almost entirely disconnected parts. In the first 70 pages we have the pure *Théorie de la croissance*—the theory of orders of greatness and smallness, the Infinitärcalcül of Paul du Bois-Reymond—or rather, we have a little of it; and there is no question that it is a fascinating subject, and that much of what M. Borel tells us is very interesting. Then we have two long chapters of analytical and arithmetical applications, and these chapters, too, contain interesting results, and a few that we had not seen before. But they are choked by a mass of work with which almost everybody who reads the book is bound to be perfectly familiar. Is this really the place for yet another account of the ordinary theory of the Gamma-function (pp. 88-104) or the elements of the theory of continued fractions (pp. 127 *et seq.*)? Surely we are entitled to expect something a little more exciting from a writer of M. Borel's reputation and originality. "Les Chapitres d'applications," says M. Borel, "ont été rédigés de manière que leurs résultats essentiels puissent être compris indépendamment du système de notation exposé dans les premiers Chapitres," and this is only too true. Whether the first chapters had been written or not, the greater part of the last two (which form more than half the book) might have been taken bodily from almost any *Traité d'Analyse*.

The fact is that the system of notation designed to represent the varying rates of increase of different functions, and developed with such detail by M. Borel in the early part of this book, has not, ingenious as it is, yet found any very serious application. It is still something of a "chinoiserie"; all the important results may be stated just as simply and clearly without it. M. Borel represents the rate of increase of x^n by n, of e^x by ω, of $\log x$ by $\frac{1}{\omega}$, of ax^n by $\frac{1}{\omega}a\omega n$, of $e^{\sqrt{(\log x)}}$ by $\omega \cdot \frac{1}{2} \cdot \frac{1}{\omega}$, of $\sqrt{\{12x^7 + 3\}}$ by

$$\frac{1}{2} \cdot \frac{1}{\omega} \cdot 3 \cdot \frac{1}{\omega} \cdot e^5 \cdot \omega \cdot 4 \cdot \omega \cdot 7,$$

and so on. He makes a distinction between the *identity* and the *equality* of such orders. Thus the order just written is equal but not identical to F_2, the order of the simple function $x^{\frac{7}{2}}$. He introduces brackets to assist him in denoting the orders of functions when they are only approximately known. Thus a function $f(x)$ such that

$$x^{-n+\epsilon}f(x) \to \infty, \quad x^{-n-\epsilon}f(x) \to 0,$$

for any positive ϵ, however small, is said to have the order (n). He studies the differentiation and integration of such orders, and so on. There is much that is interesting, but it does not seem to lead anywhere, and we wish that M. Borel, instead of dwelling on this part of the subject so long, had told us more of the *Infinitärcalcül* as originally developed by Du Bois-Reymond.

We are sorry that we cannot commend this volume with more enthusiasm. It does not maintain the standard of many previous volumes of the series. That is a very high standard, but it is the standard by which M. Borel must expect to be judged.

G. H. HARDY.

R9 *Mathematical Gazette*, 5 (1910), 210—11.

Die complexen Veränderlichen und ihre Funktionen. By G. KOWALEWSKI. Pp. ii+455. 12 m. 1911. (Teubner.)

This is a charming book, written not only with extreme clearness and precision, but also with a freshness and originality seldom to be found in books which purport to be elementary text-books. When I began reading it I was not acquainted with any of Prof. Kowalewski's writings, but I had not spent an hour over it before I went out and ordered his *Grundzüge der Differential- und Integralrechnung*, of which it is a continuation. I recommend it with confidence to all who are interested in the foundations of the theory of functions of a complex variable.

The book opens with a long chapter dealing with the algebra and geometry of complex numbers, and containing a good deal of elementary group-theory. There are sections, for example, on finite groups of linear transformations, on the equivalence of positive quadratic forms, and on the modular group.

The second chapter is short, but particularly interesting. Some classical results in *Mengenlehre* are proved, first for a sphere, and then for a plane whose points have a (1, 1) correspondence with the points of the sphere. We are then introduced to the notions of a *convex* set of points, and of the least convex set which includes a given set; and these conceptions are used for the purpose of extending the first and second mean-value theorems to integrals involving complex functions of a real variable. Finally Taylor's theorem is proved for such functions. All this work is presented in a way very novel in a text-book, and shows Prof. Kowalewski quite at his best.

The third chapter deals with the definition of functions of a complex variable, elementary properties of complex power-series, the logarithmic and exponential functions, and so on, and is on more familiar lines. In the fourth the notions of a *path*, a *rectifiable path*, and an *integral along a path* are discussed with extreme care, and applied to functions defined by power-series in z or $1/z$. The fifth is the most valuable chapter in the book, containing as it does about the best and the most complete discussion of Cauchy's Theorem that I have seen.

The theorem is first established for a rectangle, by a modification of Goursat's method due in substance to Prof. E. H. Moore. We then pass to a series of theorems concerning real functions of two real variables. A function $f(x, y)$ is said to have an *eigentliches Differential* if $\frac{\partial f}{\partial x}$, $\frac{\partial f}{\partial y}$ exist, and

$$\frac{f(x+h, y+k) - f(x, y) - h\frac{\partial f}{\partial x} - k\frac{\partial f}{\partial y}}{|h|+|k|} \to 0$$

when $|h|+|k| \to 0$. This definition, it should be observed, is practically the same as that given by Dr. W. H. Young in his papers in the *Proceedings of the London Mathematical Society*. It is then shown that if u and v have differentials, and $\frac{\partial u}{\partial y} = \frac{\partial v}{\partial x}$, then $u\,dx + v\,dy$ is a differential, and Cauchy's Theorem is deduced by what in the long run amounts to an accurate statement of the classical "double-integral" proof. After this the theorem is extended from a rectangle to any "normal region": finally a more direct proof is given which applies directly to such a region. The contents of the rest of the chapter are of a more ordinary type. But we realise the thoroughly modern character of the book when we find integral equations introduced in connection with the "Randwertaufgabe" for a circle.

I have left myself no space to say much of the last two chapters (theory of series and products, simply and doubly periodic functions, Weierstrass's and Mittag-Leffler's theorem). But this part of the book, though consistently sound and good, does not exhibit the peculiar merits of the author in as striking a way as does the earlier part. It only remains that I should wish the book every success in England as well as in Germany.   G. H. HARDY.

R10 *Mathematical Gazette*, 6 (1912), 345–6.

Estudio elemental de la prolongación analitica. Por PATRICIO PEÑALVER Y BACHILLER. (Tesis, Madrid, 1911.)

The author gives a careful account of certain well known regions of function theory. He makes no pretence to originality; but this thesis is interesting as showing an interest in modern mathematics in Spain. G. H. HARDY.

Leçons sur les principes de l'Analyse. Par R. D'ADHEMAR. Tome I. (Paris: Gauthier-Villars, 1911.)

This volume seems to me hardly worthy of the firm by whom it is published. It is slap-dash and inaccurate, and although it has some novel features, its originality is not of so striking an order as to justify an addition to the long list of *Cours* and *Traités* already on the market. Moreover, it is written in what is to me at any rate a very irritating style. French is admittedly the language best adapted for scientific purposes, and the French of the best French mathematicians is unrivalled for lucidity and charm. But it is a language which has the defects of its qualities: its lucidity can wear thin, and its nervous terseness become jerky. M. d'Adhemar's mathematical style is not like that of Picard or Goursat.

I should add a few criticisms of details to justify these remarks. M. d'Adhemar remarks in his préface: "peut-être l'ordre suivi n'est-il pas assez logique. Dans la théorie des intégrales doubles et des potentiels, je me sers de quelques propositions qui sont démontrées plus loin, dans le Chapitre X. Mais l'exposition paraît ainsi moins lourde...." And at the beginning of Chapter VII. (Les Potentiels) we are told that "dans ce Chapitre, nous admettons les règles de dérivation des séries et des intégrales, théorie qui sera faite, en détail, dans le Chapitre X." What one wants in particular is certain theorems concerning the differentiation of multiple integrals. Now these theorems are not contained in Chapter X. (or elsewhere in the book). Only simple integrals are considered there. Moreover, the discussion in Chapter X. is erroneous. M. d'Adhemar gives as sufficient conditions for the truth of the equation,

$$\phi'(a) = \frac{d}{da} \int_a^b f(x, a) \, dx^2 \int_a^b \frac{\partial f}{\partial a} dx,$$

the conditions (i) that $f'a$ is an integrable function of x, and (ii) that f is a continuous function of x and of a. "On n'a qu'à former $\phi(a+h) - \phi(a)$ et à passer à la limite." If he had taken the trouble to do this he would, of course, have found his conditions inadequate. And when he wishes to extend the result to the case of an infinite upper limit, he gives a faulty definition of the uniform convergence of an integral. A little later on he suddenly introduces a new symbol $\sim$ (without any explanation that I can find), and argues with it in the most casual way. In fact, all this part of the book is inaccurate and ill-digested. And even when it is impossible to say definitely that there is a mistake, the argument is often presented in such a form that the reader's confidence in the author is destroyed.

The best parts of the book seem to me to be the chapter on determinants and linear equations, and some of the sections dealing with integral equations. The parts about double and repeated integrals ought to be good, for the author has had the advantage of seeing the proofs of M. de la Vallée Poussin's new edition of his admirable *Cours*. It is a pity that the account which he gives of M. de la Vallée Poussin's work is not more adequate.

G. H. HARDY.

CORRECTIONS

The displayed equation in the review of d'Adhemar's book should read:

$$\phi'(\alpha) = \frac{d}{d\alpha} \int_a^b f(x, \alpha) \, dx = \int_a^b \frac{\partial f}{\partial \alpha} \, dx.$$

In the following line, in condition (i) read f'_α for $f'\alpha$.

(1) **Higher Algebra.** By CHARLES DAVISON, Sc.D. 6s. 1912. (Cambridge University Press.)

(2) **A New Algebra.** By S. BARNARD, M.A., and J. M. CHILD, B.A., B.Sc. Vol. II. Parts IV.-VI. 4s. 1912. (Macmillan.)

(3) **An Introduction to the Infinitesimal Calculus.** By H. S. CARSLAW, Sc.D. Second edition. 5s. net. 1912. (Longmans, Green & Co.)

(4) **The Calculus for Beginners.** By W. M. BAKER, M.A. 3s. 1912. (G. Bell & Sons.)

(5) **Elements of the Differential and Integral Calculus.** By W. A. GRANVILLE, Ph.D. Revised edition. 10s. 6d. 1911. (Ginn & Co.)

These five books differ widely in scope and still more widely in merit. At the same time they have something in common. They all profess to be "elementary," and they are all concerned, at one point or another, with some of the fundamental notions of analysis.

(1). The publication of this book by the Cambridge Press can only be attributed to reprehensible carelessness on the part of its expert advisers.

A reviewer who receives a book on "Higher Algebra" naturally turns first to the chapters on limits, convergence of series, and the exponential, logarithmic, and binomial series. Dr. Davison's treatment of all these subjects can only be described as hopelessly uneducated. He shows no kind of conception of the logical relations between different parts of the theory, and his definitions and proofs are not only extraordinarily slovenly, but are full of the grossest blunders. It would be waste of time to justify these remarks by a large number of criticisms of detail, but I may give a few illustrations from Chapter V. Dr. Davison defines a *series* in § 64, and a *convergent series* in § 65, both wrongly. As he never defines a *limit* at all, and postpones to § 180 any sort of explanation of what he means by a limit, this is only natural. The most important theorems in the chapter are contained in §§ 68, 73, 75, 78, and 79. In these sections Dr. Davison merely repeats the traditional blunders of English text-books of twenty years ago. In the first three cases, for example, he assumes what he professes to prove.

Nor can I say honestly that I think that Dr. Davison is very much happier when he gets away from the difficulties of limits and convergence. He is completely mistaken, for instance, in supposing that, in § 28, he has established the possibility of expressing a given rational function as a sum of partial fractions. Chapter IV., on "Complex Quantity," is a morass of confusion. I am quite unable to disentangle what the author regards as definition and what as proof. The proof (§ 164) that the arithmetic mean is greater than the geometric is unsound, as has been pointed out by a previous writer in the *Gazette*.* Finally, in § 170, Dr. Davison proves that "every rational integral equation of the nth degree has n, and only n, roots" without a word of explanation that he is assuming the existence of at least one root.

I have selected these examples more or less at random. The book is, in my opinion, a thoroughly bad one, which ought never to have been published. The fact that it appears under the *aegis* of a University Press leads me to think that I should say so with more emphasis than I should have otherwise considered necessary.

(2) Messrs. Barnard and Child's *New Algebra* seems to me a book of an altogether higher class than any other Algebra for schools that I have seen. That such a book should be produced by two authors with a wide experience of elementary teaching is a most encouraging sign of the times, and one particularly gratifying to the professional mathematicians who have protested against the superstition that accuracy is necessarily repellent and that slip-shod half-truths are all that can be interesting or intelligible to beginners.

---

*Mr. Muirhead: *vide Math. Gazette*, vol. ii. p. 283.

The authors understand what is meant by a function, or a limit. They can distinguish between a quantity and a number, between a rational and an irrational number, between a rational number and a rational function, or between a limit and a value. Their standard of accuracy is a good deal higher than that of such a well-known book as Chrystal's *Algebra*; they give a satisfactory discussion, for example, of the infinite geometric series. The gulf which separates this book from the ordinary school text-book may easily be imagined.

In spite, or rather perhaps in consequence of this, the book is bright and interesting throughout, and very seldom difficult, and it has the inestimable merit, rare indeed in a text-book, of being written in clear and decent English. The examples are numerous and well selected. In short, I feel that I can hardly recommend it too strongly to teachers of mathematics.

I have noticed very little in the way of error or obscurity. I do not like the explanation (p. 593) of what is meant by saying that "x is large" or "x is small." To say that "x is large" means *nothing*. *Statements containing* "x is large," on the other hand, may mean something. The phrase is, to use Mr. Russell's language, "incomplete": it is not a constituent of the propositions in which it occurs. I do not suggest that the authors are guilty of any real mistake, but their language seems to me confusing. My only other criticism concerns the last page of all. Here the authors are referring forward to the proof of a theorem to be given in Part VII. Exactly how much they propose to prove there, naturally I cannot say. But the proposition which they quote, regarding the rearrangement of an absolutely convergent series, does not, as ordinarily enunciated and proved, suffice for the application they propose to make of it. Here it is necessary to rearrange an absolutely convergent *double* series. Such a problem may, it is true, be reduced to a problem of a rearrangement of a simple series; but it is a rearrangement of a different and more complicated type than that contemplated in the theorem which they seem to have in mind.

(3) I have already reviewed Prof. Carslaw's excellent "Introduction" in the *Gazette* (vol. iii. p. 274). It is pleasant to find that it should have reached a second edition so quickly.

In the original edition the treatment of the logarithmic and exponential functions was open to criticism. That given in this edition is a great improvement. In spite of the authority of Prof. Carslaw and Prof. Love, I am still doubtful as to the advisability of making so much depend upon the limit of $\left(1+\dfrac{1}{n}\right)^n$, in an elementary book in which an adequate proof of the existence of the limit is out of the question. But I am prepared to believe that this method is found as clear and satisfactory as any other. I wish, however, that in the Appendix, where Prof. Carslaw now places the "older proofs of the theorems regarding the differentiation of e^x and $\log x$," he had explained more clearly exactly where the difficulties of these "older proofs" lie.

(4) I am unable to commend this book on any grounds, or to understand why it should have any prospect of competing with much better books already on the market. It is not attractively written nor, so far as I can see, particularly "practical," and the author's knowledge of the theory may be estimated in ten minutes by any competent critic. Such a critic I would refer in particular to the first four pages, to the discussion of the differentiation of x^n (pp. 9-10), to the treatment of differentials (p. 68), or that of areas (pp. 88-92).

Mr. Baker confines himself for the most part to the reproduction of other people's mistakes, but occasionally indulges in the expression of his own opinions, as when he defines an "independent variable" as "a quantity to which we may assign any value" (p. 1), or says that the differential coefficient "always exists in functions of every kind" (p. 6).

(5) "In this revised edition of Granville's *Calculus* the latest and best methods are exhibited... Those features of the first edition which contributed so much to its usefulness and popularity have been retained..." The author certainly does not err on the side of bashfulness, and invites a reviewer to judge him by the

severest standards. Still, the book, if sometimes a little disappointing, after what the preface has led us to expect, is on the whole quite a good one.

Dr. Granville does not seem to have any very consistent standard as to what may reasonably be regarded as elementary and what not. Thus, on p. 215 he quotes the fundamental theorem that a monotonic sequence tends to a limit or to infinity. The proof he regards as beyond his range. He is thus unable to establish the existence of the exponential limit. He does not even prove that $x^n \to 0$ if $|x|<1$; but this is apparently because he has not seen that any proof is needed. It is only natural, in the circumstances, that his treatment of series should be sketchy and inadequate. I may add that Theorem III. on p. 215, which asserts that the condition "$\lim_{n \to \infty} (S_{n+p} - S_n) = 0$, for all values of the integer p is *sufficient* to ensure the existence of a limit for S_n," is untrue (see Bromwich, *Infinite Series*, p. 46). Dr. Granville professes to be quoting Osgood's *Introduction to Infinite Series*, which I have not at hand; but I cannot believe that he is quoting correctly.

Another part of the book which is unsatisfactory, because the foundations have not been properly laid, is that which deals with integration as summation, the Fundamental Theorem of the Integral Calculus, and so on. On the other hand, formal proofs are sometimes given quite as difficult as those of more fundamental theorems that are omitted: I may instance that of the reversibility of two partial differentiations.

There is too much formal "bookwork." The reader is asked to regard the formula
$$\frac{d}{dx}(\text{arc } cscv) = -\frac{1}{v\sqrt{(v^2-1)}}\frac{dv}{dx}$$
as such, to "memorise" it, and to "be able to state the corresponding rule in words"! The examples are numerous, but on the whole dull and lacking in variety.

I could make many other criticisms of detail; but I do not wish to appear ungenerous to a book which, while hardly likely to excite enthusiasm, has solid merits, and is on the whole clear, readable, and reasonably accurate. To profess that I regard its methods as the "latest and best" would, however, be an exaggeration.

G. H. HARDY.

REVIEWS.

Differential and Integral Calculus. By LORRAIN S. HULBURT. Pp. xviii, 481. 1912. (Longmans, Green & Co.)

This book has many good points. It is readable and fairly accurate, and the examples are simple and interesting. The geometrical parts are the best, but even in the analytical parts a good deal is done correctly which is bungled hopelessly in many books with wide circulations: I may instance the treatment of differentials and the differentiation of x^n. The author does not pretend to be rigorous in his treatment of fundamentals, and the compromise which he attempts to set up between rigour and simplicity is often a very reasonable one. Sometimes he is less successful; and I append a few criticisms of particular passages, which might be useful if the book should reach a second edition.

P. 5. $3-(2-1)=7-(3+2)$ is certainly not an 'identity,' for it contains no variables. The author's example contradicts his own definition.

Pp. 10 *et seq.* The author seems to suggest that a discontinuity of a function is necessarily accompanied by a failure in its definition.

P. 68. The 'definition' of an 'increasing function' is not a definition at all, but a mere tautology.

Pp. 88 *et seq.* The treatment of the exponential limit is bad. The author *does* profess to prove that

$$\lim \left(1+\frac{1}{n}\right)^n = e,$$

when n is restricted to be a positive integer, but his proof is fallacious. He then tacitly assumes that his proof applies to the case in which $n \to -\infty$. He recognises that he has not proved everything, but not that he has proved nothing.

Pp. 240 *et seq.* All the discussion of the definite integral is also bad. It would have been much better to give no proof at all. The proof of the existence of a definite integral is substantially simply a proof that a certain type of area exists. If the latter proposition, which involves all the difficulties of the former, is to be assumed, nothing can be gained by not assuming the former as well. The most that can be done profitably is to give the obvious geometrical reasons for supposing the two problems to be identical.

Pp. 388 *et seq.* To define $e^{i\theta}$ as meaning $\cos\theta + i\sin\theta$, and then to 'deduce' De Moivre's theorem by assuming that $(e^{i\theta})^n = e^{ni\theta}$, is palpably absurd.

Pp. 410 *et seq.* The use of the term 'consecutive' in the treatment of envelopes is unfortunate.

I may add that I cannot regard an elementary mathematical book as a very suitable place for instruction in the rudiments of French, and that in any case I have grave doubts whether "Kō'sheé" is a very accurate phonetic rendering of the name of the great mathematician.

COMMENT

Hardy's comment on the phonetic rendering of the name Cauchy is a little severe when it is recalled that 'Corky' was the common pronunciation in Cambridge before the turn of the century; see p. 217 of E. T. Whittaker's obituary notice of A. R. Forsyth (*Obituary Notices of Fellows of the Royal Society*, vol. IV, No 11 (1942), 209-27).

R14 *Mathematical Gazette*, 7 (1914), 337.

Leçons de Mathématiques Générales. Par L. ZORETTI. Pp. xvi+753. 1914. (Gauthier-Villars.)

This is a very remarkable book, to which I should like to call the attention of schoolmasters and of all whose business it is to teach mathematics to intelligent pupils who are not mathematical specialists.

I should hardly have thought it possible to cover so much ground in a single volume. M. Zoretti assumes no knowledge in his reader beyond that of a little geometry, algebra, and trigonometry; and he carries him far beyond the limits of Part I. of the Mathematical Tripos. There are chapters, for example, on elliptic functions, Fourier's Series, partial differential equations, vectors in space, and elementary differential geometry of two and three dimensions. And in spite of the extraordinary variety of the subjects with which he deals, M. Zoretti always seems to be treating them at his leisure and to have plenty of space to spare.

Naturally M. Zoretti does not profess to give a full and rigorous treatment of fundamentals. He does not prove, for example, that a series of positive terms must converge or diverge to infinity, or establish the existence of implicit functions or of the definite integral. But he is always perfectly clear and consistent. "Ma méthode est ... admettre franchement tout ce qui présente des difficultés sérieuses, ou simplement de trop longs calculs; donner, quand c'est possible sans longueurs, soit la démonstration, soit au moins ses grandes lignes."

Herein lies the great difference between such a book as this and the majority of the English books on "Practical Mathematics" which appear in such profusion nowadays. M. Zoretti is "practical" enough. He has a most admirable chapter on methods of numerical calculation, calculating machines, and so forth, a chapter which, ignorant as I am in such matters, I read with the greatest interest: and in general he has the possibility of practical application always before him. But with him to be "practical" does not mean, as too often with English writers, merely to be ignorant of all that makes mathematics a science worth learning. For M. Zoretti is a real mathematician who has done distinguished work in the theory of functions, and knows perfectly well where the real difficulties of analysis lie.   G. H. HARDY.

Les spectres Numériques. By M. Petrovitch. With preface by M. Emile Borel. 1919. (Gauthier-Villars.)

The defect of this book is that there is nothing in it. It may seem very improbable that a book published by Gauthier-Villars, and introduced by M. Borel, should contain no proposition of interest. All that I can say is that I can find none, and that, reading between the lines of M. Borel's preface, I am inclined to suspect that that very eminent mathematician is of approximately the same opinion.

Given a series of integers, say
$$31, 17, 3, 169, 24, \ldots,$$
we can, in an infinity of ways, embody them, interspersed with zeros, in a sequence
$$31017000300169000024\ldots;$$
and such a sequence may be called a "numerical spectrum". Again, given a function $f(z)$, there may be an operation Δ such that
$$\Delta f(z) = a_0 + a_1 z + a_2 z^2 + \ldots$$
is a power series with integral coefficients. From these coefficients we may form a "spectrum", and we may, if we please, call this a "spectrum of f relative to Δ".

On these foundations M. Petrovitch builds an elaborate structure of definitions. These would be justified if some application of interest could be found for them. All that appears from the book is that M. Petrovitch has found none.

G. H. Hardy.

The Theory of the Imaginary in Geometry. By J. L. S. Hatton. Pp. vi + 216. 18s. net. 1920. (Camb. Univ. Press.)

In writing this book Prof. Hatton had a great opportunity; for the subject is one of extreme importance both logically and didactically, and there is no English treatise* in which it is dealt with adequately. The book, moreover, shows abundant evidence of thought and ingenuity. If then I am disappointed with the result, it is because I differ fundamentally from the author in my judgment as to what methods of developing the theory are, in the light of modern knowledge, the best.

Prof. Hatton's outlook is substantially that of von Staudt. The elegance of von Staudt's theory is beyond question, and it is even now of much historical importance. Von Staudt's point of view may be summarised shortly as follows. Real geometry having been established, or being taken for granted, imaginary geometry is to be grafted on to it. The process is essentially similar to that by which the theory of irrational number is grafted on to that of rational number in analysis. "Real" geometry is found to be honeycombed with exceptions and distinctions until it has become aesthetically intolerable. Homographies may or may not have double elements, straight lines may or may not meet conics; just as, in the arithmetic of rationals, $x^2 = a$ may or may not have roots. And the process by which these intolerable anomalies are banished is simply that of replacing the rational number in one case, the "real point" in the other, by some wider and more complex logical construction. In the one case the *deus ex machina* is the *section* of Dedekind, in the other the *involution*.

We begin by replacing *pairs of real points* by *involutions*, the involutions of which they are the double points. These involutions are necessarily hyperbolic, and every proposition about two real points may be translated into the new language and restated as a proposition about a hyperbolic involution. But, as soon as this step has been taken, the restriction to *hyperbolic* involutions appears to be entirely artificial; most of our propositions may be stated in a form in which they are equally true when the involutions considered are elliptic. We obtain a geometry of involutions valid whether the involutions considered have double points or not; but its propositions correspond to the propositions of the older geometry in the one case and not in the other. The older language, however, is more natural and more suggestive, and in order to preserve it we agree to speak not of an elliptic involution but of a "pair of complex points". The pair of complex points *is*, by definition, the involution. "Complex point" is an *incomplete symbol*, to use Mr. Russell's illuminating phrase.

Prof. Hatton does not seem to me, in his opening sections, to state this point of view with all the clearness that is desirable. He begins with an "Axiom": "every overlapping involution determines a pair of points . . ." To this, I think, there is an obvious and unanswerable objection. These "points" are *ex hypothesi* not "real", for if they were the axiom would be false. They are "complex"; but "complex points" have not been defined, and therefore the axiom is meaningless. Prof. Hatton is, in short, trying to state as an *axiom* what can, from the nature of the case, be only a *definition*.

The point may seem abstract and "philosophical"; and certainly it is irrelevant to the later development of the theory. Its logical importance, however, is obvious, and its pedagogical importance, I should have thought, even more so. It is sure to puzzle a student; the more intelligent he is the more certain it is that he will be puzzled; and if he ultimately decides that Prof. Hatton's axiom is devoid of meaning, I think he will be entirely in the right.

---

* No reader of Prof. Veblen's admirable *Projective Geometry* will suspect me of using "English" to include "American". I ought perhaps to except Mr. Mathews' *Projective Geometry*; but Mr. Mathews passes over these particular matters rather lightly.

In any case this is not my main point. My principal objection to Prof. Hatton's treatment is that von Staudt's procedure has long ceased to be a natural one, or one suitable for instruction of a comparatively elementary kind. It was natural only so long as it could be supposed that " real " space and " real " geometry were in some sense logically prior to, or more firmly founded than, "complex" space and geometry; and this is an exceedingly Victorian view. It is natural when we begin, and no doubt we must all pass through it; but an intelligent reader ought to have got beyond it long before he approaches the subject matter of Prof. Hatton's book. Any university student should be able to understand that the propositions of geometry are not affected by holes in a blackboard, or the presence or absence of Einstein's gravitational field; and, so soon as this is grasped, the supposed priority of " real " geometry, resting as it does on a mere naïve appeal to the reputed facts of physical experience, becomes ridiculous, and the adoption of some more abstract view imperative.

If this is granted, two courses are open to us. One is to follow Prof. Veblen, and to develop projective geometry boldly from the outset as an abstract science founded on a definite system of axioms; and this is no doubt the ideal at which we should aim. The supposed reference to real or physical space then stands out clearly as an obvious bogey, and, so far from the complex geometry appearing as a graft upon the real, we find the real geometry appearing as a specialisation of the complex.

This is the ideal: it may be, under present conditions, a difficult one. There is a natural and much easier path, and that is to base the whole theory of the imaginary in geometry upon the ordinary methods of analysis. And in this, for my part, I subscribe to the opinion of Darboux. " Il y a là " (that is to say, in the older geometrical theories) " quelque chose d'artificiel ; le développement de la théorie est nécessairement un peu compliqué. Aussi cette méthode de von Staudt n'a pu réussir à prévaloir. Il semble que, pour l'introduction et l'interprétation des imaginaires, il vaut mieux s'en tenir à la méthode analytique qui repose sur l'emploi des coordonnées rectilignes." To that I would assent whole-heartedly, with a reservation in favour of the study, later at all events, of the exceedingly beautiful axiomatic developments laid down by the modern projective geometers.

I have in any case no sympathy with attempts to treat everything, suitable or unsuitable, by synthetic methods. Elementary synthetic geometry contains a great deal that is very elegant and of very high educational value. But it has been made a fetish in English mathematical education, an excuse for mere triviality, and for ignorance of, or apathy towards, the great lines of advance of modern mathematics. Simple configurations of lines and conics, constructions of the first and second degrees, and so forth, are, after all, a very small part of mathematics, and they should be kept in their proper place.

I am afraid that I have wandered a considerable way from Prof. Hatton's book. It will be seen from what I have said that I am somewhat out of sympathy with its object, and that I regard a good deal of its undeniable ingenuity as misplaced. It is for this reason that I have not entered further into criticisms of detail, particularly with regard to Prof. Hatton's treatment of the trigonometry of the imaginary, about which a good deal might be said.

G. H. HARDY.

REVIEWS.

The psychology of invention in the mathematical field. By J. HADAMARD. Pp. xiii, 143. 13s. 6d. 1945. (Princeton University Press; Humphrey Milford)

This is, apart from one famous lecture of Poincaré *, the first attempt by a mathematician of the first rank to give a picture of his own modes of thought and those of other mathematicians. Professor Hadamard has earned the thanks of all of us for his courage in undertaking such a task, formidable from both its psychological and its mathematical difficulties, and there is one thing at any rate which he has certainly established, that the most important of the qualifications required is to be a good mathematician. He can sometimes quote illuminating remarks from " outsiders ", philosophers, poets, or even literary critics, but nothing appears more clearly from the book than the comparative futility of the speculations of psychologists who have not been mathematicians. Galton alone, of the outside authorities quoted, shows up really well; a few pages from Poincaré or Hadamard about theorems they have found themselves have ten times the value of the theories of Gall, Möbius †, Nicolle, Paulhan, Souriau, and the rest of them. One might add that " inductive " enquiries by questionnaire, such as that conducted by *L'enseignement mathématique* in 1902, have on the whole proved equally unilluminating.

We must therefore all be grateful to Professor Hadamard for this stimulating little volume, written with all the authority of one of the greatest mathematicians of the last fifty years, and with the most charming frankness concerning his own achievements, triumphs and failures alike. Indeed it is the most personal parts of the book which seem to me the most attractive. A lesser mathematician would never have explained to us how, by some odd mischance, he failed to discover " Jensen's theorem ", or how, having found the " Lorentz group " of transformations, he dismissed it as " obviously devoid of physical meaning ". A lesser mathematician may perhaps also derive some comfort from the recital of such misadventures.

I must begin, however, with two criticisms which I hope may not seem ungracious. The book is too condensed—I regret myself that it is not at least fifty per cent. longer. And I regret still more that it is not written in French: after all, there would soon have been an English translation. It is true that Professor Hadamard's English is (as anyone who has talked with him, or heard him lecture, can testify) far better than that of most Frenchmen, most even of those who have lived for years in an English-speaking country. But to write for publication to so general an audience, on a subject so packed with subtleties as this, is a test quite different from conversation, or lecturing, or the composition of a technical memoir. The result is that, while the meaning is almost always clear, and the language generally " correct ", it lacks the idiomatic force of Poincaré's French, and the whole book, especially in its earlier and more definitely psychological chapters, gives a certain impression of crabbedness and congestion. Writing in French, Professor Hadamard would naturally have been more expansive.

In this respect I find a considerable difference between the earlier and later chapters. The first part of the book is dominated by discussions of the " unconscious " and its role in discovery, an obscure subject which might daunt even an experienced psychologist, and here Professor Hadamard is wisely diffident and tentative in his conclusions. The contents of the later chapters are more miscellaneous, and include much more that is both unquestionable and easy to follow.

* Ch. 3 (' L'invention mathématique ') of *Science et méthode*.
† The neurologist, not the mathematician.

I must begin by saying something about the earlier and more difficult chapters. The main facts enumerated in them seem beyond dispute. That unconscious activity often plays a decisive part in discovery; that periods of ineffective effort are often followed, after intervals of rest or distraction, by moments of sudden illumination; that these flashes of inspiration are explicable only as the result of activities of which the agent has been unaware—the evidence for all this seems overwhelming. Poincaré's is the classical example, but a quite ordinary mathematician can recognise similar experiences. How these unconscious activities are related to those of a more normal kind, to fully conscious work or reflection on the fringe of consciousness, how they function and what is the proper language in which to describe them, are terribly difficult questions. But all of us can remember problems, even on the examination level, whose solution has come to us suddenly when we have all but forgotten them, and it is hard to believe that the rest given by forgetfulness is a sufficient explanation.

Poincaré's experiences (in the discovery of the "theta-fuchsian" functions and the transformations which govern them) were exceptional in some ways, but most of us can remember experiences similar in kind; we have only to recall the best work we have done as the result of prolonged and strenuous effort. The typical course of events is something as follows. There is a first stage of fully deliberate activity, possibly with some, but certainly with unsatisfactory results—Poincaré's were entirely negative. Then a rest, complete or partial, compulsory or deliberate, the result of other occupation or diversion to different problems, followed by a moment of sudden illumination. Then a second period of conscious effort, this time successful, in which the broad outlines of the solution become clear. Then, very likely after long delay, the final stage of what Hadamard calls "precision", in which the results are "written up" and put in order, a tiresome and subsidiary but essential process. These four periods seem the minimum, but of course there may be more; Poincaré's experiences were a good deal more elaborate, and he had at least two moments of unexpected inspiration.

The mystery lies entirely in the early stages, and first in the initial stage of "preparation", which Hadamard considers in detail in Ch. 4. It is plain that during this stage, however futile it may have seemed, we have done something essential; we have shaken up our ideas in a way which somehow makes later illumination possible. To "discover" is to combine ideas fruitfully, and we have at any rate formed a mass of combinations. These have seemed useless, and most of them have been really uninteresting; but the process of forming them has been less unfruitful than it has seemed, since it has set in motion the unconscious machinery which would never have operated without such an initial disturbance. It is a most difficult process to describe, and neither Poincaré nor Hadamard does more than suggest vague images; but the vaguest image of a mathematician may be more profitable than the theory of a psychologist who has never made a mathematical discovery in his life, and Poincaré has one of which Hadamard seems to approve. He compares the ideas which are the future elements of our combinations to "the hooked atoms of Epicurus". While our mind was inactive, these atoms were motionless; they were "hooked to the wall", but our efforts set them in motion. "After this shaking up imposed upon them by our will, they do not return to their primitive rest, but continue their dance freely", and have impacts with one another resulting in fresh combinations. In the new combinations thus formed, indirect results of the original conscious work, lie the possibilities of apparently spontaneous inspiration.

So far this is plausible enough, but we have still to *select* from among all these combinations: "invention is discernment, choice", and where and how is this choice made ? This is the most puzzling question, and I cannot feel that either Poincaré or Hadamard points at all clearly to any satisfactory reply. It seems plain that our unconscious activities must have included some process of selection, since most of our unconscious combinations never rise to our consciousness at all. Poincaré and Hadamard find the solution in the unconscious working of our aesthetic sense. " The privileged unconscious phenomena, those capable of becoming conscious, are those which affect our emotional sensibility most profoundly. . . . In what region of the mind does this sorting take place ? Surely not in consciousness, which, among all possible combinations, knows only the right ones. . . . To the unconscious belongs the most delicate and essential task, that of selecting those which satisfy our sense of beauty and are therefore most likely to be fruitful. . . ." It may be so, though I cannot say that I find it very convincing ; but I am no psychologist, and my distaste for all forms of mysticism may be prejudicing me unduly.

It is admitted that there are more commonplace explanations of some at any rate of the phenomena. Our mind, when we return to the problem, is freshened by its rest ; we have escaped from " interferences which block progress during the stage of preparation " ; we have " got rid of false leads and hampering assumptions, and can approach the problem with a more open mind " ; and it is quite natural that we should be more successful. But I am forced to agree with Hadamard that " while such explanations can be admitted in some cases, in others they are contradicted by the facts " : they do not account for experiences like that of Poincaré when he boarded the omnibus at Coutances. We seem driven to admit that " the unconscious is not merely automatic, it has tact and delicacy ", even that " it knows better how to divine than the conscious self, since it succeeds where that has failed ". But I do not *like* this kind of language, and if, with Poincaré, we begin asking " is not therefore the subliminal self superior to the conscious self ? ", then I have an uncomfortable feeling that we are rather near talking nonsense (and I gather from a remark on p. 42 that Hadamard does not altogether disagree with me).

It is something of a relief to pass to the later chapters, which are full of interesting and less controversial matter. In particular Ch. 6 (" Discovery as a synthesis ") is a long and important one. It contains a detailed discussion, with many examples, of the use of *signs*, both by mathematicians generally and by Hadamard himself ; and here I find his conclusions, though they seem to be well substantiated, rather astonishing. He may be justified in his contempt for Max Müller's view that thought is conducted almost entirely by *words*, but he goes to the other extreme ; it seems that his mind, like Galton's, is of just the opposite type. " I insist that words are entirely absent from my mind when I really think " (and here " words " are to be understood in a very wide sense, including, for example, algebraical symbols). " I use them only when dealing with easy calculations ; whenever the matter becomes difficult, they become too heavy a baggage for me. I use concrete representations, but of a quite different nature. . . . Words remain absolutely absent from my mind until I come to the moment of written or oral communication." It is an extreme view, stated with surprising emphasis, but it is supported by definite instances ; Hadamard explains, for example, his " picture " of the proof of the infinitude of the primes. And, although I should have expected his case to be exceptional, he has found, as a result of quite extensive enquiries, that most mathematicians agree with him. There were only two among those whom he consulted who confessed to belonging to the " verbal " type, Birkhoff, who " visualized algebraical symbols ", and Pólya, who seems dominated by words even more than I am myself. For my own part I must confess that I think almost entirely in words and formulae, written if possible, or visualised as printed on paper, and that for this reason I find thought almost impossible if my hands are cold and I cannot write in comfort. If Hadamard thinks with his legs ("except in the night when I cannot sleep, I never find anything except by pacing up and down the room"), then I think with my fingers. It may be a humiliating confession, but surely I have one illustrious companion in sin—did not Euler say, somewhere, that he thought with his pen ?

Another arresting discussion is that, in Ch. 7, of " different types of mathematical minds ", with special reference to the familiar contrast between the " intuitive " and " logical " types. The subject is of course one about which a great deal of nonsense has been talked. Even in 1893 we find Klein declaring that " a strong native space intuition seems to be an attribute of the Teutonic race, while the critical and purely logical sense is more developed in the Latin and Hebrew races "—and we have all heard this doctrine later in much cruder forms. It is pleasant to find that Hadamard preserves his sense of humour, and quotes French dicta of equal absurdity.

Hadamard recalls Poincaré's comparison of two pairs of famous mathematicians. Poincaré, like Hadamard, rises above the nationalistic prejudice which stakes out a claim to the gifts which sound most impressive, and chooses two pairs of mathematicians of the same nationality, Bertrand (perhaps a little overshadowed in such company) and Hermite, Riemann and Weierstrass. And here Hadamard reveals a very interesting difference of opinion. Bertrand, he agrees, " had visibly a concrete and spatial view of every question ", but it seems to him absurd to class Hermite as a " logician ", " Nothing can appear to me more directly contrary to the truth.... Methods always seemed to be born in his mind in some quite mysterious way.... I can hardly imagine a more perfect type of an intuitive mind than Hermite's." He agrees that Hermite " was not used to thinking in a concrete way " and " had a positive hatred for geometry ", and concludes that the association, so often suggested, of physics and geometry with intuition, analysis with logic, may sometimes have very little foundation. I think that one might also quote Ramanujan, an " intuitive " mathematician, surely, if ever there was one, for the same purpose.

About Riemann and Weierstrass, Hadamard agrees better with Poincaré ; and no doubt this antithesis is generally accepted and corresponds broadly to the facts. Yet even here one must be cautious about accepting such distinctions too readily. I do not know whether Hadamard has studied Siegel's analysis of Riemann's *Nachlass* on the Zeta-function. Riemann, we were told, was the outstanding example of a mathematician dominated by broad and general ideas, and here he reveals himself as a formalist quite of Ramanujan's type. It is enough to shake anybody's confidence in these facile generalisations.

Here Hadamard suggests two criteria of his own, based on his earlier discussions of the " unconscious " (while warning us that we must not expect to find them always concordant). First, we might reasonably describe a mind as " intuitive " if its original combinations of ideas are formed in a comparatively deep layer of the unconscious, as " logical " if this layer is comparatively superficial. Secondly we might (reverting to Poincaré's image of the " hooked atoms ") define the intuitive mind as one in which the initial disturbances of the atoms are notably random and scattered, the logical mind as one in which, even from the beginning, they follow comparatively narrow and convergent paths. The criteria seem plausible enough, but it is significant that, in the particular case which Hadamard considers most closely, they lead to contradictory results. Galois, he finds, was a highly " intuitive " mathematician according to the first criterion, a highly " logical " one according to the second.

I have no space to refer to more of Hadamard's many arresting discussions—the whole book is packed with provocative matter. But I cannot refrain from one final expression of satisfaction. In his last chapter he gives a short discussion of the motives which inspire research, and I am naturally delighted to find that his views agree substantially with my own. Indeed he states the case more strongly than I should have dared to do myself. That little first-rate mathematics is done with a view to immediate application; that " it seldom happens that mathematical researches are undertaken directly in view of a given practical aim "; that " practical questions are most often solved by means of existing theories ", and that the applications, however important, are usually remote in time; that, in short, research is normally inspired by " the common motive of all scientific work, the desire to know and to understand "—all this is surely (whatever the ardent young politicians of the Archimedeans may say) common ground among ninety per cent. of mathematicians, and the experiences of an Hadamard or a Hilbert merely confirm those of their humbler colleagues.

G. H. HARDY.

(b) Reviews from *Nature*

Mathematical Analysis.

(1) *The Theory of Functions of a Real Variable and the Theory of Fourier's Series.* By Prof. E. W. Hobson. Second edition, revised throughout and enlarged. Vol. 1. Pp. xvi+671. (Cambridge: At the University Press, 1921.) 45s. net.

(2) *Introduction to the Theory of Fourier's Series and Integrals and the Mathematical Theory of the Conduction of Heat.* By Prof. H. S. Carslaw. Second edition, completely revised. Vol. 1, *Fourier's Series and Integrals.* Pp. xi+323. (London: Macmillan and Co., Ltd., 1921.) 30s. net.

(3) *A Treatise on the Integral Calculus, with Applications, Examples, and Problems.* By J. Edwards. Vol. 1. Pp. xxi+907. (London: Macmillan and Co., Ltd., 1921.) 50s. net.

(1) THE first edition of Prof. Hobson's treatise fell naturally into two parts. The first five chapters were occupied with the theory of aggregates, the general theory of functions, and the theory of integration, while the last two dealt with the theory of series, and in particular with Fourier's series. It is the first five chapters which have developed into the present volume. It was inevitable that a great deal of the book would have to be rewritten, for the theory has developed very rapidly; there was a mass of recent research to be incorporated, and much of the older work has been definitely superseded. The preparation of a new edition must have been a very long and heavy piece of work, and Prof. Hobson is to be congratulated on the progress he has made with so formidable a task.

There is a singular contrast between the two great branches of the theory of functions. The complex theory has always been popular. The power of its weapons is obvious; its methods have a striking, if somewhat illusory, simplicity; and it is fascinating to investigators, to teachers, and to students alike. It is unlikely that the real theory, more abstract and in many ways more difficult, will ever be so generally attractive. Still, times have changed, very largely through the influence of Prof. Hobson himself. The theory is studied seriously even in England, and ignorance of fundamentals is no longer regarded as proof of physical insight or geometrical intuition. Prof. Hobson has every right to be satisfied with his share in this salutary revolution.

It must be admitted that there was some excuse for the conservative mathematician of twenty years ago, and his sneers at a theory which he was too lazy to try to understand. The older theory, the theory of 1900, was not only abstract and difficult, but in some ways really unattractive. There was too little simple and positive doctrine, too many intricate and irritating exceptions. Little could be proved, and the theorems which it was possible to prove were difficult to state in a terse and striking form. The theory of content in particular was obviously imperfect. The theory as a whole seemed dried up and infertile; it is easy to see now how grievously it stood in need of some refreshing storm.

All this has been changed by the rejuvenating influence of the ideas of Borel and Lebesgue. The storm has broken, and the ground has become fresh and fertile once more. There is, indeed, no other region of pure mathematics that has experienced so drastic a revolution. Prof. Hobson's book is the only English book which contains a systematic statement of the revolutionary doctrine, and it is this, above all else, that gives it its unique position.

The importance of the new theories of measure and integration is generally admitted, but their effect on the theory of functions is still very widely misunderstood. They are much more general than the older theories, and it is supposed that, being more general, they must be much more complicated and more difficult to understand. The result is that many mathematicians are too frightened to make any serious attempt to comprehend them. This attitude of panic is based on a complete misapprehension. It is not true that the new theories are much more difficult than the old. It is by no means always the most general and the most abstract that is the most difficult to understand. The trouble with the older theory lay not so much in the inherent difficulty of the subject-matter as in the complexity and clumsiness of the results. The modern theory, in acquiring generality, has acquired symmetry, terseness, and to a great extent simplicity as well. It possesses the æsthetic qualities that are characteristic of a first-rate mathematical science. Its theorems can be stated in a concise and arresting form, and make that appeal to the imagination which enables them to be mastered and remembered. It is much easier to be a master of the new theories than it was to be a master of the old, and it is also much more necessary. A young mathematician who elects to remain in ignorance of them is certain to regret his laziness or obstinacy in years when it is more difficult to learn.

It is, then, Prof. Hobson's chapters on measure (chap. 3) and integration (chaps. 6-8) that are unquestionably the most important in the book. His treatment is much more comprehensive and encyclopædic than that of any other writer. He has three serious rivals, de la Vallée Poussin, Carathéodory, and Hahn. Hahn may be disregarded for the present, as the second volume of his " Theorie der reellen Funktionen," in which the theory of integration is to be developed, has not yet appeared. The works of de la Vallée Poussin (" Cours d'analyse infinitésimale," second and third editions, 1909, 1912, 1914 ; " Intégrales de Lebesgue, fonctions d'ensembles, classes de Baire," 1916) continue to provide the best introduction to the theory. Between Carathéodory and Prof. Hobson it is unnecessary to discriminate, for both are essential for the systematic study of the subject. It is sufficient to say that there is a great deal in this volume which Carathéodory does not touch.

Chaps. 1, on number, and 4, on transfinite numbers and order-types (chap. 3 of the first edition) have not been greatly changed. We must confess that it has always been this part of the book that we like the least. Prof. Hobson often allows himself to use language which suggests the Oxford philosopher rather than the Cambridge mathematician. " The mind " maintains its position in the first sentence of chap. 1 ; " objects for thought " are " postulated " on p. 29 ; a " fundamental difference of view on a matter of Ontology " is mentioned on p. 249. We have an uneasy feeling that if one scratched the mathematician one might find the idealist, and that all these discussions, and especially those which concern the " principle of Zermelo," ought to be stated in a sharper and clearer form.

Chaps. 2 and 3 are concerned with sets of points, the theory of content and measure having very wisely been separated from the descriptive theory. The greatest difficulty is to distinguish the theorems for which Zermelo's axiom is required. We could make some criticisms of detail—we found difficulty, for example, in disentangling the proof that the measure of a measurable set satisfies the postulate (3) of p. 159, tied up as it is with the corresponding proof for the more difficult postulate (4)—but it would be ungracious to insist on such small criticisms of the most comprehensive presentation of the theory.

In chap. 5, on functions of a real variable (chap. 4 of the first edition), there are very many important additions. The ideas of absolute (p. 276) and approximate (p. 295) continuity are introduced. The treatment of functions of bounded variation (we are glad to find Prof. Hobson now adopting the ordinary language) has been materially simplified, and there is a new section (pp. 318-320) on rectifiable curves. The latter part of the chapter includes an account of some of the most recent work of Denjoy, G. C. Young, and W. H. Young concerning derivatives. Above all, there is a discussion of implicit functions, omitted somewhat unaccountably from the earlier edition. This is a most welcome addition, but we are surprised that Prof. Hobson does not state the fundamental theorem (p. 407) in its most general form. No reference to derivatives is necessary, as was made clear by Young, and a theorem more general than Prof. Hobson's is to be found in so elementary a book as the reviewer's " Pure Mathematics."

Finally, chaps. 6-8 contain the theory of integration, and it is here that we find the most that is new. These chapters are naturally far better than the corresponding parts of the first edition, both in completeness and in logical arrangement, for the first edition appeared at the awkward moment when Lebesgue's ideas were new, and the consequences of his work had not been developed to their conclusion. It may be questioned whether the space (eighty pages) devoted to the Riemann integral is not excessive, since so much of the theory is now of historical or didactic interest only ; but Prof. Hobson's object is, of course, to be complete. The importance of the Stieltjes integral is fully recognised in this edition. The last chapter (" Non-absolutely convergent integrals "), dealing as it does with the extreme limits of generalisation of which, in the hands of Denjoy and of Young, the notion of an integral has so far proved to be capable, is very heavy reading ; but to have given the first systematic account of these generalisations is in itself a most important achievement.

It is to be hoped that we shall not wait long for the appearance of the second volume, and the completion of a work which has added so much, not only to the personal reputation of the author, but to the status of English mathematics.

(2) Prof. Carslaw's book was conceived on a much less ambitious scale than Prof. Hobson's, but he too has had to rewrite it and turn one volume into two. This first volume contains pure mathematics only, and there is no reference to any physical phenomenon after the introduction. It is, in short, a treatise on analysis, restricted within certain limits, and written with a special end in view.

Prof. Carslaw confines himself quite rigidly and consistently within the limits which he has chosen.

It was necessary to have definite limits, but we do not agree entirely with his judgment in selecting them. We think that he has made them too narrow, and that he would have written a still better and more attractive book if he had allowed himself a rather wider scope. It is a very good book even as it is, for it is accurate and scholarly, it contains a mass of most interesting and important theorems which it would be difficult to find collected in an equally attractive form elsewhere, and it is written in an admirably clear and engaging style. It also contains an excellent bibliography of the subject.

Prof. Carslaw has gone too far, however, in his anxiety to eliminate the refinements of the modern theory of functions. For example, the notion of a function of bounded variation is quite explicitly and deliberately excluded (p. 207). The only functions admitted—if we confine our attention, for simplicity of statement, to bounded functions—are those which satisfy Dirichlet's famous condition; they have at most a finite number of maxima and minima within the interval considered. Now there is a serious logical objection to a treatment of Fourier's series in which this class of functions is taken as fundamental, an objection which even a physicist might feel. It is an artificial and not a natural class, since it does not form a group for the elementary operations. Neither the sum nor the product of two functions of the class is in general a function of the class; and it is difficult to see why, if a physicist is interested in two functions, he should not also be interested in their sum.

Prof. Carslaw alludes to the notion of bounded variation as "somewhat difficult," and so, no doubt, it is. But the necessary analysis, as presented, for example, by de la Vallée Poussin, is certainly not more difficult than a good deal which Prof. Carslaw includes. It is not more difficult, for example, than the second mean value theorem, or the theory of Poisson's integral, or Pringsheim's discussion of Fourier's double integral, of all of which Prof. Carslaw gives a very careful account. In any case a book may be made much easier by the inclusion of a difficult theorem, if it helps to elucidate the theorems which the book already contains.

It is inevitable that an analyst, reading a book like this, should be longing to go further all the time. No account of the theory of Fourier's series can possibly satisfy the imagination if it takes no account of the ideas of Lebesgue; the loss of elegance and of simplicity of statement is overwhelming. We recognise that it would be unreasonable to ask Prof. Carslaw for an account of the modern theories of integration. We hope, however, that, when next he has an opportunity of preparing a new edition, he will remedy the omission which we have emphasised. He should also certainly include the fundamental theorem that the Fourier constants of any integrable function tend to zero (a rather startling omission), and some account of Parseval's theorem. He would thus add greatly to the value of an already valuable book.

(3) Prof. Hobson gives us the mathematics of 1921, and Prof. Carslaw is not far behind him. Mr. Edwards's book may serve to remind us that the early nineteenth century is not yet dead. He directs our attention to "the admirable and exhaustive works of Legendre, Laplace, Lacroix, Jacobi, Serret, Bertrand, Todhunter, etc."; from which he has learnt, for example, that "a limit may be of finite, infinite, or indeterminate value," that "the processes of integration are necessarily of a tentative nature," and that any convergent series may be integrated term by term. Two proofs are offered of the last proposition. In the first it is stated to be valid "provided the series V itself, and the series V formed by the integrations of the separate terms, are both *absolutely convergent*." Mr. Edwards italicises the last condition, but we have no idea why it is inserted, for there is no pretence of making any use of it, nor is its meaning explained.

It is difficult for a reviewer to know what to say about such a book, except that it cannot be treated as a serious contribution to analysis. Twenty years ago it might have been necessary to establish the point in detail; it would be waste of time now, when the battle for accuracy has been won. There is always the danger, however, that a student who reads a textbook may suppose that the statements which it contains are true. We should therefore state explicitly that the "general theorems" asserted in this book are often false, and that, even when they are true, the arguments by which they are supported are generally invalid.

One ought, of course, to judge the book by a different standard, as a storehouse of formulæ useful for instructional purposes. Of such there is an abundance, including a good many which are seldom found in other books, and often entertaining or even important. We may mention Catalan's formula for the surface of an ellipsoid, results concerning roulettes and glissettes, the theorems of Fagnano, Burstall, Graves, MacCullagh, Schulz, and others. The book, in short, may be useful to a sufficiently sophisticated teacher, provided he is careful not to allow it to pass into his pupil's hands.

G. H. Hardy.

A Source Book in Mathematics.

A Source Book in Mathematics. By Prof. David Eugene Smith. (Source Books in the History of the Sciences, Vol. 2.) Pp. xvii + 701 + 8 plates. (New York: McGraw-Hill Book Co., Inc.; London: McGraw Hill Publishing Co., Ltd., 1929.) 25s. net.

THIS is a very entertaining volume, a surprisingly successful attempt to do what nearly all good judges would have declared to be impossible. Its aim is " to present the most significant passages from the works of the most important contributors " to mathematics " during the last three or four centuries ".

It is easy to think of a dozen excellent reasons why such an attempt is sure to be a failure. Any compilation of this kind is bound to be scrappy, and in many ways unrepresentative, and no two competent mathematicians will ever be found to agree on principles of selection. The original presentation of an important idea is usually by no means the best, either logically or for purposes of instruction. The passages selected must be translated (if the book is to sell in the United States), and, however competent the translators, a great deal of the savour of the originals must be lost in translation. In short, it is very difficult to imagine any class of students or teachers to whom the book can possibly appeal except as a curiosity, and the time and money spent on its production might be used much more profitably in other ways.

The fact remains that Prof. Smith's volume is a very definite success; it is interesting to read and pleasant to possess. The translations are accurate and vigorous, the printing and illustrations excellent, and the general level of scholarship shown by the contributors is very much higher than is usual in scientific compilations. Among them are prominent researching mathematicians such as Bateman, Bell, and Tamarkin, as well as historians like Archibald, Cajori, and Smith; and Prof. Smith, as editor, has combined the activities of his collaborators with quite remarkable skill.

It is interesting to see what mathematician scores most freely, and no one will be surprised to find that it is Gauss. Gauss has five contributions; on congruences, on the law of quadratic reciprocity, on the fundamental theorem of algebra, on regular polygons, and on the conformal mapping of surfaces. Three of these at least would probably have been unanimous selections of any editorial board. Fermat also scores five, but one is merely the famous 'marginal comment' concerning the impossibility of $x^n + y^n = z^n$; Abel and Euler each score three. There is little Newton; but the volume is limited to pure mathematics. It is surely a mistake to have omitted Cantor and Weierstrass entirely, and Riemann is represented only on the geometrical side. I should have preferred to go without Horner's method or nomography, but I can see that a great many people would disagree with me. On the other hand, I was particularly pleased to find Tchebichef's work on primes, and the early writings on non-Euclidean geometry are particularly well represented.

I may conclude by mentioning a few historical curiosities chosen at random. The first use of π occurs in " Synopsis Palmariorum Matheseos: or, a New Introduction to the Mathematics " (1706), by William Jones, who seems to have been quite a good mathematician (p. 346). The addition formulæ for the sine and cosine seem to be due to Dithmarsus and Clavius (Christopher Clavius, "Astrolabium", 1593); Clavius's 'method of prosthaphæresis' is almost an anticipation of calculation by logarithms (p. 459). Continued fractions are first found in Bombelli's " L' Algebra parte maggiore dell' aritmetica divisa in tre libri " (1572, p. 80); Cataldi (1613) gives what is practically the usual symbolism.
G. H. HARDY.

THE CONTENT OF MATHEMATICS

What is Mathematics?
An Elementary Approach to Ideas and Methods. By Richard Courant and Herbert Robbins. Pp. xix+521. (London, New York and Toronto : Oxford University Press, 1941.) 25s. net.

THIS is a thoroughly good book which deserves to run through many editions. It is not (as its title might suggest) a book on mathematical logic or philosophy : it deals not with the *nature* of mathematics but with its *content*. Its purpose is to show, not by general disquisitions but by concrete examples, drawn from almost every branch of pure mathematics, how mathematicians think and what they do.

I know only three good books with similar aims : Whitehead's "Introduction to Mathematics" in the Home University series, Rademacher and Toeplitz's "Von Zahlen und Figuren", and Dresden's "An Invitation to Mathematics" ; and Dresden's is the only one of the three which really challenges comparison. The other two books, excellent as they are, are slighter and much less systematic, and Whitehead's is written more from the point of view of a logician. It is odd that the best two books of the kind should both be American, but the public for which they are designed is no doubt largest in America.

The writing of such a book, for "beginners and scholars, students and teachers, philosophers and engineers", is a formidable task which requires an unusual combination of qualifications. The authors must be professional mathematicians, since no one else can have the reserves of knowledge and the mastery of technique which are essential ; and with these professional qualities they must combine a breadth of interest and a literary skill which very few professionals possess. Finally, they must be *honest*, and set themselves resolutely against any attempt to catch an audience by making their subject cheap.

In this respect the attitude of the authors is irreproachable. They do not pretend that their book is all easy arm-chair reading, or is likely to have a wide vogue among "tired business men". They know very well that mathematics is a serious and sometimes an exacting discipline, and that anyone who wishes to master even a little of it must be prepared to face a certain number of uncomfortable hours. They have done their best to smooth the path, and on the whole with remarkable success ; but they realize, and warn their readers honestly, that no one can learn anything worth knowing about mathematics unless he has "a certain intellectual maturity and a willingness to do some thinking on his own". Very little *knowledge* is required; no more than could be learnt from any good high-school course.

I can think of quite a number of classes of readers for whom the book is just adapted: really clever schoolboys of fifteen, freshmen undergraduates of average mathematical ability, secondary schoolmasters, university lecturers in physics and engineering, even a minority of philosophers. All of them (except the first) will find difficulties somewhere, but a book on mathematics without difficulties would be worthless. Finally, even professionals (though they will naturally skip a lot) will find the book well worth reading. Few of them will really know more than seventy per cent of the contents, and each will probably pick out for special praise the parts with which he is least familiar.

To me, for example, the best chapter is Chapter 5, on topology. Here are Euler's formula for polyhedra, the 'five colour theorem', Jordan's theorem (for polygons), Brouwer's 'fixed point theorem', and many suggestions of further developments. The proof of Brouwer's theorem, in particular, is a model of what a mathematical proof should be. Next I put Chapter 7 (on maxima and minima, with strong emphasis on the geometrical side): Schwarz's 'triangle problem', Steiner's 'road problem', the isoperimetric property of the circle, Bernoulli's investigation of the brachistochrone, and much other attractive matter. Here, on the fringes of the calculus of variations, Prof. Courant is naturally at his best as an expositor. It is a pity that the famous 'ham sandwich' problem should be a little too difficult to include: the two-dimensional analogue is solved in Chapter 6 as an application of Bolzano's theorem.

It is natural that I should find the arithmetical and analytical chapters less entertaining, but I have no serious fault to find with the authors' choice of material. There might perhaps be a little more about logic and 'fancy' algebras. The calculus comes late, in Chapter 8, but that I welcome: it will be good for 'practically minded' readers to learn that 'calculus for engineers' is nowhere near the centre of mathematics.

The chapter which I find the least satisfying is Chapter 4, on projective geometry, based as it is, like the accounts in the school-books, on ratios of lengths and other obviously metrical concepts. The geometers will agree with me, but the difficulty is fundamental, and I do not suggest that the authors are wrong in following (like Dresden before them) the course they take. The subject cannot be ignored but, treated really logically, in the manner of Veblen and Young or even of O'Hara and Ward, it becomes too bulky and formidable for a book like this. It is necessary to be rather illogical; but I think that the authors should have taken their readers a little further into their confidence, by the insertion of a few explanatory pages in which the difficulty is faced squarely. I would suggest this for the next edition: the explanations would demand great expository skill, but nothing which I should judge to be beyond the authors' powers.

I should add finally that the book is beautifully printed and the diagrams particularly well drawn.

G. H. HARDY.

(c) Reviews from *The Times Literary Supplement*

THE PHILOSOPHY OF MATHEMATICS.

The Principles of Mathematics. By Bertrand Russell, M.A., late Fellow of Trinity College, Cambridge. Vol. I. (Cambridge University Press, 12s. 6d. net.)

This book will probably arouse more interest among philosophers than among mathematicians. It is rather exceptional, in England at any rate, for a mathematician to be much interested in the philosophy of mathematics. And this is not unnatural; for that mathematics is in some sense true and valuable is generally admitted now, and never likely to be disputed. There is no reason to fear that the mathematical house has been built on sand, however deep the obscurity in which its foundations are buried. But while the constructive mathematician may remain indifferent to the questions discussed in this book, the philosopher cannot do so; he is bound to attend to any one who professes to explain the difficulties of number and quantity, infinity and continuity, time and space, matter and causality. In short, he is bound to read this book, or as much of it as he can understand—and we confess that unless he is more familiar with mathematical modes of thought than is usual among philosophers we do not altogether envy him his task.

For the book is, on the whole, exceedingly difficult. Of course a book on such a subject is bound to be difficult; but we certainly think that Mr. Russell has made it a good deal more difficult than was absolutely necessary. The truth is that in spite of its five hundred pages the book is much too short. Many chapters dealing with most important questions are compressed into five or six pages, and in some places, especially in the most avowedly controversial parts, the argument is almost too condensed to follow. And the philosopher who attempts to read the book will be especially puzzled by the constant presupposition of a whole philosophical system utterly unlike any of those usually accepted. The doctrines of this philosophy are, in Mr. Russell's opinion, "quite indispensable to any even tolerably satisfactory philosophy of mathematics." "I must leave it," he says, "to my readers to judge how far the reasoning assumes these doctrines and how far it supports them"; and this is, no doubt, the right, indeed the only possible, course, as Mr. Russell does not profess to be writing a treatise on general philosophy. But it certainly will add to the difficulties of the book for philosophical readers. The mathematician who attacks the book unfettered by any preconceived philosophical opinions will, however, be relieved to find that the conclusions of Mr. Russell's philosophy are, on the whole, much more at one with those of common sense than those of any recognized system. The principal of these conclusions are "the non-existential nature of propositions (except such as happen to assert existence) and their independence of any knowing mind" and "the pluralism which regards the world, both that of existents and that of entities, as composed of an infinite number of mutually independent entities, with relations which are ultimate, and not reducible to adjectives of their terms or of the whole which these compose." An entity, that which has being, is "any conceivable term, any possible object of thought, anything which can occur, in any proposition, true or false." Only some among entities are existents. The number 2 is an entity, my present idea of the number 2 an existent. "Every one except a philosopher," Mr. Russell laments, "can see the difference between a post and my idea of a post, but few see the difference between the number 2 and my idea of the number 2." "The number 2 is not purely mental, but an entity which may be thought of. . . . We no more create numbers than Columbus created America." And, generally, whatever we think of is something which is quite independent of our minds, but has to our minds the peculiar relation expressed by saying that we think of it. These views are on the whole taken for granted, though Mr. Russell frequently argues incidentally in favour of one or other of them, "but the fact that they allow mathematics to be true, which most current philosophies do not, is surely a powerful argument in their favour." Certainly they have enabled Mr. Russell to construct a logic of mathematics altogether in advance of any previous system; unless it be that of Professor Frege, of whom we must confess that we had never heard until Mr. Russell introduced him to our notice.

The book is divided into seven parts, of which the first differs in character from the remaining six. The latter are devoted to proving that all pure mathematics follows from twenty premisses, and that all the entities dealt with by pure mathematics can be defined in terms of those which occur in these premisses. Or, rather, Mr. Russell defines pure mathematics as all that can be deduced from these premisses, and shows that this definition includes all that is usually understood by pure mathematics, and, indeed, a good deal more. The object of Part I., on the other hand, is "a philosophical examination of the fundamental concepts accepted by mathematics as indefinable." Mr. Russell advises mathematical readers to begin with Part IV. But we would strongly recommend them not to follow his advice. The first two chapters of Part I. are essential for a proper understanding of the spirit of the book; and the early chapters of Part IV. are so unattractive that the reader who begins with them will be only too likely to end with them. After reading the first two chapters of Part I. he may turn to any chapter whose contents interest him, and read the earlier parts as necessity arises.

The first characteristic of pure mathematics, according to Mr. Russell, is that it asserts formal implications containing variables. Formal implication is a relation which holds between what Mr. Russell calls "propositional functions," and is to be distinguished from *material* implication, which holds between actual propositions. "Socrates is a man," "2 is a man," are propositions, one true, the other false; "x is a man" is neither true nor false, and therefore not a proposition; but it becomes a proposition if for x we substitute "Socrates" or "2" or any other entity. In fact "x is a man" is a propositional function. "Socrates is a man implies Socrates is mortal" asserts a material implication; "x is a man implies x is mortal, for all values of x" asserts a formal implication. But it is not a mathematical proposition, for it contains the "constant" *man*, which is not a logical constant; and the final mark of mathematics is that its propositions contain only logical constants. Mr. Russell enumerates the indefinable logical constants with which mathematics deals, and the indemonstrable propositions which must be assumed concerning them, while admitting that the number of the latter may be capable of further reduction. The philosophical analysis of the logical constants, which occupies the remainder of Part I., raises many difficult questions, especially in the case of the notion of *class*, in which Mr. Russell has discovered a strange contradiction hitherto unresolved. What propositions are taken as indemonstrable is to some extent merely a matter of convenience; and the reader may be surprised to find that the laws of contradiction and of the excluded middle are not among them. He will also be surprised by some of the results, such as that every false proposition implies every other proposition, true or false; and that every true proposition is implied by every other proposition.

In Part II. (Number) Mr. Russell begins the actual development of mathematics. Mr. Russell's view is that no further indefinables or indemonstrables are required in the whole of pure mathematics, which is consequently in a sense hardly distinguishable from symbolic logic. Every number has therefore to be defined in terms of purely logical notions. In mathematics a term is said to be definable in terms of a set of notions when it is "the only term having to certain of these notions a certain relation which itself is one of the said notions." Mr. Russell finds a definition of the number of terms in any class as follows. Two classes are *similar* when their terms can be so correlated that to each term of either class corresponds one and only one term of the other. The number of a class is the class of classes similar to the given class. Thus the number one is the class of all classes each of which is similar to the class constituted by any single term. This, as Mr. Russell admits, sounds very paradoxical. But he has certainly succeeded in showing that the entities thus defined have all the properties of numbers, and so, mathematically, fill their place in a perfectly satisfactory way. Whether the definition presupposes any philosophically more fundamental sense of number (in particular of the number one) is a more difficult question, which Mr. Russell discusses later and answers in the negative. Part III. (Quantity) is "a concession to tradition." Quantity, according to Mr. Russell, is not definable in terms of logical constants, and so "not properly a notion of pure mathematics at all." Moreover, quantity "has lost the traditional importance which it used to possess, owing to the fact that most theorems concerning it can be generalized so as to become theorems on order." This brings us to Part IV. (Order), perhaps the most constructive, and at the same time technically difficult section of the book, but it is hardly possible to give a rapid summary of the highly novel and important doctrines advocated in these two parts.

It is to Part V. (Infinity and Continuity) that most of Mr. Russell's readers will probably turn with the liveliest anticipations of interest. But the reader who is familiar with modern mathematics will find this part easier and less original than any other. For the truth is that the supposed contradictions of the infinite have been scattered once for all by the illustrious Cantor. All that has remained for Mr. Russell is to restate Cantor's doctrines with a philosophical clearness not always to be found in the writings of their originator. His statement and resolution of the classical puzzle of Achilles and the Tortoise is particularly lucid. That there are wholes containing infinitely many parts can hardly be honestly denied. The supposed contradictions of the infinite whole as such, whether stated in a purely numerical or in a spatio-temporal form, arise in the main from the assumption that the whole cannot be similar to the part, in the technical sense of similarity explained above. This is true, and can be proved, of finite wholes; but the assumption that it is therefore true of all wholes is as groundless, as Mr. Russell observes, "as the supposition formerly entertained by the inductive philosophers of Central Africa, that all men are black." Part VI. (Space) and Part VII. (Matter and Motion) conclude the book. Mr. Russell is a firm believer in absolute position in space and time—a view so much out of fashion nowadays that Chapter LI. will be read with peculiar interest. Motion is *merely* the occupation of different positions at different times, and consequently there is no such thing as a state of change, velocity and acceleration are merely mathematical fictions. Mr. Russell thus agrees with Zeno that the arrow at every moment in its flight is truly at rest. Zeno, however, drew the conclusion that therefore the world is in the same state at one time as at another; and here Mr. Russell by no means follows him. In all these discussions Mr. Russell is quite at his best.

But Mr. Russell's Dynamics does not seem to us so good; and in particular his account of causality seems definitely wrong. "Causality is the principle in virtue of which from a sufficient number of events at a sufficient number of moments, one or more events at one or more new moments can be inferred." And Mr. Russell goes on to explain that in actual dynamics, our *data* must be in the whole state of the world at two moments of time, and that from these *data* its state at any other moment can be inferred. But surely this is mathematically untrue. It is sufficient mathematically to know the positions and velocities of all the particles in the world at any moment. Mr. Russell, banishing velocity, must replace it by further *data* of positions. But to do this requires not one more set of positions, but an infinite number of such sets, for the velocity of a particle can only be calculated from a knowledge of its positions at infinitely many times. Can it be that Mr. Russell has been influenced unconsciously by the old-fashioned view that velocity is defined by two consecutive positions at two consecutive times, a view which he would, of course, be the first to repudiate? If a particle is projected vertically upwards, we can calculate its motion if we know its velocity. But we cannot calculate its motion if we only know its original position and its position at one subsequent time. Suppose that it was projected from the ground, and take the second time to be that at which it reaches the ground again. How can we tell that it has not been at rest?

It is impossible in a short review to indicate more than a few of the questions considered in this book, or to do justice to the subtlety and originality with which they are discussed; as for the principal thesis concerning the nature of mathematics, Mr. Russell seems to us to have proved his point. Cambridge is generally supposed to care much for the physical and little for the abstract side of mathematics; this book should do much to dispel so unfortunate an impression.

THE "FALL" OF EUCLID

The Thirteen Books of Euclid's Elements. By T. L. Heath. Three vols. (Cambridge University Press, 42s. net.)

Mr. Bertrand Russell has found a happy illustration of the capriciousness of posthumous fame in the person of Zeno the Eleatic. Zeno, who, if he be rightly interpreted by Mr. Russell and other modern mathematicians, was indeed the most subtle and profound of philosophers, was pronounced a juggler and a sophist by his successors, and ridiculed and contemptuously refuted for over two thousand years. In the middle of the nineteenth century the sophisms were re-established, and "made the foundation of a mathematical renaissance" by Weierstrass, a German professor who had very likely never heard of Zeno, and who was almost certainly unaware that his own work had any philosophical interest at all.

Euclid's case appears, on the face of it, to have been the opposite of Zeno's. Zeno has come to life again after two thousand years; Euclid, we are told, is at last dead, after two thousand years of an immortality that he never much deserved. He was supposed to be rigorous, and so long as it was possible to believe that, his other sins, his circumlocution, his pedantry, and his intolerable dulness, could be forgiven him. But now even his reputation for rigour has been taken away from him; logicians like Hilbert and Mr. Russell have joined forces with the Board of Education, and the schoolmasters who belong to the Mathematical Association, and Mr. Davie Mair and the apostles of squared paper, and at last they have succeeded in killing him. So we are told, and there is no doubt that in a sense it is true. The teaching of mathematics will never be dominated by Euclid again, as it was a hundred or, in the country that has longest been faithful to him, even twenty years ago; and in another twenty years it will be impossible to find a copy of the "Elements" in any public school in England, or anywhere except on the shelves of scholars and professional mathematicians.

Euclid's fall is, in a sense, final and complete. And yet, if we were to imagine him following the course of mathematics since his day, the fluctuations of his own reputation and the rise and decline of his empire in the schools, we should doubt whether the prospect of its inevitable extinction would have caused him much distress. He was, if we may trust a tradition, which it is at any rate pleasant to believe, an extremely sensible person and something of a humorist—"a man of scrupulous fairness and exemplary kindliness," "in no wise contentious, though exact, yet no braggart," tolerant of criticism, even when it was the criticism of stupidity. The modest and genial Euclid drawn by Proclus was the last person likely to have his head turned by the adulation of mathematicians less critical than himself, or to resent the gradual growth of a more discriminating appreciation of his work. Rather, we think, he would have welcomed the change which the last century has brought to his reputation among the few who are competent to understand him fully. For if it be true, in a sense, that the last century has seen the end of Euclid's reign, it is in a sense no less true of him, as of Zeno, that it is only in the last century that the full value of his work has found proper recognition.

The "Elements" is a very great book, the greatest perhaps of all mathematical books; it is no exaggeration to say, as Dr. Heath says, that it is "one of the noblest monuments of antiquity." But by no means all of it is great or even good; much of it, indeed, is positively bad; and Euclid's reputation has suffered not a little in modern times because it is precisely the parts of the

"Elements" that are least good that have attained the widest notoriety and have been drummed into the ears of every unwilling schoolboy. What is best in Euclid is just what is not for the schoolboy or the tiro. It is curious that, although geometry has generally been regarded as, above all sciences, the science of Greece, it is the purely geometrical books of the "Elements," and these only, that are open to serious reproach. Most of us have learnt Book I. by heart, and none of us should regret it, for it contains the theory of parallels, the logical formulation of which was one of Euclid's greatest individual achievements. It contains, too, Euclid's proof of the Theorem of Pythagoras, and other theorems interesting and beautiful in themselves; and yet it can only be described as on the whole a very indifferent performance. The fact is that the Greeks, for all their mathematical genius, never succeeded in formulating anything like a coherent system of definitions and postulates to serve as a foundation for their geometry. It was indeed hardly possible that any one should do so, so long as geometry was conceived as a science referring essentially to an actually existent space. To-day geometry is no longer a science of space. How does Hilbert, for instance, begin his "Grundlagen der Geometrie"? There is not a word that has any reference to space at all. "We think of three different systems of things: the things of the first system we call *points* . . . . the things of the second system we call *lines*. . . . . We conceive these 'points,' 'lines,' &c., as connected by various relations, and we denote these relations by words such as *lying on, between, parallel to, congruent with* . . : the complete and exact description of these relations follows from the *axioms* of geometry. The axioms of geometry form five groups . . ." These severely mathematical points and lines, these "entities connected by relations," "simple and secluded," as Mr. Bradley would say, "beyond belief," would never have satisfied the soul of any Greek geometer. For him points and lines were elements of the space in which we live, full of philosophical content and an integral part of the reality of experience. The result is that Euclid's definitions and postulates contain a great deal that is not mathematics at all, but indifferent popular philosophy. As De Morgan said, "there is a total absence of distinction between the various ways in which we know the meanings of things": and to mix philosophy and mathematics may be good for philosophy, but it works havoc with mathematics. Euclid's treatment of the foundations of geometry, as expounded in the definitions, postulates, and early propositions of Book I., and the corresponding portions of Books III. and XI., must, we are afraid, be pronounced a failure. "A straight line is that which lies evenly with the points on itself"—it is not even decent philosophy, certainly not serious mathematics: it is at best a confession of faith or a popular scientific illustration. Euclid's postulates, as everybody knows, are at once inadequate and redundant. The whole theory of "superposition" is pure and simple nonsense; this, as Dr. Heath points out in an interesting note, was observed by J. Peletier as long ago as 1557; it is the reason, we suppose, why its scope has been so widely extended by modern mathematical reformers. But it would be flogging a dead horse to reiterate the criticisms that have been urged against Euclid with irresistible force by the accumulated authority of modern logicians.

It is more pleasant to turn to those books of the "Elements" on which in the future Euclid's fame will rest. These, it is fairly safe to say, will be the arithmetical books (Books VII.-IX.), Book V., Book X., and (of the geometrical books) Book VI.; and there can be very little doubt that an almost unanimous consensus of mathematical opinion would follow de Morgan in assigning the first place to Book V., and the second to Book X. But if we had to single out one isolated proposition to be saved when all the rest were lost, we would choose that classical instance of the *reductio ad absurdum*, the proof (Book IX., Prop. 20) that there are infinitely many primes. The fifth Book of the "Elements" appears, in the light of modern mathematics, to have been the crowning scientific achievement of Greece. Its symbolism, judged by modern standards is clumsy: that is the first and the last of its faults. Rendered into a modern and a more convenient notation it may still stand, almost word for word, as the one and only possible theory of proportion, as irreproachable and in as little danger of supersession as when Barrow declared that "there is nothing in the whole body of the elements of a more subtile invention, nothing more solidly established and more accurately handled than the theory of proportionals." How subtle and how accurate it is is clearer now, to the mathematician familiar with the work of Weierstrass, Cantor and Dedekind, than it can possibly have been to Barrow or even to De Morgan. There is, as Dr. Heath points out, "an exact correspondence, almost coincidence, between Euclid's definition of equal ratios and the modern theory of irrationals due to Dedekind."

The doctrines expounded by Euclid in Book V. appear to be the discovery of Eudoxus. Their history and the reason for the curious position in his system assigned to them by Euclid have been discussed at length by M. Paul Tannery ("La Géometrie grecque," pp. 95–98), and his conclusions are accepted by Dr. Heath. It seems probable that the principle of similitude, the subject of Book VI., a principle no less "evident" than the postulate of parallels, was, in pre-Euclidean systems as in many modern text books of geometry, introduced much earlier and made far more fundamental than in Euclid. But this principle presupposed a theory of proportion, and the Pythagorean theory of proportion was applicable only when all the magnitudes concerned were commensurable with one another. Accordingly, the discovery of incommensurables by Pythagoras, a discovery explained and developed in the tenth Book of the "Elements," created "un véritable scandale logique" in Geometry, and made it necessary to relegate the principle of similitude to a subordinate place until such time as a theory of proportion should be developed which included incommensurables. This theory was created by Eudoxus, and for that he must rank as the greatest mathematician of antiquity. Euclid had the theory of Eudoxus before him, and it was open to him to place it in the forefront of the "Elements" and to revert to the Pythagorean order. But tradition was by Euclid's time opposed to such a course; moreover, with the instincts of a good teacher, he doubtless felt that the theory, although magnificent, was in no sense elementary; and it is for these reasons that the theory of proportion occurs in the "Elements" where it does, when it becomes essential to further progress, and not before.

The merits of Book X. are of a different kind. The substance of a good deal of it still forms and must always form an essential part of any treatise on elementary algebra. Such proofs as those of the incommensurability of $\sqrt{2}$, of the impossibility of the equation $a + \sqrt{b} = c + \sqrt{d}$, and so on, can never be superseded. But the book as a whole is less remarkable for the discoveries it expounds than for the fact that any Greek should have discovered them. It seems almost incredible that any one, with no algebraical machinery to help him, should have been able to classify all possible numbers of the form $\sqrt{(\sqrt{a} + \sqrt{b})}$ into 25 different species, and to prove that this classification is complete without being redundant, and that every number of one species is incommensurable with any number of any other species. Yet this is what Euclid has done, and the arguments against supposing that he possessed any form of algebraical symbolism seem to be overpowering. Exactly how much of the work is Euclid's own it seems impossible to say: to some extent it is certainly based upon the work of Theaetetus. But as it is precisely the kind of work the merits of which depend to a very large degree upon its presentation, to which the last touch of logical completeness and perfection of form makes all the difference, it would certainly be unfair to deny Euclid a very large share of the credit for the most startling *tour de force* recorded in the history of mathematics.

But it is time that we turned our attention from Euclid to Dr. Heath's magnificent edition. We should like to say that it had superseded all other English editions, but the truth is that there were none for it to supersede. The last complete translation was actually published in 1788, so that, as Dr. Heath observes, Books VII.–X. are "practically inaccessible to English readers in any form." We should therefore have welcomed any edition, even if it had not one-tenth of the merits of Dr. Heath's: and Euclid himself is so interesting that it would be difficult for any respectably competent editor to make him dull. But these three volumes are absolutely fascinating; they are books to take away on a holiday and to read at breakfast and lunch and tea and dinner and in bed.

Few authors, we imagine, make heavier demands on an editor than Euclid; he demands scholarship, an intimate knowledge of the history of mathematics, real expertness in the techniques of elementary geometry, and more than a slight acquaintance with modern mathematical philosophy. Dr. Heath's reputation as a scholar is high. Heiberg's text, which he follows, is admittedly by far the best, and his translation seems to us as clear and as easy to follow as any one could wish. His elaborate, historical and geometrical notes are invariably interesting, and the most difficult parts of his task, the semi-philosophical discussions which arise in connexion with the axioms of geometry and the theory of irrationals, are performed in a thoroughly competent manner. Dr. Heath is, we suppose, a scholar and a mathematician rather than a philosopher, but his philosophy is quite sufficient for his purpose and he seems to have read everything that fairly bears upon his author. As befits an editor, he is a little over-solicitous of Euclid's reputation and a little impatient of modern critics: we cannot believe that there will ever be "a return to Euclid more or less complete for the purpose of standardizing it once more." We hope not; for the reasons that we have already stated, it seems to us the very last thing that any lover of Euclid should desire.

THE NEW SYMBOLIC LOGIC.

PRINCIPIA MATHEMATICA. By A. N. WHITEHEAD and B. RUSSELL. Vol. I. (Cambridge University Press. 25s. net.)

Perhaps twenty or thirty people in England may be expected to read this book. It has many claims to be widely read; all professional mathematicians, for example, can and ought to read it; but it will have to contend with an immense mass of prejudice and misconception, and we should probably be over-sanguine if we supposed that there are half a dozen who will. It is a strange and discouraging fact that mathematicians as a class are utterly impatient of inquiries into the foundations of their own subject. Natural science has always distrusted and despised philosophy. It is hardly too much to say that any determined attempt at accurate thinking is likely to be distasteful to the average scientific mind; and this, although certainly deplorable, is not so strange as it might seem, for ingenuity and imagination, rather than accurate thought, are the ordinary weapons of science. The mathematician has no such excuse. He has received an elaborate logical training and is familiar with what is abstract and remote and unpractical; and it is not unreasonable to expect that he should have learnt to respect the truth and to tolerate, if not to sympathize with, those who, like the authors of this book, seek it with a patience and determination that no difficulties can arrest. Any such expectation will be disappointed. In England we find the authors regarded by mathematicians as amusing cranks. In France we find the great Poincaré, who has a weakness for philosophy to which we owe several most entertaining volumes, pouring contempt on *la Logistique*, and preaching a form of pragmatism as hazy and elusive as any philosopher's. Even in Germany, the home of mathematical precision, we find the successors of Cantor and Weierstrass protesting angrily that to ask really fundamental questions is an indecency and an insult to mathematics. To this sort of expert prejudice it is useless to appeal. But philosophers are more tolerant than mathematicians and more interested in fundamentals; and there are no doubt many philosophers and many laymen who would like to read this book but may be deterred from doing so by misconceptions of a different kind. To such it may be possible to give a little encouragement.

Non-mathematical readers may very naturally be frightened by an exaggerated notion of the technical difficulty of the book. The last page is a very natural place at which to open a book, and the appearance of the last page, with its crazy-looking symbolism, is appalling. And it would be silly to pretend that the book is not really difficult; some of it is very difficult indeed. It is, of course, also true that in reading it a trained mathematician enjoys a considerable advantage; some of the ideas will be more or less familiar to him; he can grasp more quickly the salient features of a symbolism; the general tone of the book is mathematical, and there are parallels and illustrations which it is difficult to appreciate without some knowledge of mathematics; in a word, the authors' minds work naturally on mathematical lines. So much the lay reader must be prepared to find. But he need not be too much discouraged. In the first place, although more knowledge of mathematics would no doubt be a great assistance to him, no profound knowledge is required. All that is wanted is some sort of general familiarity with mathematical ideas. Moreover the mathematician's advantage varies very widely from page to page, and it is just where the book is more difficult that this advantage is least; the easy parts will be particularly easy to the mathematician, the hard parts practically as hard to him as to the outsider. Finally, to be frightened by the symbolism is to run away from the most shadowy of bogeys. A mathematician can master it in an hour, and the non-mathematical reader will find in a surprisingly short time that, so far from being an additional source of difficulty, it is extraordinarily easy to understand, to read fluently, and even to use oneself. The language of modern symbolic logic, vulgarly described after its originator as "Peanese", has been developed under the authors' hands into a vehicle of astonishing flexibility and power, a real triumph of technical skill, fairly justifying their claim that it enables the mind "to construct trains of reasoning in regions of thought in which the imagination would be entirely unable to sustain itself without symbolic help." It is possible that the constant use of symbols has to some extent reacted on the authors' mastery of their own tongue. Certainly, in the explanatory portions (though there are parts that are models of lucid exposition) they are often much less clear; though sometimes, no doubt, this is due simply to the inherent difficulties of the subject-matter, and sometimes to condensation carried too far.

We trust, therefore, that no one who is prepared to recognize the value of work in logic and who appreciates the enormous importance in logic and general philosophy of mathematical concepts, *function*, *class*, *number*, *magnitude*, and so forth, will be too diffident of his ability to cope with the difficulties of this book. It is not a book that many will read right through. But it will be a foolish philosopher who will not make a serious effort to master the most essential sections. The time has passed when a philosopher can afford to be ignorant of mathematics, and a little perseverance will be well rewarded. It will be something to learn how many of the spectres that have haunted philosophers modern mathematics has finally laid to rest. And it is as bracing as a cold bath to turn from the muddy pragmatisms of current philosophy to clear-cut and dispassionate discussions in which it is recognized that words have definite meanings, and that premises imply conclusions, and that careful reasoning is the only method by which we can hope to arrive at the truth.

The first volume of "Principia Mathematica," which is all that has appeared so far, contains most of the work that is likely to be of interest to the general reader. It is true that the two later volumes will include the mathematical theories of infinity and continuity, topics which have distracted philosophers in the past and will no doubt continue to do so until philosophers make a habit of learning a little mathematics. But these theories are by now classical and easily accessible to any one who wishes to become acquainted with them. And this volume contains the part of the work most likely to excite general interest and controversy, an introduction of ninety pages written in the ordinary tongue and summarizing all that is most novel in the general doctrines of the book. There is no doubt that this introduction, dealing as it does with all the most obscure and controversial portions of the subject, is from the standpoint of the general reader too short. Its style is in places almost painfully condensed, and there are passages that are hardly intelligible until we refer forward to the detailed symbolic development which comes later. Moreover, the authors, in their desire to "avoid controversy and general philosophy," are apt to leave the reader uncertain as to what (if anything) in the way of general philosophy their whole treatment presupposes. They would probably say that the mathematical edifice is independent of the precise material with which we fill in the philosophical foundations; and no doubt in this they would be right. But they have naturally found it impossible to carry out their intention of "avoiding general philosophy" quite consistently; and the result is sometimes very puzzling. To take a definite instance—"proposition" must be regarded, for the main purposes of the book, as ultimate and unanalysable. Any analysis of the *meaning* of "proposition" is prior to mathematics. So much the reader will cheerfully accept. But it will come as an unpleasant surprise to him (at any rate if he is not familiar with Mr. Russell's "Philosophical Essays") to be told (p. 46) that "what we call a 'proposition' is not a single entity at all . . .; the phrase which expresses a proposition . . . does not have meaning in itself." It would have been better, we think, if the authors had made up their minds to "face the music," and had begun with a definitely philosophical excursus—a mere ha'porth of tar in the outfit of such a leviathan.

The main thesis of the book is the same as that of Mr. Russell's "Principles of Mathematics"—the thesis that pure mathematics involves no axioms or indefinables beyond those of formal logic. And its cardinal doctrines group themselves roughly into three divisions—the general theory of the variable and the propositional function, the theory of incomplete symbols, the doctrine of types. None of these doctrines will be found entirely novel by students of Mr. Russell's writings; all have been formulated or foreshadowed in the "Principles" or elsewhere. Here they are stated for the first time with an air of finality and as a connected whole.

The theory of propositional functions shows in a striking way how certain kinds of ultimate philosophical inquiry may be irrelevant to mathematics. It is beyond doubt that the "propositional function" is a notion of extraordinary logical importance—mathematics, one may say, is the science of propositional functions. And it is perfectly easy to recognize a propositional function when we see it; "x is x" is a propositional function; when x is determined—when for x we substitute Socrates or Plato—we obtain a proposition. But *what* is a propositional function? The question is "by no means an easy one." When we say "x is x"

> it is plain that, regarded psychologically, we have here a single judgment. But what are we to say of the object of the judgment? We are not judging that Socrates is Socrates, nor that Plato is Plato, nor any other of the definite judgments that are instances of the law of identity,

for we may be quite capable of judging "x is x" even if we have never heard of Socrates or Plato. But we cannot follow the discussion further now. Our object is to point out, on the one hand, that the reader must not expect to find this book free from ultimate doubts, and, on the other that the persistence of such doubts need not imperil the logical superstructure.

The theory of "incomplete symbols" is one of the authors' triumphs; it could hardly be clearer, nor, once understood, more consonant with common sense. It cannot be illustrated better than by the old puzzle—old, that is to say, to readers of Mr. Russell—of George IV. and Scott. George IV. wanted to know whether Scott was the author of "Waverley," and in point of fact he was. It seems, therefore, that what George IV. really wanted to know was whether Scott was Scott. For, if "the author of Waverley" is a definite object a, "Scott is the author of Waverley" means "Scott is a." And this proposition is either trivial or false; trivial if a is Scott, when "Scott is a" reduces to "Scott is Scott," and false if a is anything but Scott. On the other hand, it is perfectly obvious that "Scott is the author of Waverley" is neither trivial nor false. From this dilemma there is only one way of escape—namely, by denying that "the author of Waverley" *is* a definite object a. This is to deny that "the author of Waverley" means anything, and seems at first paradoxical. The paradox is one which soon disappears; the point is, of course, that "the author of Waverley," and "descriptions" in general, mean nothing *by themselves*, though *phrases containing them* often have a perfectly definite meaning. This idea is familiar enough to mathematicians, and in it we have in germ the whole theory of "incomplete symbols," a theory applied by the authors now not only to descriptions but also to classes and relations, though here the theory becomes a little more complicated and elusive.

With the doctrine of types we come to the most difficult, and perhaps the most controversial, theory of the book. This doctrine has been invented for the express purpose of solving a class of puzzles which has tormented generations of logicians. The classical example is the "Epimenides." In its most modern form this paradox is as follows. If I say "I am lying," then, if my statement is true, I *am* lying, and therefore it is false; and if it is false, I am *not* lying, and therefore it is true. Such paradoxes cannot (as is often supposed) be accounted for by holding up one's hands and saying "How absurd!" The doctrine of types, in which the authors find rest from all these puzzles, is in a way the least fundamental, and will probably be found the least satisfying, part of the book; and the authors are careful not to claim too much finality for their solution. There can be no doubt, too, that it does involve consequences likely to startle common sense; it forces us, for example, to believe that a whole series of common words, *true* and *false* among them, have infinitely many different meanings. And there are important points about which the authors leave us doubtful. Are there "infinite types"? Can one tell a lie "of infinite order"? Is it really true, as theologians tell us, that "the finite cannot comprehend the infinite," or is that a mere logical superstition, as De Morgan held? But of two things the authors have convinced us. One is that *some* form of some such doctrine as the doctrine of types is logically indispensable; the other is that there really *are* different meanings of "truth"—that when I say "it is true that 'x is identical with x, for all values of x'" and when I say "it is true that 'this is yellow,'" I do not in the two cases mean the same by *true*. This, we think, they show not only to follow from their premises, but to be convincing to "expert common sense"; and when we have admitted this, we have admitted that the chief paradox of their doctrine has disappeared.

It would be insulting to affix the ordinary labels of praise to a book conceived with so far-reaching an object and on so vast a scale. We may perhaps venture to pick out a minor feature of the book for commendation. It is easy to think, but hard to joke, in symbols; and this volume has not the consistent humour of the "Principles of Mathematics." Still, considering the difficulty of the medium, some of the jokes are very good. The best is that perpetrated at the expense of the law of contradiction. But it would be unfair to the circulation of the book that a reviewer should repeat them; and we leave the reader to discover them for himself.

COMMENTS

This review, and the preceding two (R22, R23), are unsigned as were, until recently, all reviews appearing in *The Times Literary Supplement*. From information provided by Mr K. Blackwell of the Bertrand Russell Archives at McMaster University, Ontario, it is clear that Hardy was the reviewer of R22 and R24. A copy of R22 forming part of P. E. B. Jourdain's correspondence with Russell bears Hardy's signature. As regards R24, a letter from Russell to Lady Ottoline Morrell in the Archives contains the passage: 'Hardy is deeply wounded because Whitehead thought the review in *The Times* was by Waterlow†—very stupid of Whitehead'. These attributions have been confirmed by the editor of the *Times Literary Supplement*.

Hardy's wounded feelings may have been assuaged by a letter, dated September 1920, received from Whitehead after the latter had read Hardy's inaugural Oxford lecture (1920, 11):

'Many thanks for your inaugural lecture, etc. I have commenced reading it though I did not get it till midnight and it is now 2 a.m. I think I ought to tell you that in the opinion of men who should know—and in mine, though you know how ill-equipped I am and how impertinent it is for me to have any opinion—you are one of the few Englishmen, now or in the past, who has a good chance of ranking finally among the great creative analysts. It seems a pity for your knowledge of this contemporary opinion to have to depend on the doubtful chance of your reading it on your tombstone.

Sincerely yours, A. N. WHITEHEAD.'

† Presumably Russell's friend, the diplomat Sydney Philip Waterlow (1878–1944), who was created K.C.M.G. in 1935.

(*d*) Review from *The Cambridge Review*

REVIEWS

Mysticism in Modern Mathematics. By Hastings Berkeley. Oxford University Press, 1910.

There is, as Mr Berkeley quite justly observes, something in the habits of mind and methods of expression of the professional mathematician that is puzzling and even repugnant to the ordinary man. Mr Berkeley is obviously a fair-minded man, and has made a very honest attempt to understand them. If he has not succeeded very well it is because he suffers, as a thinker and writer in the field of mathematical philosophy, from three fundamental disadvantages.

In the first place, he is plainly no mathematician, and it is madness for anyone but a real mathematician to write at length about the philosophy of mathematics. It is not a wide range of knowledge that is necessary, or profound originality, or great technical skill. It is that familiarity with and ease in the presence of mathematical ideas that comes only of long years spent in their company. And this is just what Mr Berkeley has not got: he does not know the language; he is a Cook's tourist in mathematical territory. If he were a mathematician he would not, for example, dispute the validity of proof by mathematical induction. He would not think that proof by *reductio ad absurdum* is a 'quibble', 'a purely verbal artifice,' in those cases where the conclusion seems as 'self-evident' as the premise. He would not confuse a 'singular point' of a surface with a 'boundary,' or say that it lies 'neither in nor out of the surface'—a confusion which leads him into an elaborate and entirely mistaken 'refutation' of a passage in one of Clifford's popular lectures. Above all, he would not have come to the extraordinary conclusion that it is 'inconsistent' to write $(+2) \times (-2) = -2^2$ because -2^2 is not itself a square. Page after page of Mr Berkeley's second part is devoted to the exhibition of this ridiculous mare's-nest.

Secondly, Mr Berkeley has not read enough. If he had he would not have taken Prof. Chrystal's *Introduction to Algebra* (no doubt a deservedly popular elementary textbook) as a serious work on the philosophy of mathematics. He might as well have gone to Todhunter for an account of the theory of the integral, or to Thomson and Tait for the philosophy of dynamics.

Thirdly, he is, as a critic, completely devoid of sympathy or imagination. His method of criticism is to select a series of passages, sometimes from such books as Mr Russell's *Principles of Mathematics*, sometimes from an article in the *Encyclopædia Britannica*, sometimes from a popular lecture by Clifford or Helmholtz, to label them indifferently 'the orthodox view' or 'the received exposition,' and to deduce from this that orthodoxy is not always consistent with itself. And as it is not every great mathematician who has had the power of expressing himself clearly and precisely—as, moreover, the philosophy of mathematics has suffered a complete revolution during the last fifty years, and there is still much that has not finally been made clear—it is not surprising that Mr Berkeley, who, to do him justice, is frequently an acute enough critic of small points, is generally able to reach the conclusion he desires.

But let us consider what Mr Berkeley has to tell us. The language of mathematics is full of conventions, fictions one might almost say, which have been adopted into the language for the sake of formal symmetry and conciseness of expression. A quadratic equation has generally two roots, but sometimes only one; in the latter case we say that it has 'two equal' roots. Lines meet unless they are parallel; in the latter case we say that they 'meet at infinity'—and so on. This is all pure convention, as every mathematician knows quite well.

Now it is easy to fall into the way of thinking, and still easier to fall into the way of expressing yourself as if you thought that a convention is more than a convention, that there really is, for example, a 'point at infinity' where parallels meet, a point real in the sense in which other points

R25 *Cambridge Review*, May 26, 1910, Literary Supplement, pp. xiii–iv.

are real. This is what Mr Berkeley calls 'mysticism'; and in so far as he points out that this tendency exists, and that some famous mathematicians seem to have been not entirely free from it, I have not a word to say against him, except that he is telling the mathematical world what it knows perfectly well already. But Mr Berkeley is not content with this. He is of opinion that this tendency to mysticism pervades and corrupts the whole of modern mathematical philosophy and that it is full of 'illusions resulting from a vicious reaction of the mode of expression upon the judgment of the reasoner.' It is to justify this thesis that he has written this book.

After a short psychological introduction, the merits of which I need not consider now, Mr Berkeley proceeds to his attack upon mathematical 'orthodoxy,' an attack which he begins with a long discussion of the 'doctrine of mathematical imaginaries.' In choosing this 'doctrine' as one of his principal objects of attack he has been peculiarly unfortunate. For if there is one part of elementary mathematical theory which, more than all others, is plain, straightforward, and uncontroversial, of merely technical importance, and absolutely devoid of all philosophical interest, it is this theory of 'imaginary quantities' or, as a mathematician would say, 'complex numbers.' A complex number is simply a pair of real numbers (a, b) which, on purely technical grounds that are no one's business but their own, mathematicians find it convenient to unite symbolically in the form $a+bi$, and to operate with in accordance with certain rules. What *does* Mr Berkeley mean when he says that a complex number 'when isolated' has 'no definite meaning'? A complex number has just as definite a meaning, is just as 'real,' just as 'interpretable,' as a 'real' number or the *Cambridge Review* or King George or Mr Berkeley. And all Mr Berkeley's talk about 'pseudo-concepts' and 'mystical delusions' is, as a criticism of current mathematical theory, simple nonsense.

It would be waste of time, therefore, to criticise Part II. of Mr Berkeley's book at length. Part III., an attack on 'metageometry,' the theory of 'varieties of space,' is at any rate more interesting. It is so rare now-a-days to find anyone who holds that space is certainly Euclidean, and so common to come across the much more irritating view that it is nonsense to ask whether space is Euclidean or not. Mr Berkeley is quite definite: the last argument in his book, he says, would 'alone be sufficient to warrant the rejection of the doctrine that different kinds of space are conceivable.'

The argument itself is a typical specimen of Mr Berkeley's methods of criticism. Riemann uses (unfortunately I grant) the term '*initial* direction' as applied to a straight line in Non-Euclidean space. If the word 'initial' were omitted his meaning would have been perfectly clear, but the word obviously came to be inserted because Riemann happened to be particularly concerned about the point from which the straight line started. Mr Berkeley, however, at once attributes to Riemann (who, after all, is known by mathematicians to have been no fool) the nonsensical doctrine that in non-Euclidean space a straight line changes its direction from point to point. Thus, he concludes, a Euclidean notion of direction is involved in the concept of a non-Euclidean space!

This is one of Mr Berkeley's three principal arguments against the metageometers. Of the other two one (pp. 206 *et seq.*, p. 256) is directed against Lobatschewsky's system of synthetic geometry, and it is sufficient comment upon it to say that, if we interchange the words 'Lobatschewsky' and 'Euclid' throughout it, it may be made to prove, with precisely the same force, that Lobatschewsky's system is 'apodeictic' and Euclid's 'a mere exercise in logic.' His third, directed specifically against Riemann and Mr Russell, is something of this sort. *Animal* (this is, of course, not the instance chosen by Mr Berkeley) is a 'qualitative' notion, and *cow*, *dog*, etc., the 'particulars subsumed under it' must therefore be 'purely qualitative.' But one cow differs from another cow in magnitude, and so *cow* has 'a quantitative aspect'; and therefore *cow* is 'at once subsumable and not subsumable' under the general notion *animal*. This, remarks Mr Berkeley, 'makes chaos of the relation of the general to the particular.' It does indeed: but the chaos is of Mr Berkeley's making.

It will be gathered from the foregoing remarks that I do not think highly of this book. It is a book, unfortunately, that may easily do a good deal of harm by worrying and confusing the minds of the large and growing circle of readers who, without professing to be philosophers or mathematicians, are interested and attracted by mathematical philosophy.

G. H. H.

COMMENTS

The author, Commander (later Captain) Hastings George Fitzhardinge Berkeley RN (1855–1934), was the author of two other works, *Wealth and Welfare* (1887) and *Japanese Letters* (1891). He entered the Royal Navy in 1869, was placed on the Retired List with the rank of Commander in 1886, and was promoted to the rank of Captain (Retired) in 1918 in recognition of services performed in the 1914–18 war. In a letter, which appeared in *The Cambridge Review* of 8 June, 1910, he disavowed some of the opinions attributed to him by Hardy in his review, but his arguments were demolished by Hardy in a subsequent letter (16 June, 1910).

6. LIST OF OTHER WRITINGS

INTRODUCTION TO THE LIST OF OTHER WRITINGS

In addition to writing well over 300 mathematical papers, Hardy was author, part-author, or editor of twelve books. Detailed information about these follows this introductory section. His first book (B1) appeared when he was a young man of 28 and his last (B12) was completed just before his death in 1947.

At least six of his books have been translated into other languages. These are listed below together with the name of the publisher, place of publication and year, where known.

Pure mathematics (B2) translated into Punjabi, Hindi, and Oriya (all published by Government of India) and into Spanish (Nigai).

Inequalities (B6) into Russian (Gosudarstv. Izdat. Inostr. Lit., Moscow, 1948) and Chinese (Science Press, Peking, 1965).

Theory of numbers (B7) into German (Oldenbourg, Munich, 1958).

A mathematician's apology (B10) into Bulgarian (Technika), Gujarati (Translation Trust), Icelandic (Gylfason), Italian (de Denato Leonardo da Vinci), Japanese (Tsurumi Shoten), Rumanian (Romana), and Swedish (Cwk Gleerup).

Fourier series (B11) into Russian (Fizmatgiz 1959, 1962) and Czech (SNTL, 1971).

Divergent series (B12) into Russian (Gosudarstv. Inostr. Lit., Moscow, 1951), with a preface and survey article by S. B. Stečkin.

The following mathematical writings are not reproduced in these volumes, nor are they included in the list of Hardy's papers.

(With A. E. Jolliffe) Examinations for mathematical scholarships at Oxford and Cambridge. *The Board of Education's special reports on educational subjects. The teaching of mathematics in the United Kingdom.* No. 31 (1912).

Prefatory note (pp. 211–12) to a paper by L. J. Rogers and S. Ramanujan: Proof of certain identities in combinatory analysis, *Proc. Cambridge Phil. Soc.* 19 (1920), 211–16.

Note (pp. 208–209) appended to a paper by Marcel Riesz: Sur le principal de Phragmén-Lindelöf, *Proc. Cambridge Phil. Soc.* 20 (1921), 205–7. Riesz's paper is an extract of a letter addressed to Hardy.

During the years 1913–1926 over forty abstracts of papers by Hardy, alone or in collaboration, appeared in the Records of the Proceedings of Meetings of the London Mathematical Society. These are published in volumes 9–24 of the second series of the Society's *Proceedings*. Some of these abstracts are reproduced in the *Collected Papers*, namely

1918, 4; 1921, 5; 1922, 8; 1924, 2–5; 1925, 6–8; 1926, 1–3.

Others are abstracts of papers that later appeared in full in the *Proceedings* or in other journals. Thus the fact that the Riemann zeta-function has infinitely many zeros on the middle line of the critical strip was first announced by Hardy in the abstract of a paper 'On the zeroes of Riemann's zeta-function' in the Records of the Proceedings of the meeting of the Society held on 12 March, 1914 (Volume 13, pp. xxix–xxx). The proof appeared first in the *Comptes Rendus* (1914, 1) and was finally further developed with Littlewood in *Acta Mathematica* (1918, 1).

In some cases it is difficult to make unique reference to the abstracts, since more than one page in the same volume may bear the same small Roman numeral; accordingly, the date of the meeting is a more reliable guide. Moreover, some abstracts are printed twice in different parts of the same or neighbouring volumes. For example, 1925, 6 appears also on pp ii–iii of Volume 23.

After the *Journal of the London Mathematical Society* was established in 1926, the Society ceased to publish abstracts of papers.

Hardy's clear and easy English style is apparent in all he wrote. It is not therefore surprising that he should have written (§29 of B10) that 'journalism is the only profession, outside academic life, in which I should have felt really confident of my chances'. It is, accordingly, quite likely that other non-mathematical articles than those mentioned below may have escaped the editors' notice.

That Hardy was interested in Bertrand Russell is evident from a perusal of B8. He also admired Russell's work on the foundations of mathematics; see the reviews R22 and R24 (pp. 851–9, of this volume). But Hardy was also interested in Russell's views on religion and in 1917 he published an article, entitled 'Mr Russell as a religious teacher' in *The Cambridge Magazine* (May 19, pp. 624–6; May 26, pp. 650–3).

C. P. Snow states in his foreword to B10 that, to show his radical allegiances, Hardy accepted public office on one occasion only. This was on 26 January 1924, when he was elected president of the National Union of Scientific Workers. He took his presidential duties seriously: In *Nature* 113 (1924), p. 746 will be found a letter from him and the General Secretary of the Union objecting to the reduction in salaries of junior staff imposed by the University Court of a certain Scottish university. He was also much concerned at the boycott of German mathematicians from international congresses; the issue of *The Scientific Worker* for 24 November, 1924, contains an article by him entitled 'The campaign against the boycott' in which he makes a vigorous protest against the exclusion of ex-enemy nations from international scientific meetings. For other articles see *The Cambridge Magazine* (May 11 & 18, 1912) and *The Scientific Worker* (March 1925).

R.A.R.

LIST OF BOOKS BY G. H. HARDY

B1. *The integration of functions of a single variable.* Cambridge Tracts in Mathematics and Mathematical Physics, 2. Cambridge University Press, 1905; 2nd edn 1916.

B2. *A course of pure mathematics.* Cambridge University Press, 1908; 2nd edn 1914; 3rd edn 1921; 4th edn 1925; 5th edn 1928; 6th edn 1933; 7th edn (revised and re-set) 1938; 8th edn 1941; 9th edn 1944, reprinted 1945, 1946, 1948; 10th edn (with index) 1952, reprinted 1955, 1958, 1960, 1962, 1967, 1975.

B3. *Orders of infinity: the 'Infinitärcalcul' of Paul Du Bois-Reymond.* Cambridge Tracts in Mathematics and Mathematical Physics, 12. Cambridge University Press, 1910; 2nd edn 1924, reprinted 1954.

B4. (with Marcel Riesz) *The general theory of Dirichlet's series.* Cambridge Tracts in Mathematics and Mathematical Physics. 18. Cambridge University Press, 1915. Reprinted 1964 by Stechert-Hafner, Inc., New York.

B5. (with P. V. Seshu Aiyar and B. M. Wilson as editors) *Collected papers of Srinivasa Ramanujan.* Cambridge University Press, 1927. Reprinted 1962 by Chelsea Publishing Company, New York.

B6. (with J. E. Littlewood and G. Pólya) *Inequalities.* Cambridge University Press, 1934; 2nd edn 1952.

B7. (with E. M. Wright) *An introduction to the theory of numbers.* Clarendon Press, Oxford, 1938; 2nd edn 1945; 3rd edn 1954; 4th edn 1960, reprinted 1962, 1965, 1968, 1971, 1975.

B8. *Bertrand Russell and Trinity.* Cambridge University Press, 1942 (privately printed). Facsimile reproduction with foreword by C. D. Broad, Cambridge University Press, 1970.

B9. *Ramanujan: twelve lectures on subjects suggested by his life and work.* Cambridge University Press, 1940. Reprinted 1959 by Chelsea Publishing Company, New York.

B10. *A mathematician's apology.* Cambridge University Press, 1940; reprinted 1941, 1948 and (with a foreword by C. P. Snow) 1967, 1969, 1973.

B11. (with W. W. Rogosinski) *Fourier series.* Cambridge Tracts in Mathematics and Mathematical Physics, 38. Cambridge University Press, 1944; 2nd edn 1950; 3rd edn 1956.

B12. *Divergent series.* Clarenden Press, Oxford, 1949.

7. OBITUARY NOTICES OF G. H. HARDY AND OTHER WRITINGS CONCERNING HIS LIFE AND WORK

OBITUARY NOTICES OF G. H. HARDY AND OTHER WRITINGS CONCERNING HIS LIFE AND WORK

The following obituary notices of Hardy by the authors listed are to be found in the literature:

Anonymous, *Nature*, 161 (1948), 797–8.
T. A. A. Broadbent, M. H. A. Newman, A. V. Hill & E. Borel, *Math. Gaz.* 32 (1948), 49–51, 98.
E. A. Milne, *Monthly Not. Roy. Astron. Soc.*, 118 (1948), 44–6.
Oskar Perron, *Jber. Bayer Akad. Wiss. München 1944/48*, (1948), 282–5.
E. C. Titchmarsh, *Obit. Notices Roy. Soc. London*, 6 (1949), 447–61.
E. C. Titchmarsh, *J. London Math. Soc.*, 25 (1950), 82–101.
T. Vijayaraghavan, *Math. Student*, 15 (1947), 121–2.
Norbert Wiener, *Bull. Amer. Math. Soc.* 55 (1949), 72–7 and 1082.

Titchmarsh's Royal Society obituary is printed with slight changes in Vol. I, pp. 1–12. Reference should also be made to the foreword by C. P. Snow to Hardy's book *A mathematician's apology* (B10).

Articles on the various aspects of Hardy's mathematical work, by the following authors, will be found in volume 25 of the *Journal of the London Mathematical Society*:

L. S. Bosanquet (Early work on divergent series), 102–6.
F. Smithies (Integral equations), 107–9.
L. J. Mordell (Diophantine approximation), 109–14.
A. E. Ingham (Additive theory of numbers), 115–19.
H. Davenport (Waring's problem), 119–25.
E. C. Titchmarsh (Riemann zeta-function and lattice-point problems), 125–8.
R. Rado (Inequalities), 129–35.
A. C. Offord (Fourier series), 136–8.

In conclusion attention is drawn to the earlier article:
H. T. H. Piaggio, Three Sadleirian professors: A. R. Forsyth, E. W. Hobson and G. H. Hardy, *Math. Gaz.* 15 (1931), 461–5.

R.A.R.

ARRANGEMENT OF THE VOLUMES

VOLUME I

I. 1 Diophantine approximation
I. 2 Additive number theory
      (*a*) Combinatory analysis and sums of squares
      (*b*) Waring's Problem
      (*c*) Goldbach's Problem
      (*d*) Inaugural Lecture (Oxford, 1920)

VOLUME II

II. 1 Multiplicative number theory (including the zeta-function)
II. 2 Other number theory
II. 3 Inequalities

VOLUME III

III. 1 Trigonometric series
      (*a*) Convergence of a Fourier series or its conjugate
      (*b*) Summability of a Fourier series or its conjugate
      (*c*) The Young–Hausdorff inequalities
      (*d*) Special trigonometric series
      (*e*) Other papers on trigonometric series
III. 2 Mean values of power series

VOLUME IV

IV. 1 Special functions
      (*a*) Zeros and asymptotic behaviour of particular integral functions
      (*b*) Taylor series and singularities
      (*c*) Orders of infinity
      (*d*) Miscellaneous
IV. 2 Theory of functions

VOLUME V

V. Integral calculus

VOLUME VI

VI. Theory of series

VOLUME VII

VII. 1 Integral equations and integral transforms
VII. 2 Miscellaneous papers
VII. 3 Questions from the *Educational Times*
VII. 4 Obituary notices by G. H. Hardy
VII. 5 Book reviews by G. H. Hardy
VII. 6 List of other writings
VII. 7 Obituary notices of G. H. Hardy and other writings concerning his life and work

LIST OF PAPERS BY G. H. HARDY

Abbreviations

N.I.C.   Notes on some points in the integral calculus
D.A.   Some problems of Diophantine approximation
P.N.   Some problems of 'Partitio Numerorum'
N.S.   Notes on the theory of series

1899

1. Question 13848, *Educational Times*, 70, 43–44.
2. Question 13917, *Educational Times*, 70, 78–79.
2a. Question 13889, *Educational Times*, 71, 26.
2b. Question 13979, *Educational Times*, 71, 61–62.
2c. Question 13964, *Educational Times*, 71, 86–87.
3. Question 14124, *Educational Times*, 71, 100–101.
4. Question 14005, *Educational Times*, 71, 111–112.
5. Question 14155, *Educational Times*, 71, 117.

VII. 3

1900

1. On a class of definite integrals containing hyperbolic functions, *Messenger of Mathematics*, 29, 25–42.   V
1a. Question 14089, *Educational Times*, 72, 49.
1b. Question 14090, *Educational Times*, 72, 57–59.
2. Question 14243, *Educational Times*, 72, 80–81.
2a. Question 14164, *Educational Times*, 72, 89–90.
2b. Question 14332, *Educational Times*, 72, 105–106.
3. Question 14271, *Educational Times*, 73, 36–37.
4. Question 14179, *Educational Times*, 73, 53–54.
5. Question 14317, *Educational Times*, 73, 61–63.
6. Question 14263, *Educational Times*, 73, 76.

VII. 3

1901

1. On Differentiation and integration under the integral sign, *Quarterly Journal of Mathematics*, 32, 66–140. (Corrected in 1915, 2.)   V
2. General theorems in contour integration, with some applications, *Quarterly Journal of Mathematics*, 32, 369–384.   V
3. N.I.C. I: On the formula for integration by parts, *Messenger of Mathematics*, 30, 185–187.   V
4. N.I.C. II: Two general convergence theorems, *Messenger of Mathematics*, 30, 187–190.   V

1901 (cont.)

5. Question 14496, *Educational Times*, 74, 37–38.
5a. Question 14511, *Educational Times*, 74, 42–43.
6. Question 14447, *Educational Times*, 74, 98–100.
7. Question 14467, *Educational Times*, 74, 111–112.
8. Question 14028, *Educational Times*, 74, 122–123.
9. Question 14369, *Educational Times*, 75, 135–136.

VII. 3

1902

1. The elementary theory of Cauchy's principal values, *Proceedings of the London Mathematical Society*, (1) 34, 16–40.   V
2. The theory of Cauchy's principal values (Second Paper: the use of principal values in some of the double limit problems of the integral calculus), *Proceedings of the London Mathematical Society*, (1) 34, 55–91.   V
3. On the Frullanian integral

$$\int_0^\infty \frac{\phi(ax^m)-\psi(bx^n)}{x} (\log x)^p \, dx,$$

   Quarterly Journal of Mathematics, 33, 113–144.   V
4. N.I.C. III: On the logarithmic criteria for the absolute convergence of an integral whose upper limit is ∞, *Messenger of Mathematics*, 31, 1–6.   V
5. N.I.C. IV: On the integral $\int^\infty \sin x \psi(x)\, dx$, *Messenger of Mathematics*, 31, 6–8.   V
6. A new proof of Kummer's series for $\log \Gamma(a)$, *Messenger of Mathematics*, 31, 31–33.   IV. 1 (d)
7. N.I.C. V: On absolutely convergent integrals of functions which are infinitely often infinite, *Messenger of Mathematics*, 31, 73–76.   V
8. N.I.C. VI: Absolute convergence of infinite multiple integrals, *Messenger of Mathematics*, 31, 125–128.   V
9. N.I.C. VII: On differentiation under the integral sign, *Messenger of Mathematics*, 31, 132–134.   V
10. N.I.C. VIII: Absolutely convergent integrals of irregular types, *Messenger of Mathematics*, 31, 177 183.   V
11. On the zeroes of the integral function

$$x-\sin x = \sum_1^\infty (-)^{n-1} \frac{x^{2n+1}}{2n+1!},$$

   Messenger of Mathematics, 31, 161–165.   IV. 1 (a)
12. Questions 1423, 2316, 3941, 4794, *Educational Times*, (2) 1, 25.
13. Question 14851, *Educational Times*, (2) 1, 58–59.
14. Question 14055, *Educational Times*, (2) 2, 41–42.

VIII. 3

1903

1. The theory of Cauchy's principal values (Third Paper: differentiation and integration of principal values), *Proceedings of the London Mathematical Society*, (1) 35, 81–107.   V
2. On the continuity and discontinuity of definite integrals which contain a continuous parameter, *Quarterly Journal of Mathematics*, 34, 28–53.   V

1903 (cont.)

3. Note on the limiting values of the elliptic modular functions, *Quarterly Journal of Mathematics*, 34, 76–86. IV. 1 (d)
4. N.I.C. IX: On the integral $\int_0^\infty \{A - \phi(\sin^2 x)\}\psi(x)\,dx$, *Messenger of Mathematics*, 32, 1–3. V
5. On the zeroes of certain integral functions, *Messenger of Mathematics*, 32, 36–45. IV. 1 (a)
6. On the integral $\int_{-\infty}^{\infty} \frac{\log(ax^2+2bx+c)^2}{\alpha x^2+2\beta x+\gamma}\,dx$, *Messenger of Mathematics*, 32, 45–50. V
7. N.I.C. X: On conditionally convergent infinite multiple integrals, *Messenger of Mathematics*, 32, 92–97. V
8. N.I.C. XI: Some conditionally convergent infinite double integrals, *Messenger of Mathematics*, 32, 159–165. V
9. N.I.C. XII: On the operation which is the inverse of double integration, *Messenger of Mathematics*, 32, 187–192. V
10. Question 14988, *Educational Times*, (2) 3, 94–95.
10a. Question 9969, *Educational Times*, (2) 4, 44–45.
11. Question 14989, *Educational Times*, (2) 4, 69–70.   VII. 3
12. Question 15019, *Educational Times*, (2) 4, 75.
13. Question 15265, *Educational Times*, (2) 4, 109–110.

1904

1. On the convergence of certain multiple series, *Proceedings of the London Mathematical Society*, (2) 1, 124–128. VI
2. A general theorem concerning absolutely convergent series, *Proceedings of the London Mathematical Society*, (2) 1, 285–290. VII. 2 (a)
3. On differentiation and integration of divergent series, *Transactions of the Cambridge Philosophical Society*, 19, 297–321. VI
4. Researches in the theory of divergent series and divergent integrals, *Quarterly Journal of Mathematics*, 35, 22–66. VI
5. A theorem concerning the infinite cardinal numbers, *Quarterly Journal of Mathematics*, 35, 87–94. VII. 2 (a)
6. Note on the function $\int_x^\infty e^{i(x^2-t^2)}\,dt$, *Quarterly Journal of Mathematics*, 35, 193–207. VII. 1 (a)
7. The asymptotic solution of certain transcendental equations, *Quarterly Journal of Mathematics*, 35, 261–282. IV. 1 (a)
8. N.I.C. XIII: On differentiation under the integral sign (continued), *Messenger of Mathematics*, 33, 62–67. V
9. The cardinal number of a closed set of points, *Messenger of Mathematics*, 33, 67–69. VII. 2 (a)
10. N.I.C. XIV: Integrals whose discontinuities are everywhere dense, *Messenger of Mathematics*, 33, 80–85. V
11. Note on divergent Fourier series, *Messenger of Mathematics*, 33, 137–144. III. 1 (b)
12. On the zeroes of two classes of Taylor series, *British Association Report*, 441–443. IV. 1 (a)

1904 (cont.)

12a. Question 9515, *Educational Times*, (2) 5, 46.
13. Question 15300, *Educational Times*, (2) 5, 61.
13a. Question 15282, *Educational Times*, (2) 5, 89–92.
14. Additional note on Question 15282, *Educational Times*, (2) 5, 113–114.   VII. 3
15. Question 15361, *Educational Times*, (2) 5, 118.
16. Question 15125, *Educational Times*, (2) 6, 31.
17. Question 15357, *Educational Times*, (2) 6, 45.

1905

1. On the roots of the equation $\frac{1}{\Gamma(x+1)} = c$, *Proceedings of the London Mathematical Society*, (2) 2, 1–7.   IV. 1 (a)
2. (With T. J. I'A. Bromwich.) Some extensions to multiple series of Abel's theorem on the continuity of power series, *Proceedings of the London Mathematical Society*, (2) 2, 161–189.   VI
3. Note in addition to a former paper on conditionally convergent multiple series. *Proceedings of the London Mathematical Society*, (2) 2, 190–191.   VI
4. On the zeroes of certain classes of integral Taylor series. Part I: On the integral function $\sum_{n=0}^{\infty} \frac{x^{\phi(n)}}{\{\phi(n)\}!}$, *Proceedings of the London Mathematical Society*, (2) 2, 332–339.   IV. 1 (a)
5. On the zeroes of certain classes of integral Taylor series. Part II: On the integral function $\sum_{n=0}^{\infty} \frac{x^n}{(n+a)^s n!}$ and other similar functions. *Proceedings of the London Mathematical Society*, (2) 2, 401–431.   IV. 1 (a)
6. A method for determining the behaviour of certain classes of power series near a singular point on the circle of convergence, *Proceedings of the London Mathematical Society*, (2) 3, 381–389.   IV. 1 (b)
7. On a class of analytic functions, *Proceedings of the London Mathematical Society*, (2) 3, 441–460.   IV. 1 (b)
8. On certain series of discontinuous functions connected with the modular functions, *Quarterly Journal of Mathematics*, 36, 93–123.   IV. 1 (d)
9. Note on an integral function, *Messenger of Mathematics*, 34, 1–2.   IV. 1 (a)
10. N.I.C. XV: On upper and lower integration, *Messenger of Mathematics*, 34, 3–6.   V
11. N.I.C. XVI: A class of conditionally convergent infinite multiple integrals, *Messenger of Mathematics*, 34, 6–10.   V
12. A generalization of Frullani's integral, *Messenger of Mathematics*, 34, 11–18, and note, p. 102.   V
13. On the zeroes of a class of integral functions, *Messenger of Mathematics*, 34, 97–101.   IV. 1 (a)
14. On certain conditionally convergent multiple series connected with the elliptic functions, *Messenger of Mathematics*, 34, 146–153.   VI
15. Question 15686, *Educational Times*, (2) 8, 74.   VII. 3
16. The expression of the double zeta function and double gamma function in terms of elliptic functions, *Transactions of the Cambridge Philosophical Society*, 20, 1–35.   IV. 1 (d)

1906

1. On Kummer's series for log $\Gamma(a)$, *Quarterly Journal of Mathematics*, 37, 49–53.   IV. 1 (d)
2. On double Fourier series, and especially those which represent the double zeta-function with real and incommensurable parameters, *Quarterly Journal of Mathematics*, 37, 53–79.   IV. 1 (d)
3. On the function $P_\rho(x)$, *Quarterly Journal of Mathematics*, 37, 146–172 (correction at end of 1906, 5).   IV. 1 (a)
4. On certain double integrals, *Quarterly Journal of Mathematics*, 37, 360–369.   V
5. On the integral function $\Phi_{a,\alpha,\beta}(x) = \sum_0^\infty \frac{x^n}{(n+a)^{\alpha n+\beta}}$, *Quarterly Journal of Mathematics*, 37, 369–378.   IV. 1 (a)
6. N.I.C. XVII: On the integration of series, *Messenger of Mathematics*, 35, 125–130.   V
7. A formula for the prime factors of any number, *Messenger of Mathematics*, 35, 145–146.   II. 1
8. N.I.C. XVIII: On some discontinuous integrals, *Messenger of Mathematics*, 35, 158–166.   V
9. Some notes on certain theorems in higher trigonometry, *Mathematical Gazette*, 3, 284–288.   V

1907

1. The continuum and the second number class, *Proceedings of the London Mathematical Society*, (2) 4, 10–17.   VII. 2 (a)
2. Some theorems connected with Abel's theorem on the continuity of power series, *Proceedings of the London Mathematical Society*, (2) 4, 247–265.   VI
3. On the singularities of functions defined by Taylor's series (Remarks in addition to a former paper), *Proceedings of the London Mathematical Society*, (2) 5, 197–205.   IV. 1 (b)
4. The singular points of certain classes of functions of several variables, *Proceedings of the London Mathematical Society*, (2) 5, 342–360.   IV. 1 (b)
5. On certain oscillating series, *Quarterly Journal of Mathematics*, 38, 269–288.   VI
6. Some theorems concerning infinite series, *Mathematische Annalen*, 64, 77–94.   VI
7. N.I.C. XIX: On Abel's lemma and the second theorem of the mean, *Messenger of Mathematics*, 36, 10–13.   V
8. Higher trigonometry, *Mathematical Gazette*, 4, 13–14.   VII. 2 (d)
9. A curious imaginary curve, *Mathematical Gazette*, 4, 14.   VII. 2 (d)
10. The line at infinity, etc., *Mathematical Gazette*, 4, 14–15.   VII. 2 (d)
11. Question 15498, *Educational Times*, (2) 12, 26–27.   VII. 3

1908

1. Generalization of a theorem in the theory of divergent series, *Proceedings of the London Mathematical Society*, (2) 6, 255–264.   VI
2. The multiplication of conditionally convergent series, *Proceedings of the London Mathematical Society*, (2) 6, 410–423.   VI
3. Further researches in the theory of divergent series and integrals, *Transactions of the Cambridge Philosophical Society*, 21, 1–48.   VI
4. (With T. J. I'A. Bromwich.) The definition of an infinite integral as the limit of a finite or infinite series, *Quarterly Journal of Mathematics*, 39, 222–240.   V

1908 (cont.)

5. Some multiple integrals, *Quarterly Journal of Mathematics*, 39, 357–375.   V
6. N.I.C. XX: On double Frullanian integrals, *Messenger of Mathematics*, 37, 96–103.   V
7. N.I.C. XXI: On a conditionally convergent multiple integral, *Messenger of Mathematics*, 37, 127–130.   V
8. N.I.C. XXII: On double Frullanian integrals (cont.), *Messenger of Mathematics*, 37, 154–161.   V
8a. Question 16041, *Educational Times*, (2) 13, 42.   }   VII. 3
9. Question 16257, *Educational Times*, (2) 13, 79–80.
10. Mendelian proportions in a mixed population, *Science* (American Association for the Advancement of Science), new series 28, 49–50.   VII. 2 (c)

1909

1. A note on the continuity or discontinuity of a function defined by an infinite product, *Proceedings of the London Mathematical Society*, (2) 7, 40–48.   VI
2. The theory of Cauchy's principal values (Fourth Paper: the integration of principal values—continued—with applications to the inversion of definite integrals), *Proceedings of the London Mathematical Society*, (2) 7, 181–208.   V
3. On an integral equation, *Proceedings of the London Mathematical Society*, (2) 7, 445–472.   VII. I (b)
4. N.I.C. XXIII: On certain oscillating cases of Dirichlet's integral, *Messenger of Mathematics*, 38, 1–8.   V
5. On certain definite integrals whose values can be expressed in terms of Bessel's functions, *Messenger of Mathematics*, 38, 129–132.   V
6. N.I.C. XXIV: Oscillating cases of Dirichlet's integral (cont.), *Messenger of Mathematics*, 38, 176–185 (correction at end of 1911, 3).   V
7. The integral $\int_0^\infty \frac{\sin x}{x} dx$, *Mathematical Gazette*, 5, 98–103.   V
8. Question 16402, *Educational Times*, (2) 16, 66.   VII. 3

1910

1. The application to Dirichlet's series of Borel's exponential method of summation, *Proceedings of the London Mathematical Society*, (2) 8, 277–294.   VI
2. The ordinal relations of the terms of a convergent sequence, *Proceedings of the London Mathematical Society*, (2) 8, 295–300.   VII. 2 (a)
3. Theorems relating to the summability and convergence of slowly oscillating series, *Proceedings of the London Mathematical Society*, (2) 8, 301–320.   VI
4. The maximum modulus of an integral function, *Quarterly Journal of Mathematics*, 41, 1–9.   IV. 2
5. On certain definite integrals considered by Airy and by Stokes, *Quarterly Journal of Mathematics*, 41, 226–240.   IV. 1 (d)
6. N.I.C. XXV: Absolutely convergent integrals of irregular types (cont.), *Messenger of Mathematics*, 39, 28–32.   V
7. The zeroes of the integral function $\sum \frac{x^{n^2}}{n^2!}$, and of some similar functions, *Messenger of Mathematics*, 39, 88–96.   IV. 1 (a)

1910 (cont.)

8. N.I.C. XXVI: On a case of term-by-term integration of an infinite series, *Messenger of Mathematics*, 39, 136–139.   V
9. To find an approximation to the large positive root of the equation $e^{e^x} = 10^{10}x^{10}e^{10^{10}x^{10}}$, *Mathematical Gazette*, 5, 333–334.   VII. 2 (d)

1911

1. Theorems connected with Maclaurin's test for the convergence of series, *Proceedings of the London Mathematical Society*, (2) 9, 126–144.   VI
2. (With S. Chapman.) A general view of the theory of summable series, *Quarterly Journal of Mathematics*, 42, 181–215.   VI
3. N.I.C. XXVII: Oscillating cases of Dirichlet's integral (cont.), *Messenger of Mathematics*, 40, 44–53.   V
4. A class of definite integrals, *Messenger of Mathematics*, 40, 53–54.   V
5. N.I.C. XXVIII: A conditionally convergent double integral, *Messenger of Mathematics*, 40, 62–69.   V
6. N.I.C. XXIX: Two convergence theorems, *Messenger of Mathematics*, 40, 87–91.   V
7. N.I.C. XXX: A theorem concerning summable integrals, *Messenger of Mathematics*, 40, 108–112.   VI
8. N.I.C. XXXI: The uniform convergence of Borel's integral, *Messenger of Mathematics*, 40, 161–165.   VI
9. Fourier's double integral and the theory of divergent integrals, *Transactions of the Cambridge Philosophical Society*, 21, 427–451.   VII. 1 (a)

1912

1. Properties of logarithmico-exponential functions, *Proceedings of the London Mathematical Society*, (2) 10, 54–90.   IV. 1 (c)
2. On the multiplication of Dirichlet's series, *Proceedings of the London Mathematical Society*, (2) 10, 396–405.   VI
3. Some results concerning the behaviour at infinity of a real and continuous solution of an algebraic differential equation of the first order, *Proceedings of the London Mathematical Society*, (2) 10, 451–468.   VII. 2 (b)
4. (With J. E. L.) Some problems of Diophantine approximation, *5th International Congress of Mathematicians, Cambridge*, 1, 223–229.   I. 1
5. Generalizations of a limit theorem of Mr. Mercer, *Quarterly Journal of Mathematics*, 43, 143–150.   VI
6. Note on Dr. Vacca's series for γ, *Quarterly Journal of Mathematics*, 43, 215–216. IV. 1 (d)
7. Note on a theorem of Cesàro, *Messenger of Mathematics*, 41, 17–22.   VI
8. N.I.C. XXXII: On double series and double integrals, *Messenger of Mathematics*, 41, 44–48.   V
9. N.I.C. XXXIII: Some cases of the inversion of the order of integration, *Messenger of Mathematics*, 41, 102–109.   VII. 1 (a)

1913

1. (With J. E. L.) The relations between Borel's and Cesàro's methods of summation, *Proceedings of the London Mathematical Society*, (2) 11, 1–16.   VI
2. (With J. E. L.) Contributions to the arithmetic theory of series, *Proceedings of the London Mathematical Society*, (2) 11, 411–478.   VI

1913 (cont.)

3. An extension of a theorem on oscillating series, *Proceedings of the London Mathematical Society*, (2) 12, 174–180. VI
4. On the summability of Fourier's series, *Proceedings of the London Mathematical Society*, (2) 12, 365–372. III. 1 (b)
5. Oscillating Dirichlet's integrals: an essay in the 'Infinitärcalcül' of Paul Du Bois-Reymond, *Quarterly Journal of Mathematics*, 44, 1–40. IV. 1 (c)
6. A theorem concerning Taylor's series, *Quarterly Journal of Mathematics*, 44, 147–60. IV. 2
7. Oscillating Dirichlet's integrals (Second Paper), *Quarterly Journal of Mathematics*, 44, 242–263. IV. 1 (c)
8. N.I.C. XXXIV: Absolutely convergent integrals of irregular types (cont.), *Messenger of Mathematics*, 42, 13–18. V
9. N.I.C. XXXV: On an integral equation, *Messenger of Mathematics*, 42, 89–93. VII. 1 (a)
10. (With J. E. L.) Tauberian theorems concerning series of positive terms, *Messenger of Mathematics*, 42, 191–192. VI
11. (With J. E. L.) Sur la série de Fourier d'une fonction à carré sommable, *Comptes Rendus*, 156, 1307–1309. III. 1 (b)

1914

1. Sur les zéros de la fonction $\zeta(s)$ de Riemann, *Comptes Rendus*, 158, 1012–1014. II. 1
2. (With J. E. L.) D.A. I: The fractional part of $n^k\theta$, *Acta Mathematica*, 37, 155–191. I. 1
3. (With J. E. L.) D.A. II: The trigonometrical series associated with the elliptic ϑ-functions, *Acta Mathematica*, 37, 193–239. I. 1
4. (With J. E. L.) Tauberian theorems concerning power series and Dirichlet's series whose coefficients are positive, *Proceedings of the London Mathematical Society*, (2) 13, 174–191. VI
5. Note on Lambert's series, *Proceedings of the London Mathematical Society*, (2) 13, 192–198. VI
6. Note in addition to a paper on Taylor's series, *Quarterly Journal of Mathematics*, 45, 77–84. IV. 2
7. A function of two variables, *Quarterly Journal of Mathematics*, 45, 85–113. IV. 1 (b)
8. N.I.C. XXXVI: On the asymptotic values of certain integrals, *Messenger of Mathematics*, 43, 9–13. V
9. N.I.C. XXXVII. On the region of convergence of Borel's integral, *Messenger of Mathematics*, 43, 22–24. VI
10. N.I.C. XXXVIII: On the definition of an analytic function by means of a definite integral, *Messenger of Mathematics*, 43, 29–33. IV. 2
11. (With J. E. L.) Some theorems concerning Dirichlet's series, *Messenger of Mathematics*, 43, 134–147. VI

1915

1. (With J. E. L.) New proofs of the prime-number theorem and similar theorems, *Quarterly Journal of Mathematics*, 46, 215–219. II. 1
2. Correction of an error, *Quarterly Journal of Mathematics*, 46, 261–262. V
3. On the expression of a number as the sum of two squares, *Quarterly Journal of Mathematics*, 46, 263–283. II. 2
4. The mean value of the modulus of an analytic function, *Proceedings of the London Mathematical Society*, (2) 14, 269–277. III. 2

1915 (cont.)

5. Proof of a formula of Mr. Ramanujan, *Messenger of Mathematics*, 44, 18–21.   V
6. N.I.C. XXXIX: Further examples of conditionally convergent infinite double integrals, *Messenger of Mathematics*, 44, 57–63.   V
7. N.I.C. XL: Some cases of term-by-term integration of an infinite series, *Messenger of Mathematics*, 44, 145–149.   V
8. N.I.C. XLI: On the convergence of certain integrals and series, *Messenger of Mathematics*, 44, 163–166.   V
9. Sur le problème des diviseurs de Dirichlet, *Comptes Rendus*, 160, 617–619.   II. 2
10. Prime numbers, *British Association Report*, 350–354.   II.1
11. Example to illustrate a point in the theory of Dirichlet's series, *The Tôhoku Mathematical Journal*, 8, 59–66.   VI
12. The definition of a complex number, *Mathematical Gazette*, 8, 48–49.   VII. 2 (d)

1916

1. The application of Abel's method of summation to Dirichlet's series, *Quarterly Journal of Mathematics*, 47, 176–192.   VI
2. Weierstrass's non-differentiable function, *Transactions of the American Mathematical Society*, 17, 301–325.   IV. 1 (d)
3. (With J. E. L.) D.A.: A remarkable trigonometrical series, *Proceedings of the National Academy of Sciences*, 2, 583–586.   I. 1
4. On Dirichlet's divisor problem, *Proceedings of the London Mathematical Society*, (2) 15, 1–25.   II. 2
5. The second theorem of consistency for summable series, *Proceedings of the London Mathematical Society*, (2) 15, 72–88.   VI
6. The average order of the arithmetical functions $P(x)$ and $\Delta(x)$, *Proceedings of the London Mathematical Society*, (2) 15, 192–213.   II. 2
7. Sur la sommation des séries de Dirichlet, *Comptes Rendus*, 162, 463–465.   VI
8. (With J. E. L.) Theorems concerning the summability of series by Borel's exponential method, *Rendiconti del Circolo matematico di Palermo*, 41, 36–53.   VI
9. (With J. E. L.) D.A.: The series $\Sigma\, e(\lambda_n)$ and the distribution of the points $(\lambda_n \alpha)$, *Proceedings of the National Academy of Sciences*, 3, 84–88.   I. 1
10. Asymptotic formulae in combinatory analysis, *Quatrième Congrès Mathématiciens, Scandinaves*, 45–53.   I. 2 (a)
11. Further remarks on the integral $\int_0^\infty \frac{\sin x}{x}\, dx$, *Mathematical Gazette*, 8, 301–303.   V

1917

1. (With S. Ramanujan) Une formule asymptotique pour le nombre des partitions de n, *Comptes Rendus*, 164, 35–38.   I. 2 (a)
2. On a theorem of Mr. G. Pólya, *Proceedings of the Cambridge Philosophical Society*, 19, 60–63.   IV. 2
3. On the convergence of certain multiple series, *Proceedings of the Cambridge Philosophical Society*, 19, 86–95.   VI
4. (With S. Ramanujan) Asymptotic formulae for the distribution of integers of various types, *Proceedings of the London Mathematical Society*, (2) 16, 112–132.   I. 2 (a)
5. N.I.C. XLII: On Weierstrass's singular integral, and on a theorem of Lerch, *Messenger of Mathematics*, 46, 43–48.   VII. 1 (a)

1917 (cont.)

6. N.I.C. XLIII: On the asymptotic value of a definite integral, and the coefficient in a power series, *Messenger of Mathematics*, 46, 70–73.   V
7. N.I.C. XLIV: On certain multiple integrals and series which occur in the analytic theory of numbers, *Messenger of Mathematics*, 46, 104–107.   V
8. N.I.C. XLV: On a point in the theory of Fourier series, *Messenger of Mathematics*, 46, 146–149.   III. 1 (a)
9. N.I.C. XLVI: On Stieltjes' 'problème des moments', *Messenger of Mathematics*, 46, 175–182.   VII. 1 (a)
10. (With J. E. L.) Sur la convergence des séries de Fourier et des séries de Taylor, *Comptes Rendus*, 165, 1047–1049.   III. 1 (a)
11. Mr. S. Ramanujan's mathematical work in England, *Journal of the Indian Mathematical Society*, 9, 30–45.   VII. 2 (e)
12. Question 14587, *Educational Times*, (3) 3, 85.   VII. 3

1918

1. (With J. E. L.) Contributions to the theory of the Riemann zeta-function and the theory of the distribution of primes, *Acta Mathematica*, 41, 119–196.   II. 1
2. (With S. Ramanujan) On the coefficients in the expansions of certain modular functions, *Proceedings of the Royal Society*, (A) 95, 144–155.   I. 2 (a)
3. Sir George Stokes and the concept of uniform convergence, *Proceedings of the Cambridge Philosophical Society*, 19, 148–156.   VII. 2 (e)
4. (With J. E. L.) On the Fourier series of a bounded function, *Proceedings of the London Mathematical Society*, (2) 17, xiii–xv.   III. 1 (b)
5. (With S. Ramanujan) Asymptotic formulae in combinatory analysis, *Proceedings of the London Mathematical Society*, (2) 17, 75–115.   I. 2 (a)
6. N.I.C. XLVII: On Stieltjes' 'problème des moments' (cont.), *Messenger of Mathematics*, 47, 81–88.   VII. 1 (a)
7. N.I.C. XLVIII: On some properties of integrals of fractional order, *Messenger of Mathematics*, 47, 145–150.   V
8. N.I.C. XLIX: On Mellin's inversion formula, *Messenger of Mathematics*, 47, 178–184.   VII. 1 (a)
9. Note on an expression of Lambert's series as a definite integral, *Messenger of Mathematics*, 47, 190–192.   IV. 1 (d)
10. On the representation of a number as the sum of any number of squares, and in particular of five or seven, *Proceedings of the National Academy of Sciences*, 4, 189–193.   I. 2 (a)

1919

1. (With J. E. L.) Note on Messrs. Shah and Wilson's paper entitled: 'On an empirical formula connected with Goldbach's theorem', *Proceedings of the Cambridge Philosophical Society*, 19, 245–254.   I. 2 (c)
2. N.I.C. L: On the integral of Stieltjes and the formula for integration by parts, *Messenger of Mathematics*, 48, 90–100.   V
3. N.I.C. LI: On Hilbert's double-series theorem, and some connected theorems concerning the convergence of infinite series and integrals, *Messenger of Mathematics*, 48, 107–112.   II. 3
4. A problem of Diophantine approximation, *Journal of the Indian Mathematical Society*, 11, 162–166.   I. 1

1920

1. (With S. Ramanujan) The normal number of prime factors of a number n, *Quarterly Journal of Mathematics*, 48, 76–92.   II. 1
2. (With J. E. L.) A new solution of Waring's problem, *Quarterly Journal of Mathematics*, 48, 272–293.   I. 2 (b)
3. Note on a theorem of Hilbert, *Mathematische Zeitschrift*, 6, 314–317.   II. 3
4. On two theorems of F. Carlson and S. Wigert, *Acta Mathematica*, 42, 327–339.   IV. 2
5. (With J. E. L.) P.N. I: A new solution of Waring's problem, *Göttinger Nachrichten* (1920), 33–54.   I. 2 (b)
6. Additional note on two problems in the analytic theory of numbers, *Proceedings of the London Mathematical Society*, (2) 18, 201–204.   II. 2
7. (With J. E. L.) Abel's theorem and its converse, *Proceedings of the London Mathematical Society*, (2) 18, 205–235.   VI
8. N.I.C. LII: On some definite integrals considered by Mellin, *Messenger of Mathematics*, 49, 85–91.   VII. 1 (a)
9. N.I.C. LIII: On certain criteria for the convergence of the Fourier series of a continuous function, *Messenger of Mathematics*, 49, 149–155.   III. 1 (a)
10. On the representation of a number as the sum of any number of squares, and in particular of five, *Transactions of the American Mathematical Society*, 21, 255–284.   I. 2 (a)
11. Some famous problems of the theory of numbers and in particular Waring's problem, Inaugural lecture, Oxford, 1920.   I. 2 (d)

1921

1. (With J. E. L.) P.N. II: Proof that every large number is the sum of at most 21 biquadrates, *Mathematische Zeitschrift*, 9, 14–27.   I. 2 (b)
2. (With J. E. L.) The zeros of Riemann's zeta-function on the critical line, *Mathematische Zeitschrift*, 10, 283–317.   II. 1
3. Note on Ramanujan's trigonometrical function $c_q(n)$, and certain series of arithmetical functions, *Proceedings of the Cambridge Philosophical Society*, 20, 263–271.   II. 2
4. A theorem concerning summable series, *Proceedings of the Cambridge Philosophical Society*, 20, 304–307.   VI
5. A convergence theorem, *Proceedings of the London Mathematical Society*, (2) 19, vi–vii.   II. 3
6. (With J. E. L.) On a Tauberian theorem for Lambert's series, and some fundamental theorems in the analytic theory of numbers, *Proceedings of the London Mathematical Society*, (2) 19, 21–29.   II. 1
7. N.I.C. LIV: Further notes on Mellin's inversion formulae, *Messenger of Mathematics*, 50, 165–171.   VII. 1 (a)

1922

1. Goldbach's Theorem, *Matematisk Tidsskrift B*, 1–16.   I. 2 (c)
2. A new proof of the functional equation for the zeta-function, *Matematisk Tidsskrift B*, 71–73.   II. 1
3. (With J. E. L.) P.N. III: On the expression of a number as a sum of primes, *Acta Mathematica*, 44, 1–70.   I. 2 (c)
4. (With J. E. L.) P.N. IV: The singular series in Waring's problem and the value of the number $G(k)$, *Mathematische Zeitschrift*, 12, 161–188.   I. 2 (b)
5. (With J. E. L.) D.A.: A further note on the trigonometrical series associated with the elliptic theta-functions, *Proceedings of the Cambridge Philosophical Society*, 21, 1–5.   I. 1
6. (With J. E. L.) D.A.: The lattice-points of a right-angled triangle, *Proceedings of the London Mathematical Society*, (2) 20, 15–36.   I. 1

1922 (cont.)

7. (With T. Carleman) Fourier's series and analytic functions, *Proceedings of the Royal Society*, (A) 101, 124–133. IV. 2
8. (With J. E. L.) Summation of a certain multiple series, *Proceedings of the London Mathematical Society*, (2) 20, xxx. I. 2 (c)
9. (With J. E. L.) D.A.: The lattice-points of a right-angled triangle, *Hamburg Abhandlungen*, 1, 212–249. I. 1
10. N.I.C. LV: On the integration of Fourier series, *Messenger of Mathematics*, 51, 186–192. III. 1 (e)
11. The theory of numbers, *British Association Report*, 90, 16–24. VII. 2 (e)

1923

1. (With J. E. L.) On Lindelöf's hypothesis concerning the Riemann zeta-function, *Proceedings of the Royal Society*, (A) 103, 403–412. II. 1
2. A chapter from Ramanujan's notebook, *Proceedings of the Cambridge Philosophical Society*, 21, 492–503. IV. 1 (d)
3. (With J. E. L.) D.A.: The analytic character of the sum of a Dirichlet's series considered by Hecke, *Hamburg Abhandlungen* 3, 57–68. I. 1
4. (With J. E. L.) D.A.: The analytic properties of certain Dirichlet's series associated with the distribution of numbers to modulus unity, *Transactions of the Cambridge Philosophical Society*, 22, 519–533. I. 1
5. (With J. E. L.) The approximate functional equation in the theory of the zeta-function with applications to the divisor-problems of Dirichlet and Piltz, *Proceedings of the London Mathematical Society*, (2) 2, 39–74. II. 1
6. N.I.C. LVI: On Fourier's series and Fourier's integral, *Messenger of Mathematics*, 52, 49–53. III. 1 (e)

1924

1. (With J. E. L.) Solution of the Cesàro summability problem for power-series and Fourier series, *Mathematische Zeitschrift*, 19, 67–96. III. 1 (b)
2. Some formulae of Ramanujan, *Proceedings of the London Mathematical Society*, (2) 22, xii–xiii. IV. 1 (d)
3. (With J. E. L.) Note on a theorem concerning Fourier series, *Proceedings of the London Mathematical Society*, (2) 22, xviii–xix. III. 1 (b)
4. (With J. E. L.) The equivalence of certain integral means, *Proceedings of the London Mathematical Society*, (2) 22, xl–xliii. VI
5. (With J. E. L.) The allied series of a Fourier series, *Proceedings of the London Mathematical Society*, (2) 22, xliii–xlv. III. 1 (b)
6. (With J. E. L.) P.N. V: A further contribution to the study of Goldbach's problem, *Proceedings of the London Mathematical Society*, (2) 22, 46–56 I. 2 (c)
7. (With J. E. L.) Abel's theorem and its converse II, *Proceedings of the London Mathematical Society*, (2) 22, 254–269. VI
8. N.I.C. LVII: On Fourier transforms, *Messenger of Mathematics*, 53, 135–142. VII. 1 (a)
9. (With E. Landau) The lattice points of a circle, *Proceedings of the Royal Society*, (A) 105, 244–258. II. 2

1925

1. (With J. E. L.) P.N. VI: Further researches in Waring's problem, *Mathematische Zeitschrift*, 23, 1–37.   I. 2 (*b*)
2. The lattice points of a circle, *Proceedings of the Royal Society*, (A) 107, 623–635.   II. 2
3. What is geometry? *Mathematical Gazette*, 12, 309–316.   VII. 2 (*e*)
4. (With J. E. L.) D.A.: An additional note on the trigonometrical series associated with the elliptic theta-functions, *Acta Mathematica*, 47, 189–198.   I. 1
5. (With J. E. L.) A theorem concerning series of positive terms, with applications to the theory of functions, *Meddelelser København*, 7, Nr. 4.   IV. 2
6. Note on a theorem of Hilbert concerning series of positive terms, *Proceedings of the London Mathematical Society*, (2) 23, xlv–xlvi.   II. 3
7. Some formulae in the theory of Bessel functions, *Proceedings of the London Mathematical Society*, (2) 23, lxi–lxiii.   IV. 1 (*d*)
8. (With E. C. Titchmarsh) Solutions of some integral equations considered by Bateman, Kapteyn, Littlewood and Milne, *Proceedings of the London Mathematical Society*, (2) 23, 1–26, and Correction ibid. 24, xxxi–xxxiii.   VII. 1 (*b*)
9. N.I.C. LVIII: On Hilbert transforms, *Messenger of Mathematics*, 54, 20–27.   VII. 1 (*a*)
10. N.I.C. LIX: On Hilbert transforms (cont.), *Messenger of Mathematics*, 54, 81–88.   VII. 1 (*a*)
11. N.I.C. LX: An inequality between integrals, *Messenger of Mathematics*, 54, 150–156.   II. 3

1926

1. A definite integral which occurs in physical optics, *Proceedings of the London Mathematical Society*, (2) 24, xxx–xxxi.   IV. 1 (*d*)
2. (With J. E. L.) Some properties of fractional integrals, *Proceedings of the London Mathematical Society*, (2) 24, xxxvii–xli.   III. 2
3. Note on the inversion of a repeated integral, *Proceedings of the London Mathematical Society*, (2) 24, l–li.   V
4. (With J. E. L.) The allied series of a Fourier series, *Proceedings of the London Mathematical Society*, (2) 24, 211–246.   III. 1 (*b*)
5. (With J. E. L.) A further note on the converse of Abel's theorem, *Proceedings of the London Mathematical Society*, (2) 25, 219–236.   VI
6. (With J. E. L. and G. Pólya) The maximum of a certain bilinear form, *Proceedings of the London Mathematical Society*, (2) 25, 265–282.   II. 3
7. (With J. E. L.) Some new properties of Fourier constants, *Mathematische Annalen*, 97, 159–209.   III. 1 (*c*)
8. (With J. E. L.) N.S. I: Two theorems concerning Fourier series, *Journal of the London Mathematical Society*, 1, 19–25.   III. 1 (*a*)
9. A theorem concerning harmonic functions, *Journal of the London Mathematical Society*, 1, 130–131.   IV. 2
10. (With J. E. L.) N.S. II: The Fourier series of a positive function, *Journal of the London Mathematical Society*, 1, 134–138.   III. 1 (*b*)
11. (With S. Bochner) Notes on two theorems of Norbert Wiener, *Journal of the London Mathematical Society*, 1, 240–244.   VII. 1 (*a*)
12. N.I.C. LXI: On the term by term integration of a series of Bessel functions, *Messenger of Mathematics*, 55, 140–144.   IV. 1 (*d*)
13. The case against the Mathematical Tripos, *Mathematical Gazette*, 13, 61–71.   VII. 2 (*e*)

1927

1. Note on Ramanujan's arithmetical function $\tau(n)$, *Proceedings of the Cambridge Philosophical Society*, 23, 675–680.   II. 2
2. (With J. E. L.) N.S. III: On the summability of the Fourier series of a nearly continuous function, *Proceedings of the Cambridge Philosophical Society*, 23, 681–684.   III. 1 (b)
3. (With J. E. L.) N.S. IV: On the strong summability of Fourier series, *Proceedings of the London Mathematical Society*, (2) 26, 273–286.   III. 1 (b)
4. (With J. E. L.) N.S. V: On Parseval's theorem, *Proceedings of the London Mathematical Society*, (2) 26, 287–294.   III. 1 (e)
5. (With A. E. Ingham and G. Pólya) Theorems concerning mean values of analytic functions, *Proceedings of the Royal Society*, (A) 113, 542–569.   IV. 2
6. (With J. E. L.) Elementary theorems concerning power series with positive coefficients and moment constants of positive functions, *Journal für Mathematik* 157, 141–158.   II. 3
7. N.I.C. LXII: A singular integral, *Messenger of Mathematics*, 56, 10–16.   VII. 1 (a)
8. N.I.C. LXIII: Some further applications of Mellin's inversion formula, *Messenger of Mathematics*, 56, 186–192.   VII. 1 (a)
9. Note on a theorem of Mertens, *Journal of the London Mathematical Society*, 2, 70–72.   II. 1
10. Note on the multiplication of series, *Journal of the London Mathematical Society*, 2, 169–171.   VI
11. (With J. E. L.) N.S. VI: Two inequalities, *Journal of the London Mathematical Society*, 2, 196–201.   II. 3

1928

1. (With J. E. L.) A theorem in the theory of summable divergent series, *Proceedings of the London Mathematical Society*, (2) 27, 327–348.   VI
2. (With A. E. Ingham and G. Pólya) Notes on moduli and mean values. *Proceedings of the London Mathematical Society*, (2) 27, 401–409.   IV. 2
3. (With J. E. L.) N.S. VII: On Young's convergence criterion for Fourier series, *Proceedings of the London Mathematical Society*, (2) 28, 301–311.   III. 1 (b)
4. (With J. E. L.) P.N. VIII: The number $\Gamma(k)$ in Waring's problem, *Proceedings of the London Mathematical Society*, (2) 28, 518–542.   I. 2 (b)
5. (With E. L.) Some properties of fractional integrals I, *Mathematische Zeitschrift*, 27, 565–606.   III. 2
6. (With J. E. L.) A convergence criterion for Fourier series, *Mathematische Zeitschrift*, 28, 612–634.   III. 1 (a)
7. N.I.C. LXIV: Further inequalities between integrals, *Messenger of Mathematics*, 57, 12–16.   II. 3
8. N.I.C. LXV: A discontinuous integral. *Messenger of Mathematics*, 57, 113–120.   IV. 1 (d)
9. A theorem concerning trigonometrical series, *Journal of the London Mathematical Society*, 3, 12–13.   III. 1 (d)
10. (With J. E. L.) N.S. VIII: An inequality, *Journal of the London Mathematical Society*, 3, 105–110.   II. 3
11. Remarks on three recent notes in the *Journal*, *Journal of the London Mathematical Society*, 3, 166–169.   II. 3
12. A formula of Ramanujan, *Journal of the London Mathematical Society*, 3, 238–240.   IV. 1 (d)
13. (With J. E. L.) N.S. IX: On the absolute convergence of Fourier series, *Journal of the London Mathematical Society*, 3, 250–253.   III. 1 (a)
14. (With J. E. L.) N.S. X: Some more inequalities, *Journal of the London Mathematical Society*, 3, 294–299.   II. 3

1929

1. (With J. E. L.) The approximate functional equations for $\zeta(s)$ and $\zeta^2(s)$, *Proceedings of the London Mathematical Society*, (2) 29, 81–97.   II. 1
2. Prolegomena to a chapter on inequalities (Presidential Address), *Journal of the London Mathematical Society*, 4, 61–78, and addenda, ibid. 5, 80.   II. 3
3. Remarks in addition to Dr. Widder's note on inequalities, *Journal of the London Mathematical Society*, 4, 199–202.   II. 3
4. (With J. E. L.) A point in the theory of conjugate functions, *Journal of the London Mathematical Society*, 4, 242–245.   III. 1 (e)
4. (With E. C. Titchmarsh) Solution of an integral equation, *Journal of the London Mathematical Society*, 4, 300–304.   VII. 1 (b)
6. An introduction to the theory of numbers, *Bulletin of the American Mathematical Society*, 35, 778–818.   VII. 2 (e)
7. N.I.C. LXVI: The arithmetic mean of a Fourier constant, *Messenger of Mathematics*, 58, 50–52.   III. 1 (e)
8. N.I.C. LXVII: On the repeated integral which occurs in the theory of conjugate functions, *Messenger of Mathematics*, 58, 53–58.   VII. 1 (a)
9. N.I.C. LXVIII: The limit of an integral mean value, *Messenger of Mathematics*, 58, 115–120.   II. 3
10. N.I.C. LXIX: On asymptotic values of Fourier constants, *Messenger of Mathematics*, 58, 130–135.   III. 1 (a)
11. (With J. E. L. and G. Pólya) Some simple inequalities satisfied by convex functions, *Messenger of Mathematics*, 58, 145–152.   II. 3
12. Mathematical proof, *Mind*, 38, 1–25.   VII. 2 (e)

1930

1. (With J. E. L.) A maximal theorem with function-theoretic applications, *Acta Mathematica*, 54, 81–116.   II. 3
2. (With E. C. Titchmarsh) Self-reciprocal functions, *Quarterly Journal of Mathematics*, 1, 196–231.   VII. 1 (a)
3. (With J. E. L.) D.A.: A series of cosecants, *Bulletin of the Calcutta Mathematical Society*, 20, 251–266.   I. 1
4. (With J. E. L.) N.S. XI: On Tauberian theorems, *Proceedings of the London Mathematical Society*, (2) 30, 23–37.   VI
5. (With E. C. Titchmarsh) Additional note on certain integral equations, *Proceedings of the London Mathematical Society*, (2) 30, 95–106.   VII. 1 (b)
6. (With J. E. L.) N.S. XII: On certain inequalities connected with the calculus of variations, *Journal of the London Mathematical Society*, 5, 34–39.   II. 3
7. Mathematics, *The Oxford Magazine*, 48, 819–21 (issue of 5 June).   VII. 2 (e)

1931

1. Some theorems concerning trigonometrical series of a special type, *Proceedings of the London Mathematical Society*, (2) 32, 441–448.   III. 1 (d)
2. (With J. E. L.) Some properties of conjugate functions, *Journal für Mathematik*, 167, 405–423.   III. 2
3. The summability of a Fourier series by logarithmic means, *Quarterly Journal of Mathematics*, 2, 107–112.   III. 1 (b)
4. (With J. E. L.) N.S. XIII: Some new properties of Fourier constants, *Journal of the London Mathematical Society*, 6, 3–9.   III. 1 (c)

1931 (cont.)

5. (With J. E. L.) N.S. XIV: An additional note on the summability of Fourier series, *Journal of the London Mathematical Society*, 6, 9–12.   III. 1 (*b*)
6. (With E. C. Titchmarsh) A note on Parseval's theorem for Fourier transforms, *Journal of the London Mathematical Society*, 6, 44–48.   VII 1 (*a*)
7. (With J. E. L.) N.S. XV: On the series conjugate to the Fourier series of a bounded function, *Journal of the London Mathematical Society*, 6, 278–281.   III. 1 (*b*)
8. (With J. E. L.) N.S. XVI: Two Tauberian theorems, *Journal of the London Mathematical Society*, 6, 281–286.   VI

1932

1. (With E. C. Titchmarsh) Formulae connecting different classes of self-reciprocal functions, *Proceedings of the London Mathematical Society*, (2) 33, 225–232.   VII. 1 (*a*)
2. On Hilbert transforms, *Quarterly Journal of Mathematics*, 3, 102–112.   VII. 1 (*a*)
3. (With J. E. L.) Some integral inequalities connected with the calculus of variations, *Quarterly Journal of Mathematics*, 3, 241–252.   II. 3
4. (With J. E. L.) Some properties of fractional integrals II, *Mathematische Zeitschrift*, 34, 403–439.   III. 2
5. (With J. E. L.) Some new cases of Parseval's theorem, *Mathematische Zeitschrift*, 34, 620–633.   III. 1 (*c*)
6. (With J. E. L.) An additional note on Parseval's theorem, *Mathematische Zeitschrift*, 34, 634–636.   III. 1 (*c*)
7. (With E. C. Titchmarsh) An integral equation, *Proceedings of the Cambridge Philosophical Society*, 28, 165–173.   VII. 1 (*b*)
8. Summation of a series of polynomials of Laguerre, *Journal of the London Mathematical Society*, 7, 138–139, and addendum, ibid. 192.   III. 1 (*e*)
9. (With J. E. L.) N.S. XVII: Some new convergence criteria for Fourier series, *Journal of the London Mathematical Society*, 7, 252–256.   III. 1 (*a*)

1933

1. (With E. C. Titchmarsh) A class of Fourier kernels, *Proceedings of the London Mathematical Society*, (2) 35, 116–155.   VII. 1 (*a*)
2. (With J. E. L.) Some more integral inequalities, *The Tôhoku Mathematical Journal*, 37, 151–159.   II. 3
3. The constants of certain inequalities, *Journal of the London Mathematical Society*, 8, 114–119.   II. 3
4. A theorem concerning Fourier transforms, *Journal of the London Mathematical Society*, 8, 227–231.   VII. 1 (*a*)

1934

1. (With J. E. L.) Theorems concerning Cesàro means of power series, *Proceedings of the London Mathematical Society*, (2) 36, 516–531.   III. 2
2. (With J. E. L.) Bilinear forms bounded in space $[p, q]$, *Quarterly Journal of Mathematics*, 5, 241–254.   II. 3
3. (With J. E. L.) Some new convergence criteria for Fourier series, *Annali Pisa*, (2) 3, 43–62.   III. 1 (*a*)
4. (With E. M. Wright) Leudesdorf's extension of Wolstenholme's theorem, *Journal of the London Mathematical Society*, 9, 38–41, and corrigendum, ibid. 240.   II. 2

1934 (cont.)

5. On the summability of series by Borel's and Mittag-Leffler's methods, *Journal of the London Mathematical Society*, 9, 153–157.   VI
6. (With J. E. L.) A problem in elementary probability, *Mathematical Gazette*, 18, 195.   VII. 2 (d)
7. The J-type and the S-type among mathematicians, *Nature*, 134, 250.   VII. 2 (e)

1935

1. Remarks on some points in the theory of divergent series, *Annals of Mathematics*, (2) 36, 167–181.   VI
2. The resultant of two Fourier kernels, *Proceedings of the Cambridge Philosophical Society*, 31, 1–6.   VII. 1 (a)
3. (With E. Landau and J. E. L.) Some inequalities satisfied by the integrals or derivatives of real or analytic functions, *Mathematische Zeitschrift*, 39, 677–695.   II. 3
4. (With J. E. L.) An inequality, *Mathematische Zeitschrift*, 40, 1–40.   II. 3
5. (With J. E. L.) The strong summability of Fourier series, *Fundamenta Mathematicae*, 25, 162–189.   III. 1 (b)
6. (With J. E. L.) N.S. XVIII: On the convergence of Fourier series, *Proceedings of the Cambridge Philosophical Society*, 31, 317–323.   III. 1 (c)
7. (With J. E. L.) N.S. XIX: A problem concerning majorants of Fourier series, *Quarterly Journal of Mathematics*, 6, 304–315.   III. 1 (c)
8. Second note on a theorem of Mertens, *Journal of the London Mathematical Society*, 10, 91–94.   II. 1
9. Some identities satisfied by infinite series, *Journal of the London Mathematical Society*, 10, 217–220.   IV. 1 (d)

1936

1. (With J. E. L.) N.S. XX: On Lambert series, *Proceedings of the London Mathematical Society*, (2) 41, 257–270.   VI
2. (With J. E. L.) Some more theorems concerning Fourier series and Fourier power series, *Duke Mathematical Journal*, 2, 354–382.   III. 1 (b)
3. (With E. C. Titchmarsh) New solution of an integral equation, *Proceedings of the London Mathematical Society*, (2) 41, 1–15.   VII. 1 (b)
4. A note on two inequalities, *Journal of the London Mathematical Society*, 11, 167–170.   II. 3

1937

1. Ramanujan and the theory of Fourier transforms, *Quarterly Journal of Mathematics*, 8, 245–254.   VII. 1 (a)
2. The Indian mathematician Ramanujan, *American Mathematical Monthly*, 44, 137–155.   VII. 2 (e)
3. (With J. E. L.) N.S. XXI: Generalizations of a theorem of Paley, *Quarterly Journal of Mathematics*, 8, 161–171.   III. 2
4. On a theorem of Paley and Wiener, *Proceedings of the Cambridge Philosophical Society*, 33, 1–5.   VII. 1 (a)
5. (With N. Levinson) Inequalities satisfied by a certain definite integral, *Bulletin of the American Mathematical Society*, 43, 709–716.   II. 3

1937 (cont.)

6. A formula of Ramanujan in the theory of primes, *Journal of the London Mathematical Society*, 12, 94–98. II. 1
7. Another formula of Ramanujan, *Journal of the London Mathematical Society*, 12, 314–318. IV. 1 (d)

1938

1. A further note on Ramanujan's arithmetical function $\tau(n)$, *Proceedings of the Cambridge Philosophical Society*, 34, 309–315. II. 2

1939

1. A note on a differential equation, *Proceedings of the Cambridge Philosophical Society*, 35, 652–653. VII. 2 (b)
2. Notes on special systems of orthogonal functions (I): The boundedness of the generalized Laguerre system, *Journal of the London Mathematical Society*, 14, 34–36. IV. 1 (d)
3. Notes on special systems of orthogonal functions (II): On functions orthogonal with respect to their own zeros, *Journal of the London Mathematical Society*, 14, 37–44.III. 1 (e)

1940

1. Notes on special systems of orthogonal functions (III): A system of orthogonal polynomials, *Proceedings of the Cambridge Philosophical Society*, 36, 1–8. IV. 1 (d)
2. Mathematics in war-time, *Eureka*, Vol. I, No. 3, 5–8. VII. 2 (e)

1941

1. (With J. E. L.) Theorems concerning mean values of analytic or harmonic functions, *Quarterly Journal of Mathematics*, 12, 221–256. III. 2
2. Note on a divergent series, *Proceedings of the Cambridge Philosophical Society*, 37, 1–8. VI
3. Notes on special systems of orthogonal functions (IV): The orthogonal functions of Whittaker's cardinal series, *Proceedings of the Cambridge Philosophical Society*, 37, 331–348. III. 1 (d)
4. A double integral, *Journal of the London Mathematical Society*, 16, 89–94. VII. 1 (a)

1942

1. Note on Lebesgue's constants in the theory of Fourier series, *Journal of the London Mathematical Society*, 17, 4–13. III. 1 (a)

1943

1. An inequality for Hausdorff means, *Journal of the London Mathematical Society*, 18, 46–50. II. 3
2. (With W. W. Rogosinski) Notes on Fourier series (I): On sine series with positive coefficients, *Journal of the London Mathematical Society*, 18, 50–57. III. 1 (d)
3. (With W. W. Rogosinski) Notes on Fourier series (II): On the Gibbs phenomenon, *Journal of the London Mathematical Society*, 18, 83–87. III. 1 (a)
4. (With J. E. L.) N.S. XXII: On the Tauberian theorem for Borel summability, *Journal of the London Mathematical Society*, 18, 194–200. VI

1944

1. (With J. E. L.) N.S. XXIII: On the partial sums of Fourier series, *Proceedings of the Cambridge Philosophical Society*, 40, 103–107.   III. 1 (c)
2. Note on the multiplication of series by Cauchy's rule, *Proceedings of the Cambridge Philosophical Society*, 40, 251–252.   VI

1945

1. (With W. W. Rogosinski) Notes on Fourier series (III): Asymptotic formulae for the sums of certain trigonometrical series, *Quarterly Journal of Mathematics*, 16, 49–58.   III. 1 (d)
2. (With N. Aronszajn) Properties of a class of double integrals, *Annals of Mathematics*, (2) 46, 220–241, and corrigendum in 47, 166.   V
3. Riemann's form of Taylor series, *Journal of the London Mathematical Society*, 20, 48–57.   VI
4. A mathematical theorem about golf, *Mathematical Gazette*, 29, 226–227.   VII. 2 (d)

1946

1. (With J. E. L.) N.S. XXIV: A curious power-series, *Proceedings of the Cambridge Philosophical Society*, 42, 85–90.   I. 1
2. (With W. W. Rogosinski) Theorems concerning functions subharmonic in a strip, *Proceedings of the Royal Society* (A), 185, 1–14.   IV. 2

1947

1. (With W. W. Rogosinski) Notes on Fourier series (IV): Summability (R_2), *Proceedings of the Cambridge Philosophical Society*, 43, 10–25.   III. 1 (b)
2. A double integral, *Journal of the London Mathematical Society*, 22, 242–247.   VII. 1 (a)

1948

1. (With J. E. L.) A new proof of a theorem on rearrangements, *Journal of the London Mathematical Society*, 23, 163–168.   II. 3

1949

1. (With W. W. Rogosinski) Notes on Fourier series (V): Summability (R_1) *Proceedings of the Cambridge Philosophical Society*, 45, 173–185.   III. 1 (b)

OBITUARY NOTICES BY G. H. HARDY

O1 S. Ramanujan, *Proceedings of the London Mathematical Society*, (2) 19 (1921), xl–lviii.
O2 ——, *Proceedings of the Royal Society* A, 99 (1921), xiii–xxix.
O3 ——, *Nature*, 105 (1920), 494–5.
O4 C. Jordan, *Proceedings of the Royal Society* A, 104 (1922), xxiii–xxvi.
O5 G. Mittag-Leffler, *Journal of the London Mathematical Society*, 3 (1928), 156–60.
O6 ——, *Proceedings of the Royal Society* A, 119 (1928), v–viii.
O7 J. W. L. Glaisher, *Messenger of Mathematics*, 58 (1929), 159–60.
O8 T. J. l'A. Bromwich, *Journal of the London Mathematical Society*, 5 (1930), 209–20.
O9 ——, *Proceedings of the Royal Society* A, 129 (1930), i–x.
O10 R. E. A. C. Paley, *Journal of the London Mathematical Society*, 9 (1934), 75–80.
O11 E. W. Hobson, *Journal of the London Mathematical Society*, 9 (1934), 225–37.
O12 ——, *Obituary Notices of Fellows of the Royal Society*, (1934), 237–49.
O13 ——, *Nature*, 133 (1934), 938–9.
O14 E. Landau (with H. Heilbronn), *Journal of the London Mathematical Society*, 13 (1938), 302–10.
O15 W. H. Young, *Journal of the London Mathematical Society*, 17 (1942), 218–37.
O16 ——, *Obituary Notices of Fellows of the Royal Society*, 4 (1943), 307–23.
O17 ——, *Nature*, 150 (1942), 227.
O18 D. Hilbert, *Journal of the London Mathematical Society*, 18 (1943), 191–2.
O19 H. L. Lebesgue, *Nature*, 152 (1943), 685.
O20 J. R. Wilton (with H. S. Carslaw), *Journal of the London Mathematical Society*, 20 (1945), 58–64.

A list of book reviews by Hardy is given in Section VII. 5, and a list of Hardy's books is given in Section VII. 6.